⑤ *2 Abbreviations for Monosaccharides and Monosaccharide Derivatives*

Ara	Arabinose	GlcNAc	*N*-Acetylglucosamine
f	furanosido	GlcUA	Glucuronic acid
Fru	Fructose	GulUA	Guluronic acid
Fuc	Fucose	Man	Mannose
Gal	Galactose	ManUA	Mannuronic acid
GalNAc	*N*-Acetylgalactosamine	p	pyranosido
GalUA	Galacturonid acid	Rha	Rhamnose
Glc	Glucose	Xyl	Xylose
GlcN	Glucosamine		

⑤ *3 Abbreviations for Amino Acid Residues*

Aba	L-α-Aminobutyric acid	Hyl	L-5-Hydroxylysine
Ala	L-Alanine	Hyp	L-4-Hydroxyproline
Arg	L-Arginine	Ile	L-Isoleucine
Asn	L-Asparagine	Leu	L-Leucine
Asp	L-Aspartic acid	Lys	L-Lysine
Cys	L-Cysteine	Met	L-Methionine
Dab	L-2,4-Diaminobutyric acid	Phe	L-Phenylalanine
DAP	Diaminopimelic acid	Pro	L-Proline
dPhe	Dehydrophenylalanine	Ser	L-Serine
Gln	L-Glutamine	Thr	L-Threonine
Glu	L-Glutamic acid	Trp	L-Tryptophan
└Glu	L-Pyroglutamic acid	Tyr	L-Tyrosine
Gly	Glycine	Val	L-Valine
His	L-Histidine		

Martin Luckner

Secondary Metabolism in Microorganisms, Plants, and Animals

Third Revised and Enlarged Edition

With 350 Figures and 83 Tables

Springer-Verlag Berlin Heidelberg New York
London Paris Tokyo Hong Kong

Professor Dr. Martin Luckner
Sektion Pharmazie der Martin-Luther-Universität
Halle-Wittenberg

Sole distribution rights for all non-socialist countries:
Springer-Verlag Berlin Heidelberg New York
London Paris Tokyo Hong Kong

ISBN 3-540-50287-4 Springer-Verlag Berlin Heidelberg New York
ISBN 0-387-50287-4 Springer-Verlag New York Berlin Heidelberg

ISBN 3-540-12771-2 2. Auflage Springer-Verlag Berlin Heidelberg New York Tokyo
ISBN 0-387-12771-2 2nd edition Springer-Verlag New York Heidelberg Berlin Tokyo

Library of Congress Cataloging-in-Publication Data
Luckner, Martin. Secondary metabolism in microorganisms, plants, and animals / Martin Luckner. –
3rd rev. and enl. ed.
p. cm. Rev. ed. of: Secondary metabolism in plants and animals. 1972. Includes bibliographical references (p.).
ISBN 0-387-50287-4 (U.S.)
1. Metabolism, Secondary. I. Luckner, Martin. Secondary metabolism in plants and animals.
II. Title.
QH521.L82   1990   574.1'33–dc20   90-9706

1st edition 1970, 2nd edition 1984

Printed in the German Democratic Republic
Typesetting, printing and binding: INTERDRUCK Graphischer Großbetrieb Leipzig
2131/3140-543210

Dedicated to my late teacher
Prof. Dr. Drs. h. c. Kurt Mothes

Contents

Contents

10   Contents

12   Contents

Preface

Many of the reactions and compounds involved in metabolism are almost identical in the different groups of living organisms. They are known as primary metabolic reactions and primary metabolic products. In addition, however, a wide variety of biochemical pathways are characteristic of only a few species of organisms, of single "chemical races", or even of a certain stage of differentiation of specialized cells. Such pathways are collectively referred to as "secondary metabolism", and the compounds formed are called "secondary products". Secondary products are frequently revealed by their color, smell, or taste. They are responsible for the flavor of most foodstuffs and beverages and for the color and fragrance of flowers and fruits. Many of them are part of the materia medica, e.g., alkaloids, cardiac glycosides, antibiotics, or compounds acting as hormones. Others are used in the industry, e.g., rubber, tannins, and cellulose.

This book treats the organization and significance of biosynthesis, storage, transformation, and degradation of the most important groups of secondary products in microorganisms, plants, and animals. It shows that the formation of secondary products is a common characteristic of specialized cells brought about by the action of special enzymes encoded by specific genetic material. It demonstrates that the biosynthesis of secondary products in most cases is without significance for the individual producer cell, but may play a decisive role in the development and function of the producer organism as a whole. The book discusses the participation of secondary products in the relationships between different types of organism, i.e., their use as repellents, attractants, pheromones, or phytoalexins, as well as their function as signal substances within the programs of gene expression and development. Last but not least the book gives an idea of the many uses of secondary products in human life and society.

The book is intended to be a textbook for advanced undergraduate and postgraduate students as well as for scientists in academics and industry, working in biochemistry, organic chemistry, biology, medicine, pharmacy, and agriculture. Basic knowledge in chemistry, biochemistry, and biology is therefore taken for granted. Space limitations forced the author to omit in most cases the means by which the given facts have been proven and discussions about the reliability of the particular results. For studies in this respect the literature listed at the end of the individual sections should be consulted.

It has been my late teacher, Professor Dr.Drs.h.c. Kurt Mothes, one of the pioneers of plant biochemistry, who has encouraged me to write this book. Its first edition appeared 1969 in German, and in the 1970s in English, Japanese, and Russian. A second edition was printed in 1984. In the preparation of this third edition the whole manuscript has been revised thoroughly to cope with the rapid development of our knowledge on the different aspects of secondary metabolism in the last years. New chapters have been included on compartmentalization and channeling in secondary metabolism at the organismic level, on colored secondary products, on the significance of secondary products as toxins in human life, and on the improvement of secondary product formation by genetic and physiological means. In general, however, the approved structure of the book and its size have been maintained. Hence, the third edition will give its readers once more a concise, up-to-date survey on all important aspects of secondary metabolism.

The author deeply acknowledges the suggestions and critical remarks of colleagues and friends who helped in the improvement of the manuscript. He would also be very grateful to receive their advice and criticism in the future. He thanks Mrs. Gudrun Reinbothe for typing the manuscript and Mrs. Ingeborg Duchek for preparation of the drawings. He is particularly indepted to Dr. Beate Diettrich for many fruitful discussions as well as for careful editing and proofreading. He is grateful to the staffs of the publishing houses VEB G. Fischer Verlag, Jena, and Springer-Verlag Berlin—Heidelberg—New York—Tokyo for rapid publication, and last but not least he thanks his family for accepting that in the last years he spent a lot of his time in preparing this manuscript.

Halle, 1989   Martin Luckner

While the enzymologist's garden is a dream of uniformity, a green meadow where the cycles of Calvin and Krebs tick round in disciplined order, the organic chemist walks in a untidy jungle of uncouthly named extractives, rainbow displays of pigments, where in every bush there lurks the mangled shape of some alkaloid, the exotic perfume of some new terpene, or some shocking and explosive polyacetylene (Bu'Lock 1961 [121]).

A General Aspects

A 1 What is Secondary Metabolism?

In addition to the reactions of primary metabolism, which are similar in all groups of living organisms and characteristic for life on earth (formation and breakdown of nucleic acids and proteins as well as their precursors, of certain carbohydrates and carboxylic acids, e. g., the intermediates of the tricarboxylic acid cycle and the ubiquitous fatty acids), a vast number of metabolic pathways lead to the formation of peculiar chemical compounds, the so-called secondary products. The most characteristic features of these substances are:
— the taxonomically restricted distribution of the individual compounds in contrast to the ubiquitous distribution of secondary metabolism in general (see below)
— the formation by specific enzymes encoded by specific genetic material (A 2)
— the compartmentalization of enzymes, precursors, intermediates, and products involved in biosynthesis, storage, and breakdown (A 3)
— the strict control of biosynthesis by regulation of enzyme amount and activity (A 4.2, A 4.3)
— the phase-dependent expression of secondary metabolism as an aspect of cell specialization brought about by the integration in the programs of differentiation and development of the producer organisms (A 4.4)
— the derivation from substances of primary metabolism on pathways leading to very diverse and in many instances complex chemical structures (D), and
— the lack of importance of secondary products for the synthesizing cell itself, but the possible significance for the producer organism as a whole (E).

The field of biosynthesis, transformation, and degradation of secondary products has been claimed mainly for didactic reasons and the frontier between primary and secondary metabolism is sometimes blurred. Several "primary" metabolites are not formed in all groups of organisms and, on the contrary, certain "secondary" products have a rather wide taxonomic distribution. In addition, there are a few cases where primary metabolic enzymes, like cysteine synthase, also form secondary products, and "secondary compounds" used as coenzymes or cosubstrates, in contrast to the definition given above, may be of direct significance for the producer cell.

The features of secondary metabolism emerged step by step during the development of chemistry, biochemistry, and biology beginning at the end of the 18th century. From 1769-1785 Scheele isolated the first pure natural products including secondary compounds, like tartaric acid (D 1.2), oxalic acid (D 4), benzoic acid (D 23.2.5), gallic acid (D 23.2.5), and uric acid (D 25.5) [589]. 1817-1820 the first alkaloids were obtained in pure state including strychnine and brucine (D 22), quinine and cinchonine (D 22.4), narcotine (D 23.1.2), emetine (D 23.1.2), piperine (D 23.3.1), and caffeine (D 25.1). 1818 Meissner coined the collective term "alkaloids" for this group of secondary products [543]. 1843 Liebig already has known 2 000 primary and secondary natural products [463] and in the following century the isolation, characterization, and synthesis of organic substances derived from natural sources was one of the most challenging tasks in organic chemistry.

With the development of biochemistry it became obviously that only some of the products occurring in nature are of prime significance in the maintainance of life. Their occurrence in living beings has been shown to be ubiquitous. In contrast other substances had only a restricted distribution and seemingly were characteristics of certain organisms. 1891 Kossel, a plant physiologist for the first time applied the designation "secondary" to this unique constituents. At a meeting of the Physiological Society of Berlin he stated [434]:

"Ebenso wie die mikroskopische Forschung dahin gelangt ist, daß sie die Zelle alles unwesentlichen Beiwerkes entkleidet hat, daß sie das Gehäuse und die in ihr aufgespeicherten Reservestoffe von den eigentlichen Trägern des Lebens zu trennen weiß, so muß auch die Chemie versuchen, diejenigen Bestandteile heraus zu sondern, welche in dem entwicklungsfähigen Protoplasma ohne Ausnahme vorhanden sind, und die zufälligen oder für das Leben nicht unbedingt nöthigen Zellstoffe als solche zu erkennen. Die Aufsuchung und Beschreibung derjenigen Atomcomplexe, an welche das Leben geknüpft ist, bildet die wichtigste Grundlage für die Erforschung der Lebensprozesse. Ich schlage vor, diese wesentlichen Bestandteile der Zelle als primäre zu bezeichnen, hingegen diejenigen, welche nicht in jeder entwicklungsfähigen Zelle gefunden werden, als sekundäre."[1]

Later the term "secondary" was adopted by Czapek [169], who in 1921 wrote with respect to the alkaloids:

"Vielleicht ist die sporadische Verbreitung dieser Basen, die Inkonstanz ihres Auftretens bei nahe verwandten Pflanzen, ein Zeichen dafür, daß es sich bei der Bildung solcher Stoffe um Prozesse handelt, die nicht jedem Zellplasma eigen sind, sondern mehr sekundären Charakter haben."[2]

---

[1]   Just as microscopy has succeeded in stripping the cell of its nonessential accessories and in separating its casing and stored reserves from the actual life carriers, chemistry must attempt to separate those compounds which always are present in protoplasm which is capable of developing and to recognize the substances which are either incidental or not absolutely necessary for life. Finding and describing those atom complexes to which life is bound is the most important basis for the investigation of life processes. I propose calling these essential components of the cell primary components, and those which are not found in every cell capable of developing, secondary components.

[2]   The sporadic distribution of these bases, the irregularity of their occurrence in closely related species, perhaps indicate that their production involves processes which are not inherent to every cell plasm, but possess a more secondary character.

It was, however, not until 1950 that the books of Paech [586] and Bonner [80] brought the term secondary products into broader use.

Several attempts have been made to avoid the separation of metabolism into a "primary" and a "secondary" field. Bonner and Galston [81] used the expression "highways and byways of metabolism", reactivating the term "Nebenprodukte des Stoffwechsels" of Sachs [663]. Zähner [838] favored the terms "general and special metabolism", Campbell [128] the phrase compounds of limited taxonomic distribution, Zenk [844], the use of the simple designation "natural products". Vanek [862] introduced the term "excessive metabolites"; Walker [796] and later on Demain [185], the designation "idiolytes" (derived from idiophase, the developmental period at which secondary products are formed in many microorganisms, A 4.1), Holzer the term luxury molecules [368] and Brattsten [93], the term "allelochemicals", because many secondary products are involved in the interactions of living organisms (E 5). All these designations, however, are not precise enough, or do not cover the field of secondary metabolism as a whole. Hence the terms "secondary products" and "secondary metabolism" are still in general use, though there is a widespread feeling that it is an unfortunate choice of words [52].

The multiplicity of the chemical structures of secondary products is enormous. In contrast to primary metabolism, which comprises several hundreds of low molecular compounds, tens of thousands of secondary substances are known today (Table 1, see also the literature cited at the end of this chapter). This num-

Table 1. Distribution of important groups of secondary products in the living world

Group of secondary products	Approximate number of known compounds	Occurrence M = microorganisms P = plants A = Animals		
Derivatives of primary metabolic sugars (D 1)		M	P	A
Secondary monosaccharides (D 1.1)		M	P	A
Sugar carboxylic acids (D 1.2)		M	P	A
Sugar alcohols (D 1.3)		M	P	A
Holosides (D 1.4.1)		M	P	A
— including polysaccharides	300	M	P	A
Heterosides (D 1.4.2)		M	P	A
Secondary products derived from intermediates of glucose degradation and gluconeogenesis (D 2)		M	P	A
Secondary products formed from propionyl CoA (D 3)		M	P	A
— including macrolide antibiotics	100			
Secondary products derived from intermediates of the tricarboxylic and glyoxylic acid cycles (D 4)		M	P	A
Acetyl CoA derivatives (D 5)		M	P	A
Acetoacetic acid and its derivatives (D 5.1)		M	P	A
Fatty acid derivatives (D 5.2)		M	P	A
Secondary fatty acids (D 5.2.1)		M	P	A
Alkanes and alkenes (D 5.2.2)		M	P	A
Alkanals and alkanols (D 5.2.3)		M	P	A
Fatty acid esters (D 5.2.4)		M	P	A
Acetylenic compounds (D 5.2.5)	750	M	P	
Eicosanoids (D 5.2.6)			P	A
Polyketides (D 5.3)	700	M	P	
— including anthraquinones	200			
Isoprenoids (D 6)		M	P	A

Group of secondary products	Approximate number of known compounds	Occurrence M = microorganisms P = plants A = Animals		
Hemiterpenes (D 6.1)		M	P	A
Monoterpenes (D 6.2)	1 000	M	P	A
Aliphatic and cyclohexanoid monoterpenes (D 6.2.1)	800	M	P	A
Iridoids (D 6.2.2)	200		P	A
Sesquiterpenes (D 6.3)	1 000	M	P	A
Diterpenes (D 6.4)	1 000	M	P	A
Sesterterpenes (D 6.5)	200	M	P	A
Pentacyclic triterpenes (D 6.6)	300	M	P	A
— including triterpene saponins	750			
Steroids (D 6.6) including	1 000	M	P	A
— Cucurbitacins (D 6.6)	50		P	
— Sterols (D 6.6.1)		M	P	A
— Steroid saponins (D 6.6.2)	250		P	
— C$_{27}$-Steroid alkaloids (D 6.6.3)	200		P	
— Bile alcohols and bile acids (D 6.6.4)				A
— Pregnane and allopregnane derivatives (D 6.6.5)			P	A
— Androstanes (D 6.6.6)			P	A
— Estranes (D 6.6.7)			P	A
— Cardiac glycosides (D 6.6.8)	150		P	A
— Vitamin D derivatives (D 6.6.9)			P	A
Carotenoids (D 6.7)	600	M	P	
Polyterpenes (D 6.8)		M	P	A
Derivatives of homoisopentenyl diphosphate (D 7)				A
Secondary products derived from amino acids on general pathways (D 8)		M	P	A
Nonprotein amino acids (D 8.1)	700	M	P	A
Amines (D 8.2)	100	M	P	A
Cyanogenic glycosides (D 8.3)	50		P	A
Glucosinolates and products of hydrolysis (D 8.4)	100		P	
Alkaloids (D 8.5)	7 000	M	P	A
Peptides (D 9)		M	P	A
Peptides formed by the degradation of proteins (D 9.1)				A
Peptides built from activated amino acids (D 9.2)		M	P	A
Dioxopiperazines (D 9.2.1)		M		
Hydroxamic acids (D 9.2.2)		M	P	
Penicillins and cephalosporins (D 9.2.3)		M		
Peptidoglycans of the bacterial cell wall (D 9.2.4)		M		
Secondary products derived from glycine, L-serine, and L-alanine (D 10)		M	P	A
Derivatives of sulfuric acid and L-cysteine (D 11)		M	P	A
Sulfinic and sulfonic acids (D 11.1)		M	P	A
S-substituted cysteine derivatives and sulfoxides (D 11.2)			P	
L-Methionine-derived secondary products (D 12)		M	P	A
Secondary products formed from S-adenosyl-L-methionine (D 12.1)		M	P	A
Secondary products originating from L-valine (D 13)		M	P	A
Secondary products formed from L-leucine (D 14)		M	P	

Group of secondary products	Approximate number of known compounds	Occurrence M = microorganisms P = plants A = Animals		
Secondary substances formed from L-isoleucine (D 15)			P	A
Secondary products formed from L-aspartic acid and L-threonine (D 16)		M	P	A
Nicotinic acid and derivatives (D 16.1)		M	P	A
Secondary products derived from L-glutamic acid, L-proline, and L-ornithine (D 17)		M	P	A
Pyrrolizidine alkaloids (D 17.1)	200		P	
Tropane alkaloids (D 17.2)			P	
Secondary products formed from L-lysine (D 18)		M	P	A
— including quinolizidines	150		P	
Secondary products derived from L-arginine (D 19)		M	P	A
Secondary products derived from L-histidine (D 20)		M	P	A
Derivatives of dehydroquinic acid, dehydroshikimic acid, shikimic acid, and chorismic acid (D 21)		M	P	
Hydroxy and amino benzoic acids formed from chorismic and isochorismic acid (D 21.1)		M	P	
Naphthoquinones and anthraquinones (D 21.2)		M	P	
Ubiquinones (D 21.3)		M	P	A
Anthranilic acid derivatives (D 21.4)		M	P	A
— including quinazolines	50	M	P	A
3-Hydroxyanthranilic acid derivatives and phenoxazines (D 21.4.1)		M	P	A
Quinoline, acridine, and benzodiazepine alkaloids (D 21.4.2)	250	M	P	
Secondary products built from L-tryptophan (D 22)		M	P	A
Indole alkylamines (D 22.1)		M	P	A
Ergoline alkaloids (D 22.2)		M	P	
β-Carbolines and related alkaloids (D 22.3)	1 200		P	A
Cinchona alkaloids (D 22.4)	35		P	
Kynurenic acid derivatives (D 22.5)		M	P	A
Secondary products derived from L-phenylalanine and L-tyrosine (phenylpropanoids) (D 23)		M	P	A
Secondary products retaining the amino nitrogen of L-phenylalanine, L-tyrosine, and L-DOPA (D 23.1)		M	P	A
Phenylethylamines and products of further transformation (D 23.1.1)		M	P	A
Tetrahydroisoquinolines (D 23.1.2)	800		P	A
Melanins (D 23.1.3)				A
Compounds formed by cleavage of the aromatic ring of L-DOPA (D 23.1.4)			P	
— including betacyanins	50	M	P	
Iodinated L-tyrosine and L-thyronine derivatives (D 23.1.5)			P	A
Cinnamic acid derivatives (D 23.2)		M	P	A
Cinnamic acids (D 23.2.1)		M	P	A
Coumarins (D 23.2.2)	1 000		P	
Lignins (D 23.2.3)			P	

2*

Group of secondary products	Approximate number of known compounds	Occurrence M = microorganisms P = plants A = Animals		
Lignans (D 23.2.4)	50		P	A
Benzoic acid derivatives (D 23.2.5)		M	P	A
Secondary products derived from cinnamic acid and malonate (D 23.3)		M	P	
Diketides derived from cinnamic acids (D 23.3.1)			P	
Triketides derived from cinnamic acids (D 23.3.2)		M	P	
— including xanthones	150		P	
Tetraktides (stilbenes and flavonoids, D 23.3.3) including	3 000	M	P	
— anthocyanins	250		P	
— chalcones	60		P	
— dihydrochalcones	10	M	P	
— flavones	350	M	P	
— flavonols	450		P	
— proanthocyanidins	50		P	
— catechins	20		P	
— flavan-3,4-diols	20		P	
— biflavonoids	65		P	
— isoflavonoids	150		P	
— aurones	20		P	
Homogentisic acid-derived hydroquinone and benzoquinone derivatives (D 23.4)		M	P	A
Porphyrin-derived secondary products (D 24)		M	P	A
Chlorophylls (D 24.1)		M	P	
Open-chain tetrapyrrols (D 24.2)		M	P	A
Cobalamins (D 24.3)		M	P	
Purine derivatives (D 25)		M	P	A
Hypoxanthine, adenine, and guanine derivatives (D 25.1)		M	P	A
Purine analogs (D 25.2)		M		
Pteridines (D 25.3)		M	P	A
Benzopteridines (D 25.4)		M	P	A
Products of purine degradation (D 25.5)		M	P	A
Thiamine (D 25.6)		M	P	
Pyrimidine derivatives (D 26)		M	P	
Antibiotics (E 5.2)	12 000	M		

ber is rapidly increasing with the refinement of analytical techniques and the investigation of a growing number of microorganisms, plants, and animals (cf. Fig. 1). There are aliphatic, carbocyclic, and heterocyclic; nitrogen, oxygen, and sulfur-containing compounds; saturated and unsaturated substances; glycosides, peptides, hydroxamic acids, and azomethines, which possess the most diverse functional groupings like hydroxy, epoxy, ester, ether, amino, nitro, and carboxylic groups (D).

At the beginning of this century the first speculations on the biosynthesis of secondary products were made (Trier 1912 [767]). In the following years they were substantiated by the synthesis of secondary products under "physiological

Number

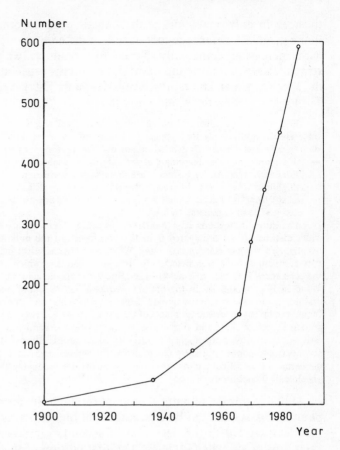

Fig. 1. Number of carotenoids with elucidated structure isolated from microorganisms, plants, and animals

conditions" (biomimetic synthesis), i. e., formation at room temperature, almost neutral pH etc. (cf. Schöpf [687]). These experiments as well as the comparison of chemical structures allowed the formulation of rules for the biosynthesis, e. g., of alkaloids (Robinson 1955 [644]), isoprenoids (Ruzicka 1959 [661]), and polyacetates (polyketides) (Birch 1962 [68]), which for the first time brought some order to the structural chaos of secondary products. In the second half of this century these ideas were further specified by feeding experiments with isotopically labelled precursors (B 1.1) and the use of enzyme preparations (B 1.2). Today they provide the base for the arrangement of nearly all known secondary products in biogenetically related groups, which are easily to survey. This type of classification is therefore used widely also within this book (cf. chapter D).

The restricted distribution of the individual secondary substances in the kingdom of living beings was first pointed out by Sachs [664] and Pfeffer [601], who demonstrated the lack of phylogenetic continuity of secondary substances such as oxalate, resins, and essential oils in plants. Usually the more chemical reactions necessary for the synthesis of a given secondary product, the more restricted is its distribution. The alkaloid nicotine, for instance, which is formed on a relatively simple biosynthetic pathway (D 16.1) occurs in many different plant species, whereas the much more complicated alkaloid brucine (D 22.3) is synthesized in only one genus of the Loganiaceae. In other words, the probability that a given secondary product occurs in different groups of organisms decreases with increasing complexity. Secondary products with complicated chemical structures are therefore important in taxonomy, especially in plant systematics at the species, genus, and family level (Hegnauer [347]; Seigler [699], Harborne and Turner [329], Waterman and Gray [804]).

There are, however, several pitfalls in the use of secondary products as markers in taxonomy, since similar compounds may appear in organisms not related to each other, and deviating substances may occur in closely related organisms. The reasons for this are:
— the selection pressure favoring distinct pattern of secondary products
    Floral anthocyanins, for instance, are more highly correlated with the class of pollinators (hummingbirds, bees, wasps, flies, etc.) than with taxonomy (E 5.5).
— and the transfer of genes which obviously occurred during evolution between taxononically unrelated -groups of organisms (A 2.1).
In addition, indications exist that the genes coding for secondary metabolic enzyme have a much wider distribution as anticipated from the occurrence of the bulk amounts of secondary products usually used as characters in taxonomy. With sensitive analytical methods small amounts of secondary compounds have been detected in cells and organisms which usually are thought to be free of the substances in question. Traces of azetidine-2-carboxylic acid (D 12.1), for instance, a compound found in large quantities in the plant *Convallaria majalis*, occur also in molasses, i. e., the syrup drained from raw sucrose obtained from sugar beets, and the alkaloid morphine (D 23.1.2) thought to be characteristic for plants of the genus *Papaver*, has been shown to occur as a minor component in lettuce, hay, and animal tissues. In addition, several secondary products are synthesized only under extreme physiological conditions. Plant cells cultivated in vitro, for instance, may build secondary compounds not known from the mother plants, and added intermediates of secondary products may be specifically transformed in organisms not producing the final products under normal physiological conditions (F 6.2.1)

At the beginning of natural product research the formation of secondary compounds was assumed to be a monopoly of higher plants. But the development of biochemistry led to the detection of secondary products also in microbial cultures and in animals (Table 1). The first edition of this book [479] was an early paper presenting evidence that similar types of secondary products are formed in microorganisms, plants, and animals and that secondary metabolism in all organisms follows the same principles.

It seems to be a consequence of the unique organization of excretion in plants, rather than of a higher metabolic potential that most known secondary products are of plant origin [541]. Animals eliminate unwanted metabolic products via kidneys, liver, and other excretory organs, and microorganisms easily release compounds into the surrounding medium. Plants, however, store most secondary products (E 1). This retention in the plant body facilitates their further transformation and is one of the reasons for the multiplicity of secondary products found in this group of organisms.

Untill the middle of this century the "absence of significance" was generally accepted as an outstanding characteristic of secondary metabolism and secondary metabolic products (cf., e. g., Paech 1950 [586]). However, already 1949 Foster has launched the idea that secondary products are "shunt" or "overflow" products, the biosynthesis of which saves the primary metabolism of the producer organisms if disturbed by nutrient imbalances causes by the depletion of certain nutrients [243]. This idea was further worked, out, e. g., by Bu'Lock [120, 121], and has its supporters even nowadays (cf. Haslam [339, 340]).

In the late fiftieth Brian [98] and Fraenkel [246] focussed attention on the significance of secondary products as allelochemicals (E 5). This aspects has been neglected in the decades before though Stahl already 1888 published an excellent paper on the deterrence of herbivores by secondary plant products [728]. 1971 the author directed interest to the usage of certain secondary products as hormones and hormone-like factors in the metabolism of microorganisms, plants, and animals [480] and in the last decades it emerged that the process of biosynthesis of secondary products as well as the secondary compounds formed serve the producer organisms in many fields (cf. chapter E). Today the known facts offer a reasonable explanation for the biosynthesis of most types of secondary products. In the 1970s it became obvious that secondary metabolism is a feature of cell specialization and that its expression during certain phases of the development of the producer organisms is due to differential gene expression [480, 485] (A 4.1). This point of view allowed for the first time the definition of secondary metabolism in terms of molecular biology [478] which has promoted a better understanding of its most interesting general aspects.

References for Further Reading

General Aspects of Secondary Metabolism

52, 68, 80, 81, 93, 98, 120, 121, 122, 128, 169, 185, 243, 246, 329, 339, 340, 347, 368, 434, 463, 478, 479, 480, 485, 539, 541, 543, 586, 589, 601, 644, 645, 661, 663, 664, 687, 697, 699, 728, 767, 796, 804, 838, 844, 862

Recent Textbooks and Survey Articles on Secondary Metabolism

120, 339, 500, 501, 762, 765, 861

Books and Journals Listing Secondary Products

16, 212, 226, 373, 404, 412, 413, 438, 452, 553, 554, 622, 722, 761, 772, 773, 791

A 2 Secondary Metabolism, a Distinct Part of General Metabolism

A 2.1 Genetics

The Genetic Material of Secondary Metabolic Pathways

A relatively large part of the genome of the producer organisms is involved in secondary product formation. Its mutation, recombination, or any other alteration may lead to qualitative and quantitative changes in the biosynthesis of secondary products.

The genome of antibiotic-producing Actinomycetes, for instance, is assumed to contain more than 6000 genes of which about 200 affect the rate of formation of 7-chlorotetracycline (D 5.3.8).

The genetic material encoding secondary metabolism is usually located in the genome. In prokaryotes contiguous DNA segments carry the genes for the enzymes forming secondary products, as well as genes governing resistance against the secondary products built and regulatory functions involved in these processes. This organization of the DNA resembles the clustering of primary metabolic genes in operons.

Neither the DNA of plastids nor that of mitochondria code for any enzymes of secondary metabolism. In a few prokaryotes, however, structural genes of secondary metabolic pathways reside in plasmids.

Examples are:
— the SCP 1 sex plasmid of *Streptomyces coelicolor* which carries at least four or five structural genes for the production of the antibiotic methylenomycin A

Methylenomycin A

— the Ti-plasmid of *Agrobacterium tumefaciens* which comprises the genes encoding octopine synthase (D 8.1) tryptophan 2'-monooxygenase, indoleacetamide hydratase (D 22), and AMP isopentenyltransferase (D 25.1) (see below)
— and the pIAA plasmid of *Pseudomonas syringae pv. sevastanoi* which houses the genes encoding tryptophan-2-monooxygenase and indoleacetamide hydratase (D 22).

Three groups of genes participate in secondary metabolism: (a) structural genes encoding secondary metabolic enzymes and other directly involved proteins; (b) genetic material regulating the expression of these structural genes, and (c) genes whose products are involved only indirectly in secondary metabolism, i. e., may control synthesis and fluxes of precursors and cosubstrates or the facilities of storage and excretion of secondary products.

Structural Genes

Mutation of structural genes usually inactivates secondary metabolic enzymes by alteration of the amino acid sequence. More rarely the gene structure is changed in such a way that the enzyme proteins are no longer synthesized. In both cases the corresponding pathway is blocked. Such structural mutants can be isolated relatively easily from microbial cultures. But similar mutants occur also in animals and higher plants, as shown by the following two examples:

a) Blockage of xanthommatin biosynthesis in artificially produced mutants of *Drosophila melanogaster*

Xanthommatin is formed by the following biosynthetic chain (D 22):

L-Tryptophan $\xrightarrow{(1)}$ L-Kynurenine $\xrightarrow{(2)}$ 3-Hydroxy-L-kynurenine $\rightarrow$ Xanthommatin

Xanthommatin-deficient mutants are devoid either of tryptophan 2,3-dioxygenase (enzyme 1) or of kynurenine 3-monooxygenase (enzyme 2).

b) Blockage of cyanogenesis (D 8.3) in "chemical races" (naturally occurring mutant strains) of *Trifolium repens*
Hydrocyanic acid is synthesized by the following reactions:

L-Valine/L-isoleucine $\xrightarrow{(1)}$ Linamarin/lotaustralin $\xrightarrow{(2)}$ HCN + glucose + acetone/methylethylketone

In noncyanogenic races either L-amino acid N-monooxygenase and/or N-hydroxyamino acid dehydrogenase belonging to enzyme group 1 or β-glucosidase (linamarase; enzyme 2) are absent.

As yet knowledge upon the structure of the genetic material involved in secondary metabolism and its usage is scarce. Some most important features known have been derived from the genes located in the T-region of the Ti-plasmid of *Agrobacterium tumefaciens* (see above). In plant cells transformed by Ti-plasmids a number of T-region-specific genes are expressed including those encoding enzymes of secondary metabolism (Fig. 2). With respect to the transcription of these genes the following general principles are of significance:

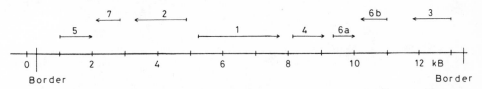

Fig. 2. Transcripts derived from the TL-region of the Ti-plasmid pTiB6S3
The transcripts are numbered according to diminishing size. The *arrows* indicate the location of the corresponding DNA and the direction of transcription. The transcripts 1–4 encode enzymes of secondary metabolism: *1* tryptophan 2-monooxygenase (D 22), *2* indoleacetamide hydratase (D 22), *3* octopine synthase (D 8.1), *4* AMP isopentenyltransferase (D 25.1)

a) The same genetic code and the same apparatus for transcription of the DNA, for processing of the transcripts and translation of the mRNA, is used as in primary metabolism
b) Both strands of the DNA may act as coding strands. Since in double-stranded DNA the strands are antiparallel RNA polymerase reads them in opposite directions (cf. the arrows in Fig. 2)
c) Transcription is initiated at specific promoter sequences. Like the promoters of primary metabolic genes these sequences may contain a CCAAT box and a TATA box
d) In eukaryotic cells the transcripts migrate from the nucleus to the cytoplasm and occur there as poly-adenylated (poly-A) mRNA
e) The frequency of transcription of the secondary metabolic genes is low. The transcripts of the T-region in Ti-transformed plant cells, for instance, represent only 0.005–0.001 % of the total poly-A mRNA present in the cells. Also, the relative abundance of transcripts of secondary mRNAs varies considerably. In Ti-transformed cells octopine mRNA has a higher concentration than AMP isopentenyltransferase mRNA, and the mRNA species encoding tryptophan 2-monooxygenase and indoleacetamide hydratase occur in even smaller quantities.

Evolution of the Genetic Material

The genetic material of secondary metabolism has developed from that of primary metabolism in the course of evolution. Most important in this respect has been the duplication of already existing genes followed by new functionalization of one or several of the multiple gene copies formed by mutation and selection (evolutionary divergence). Evidence for the derivation of secondary genetic material from that of primary metabolism has been obtained
— by determining the substrate spectrum of isoenzymes. In cases of "evolution on the way" it has been shown that one or several members of a given group of isoenzymes, in addition to their primary metabolic substrates, accept unusual substrates and catalyze a new "secondary" metabolic reaction.

In *Quisqualis indica*, for instance, two isoenzymes of cysteine synthase (D 11) have been found, one of which is able to react also with 3,5-dioxo-1,2,4-oxadiazolidine instead of H_2S forming the secondary amino acid quisqualic acid (Fig. 3).

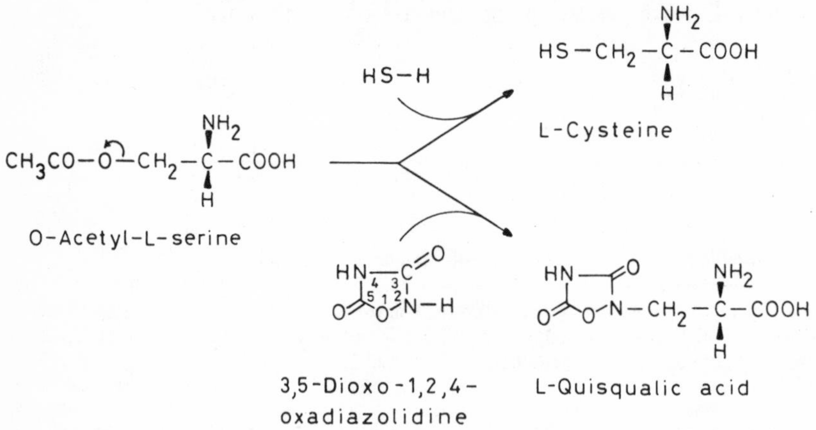

Fig.3. Reactions catalyzed by isoenzyme B of cysteine synthase from *Quisqualis indica*

— by comparison of the structures of secondary metabolic enzymes with those of primary metabolic enzymes. By this method similarities will be found even if the newly evolved protein has already lost the catalytic acitivity of the mother protein in primary metabolism during the progress of evolution.

6-Methylsalicylate synthase (D 5.3.2), for instance, shares most of the enzyme activities of the fatty acid synthase system (D 5.2), a complex of two polyfunctional enzymes active in primary metabolism, and the reaction with specific antibodies has shown it to be a very similar protein. Hence, it is likely that 6-methylsalicylate synthase has evolved from fatty acid synthase.

Gene duplication and independent evolution of the multiple gene copies formed are also the molecular basis for the derivation of new secondary pathways from already existing ones. This has been shown by comparison of amino acid sequences in protein-derived secondary products with related chemical structures and by comparison of the nucleotide sequences in the corresponding genetic material.

Probably there are phases of slow and rapid evolution of secondary genetic material. It seems to last a long time till a new function of a protein has evolved. However, if the new gene product has become of significance for the producer organism, a positive selection pressure will be originated, which in a relatively short period results in the optimization, for instance, of its catalytic or regulatory properties.

The human adenohypophysis secretes at least eight different peptide hormones (D 9.1), which may be grouped into three families according to structural similarities, Repetition of similar sequences of amino acids in the hormones suggests that duplication of genetic material followed by independent modification of the duplicated nucleotide sequences played a significant role in the evolution of these families. Comparison of the chemical structure of the above mentioned peptide hormones (and that of the corresponding mRNA) within the phylogenetic tree provides an estimate of the evolutionary rate causing diversifaction (accepted point mutations/100 nucleotides/10^8 years). This rate is remarkably variable. Values of 50–70 in the lines leading from the placental ancestor to growth hormone, human placental lactogen, or rat prolactin make these peptides some of the most rapidly evolving substances known. On the other hand, the lines leading to most nonprimate growth hormones, to whale prolactin or pig prolactin, have values of 2–7, like cytochrome C, which is generally considered to be a rather slowly evolving protein. As yet the reasons for these differences are unclear. They probably depend on the evolution of the hormone receptors, about which little is known.

Transfer of Genetic Material Encoding Secondary Metabolites between Different Types of Organisms (Lateral Gene Transfer)

In several instances secondary products with the same complicated chemical structures have been found in organisms not closely related to each other. A few examples for this chemical convergence are
— amino and guanidino cyclitols (D 1.3) which are typical products of Actinomycetes, but have also been found in several strains of *Bacillus* and *Pseudomonas*
— maytansine, an ansa macrolide (D 3), which occurs in *Nocardia* strains (Actinomycetes) and in plants of the genus *Maytenus*
— trichothecenes (D 6.3) which have been found in moulds (*Fusarium* sp.) and as constituents of the higher plant (*Baccharis megapotamica*)
— cardiac compounds (D 6.6.8) which are built in certain notrelated families of higher plants and in some beetles
— cyanogenic glycosides (D 8.3) which are synthesized in several nonrelated families of higher plants and some animals
— cephalosporins (D 9.2.3) which are formed in Actinomycetes and some fungi
— ergoline alkaloids (D 22.2) which are synthesized in several fungi and in higher plants (Convulvulaceae)
— morphine (D 23.1.2) which is a bulk compound in some *Papaver sp.* (higher plants), but is formed also in trace amounts in unrelated plant species and in animals, including man, and
— cinnamic acid which is formed by phenylalanine ammonia-lyase (PAL) in higher plants and some microorganisms (D 23.2.1).

It is likely that the genetic material encoding the pathways on which these substances are built was transferred between the different groups of producer organisms, e. g., by protoplast fusion, plasmids, phages, or other genetic vectors (natural genetic engineering).

A well known example of natural genetic engineering is the formation of crown gall tumors brought about by the transfer of genetic material (including genes encoding enzymes of secondary metabolism) form microorganisms to higher plants.

Crown galls are formed on different parts of the shoot of gymnosperms and dicotyledonous plants in response to infections with *Agrobacterium tumefaciens*. Part of a bacterial plasmid (Ti-plasmid), the T-DNA, is transferred to the plant cell and integrated in the plant chromosomal DNA. The T-DNA contains i. a. the genes for tryptophan 2'-monooxygenase and indoleacetamide hydratase transforming tryptophan to indole-3-acetic acid (D 22), of AMP isopentenyltransferase, the key enzyme of the biosynthesis of N^6-dimethylallyladenine (D 25.1), as well as the genes for opine synthases (D 8.1). It is replicated with the plant DNA and transcribed. The indole-3-acetic acid and N^6-dimethylallyladenine formed are plant hormones (E 3.1) causing unorganized division of the infected cells (crown gall formation). The opines are released in the extraplasmic space and are used as nutrients by *Agrobacterium tumefaciens* (genetic colonisation).

Also in other cases the high degree of similarity in the base sequence of the DNA or in the structure of proteins (shown by immunological techniques) strongly indicates the transfer of genetic material between different groups of organisms. Examples with respect to secondary metabolism are
— the homology of the genes encoding AMP isopentenyltransferase in *Agrobacterium tumefaciens* and *Pseudomonas sevastanoi*, and
— the structural homologies in the phenylalanine ammonia-lyase (PAL) protein from fungi and higher plants.

In most instances, however, the transfer of genetic material carrying information for the biosynthesis of secondary products from one type of organisms to an other is still a matter of debate.

It may be speculated, for example, that the biosynthesis of the cephalosporins (D 9.2.3) evolved in Actinomycetes, which form murein (D 9.2.4) as an important cell wall constituent. In these organisms the cephalosporins may control cell wall formation by inhibiting murein biosynthesis. Actinomycetes and fungi, e. g., *Penicillium* and *Cephalosporium* sp., share the same habitat, and contacts between them may be intimate enough to allow (in rare cases) transfer of the genetic material necessary for cephalosporin formation from the Actinomycetes to the fungi. After its integration in the genome the cells of the fungus form the cephalosporins as allelochemicals (E 5) which inhibit the growth of murein-synthesizing microorganisms in the neighborhood, i. e., the compounds in these organisms are no longer used as growth regulators, but as true antibiotics (E 5.2).

Genetic Regulation

Little is known about the genetics of the expression of structural genes encoding secondary metabolic enzymes. Since the principles seem to be the same as for primary metabolism, only one special case will be mentioned, the control of the activity of structural genes by transposons.

An example is the formation of the mosaic pattern on the kernels of certain maize varieties due to the presence or absence of anthocyanins (D 23.3.3) in neighboring cell clones. In maize transposons control the activity of structural genes of anthocyanin biosynthesis. The mosaic-type cultivars possess a receptor element at the locus of the affected gene(s) and a transposable element located elsewhere in the genome. Association of the transposable element with the receptor element inhibits the expression of the structural gene(s), whereas transposition away from the structural gene(s) reestablishes gene expression. There are three possibilities for transposition away from the locus:
— precise excision of the transposable element leaving the gene(s) under the control of the receptor element and giving rise to the metastable (minus)-phenotype
— excision of transposable and receptor elements generating the (+)-phenotype

— excision of transposable and receptor elements plus part of the sequence of the regulated gene(s) leaving a gene rudiment which cannot be transcribed, resulting in the stable (minus)-phenotype. The transposition proceeds in mother cells wich are still able to divide. Hence, cell clones are formed with the described phenotypes scattered over the surface of the kernels.

References for Further Reading

78, 163, 181, 201, 202, 261, 290, 350, 364, 365, 371, 372, 376, 379, 495, 537, 547, 565, 572, 598, 702, 729, 787, 800, 805, 839, 840, 867, 881

A 2.2 Enzymatics

Most reactions in secondary metabolic pathways are catalyzed by specific enzymes.[3] Spontaneous reactions not catalyzed by enzymes are significant in only a few cases, e.g., in the formation of some high molecular compounds like humic acids (D 5.3.2), melanins (D 23.1.3), and lignins (D 23.2.3), in some additions to quinones, and in cyclizations. There is also no evidence that enzymes of primary metabolism play a role in secondary metabolic pathways, i.e., secondary compounds are not side products of primary metabolism.

In the last few years a considerable number of secondary metabolic enzymes have been isolated and characterized in vitro. These enzmyes resemble in many respects those of primary metabolism. They have the same stereospecificity, catalyze the same types of reactions, show similar kinetics, have a similar affinity to substrates and inhibitors, and possess a similar regulation of their catalytic activity (C). Probably this is due to the derivation during evolution of the genetic material encöding secondary metabolic enzymes from genetic material of primary metabolism by gene duplication and diversification (A 2.1).

Secondary metabolic enzymes may be divided into two groups according to substrate specificity: (1) the enzymes forming the basic skeletons of secondary products, which usually have a high substrate specificity, and (2) those modifying the basic skeletons, e. g., certain dehydrogenases (C 2.1), monooxygenases (C 2.6), methyltransferases (C 3.3), and glycosyltransferases (C 6), which possess a relatively low substrate specificity. These enzymes cause "metabolic grids", i.e., polydimensional networks of reactions leading with different rates to the same secondary products (Fig. 4). Metabolic grids are one of the reasons for the large families of similar compounds (congeners) characteristic of most types of secondary products.

Fig. 4. Biosynthetic grid
1, 2 Early enzymes with high substrate specificity reacting only with the compounds A and B, respectively; *3, 4, 5* late enzymes with lower substrate specificity. Enzyme *5* accepts, for instance, intermediates C, E, G, and I. *Heavy line* indicates the predominant pathway leading to the product K

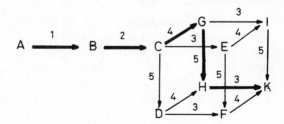

---

[3]   The use of enzyme preparations for the elucidation of secondary metabolic pathways is described in B 1.2.

As yet the biological rationale for the evolution of groups of slightly different compounds characteristic for most types of secondary products is not fully understood. Modification of the basic molecular structure may overcome feedback inhibition of the earlier enzymes (A 4.3) or make adaptation
to secondary products used as repellents more difficult (E 5.5.3). It seems most likely, however, that
many of the modifications are neutral to selection pressure and reflect the accumulation of random
mutations (E).

There is no evidence that substrate affinity is lower for enzymes of secondary
metabolism than for those of primary metabolism or that the key enzymes which
channel compounds from primary to secondary metabolic pathways have a lower
substrate affinity and therefore become active only after saturation of primary
metabolic enzymes. Enzymes, precursors, intermediates, and products in secondary metabolism are strictly compartmentalized and channeled, and the enzymes of primary and secondary metabolism may be fed from different precursor
pools (A 3).

One of the most characteristic features of (primary and) secondary metabolic
enzymes is their stereospecificity. In reactions generating or involving a chiral
center, only one enantiomer is produced or transformed. In the mutual conversion of two chiral centers one particular stereochemical mode of reaction occurs.
This stereospecificity is brought about by the chiral structure of the enzyme proteins, i. e., by the sites of substrate binding and transformation.

It is interesting that enzyme reactions of primary and secondary metabolism show the same
stereospecificity even if there is no discernible advantage of one mode of stereochemical reaction over
the other. Examples are the pyridoxal-5'-phosphate-dependent enzymes (C 5) which all have the
same stereochemistry of the coenzyme-substrate complex, though an alternative conformer formed
by 180° rotation could operate equally well. Hence, all pyridoxal-5'-phosphate-dependent enzymes
obviously did evolve from a common ancestral protein in which an arbitrary choice between the two
phases of the cofactor (and the stereochemical course of the decarboxylation; see C 5) was already
made.

The stereospecificity is not limited to reactions for which the steric course is
evident from the chemical configuration of substrate and product. Even at centers that are not chiral, the reactions are in many cases stereospecific (stereochemically cryptic reactions). The steric course of these reactions may be de-
duced by stereospecific isotope labeling of the substrates, transforming the
reaction center into a chiral center of discernible stereochemical behavior.

Studies with hydrogen isotopes make up the majority of work carried out on stereochemically
cryptic reactions. The two hydrogen atoms at a prochiral C-atom, which may be described by CHHab,
where "a" and "b" are different substituents, are distinguished by an enzyme. They are designated as
pro-R and pro-S depending on the chirality produced by substitution of the respective hydrogen
atom by deuterium or tritium.

Enzymes distinguish also between the two faces of trigonal C-atoms, most frequently encountered as part of a double-bond system. An example in this respect
is the discrimination between the A and the B sides of pyridine nucleotides by
dehydrogenases (C 2.1.1). Another example is the reduction of asymmetrically
substituted carbonyl groups, e. g., reduction of acetaldehyde by alcohol dehydrogenase (D 2), and the addition of hydrogen to carbon-carbon double bonds.

References for Further Reading

234, 322, 798

A 3 Compartmentalization and Channeling

A 3.1 Compartmentalization and Channeling at the Cellular Level

The spatial organization of cells and organelles is an important feature in the
regulation of metabolism. It is the basis for the compartmentalization of enzymes
and the channeling of precursors, intermediates, and products to the sites of metabolic activity. At the cellular level it is brought about either by membranes separating cytoplasmic and noncytoplasmic areas or by the influence of microen-
vironmental effects which form "microcompartments" within the cell, e. g., by
mutual binding of proteins, location of enzymes near each other in membranes,
and by so-called unstirred layer effects. These structural characters give rise to
high local concentrations of reactants, the protection of reactive intermediates,
and the directed transport of precursors, intermediates, and products. They favor
certain sequences of reactions (metabolic channeling) and thus facilitate the formation and function of metabolic chains.

In secondary metabolism the strict control of the intracellular location of enzymes, precursors, intermediates, and products makes possible
— the orderly synthesis of secondary products from intermediates, cosubstrates,
  etc. of primary metabolism
— high local concentrations of precursors and intermediates
— vectorial transport (channeling) of precursors and intermediates, favoring certain sequences of reactions
— facilitation of the control of metabolic chains
— accumulation of large amounts of toxic secondary products within or near the
  cell
— controlled release of stored secondary products and
— the controlled transformation and degradation of secondary compounds.

No "new" compartments have been found in cells forming secondary products, but their organelles have become adapted to the requirements of secondary metabolism. This demonstrates the flexibility of the spatial organization of
the cells and reflects the close connection between primary and secondary metabolism.

Three types of compartments are of significance in the formation of secondary
products: (a) microcompartments, (b) organelles surrounded by a single membrane, and (c) organelles surrounded by double membranes.

A 3.1.1 Microcompartments

Genetic experiments and the examination of metabolic fluxes in living cells
have shown the existence in many secondary pathways of polyfunctional proteins
and/or aggregates of functionally related enzymes bound together by noncovalent forces into highly organized structures (multienzyme complexes), which
cause metabolic channeling (Table 2).

During cell disintegration most of these entities are destroyed. Polyfunctional
proteins are split by proteases, and the weak protein-protein interactions of enzyme complexes are broken by dilution, changes in the ionic strength of the me-

Table 2. Multienzyme complexes and multifunctional proteins catalyzing reactions in secondary metabolism

Multienzyme complex (C) or multifunctional protein (P)	Catalyzed reactions; intermediates bound covalently (C) or noncovalently (N)
6-Methylsalicylate synthase (D 5.3.2)	Acetyl CoA + malonyl CoA → 6-methylsalicylic acid
P	C
Orsellinate synthase (D 5.3.2)	Acetyl CoA + malonyl CoA → orsellinic acid
P	C
Microsomal membranes containing L-amino acid N-monooxygenase, N-hydroxy amino acid dehydrogenase (oxidative decarboxylating), aldoxime dehydratase, nitrile monooxygenase (D 8.3)	L-Tyr → 4-hydroxymandelo-2′-hydroxynitrile
C	N
Bacitracin A synthetase (D 9.2)	Amino acids → bacitracin A
C/P[a]	C
Edeine A synthetase (D 9.2)	Amino acids → edeine A
C/P[a]	C
Gramicidin S synthetase (D 9.2)	Amino acids → gramicidin S
C/P[a]	C
Tyrocidine A synthetase (D 9.2)	Amino acids → tyrocidine A
C/P[a]	C
Membranes of ER containing phenylalanine ammonia-lyase and cinnamate 4-monooxygenase (D 23.2.1)	L-Phe → p-coumaric acid
C	N
Thylakoid membranes containing phenylalanine ammonia-lyase, cinnamate 4-monooxygenase and the benzoate synthetase system (D 23.2.1, D 23.2.5)	L-Phe, L-tyr → benzoic acid, 4-hydroxybenzoic acid
C	N
Thylakoid membranes containing phenylalanine ammonia-lyase and cinnamate 2-monooxygenase (D 23.2.1, D 23.2.2)	L-Phe → o-coumaric acid
C	N
Mitochondrial membranes containing the benzoate synthetase system (D 23.2.5)	p-Coumaric acid → 4-hydroxybenzoic acid
C	N
Chalcone synthase (D 23.3.3)	p-Coumaroyl CoA (caffeoyl CoA) + malonyl CoA → naringenin chalcone (eriodictyol chalcone)
P	C
Stilbene synthase (D 23.3.3)	p-Coumaryl CoA + malonyl CoA → resveratrol
P	C

[a] Complex of several multifunctional proteins.

dium, etc. These changes in the microenvironment annihilate the order of the native cytoplasm especially near membranes where protein concentration is much higher than in the cytoplasmic matrix and approaches that of pure protein crystals. The enzyme complexes and multifunctional proteins found in secondary metabolism protect unstable intermediates, e. g., in the formation of polyketides (D 5.3), and segregate competing pathways (metabolic channeling of intermediates).

Channeling of intermediates results in the phenomenon of catalytic facilitation, i. e., preferred transformation of intermediates formed by the enzyme complex, compared to that of externally added intermediates. Channeling may increase catalytic efficiency by decreased diffusion times or direct transfer of intermediates. It may also decrease the lag phase prior to steady-state production of the final product and allow a finer tuning of the overall activity by coordinate activation and inhibition.

The intermediates are either covalently bound to the multifunctional proteins or enzyme complexes, or are kept near the catalytic centers by other mechanisms.

Covalent binding of precursors and intermediates to enzyme proteins brings about a most efficient form of channeling. In secondary metabolism, this has been shown for the biosynthesis of peptide antibiotics and polyketides (Table 2).

The polyketide synthases and peptide synthetases catalyze an unusually large number of reactions. 6-Methylsalicylate synthase, which produces 6-methylsalicylic acid from acetyl CoA, malonyl CoA, and NADPH, for instance, carries out a total of 13 reactions. The peptide antibiotic synthetases activate each of the amino acids involved in the formation of the peptides by a two-step mechanism at a specific "peripheral" domain of the enzyme protein and catalyze also the linkage of the activated amino acids by sequential transpeptidation. In the case of gramicidin S synthetase, for example, there are 21 different reactions carried out by two multifunctional enzyme proteins.

In addition, metabolic channeling is originated by the proximity of enzymes located in or on membranes (Table 2). In these systems the transformation of internally produced intermediates is favored even if these compounds are not covalently bound.

Microsomal fractions of *Sorghum bicolor* transform L-tyrosine via *N*-hydroxy-L-tyrosine, 4-hydroxyphenylacetaldoxime, and 4-hydroxyphenylacetonitrile to 4-hydroxymandelonitrile. There is a very limited exchange between *N*-hydroxy-L-tyrosine or 4-hydroxyphenylacetonitrile supplied exogenously to the membrane particles and the same compounds produced in situ from labeled L-tyrosine. Similar results were obtained with thylakoid membrane preparations, chromophores of blue-green algae, chloroplasts, and the inner membranes of mitochondria, catalyzing the transformation of L-phenylalanine and L-tyrosine to o- and p-coumaric acid or to benzoic acid and 4-hydroxybenzoic acis. Cinnamic acid, formed as an intermediate in alle these metabolic pathways, is not in equilibrium with cinnamic acid added to the enzyme preparations.

The interaction of the enzymes involved may be weak, as in p-coumaric acid formation from L-phenylalanine on ER membranes or on thylakoids, or it may be so tight that intermediates added externally are hardly accepted (benzoate synthetase system in the membranes of blue-green algae or of chloroplasts of higher plants; 4-hydroxybenzoate synthetase system in mitochondria). With ER membranes which convert L-phenylalanine into o- and p-coumaric acids the extent of cooperation of the enzymes depends on the integrity of the membranes and is reduced, e.g., during membrane purification or treatment with ethylene. It is influenced also by the physiological state of the producer organism.

The enzymes belonging to a metabolic system may be attached to either side of a given membrane. Sequences spanning a membrane can sequester the products from the initial enzymes to avoid interactions like feedback inhibition (A 4.3) and maintain differences in the microenvironment needed for optimal catalysis, such as pH. Loose binding of enzymes to membranes may be brought about by ionic interactions, which are easily destroyed during purification.

References for Further Reading

148, 424, 482, 653, 654, 726

A 3.1.2 Compartments Surrounded by a Single Membrane

Table 3 presents data on the intracellular location of secondary products and secondary metabolic enzymes which demonstrate

— that many enzymes of secondary metabolism are bound to or are associated with the plasma membrane or the membranes of ER, dictyosomes, vacuoles, and microbodies

> Diverse forms of secondary metabolic enzymes may exist in different compartments if a cell or tissue forms more than one type of secondary product. From the glandular tissue of *Aesculus hippocastanum*, for instance, two species of PAL (D 23.2.1) have been isolated which are inhibited either by cinnamic acid and flavonoids or by benzoic acid derivatives, respectively. One of these forms is present in the plastids, the other in the ER. Since PAL is a key enzyme in the formation of many types of secondary products, the two PAL forms probably belong to different pathways.

— that lipophilic secondary products may accumulate in membranes, like the carotenoids, or in the cytoplasm, like rubber particles and droplets of essential oils, and
— that hydrophilic secondary compounds and some enzymes involved in secondary product formation and degradation (e. g., glucosidases, phenol oxidases, peroxidases, hemicellulases, polyuronases, glucanases, and chitinases) may

Table 3. Location of secondary metabolic enzymes and secondary products in different cell compartments

---

Space outside the plasmalemma

Enzymes: phenol oxidases (C 2.3.1), peroxidases (C 2.4), cellulase (D 1.4.1), cutinase (D 5.2.4), linamarase (D 8.3), thioglucoside glucohydrolase (D 8.4), coniferin β-glucosidase (D 23.2.3)

Secondary products: cellulose, hemicellulosic and pectine-like polymers, gums, slimes, chitin (D 1.4.1), surface lipids (alkanes, alkanols, fatty acids, wax esters, cutin, suberin, D 5.2), monoterpenoid and sesquiterpenoid essential oils (D 6.2.1 and D 6.3)[a], diterpenoids (D 6.4)[a], bile acids (D 6.6.4), triterpenes (D 6.6), sporopollenins (D 6.7), alkaloids (D 8.5)[b], cinnamic acid derivatives (D 23.2), lignins (D 23.2.3)[c], stilbenes, flavonoid aglycones and glycosides, tannins (D 23.3.3), volatile compounds, like methane (C 3.2), 3-Z-n-hexenal, 2-Z-n-hexenol (D 5.2.1), and other alcohols, aldehydes, acids, and esters, isoprene (D 6.1), amines (D 8.2), ethylene (D 12.1), secondary products used as hormones (E 3.1), neurotransmitters (E 3.2), pheromones (E 4), antibiotics (E 5.2), and flower scents (E 5.5.1)

Location/secondary metabolic enzymes, secondary products (cont.)

Plasmalemma

Enzymes: 1,4-β-glucan synthase, chitin synthase/chitin synthase zymogen (D 1.4.1)[d], edeine A synthetase (D 9.2), cyclopenase (D 21.4.2)

Secondary products: carotenoids (D 6.7)

Cytosol

Enzymes: phenol oxidases (C 2.3.1), glucosyltransferases (C 6), hydroxynitrile lyase (D 8.4), ACC synthase (D 12.1), most enzymes involved in berberine biosynthesis (D 23.1.2), chalcone synthase, chalcone isomerase (D 23.3.3)

Secondary products: rubber (D 6.8)

Tonoplast

Enzymes: peroxidases (C 2.4)

Secondary products: carotenoids (D 6.7)

Vacuole

Enzymes: peroxidases (C 2.4), β-glucosidases (C 6), inulin synthase, fructan exohydrolase (D 1.4.1), alliin lyase (D 11.2), ACC oxidase (D 12.1)

Secondary products: L-ascorbic acid (D 1.2), sorbitol (D 1.3), gentiobiose, gentianose, inulin (D 1.4.1), malic acid, citric acid, isocitric acid, oxalic acid (D 4), gibberellins (D 6.4), steroid saponins (D 6.6.2), cardiac glycosides (D 6.6.8), rubber (D 6.8), cyanogenic glycosides (D 8.3), glucosinolates (D 8.4), alkaloids (D 8.5)[b], ACC (D 12.1), capsaicins (D 23.1.1), betacyans and betaxanthins (D 23.1.4), cinnamic acid derivatives (D 23.2), coumarins (D 23.2.2), flavonoid glycosides, anthocyanins, tannins (D 23.3.3)

Endoplasmic reticulum
Enzymes, membrane-bound: peroxidases (C 2.4), alkane 1-monooxygenase, dicarboxylic fatty acid synthetase (D 5.2), steroid synthesizing and transforming enzymes (D 6.6.5), thioglucoside glucohydrolase (D 8.4), phenylalanine ammonia-lyase, cinnamate monooxygenases, phenol oxidases, O-methyltransferases (D 23.2.1 and D 23.2.2)

Secondary products, membrane-bound: carotenoids (D 6.7)

Dictyosomes

Enzymes, membrane-bound: peroxidases (C 2.4), glycan synthases (D 1.4.1), thioglucoside glucohydrolase (D 8.4), berberine bridge-forming enzyme, *(S)*-tetrahydroberberine oxidase (D 23.1.2)

Secondary products, membrane-bound: carotenoids (D 6.7); *free in the lumen:* hemicellulosic and pectinlike polymers, gums, and slimes (D 1.4.1)

---

[a] Essential oils frequently accumulate between cell wall and cuticle. Grayanotoxins and related diterpenes are toxic substances in the nectar of certain Ericaceae (E 5.5.1).

[b] Tropane alkaloid crystals (D 17.2) occur in parenchyma idioblasts was well as in root hairs of *Datura innoxia*. In *Claviceps purpurea* the ergoline alkaloids (D 22.2) are located in oil droplets.

[c] Polymerization, the final step of lignin biosynthesis, is catalyzed by peroxidases which are constituents of the cell wall. These peroxidases produce H_2O_2 from NAD(P)H involving the superoxide free radical ion as an intermediate in the complex reaction chain. An NAD-specific malate dehydrogenase produces NADH used as electron donor in the formation of H_2O_2 by the oxidation of malate.

[d] From fungi, so-called chitosomes, i. e., 105 S particles with a microvesicle-like morphology have been isolated which contain chitin synthase zymogen.

accumulate in the nonplasmic compartments separated by membranes from cytoplasm, e. g., the lumen of vacuoles, ER, and Golgi vesicles, or the extra-plasmic space, i.e., apart from the active sites of metabolism (metabolic excre-tion).

Secondary metabolic enzymes are secreted into nonplasmic compartments by the mechanisms known for primary metabolic enzymes.

Preproteins are formed in the cytoplasm containing a so-called signal squence which mediates the transport across the membrane to the nonplasmic lumen of the ER (cf. D 9.1). This signal sequence in most cases is an amino-terminal extension of the enzyme protein consisting of 15–30 amino acids. There is evidence that a common protein transport system (protein translocator) on rough ER recog-nizes the signal sequence which is cleaved during transmembrane passage releasing the enzyme pro-tein into the nonplasmic space.

Secondary products are transported from cytoplasm to nonplasmic compart-ments by membrane-integrated enzyme proteins (translocases) with high sub-strate specificity. Proton gradients act as a driving force mediating a proton anti-port. They are brought about by ATP-dependent membrane ATPases which cause the acidification of the liquid in the nonplasmic compartments (Fig. 5).

In a few cases only secondary products penetrate membranes by nonenzyme-catalyzed diffusion. Examples are certain alkaloids and other basic compounds which cross membranes in the unprotonated lipophilic form and subsequently are arrested in the hydrophilic protonated state in an acidic nonplasmic compart-

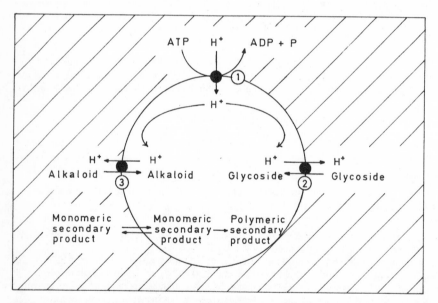

Fig. 5. Secretion of secondary products into nonplasmic compartments
Hatched area: plasm; *open circle:* nonplasmic compartment; *line between both areas:* membrane; ● en-zyme protein. *1* ATP-dependent membrane ATPase; *2* glycoside translocase; *3* alkaloid translocase
Most secondary products are secreted from plasmic compartments by a countertransport with protons as counter ions. A few groups of secondary products probably cross the membrane, separating plas-mic and nonplasmic compartments by simple diffusion, and are entrapped in the nonplasmic space by polymerization

ment (ion trap mechanism), and probably the precursors of certain polymeric compounds which in the nonplasmic compartments are transformed to nondiffusible high-molecular derivatives.

The secondary compounds or enzymes present in the extracytoplasmic space, e. g., the constituents of the cell wall, are either synthesized at the plasma membrane (like cellulose) or they are secreted into the noncytoplasmic lumen of ER or dictyosomes and transported to the extracytoplasmic space by vesicles derived from these organelles (like cell wall matrix polysaccharides and enzymes involved in lignification).

Pulse-chase experiments with living cells have shown that this vesicle-mediated transport is quite rapid. The average time of displacement of cell wall matrix polysaccharides from dictyosomes and Golgi vesicles in plant cells is about 3 min.

The secondary compounds or enzymes accumulated in the neighborhood of the plasma membrane, e. g., the constituents of the cell wall and the epidermis of plant cells, are either insoluble due to polymerization (cellulose, hemicellulose, lignin, cutin, suberin, sporopollenin) or are trapped by the sieve effect of the network of cell wall constituents. Certain enzymes, like malate dehydrogenase involved in lignification (D 23.2.3) are covalently bound to cell wall constituents. Low molecular compounds are able to penetrate the wall and may be washed out into the surroundings.

The spatial orientation of cell wall constituents arises from inhomogeneities of the plasma membrane. These may be brought about by the fusion of Golgi vesicles or ER with the membrane at distinct sites. The site of fusion seems to be directed by microtubules and other cytoskeletal structures, as well as by the certain kinds and areas of membranes.

References for Further Reading

11, 187, 398, 482, 510, 567, 654, 685, 810, 819

A 3.1.3 Double Membrane-Surrounded Compartments

Plastids

Secondary products and enzymes of secondary metabolism are common constituents of chloroplasts, which are typical organelles of plant cells (Table 4).

Lipophilic secondary compounds are found either as constituents of the membranes, e. g., chlorophylls, phycobilins, plastid quinones, gibberellic acid derivatives, or they are located in the plasm of plastids, the stroma, as lipophilic droplets, like flavonoid aglycones, plastid quinones, and carotenoids. The site of accumulation of most hydrophilic secondary products is unknown. In rooted leaves of *Nicotiana*, which accumulate large amounts of nicotine, the alkaloid was detected in the nonplasmic space between the two membranes of the chloroplast envelope and in vacuoles located within the stroma, formed by membrane convolutions. Similar vacuole-like bodies may contain peroxidases and phenol oxidases.

The secondary metabolic enzymes, which are present in chloroplasts, are syn-

Table 4. Location of secondary products and enzymes of secondary metabolism in plastids

---

Envelope

Secondary products: carotenoids (D 6.7)

Inner membranes and membrane-bound bodies

Enzymes: phenol oxidases (C 2.3.1)[a], peroxidases (C 2.4), enzymes of the violaxanthin cycle (D 6.7), phenylalanine ammonia-lyase, cinnamate monooxygenases (D 23.2.1, D 23.2.2), enzymes transform- ing *E*-cinnamic acid to benzoic acid and 4-hydroxybenzoic acid (D 23.2.1, D 23.2.5), tyrosine-3-monooxygenase, 4-hydroxyphenylpyruvate dioxygenase, homogentisate prenylase D 23, D 23.4), chalcone isomerase (D 23.3.3), enzymes converting protoporphyrin IX into chlorophylls (D 24.1)

Secondary products: carotenoids (D 6.7)[b], plastid quinones (phylloquinones, ubiquinones, plastoqui- nones, tocopherylquinones, D. 21.2, D 21.3, D 23.4)[c], flavonoid aglycones (D 23.3.3)[c], chlorophylls and precursors (D. 24.1), phycobiliproteins (D 24.2)[d]

Nonparticulate fraction[e]

Enzymes: enzymes of carotenoid biosynthesis (D 6.7), enzymes of quinolizidine alkaloid formation (D 18), enzymes converting aminolevulinic acid to protoporphyrin IX (D 24.1)

Secondary products: starch (D 1.4.1), essential oils (D 6.2.1), nicotine (D 16.1), flavonoid glyosides (D 23.3.3)

---

[a] In chloroplasts of many plants the enzyme is latent. This latency is achieved either by firm bind- ing of the enzyme to the membrane structure or by formation of an inactive complex of the ac- tive enzyme and a low molecular weight substance, probably a volatile acid.

[b] Excess amounts of carotenoids are accumulated in plastoglobuli which are found in chloroplasts as well as in chromoplasts. During synthesis of certain types of chromoplasts, tubular structures are formed from thylakoids which contain appreciable amounts of carotenoids. In other types myelin-like carotenoid-containing membrane convolutes are built. After massive accumulation of single carotenid species large membrane-covered crystals, e. g., of *β*-carotene and lycopene, may be detected.

[c] Often found in plastoglobuli.

[d] The phycobiliproteins in red and blue-green algae are aggregated to highly organized, distinctly shaped granules called phycobilisomes which are located on the outer side of the thylakoid mem- branes.

[e] Enzymes and lipophilic secondary products found in the nonparticulate fraction are probably lo- cated in the stroma. Hydrophilic secondary products may be located in the nonplasmic compart- ments, e. g., the intrathylakoidal space.

---

thesized in the cytoplasm (A 2.1). Hence, the enzymes have to penetrate the plas- tid envelope, which shows maximum permeability at early stages of greening. During this period the precursors of secondary products, such as oxaloacetic acid, succinic acid, or mevalonic acid, and even the CoA-esters of cinnamic acid derivatives may also be transported across the plastid envelope.

Several products which are known to be synthesized in plastids are eventually released into extraplastid compartments. Flavonoids or chlorogenic acid formed in chloroplasts are transported from these organelles to their storage sites in va- cuoles and the extracytoplasmic space. Quinolizidines synthesized in leaf chloro- plasts are translocated via the phloem to the roots and the maturating fruits.

In most cases the release of secondary products from the chloroplasts seems to be uncontrolled. In several instances, however, it is strictly regulated, as is the flow of the chloroplast-synthesized gibberellic acids to the cytoplasm, which is governed by phytochrome.

Evidence exists that plastids and other organelles may cooperate in the synthesis of certain secondary products.

Examples are:
— the formation of flavonoids in *Populus* gland cells, which takes place in invaginations of the plastid surface filled with cytoplasm containing ER cisternae (intraplastidal cytoplasmic areas) and
— the formation of furanocoumarins, the basic skeleton of which is built in the plastids, whereas the last steps proceed in the nonplastid part of the cells.

Nuclei and Mitochondria

Secondary products and enzymes of secondary metabolism are rare in these organelles. As yet only the following enzymes have been found in the membranes of mitochondria:
— the enzyme system cleaving the cholesterol side chain (D 6.6.5)
— steroid 11β-monooxygenase (D 6.6.5), and
— ACC oxidase (D 12.1).

References for Further Reading

267, 477, 482, 654

A 3.2 Long Distance Transport of Secondary Products in the Producer Organism

In multicellular organisms secondary products released from producing cells may be translocated to other sites where they are stored or degraded. In plants they may migrate with the xylem or phloem sap and in animals they may move with the blood or the lymph. Examples are:
— in aminals the secondary products acting as hormones (E 3.1)

Typical hormones are built in the specialized cells of a hormonal gland, are released from these cells, and are transported with the blood to the target cells, which may be located in quite another part of the body. In or at the target cells they interact with special receptors.

— in higher plants the secondary products formed in the roots, but accumulated in the shoots.

The alkaloids nicotine (D 16.1, in *Nicotiana* sp.) and L-hyoscyamine (D 17.2, in *Datura* sp.), for instance, are formed in the roots, excreted into the xylem sap and are stored in the leaves.

During transport the secondary products may be modified or degraded.

Examples are:
— the epoxidation of L-hyoscyamine (D 17.2) to L-scopolamine in stem tissue *Datura sp.* during the migration from the roots to the leaves and
— the hydroxylation of cholecalciferol (D 6.6.9) in the liver at C-25 and in the kidney at C-1 on the way from the skin where it is built to the target cells in the intestine and the kidney.

Reference for Further Reading

875

A 4 Expression and Control

A 4.1 Expression of Secondary Metabolism, a Feature of Cell Specialization

Biosynthesis and accumulation of most secondary products are limited to special cells and distinct developmental phases of the producer organisms.

Flower pigments (E 5.5.1), for instance, are formed during the development of the petals. Products responsible for the taste and smell of fruits (E 5.5.2) are synthesized during fruit maturation. Biosynthesis of alkaloids with pyridine and tropane moiety (D 16.1 and D 17.2) in several producer plants proceeds almost exclusively in root cells. Damascenine, an alkaloid formed in *Nigella damascena* (D 21.4.1), is accumulated during the maturation of special cells of the testa of the seeds. In animals specialized gland cells are formed at certain developmental stages which synthesize secondary products active as hormones or pheromones (E 3.1 and E 4) or used in defense (E 5.1).

With microorganisms it can easily be demonstrated that in many instances the formation of cells and their specialization are separated in time. Many microbial cultures pass from a phase of rapid cell division (growth phase, trophophase) to a phase of cell specialization (idiophase)[4]. During the idiophase chemical and morphological features of cell specialization are expressed, which are characteristics of the organism under investigation (idios = peculiar) (Fig. 6).

This kind of development may also occur in higher plants and animals. In these more complicated organisms the terms trophophase and idiophase, however, can be applied only to tissues or organs and not to the whole organism.

Microbial trophophase cells, meristematic cells of higher plants, and the cells of the early stages of embryo development in animals resemble each other in several respects. Their metabolism is well balanced and produces with optimum speed the compounds necessary for cell multiplication. Intracellular or extracellular signals, however, may cause cell division to slow down and already existing cells to specialize by the formation of new cytological characteristics and the synthesis of new compounds, e. g., secondary products.

The dependency of the expression of secondary metabolism on the development of the producer organism has been shown to be due to the phase-dependent formation of the enzymes synthesizing secondary compounds. Thus, the expression of secondary metabolism is the result of a differentiation process.

Cell differentiation is one of the basic features of metabolic regulation in all living organisms. It encompasses the processes by which cells become different from each other through expression of different parts of the same genetic material and has the following main aspects:
a) the triggering of differential gene expression by extracellular or intracellular factors; processes of signal reception and transformation
b) the mechanism of differential gene expression including
— gene activation and transcription of DNA into messenger RNA (mRNA) precursors,
— processing of mRNA precursors and transport of their products to the sites of protein synthesis,
— mRNA translation, and
— activation of biologically inactive proteinogens by enzymatic modification (proteinogen processing)
c) the differential breakdown of the products of gene expression
d) the coordination of individual steps of differential gene expression to differentiation programs, and

---

[4]   The terms "growthphase" and "idiophase" were coined by Bu'Lock 1965 [122].

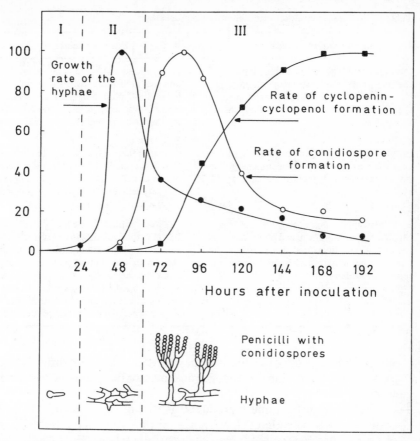

Fig. 6. Growth and cell specialization in surface cultures of *Penicillium cyclopium*
Cultures are grown on a nutrient solution containing 5% glucose, 0.12% NH_4^+, and 0.025% phosphate. Synchronization of development relative to the transition of the hyphal growth phase (trophophase) to the phase of cell specialization (idiophase) is brought about by removal of surplus nutrients, i.e., by replacement of the original culture broth 48 h after inoculation and then every 24 h by a nutrient solution containing only 20% of the original carbon and nitrogen sources and 2% of the phosphate content, respectively. Formation of the alkaloids cyclopenin and cyclopenol (D 21.4.2) as well as formation of the conidiospores begins after the growth rate has slowed down
I Germination phase; *II* trophophase; *III* idiophase

e) the coordination by intercellular interactions of the processes within individual cells (differentiation) to the developmental programs of tissues, organs, and organisms.

All partial processes offer multiple sites for specific positive or negative control by regulatory proteins and/or RNAs that are able to interact with different components of the gene expression chain. The activity of these proteins frequently depends on effectors which, as a kind of signal, carry information from inside or outside the cell to the sites of gene expression.

The proteins formed as the result of the manifold differentiation processes may be classified with respect to their biological significance for and function within the producing organism as given in Table 5. In accordance with this schedule:

— the expression of secondary product formation is an aspect of cell specializ-

Table 5. Proteins as products of differentiation processes

A. Proteins of primary metabolism	B. Specialization proteins
Proteins or their products that are typical for the cells of all or at least a large group of organisms, and that are of direct importance for the existence and reproduction of the cells themselves	Proteins that are found in only some organisms and that are frequently without direkt importance for the cells producing them. Formation and function of these proteins may have a vital function for the orgnism as a whole (E)
1. Enzymes Enzymes involved in gene expression, nucleic acid and amino acid metabolism, energy supply, synthesis and degradation of fatty acids, etc.	1. Enzymes Enzymes that enable the cell to perform special functions, e.g., *enzymes of secondary metabolism,* enzymes with a function in morphogenesis, special catabolic enzymes that degrade xenobiotics, special nutrients, etc.
2. Nonenzymatic proteins Structural proteins of the cell membrane and the protoplasm, ribosomal structural proteins, regulatory proteins, e.g., histones, acidic proteins of chromatin, etc.	2. Nonenzymatic proteins Proteins that enable the cell to perform special functions within the organisms, e.g., hemoglobin, muscle protein, storage proteins, cilial proteins and exoproteins, such as collagen blood plasma proteins, antibodies, milk proteins, and proteohormones, structural and regulatory proteins required for the specialized function of the cell, etc.

ation, and secondary metabolism may be defined as biosynthesis, transformation, and degradation of endogenous metabolic products by specialization proteins, and
— the formation of secondary products, as that of other features of cell specialization, usually is without significance for the producing cell itself, but may be of importance for the producer organism as a whole (E).

Cell division and cell specialization are not always mutually exclusive. There are instances in which specialized cells may be built de novo by so-called quantal mitoses, i.e., formation of daughter cells being determined to differ from their mother cells or from each other. These cells may already contain enzymes of secondary metabolism from the time point of their synthesis.

Conidiospores of *Penicillium cyclopium*, for instance, which are detached by the spore-forming cells, the phialides, contain the enzymes of benzodiazepine biosynthesis (D 21.4.2) as constitutive proteins (Fig. 7), whereas in the hyphae these enzymes are formed during maturation of already existing cells (Fig. 6). Thus, the enzymes must have been synthesized during conidiospore detachment. The mature conidiospores differ from the phialides in several respects, e.g., by their spherical form and their thick, rigid cell wall, by the typical spore pigment (a kind of melanin), and by the existence of the enzyme cyclopenase, which is able to transform the benzodiazepine alkaloids into quinoline derivatives (D 21.4.2). Conidiospore formation is, therefore, a typical example of a quantal cell division cycle.

References for Further Reading

119, 477, 480, 485, 573, 816

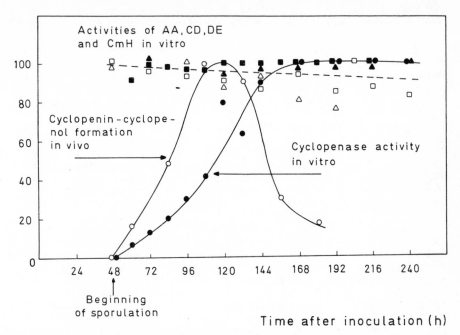

Fig. 7. In vitro activities of the enzymes of cyclopenin-viridicatin biosynthesis and rates of the formation of the alkaloids cyclopenin and cyclopenol in vivo during ripening of the conidiospores of *Penicillium cyclopium*

The enzymes of cyclopenin-cyclopenol biosynthesis (D 21.4.2) anthranilic acid adenylyltransferase (AA △), cyclopeptine dehydrogenase (CD ■), dehydrocyclopeptine epoxidase (DE □), and cyclopenin m-hydroxylase (CmH ▲) are already measurable in young, immature conidiospores. They are constitutive proteins of the spores. In contrast, the activity of cyclopenase (●), the enzyme transforming cyclopenin-cyclopenol to viridicatin-viridicatol, is not measurable in the young spores and increases during spore maturation

A 4.2 Regulation of Enzyme Amount

The basic regulatory principle in the expression of secondary product formation is the phase-dependent synthesis of secondary metabolic enzymes. Table 6 presents a selection of experimental results demonstrating that in microorganisms, higher plants, and animals the enzymatic activities of secondary metabolism appear before or during the phase of secondary product synthesis, and that expression depends on the de novo synthesis of RNA and proteins. Among the large number of methods which show the formation of RNA and proteins during the expression of new characteristics of cell differentiation, the following may be applied: (a) comparison of the rate of secondary product synthesis in vivo with the activities of the relevant enzymes in vitro, (b) application of inhibitors of gene expression, (c) radioactive labeling or immunological determination of the newly synthesized enzymes, and (d) measurement of mRNA accumulation and translation in vitro.

The beginning of secondary product formation is often directly coupled with the synthesis of the enzymes involved. One example is the biosynthesis of flavonoids in cell cultures of *Petroselinum hortense* which is triggered by irradi-

Table 6. Experimental systems in which differential gene expression in secondary metabolism has been examined

Group of secondary products	Substances whose synthesis has been investigated	Producer organism	Methods applied
Polyketides	6-Methylsalicylic acid, patulin (D 5.3.2)	*Penicillium patulum*	A, B
	Orsellinic acid (D 5.3.2)	*Aspergillus fumigatus*	B
Triterpenes	Corticosteroids (D 6.6.5.)	Mammals	B
Tetraterpenes	Carotenoids, trisporic acids (D 6.7)	*Blakeslea trispora*	B
Peptides	Peptide antibiotics (D 9.2)	*Bacillus* sp.	A, B
Amino acid derivatives	Acetylcholine (D 10)	Neoblastoma tumor cells	A
	Ethylene (D 12.1)	*Pisum sativum*	B
	Dipicolinic acid (D 18)	*Bacillus megaterium*	A, B
	Urea (D 19)	*Rana catesbeiana*	A
Anthranilic acid derivatives	Actinocin (D 21.4.1)	*Streptomyces antibioticus*	B
	Benzodiazepine alkaloids (D 21.4.2)	*Penicillium cyclopium*	A, B
Indole derivatives	Tryptamine (D 22.1)	*Catharanthus roseus* cell cultures	A, C, D
	Ergoline alkaloids (D 22.2)	*Claviceps* sp.	A
Phenylpropanoids	Hordenine (D 23.1.1)	*Hordeum vulgare*	A, B
	Betalains (D 23.1.4)	*Amaranthus, Celosia*	B
	Cinnamic acids and flavonoids D 23.2.1 and 23.3.3)	*Petroselinum hortense* cell cultures	A, B, C, D
		Sinapis alba seedlings	A, B, C
	Styrylpyrones (D 23.3.1)	*Polyporus hispidus*	A, B
	Polyphenols in pupal cuticle (D 23.2.5)	*Calliphora erythrozephala*	A, B, C, D, E

[a] Determination in vitro of an activity increase of enzymes of secondary metabolism before or at the beginning of the phase of secondary product formation.

[b] Determination of the influence of inhibitors of gene expression on the formation of secondary products or on the formation of enzymes involved in secondary metabolism.

[c] Isotope-labeling of enzymes catalyzing secondary product formation for the demonstration of their de novo formation at certain developmental phases of the producer organism.

[d] Translation of mRNA in vitro to show the phase-dependent presence of mRNA species coding for secondary metabolic enzymes.

[e] Determination of the amount of enzymes by immunological methods to demonstrate their de novo formation at specific developmental stages.

ation (Fig. 8). In this system the regulatory mechanisms have been extensively investigated with phenylalanine ammonia-lyase (PAL) (D 23.2.1) and chalcone synthase (D 23.3.3), the key enzymes of the biosynthetic chain. Determination of the amounts of the respective mRNAs by hybridization with the complementary cDNAs and the rates of enzyme biosynthesis by incorporation of labeled amino acids followed by immunoprecipitation of the enzyme proteins has demonstrated that the amounts of PAL-mRNA and chalcone synthase-mRNA increased rapidly after the beginning of irradiation, and that the rates of PAL and chalcone

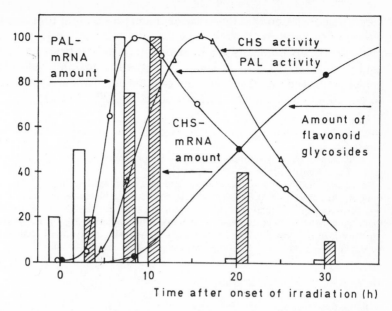

Fig. 8. Time course of events during the UV light-induced flavonoid accumulation in cell cultures of *Petroselinum hortense*
With respect to lag phases and activity changes the enzymes involved form two groups with coordinated regulation. Group *I*, which includes PAL, comprises enzymes of general cinnamic acid metabolism (D 23.2.1). Group *II*, which includes chalcone synthase (CHS), catalizes the flavone glycoside pathway (D 23.3.3)

synthase synthesis corresponded to the amount of the mRNA species. This clearly shows that in the expression of flavonoid metabolism in *Petroselinum hortense* cell cultures, as in the secondary metabolism of most other organisms, the control of transcription is the basic regulatory principle.

It is, however, not the only mechanism regulating enzyme amount. Inhibitor experiments and examination of the mRNA population present at developmental stages preceding that of secondary product formation by in vitro translation, in several instances, revealed the existence of "stable" mRNA species which are phase-dependently translated. In addition, the processing of proenzymes (enzymogens) offers a further mechanism of posttranscriptional control. In both cases the increase of enzyme activity is not directly coupled with mRNA biosynthesis.

An example of enzymogen processing is the appearance and increase of cyclopenase activity (D 21.4.2) during conidiospore maturation in *Penicillium cyclopium*. Neither process can be prevented by cycloheximide (D 5.3.8), an inhibitor of translation, in concentrations which suppress protein biosynthesis. This indicates that a proenzyme exists which is formed during an earlier stage of development. The mechanism of transformation of the cyclopenase proenzyme into active cyclopenase is still unclear. In similar cases, however, e. g., the activation of prochitin synthase in *Saccharomyces cerevisiae* and of a phenol oxidase zymogen (C 2.3.1) in the hemolymph of *Calliphora* larvae, the active enzyme is formed by proteolytic processing of a proteinogen.

In some organisms the amount of enzymes in secondary metabolism is determined not only by synthesis, but also by degradation. Again this was examined in detail with phenylalanine ammonia-lyase (PAL, D 23.2.1). Labeling experiments demonstrated that this enzyme undergoes a rapid turnover in *Petroselinum hortense* cell cultures. During induction the balance between synthesis and degradation is disturbed by an increased rate of synthesis of PAL-mRNA (see above). However, as early as about 17 h after the beginning of the induction, degradation exceeds synthesis and the enzyme activity starts to decrease, returning within about 80 h almost to the preinduction level (Fig. 8).

The reason for the decrease of PAL-mRNA biosynthesis in cell cultures of *Petroselinum hortense* is unknown. Probably product repression is involved. With seedlings of *Cucumis sativus* it has been shown that
a) *E*-cinnamic acid, the product of PAL action on L-phenylalanine (D 23.2.1), represses PAL synthesis if exogenously supplied (Table 7) and
b) the application of 2-aminooxyacetic acid, an L-phenylalanine analog, which is a powerful inhibitor of PAL, prolongs PAL activity increase and causes a higher maximum of PAL activity.
This indicates that repression by accumulating products may play a regulatory role in the cinnamic acid metabolism of *Cucumis*. In other organisms, however, e.g., in *Fagopyrum, esculentum*, this regulatory principle could not be substantiated. Here, PAL activity after illumination showed a similar time course with and without the addition of 2-aminooxyacetic acid.

References for Further Reading

314, 863

A 4.3 Regulation of Enzyme Activity

In addition to the regulation of enzyme amount, secondary metabolism may also be controlled by regulation of the activity of the enzymes involved. Evidence for this comes from the large discrepancies usually found between the relatively high enzyme activities measurable in vitro and the much lower activities in the living cell.

In some organisms the low in vivo activity has been shown to be due to limited substrate supply. One example is the low activity of cyclopenase in cultures of *Penicillium cyclopium*.

Cyclopenase transforms the benzodiazepine alkaloids cyclopenin and cyclopenol to the quinoline derivatives viridicatin and viridicatol (D 21.4.2). It is located at the inner side of the plasma membrane of the mature conidiospores. However, in spite of high activities measurable in vitro the enzyme remains inactive in vivo. Hence, in contrast to the formation and excretion of large amounts of cyclopenin and cyclopenol, the synthesis of viridicatin and viridicatol in cultures of *Penicillium cyclopium* is small. Artificial increase of membrane permeability by lowering the cellular ATP level greatly increases the rate of viridicatin-viridicatol biosynthesis because cyclopenase has now access to its substrates inside and outside the spores.

In most cases, however, the regulation of the in vivo activity of secondary metabolic enzymes is more complicated. In the hyphae of *Penicillium cyclopium* the activities of the enzymes of benzodiazepine alkaloid biosynthesis (D 21.4.2) are measurable at a relatively early stage of the mold's development, whereas alkaloid formation is expressed later (Fig. 9). The increase in the rate of biosynthesis is immediately stopped by inhibitors of gene expression, e. g., 5-fluorouracil or cycloheximide (D 5.3.8), even at developmental stages at which the in vitro measurable enzyme activities have already reached their maximum levels. This indi-

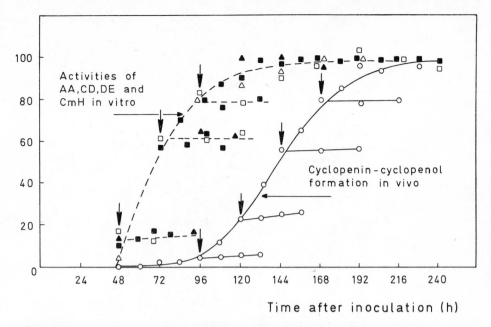

Fig. 9. Time course of the activities of enzymes of alkaloid biosynthesis and of the rates of alkaloid formation in hyphae of *Penicillium cyclopium*
The enzymes of alkaloid biosynthesis are induced in a coordinated manner by an unknown internal signal during development of the mold. However, the rates of alkaloid production increase later than the enzyme activities. The increase of both processes can be stopped by addition of inhibitors of gene expression, e. g., 6-fluorouracil or cycloheximide (cf. *arrows*), indicating that alkaloid production is controlled by the formation of a protein which is not an enzyme of alkaloid biosynthesis. Ordinate relative values: *AA* anthranilic acid adenylyltransferase (△); *CD* cyclopeptine dehydrogenase (■); *DE* dehydrocyclopeptine epoxidase (□); *CmH* cyclopenin m-hydroxylase (▲) (D 21.4.2)

cates the existence of a protein limiting the rate of alkaloid formation, which is synthesized later than the investigated enzymes.

One might speculate that intracellular channeling of precursors or cosubstrates to the site of alkaloid biosynthesis requires a specific protein which limits the production rate. In accordance with this assumption it has been shown that different pools of L-phenylalanine exist in the cytoplasm and the vacuoles of the hyphae of *Penicillium cyclopium* from which the amino acid is transported to the sites of protein and alkaloid biosynthesis. Channeling to and between these pools causes pronounced changes in the incorporation rates of L-phenylalanine into protein and alkaloids under different developmental and experimental conditions (Fig. 10).

L-Phenylalanine in the cytoplasm serves protein and alkaloid biosynthesis. Excess L-phenylalanine is accumulated in the vacuoles and can be reused later on in alkaloid formation, but not in protein biosynthesis. Hence, with respect to alkaloid formation, there are two channels for L-phenylalanine: a direct, low capacity pathway via the peripheral pool (primary labeling of alkaloids) and an indirect pathway with higher capacity from the expandable pool (secondary labeling of the alkaloids). The relative contributions of these two channels vary with the concentration of L-phenylalanine, the time of incubation, etc. Under all experimental conditions, however, in contrast to protein biosynthe-

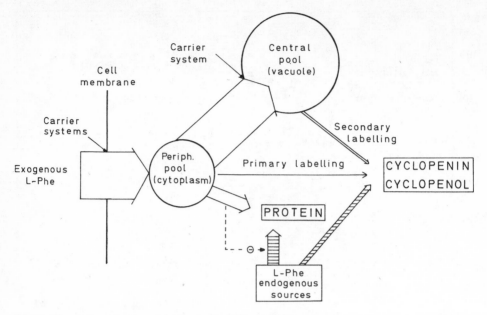

Fig. 10. Channeling of L-phenylalanine in *Penicillium cyclopium*
In the hyphal cells of *P. cyclopium* two pools exist for L-phenylalanine: a low capacity, "peripheral" pool (the cytoplasm), and a "central", expandable pool (the vacuoles). L-Phenylalanine is transported by several carrier systems through the plasma membrane and by an active transport system through the vacuole membrane. The width of the *arrows* gives an estimate of the capacity of the respective channels; - - ⊝ → negative regulatory effect

sis, about 90 % of the L-phenylalanine incorporated into the alkaloids is recruited from endogenous sources, i. e., de novo synthesis and protein degradation.

The amount of L-phenylalanine incorporated into the benzodiazepine alkaloids (D 21.4.2) and the proteins depends on the developmental stage of *P. cyclopium*. It is maximum for the proteins in the growth phase (Fig. 6) (2.7 pmol/cm^2 culture area · s, 48 h after inoculation), but zero with respect to the alkaloids. In the phase of maximum alkaloid formation, i. e., about 7–10 days after inoculation, it is only 0.4 pmol/cm^2 · s relative to the proteins, but about 10 (!) pmol/cm^2 culture area · s relative to the alkaloids. Figure 10 shows the L-phenylalanine fluxes 72 h after inoculation, i. e., at an intermediate state.

Enzyme activity in secondary metabolism may be influenced not only by the accessibility of precursors, but also by the accumulation of products (product inhibition by allosteric interaction or competition with the substrates at the binding sites). Table 7 shows results of in vitro experiments which demonstrate that product inhibition is a property of many secondary metabolic enzymes. In living cells, however, enzymes and secondary products usually are separated by compartmentalization (A 3.1). Thus, product inhibition occurs in vivo less frequently than in vitro.

One example where product inhibition obviously plays a role in living cells is ergoline alkaloid formation in *Claviceps* (D 22.2). In mycelial mats of *Claviceps*, developing in stationary batch cultures, the intracellular concentration of ergoline alkaloids and the activity of dimethylallyl diphosphate: L-tryptophan dimethylallyl transferase (DMAT synthase), measurable in homogenates, undergo fluctuations. Enzyme activity tends to decrease when the intracellular alkaloid concentration is at a maximum. This oscillation may be explained by

Table 7. Product inhibition and repression in secondary metabolism

Group of secondary products	Ezyme investigated	Inhibitors/repressors	Action I = Inhibition of activity R = Repression of synthesis
L-Ornithine derivatives	Ornithine decarboxylase, putrescine methyltransferase, methylputrescine oxidase (D 17)	Nicotine	R
Anthranilic acid derivatives	Dehydrocyclopeptide epoxidase (D 21.4.2)	Cyclopenin, cyclopenol, viridicatin, viridicatol	I
L-Tryptophan derivatives	Aromatic-L-amino acid decarboxylase (D 22.1)	N,N-Dimethyltryptamine	I
	Dimethylallyl diphosphate: L-tryptophan dimethylallyltransferase, chanoclavine-I-cyclase (D 22.2)	Agroclavine, elymoclavine, lysergic acid	I. R?
Phenylpropanoids	Phenylalanine ammonialyase (D 23.2.1)	Cinnamic acids	I. R
		Benzoic acid, flavonoids	I
	Cinnamate 4-monooxygenase (D 23.2.1)	p-Coumaric acid	I. R
	Caffeate O-methyltransferase (D 23.2.1)	5-Hydroxyferulic acid	I
	Caffeate CoA ligase (D 23.2.1)	p-Coumaric acid and caffeic acid esters of quinic acid	I
	Ferulate CoA ligase (D 23.2.1)	Naringenin	I
	Chalcone synthase (D 23.3.3)	Naringenin	I

a) accumulation of the alkaloids at periods of high enzyme activity where diffusion into the medium is limiting,

b) reversible inhibition of the enzymes by the accumulating alkaloids, in accordance with the inhibition in vitro, for instance, of DMAT synthase, the first enzyme of ergoline biosynthesis, by certain ergoline alkaloids (Table 7), and

c) decrease in the stored amount of alkaloids by release into the medium and renewed increase in enzyme activity and alkaloid biosynthesis.

References for Further Reading

238, 477, 502

A 4.4 Integration in the Programs of Differentiation and Development

The phase-dependent expression of secondary metabolism was designated in A 4.1 as one of the most important features in the control of secondary product formation. It is caused by the integration of secondary metabolism in the pro-

grams of differentiation and development. These programs, which are characteristic of all living cells

— coordinate secondary metabolism with other biochemical and morphological
  activities of the producer organism
— coordinate the different aspects of secondary product formation, e.g., synthesis of secondary metabolic enzymes, formation of the facilities necessary for
  channeling precursors and intermediates to the sites of these enzymes, and
  formation of the means necessary for release or storage of secondary products
— account for the phases of susceptibility to signals which influence secondary
  metabolism, and
— originate the determination of secondary product formation by signals which
  do not act directly on secondary metabolism.

Tightly linked with the realization of differentiation programs are the pheno-
mena of (a) competence, i.e., the stage-restricted capability of cells to respond to
particular signals by expression of distinct parts of their genetic material and (b)
determination, i.e., the commitment of cells to certain differentiation processes
or programs by prior changes of their chemical composition.

The main features of differentiation programs are:

— interrelation of the processes of differential gene expression through the action of regulatory RNAs, proteins, or effectors, which are themselves formed
  by prior processes of gene expression. The regulation may be positive or negative and may concern any of the partial processes of differential gene expression given in A 4.1
— formation of effectors either inside the cell which modify the activity of regulatory proteins and trigger sequences of differential gene expression, or outside the cell for intercellular coordination of differentiation
— multiple regulation of individual gene groups. The same regulatory RNAs,
  proteins, or effectors may exert positive or negative actions on one or several
  gene groups
— dependence on the presence of corresponding receptor proteins, i.e., on the
  cell's capacity to respond to the action of extracellular effectors
— independent operation of subprograms and regulatory interaction with other
  subprograms of the same cell or via extracellular effectors of other cells.

Except for a few simple examples, e.g., the programmed expression of viral
genetic material in bacterial host cells, the molecular organization of differentiation programs is practically unknown. This is especially true with respect to the
intercellular linkages between differentiation processes of individual cells, which
lead to the developmental programs of tissues, organs, and organisms. Present
knowledge is limited to a description of the sequence of events within differentiation programs and to a preliminary analysis of the regulatory interactions
based on observations of the influence of environmental factors or of mutations
on the programs.

An example in which integration of secondary product formation in a differentiation program has
been extensively investigated is bacterial sporulation. During formation of bacterial endospores, e.g.,
in *Bacillus sp.*, a complex sequence of morphological and biochemical characteristics is expressed
which involves at least 30 independently regulated gene groups with more than 100 genes. In addi-

Table 8. Morphological and biochemical events associated with different stages of sporulation in *Baciullus sp.*

Time scale of sporulation (h)	Stage	Morphological characteristics, formation of:	Chemical characteristics, expression of:
0	0	Vegetative cell	

| 1.5 | I | Chromatine filament | Exoprotease and ribonuclease activity
Rapid protein turnover |

| 2.5 | II | Spore septum | Alanine dehydrogenase activity
Uptake of β-alanine and panthotenic acid |

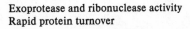

Formation of peptide antibiotics (D 9.2)

| 4.5 | III | Spore protoplast | Biosynthesis of a 70 000 dalton RNA binding protein
Alkaline phosphatase, glucose dehydrogenase and heat-resistant catalase activities |

| 6 | IV | Spore cortex | Uptake of Ca^{2+}
Ribosidase and adenosine deaminasae activities |

Formation of sulfolactic and dipicolinic acid (D 18)

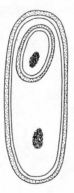

| 7 | V | Spore coat | |

Formation of a brown pigment of unknown structure

Cysteine incorporation and octanol resistance

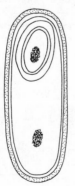

| 8 | VI | Mature spore | Alanine racemase activity and heat resistance |

---

tion to many metabolic activities and morphological characters, formation of secondary products (peptide antibiotics, dipicolinic acid, sulfolactic acid, and brown pigments of unidentified structure) is expressed. The temporal sequence of events is strictly fixed and the process of sporulation can be divided into six stages (Table 8).

The mutual regulatory interactions of sporulation processes have been investigated using mutants and gene expression inhibitors. Interruption of spore formation at certain stages prevents the expression of the cytological and biochemical characteristics of subsequent stages. This demonstrates that sporulation comprises a programmed sequence of dependent processes. Some steps are necessary for the discharge of the program, and others are not. Secondary product formation, for instance, may be eliminated by mutation without impairing the expression of the subsequent steps of sporulation. It thus appears that at least in the artificial milieu of the laboratory none of the secondary products shown in Table 8 is essential to spore formation.

In the following sections some general principles are discussed, which show the interrelations between the control of secondary metabolism and other metabolic activities of the producer organisms.

References for Further Reading

477, 485

A 4.4.1 The Influence of Nutrients

Suppression of secondary product formation by excess nutrients, especially by glucose and other easily degradable carbon sources, but also by nitrogen-containing compounds and phosphate, is a general phenomenon in microbial cultures. Suppression by excess nutrients has been found in the biosynthesis of certain antibiotics, e. g., streptomycin and neomycin C (D 1.3), bacitracin A and enniatin B (D 9.2), penicillins and cephalosporins (D 9.2.3), chloramphenicol (D 21.1), actinomycins (D 21.4.1), as well as in the formation of polyketides (D 5.3), gibberellins (D 6.4), and alkaloids, e. g., benzodiazepines (D 21.4.2), and ergolines (D 22.2).

Usually the suppression of secondary product formation is accompanied by the suppression of other characteristics of cell specialization (such as conidiospore formation in *Penicillium cyclopium*), indicating a general influence of nutrient supply on cell specialization.

The mechanisms by which nutrients control secondary metabolism are unknown in most cases. However, it has been demonstrated that D-glucose and other rapidly used carbon sources repress, for instance, the formation of *N*-acetylkanamycin amidohydrolase, which is thought to be the final enzyme in kanamycin A biosynthesis (D 1.3) and phenoxazinone synthase, an enzyme required in actinomycin biosynthesis (D 21.4.1).

The report that cAMP relieves glucose repression of *N*-acetylkanamycin amidohydrolase in *Streptomyces kanamyceticus*, a prokaryote (!), indicates that the repression mechanism resembles that of different catabolic enzymes in bacteria, which proceeds via the inhibition of adenylate cyclase, the enzyme that converts ATP to cAMP (D 25.1). As a consequence, the concentration of cAMP decreases and the transcription by RNA polymerase of operons subjected to cAMP control is inhibited (catabolite repression). In eukaryotes, however, catabolite repression could not be demonstrated. In *Penicillium cyclopium*, for instance, glucose suppression of benzodiazepine alkaloid biosynthesis cannot be overcome by administration of cAMP or cAMP derivatives.

It is of interest that glucose suppression of alkaloid metabolism in *P. cyclopium* (D 21.4.2) is overcome by the morphological organization of the mycelium. In mycelial mats, growing at the surface of solid or liquid media, i.e., under natural growth conditions, cell specialization, as indicated by alkaloid production and conidiospore formation, is not suppressed by high glucose concentrations. This is because only one side of the mat has contact with the medium and the compact cell layer slows the permeation of nutrients to the other cells.

In higher plants and animals the influence of nutrients on secondary metabolism and other features of cell specialization is of minor significance. However, the response to nutrients of in vitro cultivated plant and animal cells, i.e., cells grown under conditions which cause a much lower level of morphological organization, resembles that of microbial cells. This indicates that it is the morphological organization of plants and animals which overcomes the nutrient control on cell specialization, including expression of secondary metabolism.

References for Further Reading

7, 184, 260, 873, 874

A 4.4.2 The Action of Signals

A large number of chemical and physical signals mediate the interaction of cells with their surrounding, i.e., the adaptation of cells to the metabolic and functional requirements imposed by the milieu or by adjacent cells. Many of the signals influencing secondary metabolism are by themselves secondary products (Table 9). They may be divided into substrate-like and nonsubstrate-like signals.

Substrate-like signals have a direct relation to the biological function of the proteins whose synthesis they act upon. They include substrates or products of enzymes or enzymatic chains, prosthetic groups of proteins, etc. Substrate-like signals generally influence the synthesis of a restricted number

Table 9. Secondary products acting as signals in the expression of secondary metabolic pathways

Signal	Action S = substratelike N = nonsubstrate-like	Secondary products whose synthesis is influenced
Glucose heptamer elicitor (D 1.4.1)	N	Phytoalexins (different groups of secondary products and enzymes, E 5.4)
β-Ionone (D 6.3)	S	Carotenoids (D 6.7)
Gibberellic acid (D 6.4)	N	Betalains (D 23.1.4), cinnamic acids (D 23.2.1), flavonoids (D 23.3.3)
Trisporic acids (D 6.7)	S	Trisporic acids (D 6.7)
Peptides (D. 9.2)	N	Hormones (different groups of secondary products in animals and human beings, E. 3.1)
Ethylene (D 12.1)	N	Cinnamic acids (D 23.2.1), flavonoids (D 23.3.3.)
Indole-3-acetic acids (D 22)	N	Anthraquinones (D 5.3.6), ethylene (D 12.1), cinnamic acids (D 23.2.1), flavonoids (D 23.3.3)
Cinnamic acids (D 23.2.1)	S	Cinnamic acids (D 23.2.1), flavonoids (D 23.3.3)
Phytochrome (D 24.2)	N	Carotenoids (D 6.7), cinnamic acids (D 23.2.1), flavonoids (D 23.3.3)
Cytokinins (D 25.1)	N	Cardiac glycosides (D 6.6.8)

of proteins that have the same biological functions in very different organisms, though their detailed modes of action may differ from one organism to the other.

Nonsubstratelike signals have no direct relationship to the biological functions of the proteins whose synthesis they control. They include the light-phytochrome system in plants (D 24.2), cAMP (D 25.1), the plant and animal hormones (E 3.1), and similar systems. The particular effect of these substances is often restricted to closely related groups of organisms and, within one organism, to a limited number of cells, the so-called target cells. Depending on the stage of differentiation, cells of different tissues may respond to the same nonsubstratelike effector with totally different changes in the pattern of protein synthesis.

The classification into substratelike and nonsubstratelike signals is, however, not a strict one and does not imply a principal difference in the mode of action.

In most cases the signals influence differential gene expression. Their molecular mode of action, however, is still unknown. A detailed analysis is complicated due to the following facts:

a) in nearly all instances the actual regulatory effector governing the expression of the genes in question is unknown. The signals present in the extracellular milieu either are chemically transformed to yield the intracellular effector or they are only the first link in a signal transformation chain causing an increase or decrease in the cellular concentration of the actual effector, which reacts with the respective promotor region at the DNA. The latter situation is found with some animal hormones, with signals effective in C- and N-catabolite repression in bacteria or with the action of light on the reversible transformation of phytochrome (D 24.2) in plants, and

b) the signals frequently influence a multiplicity of gene expressions, particularly those acting on differentiation programs.

Only a few substrate-like signals seem to influence secondary metabolism directly.

One example is probably cinnamic acid (D 23.2.1), which represses the formation of phenylalanine ammonia-lyase (PAL) and induces hydroxycinnamoyl CoA: quinate hydroxycinnamoyltransferase, an enzyme of chlorogenic acid biosynthesis.

Most signals show indirect action.

The plant hormone gibberellic acid (D 6.4; a nonsubstrate-like signal), for instance, reduces the biosynthesis of amaranthine, a betalain (D 23.1.4), by limitation of precursor supply, an effect normalized by the administration of L-tyrosine or L-DOPA.

Furthermore, the plant hormones and most microbial signals affect not only secondary metabolism but also a wide variety of other metabolic activities.

Usually the response of the target cells depends on their state of differentiation (competence). Since secondary metabolism is integrated into differentiation programs, the phase during which secondary metabolism may be influenced is not necessarily linked directly to the stage of secondary product formation.

The following microbial signal substances, for instance, are active only if added at the beginning of the growth phase, i.e., at a developmental stage before secondary product formation is actually expressed (determination phase):

— the A-factor, which triggers streptomycin formation (D 1.3) in *Actinomyces streptomycini*
— L-phenylalanine and the P-factor, which increase benzodiazepine alkaloid biosynthesis (D 21.4.2) in *Penicillium cyclopium*
— L-tryptophan, which influences ergoline alkaloid biosynthesis (D 22.2) in *Claviceps* and
— L-methionine and L-norleucine, which stimulate cephalosporin formation (D 9.2.3) in *Cephalosporium acremonium*.

Phases of susceptibility, in which secondary metabolism is influenced by appropriate signals, have also been detected in higher plants. The induction of anthocyanin biosynthesis (D 23.3.3) by phytochrome in the epidermal cells of the cotyledons of *Sinapis alba*, for instance, begins only 27 h after sowing, despite the fact that the phytochrome system per se is fully functional immediately after sowing. Competence with respect to anthocyanin formation is determined endogenously, i.e., by the state of differentiation of the epidermis cells in which the anthocyanins are formed and accumulated.

References for Further Reading

477, 573

A 4.4.3 Coordinated Enzyme Expression

Frequently the enzymes participating in a secondary pathway are formed in a coordinated manner leading to constant proportions of enzyme activities (cf. Figs. 8 and 9). Genetic and biochemical experiments in bacteria have shown that the genetic material of such pathways may be grouped into regulatory units (operons, regulons). In eukaryotic cells, however, the genetic and detailed biochemical evidence for the coordinated expression of gene groups is scarce (A 2.1).

The coordinated expression of enzymes belonging to secondary pathways has been examined in detail in the flavonoid metabolism of higher plants.

In cell cultures of *Petroselinum hortense* after light induction the enzymes of flavonoid biosynthesis fall into two groups according to their in vitro measurable activities. Group I comprises the enzymes of general cinnamic acid metabolism: phenylalanine ammonia-lyase (PAL), cinnamate 4-monooxygenase, and p-coumarate CoA ligase (D 23.2.1), which produce the precursors of different groups of secondary phenylpropanoids. Group II is formed from the enzymes of the special pathway leading from the cinnamic acid derivatives to the flavonoids themselves, e.g., chalcone synthase, chalcone isomerase, flavanone oxidase, malonyl, glucosyl, and apiosyltransferases (D 23.3.3). Group I enzymes and group II enzymes differ with respect to their lag phases and the time points of maximal activity (Fig. 8).

In the plant itself, however, another pattern of control is established. In cotyledons and leaves of *Petroselinum hortense* seedlings the regulatory behavior of PAL and several group II enzymes indicate interdependence. Furthermore, PAL is induced by light, whereas p-coumarate CoA ligase, another group I enzyme, is not. These results reflect the flexibility in the integration of genes into groups with coordinated expression in the eukaryotic cell.

References for Further Reading (see A 4.4.2)

A 4.4.4 Sequential Gene Expression

The step-by-step induction of enzymes which belong to one metabolic pathway by the products of preceding enzymes (sequential gene expression) is a characteristic of catabolic pathways in bacteria. Although sequential formation of secondary products and secondary metabolic enzymes in many plants and microorganisms has been observed, the direct influence of secondary products on the expression of enzymes catalyzing their further transformation has been demonstrated in very few organisms only.

An example is probably the light-induced chlorogenic acid biosynthesis (D 23.2.1) in *Solanum tuberosum*. In potato tuber disks radiation causes an increase in phenylalanine ammonia-lyase (PAL) activity, followed by an increase in hydroxycinnamoyl CoA: quinate hydroxycinnamoyltransferase (HQT) activity and the accumulation of chlorogenic acid. Since the feeding of cinnamic acid to potatoes increases HQT activity, the cinnamic acid produced endogenously by PAL may also induce the enzymes of chlorogenic acid biosynthesis. In *Fagopyron esculentum*, however, indications for such a feedforward control of chlorogenic acid biosynthesis could not be demonstrated.

References for Further Reading (see A 4.4.2)

A 5 Transformation and Degradation of Secondary Products

For a long time metabolic stability was thought to be a characteristic of secondary products. Recent experiments, however, have demonstrated that many secondary substances are transformed or are even degraded to compounds of primary metabolism. Three types of secondary compounds may be distinguished with respect to their fate in metabolism: (a) truly metabolically inert end products, (b) products stable only at a given physiological or developmental state, and (c) substances undergoing a continuous turnover.

A 5.1 Pathways

Many secondary products are metabolic intermediates. Of importance in further transformation are:

a) modifications, e.g., by substitution, reduction, oxidation, etc., cf.

- the epoxidation of dehydrocylopeptine to cyclopenin (D 21.4.2),
- the methylation of serotonin to 5-hydroxytryptamine (D 22.1), and
- the hydroxylation of p-coumaric acid to caffeic acid (D 23.2.1)

b) transformation to other types of secondary products, cf.

- the rearrangement of the benzodiazepine alkaloid cyclopenin to the quinoline derivative viridicatin (D 21.4.2),
- the cleavage of the porphyrin ring system with the formation of open-chain tetrapyrrols (D 24.2), and
- the incorporation of cinnamoyl CoA esters into flavonoids (D 23.3.3)

c) degradation to primary metabolic substances, e.g., formic, acetic, propionic, pyruvic, and succinic acids, and finally to CO_2, cf.

- the cleavage of polyhydroxybutyric acid to acetyl CoA (D 5.1),
- the splitting of kynurenic acid to 2-oxoglutaric acid and L-aspartic acid (D 22.5), and
- the degradation of flavonoids to CO_2, acetic acid, and succinic acid (D 23.3.3).

In several instances polymeric substances arise during the transformation of secondary products, such as polymeric carbohydrates (D 1.4.1), humic acidelike polyphenols (D 5.3.2), sporopollenins (D 6.7), rubber (D 6.8), muramin (D 9.2.4), polymeric products derived from 3-hydroxyanthranilic acid (D 21.4.1), melanins (D 23.1.3), and lignins (D 23.2.3). Many of these compounds are formed by oxidative polymerization catalyzed by phenol oxidases (C 2.3.1) and peroxidases (C 2.4). The periplasmic space, e.g., the cell wall of plants, is the preferred site of the formation of these macromolecules.

Degradation of secondary products to primary metabolic substances usually involves metabolic enzymes with relatively low substrate specificity, like demethylating monooxygenases (C 2.6.4) and glucosidases. However, there are also highly specific enzymes like dioxygenases cleaving aromatic rings (C 2.5) and enzymes splitting aliphatic carbon chains (cf. D 5.1 and D 22).

References for Further Reading (see A 5.3)

A 5.2 Regulation

Transformation and degradation are subject to the same principles of control as biosynthesis (regulation by enzyme amount, enzyme activity, compartmentalization, etc.). Expression of the reactions transforming and degrading secondary products is either part of the same differentiation programs which also govern biosynthesis, or is integrated in other programs. If the enzymes catalyzing biosynthesis and transformation/degradation are expressed within the same program, they appear in the same cells and at more or less the same time. As a result the products formed may undergo a continuous turnover, even if there is no change in the absolute amount. This can be shown by inhibition of biosynthesis or degradation as well as by labeling of the secondary products with radioactive precursor in pulse-chase experiments (B 3.3, Fig. 11).

The biological half-life of secondary products which are simultaneously synthesized and transformed and/or degraded is controlled by the activity of the enzymes involved which may depend on many internal or external factors. It varies greatly from a few minutes to a couple of months in different organisms and may also be variable in the same organism (Table 10). Even the half-lives of closely related substances may differ greatly depending on their location and derivation. For several substances more than one pool with different turnover rates exists in the same organism.

The accumulation of products undergoing turnover depends on the rates of synthesis and of transformation/degradation. These may be regulated independently of each other. Since many endogenous and exogenous factors affect synthesis and degradation, the amount of secondary products stored may vary with the developmental state, the season, the climate, and even the time of the day.

An example is the concentration of the isoquinoline alkaloids morphine, codeine, and thebaine (D 23.1.2) in the latex of *Papaver somniferum*. The concentration of morphine peaks in the morning, that of codeine, at noon, and the concentration of thebaine gradually increases from 6 a.m. until the late evening.

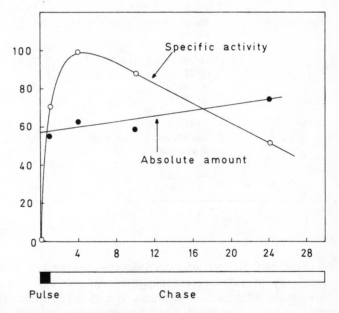

Fig. 11. Turnover of chlorogenic acid in leaf disks of *Xanthium pennsylvanicum*
[U- ^{14}C]-L-Phenylalanine has been applied to the upper surface of leaf disks for 1 h (pulse). Then the disks were rinsed with water and placed on moist filter paper exposed to light (chase). At the time points indicated the absolute amount and the specific activity of chlorogenic acid were determined.
The curves show that after administration of ^{14}C-phenylalanine the specific activity of chlorogenic acid first increases rapidly and then decreases with a half-life of less than 20 h in spite of a slight increase in its absolute amount. This indicates, on the one hand, steady synthesis of the new unlabeled compound during the chase period and, on the other hand, continuous degradation of labeled and unlabeled chlorogenic acid

Table 10. Half-life of some secondary products

Group of substances	Individual compounds	Half-life	Organisms/organ
Isoprenoids	1-Menthol and other constituents of the essential oil D 6.2.1)	Several h	*Mentha piperita*
	Monoterpenes (D 6.2.1) and diterpenes (D 6.4)	170 d 46 h	*Pinus sylvestris,* cortex, needles
	Marrubiin (D 6.4)	24 h	*Marrubium vulgare*
	α-Tomatine (D 6.6.3)	6 d	*Solanum esculentum*, fruits
	Corticosteroids (D 6.6.5)	1-1.5 h	Human beings
Cyanogenic glycosides	Dhurrin (D 8.3)	10 h	*Sorghum*, seedlings
Alkaloids	Nicotine (D 16.1)	22 h	*Nicotiana tabacum*
	Ricinine (D 16.1)	4 h[a]	*Ricinus communis*
	Quinolizidines (D 18)		*Lupinus polyphyllus*
		5 h	Leaves
		2.5 h	Cell cultures
	Gramine (D 22.1)	80 h	*Hordeum vulgare*
	Morphine (D 23.1.2)	7.5 h	*Papaver somniferum*
Cinnamic acid derivatives	Chlorogenic acid (D 23.2.1)	20 h	*Xanthium pennsylvanicum,* leaves
	Coniferin (D 23.2.3)	60-120 h	*Picea abies*, seedlings
Flavonoids	Kaempferol and quercetin glycosides (D 23.3.3)	7—12 d	*Cicer arietinum*, leaves
	Delphinidin glycosides (D 23.3.3)	25-31 h	*Petunia hybrida*, corolla
	Biochanin A (D 23.3.3)	25-320 h	*Cicer arietinum*, several organs, cell cultures
	Formononetin (D 23.3.3)	72 h	*Cicer arietinum*, roots
	Kaempferol, quercetin and isorhamnetin (D 23.3.3)		*Cucurbita maxima*, seedlings
	— glucosides	30-36 h	
	— biosides	48 h	
	— triosides	None	

[a]   Later increasing to 6-7 d.

Synthesis, storage, and degradation of secondary products may go on simultaneously in different compartments of one cell. This needs the continuous movement of secondary products in and out of the storage compartments, e. g., the vacuoles. Intracellular compartmentalization is therefore especially important in the regulation of the amount of secondary compounds undergoing continuous turnover. In several instances, however, the synthesis of secondary products and their further transformation/degradation are part of different programs of cell differentiation. In this case biosynthesis and transformation and/or degradation are separated in space and/or time.

This may be seen for instance

— in the biosynthesis of benzodiazepine alkaloids in the hyphae of *Penicillium*

cyclopium and their transformation to quinoline derivatives in the conidio-spores (D 21.4.2),
— in the formation of porphyrins in the erythrocytes of animals and their subsequent transformation to open-chain tetrapyrrols in the liver (D 24 and D 24.2),
— in the formation of chlorophylls (D 24.1) in young leaves and their degradation during leaf senescence, and
— in the different secondary products used for the storage of carbon and nitrogen in the seeds of plants, e.g., starch, hemicelluloses, different types of lipids, nonprotein amino acids, and so on (E 2.2), which are formed during seed maturation and degraded during seed germination.

References for Further Reading (see A 5.3)

A 5.3 The Significances of Degradation

Secondary substances present in large quantities, e. g., the compounds used as carbon and nitrogen storage in plant seeds or the chlorophylls in leaves, may furnish substantial amounts of energy and compounds of primary metabolism if degraded during certain developmental stages. Usually, however, secondary products are formed in quantities too small to be important in this respect. In these cases transformation and degradation may serve to adapt secondary product accumulation to changing needs or to remove secondary products from the producer (E 1). In many cases, however, the significance of the transformation and degradation of secondary products is still an open question.

References for Further Reading

32, 646, 700, 799, 843

B Methods Used in the Investigation of Secondary Metabolism

B 1 Examination of Secondary Metabolic Pathways

B 1.1 Use of Isotopically Labeled Compounds (Tracer Technique)

Until the mid-1940s the metabolic fate of a compound could be followed only with great difficulty. "Biosynthetic" studies often were limited to model experiments in unbiological systems, which served to demonstrate the chemical feasibility of certain reactions under "physiological conditions". This situation changed drastically in the 1950s when radioactive isotopes became readily available. The use of the so-called tracer technique, i. e.,

— administration ("feeding") of putative precursors labeled with isotopes at one or more specific positions to organisms producing secondary products,
— isolation of the secondary compounds after a suitable period of time, and
— determination whether they contain any of the isotopes enriched in the administered precursor

increased our knowledge on the pathways of secondary product formation almost explosively. In most cases the first step was to establish precursor-product relations by using easily available compounds of primary metabolism in tracer experiments. Today research concentrates on the detection of intermediates and on the elucidation of mechanistic details of biosynthetic pathways using multiple and stereospecific labeling techniques.

The Isotopes Used

The isotopes most important in biosynthetic studies are ^{2}H (deuterium), ^{3}H (tritium), ^{13}C, ^{14}C, ^{15}N, ^{18}O, ^{32}P, and ^{35}S. They may be introduced in a compound either during chemical or biochemical synthesis, or like ^{3}H, by isotope exchange into the already present nonisotopically labeled compounds. The radioactive isotopes ^{3}H, ^{14}C, ^{32}P, or ^{35}S may be detected by special counters; the stable isotopes, by the methods listed in Table 11.

The Rate of Incorporation

After administration of isotopically labeled compounds the substance in which the labeled precursor (or more precisely, the labeled part of the precursor) is incorporated and the extent of the incorporation rate may be determined.

The incorporation rate can be calculated from radioactivity whose units are the becquerel (1 Bq = 1 disintegration/s) or the curie (1 Ci = $3.7 \cdot 10^{10}$ disintegrations/s) or from the increase in the

Table 11. Spectroscopic techniques used for the detection of stable isotopes in biosynthesis studies

Isotope	Emission spectroscopy	ORD spectroscopy[a]	Mass spectrometry	NMR spectroscopy[a]
^2H		+	+	+[c]
^{13}C			+	+[d]
^{15}N	+[b]		+	
^{18}O			+	

[a] Sensitivity relatively low.
[b] After combustion to nitrogen gas. It therefore does not allow determination of the position of label if there is more than one nitrogen atom per molecule.
[c] Used primarily in mechanistic studies with stereospecific labeling.
[d] Most important method in the analysis of tracer experiments with ^{13}C.

natural isotopic concentrations, given in atoms % excess in the precursor and the product. Both absolute and specific incorporation rates can be calculated and usually are expressed as a percentage:

a) The specific incorporation rate (specific radiochemical yield) is given by the equation:

$$\frac{\text{Specific activity (or atoms \% excess) of the product}}{\text{Specific activity (or atoms \% excess) of the precursor}}$$

The specific activity usually is given in MBq/mmol or mCi/mmol.

A parallel quantity is the dilution of the precursor. Here the specific activity (or the atoms % excess) of the precursor is set equal to 1:

$$\text{Dilution} = 1 : \frac{\text{Specific activity (or atoms \% excess) in the precursor}}{\text{Specific activity (or atoms \% excess) in the product}}$$

b) The absolute incorporation rate is given by the equation:

$$\frac{\text{Specific activity (or atoms \% excess) in the product} \cdot \text{mol of product}}{\text{Specific activity (or atoms \% excess) in the precursor} \cdot \text{mol of precursor}}$$

The absolute incorporation rate in which the amount of administered precursor is correlated with the amount incorporated into the products is usually less reliable than the specific incorporation rate or the dilution of the precursor. It depends, to a great extent, on how much product is synthesized during the experiment and is therefore usually subject to wide variation. In addition, its calculation requires an exact determination of the total amount of product formed, a condition which usually cannot be satisfied.

The specific incorporation rate (or the dilution of the precursor) gives the amount of product formed from the labeled precursor relative to that formed from the pool of endogenous precursor. Thus, for a specific incorporation rate of 0.1 % (or a dilution of 1:1,000), one product molecule in a thousand is formed from the isotopically labeled precursor. The specific incorporation rate is therefore dependent on the ratio of incorporation of endogenous to administered precursor molecules. It is influenced by the absolute rate of synthesis during the experiment only when this ratio is altered. It is easier to determine the specific activity of the product than its absolute amount since only a small portion of the product has to be isolated.

The specific incorporation rate permits conclusions regarding the closeness of relationship between the product and the precursor since it is usually higher if the precursor is transformed to the product in one or a few steps. This correlation must, however, not be overrated since factors such as the size of the endogenous pools of precursors and intermediates and the permeability of the cell to the precursor also play an important role.

Feeding Techniques

Biosynthetic investigations can be carried out on whole organisms or parts of organisms, e.g., certain organs, tissues, or cells. The compounds to be administered are usually injected into animals and plants. In the case of higher plants, they are also fed via the roots (in hydroculture) or by a cotton thread dipped into the precursor solution and passing through the shoot. The cut end of shoots may be dipped directly into the precursor solution and pieces of tissue may be floated on or are shaken with the precursor solution.

The specific incorporation rates in these types of experiments are usually low. In higher plants they are frequently in the range of 0.01%. The low rates are mainly due to loss of precursors to competing reactions during transport to the sites of biosynthesis and to dilution of labeled precursors by large pools of unlabeled precursors and of the labeled products by unlabeled products already present at the beginning of the experiment.

In microorganisms the specific incorporation rates may be much higher because precursor transport plays a less significant role and the experimental conditions can be controlled in such a way that the product present at the end of the experiment has been synthesized only during the course of the experiment itself.

The experimental difficulties of experiments with higher plants and animals are smaller if cell cultures are used. Cell cultures can be fed like cultures of microorganisms, so that the incorporation rates can be much higher than for whole plants and animals, or isolated organs. They are, however, not universally applicable for the investigation of secondary metabolism. Cultured plant cells, for instance, are in many cases in a meristematic, unspecialized state and since expression of secondary metabolism is a feature of cell specialization (A 4.1), they do not synthesize considerable amounts of secondary products. However, with those cell cultures synthesizing secondary compounds tracer experiments and investigations with isolated enzymes have been performed with great success.

Since an organism's capacity to produce certain compounds may change in the course of its development (A 4.1), the physiology of secondary product formation must always be studied before biosynthetic investigations are undertaken.

Whether the product is actually synthesized during the period of the experiment may be detected in control experiments by feeding [U-^{14}C]glucose, $^{14}CO_2$, or similar substances which enter primary metabolism easily. If there is synthesis, the isolated secondary product will be labeled after a certain delay since it is ultimately formed from the products of primary metabolism. If no labeled secondary products can be detected after administration of these unspecific precursors, feeding with specific precursors is unlikely to be useful.

The material used for the experiments should not contain any other organisms (axeny). This excludes, for instance, the transformation of the labeled precursor by contaminating microorganisms. Preparation of microbial cultures free of other organisms is not difficult. Recently, methods have been developed which enable also the growth of higher plants, e.g., seedlings, or parts of higher plants, e.g., roots, as well as the growth of animals, under axenic conditions.

Isotope Distribution in the Products

To demonstrate a direct biogenetic relationship between the administered precursor and the isolated product it is necessary to determine the isotope distribution of the latter. If there is a direct biogenetic relation, i.e., if the administered compound is a direct precursor of the product, only those atoms will be labeled in the product that correspond directly to the labeled positions of the precursor. If, however, the precursor first enters the pathways of general metabolism, and the product is synthesized from conversion products, most of the atoms will be labeled. This smearing (randomization) of labeling will be greater the longer the time between feeding and extracting the material. Hence experiments in which the pertinent isotope distribution is not localized do not usually stand up to rigorous criticism.

In many cases chemical degradation is necessary for determining the distribution of radioactive isotopes in the products.

The degradation of L-tropic acid will be discussed as an example (Fig. 12). L-Tropic acid originates from L-phenylalanine by an intramolecular shift of the carboxy group (D 23). To determine the isotope content in the individual carbon atoms of the side chain, L-tropic acid is first oxidatively converted to benzoic acid which is then decarboxylated. This separates carbon atom 2 from the other carbon atoms as CO_2. Conversion of L-tropic acid to atropic acid which is then decarboxylated releases carbon atom 1 as CO_2. The methylene group can, in addition, be cleaved off by a periodate oxidation so that carbon atom 3 is removed as formaldehyde. The isotope content of each degradation product can be determined to give the isotope distribution within the L-tropic acid side chain.

Degradation for the localization of label can be avoided by examining the incorporation of stable isotopes with ORD spectroscopy, but especially with NMR and to a certain extent mass spectrometry. The position of signals or the weight of fragments containing the isotopes may allow their localization. However, these detection methods usually are relatively insensitive and can be used only with substances containing a high percentage of the stable isotope.

If for instance a ^{13}C-labeled precursor is incorporated into a secondary product, the typical NMR signal should be enhanced at least by 30–40%. As the natural content of ^{13}C is 1.11%, the incorporation should be at least 0.5% when the precursor is enriched 90% in the analogous carbon atom.

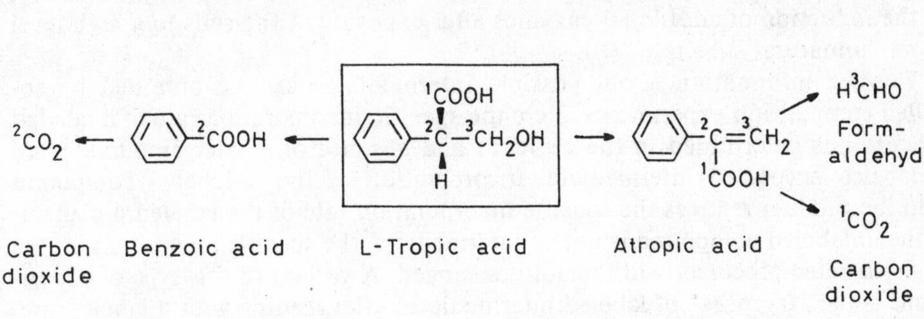

Fig. 12. Degradation of L-tropic acid yielding separately the carbon atoms 1–3 of the side chain

The Use of Multiple Labeled Precursors

Since only the metabolic fate of the labeled portion of the precursor molecule can be determined, it is necessary to label all the important atoms. This is achieved in most cases in separate experiments, in order to follow the behavior of a particular atom in metabolism. For certain investigations, however, multiple labeling of the precursors is necessary. Either different atoms of the same element, or different elements participating in the structure, may be labeled with isotopes.

Experiments which will demonstrate, for instance, the incorporation of the C-skeleton and the amino group of amino acids as intact entity into an alkaloid need the labeling of the nitrogen atom of the amino group as well as the carbon atoms as an internal standard. In the case of direct incorporation, the ratio of the amount of label in both atom species in the precursor and the isolated alkaloid must remain the same.

It is of special interest that the relative insensitivity of ^{13}C-NMR analysis can be overcome by labeling two contiguous C-atoms with ^{13}C or adjacent C and N atoms with ^{13}C and ^{15}N, respectively (coupling analysis). Taking into consideration the natural abundance of ^{13}C and ^{15}N, the probability that two contiguous C-atoms contain ^{13}C is only about 0.012 3 %, and that a ^{13}C-atom occurs adjacent to a ^{15}N-atom is only 0.004 %. Hence after feeding of compounds highly enriched (>90 %) with ^{13}C—^{13}C or ^{13}C—^{15}N, incorporation can be measured even if the rates are less than 0.1 %, i. e., about 10-20 times the background caused by the natural content. Contiguous ^{13}C—^{13}C or ^{13}C—^{15}N groupings can be determined by measuring the satellite peaks in the neighborhood of the corresponding ^{13}C-peaks, caused by spin-spin coupling. However, two disadvantages remain (a) the large amount of product necessary for NMR analysis (about 5 mg) and (b) the length of time necessary for the accumulation of spectra with satellite peaks due to the long relaxation periods.

Elucidation of Pathways

For the elucidation of secondary metabolic pathways it is necessary to test the incorporation of potent intermediates. If these are converted o the product they are probably involved also in normal metabolism. In certain cases, however, "unusual" molecules may also be incorporated into secondary products. This may be due to low substrate specificity of the enzymes involved (A 2.2) or, more rarely, to the induction of additional enzymes after exposure of the cells to a high level of an "unnatural substrate" (E).

Further information about possible intermediates can be obtained by so-called competition experiments. Here the specific incorporation rate of a labeled precursor is determined in the presence and absence of greater amounts of an unlabeled suspected intermediate. Incorporation of the unlabeled compound into the product reduces the specific incorporation rate of the labeled precursor. If the unlabeled compound is not an intermediate, the specific incorporation rate of the labeled precursor will remain unchanged. A variant of this type of experiment is the "trapping" of labeled intermediates after feeding with a labeled precursor and an excess of the unlabeled suspected intermediates. If the labeled intermediates formed from the labeled precursor mix with the pool of unlabeled administered intermediates, the whole population of molecules becomes labeled. If at the end of the experiment no isotopic label is found in the added unlabeled compounds, it may be assumed that they are not intermediates.

Competition and trapping experiments make it possible to study hypothetical intermediates even if they are not available in an isotopically labeled form. The experiments fail, however, if endogenous intermediates are strictly channeled and do not mix with compounds administered from outside (A 3).

References for Further Reading

110, 218, 219, 235, 500, 545, 580, 602, 626, 689, 785

B 1.2 Use of Enzyme Preparations

In tracer experiments (B 1.1) administered substances have to pass many barriers before reaching the site of the biosynthesis orf secondary products. They may be metabolized in nonproducing cells and, due to intracellular compartmentalization, in other parts of the producer cell than the natural precursor. Hence, the incorporation rates are often low and in some instances misleading.

Most of the problems caused by compartmentalization may be avoided if homogenates, certain fractions of homogenates, or enzyme preparations are used instead of whole cells. To make a homogenate the integrity of cells and tissues is destroyed by chemical or physical methods, e.g., by the action of enzymes, ultrasound, grinding, or freezing. The resulting mixture, the homogenate, contains the enzymes of secondary metabolism as well as the other cell constituents. More or less pure preparations of secondary metabolic enzymes may be prepared from the homogenate by centrifugation, precipitation, chromatography, etc. However, in vitro experiments with enzymes of secondary metabolism are still difficult because these enzymes are usually present only in small quantities and they often must be isolated from cells containing large amounts of secondary products which after disintegration of their sites of storage (A 3) may denature proteins and inactivate enzymes.

Investigations with enzyme preparations are of importance in the elucidation of the details of secondary metabolic pathways. Each step can be examined independently from the metabolic grid of the producer cell, i.e., without interference with other transformations of the administered precursor or the products formed. The reaction in question can be examined by following the transformation of the substrates as well as the synthesis of the products and the consumption of cosubstrates, e.g., NADH or NADPH. The sensitivity of the determination of enzyme activities may be improved by sophisticated methods using other enzymes or isotopes.

For example, cultures of the mold *Penicillium cyclopium* form the epoxide cyclopenin from dehydrocyclopeptine by a monooxygenase (dehydrocyclopeptine epoxidase, D 21.4.2). However, in spite of a high rate of epoxidation in vivo the epoxidase activity measurable in vitro is low. Therefore, radioactively labeled dehydrocyclopeptine has been used as substrate in the in vitro test. To improve separation of the product cyclopenin from excess substrate, cyclopenin has been transformed into viridicatin by the enzyme cyclopenase. After chromatographic purification viridicatin has been oxidize to 2-aminobenzophenone which unlike viridicatin can be quantitatively eluted from the adsorbent used in the chromatographic separation and quantified by counting the radioactivity. By this method less than 0.5 pkat dehydrocyclopeptine epoxidase per test have been determined precisely.

Before starting enzymatic work it is necessary to have basic information about the pathway in question, i. e., about the intermediates and types of reactions which are involved. This knowledge may come from tracer experiments (B 1.1), which therefore are usually a prerequisite of successful enzymatic work, and from general experience on reactions of secondary product formation, i. e., on well-founded knowledge of the biochemistry of secondary metabolism.

References for Further Reading

97, 376

B 2 Investigations with Mutants

The use of mutants is of special importance in biosynthetic research on microorganisms. Their small size, the comparatively simple constitution of their genetic material, e.g., the frequently found haploidy, as well as their saprophytic nutrition and fast multiplication, allow the artificial production of large numbers of mutants, e.g., by treatment with UV radiation and chemical agents.

Studies with mutants of higher plants and animals are more difficult because the cells of these organisms are usually diploid, triploid, or polyploid, i. e., every cell contains two, three, or several sets of identical chromosomes. Mutations in which usually one gene is altered are, therefore, noticeable only when the mutant gene is dominant, i. e., it expresses itself in the presence of nonmutated genes. If the gene is recessive, the mutation can be recognized only if, by breeding and crossing, cells are obtained which in all homologues genes are mutated. Moreover, plants and animals are multicellular organisms. Therefore, a certain amount of tissue, or a whole organ, must be treated with mutagens. If a mutation takes place in a certain cell, it may only be noticed if this cell is capable of division and generates a large number of cells with altered genetic material.

There are two types of mutants which are of significance in the examination of secondary metabolism:

a) Mutants of structural genes encoding enzymes involved in synthesis, transformation, or degradation of secondary products (A 2.1)

Blocking a structural gene that encodes a secondary metabolic enzyme depresses synthesis of the product and frequently causes accumulation of its substrate. Since in most metabolic pathways the starting material is converted to the end product through intermediary stages, the reaction chain may be blocked at different places. The number and location of blocks within the metabolic chain can be determined by "cosynthesis", i. e., normalization of product formation if mutants

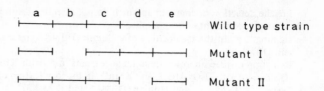

Fig. 13. Scheme representing the use of cosynthesis for the elucidation of secondary metabolic pathways

In the wild-type strain enzymes *a–e* form a secondary product. In mutants *I* and *II* different enzymes are blocked by mutation. Both mutants grown separately will not produce the secondary product formed in the wild-type strain. They will build this product, however, if cultivated together, if the product of enzyme *c* is released from mutant *II* and is able to enter the cells of mutant *I*

with different blocks are grown together (Fig. 13). This method gives an estimate of the number of structural genes and enzymes involved in a metabolic chain.

By feeding compounds which "normalize" synthesis of secondary products in the blocked mutants and by isolation of the substances accumulated in the particular mutants, it is possible to elucidate the metabolic pathway in question. Difficulties may arise, however, if the intermediates themselves fail to accumulate and modified derivatives are built on pathways not observed in the wild-type strain.

b) Mutants of regulatory regions of the genetic material

The rate of secondary product formation may be influenced by the mutation of many different parts of the genetic material of the producer organism. In most cases these mutations act indirectly, e.g., by influencing the flow of precursors, cosubstrate biosynthesis, or differentiation programs (pleiotropic effects). Mutations in the genetic material directly controlling the expression of secondary metabolism are rare (A 2.1).

References for Further Reading

372, 376, 513, 747

B 3 Methods Used in the Investigation of Compartmentalization and Channeling

B 3.1 Cytochemical Methods

In the last century it has become known from light microscopic studies that colored secondary products are located in distinct compartments of living cells, e.g., anthocyanins and anthraquinones in vacuoles, and carotenoids in chromoplasts. Since then the use of UV fluorescence microscopy or electron microscopy as well as different methods of staining and precipitation have led to the localization of many other secondary products (A 3).

For the detection of compounds which do not directly absorb visible light, UV, or electrons, the following procedures have been applied:
— staining, e.g., in the localization of cellulose (D 1.4.1), lignins (D 23.2.3), and tannins (D 23.2.5 and D 23.3.3)
— formation of precipitates with suitable reagents, e.g., with KBiI$_4$ in the localization of alkaloids (D 8.5), and with OsO$_4$, UO$_2^{2+}$, and KMnO$_4$ in the localization of essential oils (D 6.2.1 and D 6.3), flavonoids (D 23.3.3), and tannins (D 23.2.5 and D 23.3.3)
— chemical modification of the compounds followed by the formation of precipitates, e.g., oxidation with periodic acid in the localization of carbohydrates (D 1.4.1)
— reaction with peroxidase-labeled or fluorescent antibodies, e.g., in the localization of alginic acid (D 1.4.1), or with antibodies labeled with gold granules, e.g., in the localization of mannan (D 1.4.1)
— reaction with gold-labeled concanavalin A, e.g., in the localization of yeast mannan and other branched polysaccharides (D 1.4.1), or reaction with wheat germ agglutinin, e.g., in the localization of chitin (D 1.4.1)
— labeling with radioactive isotopes and autoradiographic examination, e.g., in the localization of lignin (D 23.2.3), of cell wall polysaccharides (D 1.4.1), of cyanogenic glycosides (D 8.3), and of alkaloids (D 8.5), or by use of X-ray microanalysis, e.g., in the localization of fucan (D 1.4.1)
— examination before and after specific enzymatic degradation or selective extraction of secondary products, e.g., in the localization of carbohydrates (D 1.4.1).

In rare cases, where enzymes of secondary metabolism form insoluble or easily detectable products their intracellular location may also be studied cytochemically (localization of thioglucosidase, phenol oxidases, and peroxidases).

Artifacts may be caused by the migration of proteins and low molecular compounds within the cell because of the increased membrane permeability after cell death. In most cases, therefore, it is necessary to use appropriate fixation techniques. Enzymes, for instance, can be bound to particulate cell constituents by glutaraldehyde. The fixation of low molecular weight substances, however, is still difficult. In the case of hydrophilic compounds, diffusion after cell death may be prevented either by rapid freezing and subsequent exchange of the cell water by a suitable organic solvent, or by freeze-drying followed by embedding in plastic.

Reference for Further Reading

482

B 3.2 Examination of Isolated Cell Constituents

In contrast to the microscopic in situ methods, disintegration of cells necessarily destroys certain cell structures, such as the endoplasmic reticulum (ER). Secondary products stored in the vacuole and the extracytoplasmic space may then come into contact with cytoplasmic or membrane-bound enzymes and inactivate them. In addition, homogenization of the rigid cell wall may produce fragments which demolish fragile organelles, such as the plasma and vacuole membranes. These latter difficulties may be overcome by the preparation of protoplasts, which make it possible to break "cells" more gently, and allow the easy isolation of vacuoles from microbial or plant cells. Also, the addition of compounds which bind phenolic substances may reduce enzyme inactivation and protect organelles. One serious problem is that substances, which in vivo are associated with cell structures, are solubilized and then absorbed randomly onto other particulate cell constituents. The cell wall, for instance, may bind cytoplasmic proteins by absorption, ionic linkages, or by coprecipitation with polyphenols. It may also release enzymes at unsuitable ionic strength and/or pH.

After cell disintegration, organelles are usually separated by differential centrifugation. But it is difficult to obtain really pure organelles in the native state by this method. Preparations are often contaminated by unwanted cell constituents, fragments of other organelles, or parts of the cytoplasm. Relatively pure samples can usually be obtained only by further fractionation, preferably on density gradients. However, the compounds used to form such gradients may cause changes in membrane structures. These changes are often accompanied by permeability changes, increase or decrease in the activity of enzymes, and release of compounds attached to the surface of membranes.

In some instances, cell fractionation has been aided by specific labeling of organelles. The density of the rough ER may be increased by prior incubation with lead acetate and glucose-6-phosphate. The latter compound gives rise to inorganic phosphate which is precipitated as the lead salt directly at the site of glucose-6-phosphatase which is characteristic of the rough ER. Reaction with iodinated tetrazolium salts increases the density of mitochondria and favors their separation from microbodies. Labeling of membrane proteins with ^{125}I by treatment with lactoperoxidase may increase the specific weight of the plasma membrane and the envelope membranes of isolated cell organelles. Coating membranes with concanavalin A can enhance their rigidity, as was shown by the fact that after protoplast formation, the coated plasma membrane of *Saccharomyces cerevisiae* was separated from the other cell membranes practically in one piece. For strengthening the binding of proteins to organelles and for stabilization of organelles, e.g., Golgi vesicles, fixation with aldehydes, compounds used for the same purpose in cytochemistry or imidoesters may be helpful.

Whether the occurrence of an enzyme in a particulate fraction really represents its in vivo location can be substantiated by the following properties: (a) close binding to membrane lipids, cf. cyclopenase (D 21.4.2), which is a lipoprotein of the plasma membrane; (b) cooperation with other enzymes, cf. the catalytic facilitation in the biosynthesis of cyanogenic glycosides, cinnamic and benzoic acids (A 3.1.1); and (c) in situ examination by cytochemical methods, cf. the localization of phenol oxidases (C 2.3.1), peroxidases (C 2.4), and thioglucoside glucohydrolase (D 8.4).

Reference for Further Reading

482

B 3.3 Tracer Experiments with Living Cells

Biochemical tracer experiments on compartmentalization and channeling in living cells take into consideration the cellular "microenvironment" as well as the interrelations between enzymes, low molecular compounds, and other cell constituents. Such experiments are especially useful for studying the dynamic aspects of secondary metabolism. However, they have the drawback of not allowing the spatial localization of existing pools and channels of the metabolites.

Thus, they give reliable results only if they are combined with cytochemical experiments or with examination of isolated cell constituents.

There are two basic types of experiments using tracer molecules. In one, tracers are applied in "pulse" experiments and the pattern of labeling in the tracer-derived products is determined over time. In the other, the fate of labeled compounds is examined after a "chase" period following the initial pulse.

Typical examples are experiments in which CO_2 has been administered to intact leaves in competition with mevalonic acid for localization of the site of plastid quinone biosynthesis, and experiments with intact mycelia of *Penicillium cyclopium*, where the channeling of the precursor L-phenyl-alanine and its incorporation into the alkaloids of the cyclopenin-viridicatin group and into proteins have been studied (Fig. 10).

Reference for Further Reading

482

C Important Groups of Secondary Metabolic Enzymes

In the last years an increasing number of secondary metabolic enzymes has been purified and characterized. The most important groups are discussed in the following section.

C 1 Activating Enzymes

C 1.1 Phosphotransferases

Phosphorylated compounds, e.g., the esters, amides, and anhydrides of phosphoric acid, participate in many secondary reactions. High-energy and low-energy phosphates may be distinguished. The former liberate up to 62 kJ mol^{-1} if the bond between the phosphate residue and the acceptor molecule is hydrolyzed, the latter set free about 9–21 kJ mol^{-1} (Table 12).

Table 12. Energy liberated during the hydrolysis of posphorylated compounds under standard conditions

Group of compounds	Representative, reaction of hydrolysis	Energy liberated [kJ mol^{-1}][a]
High-energy phosphates	Phosphoenol pyruvate (D 2) → Pyruvate + Phosphate	62
	Creatine phosphate (D 10) → Creatine + Phosphate	43
	Acetyl phosphate (D 5) → Acetate + Phosphate	42
	ATP (D 25) → AMP + Disphosphate	36
	L-Arginine phosphate (D 19) → L-Arginine + Phosphate	32
	ATP (D 25) → ADP + Phosphate	31
Low-energy phosphates	Aldose-1-phosphates (C 6) → Aldose + Phosphate	21
	Aldose-6-phosphates (C 6) → Aldose + Phosphate	14
	Phosphate esters → Alcohol + Phosphate	9

[a] Determined under standard conditions. Because these deviate from the conditions within the cell, the figures give only an indication of the values occurring in vivo.

The energy content of the phosphate bond depends considerably on the extend of the disturbance of the electron resonance in the phosphorylated compound. In phosphates with low-energy content the resonance of the phosphate anion, which depends on the mutual convertibility of the mesomeric structures I-IV:

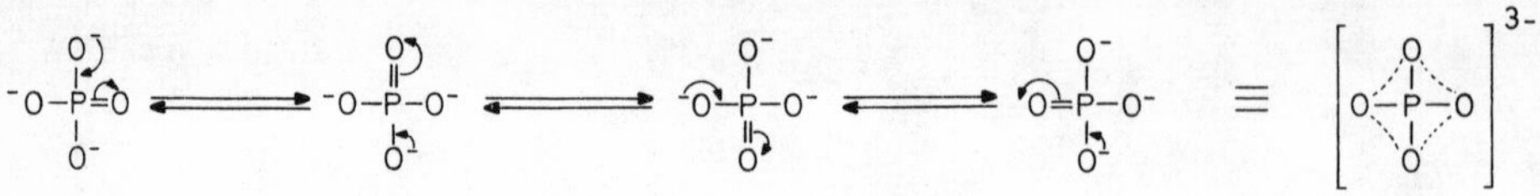

$$\text{I} \qquad \text{II} \qquad \text{III} \qquad \text{IV}$$

is reduced, because one of these structures is excluded from resonance:

$$\text{I} \qquad \text{II} \qquad \text{III}$$

In the case of energy-rich phosphates electron resonance is also hindered in the acceptor molecule. For example

— in acyl phosphates, e.g., acetyl phosphate (D 5) the resonance of the carboxy group is disturbed:

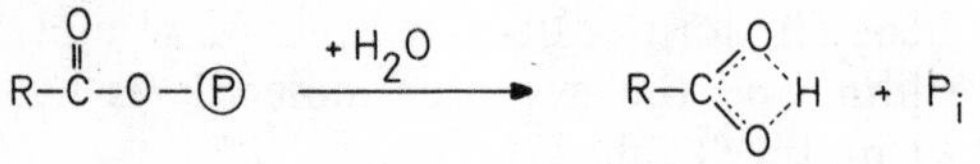

Acyl phosphate   Carboxylic
                        acid

— in guanidine phosphates, e.g., creatine phosphate (D 10) and L-arginine phosphate (D 19) that of the free guanidine group is hindered:

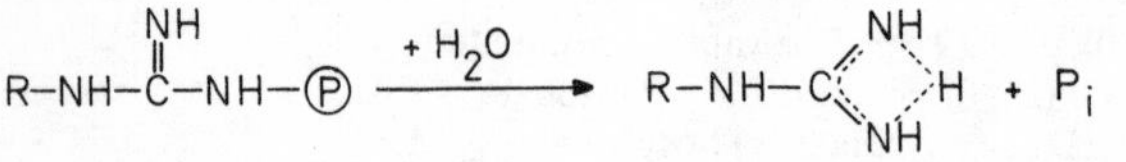

Guanidine   Guanidine
phosphate

— and in phosphoric acid anhydrides, like ATP the resonance of the acceptor phosphate group is diminished:

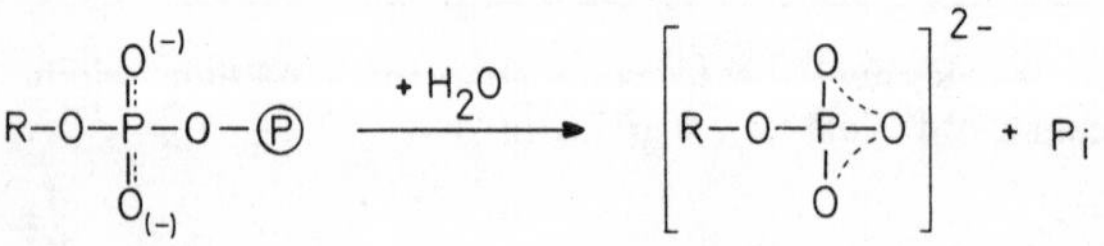

Phosphoric acid   Monophosphate
anhydrides

The decrease of resonance causes bond polarization. The stronger this polarization, the easier electrophilic or nucleophilic substitution of a substance, or the greater its capacity to substitute. It is, therefore, reactive and is termed "activated". Activated compounds are, for example, acyl phosphates and acyl AMP derivatives (C 1.2), sugar phosphates (C 6), and isopentenyl diphosphate (D 6).

Phosphotransferases catalyze the transfer of a phosphate group from one compound to another according to the following equation:

$$R_1X-\text{(P)} + R_2X-H \rightleftharpoons R_1X-H + R_2X-\text{(P)}$$

If the phosphate bond formed has approximately the same energy content as the cleaved phosphate bond, R_1X—(P) and R_2X—(P) remain in equilibrium with each other. If the energy content of the newly formed bond is much smaller, the transphosphorylation is practically irreversible. Consideration of the energy content of the numerous phosphorylated compounds present in the cell is therefore important for the understanding of possible reactions, their directions, and rates. Relatively energy-rich phosphorylated intermediates, like aldose 1-phosphates (C 6) or isopentenyl diphosphate (D 6) may be built from high-energy phosphates like ATP by means of phosphotransferases. They are precursors of secondary products. During their further transformation, which is usually accompanied with the release of the phosphate residue, sufficient energy is liberated to provide favorable equilibrium conditions for the synthesis of new bonds.

References for Further Reading

85, 90, 538

C 1.2 Acid-Thiol Ligases and CoA-Transferases

Thioesters of coenzyme A (D 11) are important intermediates in carboxylic acid metabolism. They are formed by acid-thiol ligases (thiokinases) either via acyl phosphates:

$$R-C\underset{OH}{\overset{O}{<}} + ATP \longrightarrow R-C\underset{O-\text{(P)}}{\overset{O}{<}} + ADP \;;\; R-C\underset{O-\text{(P)}}{\overset{O}{<}} + HSCoA \longrightarrow R-C\underset{S-CoA}{\overset{O}{<}}$$

or via acyl AMP derivatives:

$$R-C\underset{OH}{\overset{O}{<}} + ATP \longrightarrow R-C\underset{O-AMP}{\overset{O}{<}} + \text{(P)(P)} \;;\; R-C\underset{O-AMP}{\overset{O}{<}} + HSCoA \longrightarrow R-C\underset{S-CoA}{\overset{O}{<}}$$

CoA-transferases catalyze an exchange of the acyl group according to the following equation:

$$R_1-COOH + R_2-C\underset{S-CoA}{\overset{O}{<}} \rightleftharpoons R_1-C\underset{S-CoA}{\overset{O}{<}} + R_2-COOH$$

In the thioester grouping the resonance of the carboxy groups is disturbed and the CO-grouping has a considerable carbonyl character (as is the case in acyl phosphates, C 1.1).

Because of this "activation", the oxygen atom carries a negative, and the carbon atom a positive, fractional charge. The positively charged C-atom attracts electrons from the α-carbon atom causing weakening of the C—H bonds at this C-atom. Thus, thioesters may be easily substituted electrophilically at the α-carbon atom and nucleophilically at the carbon atom of the carbonyl group:

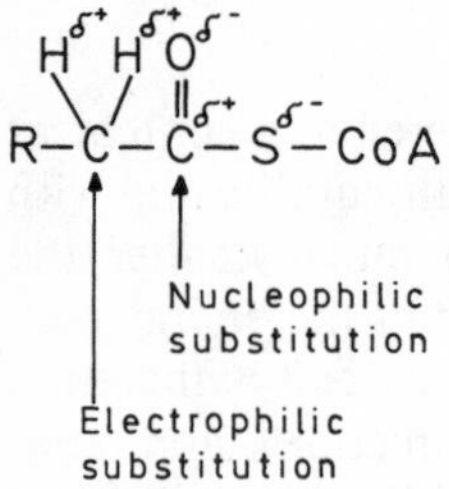

Nucleophilic
substitution

Electrophilic
substitution

Reference for Further Reading

197

C 2 Oxidoreductases and Oxygenases

Oxidoreductases (dehydrogenases or oxidases) catalyze the addition or the removal of hydrogen or electrons. Oxygenases bring about the incorporation of oxygen which originates from molecular oxygen.

C 2.1 Dehydrogenases Containing Pyridine Nucleotides

C 2.1.1 Mechanism of Reaction

In a large number of dehydrogenations the hydrogen of the substrate is transferred to the pyridine nucleotides nicotinamide adenine dinucleotide (NAD^+) or nicotinamide adenine dinucleotide phosphate ($NADP^+$) (D 16.1) according to the following equation:

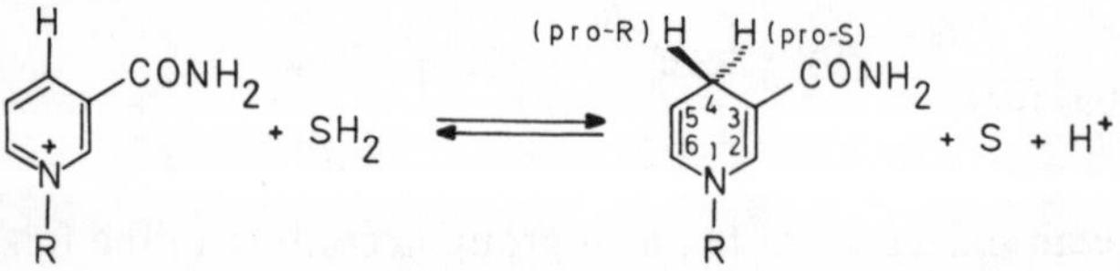

In the case of dehydrogenation in addition to a proton, a hydride ion is removed from the substrate under strict stereospecific control and is linked to the pyridine ring of nicotinamide at position 4. This causes formation of a center of prochirality at the pyridine ring (A 2.2). The hydrogen atom located at position 4 above the plane of the molecule (pro-R position) is known as H_A and the one below the plane (pro-S position) as H_B. H_A and H_B are not equivalent with respect to their transferability by dehydrogenases to substrates. According to the stereospecificity of the particular apoenzyme present, one or the other hydrogen atom participates in the oxidoreductions (Table 13). The reactions catalyzed by pyridine nucleotide-dependent dehydrogenases in most cases are reversible.

Table 13. Stereospecificity of dehydrogenases with respect to position 4 of the pyridine nucleus of NAD^+ and $NADP^+$

Transfer of hydrogen from			
H_A		H_B	
Alcohol dehydrogenase (alcohol: NAD^+ oxidoreductase, D 2)	Shikimate dehydrogenase (shikimate: $NADP^+$ oxidoreductase, D 21)	Glucose-6-phosphate dehydrogenase (D-glucose-6-phosphate: $NADP^+$ oxidoreductase, D 1.2)	Glutamate dehydrogenase (L-glutamate: NAD^+ oxidoreductase, C 2.1.2)
Hydroxymethylglutaryl CoA reductase (mevalonate: $NADP^+$ oxidoreductase, D 6)	Cyclopeptine dehydrogenase (cyclopeptine: $NAD(P)^+$ oxidoreductase, D 21.4.2)	γ-Coniceine reductase (D 5.3.2)	β-Hydroxysteroid dehydrogenase (3 [or 17] β-hydroxysteroid: $NAD(P)^+$ oxidoreductase, D 6.6.6)
		Squalene synthase (D 6)	

References for Further Reading

392, 445

C 2.1.2 Oxidative Deamination of Amino Acids

A group of special $NAD(P)^+$-dependent dehydrogenases catalyzes the oxidative elimination of the NH_2-group from α-amino acids with the formation of 2-oxo acids. The most important enzyme of this group is glutamate dehydrogenase (Table 13). 2-Oxoglutaric acid and ammonia are formed. An imino acid which decomposes spontaneously is built as intermediate:

$$\underset{\text{L-Glutamic acid}}{\overset{NH_2}{\underset{\text{COOH}}{CH_2-CH_2-\underset{H}{\overset{|}{C}}-COOH}}} + NAD^+ \rightleftharpoons \underset{\text{Imino acid}}{\overset{NH}{\underset{\text{COOH}}{CH_2-CH_2-\overset{\|}{C}-COOH}}} + NADH + H^+$$

$$\underset{\text{Imino acid}}{\overset{NH}{\underset{\text{COOH}}{CH_2-CH_2-\overset{\|}{C}-COOH}}} + H_2O \rightleftharpoons \underset{\text{2-Oxoglutaric acid}}{\overset{O}{\underset{\text{COOH}}{CH_2-CH_2-\overset{\|}{C}-COOH}}} + NH_3$$

The NH_2-group of amines and other amino acids may be transferred by trans-amination (C 5) to 2-oxoglutaric acid. Hence, most of the ammonia liberated by the degradation of amino acids and amines is formed by glutamate dehydrogen-ase.

References for Further Reading (see C 2.1.1)

C 2.2 Flavin Enzymes

C 2.2.1 Mechanism of Reaction

The flavin enzymes contain the riboflavin derivatives flavin mononucleotide (FMN) or flavin adenine dinucleotide (FAD) (D 25.4) as prosthetic groups. They act as the carrier of hydrogen or electrons and participate in a large number of oxidoreductions. Flavin enzymes may be oxidized or reduced in a one-step me-chanism (Fig. 14, pathway A) or a two-step reaction (Fig. 14, pathway B). In the case of the one-step mechanism the addition and elimination of a hydride ion plays an important role. In the two-step transformation a flavin radical appears as intermediate.

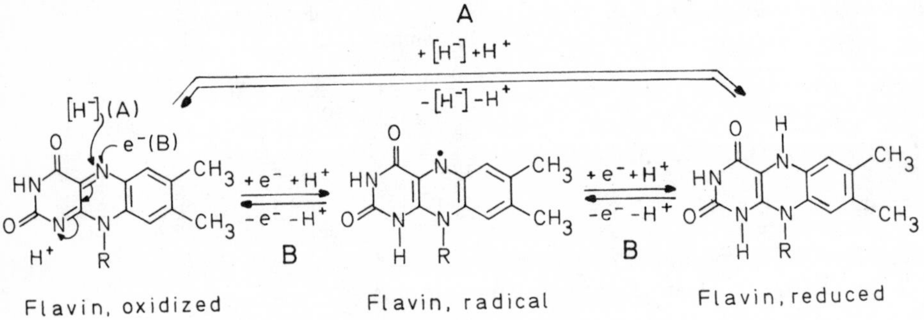

Fl avin, oxidized   Fl avin, radical   Fl avin, reduced

Fig. 14. Oxidoreduction of flavins by one- and two-step reactions

References for Further Reading

86, 352

C 2.2.2 Flavin Enzymes in Secondary Metabolism

Important types of reactions catalyzed by flavin enzymes are:

a) The oxygen-linked dehydrogenation of substrates, like D-glucose, amino acids, or xanthines

In addition to glutamate dehydrogenase (C 2.1.2), two amino acid oxidases which contain FAD have been isolated from animals and microorganisms. One type oxidatively deaminates D-amino acids (D 8.1) and the other type L-amino acids with the formation of an imino acid and $FADH_2$. The reduced coenzyme is reoxidized by means of oxygen and the formation of H_2O_2:

$$\text{Amino acid} + \text{FAD-Enzyme} \rightarrow \text{Imino acid} + \text{FADH}_2\text{-Enzyme}$$
$$\text{Imino acid} + \text{H}_2\text{O} \rightarrow 2\text{-Oxo acid} + \text{NH}_3$$
$$\text{FADH}_2\text{-Enzyme} + \text{O}_2 \rightarrow \text{FAD-Enzyme} + \text{H}_2\text{O}_2$$

b) The NAD(P)^+-linked dehydrogenation of substrates, like dihydrolipo-amide by dihydrolipoamide reductase (C 2.7), or glutathione (D 9.2) by gluta-thione reductase according to the equations:

$$\text{Dihydrolipoamide} + \text{NAD}^+ \rightleftarrows \text{Lipoamide} + \text{NADH} + \text{H}^+$$
$$2\,\text{Glutathione} + \text{NAD(P)}^+ \rightleftarrows \text{Glutathione, oxidized} + \text{NAD(P)H} + \text{H}^+$$

c) The cytochrome-linked dehydrogenation of the initial members of the membrane-bound electron transport chain of certain monooxygenases (C 2.6.1) and

d) the monooxygenation of aldehydes with the concomitant emission of light (D 25.4).

References for Further Reading (see C 2.2.1)

C 2.3 Copper-Containing Oxidases

C 2.3.1 Phenol Oxidases

Phenol oxidases form two groups: catechol oxidases and laccases.

Catechol Oxidases

Catechol oxidases hydroxylate monophenols to diphenols and oxidize diphenols to o-quinones:

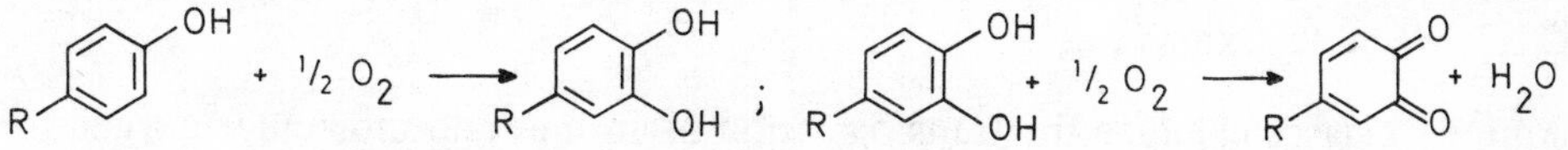

Both reactions need molecular oxygen. They are coupled to each other, if monophenols are the substrates. In contrast to the laccases, described in the fol-lowing, the copper atom(s) in the enzyme molecule probably do not change the charge during catalysis.

Catechol oxidases are widespread in nature. They are named according to their most important substrates as monophenol oxidases, polyphenol oxidases, phenolases, DOPA oxidases, cresolases, tyrosinases, etc. The specificity of most catechol oxidases is rather broad.

Laccases

Laccases do not posses hydroxylating properties. They oxidize o-diphenols and p-diphenols by a radicalic mechanism. Enzymes of this type were first obtained from the Japanese lack tree *Rhus vernicifera*. Laccases contain 4 atoms of copper: $2\,\text{Cu}^+$ ions and $2\,\text{Cu}^{2+}$ ions. One of the latter is responsible for the blue color of the enzymes. This Cu^{2+} ion is reduced by the substrate to Cu^+, i.e., it has the properties of an electron carrier (as is the case with iron in dioxygenases and some monooxygenases, C 2.5 and C 2.6). Hence, laccases are capable of producing radicals from phenols according to the following equation:

$$E-Cu^{2+} + S-OH \longrightarrow E-Cu^+ + S-O^{\bullet} + H^+$$

The bivalent copper of the prosthetic group of the enzymes is subsequently regenerated by molecular oxygen:

$$2\ E-Cu^+ + \tfrac{1}{2}\ O_2 + 2\ H^+ \longrightarrow 2\ E-Cu^{2+} + H_2O$$

The phenoxy radicals are transformed by reactions which are not enzymatically mediated. The radical may shift in the molecule, may react with another radical (oxidative coupling, radicalic coupling), may interact with double bonds, etc.

The quinoid products of phenol oxidases undergo self-polymerization or condensation with compounds carrying, e.g., —NH$_2$, —NH—, —OH, or —SH groups, i. e., carbohydrates, proteins, etc. (cf. the formation of eumelanins, D 23.1.3). Of significance are the phenol oxidase-dependent browning reactions in damaged, infected, or senescent plant tissues. In these reactions many groups of phenolic substances, e.g., flavonoids (D 23.3.3), are polymerized, which under "normal" physiological conditions do not undergo this type of reaction.

References for Further Reading

86, 520

C 2.3.2 Amine Oxidases

Amine oxidases catalyze the transformation of amines and molecular oxygen to aldehydes, ammonia, and H_2O_2:

$$R-CH_2-NH_2 + O_2 + H_2O \longrightarrow R-C\overset{H}{\underset{O}{\diagdown}} + NH_3 + \tfrac{1}{2}\ H_2O_2$$

They also react with monomethylated amines with the formation of methylamine (D 8.2). Mono-, di-, and polyamine oxidases may be distinguished according to the preferred substrate. The substrate specificity of the enzymes, however, is relatively low.

References for Further Reading

530, 716, 717

C 2.4 Peroxidases

C 2.4.1 Dehydrogenations and Hydroxylations Catalyzed by Peroxidases

Peroxidases catalyze the reaction of substrates with hydrogen peroxide. Most of the plant peroxidases, e.g., the well-investigated peroxidase of horseradish, contain ferriprotoporphyrin IX (hemin, D 24) as coenzyme. The prosthetic groups of the animal peroxidases are similar. Up to now only a few peroxidases are known whose prosthetic group has a fundamentally different structure. One example is a flavin enzyme from *Streptococcus faecalis*.

The peroxidases first break down hydrogen peroxide to form hydroxyl radicals. Then they cleave either two hydrogen radicals from the substrate:

$$SH_2 + OH^\bullet \longrightarrow SH^\bullet + H_2O$$
$$SH^\bullet + OH^\bullet \longrightarrow S + H_2O$$
$$\overline{SH_2 + 2\,OH^\bullet \longrightarrow S + 2\,H_2O}$$

or one hydroxyl radical replaces a hydrogen radical of the substrate which reacts with the second hydroxyl radical to form water:

$$SH_2 + OH^\bullet \longrightarrow SHOH + H^\bullet$$
$$H^\bullet + OH^\bullet \longrightarrow H_2O$$
$$\overline{SH_2 + 2\,OH^\bullet \longrightarrow SHOH + H_2O}$$

The substrate radicals obtained as intermediates may react spontaneously with each other or with other compounds.

Similar products are formed by the action of peroxidases and laccases (C 2.3.1) on phenols. In both cases phenoxy radicals are built which may either condense with each other or may be added to unsaturated compounds forming irregular, high molecular polymers (see the biosynthesis of lignins, D 23.2.3).

Peroxidases have low substrate specificity.

References for Further Reading

84, 552

C 2.4.2 Halogenation by Halogenoperoxidases

Halogenoperoxidases catalyze the substitution of a substrate hydrogen atom by a halogen atom according to the following equation:

$$SH + Halogen^- + H_2O_2 \longrightarrow S-Halogen + H_2O + OH^-$$

They have many properties in common with normal peroxidases (C 2.4.1) and also contain ferriprotoporphyrin IX (D 24) as prosthetic group.

With respect to halogen anions the enzymes possess little substrate specificity. They react with chloride as well as bromide and iodide ions. These halogen anions are probably first oxidized to radicals by means of hydroxyl radicals before substituting the substrate:

$$OH^{\cdot} + Halogen^{-} \longrightarrow Halogen^{\cdot} + OH^{-}$$

$$SH + Halogen^{\cdot} \longrightarrow S-Halogen + H^{\cdot}$$

$$H^{\cdot} + OH^{\cdot} \longrightarrow H_2O$$

Fluoride ions are not used by halogenoperoxidases. This agrees with the fact that H_2O_2 (or OH-radicals) are not capable of oxidizing fluorine.

References for Further Reading

557, 558

C 2.5 Dioxygenases

Dioxygenases introduce both atoms of an oxygen molecule into substrates. In most instances the two oxygen atoms react with one substrate molecule (intramolecular dioxygenases):

$$S + O_2 \longrightarrow SO_2$$

Some dioxygenases, however, incorporate one atom each of the oxygen molecule into different molecules of the same substrate:

$$2 S + O_2 \longrightarrow 2 SO$$

or into two different substrate molecules (intermolecular oxygenases):

$$S + S' + O_2 \longrightarrow SO + S'O$$

In the last case one of the two substrates is invariably 2-oxoglutaric acid. 2-Oxoglutaric acid is converted to succinic acid by incorporation of one atom of oxygen with concomitant decarboxylation:

$$HOOC-CH_2-CH_2-CO-COOH \longrightarrow HOOC-CH_2-CH_2-COOH + CO_2$$

2-Oxoglutaric acid   Succinic acid

A similar reaction is catalyzed by 4-hydroxyphenylpyruvate dioxygenase (D 23). The substrate, 4-hydroxyphenylpyruvic acid, possesses an oxo grouping in the side chain. The oxygen molecule probably reacts first with this carbonyl group under simultaneous decarboxylation. The other oxygen atom is added to the aromatic ring and appears in the newly formed hydroxyl group of the product, homogentisic acid (Fig. 15).

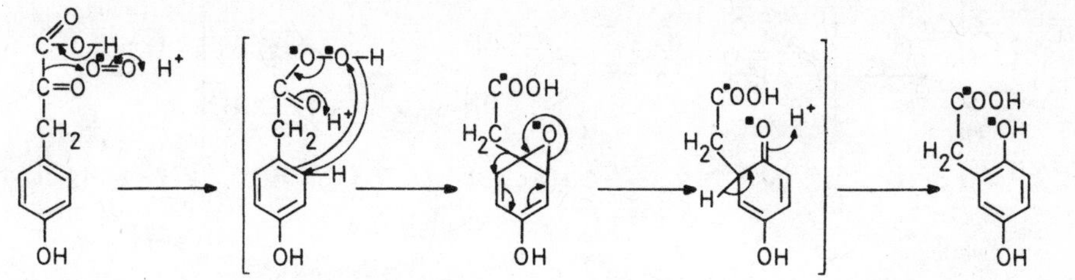

4-Hydroxyphenyl-pyruvic acid   **Homogentisic acid**

Fig. 15. Formation of homogentisic acid from 4-hydroxyphenylpyruvic acid by 4-hydroxyphenylpyruvate dioxygenase

Nearly all dioxygenases contain iron or copper, which in most cases is a direct constituent of the enzyme protein. Iron may also be part of ferroprotoporphyrin IX (heme, D 24; Table 14). The metal activates the oxygen linked to the enzyme, a process which is associated with polarization of the complex. In the case of the iron-containing enzymes this may be formulated as follows:

$$E-Fe^{2+} + O_2 \rightleftharpoons E-Fe^{2+}-O-O \rightleftharpoons E-Fe^{3+}-O-O^-$$

The enzyme-oxygen complex adds to the substrate with the elimination of a proton and the E—Fe^{2+}-complex is regenerated.

Dioxygenases frequently break C=C-bonds with the formation of two carbonyl groups. Cyclic peroxides may be intermediates. Catechol 2,3-dioxygenase (metapyrocatechase), for instance, cleaves the aromatic ring of catechol next to a hydroxyl group (extradiol cleavage, Fig. 16). Catechol 1,2-dioxygenase (pyrocatechase) splits the ring of catechol between the hydroxylated C-atoms (intradiol cleavage).

Table 14. Dioxygenases and their cofactors

Enzyme	Type of reaction catalyzed	Cofactors
Metapyrocatechase (C 2.5)	Ring cleavage	Fe^{2+}
Pyrocatechase (C 2.5)	Ring cleavage	Fe^{3+}
Lipoxygenase (D 5.2.1)	Formation of an open peroxide	Fe^{2+}
Prostaglandin cyclooxygenase (D 5.2.6)	Formation of a cyclic peroxide	Heme, L-tryptophan, glutathione
Cysteamine dioxygenase (D 11.1)	Oxygenation of sulfur	Fe^{3+}, sulfide
Cysteine dioxygenase (D 11.1)	Oxygenation of sulfur	Fe^{2+}, NAD(P)H
Tryptophan 2,3-dioxygenase (D 22)	Ring cleavage	Heme
Kynurenate 7,8-hydroxylase (D 22.5)	Dihydroxylation	Pyridine nucleotide-linked flavoproteins
4-Hydroxyphenylpyruvate dioxygenase (D 23)	Hydroxylation, oxidative decarboxylation	Fe^{2+}, L-ascorbic acid, requires 2-oxoglutaric acid

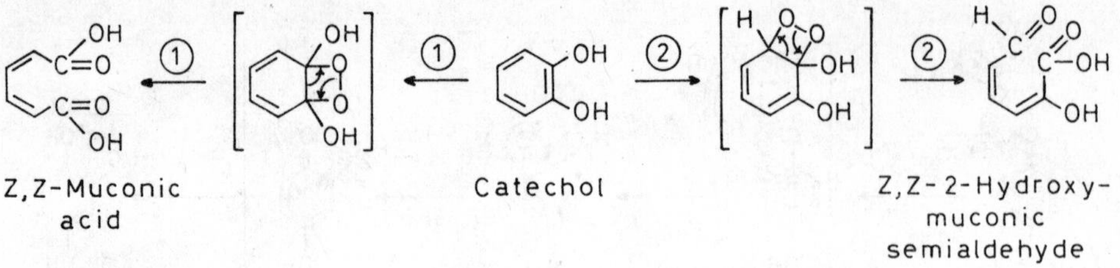

Fig. 16. Splitting of catechol by dioxygenases
1 Pyrocatechase; *2* metapyrocatechase

Luciferases are an interesting group of dioxygenases, which produce bioluminescence on reaction with special substrates (luciferins). Luciferins may have
diverse structures (D 11, D 24.2, D 25.4, and Fig. 17 for Renilla and Cypridina luciferin). Luminescence is a consequence of the oxygenation (formation of dioxethanone derivatives) and subsequent decarboxylation of luciferins.

References for Further Reading

86, 118, 343, 393, 442

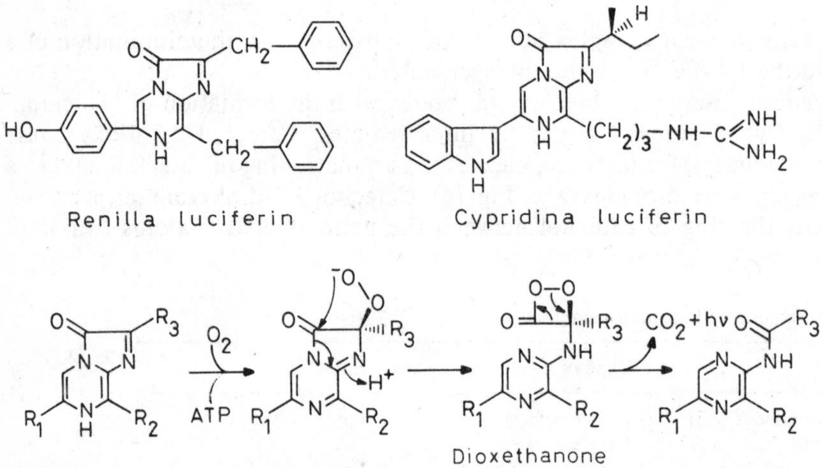

Fig. 17. Action of luciferases (Renilla luciferin 2-monooxygenase and Cypridina luciferin 2-mono-
oxygenase) on Renilla and Cypridina luciferin

C 2.6 Monooxygenases

C 2.6.1 Structure and Mechanism of the Reaction

Monooxygenases (mixed-function oxygenases) interact with molecular oxygen.
They introduce one atom of an oxygen molecule into the substrate (oxygenation), while the second oxygen atom reacts with a reduced cosubstrate forming

water (oxidation). The term "mixed-function" oxygenases reflects this double function. By the so-called external monooxygenases NADH, NADPH (D 16.1), L-ascorbic acid (D 1.2), tetrahydropteridines (D 25.3), etc. are used as cosubstrates according to the equation:

$$S + O_2 + RH_2 \longrightarrow SO + H_2O + R$$

$$RH_2 = \text{reduced cosubstrate}, \quad R = \text{oxidized cosubstrate}$$

A few monooxygenases split hydrogen from the substrate itself (internal monooxygenases):

$$SH_2 + O_2 \longrightarrow SO + H_2O$$

Examples of this group of enzymes are the amino acid oxygenases (C 2.6.3).

The hydrogen that is cleaved from the cosubstrates may be transferred in different ways to the oxygenating enzyme. In the simplest case it is taken over by a flavin enzyme which also carries out the oxygenation (Fig. 18). Enzymes of this type are, for instance, dehydrocyclopeptine epoxidase and cyclopenin m-hydroxylase from *Penicillium* (D 21.4.2).

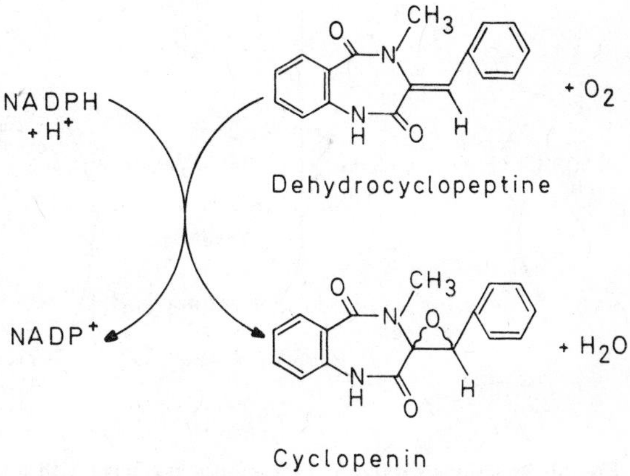

Fig. 18. Reaction catalyzed by a monooxygenase without an electron transport chain (dehydrocyclopeptine epoxidase, D 21.4.2)

More complicated are enzymes which dehydrogenate the cosubstrate by a separate protein which itself interacts with the oxygenating enzyme. An example of this type is phenylalanine 4-monooxygenase (Fig. 19, see also D 23).

The most complex monooxygenases, however, are the membrane-integrated enzyme complexes occurring, e.g., in liver and adrenocortex of mammals, which are involved in steroid hydroxylation (D 6.6.5). These complexes possess special electron transport chains with cytochrome P_{450} (D 24) at the terminal site (Fig. 20). The substrate is linked to oxidized cytochrome P_{450}. The cytochrome-substrate complex is reduced by an enzyme which has a "nonheme-iron" as prosthetic group, e.g., adrenodoxin. This protein reacts with a flavin enzyme which in turn oxidizes the cosubstrate, e.g., NADPH. The reduced cytochrome-substrate complex interacts with molecular oxygen and breaks down with the liberation of the oxygenated substrate and the formation of oxidized cytochrome.

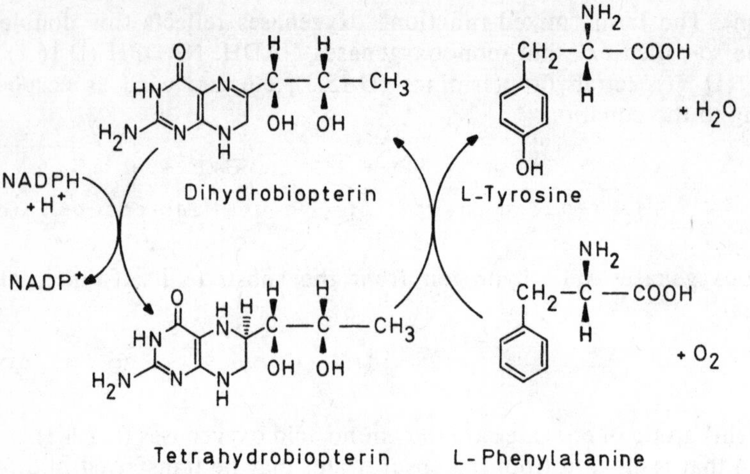

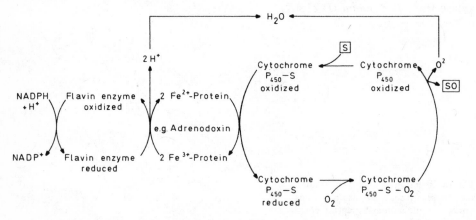

Fig. 19. Reactions carried out by a monooxygenase with a simple electron transport chain (phenylalanine 4-monooxygenase, D 23)

Fig. 20. Reactions catalyzed by a monooxygenase with a complex electron transport chain
S Substrate; *SO* oxygenated substrate

The most important types of reactions catalyzed by monooxygenases are discussed in the following.

References for Further Reading

70, 555, 579

C 2.6.2 Oxygenation of Amines and Thio Compounds

A number of monooxygenases are capable of oxygenating amines and thio compounds to hydroxy derivatives and oxides, respectively. The following types of reactions are involved:

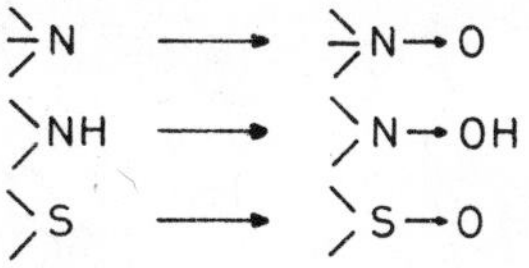

$$\text{>N} \longrightarrow \text{>N} \rightarrow \text{O} \qquad (D\,9.2.1)$$

$$\text{>NH} \longrightarrow \text{>N} \rightarrow \text{OH} \qquad (D\,9.2.2)$$

$$\text{>S} \longrightarrow \text{>S} \rightarrow \text{O} \qquad (D\,11.2)$$

References for Further Reading (see C 2.6.5)

C 2.6.3 Amino Acid Oxygenases

Amino acid oxygenases are internal monooxygenases (C 2.6.1). They catalyze the following reaction:

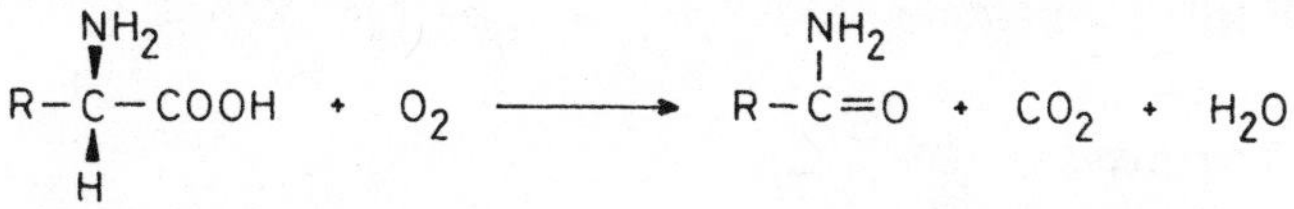

$$\underset{H}{\overset{NH_2}{R-C-COOH}} + O_2 \longrightarrow \overset{NH_2}{R-C=O} + CO_2 + H_2O$$

L-Amino acid (C_n)	Acid amide (C_{n-1})	R
L-Arginine	4-Guanidino-butyramide	$H_2N-\underset{NH}{\overset{\|\|}{C}}-NH-(CH_2)_3-$
L-Lysine	5-Amino-valeramide	$H_2N-(CH_2)_4-$
L-Tryptophan	Indole-3-acetamide	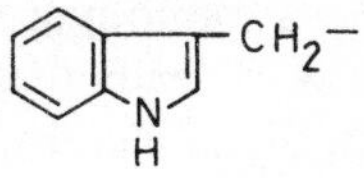 (indole)$-CH_2-$

References for Further Reading (see C 2.6.5)

C 2.6.4 Hydroxylation of Tetragonal Carbon Atoms and Oxidative Demethylation

Hydroxylation of tetragonal carbon atoms by monooxygenases replaces stereospecifically a hydrogen atom by a hydroxyl group. Nearby hydrogen atoms do not undergo any change in position or configuration. The monooxygenases directly attack the electrons of the C—H bond:

$$\underset{H}{\overset{H}{R-C-H}} \longrightarrow \underset{H}{\overset{H}{R-C-OH}}$$

The oxidative elimination of *O*-methyl and *N*-methyl groups is initiated in this way by hydroxylation of the CH_3-group. The resulting hydroxy derivatives decompose spontaneously to amino or hydroxy compounds and formaldehyde:

$$\text{>N--CH}_3 \longrightarrow \text{>N--CH}_2\text{--O--H} \rightleftharpoons \text{>N--H} + HCHO$$

$$\text{--O--CH}_3 \longrightarrow \text{--O--CH}_2\text{--O--H} \rightleftharpoons \text{--O--H} + HCHO$$

The hydroxylated primary products can be detected only when they become stabilized, e.g., by glycosylation. The formaldehyde formed may react spontaneously with different organic compounds, e.g., L-arginine, L-lysine, L-histamine, proteins, and nucleic acids. As products again hydroxymethylderivatives are built which may be transformed to more stable methyl and formyl derivatives. Different *N*-guanyl-hydroxymethyl derivatives of L-arginine (D 19) were found as normal constituents of human blood and urine built by the spontaenous reaction of L-arginine with free formaldehyde.

References for Further Reading (see C 2.6.5)

C 2.6.5 Hydroxylation of Unsaturated or Aromatic Compounds and the NIH-Shift

Epoxides are the first reaction products when monooxygenases attack C=C bonds. In the case of aromatic substrates the epoxides are very unstable and may be further transformed by the reactions shown in Fig. 21.

Pathway I demonstrates that by the addition of water a dihydrodiol is formed, which is either dehydrogenated to a dihydroxy derivative (D 22.5), or is transformed to a monohydroxylated compound with the elimination of water.

Pathway II leads directly to the monohydroxylated compounds. The epoxide is opened by an electrophilic agent, e.g., a proton, and the formed cationic intermediate stabilizes with the elimination of a proton (see transformation of squalene epoxide, D 6.6).

Pathway III shows the degradation of the epoxide by compounds containing an SH-group, e.g., glutathione (D 9.2). Thioethers are formed with or without elimination of water.

A shift of constituents may occur in pathway II. This shift was observed first by scientists of the National Institute of Health and thus is termed "NIH-shift". Some reactions which are accompanied by an NIH-shift are the hydroxylations of L-tryptophan (Fig. 255), L-phenylalanine (Fig. 267), and cinnamic acid (Fig. 296). During formation of 5-hydroxy-L-tryptophan, a hydrogen atom is shifted from position 5 to position 4 and during formation of L-tyrosine or p-coumaric acid the hydrogen atom of position 4 is shifted to the equivalent positions 3 and 5 (Fig. 22).

The NIH-shift is based on the translocation of a hydride ion or another substituent in its anionic form, which is located at the hydroxylated *C*-atom, e.g., a

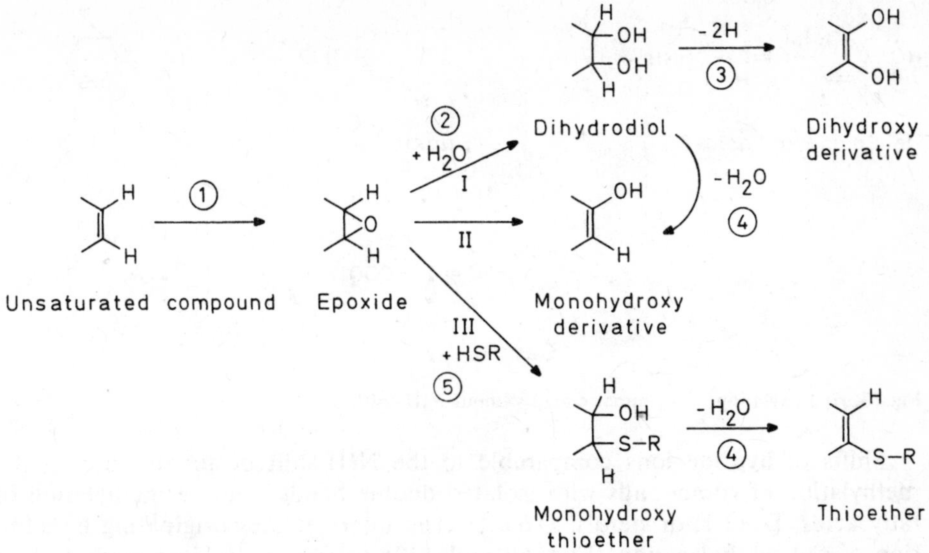

Fig. 21. Reactions of epoxides formed by monooxygenases from unsaturated substrates
1 Monooxygenase; 2 hydratase; 3 dehydrogenase; 4 dehydratase; 5 glutathione S-epoxidetransferase

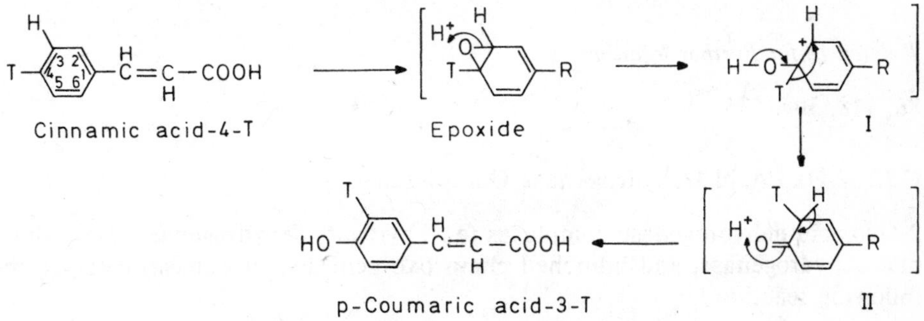

Fig. 22. NIH-shift during hydroxylation of cinnamic acid

chloride or a bromide anion, a negatively charged alkyl group, or another carbon chain as in homogentisic acid biosynthesis (Fig. 15). It takes place when the intermediate I in Fig. 22 is stabilized by the shift of electrons.

In contrast in caffeic acid biosynthesis, the intermediate I is stabilized by the loss of a proton (formation of intermediate II) and the hydrogen (or tritium) in o-position without any shift (Fig. 23).

The group that is most loosely bound is preferentially eliminated from intermediate II. In the presence of both tritium and a hydroxyl group (in the formation of caffeic acid-5-T), the tritium atom is eliminated, while in the presence of tritium and hydrogen (formation of p-coumaric acid-3-T), the hydrogen atom is preferentially removed, since the C—T bond is considerably stronger than the C—H bond.

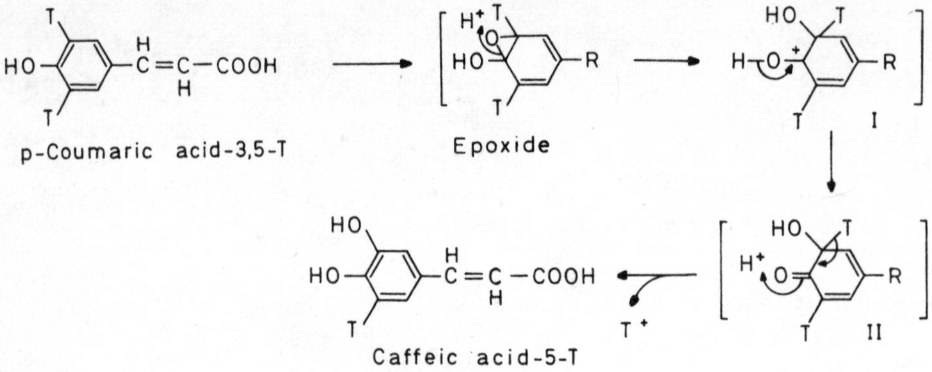

Fig. 23. Hydroxylation of p-coumaric acid without NIH-shift

Shifts of hydride ions comparable to the NIH-shift occur also during the methylation of compounds with isolated double bonds (see the methylation of fatty acids, D 5.2.1; or sterols, D 6.6.1). The intermediates originating by addition of the methyl cation (Figs. 69 and 129) correspond to intermediate I in Fig. 22 and are stabilized according to the rules mentioned above. Corresponding intermediates and a hydride ion shift may also be expected during substitution of products by the isopentenyl cation (D 6), but so far have not been detected.

References for Further Reading

86, 118, 307, 575

C 2.7 2-Oxo Acid Dehydrogenase Complexes

2-Oxo acid dehydrogenase complexes (e.g., pyruvate dehydrogenase, 2-oxoglutarate dehydrogenase, and branched chain oxo acid dehydrogenase) catalyze the following reaction:

$$R-CO-C\underset{OH}{\overset{O}{\diagup}} + CoASH + NAD^+ \longrightarrow R-C\underset{S-CoA}{\overset{O}{\diagup}} + CO_2 + NADH + H^+$$

Three enzymes forming a relatively stable complex participate in the overall process (Fig. 24): 2-oxo acid dehydrogenase, dihydrolipoamide acyltransferase, and the flavoprotein dihydrolipoamide reductase. Thiamine diphosphate (D 25.6), lipoic acid (Table 26), and coenzyme A (D 11) are involved. Acids shortened by one C-atom and activated by binding to CoA are the final products. The activated bond is formed during the dehydrogenation of the "activated aldehyde" by formation of the acylated lipoic acid (Fig. 25). The last steps are transfer of the acyl moiety to coenzyme A and regeneration of the lipoic acid.

Reference for Further Reading

834

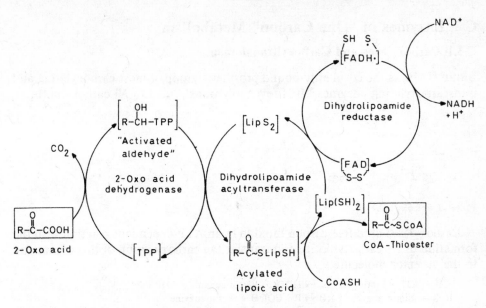

Fig. 24: Reactions involved in the dehydrogenation of 2-oxo acids
TPP thiamine diphosphate; *LipS$_2$* and *Lip(SH)$_2$* lipoyl moiety and its reduced form, respectively;
CoASH coenzyme A

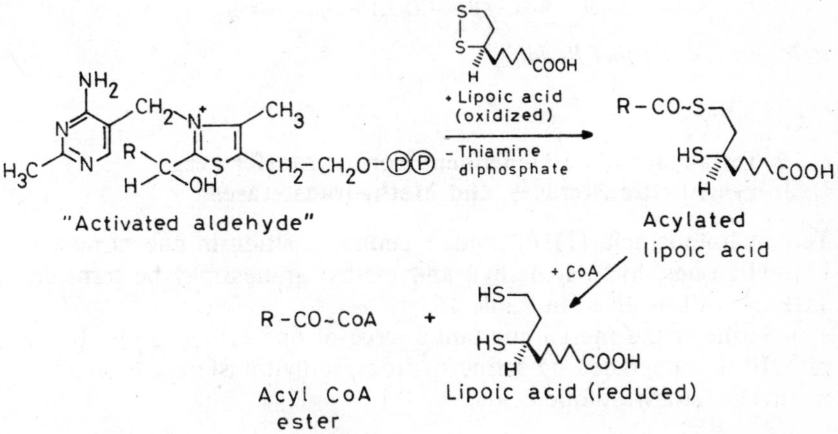

Fig. 25. Formation of acyl CoA esters from "activated aldehydes"

C 3 Enzymes of "One Carbon" Metabolism

C 3.1 Carboxylases and Carboxyltransferases

Biotin (D 10) is the covalently-bound prosthetic group of most carboxylating and transcarboxylating enzymes.[5] It is reversibly transformed to N^1-carboxybiotin:

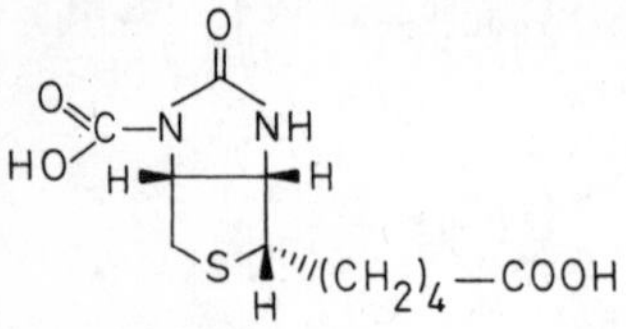

N^1-Carboxybiotin

Carboxylases bind free carbon dioxide in an ATP-dependent reaction with the formation of N^1-carboxybiotin, and transfer the carboxy group of this compound to the acceptor molecule:

$$CO_2 + ATP + \text{Biotin-enzyme} \rightarrow HOOC\text{—Biotin-enzyme} + ADP + Pi$$
$$HOOC\text{—Biotin-enzyme} + RH \rightarrow R\text{—COOH} + \text{Biotin-enzyme}$$

Carboxyl transferases catalyze the transfer of a carboxy group from a donor via N^1-carboxybiotin to an acceptor:

$$R_1\text{—COOH} + \text{Biotin-enzyme} \rightarrow R_1H + HOOC\text{—Biotin-enzyme}$$
$$HOOC\text{—Biotin-enzyme} + R_2H \rightarrow R_2\text{—COOH} + \text{Biotin-enzyme}$$

References for Further Reading

89, 745, 828

C 3.2 Tetrahydrofolate-Dependent Formyltransferases, Hydroxymethyltransferases, and Methyltransferases

Tetrahydrofolic acid (THF) holds a central position in one-carbon metabolism. Formyl groups, hydroxymethyl, and methyl groups may be transferred via the THF derivatives given in Table 15.

L-Serine is the most important source of one-carbon units. In the presence of THF it is degraded by serine hydroxymethyltransferase to glycine and 5,10-methylenetetrahydrofolic acid:

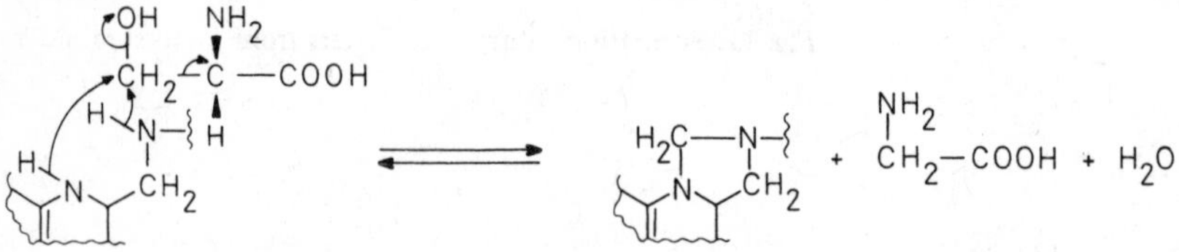

Tetrahydrofolic L-Serine   5,10-Methylene-   Glycine
       acid   tetrahydrofolic acid

---

[5]   In the posttranslational conversion of glutamyl to 4-carboxyglutamyl residues (D 17), however, vitamin K-dependent carboxylases are involved (D 21.2).

5,10-Methylenetetrahydrofolic acid can be transformed to the other cosubstrates by the reactions given in Fig. 26. On hydrolysis of 5,10-methylenetetrahydrofolic acid formaldehyde is formed. The compound therefore may be called "activated formaldehyde". Formic acid and methanol are derived from the other cosubstrates (Table 15). It is unknown, however, whether the formic acid released as an alarm pheromone and repellent by certain ants (E 4), or formic acid and methanol occurring in some plants are synthesized in this way.

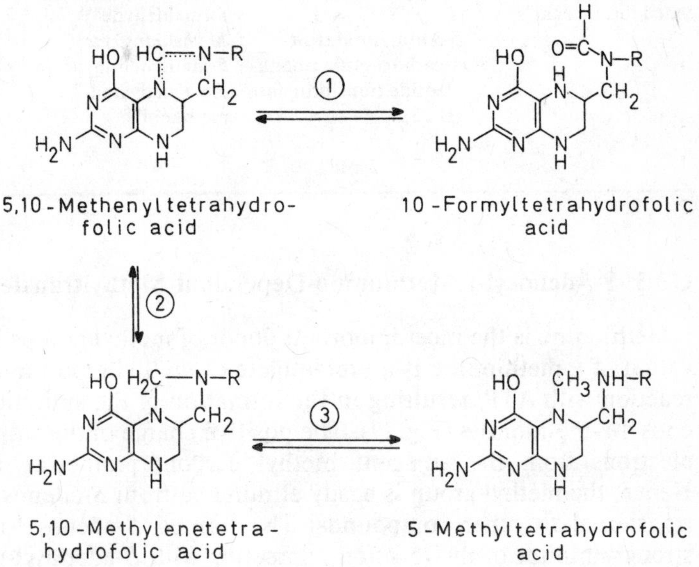

Fig. 26. Mutual conversion of tetrahydrofolic acid derivatives
1 Methenyltetrahydrofolate cyclohydrolase; *2* methylenetetrahydrofolate dehydrogenase; *3* 5,10-methylenetetrahydrofolate reductase

5-Methyltetrahydrofolic acid is a donor of methyl groups in selected reactions, e.g., in the formation of L-methionine (D 12). An intermediate is methylcobalamin (D 24.3). In the formation of methane (CH_4) by archaebacteria the methyl group of methylcobalamin is transferred by a methyltransferase to "coenzyme M" ($HS—CH_2—CH_2—SO_3H$, 2-mercaptoethane sulfonic acid) yielding methylcoenzyme M ($CH_3—S—CH_2—CH_2—SO_3H$, 2-methylmercaptoethane sulfonic acid). Methylcoenzyme M is reductively split to coenzyme M and to methane. The methylcoenzyme M reductase comprises an Ni-containing porphyrin (D 24) as coenzyme. Molecular hydrogen (H_2) acts as hydrogen source.

Methane is formed in anaerobic environments, e.g., in fresh water or marine sediments, flooded soils, anaerobic sludges, but also in the animal digestive system, from complex organic compounds such as carbohydrates, proteins, and lipids. In the rumen, methane appears to be almost exclusively produced by the reduction of CO_2. In nongastrointestinal habitats methane appears to be built chiefly by cleaving acetate into CO_2 and CH_4 and secondarily by the reduction of CO_2.

References for Further Reading

49, 88, 131, 153, 161, 162, 493, 755

Table 15. Tetrahydrofolate-dependent enzymes

Enzyme group/ cosubstrates	Representatives	Enzyme group/ cosubstrates	Representatives
Formyltransferase/ 5,10-methenyltetrahydrofolate, 10-formyltetrahydrofolate ("activated formic acid")	Glycinamide ribonucleotide transformylase (D 25)	*Hydroxymethyltransferases/* 5,10-methylenetetrahydrofolate ("activated formaldehyde")	Serine hydroxymethyltransferase (D 10)
	5-Aminoimidazole-4-carboxamide ribonucleotide transformylase (D 25)	*Methyltransferase/* 5-methyltetrahydrofolate ("activated methanol")	Tetrahydropteroylglutamate methyltransferase (D 12)
			Thymidylate synthase (D 26)

C 3.3 *S*-Adenosyl-ʟ-Methionine-Dependent Methyltransferases

ʟ-Methionine is the most important donor of methyl groups in metabolism. Activation of ʟ-methionine is a prerequisite of methyl group transfer. It proceeds by reaction with ATP, resulting in the formation of the sulfonium compound *S*-adenosyl-ʟ-methionine (Fig. 27). The positive charge of the sulfonium atom attracts electrons from the adjacent methyl carbon, promoting nucleophilic attack. Hence, the methyl group is easily eliminated from *S*-adenosyl-ʟ-methionine and transferred to other compounds. The formed *S*-adenosyl-ʟ-homocysteine is a strong inhibitor of the reaction competing with *S*-adenosyl-ʟ-methionine. *S*-Adenosyl-ʟ-homocysteine inhibition probably is a general mechanism for the regulation of methyltransferase activities. In vivo *S*-adenosyl-ʟ-homocysteine may be

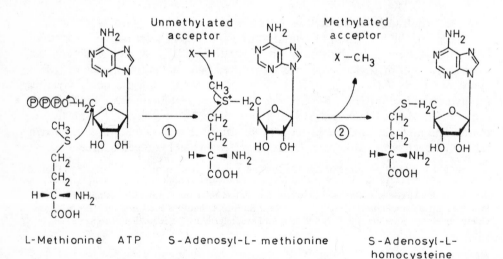

L-Methionine   ATP   S-Adenosyl-L-methionine   S-Adenosyl-L-homocysteine

Fig. 27. Synthesis and transformation of *S*-adenosyl-ʟ-methionine
1 Methionine adenosyltransferase; *2* methyltransferase

Table 16. Secondary compounds whose methyl groups originate form S-adenosyl-L-methionine

a) *Compounds with O-methyl-groups*	c) *Compounds with C-methyl groups*
Pectins (D 1.4.1)	L-Mycarose, D-cladinose (D 1.1)
Colchicine (D 23.1.2)	Tuberculostearic acid (D 5.2.1)
Ferulic acid, sinapic acid (D 23.2.1)	Ergosterol (D 6.6.1)
Lignins (D 23.2.3)	Cobalamins (D 24.3)
b) *Compounds with N-methyl groups*	d) *Compounds with S-methyl groups*
Betaines (D 8.1)	S-Methyl-L-cysteine (D 11.2)
Nicotine (D 16.1)	S-Methyl-L-methionine (D 12)
Gramine (22.1)	Dimethyl-β-propiothetin (D 12)
1-Methyladenine (D 25.1)	5'-Methylthioadenosine (D 12.1)

cleaved to form adenosine and L-homocysteine by S-adenosylhomocysteine hydrolase. S-Adenosyl-L-methionine-dependent methyltransferases show a surprisingly strict specificity toward only one class of substrates or to a limited number of closely related compounds, and to specific positions in the substrate molecule. A few compounds formed by methylation with S-adenosyl-L-methionine are given in Table 16. During transmethylation all hydrogen atoms of the methyl group are transferred to the acceptor molecule. (However, one hydrogen atom is eliminated during C-methylation of unsaturated compounds in a reaction following the methyl transfer, Figs. 69 and 129.) The methyl transfer has been shown to occur with overall inversion of the methyl group:

$$H-\underset{D}{\overset{T}{C}}-\overset{+}{\underset{\text{Adenosine}}{S}}-CH_2-CH_2-\underset{H}{\overset{NH_2}{C}}-COOH \longrightarrow H-\underset{D}{\overset{T}{C}}-\text{Acceptor}$$

Thus, an S_N2 transition state including donor and acceptor is likely.

References for Further Reading

131, 142, 613, 775, 821

C 4 2-Oxo Acid Decarboxylases

Enzymes which degrade 2-oxo acids to CO_2 and aldehydes contain thiamine diphosphate (D 25.6) as coenzyme. The oxo acid is added in a reversible reaction to carbon atom 2' of the thiazole ring of thiamine diphosphate giving a chiralic 2-hydroxy acid derivative. This compound is decarboxylated and split to an aldehyde and thiamine diphosphate. Figure 28 shows this sequence of reactions for the enzyme pyruvate decarboxylase which splits pyruvate to acetaldehyde and CO_2.

References for Further Reading

868, 869

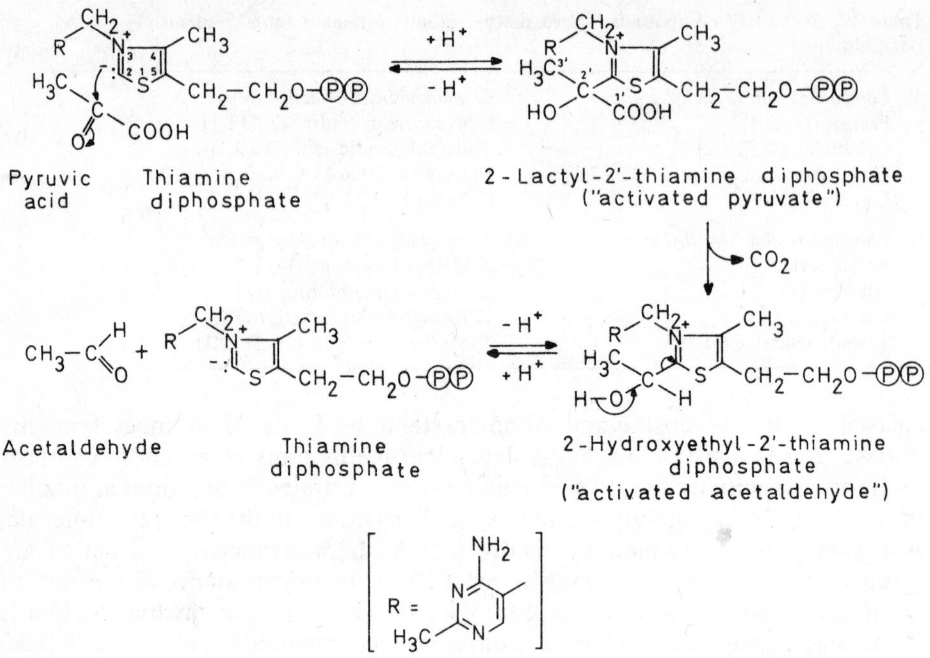

Fig. 28. Formation of acetaldehyde from pyruvic acid by pyruvate decarboxylase

C 5 Aminotransferases and Amino Acid Decarboxylases

The prosthetic group of these enzymes is usually pyridoxal-5'-phosphate (D 2), a compound forming Schiff bases (azomethines) with amino acids and amines (Fig. 29). The Schiff base provides a conjugated system of double bonds extending from the reaction site to the strongly electrophilic pyridine nitrogen atom, and aiding the displacement of a pair of electrons adjacent to the α-carbon of the amino acid. Depending on the type of enzyme, either a proton is eliminated, as is the case in transamination of amino acids and amines (Fig. 29, pathway A), or the carboxyl group is lost, as is the case in the formation of amines from amino acids (Fig. 29, pathway B).

Aminotransferases catalyze a binary mechanism illustrated by the following equations:
a) Amino acid$_1$ (amine$_1$) + Pyridoxal-5'-phosphate-enzyme $\rightleftarrows$
   Oxo acid$_1$ (aldehyde$_1$) + Pyridoxamine-5'-phosphate-enzyme
b) Oxo acid$_2$ (aldehyde$_2$) + Pyridoxamine-5'-phosphate-enzyme $\rightleftarrows$
   Amino acid$_2$ (amine$_2$) + Pyridoxal-5'-phosphate-enzyme

The decarboxylation of amino acids proceeds with net retention of the configuration at C-2:

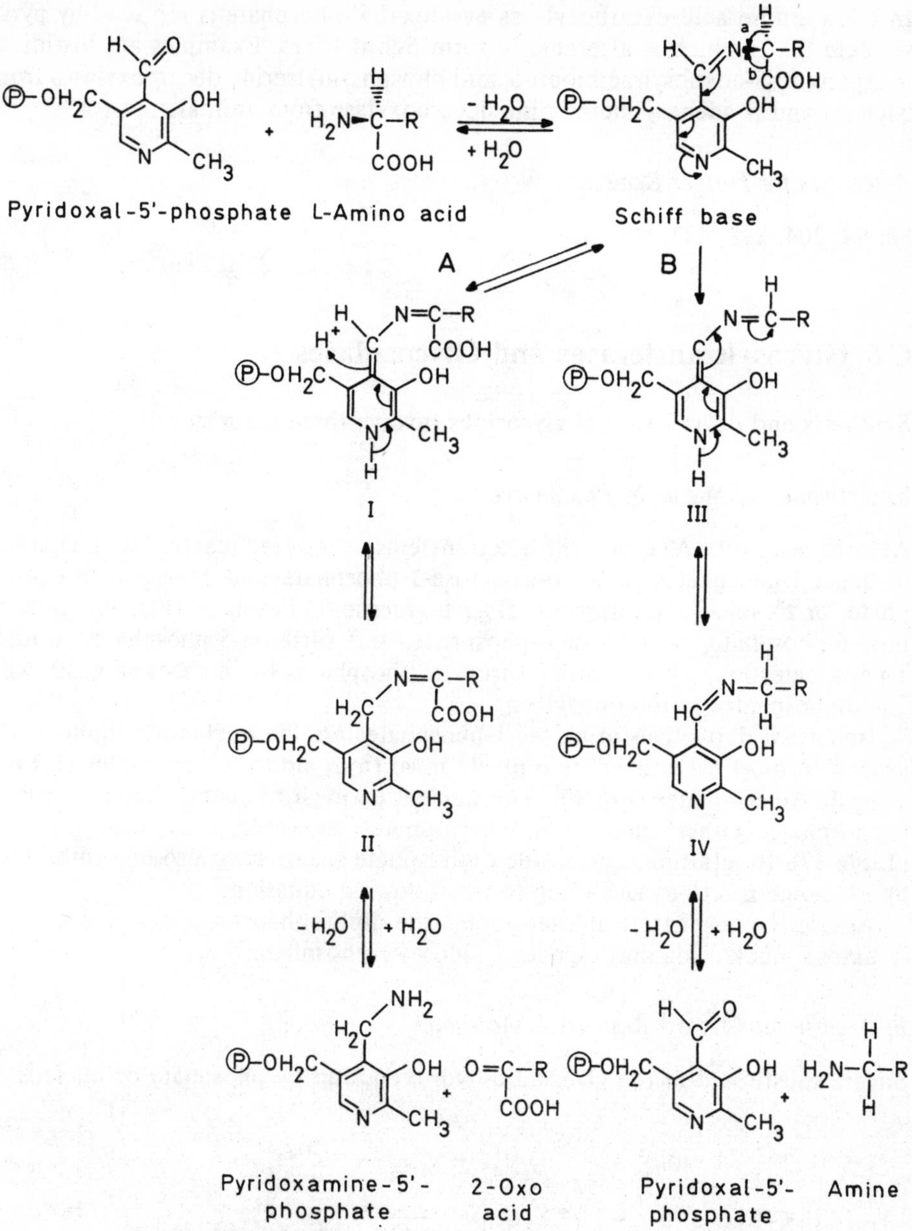

Fig. 29. Transamination and decarboxylation of amino acids by pyridoxal-5'-phosphate-containing enzymes

In a few amino acid decarboxylases pyridoxal-5'-phosphate is replaced by pyruvic acid (D 2) which is also able to form Schiff bases. Examples are histidine, α-aspartate, S-adenosylmethionine, and phosphatidylserine decarboxylases from bacteria and S-adenosylmethionine decarboxylase from animals.

References for Further Reading

88, 94, 204, 242, 433, 505, 628

C 6 Glycosyltransferases and Glycosidases

Synthesis and degradation of glycosides involve three main steps:

a) Activation of Sugars by Phosphorylation

Aldoses react with ATP in a phosphotransferase-catalyzed reaction (C 1.1) either to sugar-1-phosphates, e. g., D-galactose-1-phosphate and L-arabinose-1-phosphate, or to sugar-ω-phosphates, e. g., D-glucose-6-phosphate (Fig. 30), D-mannose-6-phosphate, D-galactose-6-phosphate, and D-ribose-5-phosphate. Aldose-1-phosphates may also be formed from ω-phosphates by mutases (Fig. 30, with 1,ω-diphosphates as intermediates).

Important derivatives of aldose-1-phosphates are the nucleoside diphosphate sugars. In most cases they are built de novo from aldose-1-phosphates and nucleoside triphosphates (Fig. 30). The nucleotide most frequently used is uridine triphosphate. Other nucleoside triphosphates, however, may react as well (Table 17). In addition, nucleoside diphosphate sugars may also be synthesized by exchange reactions according to the following equation:

Aldose$_1$-1-phosphate + aldose$_2$-nucleoside diphosphate $\rightleftarrows$
aldose$_1$-nucleoside diphosphate + aldose$_2$-1-phosphate

b) Transfer of Glycosyl Residues to Acceptors

Sugars substituted at the glycosidic hydroxy group by phosphate or nucleoside

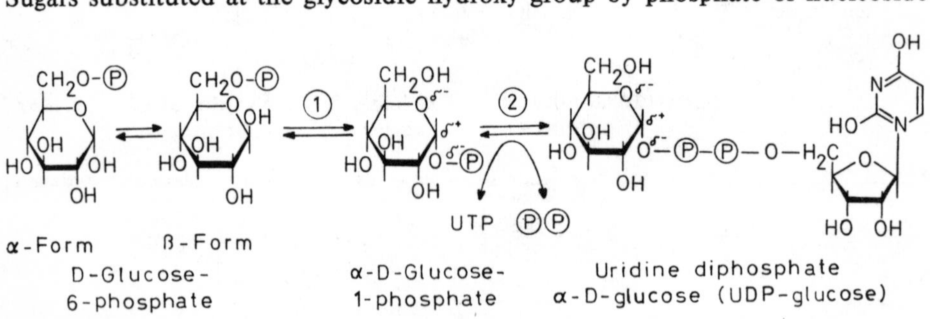

Fig. 30. Formation of "activated" glucose (UDP-glucose)
1 Phosphoglucomutase; *2* glucose-1-phosphate uridylyltransferase

Table 17. Important sugar nucleotides found in living beings

Nucleotide	Sugar moiety	Nucleotide	Sugar moiety
UDP	D-Glucose, D-galactose, D-fructose, D-xylose, L-arabinose, D-glucuronic acid, D-galacturonic acid, L-rhamnose, N-acetyl-D-glucosamine, N-acetyl-D-galactosamine	GDP	D-Glucose, D-mannose, L-fucose
		ADP	D-Glucose
		TDP	D-Glucose, D-galactose, L-rhamnose
		dUDP	D-Glucose, D-galactose

diphosphate residues are "activated" due to the disturbance of resonance between their α- and β-forms (Fig. 30) and the dislocation of electrons (C 1.1).

The "activated" sugars have a positive fractional charge at the glycosidic C-atom, which in the nucleoside diphosphate derivatives is even larger than in the sugar-1-phosphates. They easily undergo nucleophilic substitution at the glycosidic C-atom (glycosyl transfer). The most important glycosyl acceptors are hydroxylated compounds, amines, substances with SH-groups, and products with acidic C-atoms corresponding to the classes of glycosides mentioned in D 1.4. Most glycosides are formed with nucleoside diphosphate sugars as glycosyl donors.

Aldose-1-phosphates and nucleoside diphosphate sugars possess the α-configuration at the glycosidic C-atom. Most of the products formed by glycosyltransferases, however, show the β-configuration. In the biosynthesis of these compounds a Walden conversion occurs at the glycosidic C-atom during the nucleophilic attack by the glycosyl acceptor molecule.

c) Hydrolysis of Glycosides by Glycosidases

Glycosidases mediate the nucleophilic attack of a hydroxyl ion at the glycosidic C-atom according to the following equation:

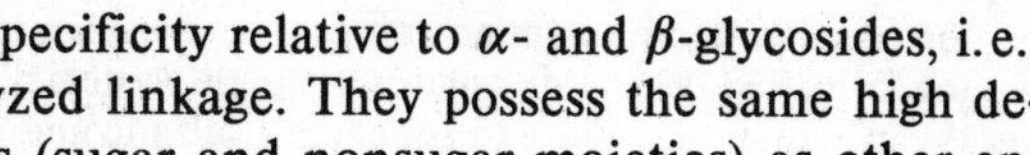

Glycoside   Sugar   Aglycone

Though the reaction of glycosides with water is reversible, hydrolysis is the favored reaction.

Glycosidases show an absolute specificity relative to α- and β-glycosides, i.e., to the configuration of the hydrolyzed linkage. They possess the same high degree of specificity to the substrates (sugar and nonsugar moieties) as other enzyme groups involved in secondary metabolism.

References for Further Reading

65, 194, 227, 362, 679

7*

D Structure, Biosynthesis, and Metabolism of Secondary Products in Microorganisms, Plants, and Animals

Secondary products are derived from compounds of primary metabolism and in this chapter they are arranged according to their primary metabolic precursors. A survey of the relations between primary and secondary metabolism is given in Fig. 31. The exact branch points of secondary metabolic pathways from primary metabolism are indicated at the beginning of the following sections.

In comparison to primary metabolism the biosynthesis of secondary products is more complex and the number of secondary compounds is much larger than that of primary metabolic substances. This is caused by a combination of the following principles:

a) Occurrence of Many Different Types of Chemical Reactions

These reactions resemble, on the one hand, those of primary metabolism (group a) or are unique for secondary metabolism (group b). The enzymes catalyzing the reactions evolved from primary metabolic enzymes either by altering the substrate specificity (group a enzymes) or the reaction specificity (group b enzymes).

Reactions resembling those occurring in primary metabolism are:
— the transformation of carboxylic acids to CoA esters by acid-thiol ligases and CoA-transferases (C 1.2)
— hydrogenations and dehydrogenations catalyzed by dehydrogenases (C 2.1, C 2.2)
— reactions of one carbon metabolism brought about by carboxylases, carboxyltransferases, and tetrahydrofolate-dependent enzymes (C 3)
— decarboxylations and aminations/deaminations catalyzed by pyridoxal-5′-phosphate-dependent enzymes (C 5), and
— the formation and breakdown of glycosides by glycosyltransferases and glycosidases (C 6).

Examples of reactions unique to secondary metabolism are:
— the oxidative coupling of phenols by phenol oxidases (C 2.3.1)
— halogenations catalyzed by halogenoperoxidases (C 2.4.2)
— the reduction of double bonds to triple bonds (D 5.2.5)
— the transformation of amino groups to hydroxylamino and cyano groupings (D 8.3)
— the formation of peptide bonds via the thiolesters of amino acids (D 9.2)
— the oxidation of sulfides to sulfoxides, sulfinic and sulfonic acids (D 11.1, D 11.2)

Of special interest are polymerization reactions in which the intermediates (in contrast, e. g., to fatty acid biosynthesis in primary metabolism, D 5.2) are not fully processed before the next monomer is added to the growing chain. The compounds formed thus contain several reactive centers and may be transformed in different ways.

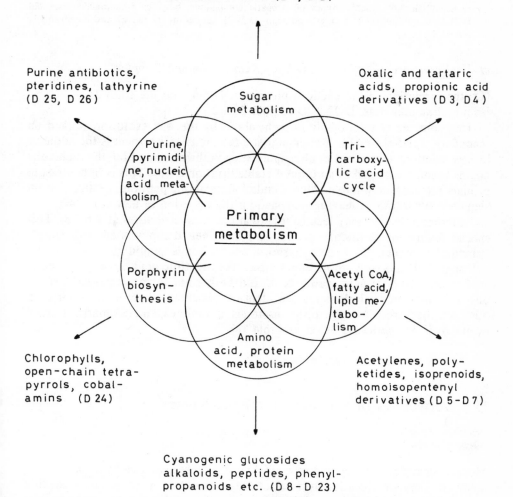

Secondary monosaccharides,
sugar carboxylic acids,
heterosides etc. (D 1, D 2)

Purine antibiotics,
pteridines, lathyrine
(D 25, D 26)

Oxalic and tartaric
acids, propionic acid
derivatives (D 3, D 4)

Sugar
metabolism

Purine,
pyrimidi-
ne, nucleic
acid meta-
bolism

Tri-
carboxy-
lic acid
cycle

Primary
metabolism

Porphyrin
biosyn-
thesis

Acetyl CoA,
fatty acid,
lipid me-
tabo-
lism

Amino
acid, protein
metabolism

Chlorophylls,
open-chain tetra-
pyrrols, cobal-
amins   (D 24)

Acetylenes, poly-
ketides, isoprenoids,
homoisopentenyl
derivatives (D 5-D 7)

Cyanogenic glucosides
alkaloids, peptides, phenylpropanoids etc. (D 8-D 23)

Fig. 31. Derivation of secondary products from the different areas of primary metabolism

Examples are
— the large numbers of compounds derived from the highly reactive polyoxo acid intermediates
  formed as the first step in polyketide biosynthesis (D 5.3), and
— the many secondary products built from the prenyl diphosphates, which are reactive intermediates in isoprenoid biosynthesis (D 6).

b) The Use of a Multitude of Primary Metabolic Compounds as Precursors and the Formation of Secondary Products by Combining Precursors from Different Fields of Primary Metabolism

Examples for the formation of secondary products from more than one precursor type are:
— the glycosylation of many groups of secondary products with the formation of heterosides
  (D 1.4.2)

— the biosynthesis of fatty acids and polyketides with unusual starter molecules, like propionyl CoA, isobutyryl CoA, aleprolic acid CoA ester, anthranilyl CoA, or p-coumaroyl CoA (D 5.2.1, D 5.3), and

— the prenylation of different types of aromatic compounds, e. g., of 4-hydroxybenzoic acid (D 21.3), quinolines (D 21.4.2), L-tryptophan (D 22.2), coumarins (D 23.2.2), and homogentisic acid (D 23.4).

c) The Occurrence of Pathways which are Rather Long and Complex

(Cf. the formation of the cyclic tripterpenes, D 6.6, and the iridoid indole and isoquinoline alkaloids, D 22.3, D 22.4, D 23.1.2.)

This complexity is probably brought about by the way evolution worked on secondary metabolism (E). A characteristic feature in this respect is the addition of new reactions to the end of already existing pathways, by which the metabolic chains became enlarged in an unpredictable, irrational way. This is in contrast to primary metabolism where no "loose ends" of metabolic pathways exist and every sequence fits to others leaving no space for the insertion of new reactions.

Different pathways may lead to identical chemical structures (Table 18). This special feature of secondary metabolism has been demonstrated with simple structural elements, like hydroxyl groups, double bonds, peptide bonds, and phenyl rings, but also with whole compounds like nicotinic acid, urea, p-coumaric acid, protocatechuic acid, gallic acid, and berberine. In most cases, however, only certain structural elements are identical, whereas others differ from each other (cf. the formulas of the anthraquinones, quinolines, and coumarins formed on the different pathways listed in Table 18).

D 1 Derivatives of Primary Metabolic Sugars

Sugar Chemistry

Monosaccharides are polyhydroxy aldehydes (aldoses) or polyhydroxy ketones (ketoses). They are classified as trioses, tetroses, pentoses, hexoses, etc. according to the number of carbon atoms. With the exception of dihydroxyacetone, monosaccharides contain one or more chiral C-atoms. Most monosaccharides carry an oxygen-containing group at each carbon atom. However, there are also deoxy sugars which have one or several oxygen-free carbon atoms (Fig. 32).

Monosaccharides with the configuration of D-glyceraldehyde (Fig. 32) at the chiral C-atom furthest away from the carbonyl function belong to the group of D-sugars, e. g., D-ribose, D-glucose, D-fructose. Their counterparts are the L-sugars, e. g., L-rhamnose, L-fucose, L-streptose (Fig. 33).

Monosaccharides exist as cyclic hemi-acetals, called lactols, rather than in the oxo-form given in Figs. 32 and 33. Intramolecular substitution of the carbonyl C-atom by one of the hydroxy groups yields tetrahydropyrane derivatives (pyranoses, shown for glucose in Fig. 34) or tetrahydrofurane derivatives (furanoses, shown for D-fructose-6-phosphate in Fig. 37). This cyclization is accompanied by

Table 18. Different metabolic pathways to similar chemical structures

---

Hydroxyl groups:
Introduction by monooxygenases (C 2.6.4) or formation by the hydratation of unsaturated compounds (C 2.6.5)

C=C double bonds:
Derivation form saturated compounds by dehydrogenases (C 2) or building by dehydration of hydroxylated substances

Peptide bonds:
Synthesis via aminoacyl-tRNAs on ribosomes (D 9.1) or formation from activated amino acids with the participation of simple enzymes or enzyme systems (D 9.2)

Phenyl rings:
Synthesis by dehydration of dehydroshikimic acid, chorismic acid, isochorismic acid, or prephenic acid (D 21, D 21.1, D 23), derivation by ring closure from polyines (D 5.2.5), or formation by the elimination of formaldehyde from sterane derivatives (synthesis of estranes, D 6.6.7)

Nicotinic acid:
Biosynthesis from aspartic acid and a glycerol-derived C_3-unit (D 16.1) or derivation from 3-hydroxyanthranilic acid (D 22)

Urea:
Formation from L-ornithine (D 19) or on the glycine-allantoin cycle (D 25.5)

p-Coumaric acid:
Synthesis by phenylalanine ammonia-lyase or tyrosine ammonia-lyase from L-tyrosine (D 23.2.1), derivation from cinnamic acid by cinnamate monooxygenase (D 23.2.1), formation from 4-hydroxyphenyl pyruvic acid via 4-hydroxyphenyl lactic acid (D 23.2.1)

Protocatechuic acid, gallic acid:
Derivation from dehydroshikimic acid (D 21) or formation from dihydrocinnamic acids (D 23.2.5)

Anthraquinones:
Synthesis on the polyketide pathway (D 5.3.6) or derivation from isochorismic acid (D 21.2)

Quinolines:
Formation with participation of anthranilic acid by reaction with acetyl CoA or a 3-oxo acid (D 21.4.2), or by rearrangement of benzodiazepines (D 21.4.2). Synthesis from indole derivatives after splitting the bond between N-1 and C-2 (*Chinchona* alkaloids, D 22.4) or after splitting the bond between C-2 and C-3 (kynurenic acid derivatives, D 22.5)

Coumarins:
Formation from o-coumaric acid derivatives (D 23.2.2) or derivation from flavonoids (D 23.3.3)

---

the formation of a new chiral center and a new hydroxy group, the glycosidic hydroxyl. Monosaccharides differing in the stereochemistry of this center are called anomeric and are distinguished by the Greek letters α and β, see the formulas of α-D- and β-D-glucose (Fig. 34). Using the Fischer projection, i. e., the arrangement of the carbon chain of monosaccharides in such a way that
— the bonds between the carbon atoms are behind the plane and those to the left-hand and right-hand substituents are in front of the plane, and
— the C-atom with the highest state of oxidation is at the top of the molecule
in the α-form, the glycosidic (anomeric) hydroxyl group has the same configuration as the hydroxyl group determining the D- or L-configuration. In the β-form

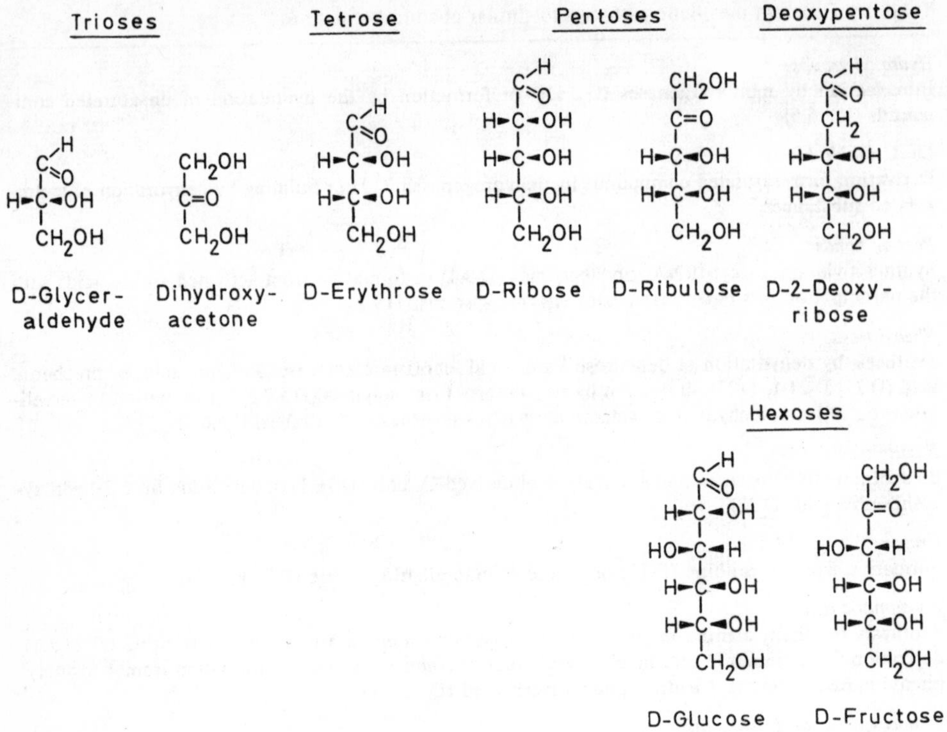

Trioses **Tetrose** **Pentoses** **Deoxypentose**

D-Glycer-aldehyde Dihydroxy-acetone D-Erythrose D-Ribose D-Ribulose D-2-Deoxy-ribose

Hexoses

D-Glucose D-Fructose

Fig. 32. Monosaccharides of primary metabolism

the hydroxyl group points in the opposite direction. Both anomeric forms of a given monosaccharide equilibrate in solution via the oxo-form.

The six-membered ring of pyranoses takes on the chair conformation. Possible conformers of α- and β-D-glucose are given in Fig. 34. Several conformers of a sugar may coexist and are easily converted into each other. For convenience in this book therefore the oxo-form or more frequently the Haworth projection of sugar molecules is used, i.e., the sugar molecule is represented by a flat ring with the oxygen atom in back, as is shown in Fig. 34 for α- and β-D-glucose.

Formation of Sugars in Primary Metabolism

Three metabolic pathways are of significance in the biosynthesis, interconversion, and degradation of sugars in primary metabolism:

— the Embden-Meyerhoff-Parnas pathway (D 2) on which D-glucose, D-fructose, D-glyceraldehyde, and dihydroxyacetone phosphate are synthesized and degraded

— the pentosephosphate cycle on which C_3-, C_4-, C_5-, C_6-, and C_7-sugars are interconverted (Fig. 35) and

— the deoxygenation of ribose at the level of the nucleotides.

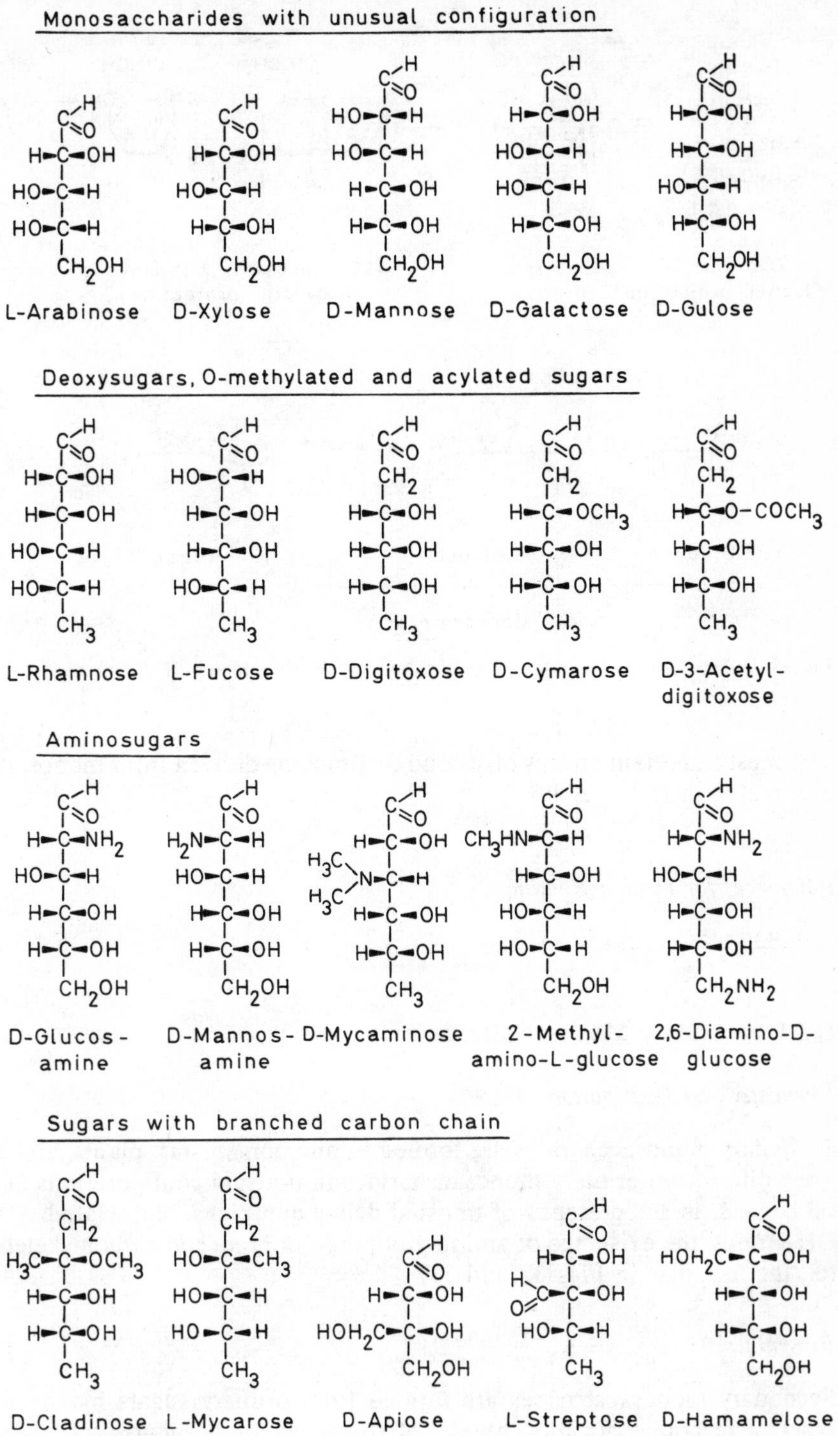

Monosaccharides with unusual configuration

L-Arabinose   D-Xylose   D-Mannose   D-Galactose   D-Gulose

Deoxysugars, O-methylated and acylated sugars

L-Rhamnose   L-Fucose   D-Digitoxose   D-Cymarose   D-3-Acetyl-digitoxose

Aminosugars

D-Glucos-amine   D-Mannos-amine   D-Mycaminose   2-Methyl-amino-L-glucose   2,6-Diamino-D-glucose

Sugars with branched carbon chain

D-Cladinose   L-Mycarose   D-Apiose   L-Streptose   D-Hamamelose

Fig. 33. Secondary monosaccharides

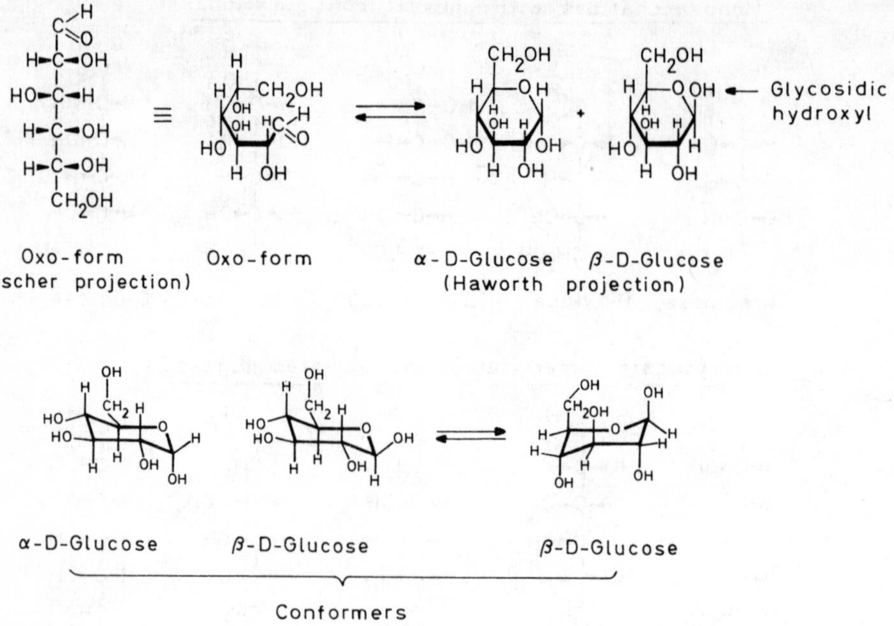

Oxo-form   Oxo-form   α - D - Glucose   β - D - Glucose
(Fischer projection)   (Haworth projection)

α - D - Glucose   β - D - Glucose   β - D - Glucose

Conformers

Fig. 34. Different forms and projections of the molecule of D-glucose

The most important groups of secondary products derived from monosaccharides are treated in D 1.1-D 1.4.

References for Further Reading

616, 829

D 1.1 Secondary Monosaccharides

Chemistry and Distribution

Secondary monosaccharides are formed in microorganisms, plants, and animals. They differ from primary monosaccharides in unusual configurations at the chiral centers, in the presence of unusual deoxy groupings, methylated or acylated OH-groups, the existence of amino groupings, of branched carbon skeletons, etc. (cf. the formulas in Figs. 32 and 33).

Biosynthesis

Secondary monosaccharides are formed from primary sugars by the following types of reactions (for biosynthesis of some secondary monosaccharides by decarboxylation of uronic acids, see D 1.2):

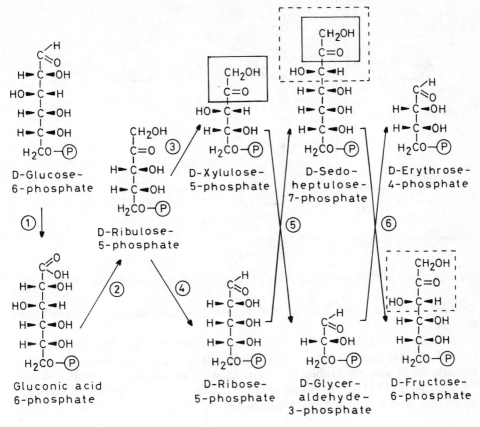

Fig. 35. Interconversion of sugars by reactions of the pentosephosphate cycle
1 Glucose-6-phosphate dehydrogenase, glucose lactonase; *2* 6-phosphogluconate dehydrogenase; *3* ribulose-5-phosphate 3-epimerase; *4* ribulose-5-phosphate isomerase; *5* transketolase; *6* transaldolase

a) *Epimerization.* Epimerases change the configuration of chiralic *C*-atoms. Examples are the epimerization at C-2 and C-4 of D-glucose in the formation of D-mannose and D-galactose, respectively (cf. the formulas in Figs. 32 and 33).

b) *Deoxidation.* 6-Methylpentoses, e. g., L-rhamnose (Fig. 33), originate from the corresponding hydroxylated monosaccharides.

For example, *d*TDP-L-rhamnose is formed from *d*TDP-D-glucose by the pathway given in Fig. 36. *d*TDP-D-glucose oxidoreductase, an enzyme with a tightly bound pyridine nucleotide, transfers a hydrogen atom from C-4 to C-6 of the substrate. The TDP-4-oxo-6-deoxyglucose formed as an intermediate undergoes epimerization (probably via the enol-form) and finally is reduced by NADH.

2-Deoxyaldoses, e. g., D-digitoxose and D-cymarose (Fig. 33) may be formed by direct reduction at the level of nucleotides by pyridine nucleotide-dependent dehydrogenases (C 2.1) in analogy to the biosynthesis of D-2-deoxyribose from D-ribose.

c) *Methoxylation.* Formation of sugars with methoxy groups, e. g., D-cymarose

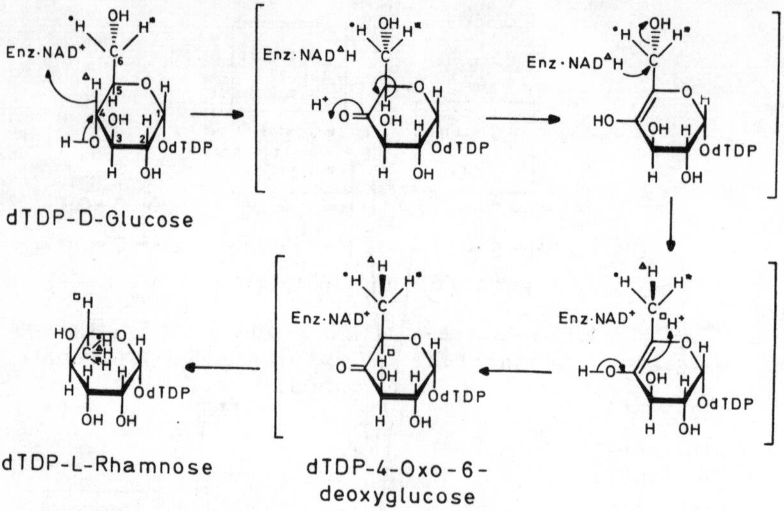

Fig. 36. Synthesis of *d*TDP-rhamnose from *d*TDP-glucose by *d*TDP-D-glucose oxidoreductase

(Fig. 33), is catalyzed by *S*-adenosyl-L-methionine-dependent methyltransferases (C 3.3).

d) Acylation. Acylated OH-groups, e. g., in D-3-acetyldigitoxose (Fig. 33), are formed by acyl transferases. Sugar sulfate esters, e. g., L-galactose-6-sulfate, a constituent of agarose (Table 21) are built with adenosine-3'-phosphate-5'-sulfatophosphate (D 11) as the donor of the sulfate group. They are precursors of anhydro sugars, e. g., 3,6-anhydro-L-galactose:

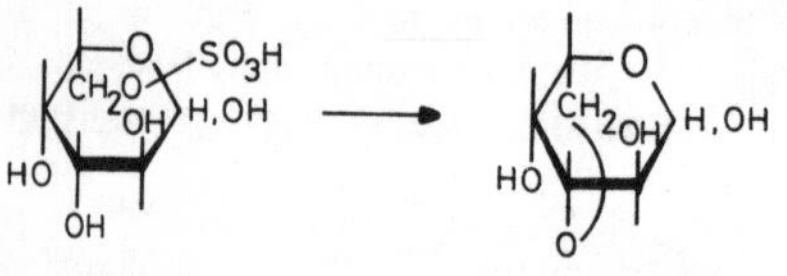

L-Galactose-6-   3,6-Anhydro-L-
sulfate   galactose

e) Amination. Amino sugars carrying an amino group in place of a hydroxyl group are formed from the corresponding ketoses. Ammonia or glutamine act as amino group donors.

Of significance are D-glucosamine and D-mannosamine (Fig. 33). D-Glucosamine-6-phosphate is formed from D-fructose-6-phosphate on the pathway given in Fig. 37. It is subsequently transformed to *N*-acetylglucosamine-6-phosphate and *N*-acetylmuramic acid. Synthesis of D-glucosamine-6-phosphate includes an intramolecular oxidoreduction at positions 1 and 2 transforming the ketose to an aldose derivative.

N-Acetyl-D-glucosamine is the precursor of chitin in fungi and animals, e. g., insects (D 1.4.1). Together with *N*-acetylmuramic acid it is a constituent of the cell wall of bacteria (D 9.2.4). *N*-Acetyl-D-mannosamine-6-phosphate is formed

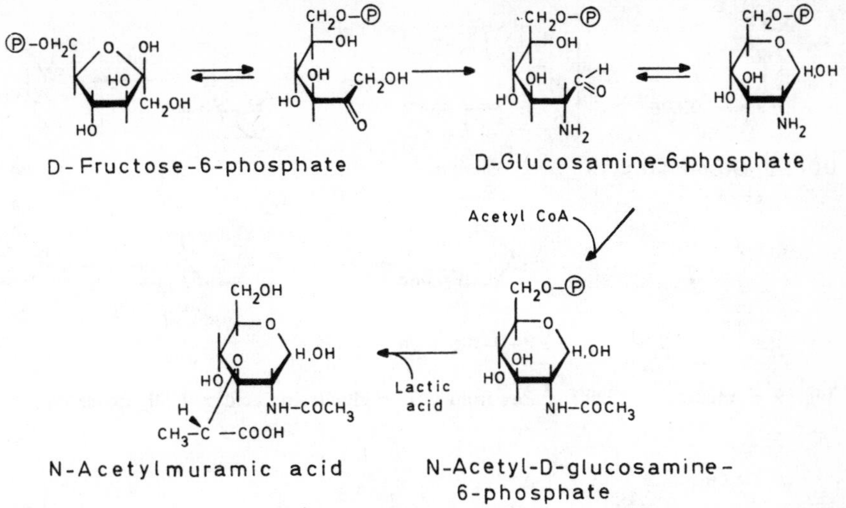

Fig. 37. Formation and conversion of D-glucosamine derivatives

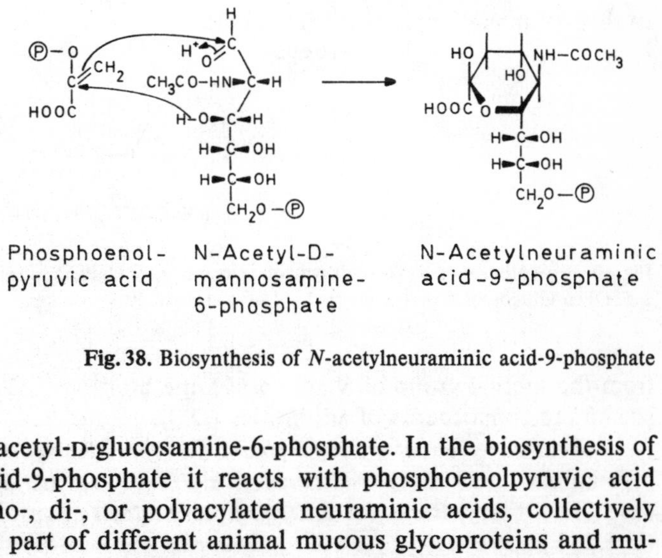

Fig. 38. Biosynthesis of N-acetylneuraminic acid-9-phosphate

by epimerization of N-acetyl-D-glucosamine-6-phosphate. In the biosynthesis of N-acetylneuraminic acid-9-phosphate it reacts with phosphoenolpyruvic acid (Fig. 38). Various mono-, di-, or polyacylated neuraminic acids, collectively termed sialic acids, are part of different animal mucous glycoproteins and mucins, which occur, for instance, in saliva, sperm, and vaginal secretion. In microorganisms and animals they are constituents of glycolipids present in cell membranes.

The dimethylamino sugar mycaminose is a constituent of macrolide antibiotics synthesized in Actinomycetes (D 3). 2-Methylamino-L-glucose and 2,6-diamino-D-glucose are building blocks of amino and guanidino cyclitol antibiotics (D 1.3). Also the polysaccharides heparin, chondroitin sulfate, and hyaluronic acid (Table 21) formed in animals contain amino sugars.

f) C-Methylation. The additional methyl groups of certain sugars with a branched carbon chain, e. g., D-cladinose and L-mycarose (Fig. 33), are derived

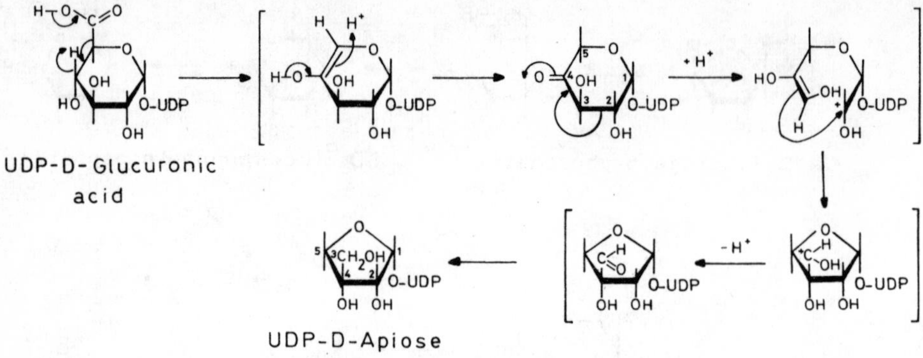

Fig. 39. Formation of UDP-D-apiose from UDP-D-glucuronic acid by UDP-apiose synthase

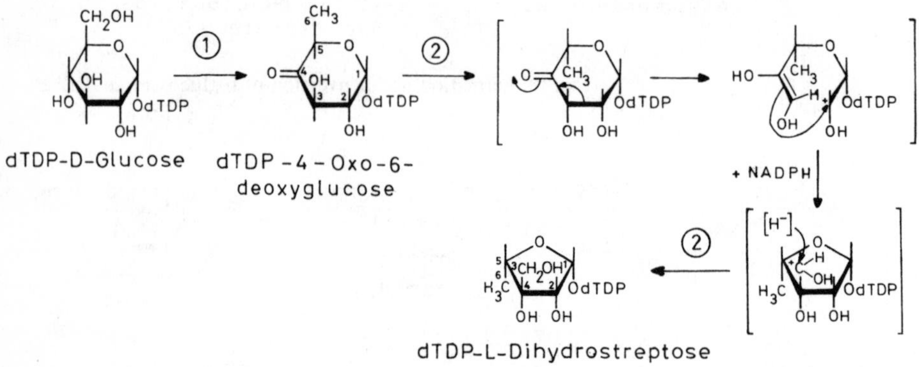

Fig. 40. Biosynthesis of dTDP-L-dihydrostreptose from dTDP-D-glucose
1 dTDP-D-Glucose-4,6-dehydratase; *2* dTDP-L-dihydrostreptose synthase

from the methyl group of *S*-adenosyl-L-methionine (C 3.3). These *C*-methylated sugars are constituents of antibiotics (D 3).

g) Rearrangement of the sugar chain. The "additional" CH_2OH- or CHO-groups of D-apiose and L-streptose, respectively (Fig. 33), are derived from C-3 of glucose. The biosynthesis of UDP-D-apiose starts from UDP-D-glucuronic acid (D 1.2; Fig. 39).

The synthesis of dTDP-L-dihydrostreptose begins with the formation of dTDP-4-oxo-6-deoxyglucose (Fig. 40, see also Fig. 36).

Apiose is a constituent of the flavonoid apiin (D 23.3.3). L-Streptose is part of the antibiotic streptomycin (D 1.3).

Significance

Several secondary monosaccharides have a sweet taste. If present in higher concentration they attract animals involved, for instance, in pollination or seed dispersal (E 5.5.1 and E 5.5.2) as well as human beings (F 1).

Transformation

Secondary monosaccharides are precursors of carboxylic acids (D 1.2), sugar alcohols (D 1.3), and glycosides (D 1.4).

References for Further Reading

227, 289, 421, 471, 669, 829

D 1.2 Sugar Carboxylic Acids

Chemistry and Distribution

Aldose-derived carboxylic acids are synthesized in microorganisms, plants, and animals. With respect to the location of the carboxyl group within the molecule, aldonic acids (onic acids), uronic acids, and glycaric acids (sugar dicarboxylic acids, aldaric acids) may be distinguished (Fig. 41). Aldonic acids cyclize easily to the corresponding γ-lactones. Glycaric acids may form dilactones.

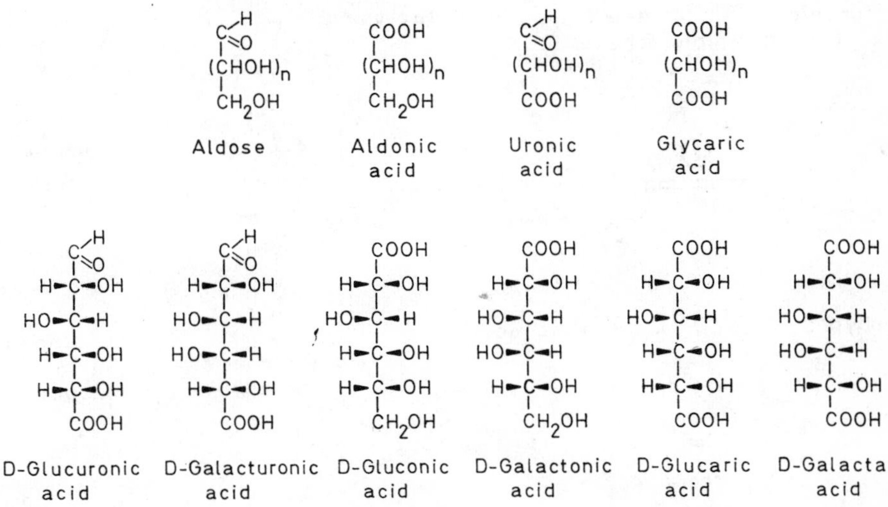

Fig. 41. Carboxylic acids derived from aldoses

Biosynthesis

Pyridine nucleotide-dependent dehydrogenases (C 2.1) transform aldose derivatives to uronic acids, e.g., UDP-D-glucose to UDP-D-glucuronic acid and in animals also to aldonic acids, e. g., D-glucose-6-phosphate to D-gluconic acid-6-phosphate (glucose-6-phosphate dehydrogenase).

In microorganisms and plants, however, aldonic acids, e. g., D-gluconic acid and D-galactonic acid are synthesized by flavin-containing oxidases, e.g., glucose oxidase (C 2.2), using O_2 and forming H_2O_2 as a side product. Glycaric acids are

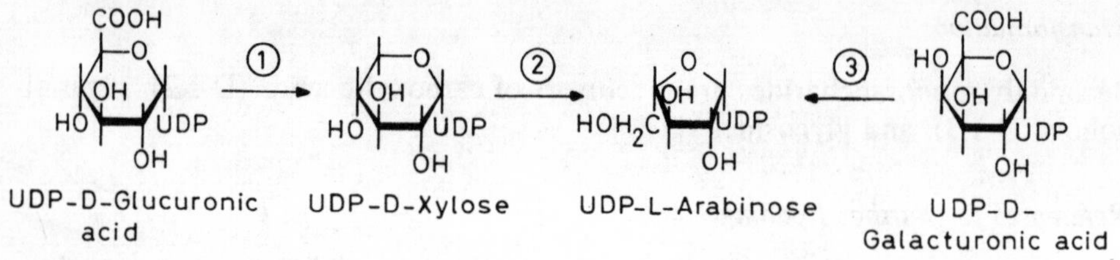

Fig. 42. Decarboxylation of D-uronic acids
1 UDP-D-Glucuronate decarboxylase; *2* UDP-D-xylose-4-epimerase; *3* UDP-D-galacturonate decarboxylase

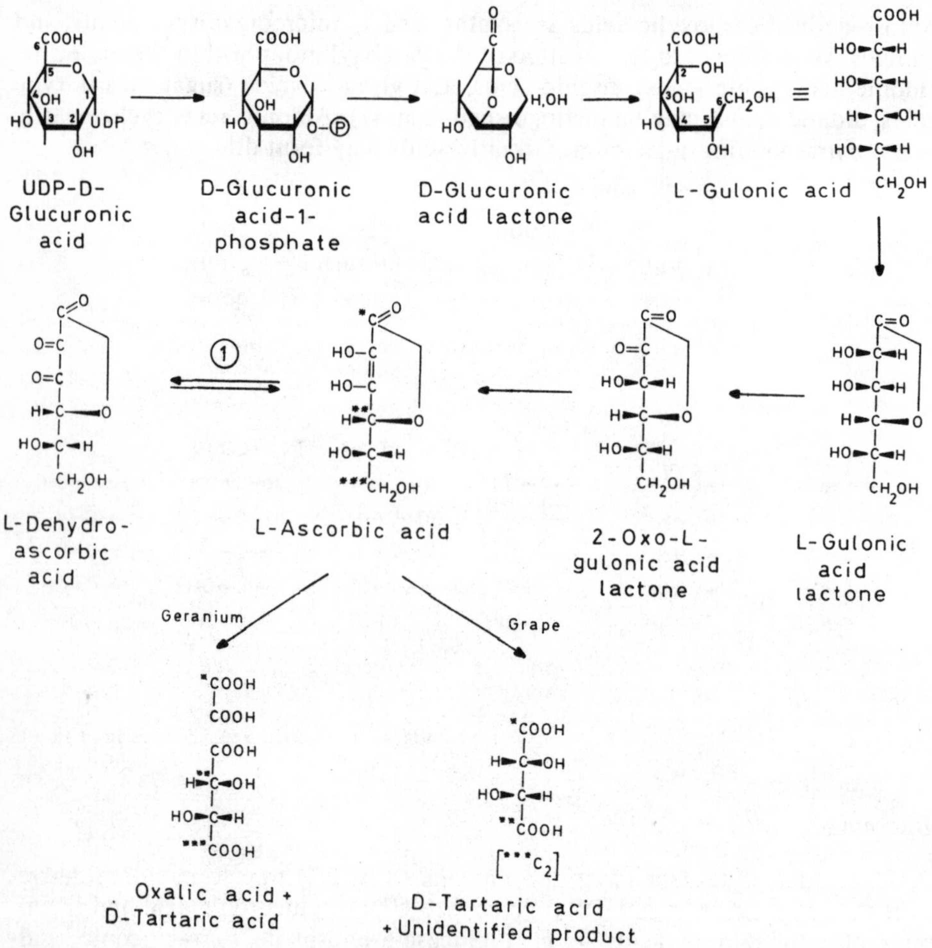

Fig. 43. Synthesis and degradation of L-ascorbic acid
Because C-6 of D-glucuronic acid becomes C-1 of L-gulonic acid, numbering of the C-atoms of L-ascorbic acid is reverse to that of D-glucuronic acid. *1* Certain monooxygenases which use L ascorbic acid as cosubstrate

built by similar enzymes from uronic acids, e.g., D-glucaric acid is formed from D-glucuronic acid and D-galactaric acid from D-galacturonic acid.

Significance

Heterosidic uronides are frequently built in animals from nucleotide diphosphate uronic acids (cf. C 6). Formation of uronides increases water solubility and is a prerequisite for the excretion of many secondary products and other compounds (E 1).

Transformation of uronic acids

a) Nucleotide diphosphate uronic acids (cf. C 6) are precursors of holosidic polymers, e.g., pectin, alginic acid, hyaluronic acid, and chondroitin sulfuric acids (D 1.4.1).

b) UDP-D-Xylose and UDP-L-arabinose (C 6) are formed by decarboxylation of UDP-D-glucuronic acid and UDP-D-galacturonic acid, respectively (Fig. 42).

Oxo sugar derivatives probably occur as intermediates since the reaction is NAD^+-dependent. The pentoses formed may be polymerized to yield L-arabanes and D-xylanes which are important cell wall constituents of plants (D 1.4.1). Together with polyuronides (see above) they form the group of hemicelluloses, in which cellulose fibrils (D 1.4.1) are embedded.

c) L-Ascorbic acid is derived in animals from D-glucuronic acid via L-gulonic acid (Fig. 43). In plants L-ascorbic acid is derived from D-glucose on a pathway still unknown. L-Ascorbic acid is a cosubstrate of certain monooxygenases (C 2.6.1). During the oxygenation of the substrate it is transformed to dehydroascorbic acid. L-Ascorbic acid is a vitamin for humans (E 2.1, vitamin C, necessary intake about 50 mg/day).

D-Tartaric acid and oxalic acid are built from L-ascorbic acid in higher plants. They are formed by oxidative cleavage of the ascorbic acid molecule (Fig. 43).

References for Further Reading

421, 468, 470, 471, 616, 829

D 1.3 Sugar Alcohols

Chemistry and Distribution

Sugar alcohols are polyols with an aliphatic (alditols) or cyclic carbon chain (cyclitols). According to the number of hydroxy groups present in the molecule, triols, tetraols, pentaols, hexaols, etc. may be distinguished (Fig. 44). The most important cyclic polyols possess a six-membered carbocyclic ring with a hydroxy group at each C-atom (inositols). However, there may also be cyclitols with oxygen-free carbon atoms, e.g., proto-quercitol, or with double bonds, e.g., conduritol. In addition, several cyclitols have methylated hydroxy groups, e. g., D-pinitol. Sugar alcohols occur in microorganisms, plants, and animals.

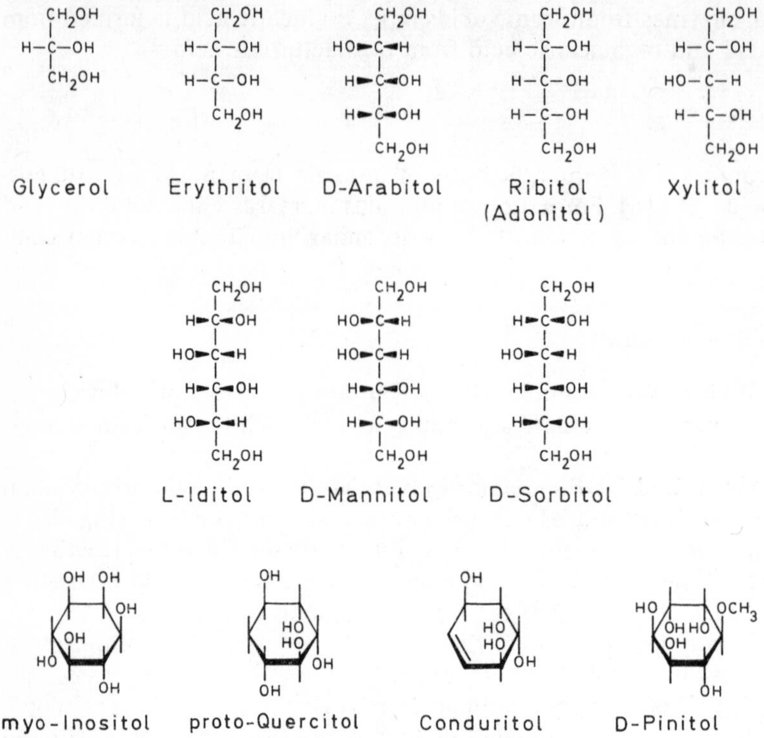

Glycerol   Erythritol   D-Arabitol   Ribitol (Adonitol)   Xylitol

L-Iditol   D-Mannitol   D-Sorbitol

myo-Inositol   proto-Quercitol   Conduritol   D-Pinitol

Fig. 44. Aliphatic and cyclic sugar alcohols

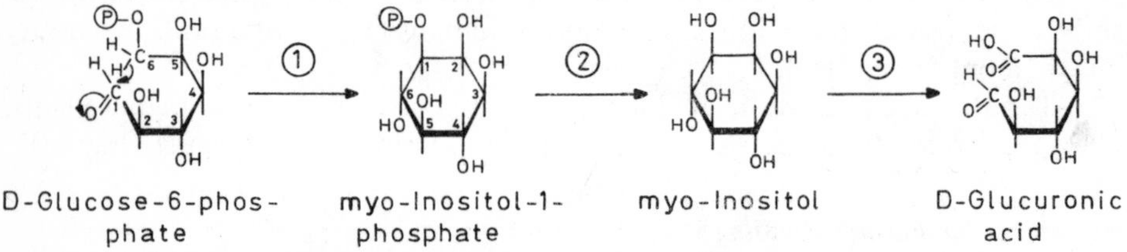

D-Glucose-6-phosphate   ① →   myo-Inositol-1-phosphate   ② →   myo-Inositol   ③ →   D-Glucuronic acid

Fig. 45. Biosynthesis and degradation of myo-inositol
1 myo-Inositol-1-phosphate synthase; *2* 1L-myo-inositol-1-phosphatase; *3* myo-inositol oxygenase

Table 19. Monosaccharides and related sugar alcohols

Sugars	Sugar alcohols	Sugars	Sugar alcohols
D-Glyceraldehyde	Glycerol	L-Ribulose and D-ribulose	Ribitol (adonitol)
D-Erythrose	Erythritol	D-Glucose, L-sorbose, and D-fructose	D-Sorbitol (D-glucitol)
D-Xylulose and L-xylulose	Xylitol		
D-Xylulose and D-ribulose	D-Arabitol	L-Sorbose	L-Iditol
		D-Fructose	D-Mannitol

Biosynthesis

Alditols are formed on the reduction of aldoses or ketoses by pyridine nucleotide-dependent dehydrogenases (aldose reductases, ketose reductases, C 2.1). Since this reaction is reversible, aldoses and ketoses, e. g., D-glucose, L-sorbose, and D-fructose, may be converted into each other via the common polyol D-sorbitol (Table 19).

Inositols are formed from glucose-6-phosphate by an intramolecular aldol reaction (Fig. 45). They are degraded to uronic acids. From myo-inositol several other inositols are built by isomerization (probably via the corresponding oxo inositols) and methylation. Their degradation yields unusual uronic acids and after decarboxylation (D 1.2) unusual pentoses which are frequently found as constituents of plant cell walls.

Significance

Sugar alcohols have a sweet taste and attract animals (E 5.5.2) and humans (F 1). D-Sorbitol serves as a sweetener for diabetics (F 2). Glycerol and D-sorbitol act as cryoprotectors in insects (E 2.2). Glycerol is an osmotic in halophilic algae (E 2.2). D-Sorbitol is the means of transport of the photosynthate from the leaves to the fruits in apple trees.

Myo-inositol promotes the growth of molds (E 2.2) and probably has vitamin effects in human beings. Phytic acid (myo-inositol hexaphosphate) is an important store of phosphate in plant cells, e. g., in pollen and seeds. In wheat endosperm the phosphate residues of phytic acid may be transferred to ADP by a phosphotransferase with the formation of ATP. In germinating pollen the enzyme phytase hydrolyzes phytic acid to myo-inositol and inorganic phosphate (E 2.2). Galactinol (1-O-α-D-galactopyranosyl-myo-inositol) is formed from myo-inositol and UDP-D-galactose (C 6) by a galactosyltransferase. Galactinol is widespread in higher plants and, like the sugar nucleotides, acts as a glycosyl donor (E 2.2).

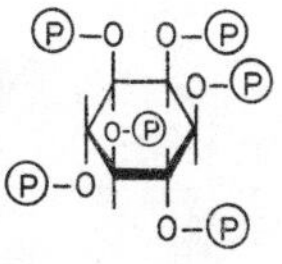

Phytic acid 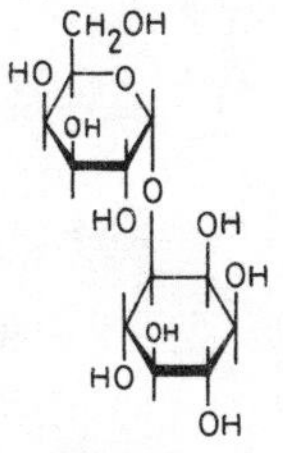   Galactinol

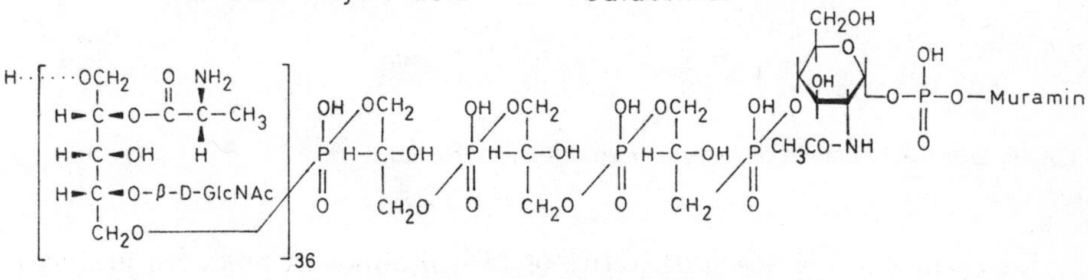

Teichoic acid from the cell wall of Staphylococcus aureus

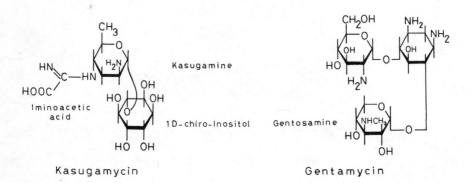

Fig. 46. Structural formulas of some aminoglycoside antibiotics

Kasugamycin (Fig. 46), a derivative of 1*D*-chiro-inositol produced in *Strepto-myces kasugaensis*, is used as an antifungal antibiotic in agriculture (E 5.2).

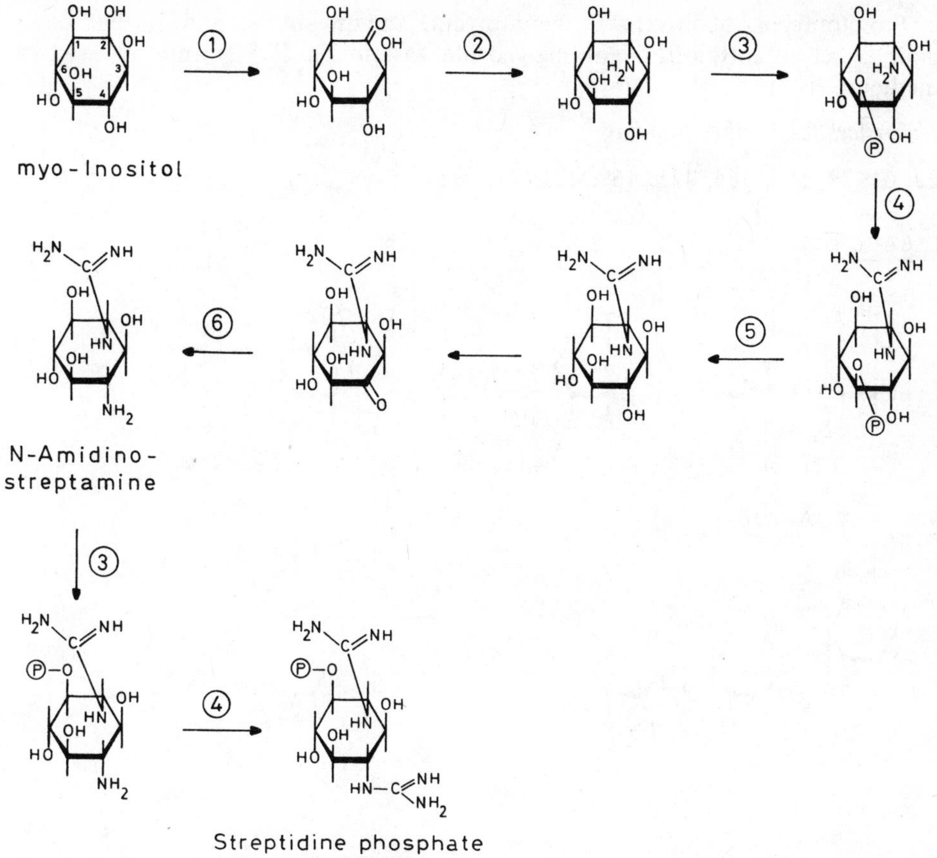

myo-Inositol

N-Amidino-
streptamine

Streptidine phosphate

Fig. 47. Biosynthesis of streptidine phosphate
1 myo-Inositol: NAD 2-oxidoreductase; *2* L-glutamine: oxo-scylloinositol amidinotransferase; *3* ATP: inosamine phosphotransferase(s); *4* L-arginine: inosaminephosphate amidinotransferase(s); *5* 1-guanidino-1-deoxy-scyllo-inositol-4-phosphate phosphorylase; *6* L-alanine: 1D-1-guanidino-1-deoxy-3-oxo-scyllo-inositol aminotransferase

Transformation

a) Teichoic acids are constituents of the cell wall of bacteria. They are attached to the murein sacculus (D 9.2.4). Teichoic acids consist of chains of ribitol and glycerol moieties joined by phosphate groups. This core carries glycosyl and D-alanyl residues.

b) Amino and Guanidino Cyclitols. Streptidine, the aglycone of the guanidino cyclitol, streptomycin (Fig. 46), and similar amino cyclitols formed in Actinomycetes, e.g., deoxystreptamine, the aglycone of neomycin C, gentamycin, and kanamycin A, are derived from cyclitols. Streptidine is formed from myo-inositol (Fig. 47). The amino groups are derived from L-glutamine. L-Arginine serves as donor of the guanidino groups. In the formation of streptomycin, nucleoside diphosphate streptose and 2-methylamino-L-glucose (D 1.1) are linked with streptidine by glycosyltransferases (C 6).

Streptomycin, neomycin C, gentamycin, kanamycin A, and similar compounds act as antibiotics (aminoglycoside antibiotics, E 5.2) and are used in medicine (F 2).

References for Further Reading

22, 63, 72, 160, 354, 421, 469, 471, 778, 829

DISACCHARIDES

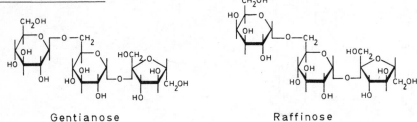

TRISACCHARIDES

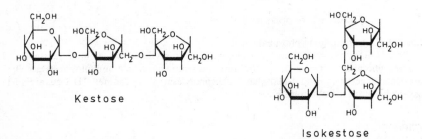

TETRASACCHARIDE

Fig. 48. Structural formulas of several oligosaccharides

D 1.4 Glycosides

Chemistry and Distribution

Glycosides are sugar derivatives in which the glycosidic hydroxy group of sugar hemi-acetals (lactols, D 1) is substituted. They are formed in microorganisms, plants, and animals. In the so-called holosides (D 1.4.1) the substituent is the hydroxy group of another sugar moiety. In the heterosides (D 1.4.2) the nucleophilic grouping of a nonsugar residue reacts as substituent. With respect to the sugars involved, glucosides, galactosides, mannosides, ribosides, etc. can be distinguished.

The chemical designation of holosides starts with the terminal sugar moiety whose glycosidic hydroxyl is substituted and ends with the sugar residue which has the free glycosidic hydroxyl. Examples are:

maltose = α-D-glucopyranosyl-(1→4)-D-glucopyranose
lactose = β-D-galactopyranosyl-(1→4)-D-glucopyranose (Fig. 48)

The designation of holosides with no free glycosidic hydroxyl ends with the suffix -ide:

sucrose = α-D-glucopyranosyl-(1→1)-β-D-fructofuranoside (Fig. 48)

The name of heterosides begins with the description of the aglycone:

glucofrangulin A (D 5.3.6) = emodine-6-O-α-L-rhamnopyranosido-8-O-β-D-glucopyranoside

For convenience the names may be abbreviated (see list of abbreviations):

maltose = D-glc-p-α(1,4)-D-glc-p
lactose = D-gal-p-β(1,4)-D-glc-p
sucrose = D-glc-p-α(1,1)β-D-fru-f
glucofrangulin A = emodine-6-O-α-L-rha-p-8-O-β-D-glc-p

Biosynthesis

Aldose-1-phosphates are the primary glycosides. They are built by the reaction of aldoses with ATP or by isomerization of sugar-ω-phosphates. They may act as glycosyl donors in the synthesis of other glycosides either directly or after transformation to nucleoside diphosphate sugars (C 6).

D 1.4.1 Holosides

Chemistry and Distribution

Holosides occur in all groups of organisms. They are glycosidic compounds completely composed of sugar moieties. According to the number of sugar residues, oligosaccharides (di-, tri-, tetrasaccharides, etc., Fig. 48 and Table 20) and polysaccharides may be distinguished. Polysaccharides built completely from one type of sugar, e. g., D-glucose, D-galactose, or D-mannose are called glucans, galactans, and mannans, respectively. In addition, there are complex polysaccharides formed from different sugar units like agarose, heparin, and xanthan (Table 21).

With respect to the configuration at the glycosidic carbon atom (D 1), α- and β-glycosides may be distinguished. In most holosides the glycosidic hydroxyl of

Table 20. Di- and oligosaccharides

Compound	Composition	Occurrence
Disaccharides		
Maltose	D-glc-p-α(1,4)D-glc-p	Plants
Trehalose	D-glc-p-α(1,1)α-D-glc-p	Mushrooms, insects
Lactose	D-gal-p-β(1,4)-D-glc-p	Animals (milk)
Sucrose	D-glc-p-α(1,2)β-D-fru-f	Plants
Melibiose	D-gal-p-α(1,6)-D-glc-p	Plants
Turanose	D-glc-p-α(1,3)-D-fru-p	Honey
Primverose	D-xyl-p-β(1,6)-D-glc-p	Plants
Vicianose	L-ara-p-α(1,6)-D-glc-p	Plants
Rutinose	L-rha-p-α(1,6)-D-glc-p	Plants
Gentiobiose	D-glc-p-β(1,6)-D-glc-p	Plants
Trisaccharides		
Gentianose	D-glc-p-β(1,6)-D-glc-p-α(1,2)β-D-fru-f	Plants
Raffinose	D-gal-p-α(1,6)-D-glc-p-α(1,2)β-D-fru-f	Plants, honey
Melezitose	D-glc-p-α(1,3)-D-fru-f-β(2,1)α-D-glc-p	Plants (nectar)
Kestose	D-glc-p-α(1,2)β-D-fru-f-(6,2)β-D-fru-f	Plants
Isokestose	D-glc-p-α(1,2)β-D-fru-f-(1,2)β-D-fru-f	Plants
Tetrasaccharide		
Stachyose	D-gal-p-α(1,6)-D-gal-p-α(1,6)-D-glc-p-α(1,2)β-D-fru-f	Plants
Pentasaccharide		
Verbascose	D-gal-p-α(1,6)-D-gal-p-α(1,6)-D-gal-p-α(1,6)-D-glc-p-α(1,2)β-D-fru-f	Plants

Table 21. Properties and occurrence of homoglycans and heteroglycans

Compound	Constituent (terminal group)	Connection [branching]	Molecular weight	Occurrence
Amylose	D-glc-p	α(1,4)	50,000	Plants (starch)
Amylopectin	D-glc-p	α(1,4) [α(1,6)]	500,000 to 1 million	Plants (starch)
Glycogen	D-glc-p	α(1,4) [α(1,6)]	1 to 15 million	Molds, animals
Cellulose	D-glc-p	β(1,4)	300,000 to 500,000	Plants (cell wall)
Lichenin	D-glc-p	β(1,4) [β(1,3)]	25,000	Plants
Laminarin	D-glc-p (Mannitol)	β(1,3) [β(1,6)]	5,000	Seaweeds
Dextran	D-glc-p	α(1,6) [α(1,4), α(1,3), α(1,2)]	Several million	Bacteria
Inulin	D-fru-f (D-glc-p)	β(1,2)	5,000	Plants (carbohydrate reserve)
Levan	D-fru-f (D-glc-p)	β(2,6)	10 million	Bacteria, Poaceae
Mannan	D-man-p	β(1,4)	100,000	Plants
Galactan	D-gal-p	β(1,4)		Plants
Xylan	D-xyl-p	β(1,4)	10,000	Plants

Compound	Constituent (terminal group)	Connection [branching]	Molecular weight	Occurrence
Araban	L-ara-f	α(1,5) [α(1,3)]	6,000	Plants
Fucan	L-fuc-f-4-sulfate	α(1,2)		Seaweeds
Chitin	D-glcNAc-p	β(1,4)	200,000	Fungi, arthropods, insects
Pectin	D-galUA-p	α(1,4)	50,000	Plants
Agarose	D-gal-p, L-gal-p, 3,6-anhydro- L-gal-p, sulfate groups	β(1,4) α(1,3)	100,000	Algae (constituent of agar)
Alginic acid	D-manUA-p, L-gulUA-p	β(1,4) β(1,4)	12,000 to 120,000	Brown algae, microorganisms
Hyaluronic acid	D-glcUA-p, D-glcNAc-p	β(1,4) β(1,3)	200,000 to 500,000	Animals
Chondroitin sulfuric acid	D-glcUA-p, D-galNAc-p, sulfate groups	β(1,4) β(1,3)	250,000	Animals (cartilage, collagen)
Heparin	D-glcUA-p- 2-sulfate, D-glcN-p- N-sulfonate	α(1,4) α(1,4)	16,000	Animals (inhibits blood coagulation)
Murein	D-glcNAc-p, N-acetyl- muramic acid, amino acids	β(1,4)		Bacteria (cell wall)
Xanthan	D-glc-p D-man-p, pyruvic acid, D-glucosyl- uronic acid, D-manAc-p	β(1,4) β(1,4) β(1,2) α(1,3)		*Xanthomonas campestris* (exopolysaccharide)

one sugar unit is substituted by one of the alcoholic hydroxyls of a second sugar moiety (see the formula of lactose, Fig. 48). But in a few instances the glycosidic hydroxyls of both partners may react with each other (see the formulas of sucrose and trehalose, Fig. 48).

Biosynthesis

The formation of holosides is catalyzed by glycosyltransferases (C 6). Nucleoside diphosphate sugars are of special significance as glycosyl donors. In addition, aldose-1-phosphates, oligo- and polysaccharides, as well as certain heterosides may donate glycosyl residues (Table 22). Growing holoside chains are elongated by one carbohydrate unit at a time. In most cases the new glycosyl residues are added at the nonreducing end of the chain. Branchings in the chain of sugar

Table 22. Formation of some low molecular glycosides

Glycosyl acceptor	Glycosyl donor	Glycoside formed
β-D-Fructose	α-D-Glucose-1-phosphate; UDP-D-Glucose[a]	Sucrose
β-D-Fructose-6-phosphate	UDP-D-Glucose	Sucrose-6-phosphate[b]
Purines, pyrimidines	α-D-Ribose-1-phosphate	Nucleosides
Aglycones with different structures	UDP-D-Glucose	Glucosides
	UDP-D-Glucuronic acid	Glucuronides
Phosphate	Sucrose	Glucose-1-phosphate
Sucrose	UDP-D-Galactose	Raffinose
Raffinose	Galactinol[c]	Stachyose
Stachyose	Galactinol[c]	Verbascose
Sucrose	Sucrose	Isokestose

[a] The reverse reaction is important in the breakdown of sucrose. The UDP-glucose formed may be used in the synthesis of starch.

[b] After hydrolysis of sucrose-6-phosphate to sucrose, the reaction is no longer reversible. Thus, the accumulation of large amounts of sucrose is possible, e. g., in sugarcane and sugar beet.

[c] Galactinol (D 1.3) is a glucosyl donor in higher plants.

Table 23. Heteroglycans present in gum exudates from plants

Plant source (commercial product)	Composition of the gum
Acacia sp. (Gum arabic)	Arabinogalactan consisting of a branched, heavily substituted β-D-galactan in which $1 \rightarrow 3$ linkages predominate over $1 \rightarrow 6$ linkages. Single units or short chains of sugar residues are attached to the galactan core comprising L-arabinose, L-rhamnose, and D-glucuronic acid
Astragalus gummifer (Gum tragacanth)	Galacturonorhamnan consisting of highly branched molecules in which the chains of $1 \rightarrow 4$-linked acids are interrupted at intervals by $1 \rightarrow 2$-linked L-rhamnose units

moieties, e. g., in the formation of glycogen and amylopectin, are brought about by transglycosylation.

The degradation of holosides is carried out by hydrolyzing enzymes (glycosidases) or phosphorylating enzymes.

Hydrolases are:
— β-amylase, an exohydrolase splitting off maltose from the nonreducing end of amylose and amylopectin
— α-amylase, an endohydrolase forming oligosaccharides from starch, which may be further degraded to maltose
— isoamylase (debranching enzyme), which splits $\alpha(1,6)$-bonds in amylopectin
— fructosan hydrolase, which degrades fructosans, e. g., inulin
— cellulases splitting celluloses
— $\beta(1,3)$-glucanases, xylosidases, and pectinases, which degrade the hemicelluloses and pectins present in the cell wall of plants, and
— chitinase which degrades chitin, a cell wall polymer of fungi and a component of the insect exoskeleton.

A phosphorylase catalyzes the degradation of glycogen. In the presence of inorganic phosphate it splits off glucose units as glucose-1-phosphate from the nonreducing end of the molecule.

Significance

Oligosaccharides and polysaccharides produced in plants are the main nutrients for heterotrophic microorganisms and animals. Several oligosaccharides have a sweet taste and attract animals involved in pollination or seed dispersal (E 5.5.1 and 5.5.2), as well as human beings (F 1). Polysaccharides serve as structural elements and a mechanical barrier to invading organisms, especially in plant or microbial cell walls and the exoskeleton of insects (E 2.2, E 5.4). Others are storage products, forming a reserve of sugar moieties (E 2.2).

Plant cell walls consist of cellulose fibers embedded in a covalently cross-linked matrix of hemicelluloses such as arabans and galactans (Table 21), arabinogalactans, arabinoxylans, galacturonoxylans, and 4-*O*-methylglucuronoxylans. In lignified cells the fibers are embedded in a lignin matrix (D 23.2.3). Cell walls of algae contain polymannuronic and polyguluronic acids (Phaeophyta) as well as sulfated galactans which comprise a considerable amount of 3,6-anhydrogalactosyl residues.
Examples of reserve polysaccharides are
— starch, which is produced in plants and consists of 10–30 % amylose and 70–80 % amylopectin,
— glycogen, an animal and microbial glucan, which resembles amylopectin, but is more highly branched, and
— levans and inulins, found in plants (grasses, Compositae, and Boraginaceae).

When injured certain plants synthesize protective gums (E 2.2) which consist of highly branched polysaccharides with complex structure (Table 23).

Polysaccharides, e. g., cellulose, starch, agar, alginates, and other microbial polysaccharides forming slimes and gels, gum arabic, and gum tragacanth are used in technology (F 5). Starch serves as an important food source for human beings. Inulin is used as a diet for diabetics. Lactose is applied as an mild laxative in medicine (F 2).

Transformation

Branched oligosaccharides with elicitor activity (A 4.4.2) are formed by the degradation of fungal cell wall glucans containing $\beta(1,3)$- and $\beta(1,6)$-linked glucosyl residues. The most simple elicitor is a heptamer of the following structure:

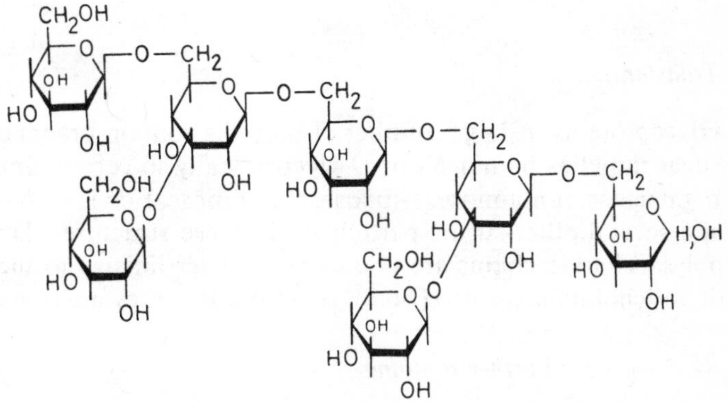

Glucose heptamer elicitor

References for Further Reading

17, 18, 61, 109, 125, 175, 182, 192, 193, 195, 228, 253, 356, 421, 437, 461, 466, 471, 515, 550, 614, 615, 616, 617, 665, 730, 734, 744, 754, 837

D 1.4.2 Heterosides

Chemistry and Distribution

Heterosides are formed in microorganisms, plants, and animals. They are composed of one or more sugar residues and a nonsugar constituent, the genin or aglycone. Depending on the functional group of the genin which reacts with the sugar, *O-*, *N-*, *S-*, and *C*-heterosides may be distinguished (Table 24). In most heterosides a relatively short, straight chain of sugar moieties is linked to the genin. In only a few groups, e. g., in steroid saponins and alkaloids (D 6.6.2 and D 6.6.3) as well as in glycoproteins, the chain of sugar moieties is branched. Sugars of primary metabolism, e. g., D-glucose, D-ribose, and D-deoxyribose participate in the formation of many heterosides. There are, however, also groups of glycosides whose biosynthesis involves strange secondary monosaccharides. Examples are the amino and guanidino cyclitol antibiotics (D 1.3), the macrolide antibiotics (D 3), the pregnane derivatives (D 6.6.5), and the cardiac glycosides (D 6.6.8). This book treats the different groups of heterosides in the sections pertaining to their respective aglycones.

Biosynthesis

Heterosides are built by glycosyltransferases (C 6). In most cases nucleoside diphosphates, in some instances other energy-rich glycosides, e. g., sugar 1-phosphates, galactinol (D 1.3), etc., act as glycosyl donors.

Significance

Formation of heterosides increases the water solubility of most aglycones and frequently is a prerequisite for accumulation or release of secondary products (E 1).

Transformation

Glycoproteins. A large number of proteins contain branched chains of up to 100 sugar moieties bound *N-* or *O*-glycosidically to certain amino acids. D-Glucose, D-galactose, D-mannose, L-fucose, L-arabinose, D-xylose, *N*-acetyl-D-glucosamine, and several other sugars participate in these structures. During biosynthesis the polysaccharide chains are bound by an ester linkage to the diphosphate residue of dolicholdiphosphate (D 6.8). In a final step they are transferred to the protein.

References for Further Reading

362, 421, 459, 471

Table 24. Important groups of heterosides

Group of compounds/ representatives	Frequently occurring sugar residues	Occurrence M = microorganisms P = plants A = animals		
O-Glycosides				
Amino and guanidino cyclitols (strep- tomycin, neomycin, kanamycin A, gentamycin, D 1.3)	Branched chain sugars, amino sugars	M		
Macrolides (erythromycins, methy- mycin, nystatin, D 3)	Branched chain sugars, amino sugars	M		
Anthrone and anthraquinone deriva- tives (glucofrangulin A, D 5.3.6)	D-Glucose, L-rhamnose, D-apiose	M	P	
Steroid saponins and alkaloids (dios- cin, α-tomatine, D 6.6.2 and D 6.6.3)	D-Glucose, D-galactose, L-rhamnose, D-xylose, L-arabinose		P	
Estranes (estrone-3-*O*-β-D-glucuro- nide, D 6.6.7)	D-Glucuronic acid			A
Cardiac glycosides (lanatoside C, pur- pureaglycoside A, D 6.6.8)	Deoxysugars, *O*-methylated and ace- tylated sugars, D-glucose		P	A
Cyanogenic glycosides (linamarin, D 8.3)	D-Glucose		P	A
Cinnamic acid derivatives (couma- rinic acid-β-D-glucoside, D 23.2.2)	D-Glucose, L-rhamnose, D-fructose, gentiobiose		P	
Flavonoids (rutin, naringin, apiin, D 23.3.3)	D-Glucose, D-galactose, L-rhamnose, L-arabinose		P	
S-Glycosides				
Glucosinolates (glucobrassicin, allyl- glucosinolate, D 8.4)	D-Glucose		P	
N-Glycosides				
Imidazoles (imidazole acetic acid, D 20)	D-Ribose			A
5,6-Dimethylbenzimidazol (cobal- amins, D 24.3)	D-Ribose	M	P	
Purines (cAMP, ppGpp, cordycepin, D 25.1)	D-Ribose, D-deoxyribose	M	P	A
Pyrrolopyrimidines (toyocamycin, tubercidin, D 25.2)	D-Ribose	M		
C-Glycosides				
Anthrones (aloin, D 5.3.6)	D-Glucose		P	
Flavonoids (vitexin, D 23.3.3)	D-Glucose		P	

D 2 Secondary Products Derived from Intermediates of Glucose Degradation and Gluconeogenesis

Glucose Degradation and Gluconeogenesis

The most important pathway for the degradation of D-glucose is the Embden-Meyerhof-Parnas (EMP) pathway, which leads via fructose-1,6-diphosphate to pyruvic acid (Fig. 49). The pyruvic acid may be transformed to acetyl CoA (D 5) which enters the tricarboxylic acid cycle (D 4).

Gluconeogenesis (resynthesis of glucose) starts with phosphoenolpyruvic acid

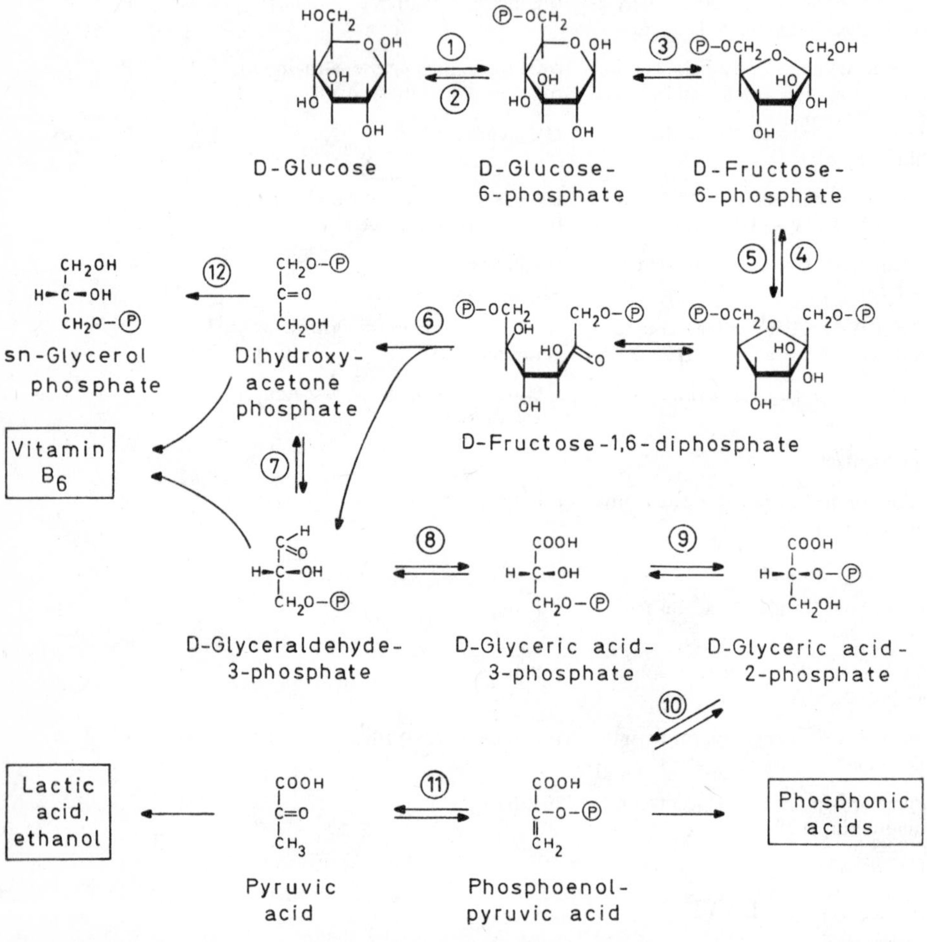

Fig. 49. Degradation of glucose on the Embden-Meyerhof-Parnas pathway
1 Hexokinase; *2* glucose-6-phosphatase; *3* glucosephosphate isomerase; *4* fructose-1,6-bisphosphatase; *5* 6-phosphofructokinase; *6* fructose-bisphosphate aldolase; *7* triosephosphate isomerase; *8* glyceraldehydephosphate dehydrogenase; phosphoglycerate kinase; *9* phosphoglyceromutase; *10* enolase; *11* pyruvate kinase; *12* dihydroxyacetone phosphate reductase

(PEP) and follows the pathway of D-glucose degradation in the reverse direction. Phosphoenolpyruvic acid is formed from oxaloacetic acid by PEP-carboxykinase:

$$\underset{\text{Oxaloacetic acid}}{\overset{O}{\underset{HO}{\diagdown}}C-CH_2-CO-C\overset{\diagup O}{\diagdown}{OH}} \;+\; GTP \;\rightleftharpoons\; \underset{\text{Phosphoenolpyruvic acid}}{CH_2=C\overset{O-\textcircled{P}}{\underset{}{|}}-C\overset{\diagup O}{\diagdown}{OH}} \;+\; GDP \;+\; CO_2$$

Oxaloacetic acid is an intermediate of the tricarboxylic acid cycle (D 4), but may also be built from pyruvic acid, either directly by pyruvate carboxylase:

$$\underset{\text{Pyruvic acid}}{CH_3-\overset{O}{\overset{\|}{C}}-C\overset{\diagup O}{\diagdown}{OH}} + ATP + CO_2 + H_2O \longrightarrow \underset{\text{Oxaloacetic acid}}{\overset{O}{\underset{HO}{\diagdown}}C-CH_2-\overset{O}{\overset{\|}{C}}-C\overset{\diagup O}{\diagdown}{OH}} + ADP + P_i$$

or via L-malic acid, another intermediate of the tricarboxylic acid cycle (D 4):

$$\underset{\text{Pyruvic acid}}{CH_3-\overset{O}{\overset{\|}{C}}-C\overset{\diagup O}{\diagdown}{OH}} + CO_2 + NADPH + H^+ \longrightarrow \underset{\text{L-Malic acid}}{\overset{O}{\underset{HO}{\diagdown}}C-CH_2-\overset{OH}{\underset{H}{\overset{|}{C}}}-C\overset{\diagup O}{\diagdown}{OH}} + NADP^+$$

Biosynthesis of Secondary Products

Secondary products are derived from pyruvic acid and the intermediates of glucose degradation and resynthesis in microorganisms, plants, and animals (Fig. 49). The most important are:

a) Pyridoxyl derivatives. Pyridoxin (pyridoxol), pyridoxal, and pyridoxamine are pyridines synthesized in plants and most microorganisms. In bacteria obviously two biosynthetic pathways exist. In the main route probably acetaldehyde or a biological equivalent, dihydroxyacetone phosphate, and D-glyceraldehyde-3-phosphate act as precursors (Fig. 50). In the minor route C-6 and N-1 are formed from a glycine-derived C-N unit. Pyridoxin, pyridoxal, and pyridoxamine are vitamins (vitamins B_6) for humans and animals (E 2.1). They are transformed to pyridoxal-5'-phosphate, the coenzyme of aminotransferases and amino acid decarboxylases (C 5) as well as of several other enzymes.

b) Glycerol derivatives. Reduction of D-glyceraldehyde-3-phosphate yields sn-glycerol-3-phosphate (L-glycerol-3-phosphate), a precursor of fatty acid esters (glycerides, D 5.2.4). Glycerol glycosides, e. g., lilioside C, and glycerol cinnamic acid esters (Table 48) occur in higher plants.

Lilioside C

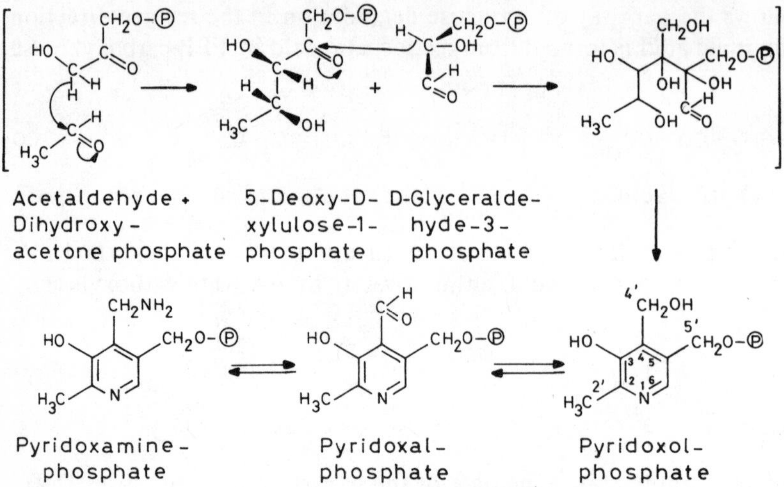

Acetaldehyde +   5-Deoxy-D-   D-Glyceralde-
Dihydroxy-   xylulose-1-   hyde-3-
acetone phosphate phosphate   phosphate

Pyridoxamine-   Pyridoxal-   Pyridoxol-
phosphate   phosphate   phosphate

Fig. 50. Probable pathway of biosynthesis of pyridoxyl derivatives

c) Lactic acid. Reduction of pyruvic acid yields D- or L-lactic acid depending on the stereospecificity of the enzyme lactate dehydrogenase (Fig. 51). L-Lactic acid is formed for example in animal muscle cells under high rates of glucose degradation which create anaerobic conditions. The accumulated L-lactic acid causes musclecramps. The D- or L-lactic acid synthesized by some microorganisms is used in the processing of preserved food, e. g., sauerkraut.

d) Ethanol. Decarboxylation of pyruvic acid yields acetaldehyde (Fig. 51), which under anaerobic conditions, i. e., alcoholic fermentations, is reduced to ethanol. Ethanol is a frequently used stimulant of human beings (F 3) and an important fuel (F 5).

e) L-Alanine-3-phosphonic acid and 2-aminoethylphosphonic acid. Phosphonic acids contain a phosphate group attached to the carbon skeleton by the stable C-P bond. They are formed, for instance, by protozoa living in the stomach of ruminants, but also in other animals, e. g., Coelenterata, Mollusca, and Vertebrata.

Phosphonic acids are synthesized from phosphate esters by intramolecular rearrangement (Fig. 52). Ciliatin (2-aminoethylphosphonic acid) is a constituent of phosphonocephalin, which corresponds structurally to cephalin (D 5.2.4), and is incorporated into membranes. Ciliatin may also be methylated to the N-trimethyl derivative, which corresponds to choline.

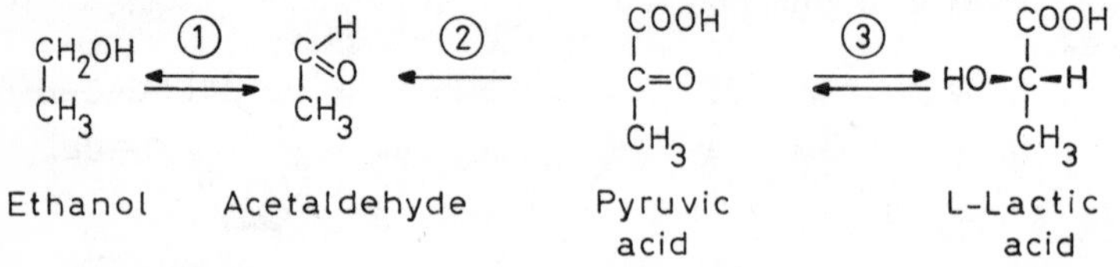

Ethanol   Acetaldehyde   Pyruvic   L-Lactic
                                acid   acid

Fig. 51. Formation of L-lactic acid and ethanol from pyruvic acid
1 Alcohol dehydrogenase (C 2.1); *2* pyruvate decarboxylase (C 4); *3* lactate dehydrogenase

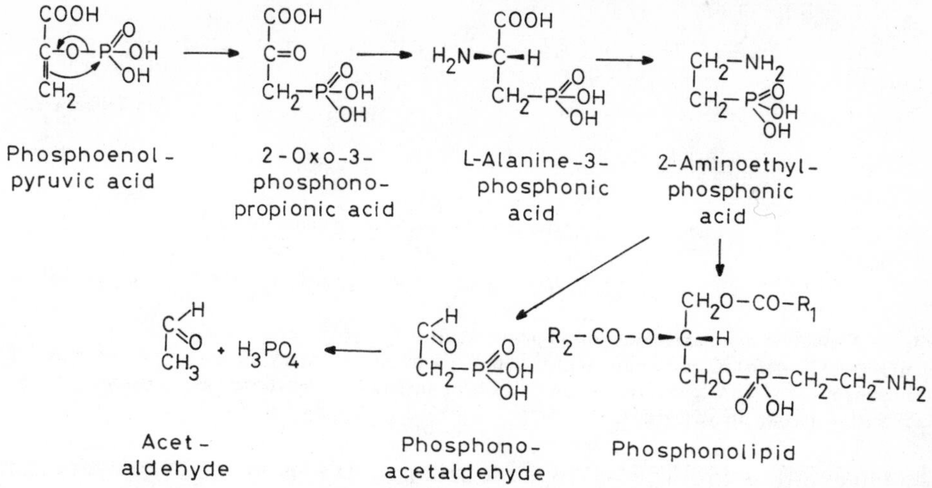

Fig. 52. Biosynthesis and degradation of L-alanine-3-phosphonic acid

Specialized microorganisms are able to degrade 2-aminoethylphosphonic acid to phosphonoacetaldehyde which is split to acetaldehyde and phosphoric acid.

References for Further Reading

259, 407

D 3 Derivatives of Propionyl CoA

Biosynthesis and Degradation of Propionyl CoA

Propionyl CoA is built from succinyl CoA, an intermediate of the tricarboxylic acid cycle (D 4), via *(S)*-methylmalonyl CoA (Fig. 53), or directly from methyl-malonyl CoA, a product of L-valine degradation (D 13).

Propionyl CoA is carboxylated by a biotin-containing enzyme (C 3.1) to me-thylmalonyl CoA.

Chemistry and Distribution

Propionyl CoA is an activated compound (C 1.2) and may be substituted easily at the carbonyl *C*-atom by nucleophilic groups and at the α-carbon atom by elec-trophilic groupings. It is a precursor of secondary products in microorganisms and animals.

Secondary Products Formed from Propionyl CoA

a) Propionic acid is derived from propionyl CoA by transfer of the CoA residue to another compound or by hydrolysis. It is degraded via acrylic acid and L-lactic

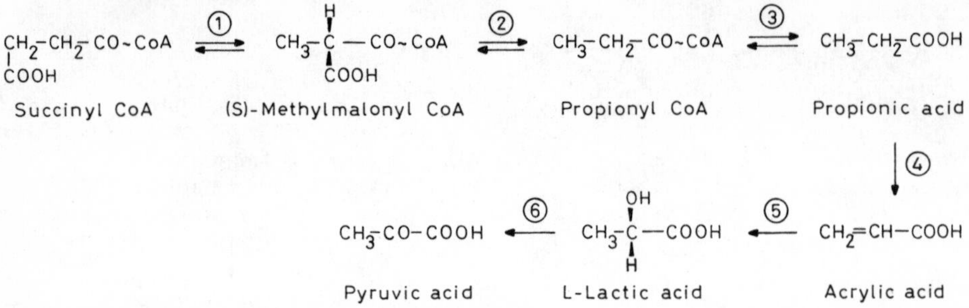

$$\underset{\underset{\text{Succinyl CoA}}{\overset{\displaystyle CH_2-CH_2-CO\sim CoA}{\underset{\displaystyle COOH}{|}}}}{} \overset{①}{\rightleftarrows} \underset{\underset{\text{(S)-Methylmalonyl CoA}}{\overset{\displaystyle CH_3-\overset{H}{\underset{|}{C}}-CO\sim CoA}{\underset{\displaystyle COOH}{|}}}}{} \overset{②}{\rightleftarrows} \underset{\text{Propionyl CoA}}{CH_3-CH_2-CO\sim CoA} \overset{③}{\rightleftarrows} \underset{\text{Propionic acid}}{CH_3-CH_2-COOH}$$

$$\downarrow ④$$

$$\underset{\text{Pyruvic acid}}{CH_3-CO-COOH} \overset{⑥}{\leftarrow} \underset{\text{L-Lactic acid}}{CH_3-\overset{OH}{\underset{H}{\overset{|}{\underset{|}{C}}}}-COOH} \overset{⑤}{\leftarrow} \underset{\text{Acrylic acid}}{CH_2\!=\!CH-COOH}$$

Fig. 53. Formation and degradation of propionic acid
1 (R)-Methylmalonyl CoA mutase (coenzyme B$_{12}$-dependent), methylmalonyl CoA racemase; *2* (S)-methylmalonyl CoA carboxyltransferase; *3* CoA transferase; *4* dehydrogenase; *5* reaction not established; *6* lactate dehydrogenase

acid to pyruvic acid (Fig. 53). Propionic acid is a sex pheromone of monkeys and human being (E 4).

b) Propionic acid esters occur in the essential oils of many plants. They have a characteristic smell.

c) Polymethylated fatty acids have been found in microorganisms and animals. They either are synthesized completely from propionyl CoA and methylmalonyl CoA by the pathway described in D 5.2 for normal fatty acids, or may contain acetyl CoA/malonyl CoA-derived parts (Table 25). Polymethylated fatty

Table 25. Fatty acids fully or partly synthesized from propionic acid units

Formula	Occurrence	No. of Precursor molecules Acetate	Propionate
$CH_2-CH_2-\overset{CH_3}{\underset{H}{\overset{|}{\underset{|}{C}}}}-COOH$, CH_3 2-Methyl valeric acid	Ascaris lumbricoides	–	2
$CH_3-(CH_2)_{19}-\overset{H}{\underset{CH_3}{\overset{|}{\underset{|}{C}}}}-CH_2-\overset{H}{\underset{CH_3}{\overset{|}{\underset{|}{C}}}}-CH_2-\overset{H}{\underset{CH_3}{\overset{|}{\underset{|}{C}}}}-CH_2-\overset{H}{\underset{CH_3}{\overset{|}{\underset{|}{C}}}}-COOH$ C$_{32}$-Mycocerosic acid	Mycobacteria	10	4
$CH_3-(CH_2)_{17}-\overset{CH_3}{\underset{H}{\overset{|}{\underset{|}{C}}}}-CH_2-\overset{CH_3}{\underset{H}{\overset{|}{\underset{|}{C}}}}-\overset{H}{\underset{CH_3}{\overset{|}{\underset{|}{C}}}}\!=\!C-COOH$ C$_{27}$-Phthienoic acid	Mycobacteria	9	3
$CH_3-CH_2-\overset{H}{\underset{CH_3}{\overset{|}{\underset{|}{C}}}}-CH_2-\overset{H}{\underset{CH_3}{\overset{|}{\underset{|}{C}}}}-CH_2-\overset{H}{\underset{CH_3}{\overset{|}{\underset{|}{C}}}}-CH_2-\overset{H}{\underset{CH_3}{\overset{|}{\underset{|}{C}}}}-COOH$ 2,4,6,8-Tetramethyldecanoic acid	Anser anser domesticus	1	4
$CH_2-CH_2-\overset{H}{\underset{CH_3}{\overset{|}{\underset{|}{C}}}}-CH_2-\overset{H}{\underset{CH_3}{\overset{|}{\underset{|}{C}}}}-CH_2-\overset{H}{\underset{CH_3}{\overset{|}{\underset{|}{C}}}}-CH_2-\overset{H}{\underset{CH_3}{\overset{|}{\underset{|}{C}}}}-COOH$, CH_3 2,4,6,8-Tetramethylundecanoic acid	Anser anser domesticus	–	5

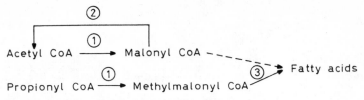

Fig. 54. Synthesis of methyl fatty acids in birds
1 Carboxylase; *2* decarboxylase; *3* fatty acid synthase

acids are precursors of the corresponding alkanols (see D 5.2.3) and esters (see D 5.2.4) found, e. g., in the uropygial gland secretions of birds.

The uropygial gland of birds, which produces polymethylated fatty acids, contains a carboxylase which forms malonyl CoA from acetyl CoA and methylmalonyl CoA from propionyl CoA. In addition, there is a decarboxylase, which degrades malonyl CoA, drastically lowering its concentration. Thus, in the formation of fatty acids methylmalonyl CoA is preferred though the fatty acid synthase of the gland reacts much faster with malonyl CoA than with methylmalonyl CoA (Fig. 54).

d) Macrolides, i. e., lactones with a macrocyclic ring more or less regularly substituted with methyl groups, are formed in *Streptomyces* sp. They may be divided in nonpolyenes, e. g., erythromycins (Fig. 55) and methymycin, and in polyenes, e. g., nystatin and amphotericin B (Fig. 56). The polyenes have a chromophoric system of conjugated double bonds and therefore show strong absorption in UV and visible light. The lactone ring of the polyene macrolides comprises 26–38 C-atoms, whereas that of the nonpolyenes is much smaller, cf. the 14-membered ring of erythromycins. Both types of lactones are coupled to rare sugars (D 1.1), acyl residues, etc.

In analogy to fatty acid and polyketide biosynthesis (D 5.2 and D 5.3) it is assumed that macrolides are derived from a starter molecule, e. g., acetyl CoA or propionyl CoA, and several malonyl and/or methylmalonyl CoA units as well as ethylmalonyl CoA moieties (derived from butyric acid, D 5.1). These building blocks are linked together and transformed by polyfunctional proteins. The fundamental skeletons may then be completed by reaction with other precursors.

The macrolides possess antibiotic properties (E 5.2) and are used in medicine (F 2).

e) Ansa macrolides (ansamycins), like rifamycin B and maytansine (Fig. 56), possess a small aromatic nucleus derived from benzene or naphthene, and a second large ring built from a branched-chain fatty acid, which is formed in part from propionic acid residues. The rifamycins are antibiotics built in *Nocardia* sp. (E 5.2). Rifampicin, a semisynthetic derivative is used in medicine. Maytansine has antileukemic activity. It is synthesized in Actinomycetes and in plants of the genus *Maytenus* (Celastraceae).

f) Daunomycin (daunorubicin) and adriamycin (Fig. 56) are anthracycline derivatives formed in *Streptomyces peucetius*. The anthracycline ring system is of polyketide origin (D 5.3). It is built with propionyl CoA as starter and nine malonyl CoA residues, with loss of the terminal carboxy group. Daunomycin and adriamycin are antibiotics (E 5.2). They possess antineoplastic activity and are used in the treatment of leukemia (F 2).

g) The piericidins (Fig. 55) are pyridine derivatives built in Actinomycetes. It

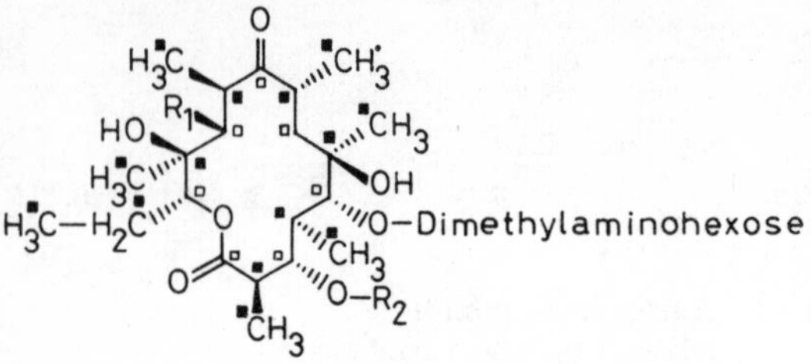

O—Dimethylaminohexose

Aminohexose

Methymycin

Erythromycin A R_1=OH, R_2=L-Cladinose
Erythromycin B R_1=H, R_2=L-Cladinose
Erythromycin C R_1=OH, R_2=L-Mycarose

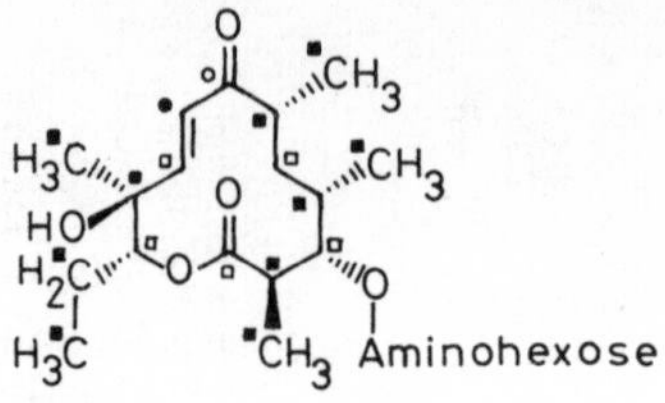

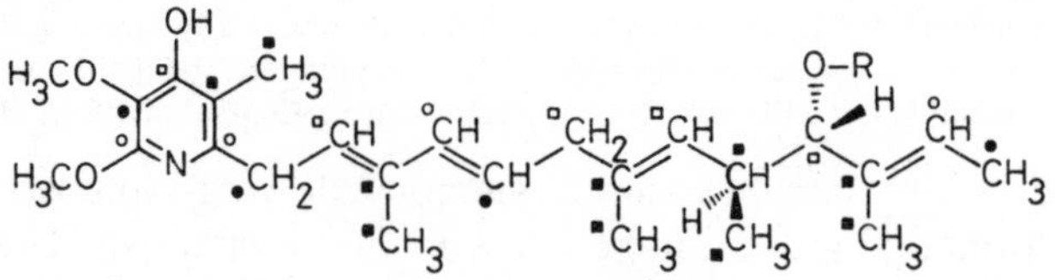

Piericidin A (R=H)
Piericidin B (R=CH$_3$)

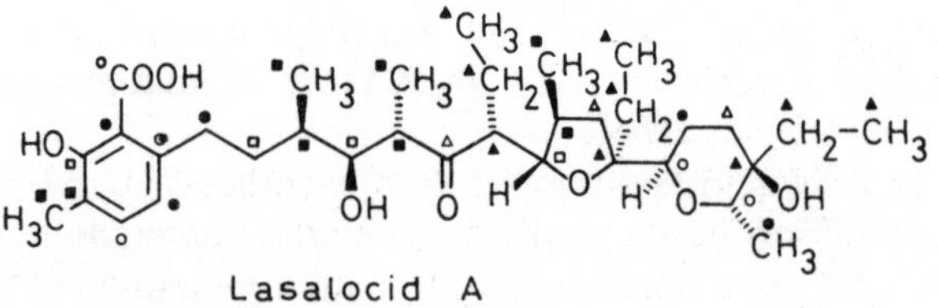

Lasalocid A

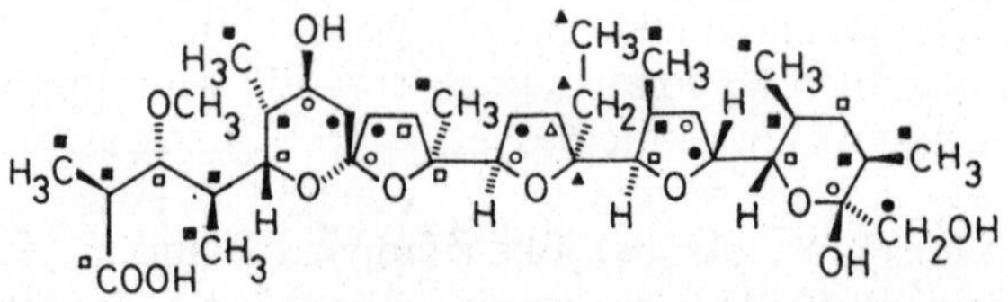

Monensin A

Fig. 55. Incorporation of acetic, propionic, and butyric acids in macrolide, pyridine, and polyether antibiotics
Acetic acid (CH$_3$—COOH, ●—o), propionic acid (CH$_3$—CH$_2$—COOH, ■—■—□), butyric acid (CH$_3$—CH$_2$—CH$_2$—COOH, ▲—▲—▲—△)

is assumed that they are synthesized via a polyoxo acid, which condenses with ammonia in a reaction resembling the formation of coniine (D 5.3.2).

h) Polyether antibiotics (E 5.2), e. g., lasalocid A and monensin A (Fig. 55) have been found in Streptomycetes. They are formed with polyketides as intermediates. Acetyl CoA or malonyl CoA (D 5.3) act as starter. Chain elongation

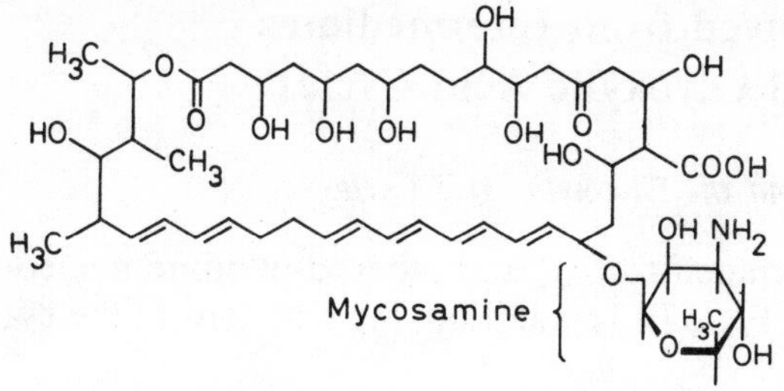

Nystatin

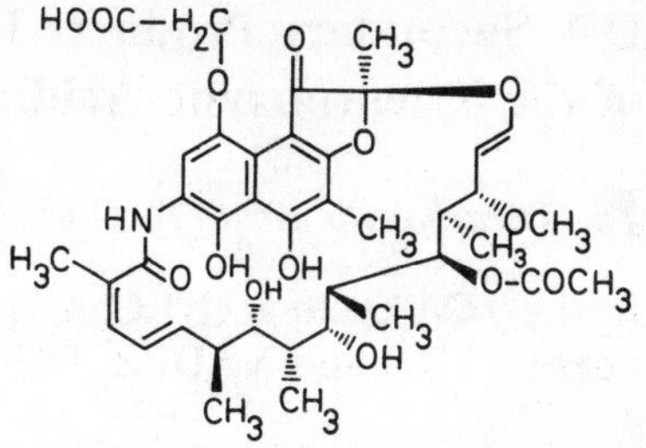

Rifamycin B

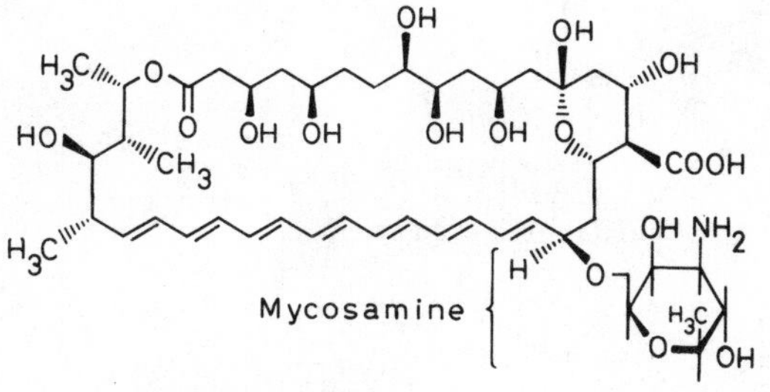

Amphotericin B

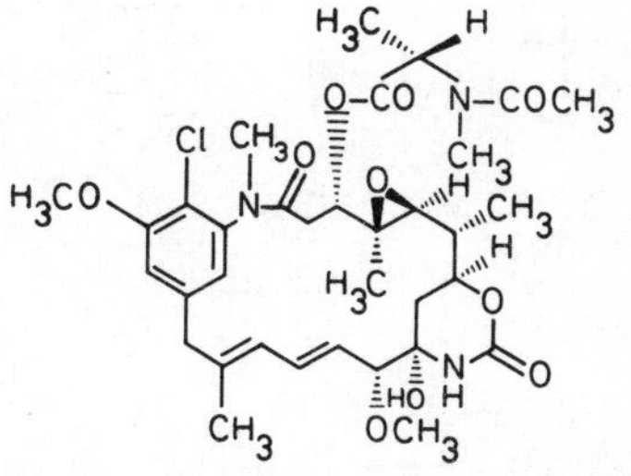

Maytansine

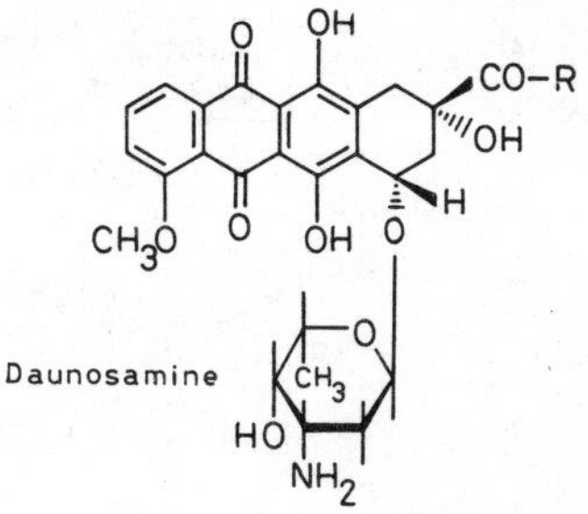

Daunomycin (R = –CH$_3$)
Adriamycin (R = –CH$_2$OH)

Fig. 56. Structural formulas of several macrolide and anthracycline antibiotics

proceeds with malonyl CoA, methylmalonyl CoA, and/or ethylmalonyl CoA. Polyether antibiotics are used as coccidiostatics and ergotropics (F 2).

References for Further Reading

179, 281, 286, 386, 403, 577, 708, 737, 783

D 4 Secondary Products Derived from Intermediates of the Tricarboxylic Acid and Glyoxylic Acid Cycles

The Tricarboxylic Acid (TCA) Cycle and the Glyoxylic Acid Cycle

In the TCA cycle acetyl CoA is degraded to CO_2 and reduced pyridine nucleotides (NADH and NADPH, D 16.1, Fig. 57). In addition, the reactions of the cy-

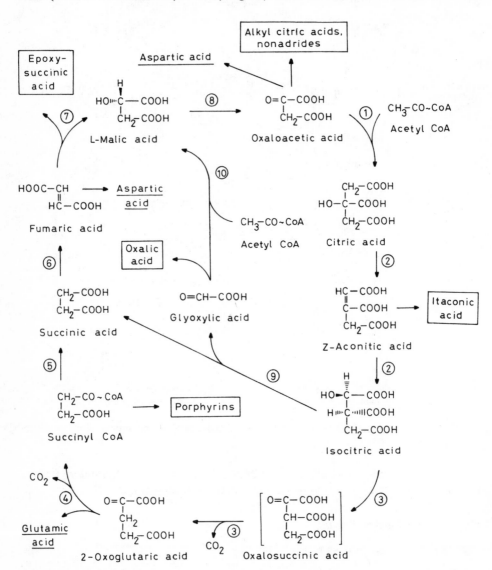

Fig. 57. Secondary products derived from intermediates of the tricarboxylic and glyoxylic acid cycles
1 Citrate synthase; *2* aconitate hydratase; *3* isocitrate dehydrogenase; *4* oxoglutarate dehydrogenase; *5* succinyl-CoA synthetase; *6* succinate dehydrogenase; *7* fumarate hydratase; *8* malate dehydrogenase; *9* isocitrate lyase; *10* malate synthase

cle supply intermediates for the biosynthesis of the amino acids L-glutamic acid (D 17) and L-aspartic acid (D 16). The carbon chains withdrawn from the cycle are replaced by degradation of isocitric acid to succinic and glyoxylic acid and the formation of malic acid from the latter. The TCA cycle "short-circuited" by these reactions is known as the glyoxylic acid cycle.

In certain species and special types of cells intermediates of the TCA cycle are produced in excess. Citric acid, for instance, is accumulated in large amounts in cultures of certain microorganisms and in the fruits of lemon. It is used in the food industry. Malic acid accumulates in the guard cells of plants. It causes swelling of the protoplast and opening of the stomata.

Groups of Secondary Products Formed from Cycle Intermediates

Secondary products are derived from intermediates of the TCA and glyoxylic acid cycles in microorganisms, plants, and animals. They are either carboxylic acids, e. g., agaricic, epoxysuccinic, and meso-tartaric acid, or lactones like the nonadrides.

a) Alkylcitric acid derivatives are formed from oxyloacetic acid in several fungi. Compounds of the agaricic acid type are built with the participation of a fatty acid CoA ester, e. g., stearyl CoA. The reaction resembles the formation of citric acid:

$$CH_3-(CH_2)_{15}-CH_2-CO \sim CoA \ + \ \begin{matrix} O=C-COOH \\ | \\ CH_2-COOH \end{matrix} \longrightarrow \ \begin{matrix} CH_3-(CH_2)_{15}-CH-COOH \\ | \\ HO-C-COOH \\ | \\ CH_2-COOH \end{matrix}$$

$$\quad \text{Stearyl CoA} \qquad\qquad \begin{matrix}\text{Oxaloacetic} \\ \text{acid}\end{matrix} \qquad\qquad \text{Agaricic acid}$$

n-Butylcitric acid is the precursor of the nonadrides, compounds with a nine-membered carbocyclic ring and two five-membered anhydride rings. Different types of condensation yield glaucanic and byssochlamic acid, respectively (Fig. 58).

b) Itaconic acid is formed from Z-aconitic acid by decarboxylation. It is used as additive in the production of synthetic fibers.

$$\begin{matrix} CH_2 \\ \parallel \\ C-COOH \\ | \\ CH_2-COOH \end{matrix}$$

$$\text{Itaconic acid}$$

c) *E*-Epoxysuccinic acid is formed from fumaric acid accumulating as the result of *Rhizopus* infections in almond trees (E 5.4). *E*-Epoxysuccinic acid is toxic. It is degraded to meso-tartaric acid and oxaloacetic acid (Fig. 59).

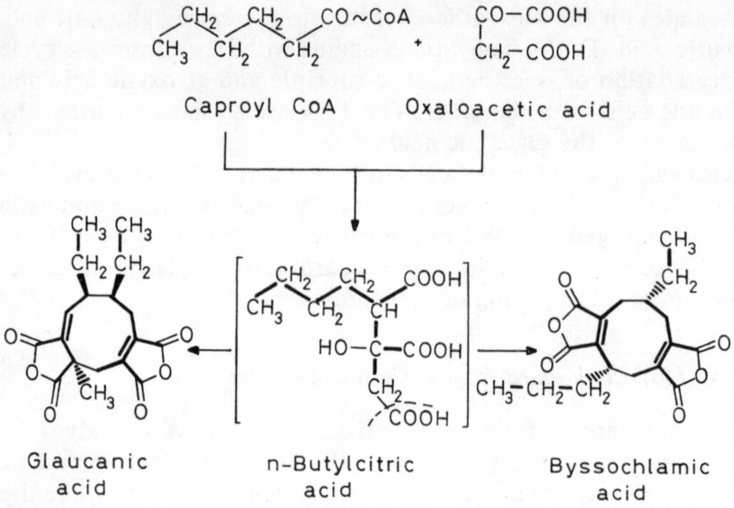

Caproyl CoA   Oxaloacetic acid

Glaucanic   n–Butylcitric   Byssochlamic
acid   acid   acid

Fig. 58. Possible pathway for the biosynthesis of nonadrides

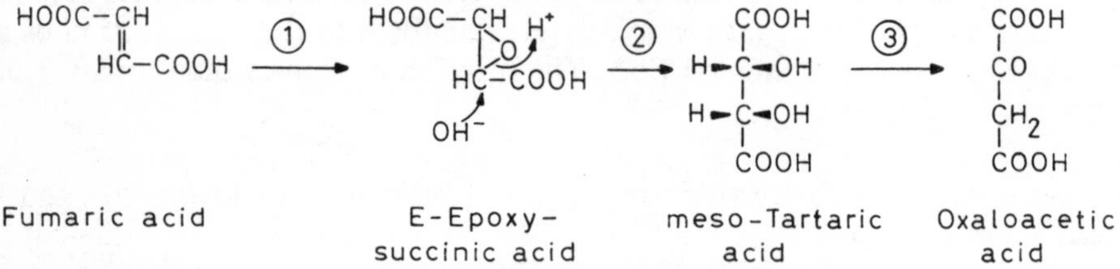

Fumaric acid   E-Epoxy-   meso-Tartaric   Oxaloacetic
                    succinic acid   acid   acid

Fig. 59. Formation and degradation of *E*-epoxysuccinic acid
1 Fumarate monooxygenase; *2* E-epoxysuccinate hydratase; *3* tartrate dehydratase

d) Oxalic acid is synthesized from glyoxylic acid by glycolate oxidase:

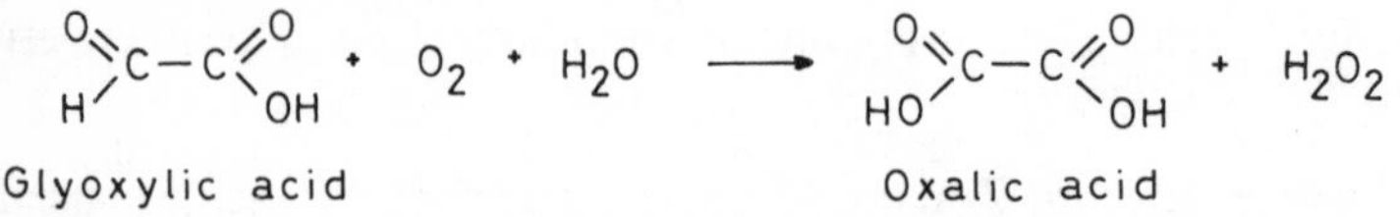

Glyoxylic acid   Oxalic acid

Glyoxylic acid is either derived from isocitric acid (Fig. 57) or, in plants, from glycolic acid formed in photosynthesis. The oxidation of glycolic acid to glyoxylic acid is also catalyzed by glycolate oxidase:

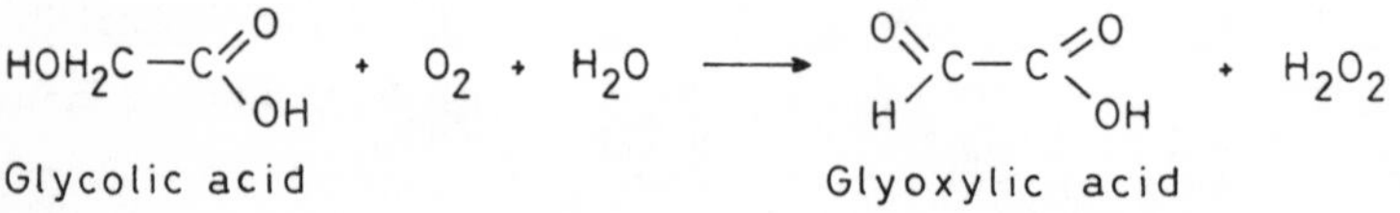

Glycolic acid   Glyoxylic acid

Oxalic acid and calcium oxalate, a store for excess Ca^{2+} (E 2.2) and a repellent for potential predators (E 5.5.3), accumulate frequently in plants and microbial cultures. Oxalyl derivatives of amino acids occur in different groups of organisms (D 8.1). *O*-Oxalyl-L-homoserine is involved in the biosynthesis of L-methionine in microorganisms (D 12).

Oxalic acid is decarboxylated to formic acid which may enter one-carbon metabolism (C 3.2). Oxalate oxidase degrades oxalic acid to CO_2:

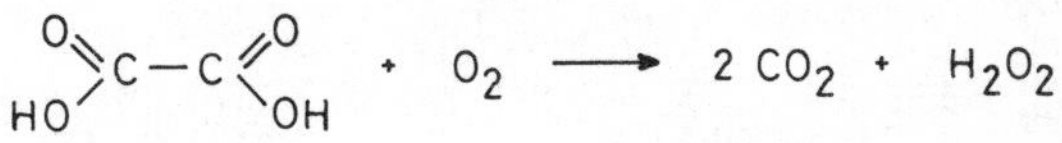

$$\underset{HO}{\overset{O}{\diagdown}}C-C\underset{OH}{\overset{O}{\diagup}} + O_2 \longrightarrow 2\,CO_2 + H_2O_2$$

References for Further Reading

136, 360, 525, 790

D 5 Compounds Formed from Acetyl CoA

Biosynthesis and Transformation of Acetyl CoA in Primary Metabolism

The main sources of acetyl CoA in primary metabolism are
— the oxydative decarboxylation of pyruvic acid (C 2.7), which proceeds via activated acetaldehyde and acetyllipoic acid
— the thiolytic cleavage of fatty acids (D 5.2), and
— the activation of acetic acid by reaction with ATP and CoA with the formation of acetyl phosphate or acetyl-AMP (C 1.2) as intermediates.

The most important pathways of primary metabolism for the transformation of acetyl CoA are the tricarboxylic acid cycle (D 4) and the biosynthesis of fatty acids (D 5.2).

Acetyl CoA-Derived Secondary Products

Acetyl CoA, "activated acetic acid", is a precursor of secondary products in microorganism, plants, and animals. It is easily substituted nucleophilically at the carbonyl group and electrophilically at the methyl group (C 1.2). Of special interest is its transformation to malonyl CoA by acetyl CoA carboxylase (C 3.1):

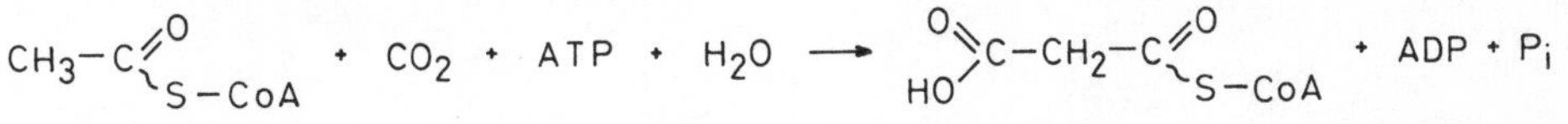

$$CH_3-C\underset{S-CoA}{\overset{O}{\diagup}} + CO_2 + ATP + H_2O \longrightarrow \underset{HO}{\overset{O}{\diagdown}}C-CH_2-C\underset{S-CoA}{\overset{O}{\diagup}} + ADP + P_i$$

Acetyl CoA   Malonyl CoA

The following groups of secondary products are derived from acetyl CoA and malonyl CoA:

a) Acetic acid, CH_3COOH, is formed from activated acetic acid derivatives, e. g., acetyl CoA and acetyl phosphate, by hydrolysis, but may also be derived from acetyl esters by esterases or by the oxidation of ethanol (D 2). It is part of the sex pheromones of monkeys and human beings (E 4).

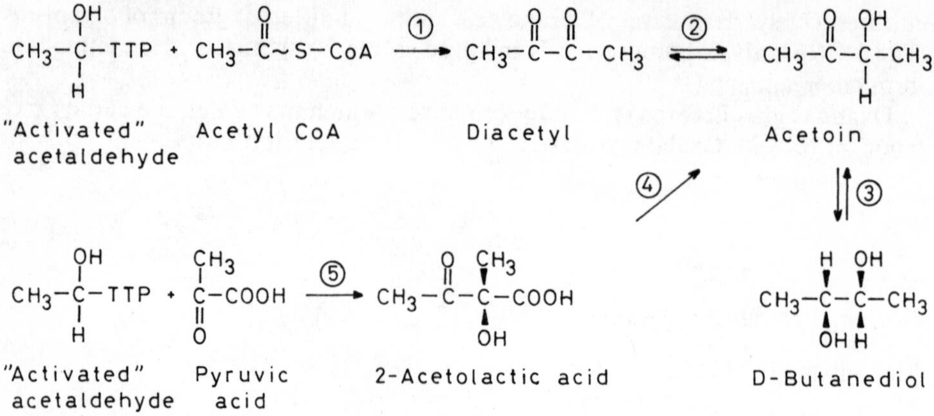

Fig. 60. Formation of diacetyl and acetoin
1 Diacetyl synthase; *2* acetoin dehydrogenase; *3* D-(−)-butanediol dehydrogenase; *4* acetolactate dehydrogenase; *5* acetolactate synthase

b) Fluoroacetic acid, CH_2FCOOH, is a toxic constituent of certain plants. It is a potent inhibitor of citrate synthase, the key enzyme of the tricarboxylic acid cycle (D 4). Fluoroacetic acid repels potential predators (E 5.5.3).

c) Trifluoroacetic acid, CF_3COOH, is a constituent of *Dichapetalum* sp. It is toxic to animals (E 5.5.3).

d) Malonic acid, $HOOC—CH_2—COOH$, is formed from malonyl CoA in some microorganisms and plants. As a succinic acid analog it inhibits succinate dehydrogenase (D 4). Malonic acid is toxic and repels potential predators (E 5.5.3). Malonic acid esters occur rather widespread in several groups of secondary products (E 1).

e) Diacetyl and acetoin are formed in microorganisms, e. g., in *Streptococcus* and *Leuconostoc* sp., from acetyl CoA and "activated" acetaldehyde or from pyruvic acid (Fig. 60). The compounds cause the characteristic flavor of butter.

f) Acetoacetic acid and fatty acid derivatives as well as polyketides are treated in D 5.1-D 5.3.

References for Further Reading

30, 345

D 5.1 Acetoacetic Acid and its Derivatives

Chemistry and Distribution

Acetoacetic acid, CH_3COCH_2COOH, is an aliphatic 3-oxo acid. Derivatives are formed by reduction, activation of the carboxylic group, and decarboxylation. They occur in microorganisms, plants, and animals.

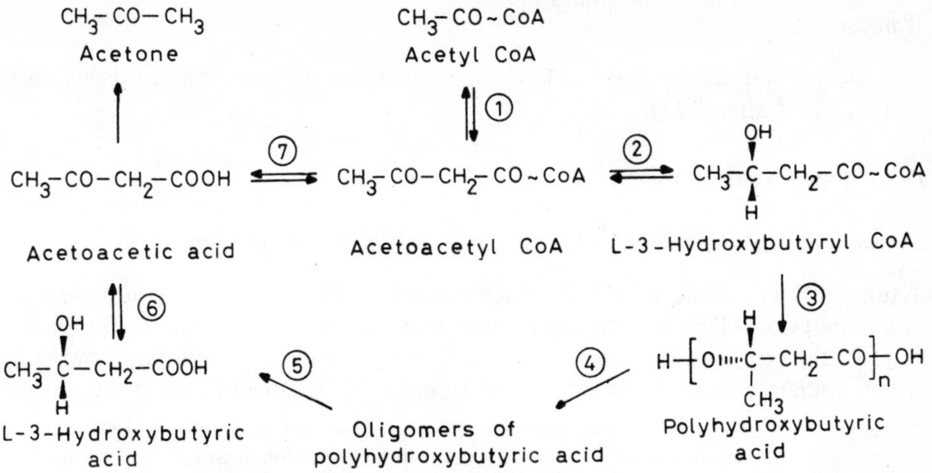

CH₃–CO–CH₃ — written as $CH_3-CO-CH_3$

$$CH_3-CO-CH_3 \qquad CH_3-CO\sim CoA$$

Acetone   Acetyl CoA

$$CH_3-CO-CH_2-COOH \overset{\textcircled{7}}{\rightleftharpoons} CH_3-CO-CH_2-CO\sim CoA \overset{\textcircled{2}}{\rightleftharpoons} CH_3-\underset{H}{\overset{OH}{C}}-CH_2-CO\sim CoA$$

Acetoacetic acid   Acetoacetyl CoA   L-3-Hydroxybutyryl CoA

$$CH_3-\underset{H}{\overset{OH}{C}}-CH_2-COOH \overset{\textcircled{5}}{\longleftarrow} \qquad \overset{\textcircled{4}}{\longrightarrow} \qquad H-\left[O\text{\tiny{IIII}}\underset{CH_3}{\overset{H}{C}}-CH_2-CO\right]_n OH$$

L-3-Hydroxybutyric   Oligomers of   Polyhydroxybutyric
acid   polyhydroxybutyric acid   acid

Fig. 61. Formation of secondary products from acetoacetyl CoA
1 3-Oxothiolase; *2* acetoacetyl CoA reductase; *3* 3-hydroxybutyryl CoA polymerase; *4* polyhydroxy-butyrate depolymerase; *5* hydrolase; *6* 3-hydroxybutyrate dehydrogenase; *7* CoA-transferase

Biosynthesis

Acetoacetyl CoA is built from two molecules of acetyl CoA. Free acetoacetic acid is either formed from acetoacetyl CoA or from L-3-hydroxybutyric acid (Fig. 61). Acetoacetic acid decarboxylates spontaneously to acetone. Reduction of acetoacetyl CoA yields L-3-hydroxybutyryl CoA which may be transformed to a polymeric derivative (Fig. 61), to butyric acid, or to butanol (Fig. 62).

Significance

Polyhydroxybutyric acid is a storage for excess carbon in many microorganisms (E 2.2). It may be used in the production of plastics (F 5). Acetoacetic acid, acetone, and L-3-hydroxybutyric acid are constituents of the urine of patients with a pathologically high blood sugar level (diabetes mellitus) (E 1). Butyric acid, butanol, and acetone are products of microbial fermentations.

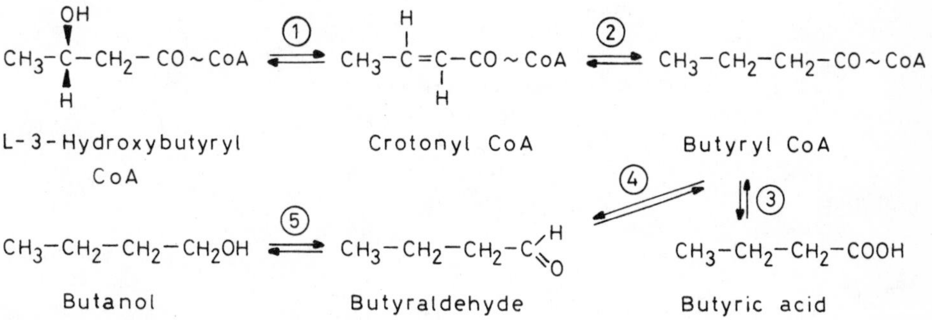

$$CH_3-\underset{H}{\overset{OH}{C}}-CH_2-CO\sim CoA \overset{\textcircled{1}}{\rightleftharpoons} CH_3-\underset{H}{\overset{H}{C}}=C-CO\sim CoA \overset{\textcircled{2}}{\rightleftharpoons} CH_3-CH_2-CH_2-CO\sim CoA$$

L-3-Hydroxybutyryl   Crotonyl CoA   Butyryl CoA
CoA

$$CH_3-CH_2-CH_2-CH_2OH \overset{\textcircled{5}}{\rightleftharpoons} CH_3-CH_2-CH_2-C\overset{H}{\underset{O}{<}} \qquad \overset{\textcircled{4}}{\rightleftharpoons} \quad \text{\textcircled{3}} \quad CH_3-CH_2-CH_2-COOH$$

Butanol   Butyraldehyde   Butyric acid

Fig. 62. Formation of butyric acid and butanol
1 Enoyl CoA hydratase (crotonase); *2* butyryl CoA dehydrogenase; *3* CoA-transferase; *4* butyralde-hyde dehydrogenase; *5* butanol dehydrogenase

Transformation

Acetoacetic acid (or acetoacetyl CoA) is a precursor of tropane and Punica alkaloids (D 17.2 and D 18).

D 5.2 Fatty Acid Derivatives

Biosynthesis and Degradation of Fatty Acids in Primary Metabolism

Saturated fatty acids are built from acetyl CoA as a starter and several molecules of malonyl CoA (D 5) by fatty acid synthase, as is shown schematically in Fig. 63. In yeast the enzymes involved are constituents of two polyfunctional proteins. It is still unclear whether similar polyfunctional enzymes exist in other organisms.

The synthesis of fatty acids begins with the linkage of acetyl CoA (or another acyl CoA), to the SH-group of a peripheral subunit of the enzyme protein with the elimination of the CoA residue.

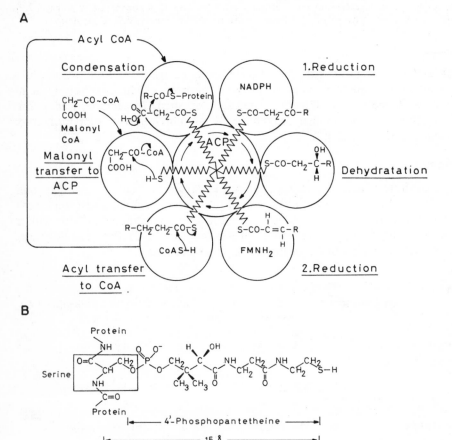

Fig. 63. Fatty acid synthase shown schematically. B shows the 4-phosphopanthetheine residue of the acyl-carrier protein (ACP) in detail

Simultaneously a molecule of malonyl CoA is bound to another domain, the acyl carrier protein (ACP), again with the loss of CoA. The malonyl residue becomes attached to the SH-group of a 4'-phospho-pantetheine grouping (D 11) linked to a serine residue of the peptide chain. The ACP-bound malonyl group substitutes the acetyl residue nucleophilically at the carbonyl group (C 1.2) releasing it from the enzyme protein with the formation of acetoacetyl-ACP.

Reduction of the ACP-bound acetoacetate yields D-3-hydroxybutyryl-ACP from which crotonyl-ACP is formed by the elimination of water. By the subsequent reduction butyryl-ACP is obtained from which the butyryl residue is transferred to CoA, releasing the SH-group of pantetheine which again reacts with a molecule of malonyl CoA. Butyryl CoA, like acetyl CoA, may react with the peripheral, acyl-binding SH-group of the synthase and subsequently with the malonyl residue. In the following reactions caproic acid is formed from which caprylic, capric, lauric, myristic, palmitic, stearic, and arachidonic acid (Table 26) are built. The specificity of the enzymes involved as well as other factors, such as compartmentalization stop chain elongation at about 10-20 C-atoms.

Unsaturated fatty acids arise on two independent pathways:

a) In animals, plants, and fungi double bonds are introduced into the CoA esters or ACP derivatives of saturated fatty acids. Stearyl ACP, for example, is converted to oleyl ACP in plants by stearyl-ACP desaturase. Desaturation proceeds by a concerted syn-elimination of a pair of neighboring pro-R hydrogen atoms without involvement of oxygenated intermediates:

b) In bacteria the second reduction in one of the cycles of fatty acid synthesis (Fig. 63) is omitted. The unsaturated fatty acid is converted directly to the CoA ester and may be further elongated without loss of the double bond.

Fatty acids may be degraded in three different ways of which the
β-oxidation is most widespread:

a) β-Oxidation. The β-oxidation begins with the formation of acyl CoA derivatives (C 1.2). These are transformed to the 3-oxo acid CoA esters, which are cleaved thiolytically by reaction with CoA to form acetyl CoA and an acyl CoA shorter than the starting compound by two carbon atoms (Fig. 64).

Saturated fatty acids with an even number of C-atoms are degraded totally to acetyl CoA by repetition of these reactions. Unsaturated fatty acids have to be converted during the course of degradation to compounds which can be attacked by the enzymes of β-degradation, e. g., by isomerization, shift of double bonds, addition of water, etc. (Fig. 65).

b) α-Oxidation. Free fatty acids may be degraded by flavin enzymes (C 2.2.1) to fatty acids shorter by one carbon atom (Fig. 66). These enzymes probably remove a hydrogen radical from the fatty acid (formation of intermediate I) and then add on OOH-radical with synthesis of an L- or D-2-hydroperoxy fatty acid. D-2-Hydroperoxy fatty acids may undergo reduction to D-2-hydroxy fatty acids.

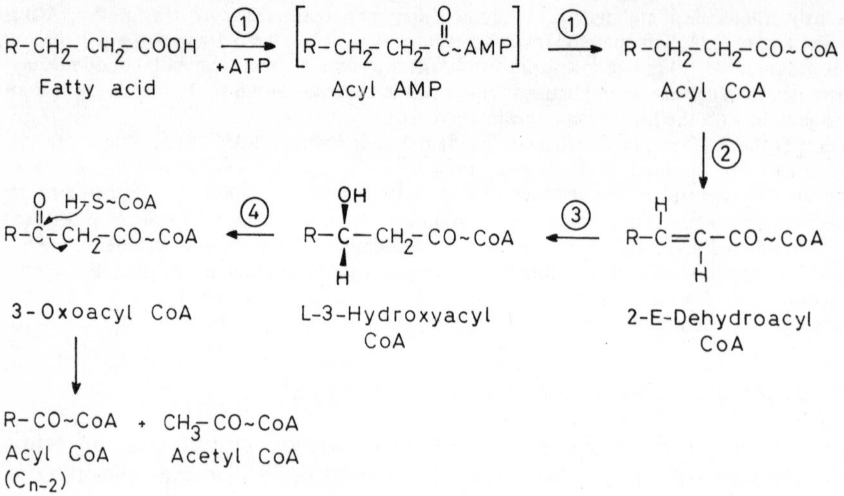

R-CH$_2$-CH$_2$-COOH $\xrightarrow[\text{+ATP}]{①}$ $\left[\text{R-CH}_2\text{-CH}_2\text{-}\overset{\overset{\textstyle O}{\|}}{C}\text{~AMP} \right]$ $\xrightarrow{①}$ R-CH$_2$-CH$_2$-CO~CoA

Fatty acid   Acyl AMP   Acyl CoA

$\downarrow ②$

R-$\overset{\overset{\textstyle O}{\|}}{C}$-CH$_2$-CO~CoA $\xleftarrow{④}$ R-$\overset{\overset{\textstyle OH}{|}}{\underset{\underset{\textstyle H}{|}}{C}}$-CH$_2$-CO~CoA $\xleftarrow{③}$ R-$\overset{\overset{\textstyle H}{|}}{C}$=$\overset{\underset{\textstyle H}{|}}{C}$-CO~CoA

H~S~CoA

3-Oxoacyl CoA   L-3-Hydroxyacyl   2-E-Dehydroacyl
                              CoA   CoA

$\downarrow$

R-CO~CoA + CH$_3$-CO~CoA
Acyl CoA   Acetyl CoA
(C$_{n-2}$)

Fig. 64. β-Oxidation of saturated fatty acids
1 Acyl-CoA synthetase; *2* acyl-CoA dehydrogenase; *3* enoyl-CoA hydratase; *4* 3-hydroxyacyl-CoA dehydrogenase; *5* acetyl-CoA acyltransferase

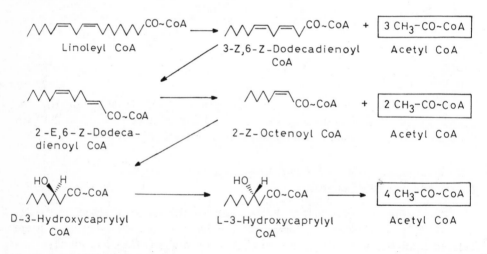

Linoleyl CoA

3-Z,6-Z-Dodecadienoyl
CoA

2-E,6-Z-Dodeca-
dienoyl CoA

2-Z-Octenoyl CoA

D-3-Hydroxycaprylyl
CoA

L-3-Hydroxycaprylyl
CoA

+ 3 CH$_3$-CO~CoA
Acetyl CoA

+ 2 CH$_3$-CO~CoA
Acetyl CoA

4 CH$_3$-CO~CoA
Acetyl CoA

Fig. 65. β-Oxidation of linoleyl CoA

L-2-Hydroperoxy fatty acids are decarboxylated to CO_2 and an aldehyde. The aldehyde is dehydrogenated by an NAD$^+$-dependent dehydrogenase (C 2.1) to the corresponding fatty acid shorter by one carbon atom, which may undergo the same series of reactions.

α-Oxidation plays an important role in the formation of alkanes (D 5.2.2) and acetylenic compounds (D 5.2.5).

c) ω-Oxidation. In this type of oxidation the terminal methyl group is hydroxylated by a monooxygenase (C 2.6.4) and further on oxidized to a carboxylic

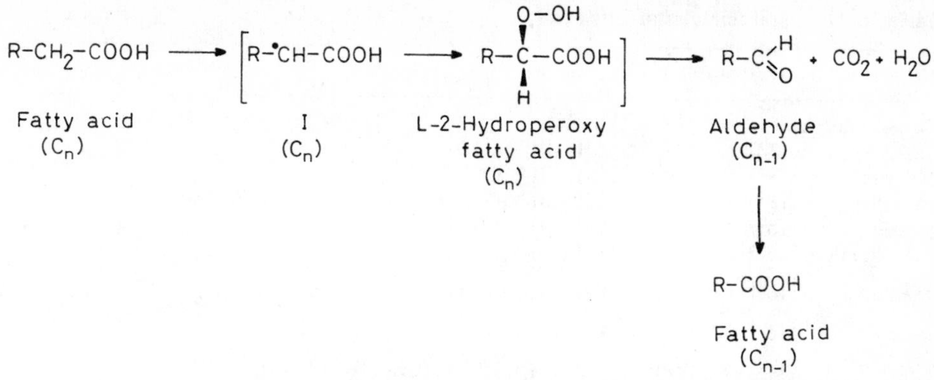

$$R-CH_2-COOH \longrightarrow \left[R-\overset{\bullet}{C}H-COOH \longrightarrow R-\underset{H}{\overset{O-OH}{\underset{|}{\overset{|}{C}}}}-COOH \right] \longrightarrow R-C\overset{H}{\underset{O}{\diagdown}} + CO_2 + H_2O$$

| Fatty acid (C_n) | I (C_n) | L-2-Hydroperoxy fatty acid (C_n) | Aldehyde (C_{n-1}) |

$$\downarrow$$

$$R-COOH$$

Fatty acid
(C_{n-1})

Fig. 66. α-Oxidation of fatty acids

acid group. The dicarboxylic acids formed are precursors of cutins and suberins (D 5.2.4).

Alkanes (D 5.2.2) are also attacked by the enzymes of ω-oxidation. The fatty acids formed may be degraded by the enzymes of β-oxidation.

Microorganisms which oxidize alkanes are of great economic importance in the production of fodder from petroleum and natural gas. A mysterious disintegration of roads in some districts of Western Australia has been attributed to the degradation of the petroleum portion of the asphalt by bacteria.

Chemistry of Fatty Acids

Most fatty acids are carboxylic acids with an unbranched, even-numbered carbon chain. They may be saturated or may contain one or more double bonds which usually have the Z-configuration. In the polyunsaturated compounds the double bonds are separated by methylene groups (Table 26).

The following shorthand designation may be used to describe the chemical structure of fatty acids: the number of carbon atoms in the chain separated by a colon from the number of unsaturated bonds is followed by a set of parentheses containing the numbers of the double-bond positions and the letters E, Z, or a to indicate whether a bond is Z-olefinic, E-olefinic, or acetylenic. The position of double bonds is numbered beginning with the carboxyl group (cf. Table 26).

Significance

Butyric acid is a sex pheromone of vertebrates, including humans (E 4). Lipoic acid acts as a cosubstrate of certain oxidoreductases and is involved in the transfer of acyl groups (E 2.1, C 2.7). Linolenic acid is a vitamin for human beings (vitamin F, E 2.1).

Formation of Secondary Products

The fatty acids of primary metabolism are precursors of secondary products in microorganisms, plants, and animals. The most important compounds are treated in D 5.2.1–D 5.2.6.

Table 26. Structural formulas of fatty acids

Name	Symbol	Structure
Common fatty acids		
Lauric	12:0	$CH_3(CH_2)_{10}COOH$
Myristic	14:0	$CH_3(CH_2)_{12}COOH$
Palmitic	16:0	$CH_3(CH_2)_{14}COOH$
Stearic	18:0	$CH_3(CH_2)_{16}COOH$
Arachidonic	20:0	$CH_3(CH_2)_{18}COOH$
Palmitoleic	16:1 (9Z)	$CH_3(CH_2)_5CH\overset{Z}{=\!=}CH(CH_2)_7COOH$
Oleic	18:1 (9Z)	$CH_3(CH_2)_7CH\overset{Z}{=\!=}CH(CH_2)_7COOH$
Linoleic	18:2 (9Z, 12Z)	$CH_3(CH_2)_4CH\overset{Z}{=\!=}CHCH_2CH\overset{Z}{=\!=}CH(CH_2)_7COOH$
α-Linolenic	18:3 (9Z, 12Z, 15Z)	$CH_3CH_2CH\overset{Z}{=\!=}CHCH_2CH\overset{Z}{=\!=}CHCH_2CH\overset{Z}{=\!=}CH(CH_2)_7\!-\!COOH$
γ-Linolenic	18:3 (6Z, 9Z, 12Z)	$CH_3(CH_2)_4CH\overset{Z}{=\!=}CHCH_2CH\overset{Z}{=\!=}CHCH_2CH\overset{Z}{=\!=}CH(CH_2)_4\!-\!COOH$
Unusual fatty acids		
Acetic	2:0	CH_3COOH
Butyric	4:0	$CH_3(CH_2)_2COOH$
Valeric	5:0	$CH_3(CH_2)_3COOH$
Caproic	6:0	$CH_3(CH_2)_4COOH$
Caprylic	8:0	$CH_3(CH_2)_6COOH$
Capric	10:0	$CH_3(CH_2)_8COOH$
Lignoceric	24:0	$CH_3(CH_2)_{22}COOH$
Petroselinic	18:1 (6Z)	$CH_3(CH_2)_{10}CH\overset{Z}{=\!=}CH(CH_2)_4COOH$
Tariric	18:1 (6a)	$CH_3(CH_2)_{10}C\!\equiv\!C(CH_2)_4COOH$
Sterculic		$CH_3(CH_2)_7\,C\!=\!C(CH_2)_7COOH$

$$Sterculic:\quad CH_3(CH_2)_7\,\underset{\diagdown\;\diagup}{\overset{}{C=C}}(CH_2)_7COOH$$
$$\underset{CH_2}{}$$

Name	Symbol	Structure
Chaulmoogric		
Ricinoleic		$CH_3(CH_2)_5CHOHCH_2CH\overset{Z}{=\!=}CH(CH_2)_7COOH$
Vernolic		$CH_3(CH_2)_4CH\!-\!CHCH_2CH\overset{Z}{=\!=}CH(CH_2)_7COOH$
9-Oxo-2-E-decenoic		$CH_3CO(CH_2)_5CH\overset{E}{=\!=}CHCOOH$
α-Eleostearic		$CH_3(CH_2)_3CH\overset{E}{=\!=}CHCH\overset{E}{=\!=}CHCH\overset{Z}{=\!=}CH(CH_2)_7COOH$
Lipoic		

Chaulmoogric:
$$\begin{array}{c} HC\!=\!CH \\ |\qquad\diagdown H \\ H_2C\!-\!CH_2\;C\cdots(CH_2)_{12}COOH \end{array}$$

Vernolic (epoxide):
$$CH_3(CH_2)_4CH\!-\!CHCH_2CH\overset{Z}{=\!=}CH(CH_2)_7COOH$$
$$\underset{O}{\diagdown\;\diagup}$$

Lipoic:
$$\begin{array}{c} S\!-\!\!-\!S \\ |\qquad| \\ H_2C\quad C\cdots CH_2CH_2CH_2CH_2COOH \\ \diagdown\; \underset{H_2}{C}\;\diagup\quad H \end{array}$$

References for Further Reading

73, 146, 256, 257, 308, 309, 335, 516, 740, 741

D 5.2.1 Secondary Fatty Acids

Chemistry and Distribution

Secondary fatty acids differ from common fatty acids in chain length, additional functional groups, e.g., hydroxy and hydroperoxy groups, or other structural elements (branches in the carbon chain, presence of cyclopropane or cyclopentene rings, etc.). They occur in microorganisms, plants, and animals.

Biosynthesis

Different mechanisms cause the formation of secondary fatty acids (Fig. 67):

a) Participation of unusual starter molecules in fatty acid biosynthesis. Instead of acetyl CoA (D 5.2) other acyl CoA derivatives may serve as starters, e.g.,
— propionyl CoA (D 3) yielding fatty acids with an odd number of carbon atoms
— isobutyryl CoA (D 13), isovaleryl CoA (D 14), and 2-methylbutyryl CoA (D 15) causing synthesis of fatty acids with a branch in the carbon chain near the terminal methyl group, or
— the CoA ester of aleprolic acid (D 17) giving rise to the formation of cyclopentenyl fatty acids, e.g., chaulmoogric acid.

b) Extension of the carbon chain. The carbon chain of saturated or unsaturated fatty acids may be extended by one or several C_2-units by the enzymes of

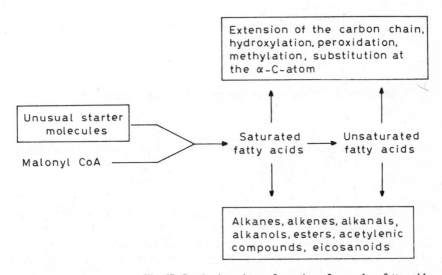

Fig. 67. Synthesis and transformation of secondary fatty acids

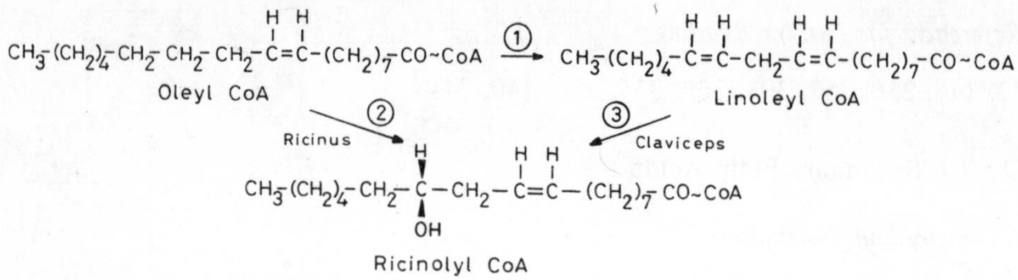

$$CH_3-(CH_2)_4-CH_2-CH_2-CH_2-\overset{H}{\underset{}{C}}=\overset{H}{\underset{}{C}}-(CH_2)_7-CO\sim CoA \xrightarrow{\text{①}} CH_3-(CH_2)_4-\overset{H}{\underset{}{C}}=\overset{H}{\underset{}{C}}-CH_2-\overset{H}{\underset{}{C}}=\overset{H}{\underset{}{C}}-(CH_2)_7-CO\sim CoA$$

Oleyl CoA   ② Ricinus   ③ Claviceps   Linoleyl CoA

$$CH_3-(CH_2)_4-CH_2-\overset{H}{\underset{OH}{C}}-CH_2-\overset{H}{\underset{}{C}}=\overset{H}{\underset{}{C}}-(CH_2)_7-CO\sim CoA$$

Ricinolyl CoA

Fig. 68. Formation of ricinolyl CoA
1 Dehydrogenase; *2* monooxygenase; *3* hydratase

the fatty acid elongation system. This enzyme system catalyzes reactions analogous to those of fatty acid synthase (D 5.2). The chain elongation proceeds with acetyl CoA in animals and with malonyl CoA in plants.

From linolenic acid bishomo-γ-linolenic acid (eicosatrienoic acid) is formed which may be desaturated to arachidonic acid (eicosatetraenoic acid) (Fig. 79) and eicosapentaenoic acid. Very long-chain fatty acids are constituents of waxes (see Table 29).

c) Hydroxylation of fatty acids. The hydroxy groups present in several fatty acids are formed either
— by the hydroxylation of CH_2-groups by monooxygenases (C 2.6.4) or
— by the action of hydratases adding water to —CH=CH— groups.

Hence, ricinolyl CoA may be built either from oleyl CoA or from linoleyl CoA (Fig. 68).

d) Peroxidation of fatty acids. Lipoxygenases (C 2.5) introduce stereospecifically hydroperoxy groups into polyunsaturated fatty acids according to the following equation:

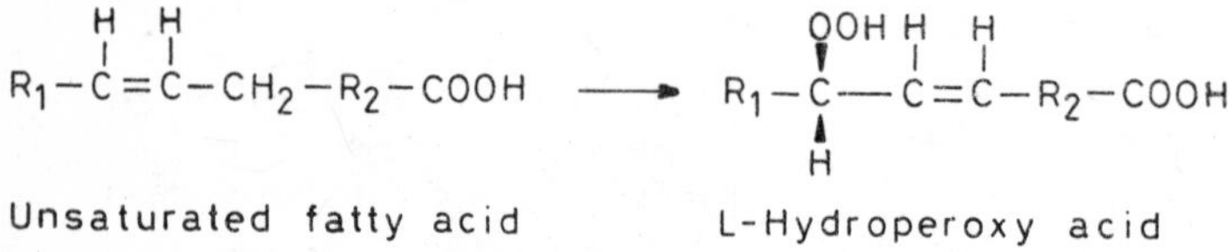

$$R_1-\overset{H}{\underset{}{C}}=\overset{H}{\underset{}{C}}-CH_2-R_2-COOH \longrightarrow R_1-\overset{OOH}{\underset{H}{C}}-\overset{H}{\underset{}{C}}=\overset{H}{\underset{}{C}}-R_2-COOH$$

Unsaturated fatty acid   L-Hydroperoxy acid

Hydroperoxide lyases cleave the peroxidic fatty acids to form aldehydes:

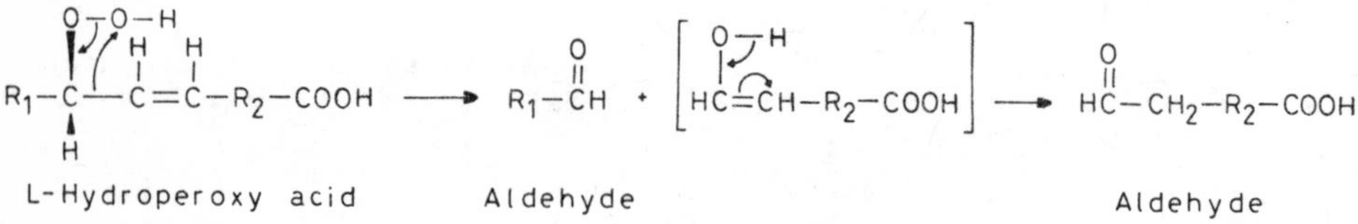

$$R_1-\overset{O-O-H}{\underset{H}{C}}-\overset{H}{\underset{}{C}}=\overset{H}{\underset{}{C}}-R_2-COOH \longrightarrow R_1-\overset{O}{\underset{}{C}}H + \left[HC\overset{O-H}{=}CH-R_2-COOH \right] \longrightarrow HC\overset{O}{\underset{}{=}}CH_2-R_2-COOH$$

L-Hydroperoxy acid   Aldehyde   Aldehyde

From linoleic and linolenic acid (Table 26) volatile C_6-aldehydes (n-hexanal and 3-Z-n-hexenal) are formed in this way. The characteristic smell of green leaves and grass is due to 2-Z-n-hexenal (leaf aldehyde) and the corresponding alcohol (leaf alcohol). They may be formed from the 3-Z-derivatives by isomerization.

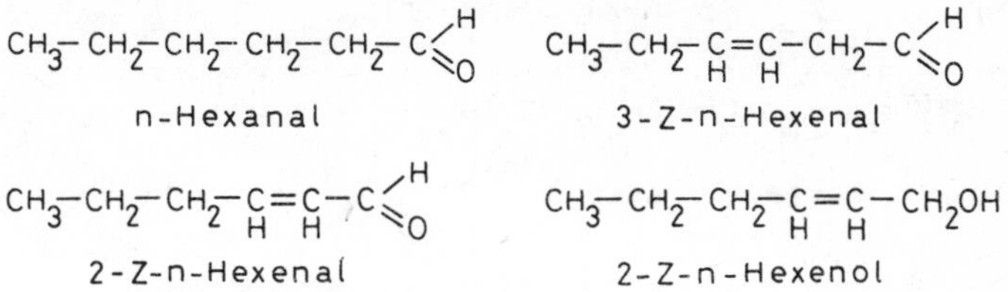

$$CH_3-CH_2-CH_2-CH_2-CH_2-C\overset{H}{\underset{O}{\Big\langle}}$$

n-Hexanal

$$CH_3-CH_2-C=C-CH_2-C\overset{H}{\underset{O}{\Big\langle}}$$
$${}^{H}{}^{H}$$

3-Z-n-Hexenal

$$CH_3-CH_2-CH_2-C=C-C\overset{H}{\underset{O}{\Big\langle}}$$
$${}^{H}{}^{H}$$

2-Z-n-Hexenal

$$CH_3-CH_2-CH_2-C=C-CH_2OH$$
$${}^{H}{}^{H}$$

2-Z-n-Hexenol

e) **Methylation of fatty acids.** Methylation of unsaturated fatty acids by
S-adenosyl-L-methionine (C 3.3) takes place in microorganisms and plants. It
yields a cationic intermediate which may be stabilized
— either by shift of a hydride ion and elimination of a proton (formation of
  methylene derivatives, which subsequently may be reduced by a dehydrogenase, Fig. 69; see also the formation of sterols, D 6.6.1), or
— by the addition of a methyl group to the cationic C-atom (formation of fatty
  acids with cyclopropane ring, e.g., lactobacillic acid, Fig. 70, and sterculic
  acid, Table 26).

f) **Substitution at the α-C-atom.** The α-C-atom of activated fatty acids, e.g.,
the CoA ester, ist easily substituted nucleophilically (C 1.2). (+)-Corynomycolic
acid is formed in this way in mycobacteria from two molecules of palmityl CoA
(Fig. 71).

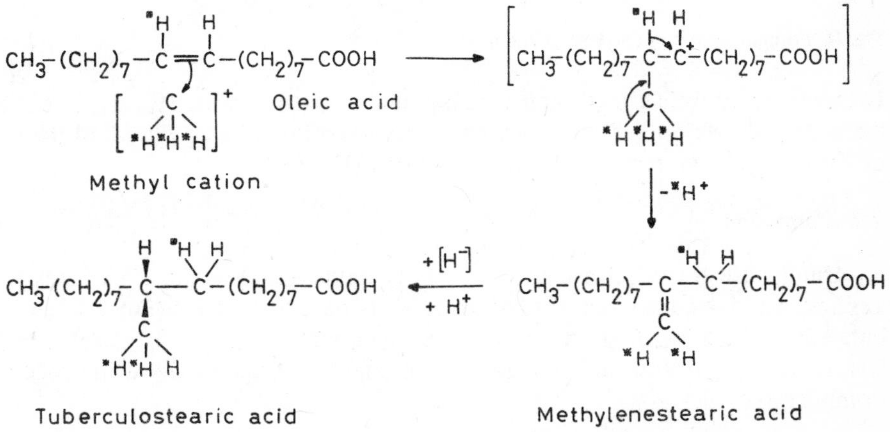

Fig. 69. Biosynthesis of tuberculostearic acid

Significance

Valeric acid, $CH_3(CH_2)_3COOH$, and 9-oxo-2-E-decenoic acid (Table 26) act as
sex pheromones in the sugar beet wireworm and the honeybee queen, respectively (E 4). Myrmicacin, $CH_3(CH_2)_6CHOHCH_2COOH$, is produced by leaf-cutting ants to prevent the germination of undesirable fungal spores in the nest
(E 5.1). Ricinoleic acid (Table 26) is the active principle of castor oil used as a
laxative (F 2). It is set free on hydrolysis of the oil in the intestine by means of
lipases. Capric acid, present, e.g., in the seed fat of Ulmaceae, Lauraceae, and

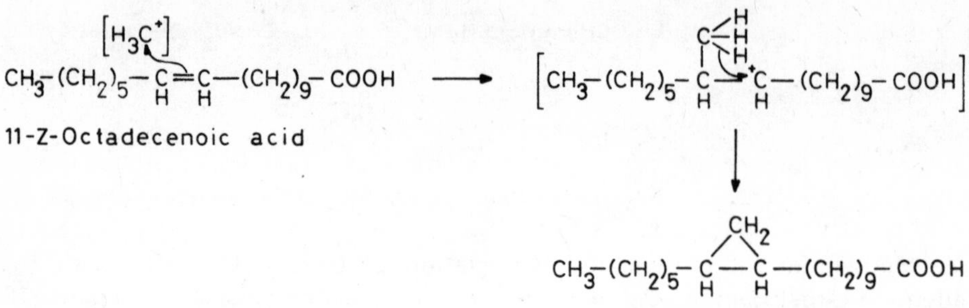

$$[H_3C^+]$$

$$CH_3-(CH_2)_5-C\overset{H}{=}C-(CH_2)_9-COOH \longrightarrow \left[CH_3-(CH_2)_5-\overset{H}{\underset{H}{C}}\overset{CH_2}{\cdots}\overset{+}{C}-(CH_2)_9-COOH \right]$$

11-Z-Octadecenoic acid

$$CH_3-(CH_2)_5-\underset{H}{C}\overset{CH_2}{\diagdown}\underset{H}{C}-(CH_2)_9-COOH$$

Lactobacillic acid

Fig. 70. Formation of lactobacillic acid

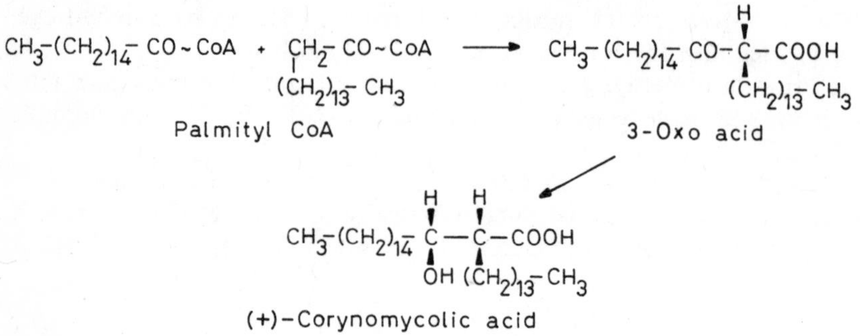

$$CH_3-(CH_2)_{14}-CO\sim CoA + \underset{(CH_2)_{13}-CH_3}{CH_2-CO\sim CoA} \longrightarrow CH_3-(CH_2)_{14}-CO-\underset{(CH_2)_{13}-CH_3}{\overset{H}{C}}-COOH$$

Palmityl CoA

3-Oxo acid

$$CH_3-(CH_2)_{14}-\underset{OH}{\overset{H}{C}}-\underset{(CH_2)_{13}-CH_3}{\overset{H}{C}}-COOH$$

(+)-Corynomycolic acid

Fig. 71. Formation of (+)-corynomycolic acid

Lythraceae, inhibits growth and development of many plants (E 2.2). The highly toxic sterculic acid (Table 26), which comprises up to 70 % of the fat of the seeds of *Sterculia foetida*, repels potential predators (E 5.5.3).

Transformation

a) Amides are formed from activated fatty acids (usually the CoA esters) and amines. Amides composed of olefinic or acetylenic fatty acids (D 5.2.5) and iso-butylamine (D 8.2), Δ^1-pyrroline (D 17), or piperidine (D 18), e.g., 2,3-dihydro-stearic acid piperidide and hexadeca-7-Z-en-10-ynoic acid pyrrolide, occur in Compositae.

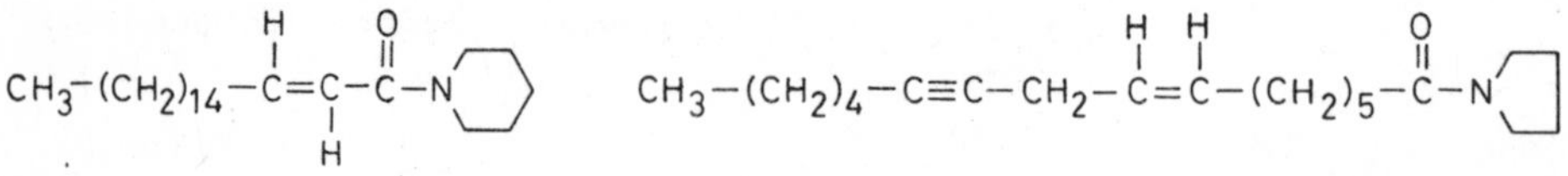

$$CH_3-(CH_2)_{14}-\overset{H}{\underset{H}{C}}=C-\overset{O}{\overset{\parallel}{C}}-N\diagup$$

2-E-Dehydrostearic acid piperidide

$$CH_3-(CH_2)_4-C\equiv C-CH_2-\overset{H}{C}=\overset{H}{C}-(CH_2)_5-\overset{O}{\overset{\parallel}{C}}-N\diagup$$

Hexadeca-7-Z-en-10-ynoic acid pyrrolide

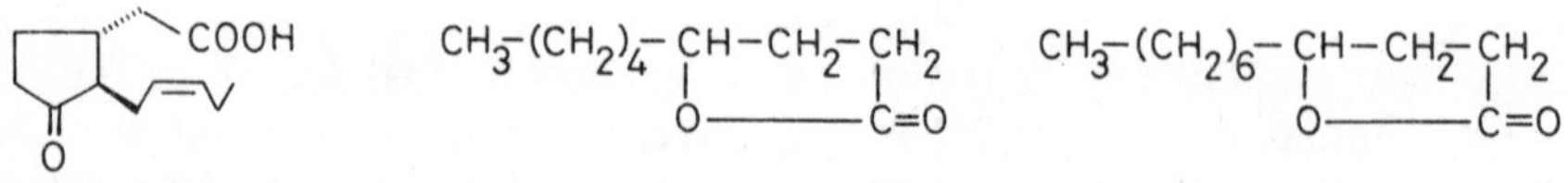

(-)-Jasmonic acid

$$CH_3-(CH_2)_4-\underset{O\rule{1cm}{0.4pt}}{CH-CH_2-CH_2}\underset{C=O}{}$$

α-Nonalactone

$$CH_3-(CH_2)_6-\underset{O\rule{1cm}{0.4pt}}{CH-CH_2-CH_2}\underset{C=O}{}$$

Undecalactone

b) Lactones are derived from γ-hydroxylated fatty acids. α-Nonalactone and undecalactone dominate the smell of coconuts and peaches, respectively (F 1).

c) (−)-Jasmonic acid and its derivatives, a group of plant hormones (E 3.1), are built from unsaturated fatty acids.

d) Alkanes, alkenes, alkanals, and alkanols (D 5.2.2 and D 5.2.3) as well as fatty acid esters (D 5.2.4), acetylenes (D 5.2.5), and eicosanoids (D 5.2.6) are derived from different types of fatty acids.

References for Further Reading

308, 309, 740, 759, 792

D 5.2.2 Alkanes and Alkenes

Chemistry and Distribution

The n-alkanes are hydrocarbons with a straight carbon chain. The iso-alkanes and anteiso-alkanes contain a methyl branch near one of the terminal groups of the molecule:

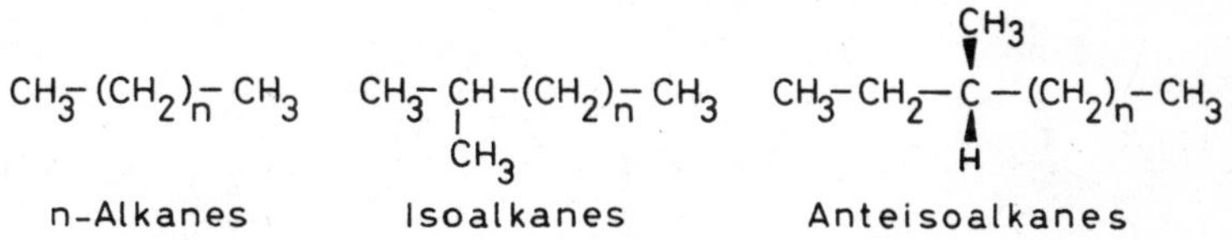

$$CH_3-(CH_2)_n-CH_3 \qquad CH_3-\underset{\underset{CH_3}{|}}{CH}-(CH_2)_n-CH_3 \qquad CH_3-CH_2-\overset{\overset{CH_3}{|}}{\underset{\underset{H}{|}}{C}}-(CH_2)_n-CH_3$$

n-Alkanes   Isoalkanes   Anteisoalkanes

Alkenes are unsaturated and possess one or several double bonds. Alkanes and alkenes are formed in microorganisms, plants, and animals. Often they contain more than 30 C-atoms.

Biosynthesis

Alkanes and alkenes are formed by decarboxylation of the corresponding saturated and unsaturated fatty acids (D 5.2 and 5.2.1). Probably the mechanism of

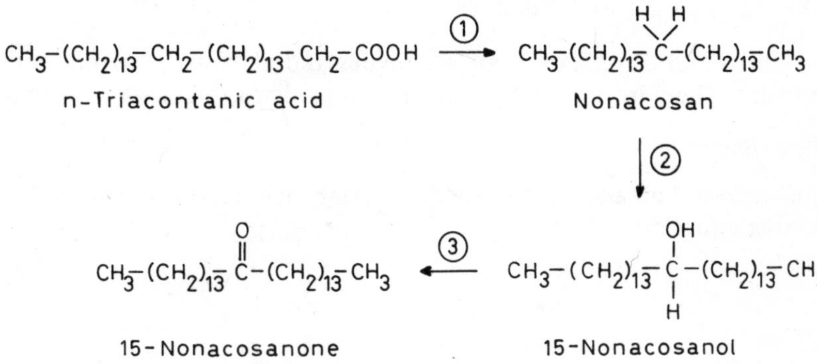

$$CH_3-(CH_2)_{13}-CH_2-(CH_2)_{13}-CH_2-COOH \xrightarrow{\text{①}} CH_3-(CH_2)_{13}-\overset{\overset{H}{\diagup}\overset{H}{\diagdown}}{C}-(CH_2)_{13}-CH_3$$

n-Triacontanic acid   Nonacosan

$$\downarrow \text{②}$$

$$CH_3-(CH_2)_{13}-\overset{\overset{O}{\|}}{C}-(CH_2)_{13}-CH_3 \xleftarrow{\text{③}} CH_3-(CH_2)_{13}-\overset{\overset{OH}{|}}{\underset{\underset{H}{|}}{C}}-(CH_2)_{13}-CH_3$$

15-Nonacosanone   15-Nonacosanol

Fig. 72. Synthesis and transformation of nonacosan
1 Decarboxylase system; *2* monooxygenase; *3* dehydrogenase

α-oxidation (D 5.2) is involved, and 2-hydroxy acids are intermediates. The alkanes and alkenes may be subsequently modified by hydroxylation and further dehydrogenation (Fig. 72).

Significance

Alkanes and alkenes containing about 30 C-atoms are constituents of plant cuticular waxes and the surface lipids of microorganisms and animals (D 5.2.4). These waxes play an important role in the hardiness and in the water economy of plants (E 2.2 and 5.4). Undecane, $CH_3(CH_2)_9CH_3$, tridecane, $CH_3(CH_2)_{11}CH_3$, and pentadecane, $CH_3(CH_2)_{13}CH_3$, are alarm pheromones of ants. 7,8-Epoxy-2-methyl-octadecane is a sex pheromone of *Lymantria dispar*. In brown algae fucoserraten ($CH_3CH_2CH\overset{Z}{=}CHCH\overset{E}{=}CHCH=CH_2$), ectocarpen, and multifiden are sex pheromones (E 4).

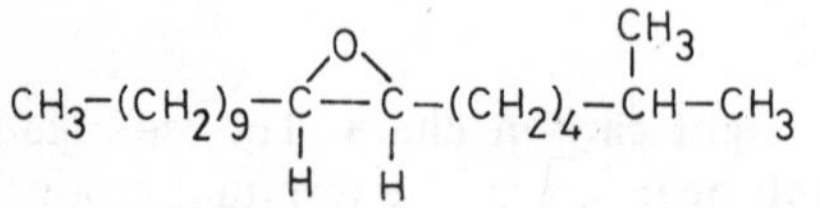

CH₃-(CH₂)₉-C—C-(CH₂)₄-CH-CH₃

7,8-Epoxy-2-methyl-octadecane 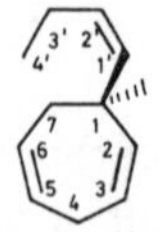   Ectocarpen 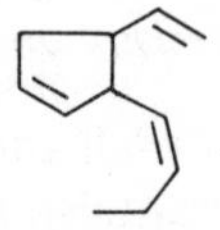   Multifiden

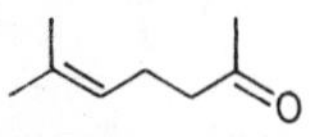

6-Methylhept-5-en-2-one 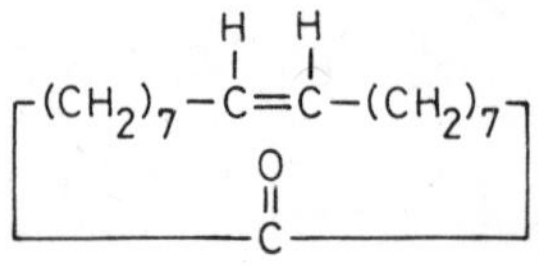   Civetone 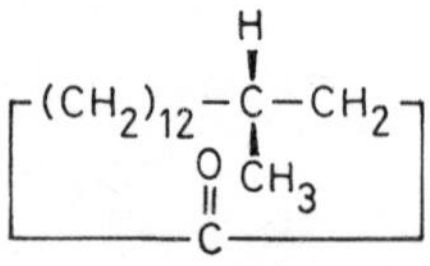   Muscone

6-Methylhept-5-en-2-one stimulates the outgrowth of rust spores (E 2.2). Civetone and muscone, cyclic ketones, which obviously are derived from fatty acids, act as sex pheromones in the civet cat and the musk deer, respectively (E 4).

References for Further Reading

706, 740

D 5.2.3 Alkanals and Alkanols

n-Alkanals and n-alkanols are aldehydes and alcohols structurally derived from alkanes. They are synthesized in microorganisms, plants, and animals.

Biosynthesis

Alkanals and alkanols are formed from fatty acid CoA esters according to the following equation:

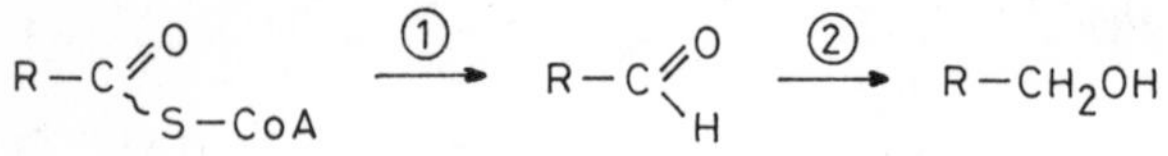

Fatty acid CoA ester   Alkanal   Alkanol

1 Acyl CoA reductase; *2* aldehyde reductase

Significance

Bombykol, $CH_3(CH_2)_2CH\overset{Z}{=\!\!=}CHCH\overset{E}{=\!\!=}CH(CH_2)_8CH_2OH$,   and   bombykal,
$CH_3(CH_2)_2CH\overset{Z}{=\!\!=}CHCH\overset{E}{=\!\!=}CH(CH_2)_8CHO$, are sex pheromones of *Bombyx mori.* 3-*Z*-6-*Z*-8-*E*-dodecatriene-1-ol, $CH_3(CH_2)_2CH\overset{E}{=\!\!=}CHCH\overset{Z}{=\!\!=}CHCH_2CH\overset{Z}{=\!\!=}CHCH_2CH_2OH$, is a trail pheromone of termites (E 4). Nonanal stimulates the germination of rust spores (E 2.2). 2-*E*-6-*Z*-Nonadienal, $CH_3CH_2CH\overset{Z}{=\!\!=}CH(CH_2)_2CH\overset{E}{=\!\!=}CHCHO$, is the flavor principle of cucumber (F 1). 1-Triacontanol, $CH_3(CH_2)_{28}CH_2OH$, acts as a growth-stimulating compound in several plants (E 2.2). Alkanols present in the lipophilic layer, covering many microorganisms and plants, participate in the resistance to microbial attack (E 5.4).

Transformation

Alkanols are constituents of wax esters (D 5.2.4)

Alkanals are precursors of ether lipids. The most important groups are discussed in the following:

a) Sphingolipids contain 4-sphingenin (sphingosine) (Fig. 73), or a related substance as the central structural element. They may be classified as sphingomyelins, cerebrosides, and gangliosides according to the substituents adjacent to the basic skeleton. Representatives of these classes of substances occur in all groups of organisms. They are constituents of the lipid portion of the cell membrane and the membranes of cell organelles. The gangliosides participate in the functions of the central nervous system of human beings and animals.

4-Sphingenin originates from palmitoyl CoA and L-serine (Fig. 73). In the formation of sphingolipids 4-sphingenin is acylated at the NH$_2$-group by a fatty acid CoA ester, forming *N*-acyl sphingenins (ceramides). These may further be substituted by CDP-choline (D 10) to form sphingomyelins. In the biosynthesis of cerebrosides the ceramides are glycosidically linked with a sugar (galactose or glucose) which in the case of sulfatides is further esterified with a sulfate group at position 3. In the case of mucolipids, e.g., the gangliosides, the ceramides are linked at the α-position to several sugars, e.g., D-glucose, D-galactose, *N*-acetyl-D-glucosamine, *N*-acetyl-D-galactosamine, and L-fucose, as well as to *N*-acetylneuraminic acids (sialic acids, D 1.1).

The mucolipids may combine with proteins to yield lipoproteins, proteolipids, and phosphatidolipids. Defects in the degradation of gangliosides in human beings cause the syndrome of sphingolipidoses, congenital disorders, which are fatal in the first few years of life.

b) Plasmalogens are constituents of the cell membranes of microorganisms, plants, and animals in which sn-glycerol-3-phosphate (L-glycerol-3-phosphate, D 2) is attached

— at the phosphate group to choline, ethanolamine, L-serine, etc.
— at position 2 to a fatty acid residue, and
— at position 1 to an aliphatic α,β-unsaturated alkane via an ether group.

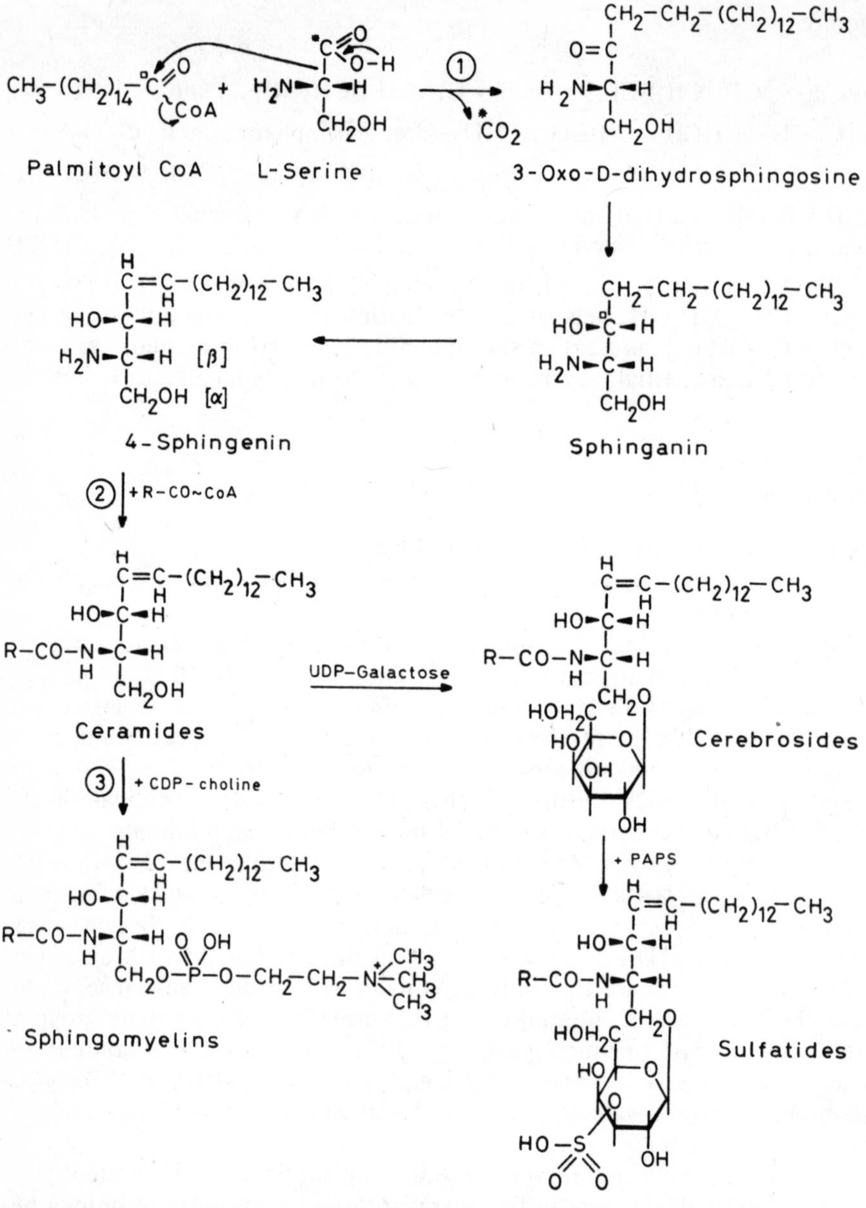

Fig. 73. Formation of sphingomyelins, cerebrosides, and sulfatides
1 Serine palmitoyltransferase; *2* sphingosine acyltransferase; *3* ceramide cholinephosphotransferase

The α,β-unsaturated aliphatic carbon chain is probably derived from an alkanal, e.g., palmitic aldehyde (Fig. 74).

c) PAF (platelet-activating factor) activates thrombocytes and is probably a mediator in allergy and inflammation. It occurs in humans and higher animals.

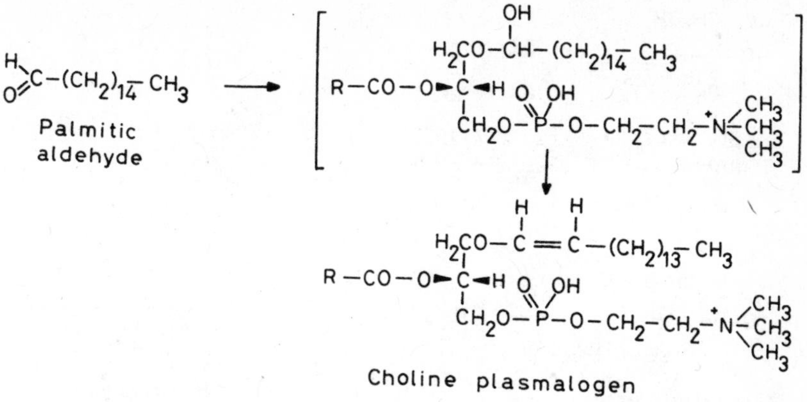
Palmitic aldehyde
Choline plasmalogen

Fig. 74. Biosynthesis of choline plasmalogen

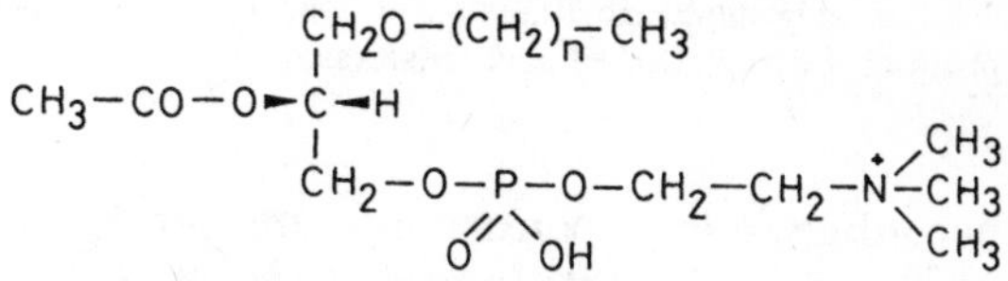

Platelet activating factor
n = 15, 17

References for Further Reading

317, 408, 498, 706, 740

D 5.2.4 Fatty Acid Esters

Chemistry and Distribution

Fatty acid esters contain the structure $R_1CO—OR_2$. Thereby R_1 is a fatty acid residue, whereas R_2 may be the residue of any one of the many alcohols found in nature, e.g., of glycerol, of the alcohols derived from fatty acids by reduction, etc. Fatty acid esters occur in microorganisms, plants, and animals.

Biosynthesis

The ester group is formed by the reaction of an activated fatty acid (usually the CoA ester) with an alcohol:

$$R_1-C(=O)-S-CoA + HO-R_2 \longrightarrow R_1-C(=O)-O-R_2 + HS-CoA$$

The following classes of esters may be distinguished:

a) Glycerides: Triacylglycerols (triglycerides), designated as fats or fatty oils

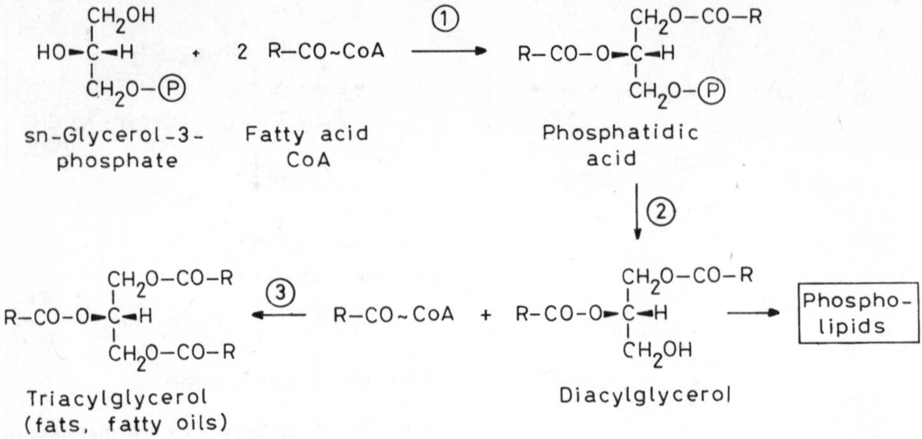

$$\underset{\substack{\text{sn-Glycerol-3-}\\\text{phosphate}}}{\begin{array}{c}CH_2OH\\ |\\ HO-C-H\\ |\\ CH_2O-\textcircled{P}\end{array}} + \underset{\substack{\text{Fatty acid}\\\text{CoA}}}{2\ R-CO\sim CoA} \xrightarrow{\ \textcircled{1}\ } \underset{\substack{\text{Phosphatidic}\\\text{acid}}}{\begin{array}{c}CH_2O-CO-R\\ |\\ R-CO-O-C-H\\ |\\ CH_2O-\textcircled{P}\end{array}}$$

$$\Big\downarrow \textcircled{2}$$

$$\underset{\substack{\text{Triacylglycerol}\\\text{(fats, fatty oils)}}}{\begin{array}{c}CH_2O-CO-R\\ |\\ R-CO-O-C-H\\ |\\ CH_2O-CO-R\end{array}} \xleftarrow{\ \textcircled{3}\ } R-CO\sim CoA + \underset{\substack{\text{Diacylglycerol}}}{\begin{array}{c}CH_2O-CO-R\\ |\\ R-CO-O-C-H\\ |\\ CH_2OH\end{array}} \longrightarrow \boxed{\begin{array}{c}\text{Phospho-}\\\text{lipids}\end{array}}$$

Fig. 75. Formation of fats and fatty oils
1 Glycerophosphate acyltransferase, 1-acylglycerophosphate acyltransferase; *2* phosphatidate phosphatase; *3* diacylglycerol acyltransferase

according to their consistency, are formed in all groups of organisms. Phosphatidic acid is the key intermediate. It is built by the acylation of sn-glycerol-3-phosphate (L-glycerol-3-phosphate, D 2) by two molecules of fatty acid CoA ester. Then the phosphate ester bond is hydrolyzed and a further molecule of fatty acid CoA ester reacts to give the triacylglycerol (Fig. 75).

The naturally occurring triglycerols are mixtures. They are solid at room temperature if the content of saturated fatty acids like stearic acid and palmitic acid, which have a high melting point, is high, and they are liquid if large amounts of unsaturated fatty acids, e.g. oleic, linoleic, and linolenic acids (Table 26), are present. Lipases degrade glycerides to glycerol and free fatty acids.

b) Diacylglycerols in the free or phosphorylated form (phosphatidic acids) are precursors of different groups of phospholipids that are important membrane constituents (Table 27). In addition, the phosphatidylinositols participate in the regulation of cell metabolism.

Phospholipids are formed by the following reactions:
— phosphatidylcholines: diacylglycerols + cytidine diphosphate choline (D 10) → phosphatidylcholines (lecithins) (reaction catalyzed by choline phosphotransferase)
— phosphatidylethanolamines: diacylglycerols + cytidine disphosphate ethanolamine (D 10) → phosphatidylethanolamines (cephalins) (reaction catalyzed by ethanolamine phosphotransferase)
— phosphatidylserines: phosphatidic acids + cytidine triphosphate → cytidine diphosphate diacylglycerol + diphosphate (reaction catalyzed by phosphatidate cytidyl transferase); cytidine diphosphate diacylglycerol + L-serine → phosphatidyl-L-serine (reaction catalyzed by cytidyldiphosphate-serine-O-phosphatidyltransferase)
— phosphatidylglycerols: cytidine diphosphate diacylglycerol + sn-glycerol-3-phosphate (D 2) → phosphatidylglycerol phosphate → phosphatidylglycerol

Table 27. Structure of different types of glycerol-derived phospholipids

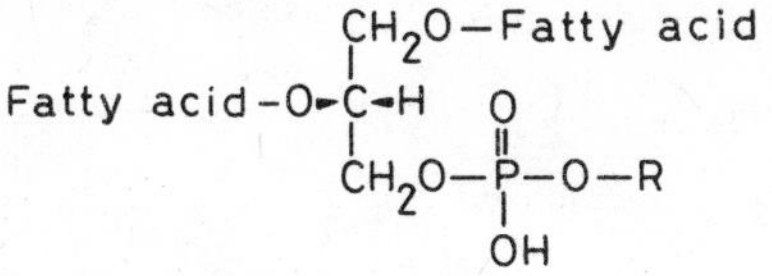

$$CH_2O-Fatty\ acid$$
$$Fatty\ acid-O-C-H \qquad O$$
$$CH_2O-P-O-R$$
$$OH$$

Group of compounds	R
Phosphatidylcholines	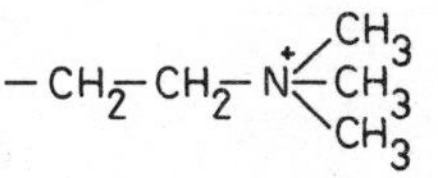 $-CH_2-CH_2-\overset{+}{N}\underset{CH_3}{\overset{CH_3}{<}}CH_3$
Phosphatidylethanolamines	$-CH_2-CH_2-NH_2$
Phosphatidylserines	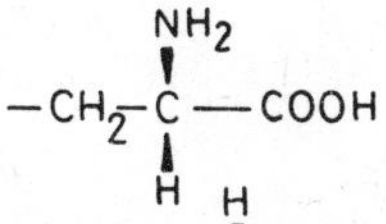 $-CH_2-\overset{NH_2}{\underset{H}{C}}-COOH$
Phosphatidylglycerols	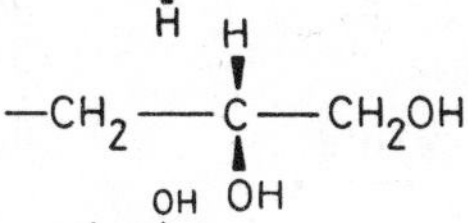 $-CH_2-\overset{H}{\underset{OH}{C}}-CH_2OH$ (OH)
Phosphatidylinositols	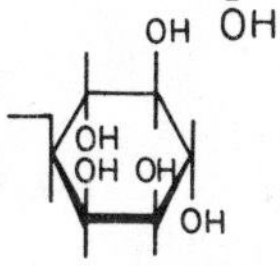 (inositol ring structure with OH groups)

(reaction catalyzed by cytidyldiphosphate-glycerol phosphatidyltransferase and a phosphatase). Bisphosphatidylglycerols (cardiolipids) are formed from phosphatidylglycerol and a second molecule of cytdine diphosphate diacyl glycerol

— phosphatidylinositols: cytidine disphosphate diacylglycerol + myo-inositol (D 1.3) → phosphatidylinositol (reaction catalyzed by diphosphate-diglyceride-inositol phosphatidyltransferase).

Glycerol-derived glycolipids (Table 28) are important building blocks of chloroplast membranes. They are formed from diacylglycerol and the corresponding uridine diphosphate sugars (reaction catalyzed by uridine diphosphate glycosyltransferases).

c) Cyanolipids have an alcoholic core, derived from L-leucine (D 14), which is esterified with fatty acids, mainly oleic acid and unsaturated and saturated C_{20}-acids. Those cyanolipids forming cyanohydrins on hydrolysis of the ester

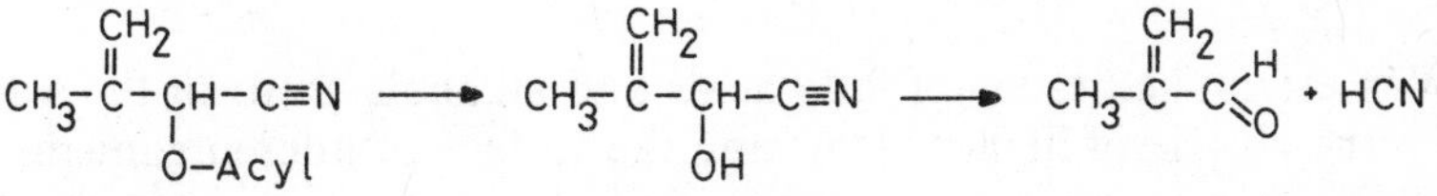

$$CH_3-\overset{CH_2}{\underset{O-Acyl}{C}}-CH-C\equiv N \longrightarrow CH_3-\overset{CH_2}{\underset{OH}{C}}-CH-C\equiv N \longrightarrow CH_3-\overset{CH_2}{C}-C\overset{H}{\underset{O}{<}} + HCN$$

Cyanolipid   Cyanohydrin   Aldehyde   Hydrocyanic
                                                                        acid

Fig. 76. Hydrolysis of cyanolipids with formation of hydrocyanic acid

Table 28. Structure of glycerol-derived glycolipids

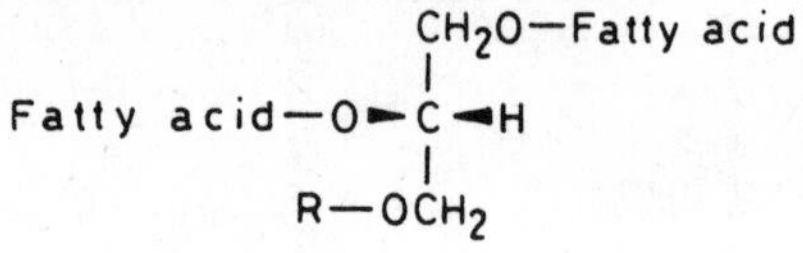

Group of compounds	R
Diacylgalactosylglycerols	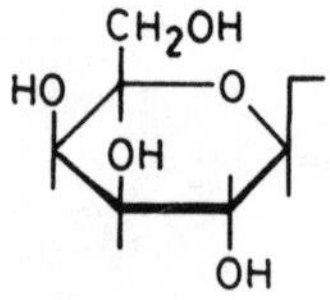
Diacylgalabiosylglycerols	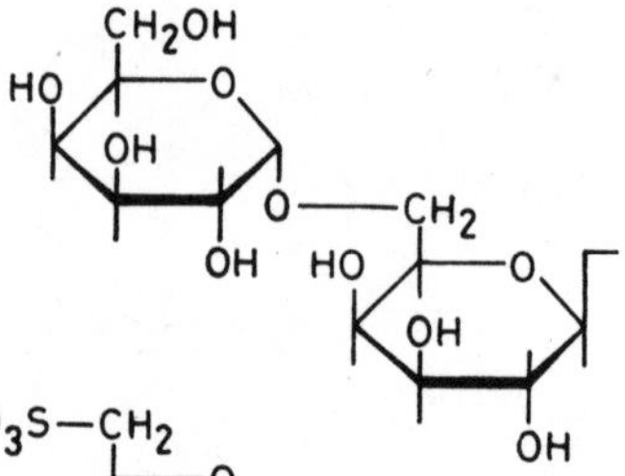
Diacylsulfoquinovosylglycerols	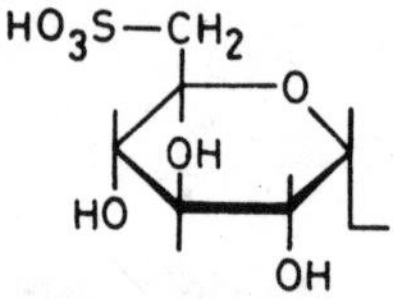

bond are able to liberate hydrocyanic acid (Fig. 76). This reaction resembles the degradation of cyanogenic glycosides (D 8.3).

d) 3-Acetoxyfatty acids (3-acetoxypalmitic, stearic, and -arachidic acid) occur in the floral glands of several Angiospermae either in the free state or as glycerides. They are collected by bees during pollination as prey for the larvae (E 5.5.1).

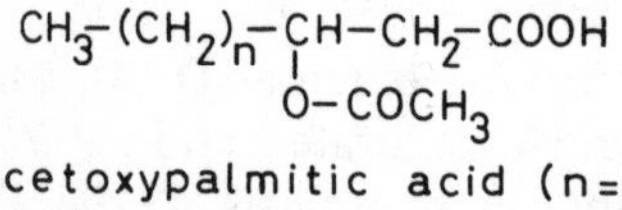

$$CH_3-(CH_2)_n-CH-CH_2-COOH$$
$$O-COCH_3$$

3-Acetoxypalmitic acid (n=12)
3-Acetoxystearic acid   (n=14)
3-Acetoxyarachidic acid (n=16)

e) Wax esters, i.e., esters of fatty acids and alkanols occur in the waxlike deposits on the cuticle of higher plants and the surface of microorganisms and animals.

In these surface lipids the wax esters are accompanied by n-alkanes, isoalkanes, anteisoalkanes, alkenes, and derivatives of these compounds (D 5.2.2), homologous series of very long-chain fatty acids

Table 29. Components of plant cuticular waxes

Type of compounds	Structural formula	Usual range of chain length
n-Alkanes	$CH_3(CH_2)_nCH_3$	$C_{25}-C_{35}$
Isoalkanes	$CH_3\overset{\displaystyle CH_3}{\underset{\displaystyle |}{CH}}(CH_2)_nCH_3$	$C_{25}-C_{35}$
Alkenes	$CH_3(CH_2)_nCH{=}CH(CH_2)_mCH_3$	$C_{17}-C_{33}$
Monoketones	$CH_3(CH_2)_n\overset{\displaystyle O}{\overset{\displaystyle \|}{C}}(CH_2)_mCH_3$	$C_{24}-C_{33}$
β-Diketones	$CH_3(CH_2)_n\overset{\displaystyle O}{\overset{\displaystyle \|}{C}}CH_2\overset{\displaystyle O}{\overset{\displaystyle \|}{C}}(CH_2)_mCH_3$	$C_{31}-C_{33}$
Primary alcohols	$CH_3(CH_2)_nCH_2OH$	$C_{12}-C_{36}$
Secondary alcohols	$CH_3(CH_2)_n\overset{\displaystyle OH}{\underset{\displaystyle |}{CH}}(CH_2)_mCH_3$	$C_{20}-C_{33}$
Fatty acids	$CH_3(CH_2)_nCOOH$	$C_{12}-C_{36}$
Wax esters	$CH_3(CH_2)_n\overset{\displaystyle O}{\overset{\displaystyle \|}{C}}{-}O(CH_2)_mCH_3$	$C_{30}-C_{60}$
ω-Hydroxy acids	$\overset{\displaystyle OH}{\underset{\displaystyle |}{CH_2}}(CH_2)_nCOOH$	$C_{10}-C_{34}$

(D 5.2.1), ω-hydroxy acids (D 5.2), and alkanols (D 5.2.3) (Table 29). Within the homologous series the fatty acids usually have an even number of C-atoms, while most alkane derivatives have an uneven number.

Spermaceti obtained from the head of the sperm whale, *Physeter macrocephalus*, consists mainly of cetylpalmitate, $CH_3(CH_2)_{14}CO{-}O(CH_2)_{15}CH_3$. Beeswax, which is secreted from specialized areas of the abdomen of *Apis mellifera*, the honeybee, and is used in the construction of the honeycomb, contains more than 50% myricylpalmitate, $CH_3(CH_2)_{14}CO{-}O(CH_2)_{29}CH_3$. Carnauba wax obtained from the leaves of *Copernicia cerifera* (Palmae) consists chiefly of myricylcerotate, $CH_3(CH_2)_{24}CO{-}O(CH_2)_{29}CH_3$.

Three mechanisms are involved in the formation of the ester groups of waxes:

— most important is the reaction of fatty acid CoA esters with alcohols catalyzed by alcohol: acyl CoA transacylase according to the following equation:

$$R_1-C\!\!\begin{array}{c}\nearrow O\\\searrow S-CoA\end{array} \quad + \quad HO-R_2 \quad \longrightarrow \quad R_1-C\!\!\begin{array}{c}\nearrow O\\\searrow O-R_2\end{array} \quad + \quad HS-CoA$$

— the reversal of saponification, i.e., the direct esterification of free fatty acids and free alcohols:

$$R_1-C\!\!\begin{array}{c}\nearrow O\\\searrow OH\end{array} \quad + \quad HO-R_2 \quad \rightleftharpoons \quad R_1-C\!\!\begin{array}{c}\nearrow O\\\searrow O-R_2\end{array} \quad + \quad H_2O$$

— and the transfer of acid groups from sphingolipids, lecithin, and related compounds to the alcohols:

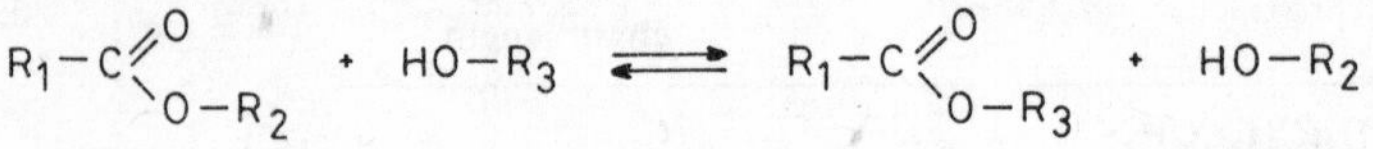

$$R_1-C\overset{O}{\underset{O-R_2}{\diagdown}} \;+\; HO-R_3 \;\rightleftharpoons\; R_1-C\overset{O}{\underset{O-R_3}{\diagdown}} \;+\; HO-R_2$$

f) Cutin and suberin. Cutin and suberin are polymeric substances covering the above and underground parts of plants. They are formed mainly from fatty acid derivatives, but contain also phenolic compounds. Though similar in their fundamental structure, the polymers differ considerably in their composition (Table 30). Some of the major fatty acid monomers of cutin and suberin are listed in Table 31. They are formed from palmitic and oleic acid by the following types of reaction:
— ω-hydroxylation (D 5.2)
— midchain hydroxylation (C 2.6.4)
— epoxygenation and hydration of the epoxide ring (C 2.6.5), and
— oxidation of the ω-hydroxyl group.

The CoA esters of the acids are the substrates of transacetylase(s) forming an interesterified, rigid meshwork to which different amounts of cinnamic acid derivatives are bound.

Cutin and suberin are embedded in a matrix of wax. They are degraded by exoenzymes with esterase activity (cutinases). After the decay of plants most of the cutin and suberin is hydrolyzed by soil bacteria and fungi. Degradation of both types of compounds also plays a role in the penetration of plants by phytopathogenic fungi (E 5.4).

Table 30. Composition of cutin and suberin

Monomer	Cutin	Suberin
Dicarboxylic acids	Minor	Major
In-chain substituted acids	Major	Minor
Phenolics, e. g., p-coumaric acid and ferulic acid	Low	High
Very long chain (C_{20}-C_{26}) acids	Rare	Common
Very long chain alcohols	Rare	Common

Table 31. Characteristic acid monomers of cutin and suberin

Name	Structural formula
Cutin	
10,16-Dihydroxypalmitic acid	$CH_2OH-(CH_2)_5-CHOH-(CH_2)_8-COOH$
threo-9,10,18-trihydroxystearic acid	$CH_2OH-(CH_2)_7-CHOH-CHOH-(CH_2)_7-COOH$
Suberin	
18-Hydroxyoleic acid	$CH_2OH-(CH_2)_7-CH\overset{z}{=\!\!=}CH-(CH_2)_7-COOH$
16-Hydroxypalmitic acid	$CH_2OH-(CH_2)_{14}-COOH$
Tetradecane dicarboxylic acid	$HOOC-(CH_2)_{14}-COOH$

Significance

Fats and fatty oils are the most important means of fatty acid storage (E 2.2). Phospholipids, e. g., lecithins and cephalins, and glycolipids are widespread membrane constituents (E 2.2). Cyanolipids serve as a store of fatty acids and nitrogen in the seeds of Sapindaceae (E 2.2). Wax esters as well as cutin and suberin, covering the above and underground parts of plants, make the surface of most organisms impermeable to water and form a screen against infections (E 2.2, and E 5.4). The wax layer of glaucous fruits, which highly reflects UV light, increases the visibility of fruits and hence their signal character for birds involved in seed dispersal (E 5.5.2, and E 5.6).

The oak-leaf-roller moth uses 11-E- and 11-Z-tetradecenyl acetate, $CH_3CH_2CH{=}CH(CH_2)_{10}O{-}COCH_3$, as sex pheromones. Hexadecanyl acetate, $CH_3(CH_2)_{15}O{-}COCH_3$, is used for the same purpose by the butterfly *Lycorea ceres ceres* (E 4). Ethyl 2-E-4-Z-decadienoate, $CH_3(CH_2)_4CH\overset{Z}{=}CHCH\overset{E}{=}CHCO{-}OC_2H_5$, causes the characteristic smell of pears (F 1). Amyl acetate, $CH_3(CH_2)_3CH_2O{-}COCH_3$, and amyl propionate, $CH_3(CH_2)_3CH_2O{-}COCH_2CH_3$, are constituents of the smell of bananas, ethyl 2-methylbutyrate, $CH_3CH_2CH(CH_3)CO{-}OC_2H_5$, dominates the smell of apples (F 1). Castor oil, linseed oil, spermaceti, beeswax, and other fatty oils and waxes are used in pharmacy (F 2) and industry (F 5). Cork, the light, flexible outer bark of the cork oak, consists mainly of suberin and is used in the tobacco industry (cork-tipped cigarettes), in the bottling of wine and other beverages (cork stoppers), as insulating material, etc. (F 5). Beeswax and carnauba wax are used in technology (F 5).

References for Further Reading

28, 63, 72, 87, 140, 211, 308, 309, 366, 512, 516, 523, 532, 706, 740

D 5.2.5 Acetylenes

Chemistry and Distribution

Acetylenic bonds (trible bonds, $-C{\equiv}C-$) are the characteristic features of acetylenic compounds. In addition double bonds, as well as oxygen-, sulfur-, and nitrogen-containing groups may be present. Acetylenic secondary products may be aliphatic, alicyclic, and/or heterocyclic (Table 32). Several hundred acetylenic compounds have been found in fungi, e.g., Basidiomycetes, algae, and higher plants, e.g., Compositae and Umbelliferae.

Biosynthesis

Acetylene derivatives are formed from fatty acids. In higher plants linoleic acid is the most common precursor. The acetylenic groups are built from double

Table 32. Naturally occurring acetylene derivatives

Name	Structure	Occurrence
Diatrene	$HOOC{-}CH{=}CH{-}(C{\equiv}C)_3{-}CH_2OH$	Fungi
Laurencin	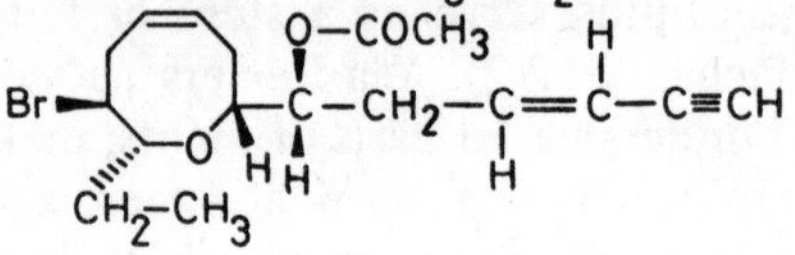	Laurentia glandulifera (alga)
Artemisia ketone	$CH_3{-}(C{\equiv}C)_3{-}CH{=}CH{-}CH_2{-}CH_2{-}CO{-}CH_2{-}CH_3$	Anthemideae
Falcarinone	$CH_3{-}(CH_2)_6{-}CH{=}CH{-}CH_2{-}(C{\equiv}C)_2{-}CO{-}CH{=}CH_2$	Araliaceae, Ammiaceae
Matricaria lactone	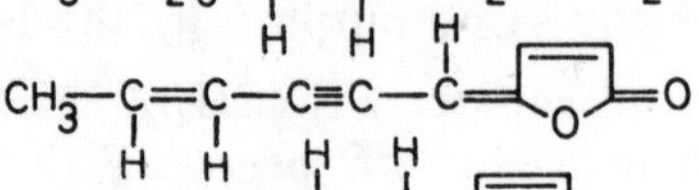	Asteraceae
Artemisia lactone	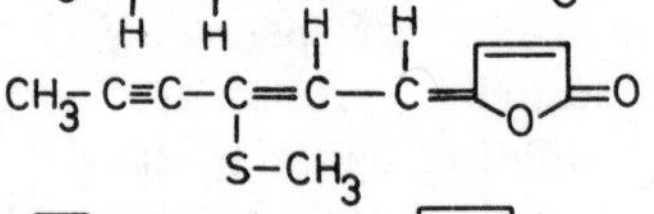	Asteraceae
Carlina oxide	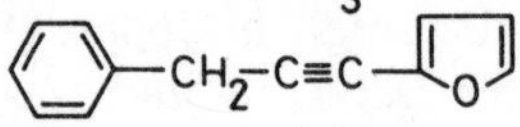	Carlina acaulis
	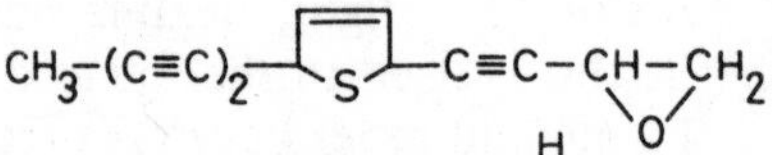	Echinops species
	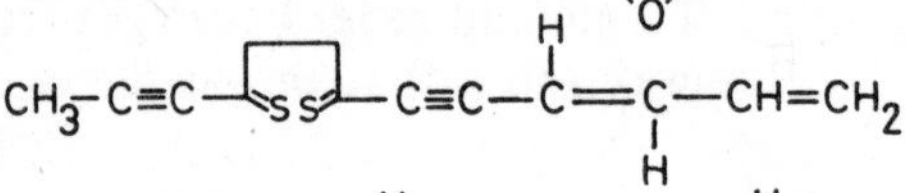	Echinops species
Safynol	$HO{-}CH_2{-}CHOH{-}CH{=}CH{-}(C{\equiv}C)_3{-}CH{=}CH{-}CH_3$	Asteraceae
Wyerone	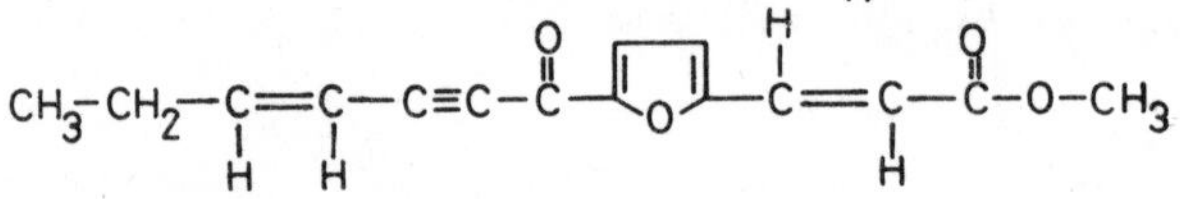	Leguminosae
Phenyl-heptatriyne	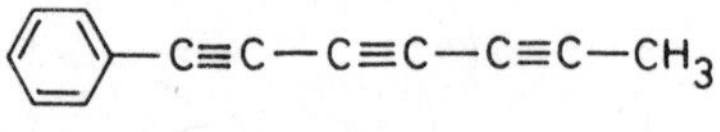	Asteraceae

bonds by dehydrogenation (Fig. 77). The following modifications play an important role in the biosynthesis of the acetylenic compounds found in nature:

Shortening of the carbon chain: Most acetylenes are devoid of the carboxylic group of the precursor fatty acid, which is removed by decarboxylation. The carbon chain may also be shortened by *α*- and *β*-oxidations (D 5.2).

The formation of allenic groups: The $—C{\equiv}C—CH_2$-group may be converted to $—CH{=}C{=}CH—$ by isomerases. Depending on the particular enzyme both possible absolute configurations may be formed:

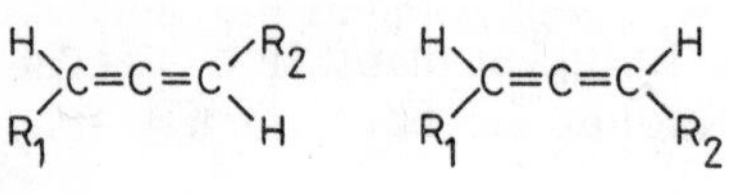

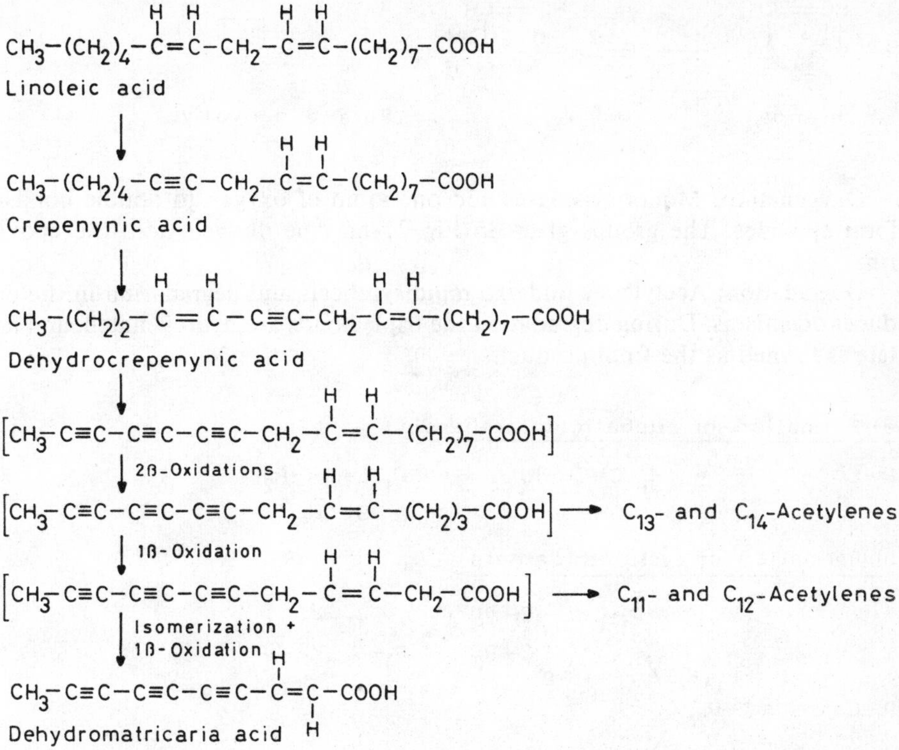

$$CH_3-(CH_2)_4-\overset{\overset{H}{|}}{C}=\overset{\overset{H}{|}}{C}-CH_2-\overset{\overset{H}{|}}{C}=\overset{\overset{H}{|}}{C}-(CH_2)_7-COOH$$

Linoleic acid

$$\downarrow$$

$$CH_3-(CH_2)_4-C\equiv C-CH_2-\overset{\overset{H}{|}}{C}=\overset{\overset{H}{|}}{C}-(CH_2)_7-COOH$$

Crepenynic acid

$$\downarrow$$

$$CH_3-(CH_2)_2-\overset{\overset{H}{|}}{C}=\overset{\overset{H}{|}}{C}-C\equiv C-CH_2-\overset{\overset{H}{|}}{C}=\overset{\overset{H}{|}}{C}-(CH_2)_7-COOH$$

Dehydrocrepenynic acid

$$\downarrow$$

$$\left[CH_3-C\equiv C-C\equiv C-C\equiv C-CH_2-\overset{\overset{H}{|}}{C}=\overset{\overset{H}{|}}{C}-(CH_2)_7-COOH\right]$$

$\downarrow$ 2β-Oxidations

$$\left[CH_3-C\equiv C-C\equiv C-C\equiv C-CH_2-\overset{\overset{H}{|}}{C}=\overset{\overset{H}{|}}{C}-(CH_2)_3-COOH\right] \longrightarrow C_{13}\text{- and } C_{14}\text{-Acetylenes}$$

$\downarrow$ 1β-Oxidation

$$\left[CH_3-C\equiv C-C\equiv C-C\equiv C-CH_2-\overset{\overset{H}{|}}{C}=\overset{\overset{H}{|}}{C}-CH_2-COOH\right] \longrightarrow C_{11}\text{- and } C_{12}\text{-Acetylenes}$$

$\downarrow$ Isomerization +
1β-Oxidation

$$CH_3-C\equiv C-C\equiv C-C\equiv C-\overset{\overset{H}{|}}{C}=\overset{\underset{H}{|}}{C}-COOH$$

Dehydromatricaria acid

Fig. 77. Formation of different groups of acetylenic compounds in plants

E-Z-Isomerization: The double bonds present in most acetylenes may be isomerized enzymatically. *E* and *Z* isomers occur frequently together.

Addition: The reaction of different chemical groups with the triple bond of acetylene derivatives results in the following structures:
— Aromatic rings may be formed by joining the terminal part of the molecule to the triple bond:

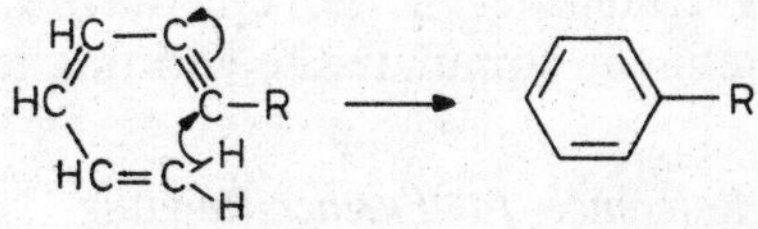

— Thioethers are synthesized by adding H$_2$S to the triple bond with or without subsequent alkylation. Cyclic thio derivatives are formed from dithio compounds by the elimination of sulfur or directly from monothiolated compounds (Fig. 78). Sulfur from L-cysteine is incorporated. However, the actual donor of the sulfhydryl group is unknown.
— Furane rings are formed by addition of oxygen-containing groups to the triple bonds. The precursors are inenols:

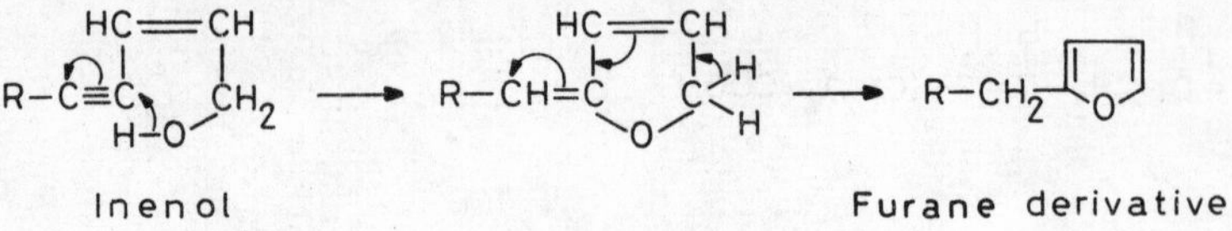

Inenol   Furane derivative

Oxygenation: Monooxygenases add one atom of oxygen to double bonds to form epoxides. The groups given in Fig. 21 may be derived from the epoxide ring.

Degradation: Acetylenes undergo rapid synthesis and degradation in the producer organisms. During degradation the triple bonds are hydrogenated, and acetate is formed as the final product.

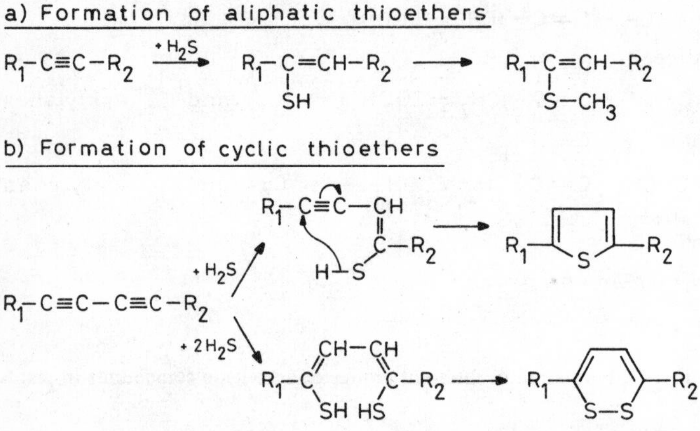

a) Formation of aliphatic thioethers

b) Formation of cyclic thioethers

Fig. 78. Biosynthesis of thio derivatives from acetylenic compounds

Significance

Most acetylenes are highly toxic to animals. Their formation is of ecological significance in deterring predators (E 5.5.3). Safynol and wyerone (Table 32) act as phytoalexins (E 5.4). Phenylheptatriyne disturbs membrane functions (E 5.4). 1-Tridecene-3,5,7,9,11-pentayne and *E*- and *Z*-trideca-1,11-diene-3,5,7,9-tetrayne stimulate the germination of rust spores (E 5.4).

References for Further Reading

79, 309, 394

D 5.2.6 Eicosanoids

Chemistry and Distribution

Eicosanoids are unsaturated C_{20} acids. They are aliphatic (leucotrienes) or contain either a five-membered carbocyclic ring (prostaglandins, prostacyclins) or a six-membered *O*-heterocyclic ring (thromboxanes). Hydroxy, oxo, and epoxy

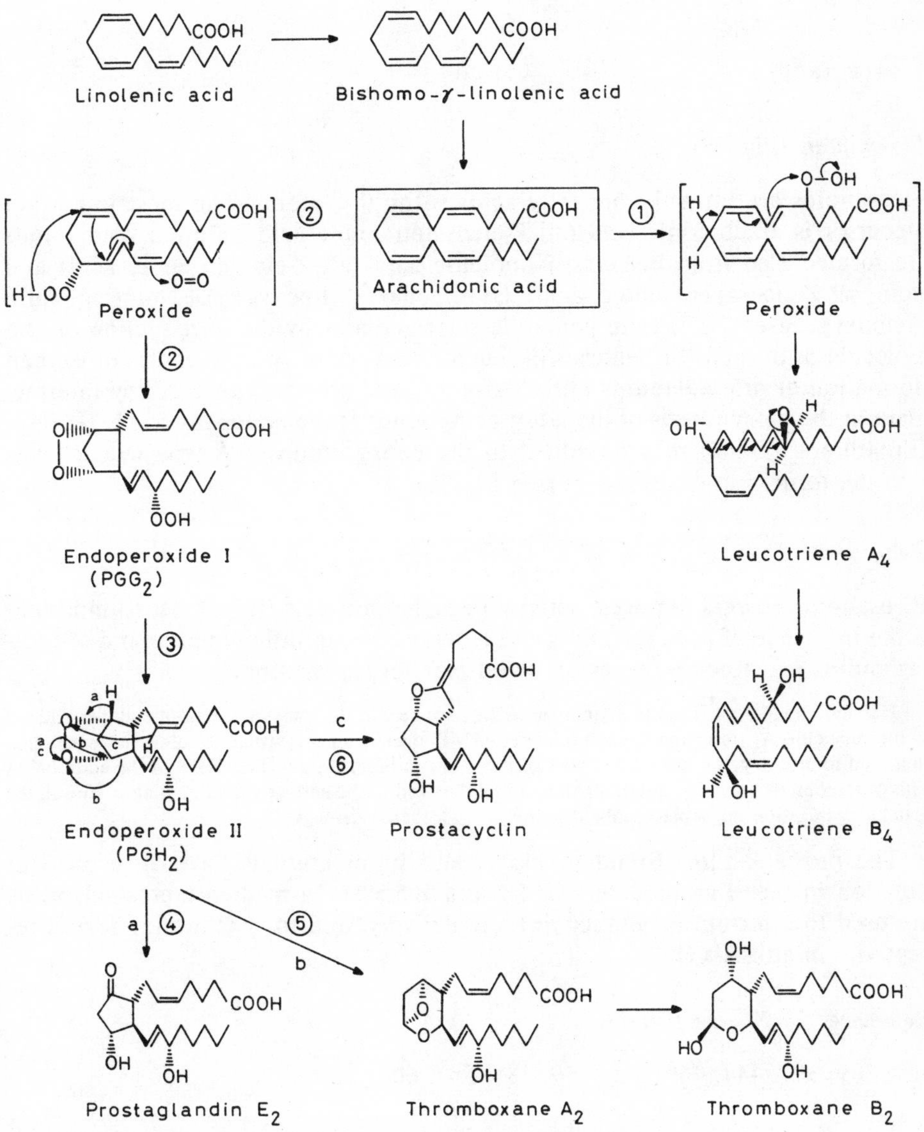

Fig. 79. Conversion of arachidonic acid to eicosanoids
1 Lipoxygenase; *2* prostaglandin cyclooxygenase; *3* reductase; *4* prostaglandin synthase; *5* thromboxane synthase; *6* prostacyclin synthase

groups may be present. Eicosanoids are produced in higher animals. Prostaglandins have also been found in invertebrates. The coral *Plexaura homomalla* contains several percent of prostaglandin A_2 or the epimeric 15*(R)*-prostaglandin A_2 acting as feeding deterrents (E 5.1).

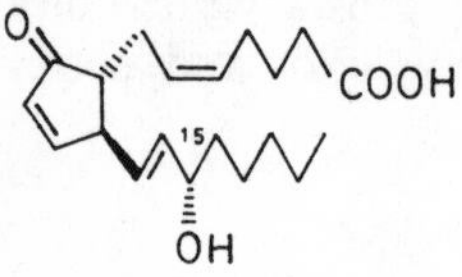

Prostaglandin A_2 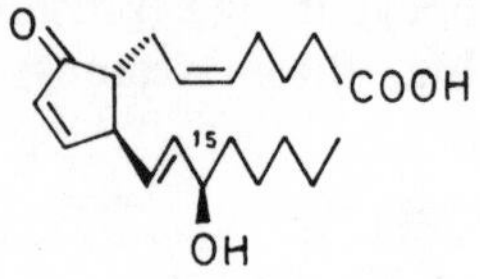   15(R)-Prostaglandin A_2

Biosynthesis (Fig. 79)

Eicosanoids are formed from fatty acids with 20 C-atoms. The most important precursor is arachidonic acid (all-*Z*-eicosatetraenoic acid). Similar compounds are formed also from bishomo-γ-linolenic acid (all-*Z*-eicosatrienoic acid) and from all-*Z*-eicosapentaenoic acid. Dioxygenases (lipoxygenase, prostaglandin cyclooxygenase, C 2.5) form peroxidic intermediates by the introduction of one molecule of oxygen (biosynhesis of leucotrienes) or of two molecules of oxygen (formation of prostaglandins, thromboxanes, and prostacyclins). A key intermediate in the biosynthesis of the latter compounds is the endoperoxide II (PGH₂). Glutathione (D 9.2) may be added to the epoxy group of A-type leucotrienes with the formation of thioesters (see Fig. 21).

Significance

Eicosanoids possess hormone activity in higher animals (E 3.1). Most important is the influence of prostaglandins and leucotrienes on inflammation and of prostaglandins and thromboxanes on blood platelet aggregation.

Platelet homeostasis is maintained by a delicate balance in the transformation of endoperoxide II to thromboxane A_2 (reducing the intraplatelet cAMP level, a process which is followed by aggregation) and to prostacyclin (doing the reverse). Prostacyclin biosynthesis takes place in the endothelial cells of arteries, whereas the thromboxanes are synthesized in the collagen. Thus, collagen-stimulated platelet aggregation takes place only at points of endothelial damage.

The prostaglandins found in plants and lower animals have an ecological function in repelling predators (E 5.1 and E 5.5.3). In medicine prostaglandins are used to interrupt pregnancy, initiate delivery, and to synchronize sexual receptivity in animals (F 2).

References for Further Reading

308, 309, 397, 447, 449, 516, 569, 585, 587, 606, 771

D 5.3 Polyketides Formed with Participation of Malonyl CoA

Chemistry and Distribution

Polyketides (acetogenins) show a regular pattern of oxygen-substituted and oxygen-free C-atoms at least in part of the molecule. They are derivatives of (—CHR —CO)$_n$-polyoxo acids. In the acetyl/malonyl CoA-derived compounds is R = —H. However, there are also substances formed from propionyl/methylmalonyl CoA and butyryl/ethylmalonyl CoA (R = —CH$_3$ and —C$_2$H$_5$, respectively, D 3). Most polyketides are compounds with carbocyclic, O- and N-heterocyclic rings. The regular alternation of oxygenated and oxygen-free C-atoms is frequently modified by the introduction of further hydroxy groups, by reduction, etc. Additional substituents, e.g., chlorine and bromine atoms, methyl or dimethylallyl groups may be present.

Most polyketides are synthesized in microorganisms. A few are built in higher plants.

Biosynthesis

Polyketides are formed by multifunctional proteins (A 3.1.1), e.g., 6-methylsalicylate synthase (D 5.3.2) and chalcone synthase (D 23.3.3), from a starter molecule, in most cases acetyl CoA, but also malonyl CoA, malonamido CoA, anthranilyl CoA, cinnamoyl/p-coumaroyl CoA, or 4-hydroxybenzoyl CoA. These starter molecules react with several molecules of malonyl CoA, methylmalonyl CoA, or ethylmalonyl CoA (extender units) with loss of the free carboxy group of the latter. According to the number of acid molecules involved in the formation of the individual polyketides, di-, tri-, tetraketides, etc. may be distinguished (Table 33).

The synthesis of polyketides resembles the formation of fatty acids by fatty acid synthase in terms of the precursors used and the mode of their alignment (D 5.2). In contrast to the formation of fatty acids, however, the biosynthesis of polyketides proceeds without obligate reduction of intermediates. Most probably polyoxo acids are formed as the first step which are unstable and have not yet been detected in the free state. They seem to be attached to the core unit of the enzyme complex, stabilized by hydrogen bonding or by chelation of their semi-enolates with metal ions held by the enzymes.

Some of the bound oxo groups may be reduced by pyridine nucleotide-dependent dehydrogenases (C 2.1, see the formation of 6-methylsalicylic acid, D 5.3.2), and some of the activated CH$_2$-groups may be alkylated by S-adenosyl-L-methionine (C 3.3, see the biosynthesis of tetracyclines, D 5.3.8), or may be substituted by dimethylallyldiphosphate (D 6, see the formulas of lupulone and humulone, D 5.3.1).

The spatial arrangement of the polyoxo acids leads to cyclizations, either by aldol condensation, i.e., the reaction of carbonyl groups with acidic CH$_2$-groups, or by Claisen condensation, i.e., reaction of the ester group at the head of the molecule with one of the acidic CH$_2$-groups. The actual type of condensation depends on the nature of the polyketide synthetase in question and the spatial arrangement of the polyoxo acid directed by the enzyme. Thus, several cyclic

Table 33. Groups of polyketides formed with the participation of malonyl CoA

Poly-ketide families	Starter						
	Propionyl CoA (D 3)	Acetyl CoA (D 5)	Malonyl CoA (D 5)	Malonamido CoA (D 5.3.8)	Anthranilyl CoA (D 21.4)	Cinnamoyl/p-coumaryl CoA (D 23.2.1)	p-hydroxybenzoyl CoA (D 23.2.5)
Diketides						Piperine, paracotoin (D 23.3.1)	
Triketides		Phloroglucinol, triacetic acid lactone (D 5.3.1)				Kawain, hispidin (D 23.3.2)	
Tetraketides		6-Methylsalicylic acid, acetophenones, lecanoric acid, patulin, coniine, gallic acid (D 5.3.2)			Acridone alkaloids (D 21.4.2)	Stilbene derivatives, flavonoids (D 23.3.3)	Gentisein (D 23.2.5)
Pentaketides		Eugenone, citrinin, sepedonin, coumarin derivatives (D 5.3.3)					
Hexaketides		Plumbagin (D 5.3.4)		Cycloheximide (D 5.3.8)			
Heptaketides		Griseovulvin (D 5.3.5)					
Octaketides		Anthraquinones, ergochromes (D 5.3.6)					
Nonaketides			Tetracyclines (D 5.3.8)				
Decaketides	Anthracyclines (D 3)	Aflatoxins (D 5.3.7)					

structures may be formed from one polyoxo acid. The great variety of polyketides found in nature is based on the different starter molecules, the varying number (and chemical structure) of the extender units involved in the formation of the polyoxo acid, different modes of cyclization, and a wide variety of transformations of the basic skeletons.

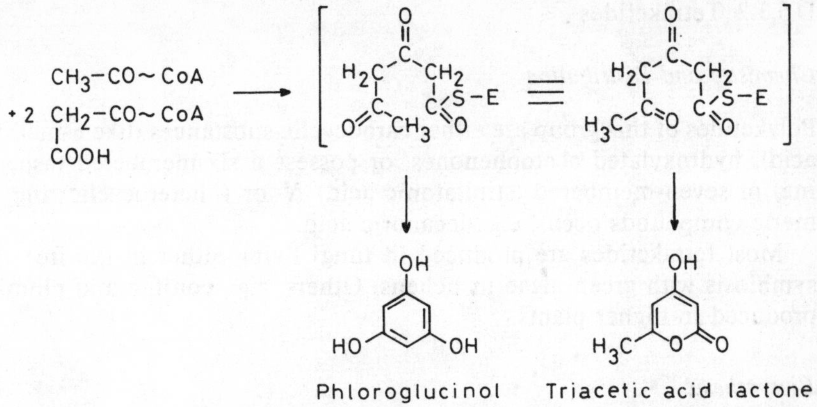

Fig. 80. Biosynthesis of phloroglucinol and triacetic acid lactone

References for Further Reading

708, 808

D 5.3.1 Triketides

Chemistry, Biosynthesis, Significance

The most important triketides are phloroglucinol and its derivatives (found in higher plants and brown algae) and the pyrone triacetic acid (built in microorganisms). Triketides are formed by the pathway given in Fig. 80. Phloroglucinol may be glucosylated (cf. the structure of phlorin), substituted by isopentenyl residues (cf. the formulas of lupulone and humulone, the bitter principles of hop cones used in brewing beer, F 1), or condensed by dehydrogenation to the so-called phlorotannins. A representative of the latter group with relatively low molecular weight is tetraphloretol.

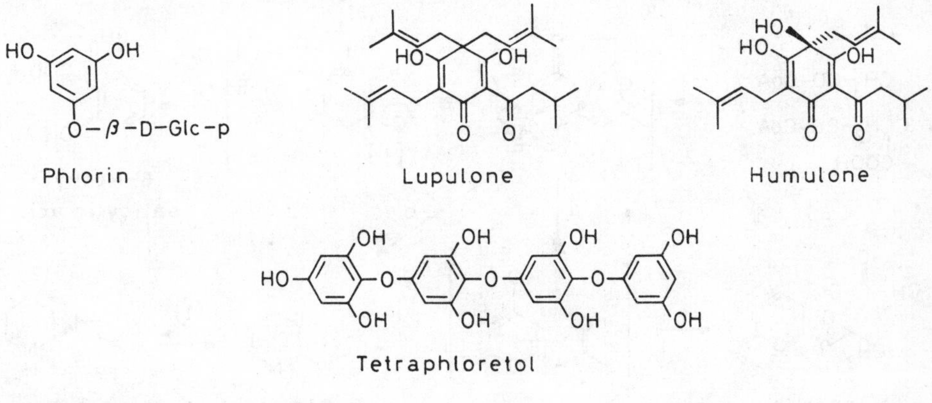

Phlorin   Lupulone   Humulone

Tetraphloretol

Reference for Further Reading

708

D 5.3.2 Tetraketides

Chemistry and Distribution

Polyketides of this group are either carbocyclic substances (like 6-methylsalicylic acid), hydroxylated acetophenones, or possess a six-membered (asperlin, coniine) or seven-membered (stipitatonic acid) *N*- or *O*-heterocyclic ring. Often dimeric compounds occur, e.g., lecanoric acid.

Most tetraketides are produced in fungi living either in the free state or in symbiosis with green algae in lichens. Others, e.g., coniine and plumbagin, are produced in higher plants.

Biosynthesis

A. Cyclization. Cyclization of the polyoxo acid intermediate may proceed in different ways (Fig. 81).

Formation of 6-methylsalicylic acid (6-MSA, Fig. 82). 6-MSA synthase, a polyfunctional protein (A 3.1.1), forms 6-MSA from acetyl CoA, malonyl CoA, and NADPH. 6-MSA possesses two types of SH-groups to which the intermediates are bound. The SH-groups are designated as central and peripheral in analogy to those of fatty acid synthase (D 5.2). A 4'-phosphopantetheine moiety has been identified as part of the enzyme. It is thought that acetyl CoA is bound to the peripheral SH-group and malonyl CoA to the central SH-group. The reduction probably proceeds at the triacetic acid level before dehydration and the third condensation with malonyl CoA. The last stage seems to be the release of 6-MSA from the enzyme protein.

In similar types of reactions orsellinic acid (Fig. 86) and 3-methylorsellinic acid (Fig. 85) are built.

Formation of asperlin and coniine. The biosynthesis of these compounds includes reduction of the polyoxo acid which is followed by the formation of a

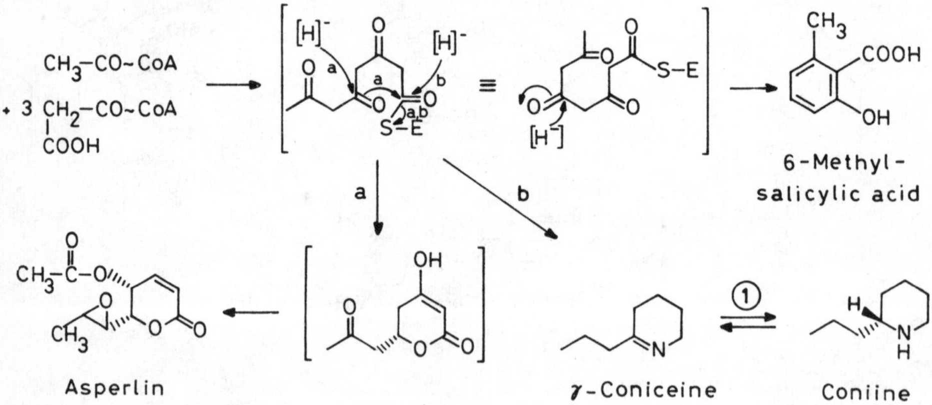

Fig. 81. Synthesis of different types of polyketides with eight C-atoms
1 γ-Coniceine reductase

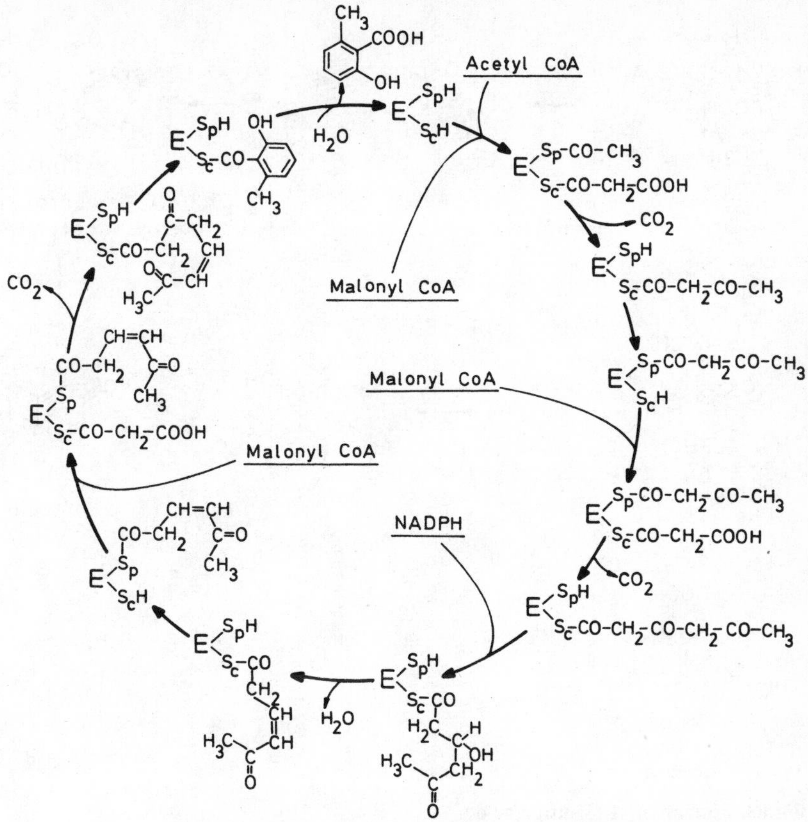

Fig. 82. Hypothetical scheme of 6-methylsalicylic acid biosynthesis
E Enzyme; S_pH peripheral SH-group; S_cH central SH-group

heterocyclic ring. γ-Coniceine and coniine are easily converted into each other (Fig. 81).

Formation of acetophenones (Fig. 83). Condensation between C-1 and C-6 of the polyoxo acid yields a hydroxylated acetophenone derivative. The methyl group in position 5′ originates from S-adenosyl-L-methionine (D 12.1). It is introduced into the molecule before cyclization.

B. Further modification. The compounds formed by cyclization may undergo further reactions. Of special interest are:

— the formation of esters (depsides), cf. the biosynthesis of lecanoric acid from two molecules of orsellinic acid:

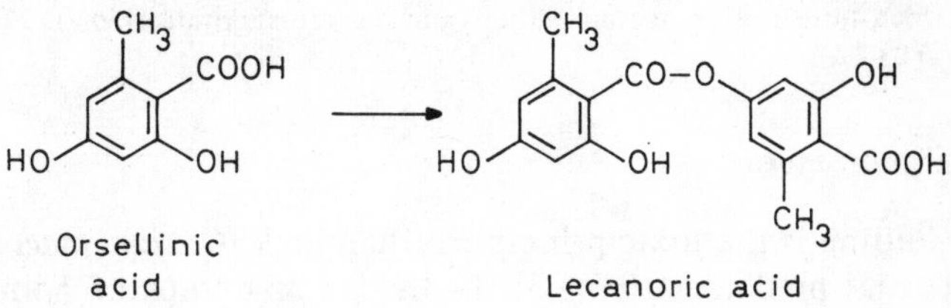

Orsellinic
acid

Lecanoric acid

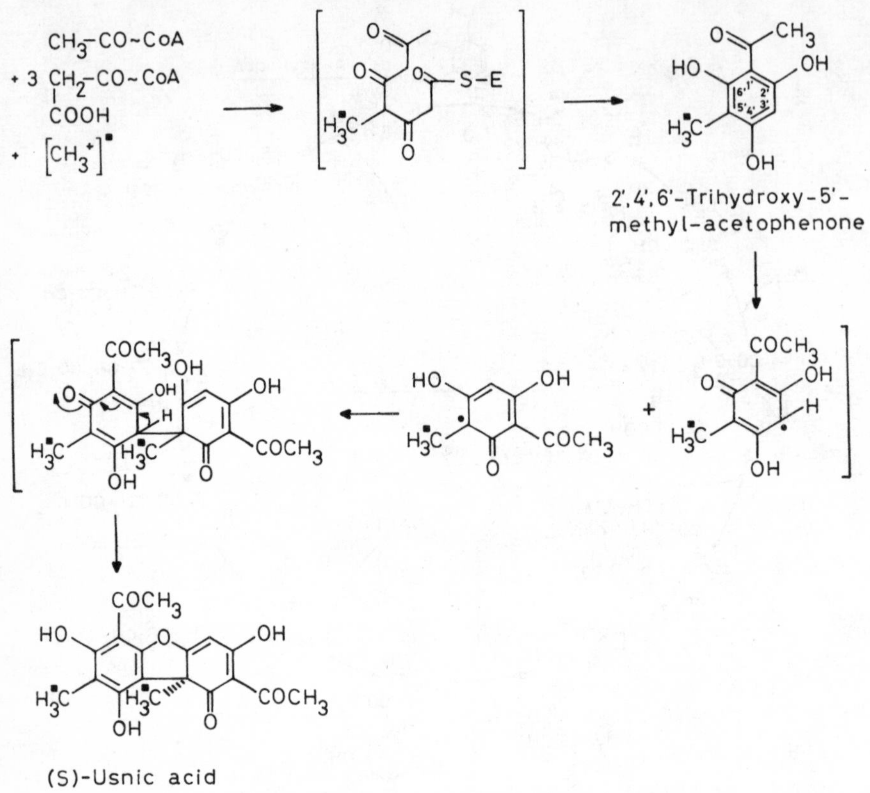

Fig. 83. Formation of *(S)*-usnic acid

— coupling via radicals produced by phenol oxidases (C 2.3.1) or peroxidases
  (C 2.4.1),   see   the   formation   of   *(S)*-usnic   acid   from
  2',4',6'-trihydroxy-5'-methylacetophenone (Fig. 83)
— the oxidative splitting of the aromatic ring by dioxygenases (C 2.5) followed
  by recyclization, see the synthesis of patulin from 6-methylsalicylic acid
  (Fig. 84)
— ring enlargement, see the transformation of 3-methylorsellinic acid to stipitatonic acid (Fig. 85)
— methylations, prenylations, decarboxylations, hydroxylations, oxidations, and
  reductions, see the synthesis of patulin (Fig. 84) and the formation of phenolic
  compounds by soil fungi (Fig. 86).

It is of interest that fungi synthesize gallic acid from orsellinic acid in contrast to other organisms
which form this compound either from dehydroshikimic acid (D 21) or cinnamic acid derivatives
(D 23.2.5).

Significance

Coniine is the toxic principle of hemlock *(Conium maculatum)* (F 4). It repels potential predators (E 5.5.3). In the pitcher traps of *Sarracenia* the alkaloid para-

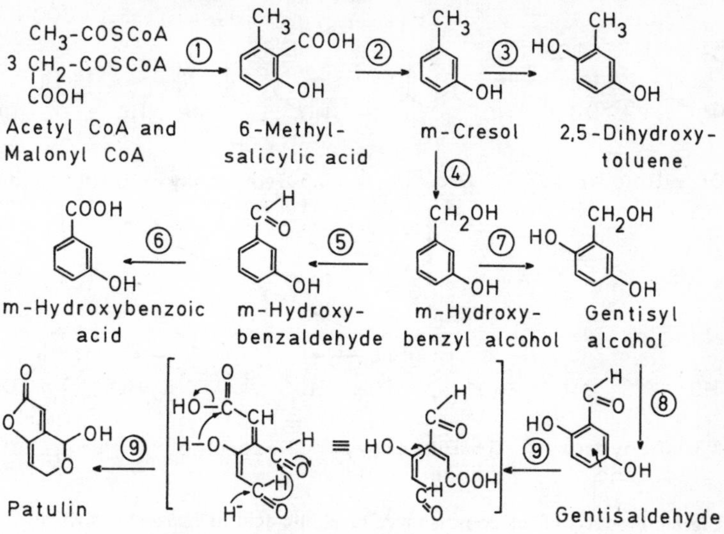

Fig. 84. Transformation of 6-methylsalicylic acid to patulin
1 6-Methylsalicylate synthase; *2* 6-methylsalicylate decarboxylase; *3* m-cresol 6-monooxygenase; *4* m-cresol methylmonooxygenase; *5* m-hydroxybenzyl alcohol dehydrogenase; *6* m-hydroxybenzaldehyde dehydrogenase; *7* m-hydroxybenzyl alcohol 6-monooxygenase; *8* gentisyl alcohol dehydrogenase; *9* patulin synthase

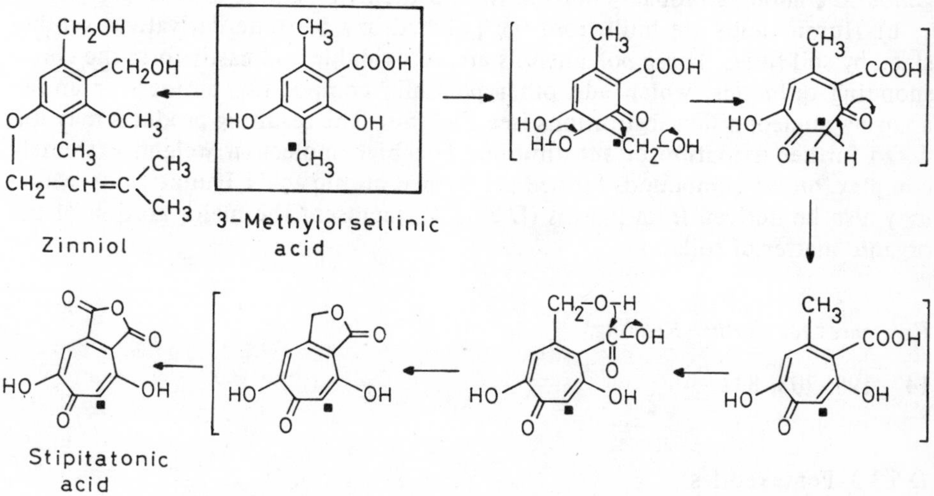

Fig. 85. Possible pathway of stipitatonic acid formation
The C-atom marked by ■ is derived from the methyl group of L-methionine

lyzes captured insects. Patulin is a phytotoxin (F 4). 6-Methylsalicylic acid acts as an ant trail pheromone (E 4). (S)-Usnic acid and other lichen phenolics act as radiation screen protecting the photosynthetic pigments of the photobiont algal cells.

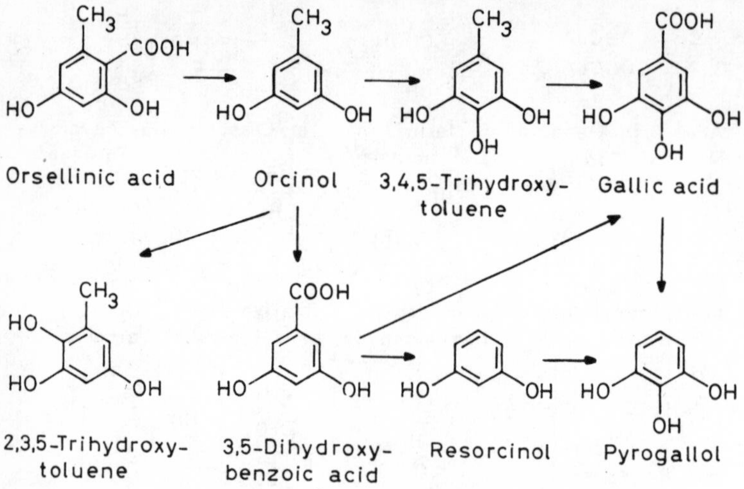

Fig. 86. Products of the conversion of orsellinic acid in *Epicoccum nigrum*

Transformation

a) Zinniol, a phytotoxic compound (E 5.4) produced in plant pathogens of the genus *Alternaria*, is probably derived from 3-methylorsellinic acid (Fig. 85).

b) Humic acids are built from the polyhydroxy benzene derivatives synthesized by soil fungi. These polyphenols are autoxidable and easily form the corresponding quinones, which add other phenolic compounds, amines, or amino acids by nucleophilic substitutions (see Fig. 160). The resulting products may undergo further oxidation or substitution. The high molecular weight, extremely complex, brown compounds formed are termed humic acids. Humic acids, which may also be derived from lignins (D 23.2.3), represent the major portion of the organic matter of soils.

References for Further Reading

147, 198, 708, 841

D 5.3.3 Pentaketides

Chemistry and Distribution

Pentaketides are built in molds, e.g., *Aspergillus* and *Penicillium* sp. They are either monocyclic compounds, e.g., eugenone and Ceratocystis toxins, or bicyclic substances, e.g., citrinin. Most pentaketides possess a six-membered carbocyclic ring. A few compounds have a seven-membered ring (tropolones, e.g., sepedonin).

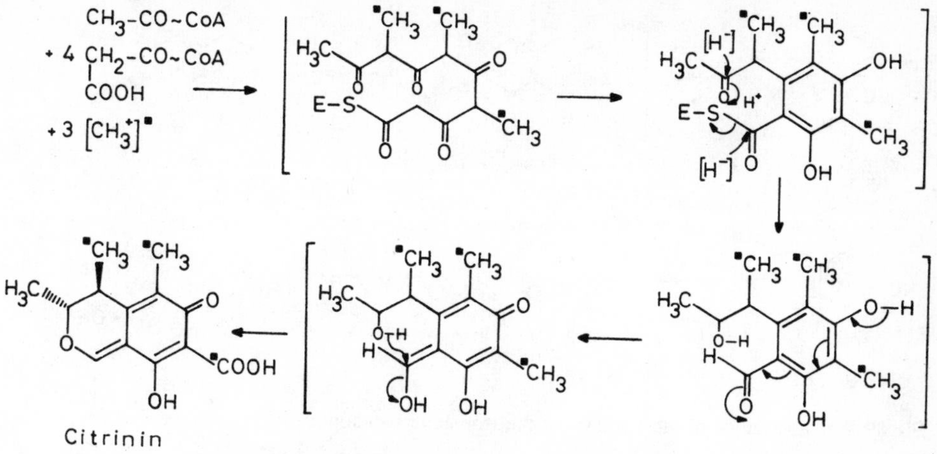

Fig. 87. Biosynthesis of citrinin
The C-atoms marked by ■ are derived from the methyl group of L-methionine

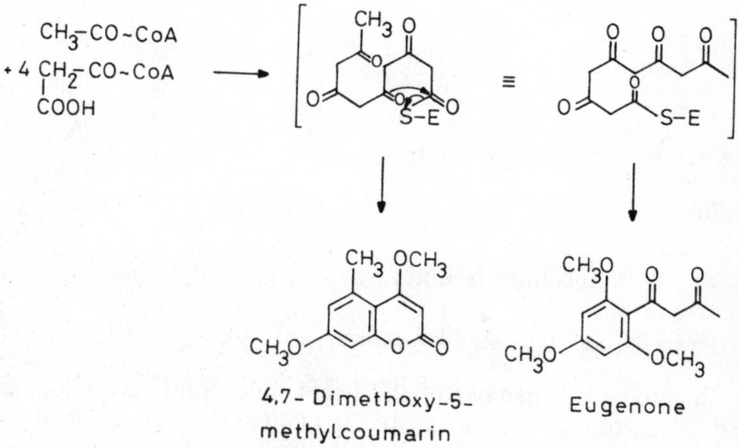

4,7-Dimethoxy-5-methylcoumarin

Eugenone

Fig. 88. Formation of polyketide-derived coumarins of the eugenone type

Biosynthesis

The corresponding pentaoxo acid may cyclize in different ways (Figs. 87 and 88). Transformation of the cyclic compounds is common. Of special interest is the ring enlargement during the formation of tropolones (Fig. 89).

Significance

Sepedonin is involved in the uptake and transport of Fe^{3+} (E 2.2). Citrinin is a carcinogenic toxin produced in several molds (F 4). The *Ceratocystis* toxins cause the Dutch elm disease (E 5.4).

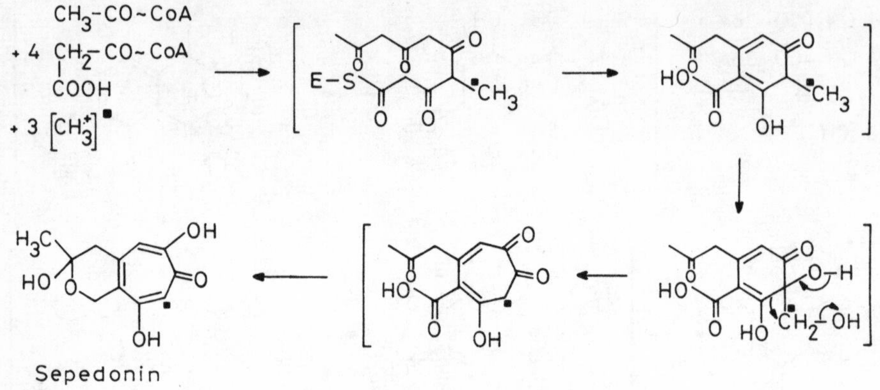

Fig. 89. Probable mode of biosynthesis of the tropolone sepedonin

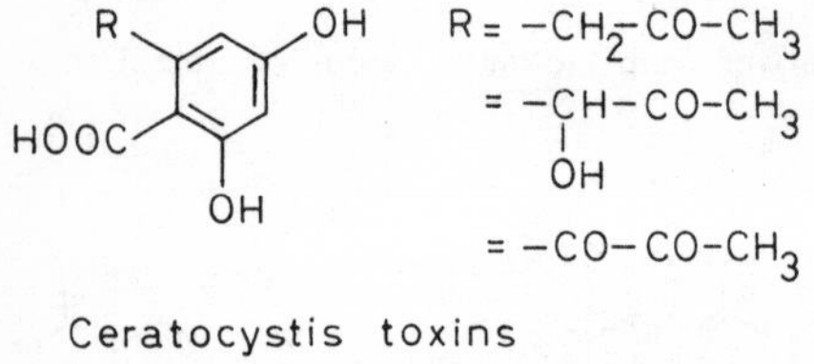

Ceratocystis toxins

Reference for Further Reading

708

D 5.3.4 Hexaketides (Plumbagin)

Chemistry, Distribution, Biosynthesis

An interesting member of the hexaketide family is plumbagin, a naphthoquinone derivative, formed in higher plants of the genus *Plumbago*. In contrast to many other naphthoquinones (D 21.2), plumbagin is a polyketide. Its formation resembles anthraquinone biosynthesis (D 5.3.6, Fig. 90).

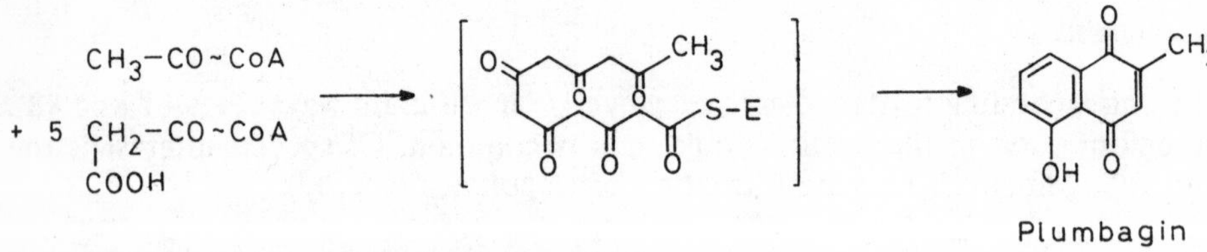

Fig. 90. Probable pathway of plumbagin biosynthesis

Reference for Further Reading

708

D 5.3.5 Heptaketides (Griseofulvin)

Chemistry and Distribution

A representative of the heptaketides is griseofulvin, a chlorinated cumarane-2-spiro-cyclohexane derivative formed in *Penicillium* sp.

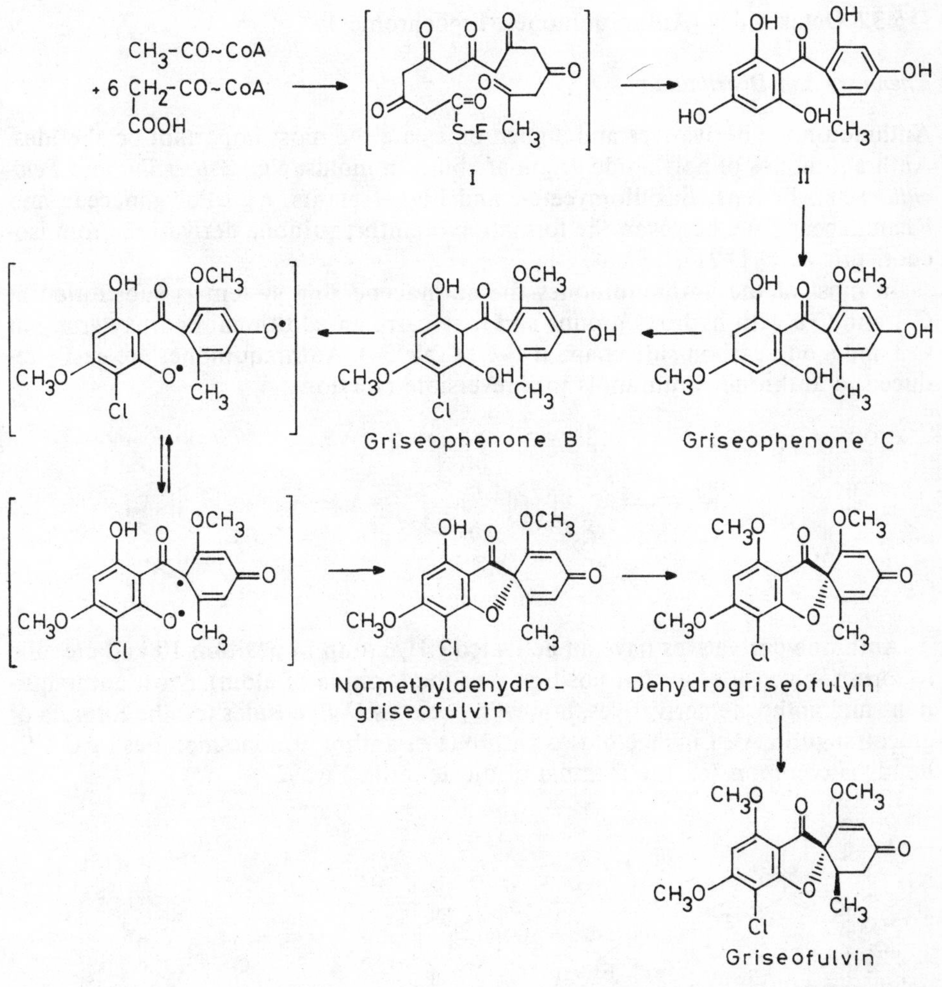

Fig. 91. Biosynthesis of griseofulvin

Biosynthesis (Fig. 91)

In the biosynthesis of griseofulvin the polyoxo acid I cyclizes to the benzophenone derivative II, which is methylated (C 3.3) and chlorinated (C 2.4.2). Attack of a phenol oxidase (C 2.3.1) or of peroxidase (C 2.4.1) yields a biradical which is the key intermediate in the closure of the O-heterocyclic ring.

Significance

Griseofulvin is an antifungal antibiotic (E 5.2) used in medicine (F 2).

References for Further Reading

708, 790

D 5.3.6 Octaketides (Anthraquinones, Ergochromes)

Chemistry and Distribution

Anthraquinone derivatives and ergochromes are the most important octaketides. Anthraquinones of polyketide origin are built in molds, e.g., *Aspergillus* and *Penicillium* sp., lichens, Basidiomycetes, and higher plants, e.g., Polygonaceae and Rhamnaceae (see, however, the formation of anthraquinone derivatives from isochorismic acid, D 21.2).

In most of the anthraquinones the anthracene ring system is substituted at C-1 and C-8 with hydroxy groups and may carry an additional hydroxy group at C-3 and a one-carbon side chain at C-6 (Table 34). Anthraquinones are easily reduced to anthrones/anthranols in a reversible reaction:

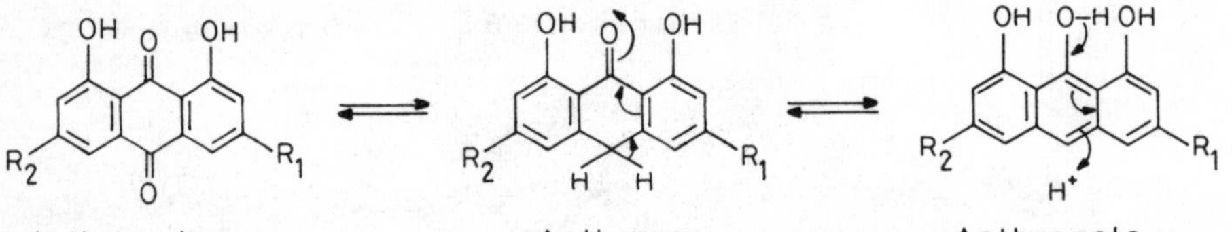

Anthraquinones   Anthrones   Anthranols

Anthrone derivatives have an activated CH_2-group in position 10 and are able to form *C*-glycosides at that position (see the formula of aloin). Most anthraquinone and anthrone derivatives, however, occur as *O*-glycosides (cf. the formula of glucofrangulin A). Linkage of two anthrone or anthraquinone moieties by C—C-bonds is common (cf. the formula of dicatenarin, Fig. 92).

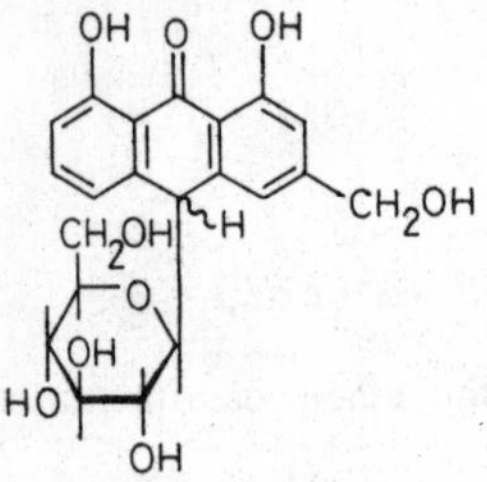

Aloin 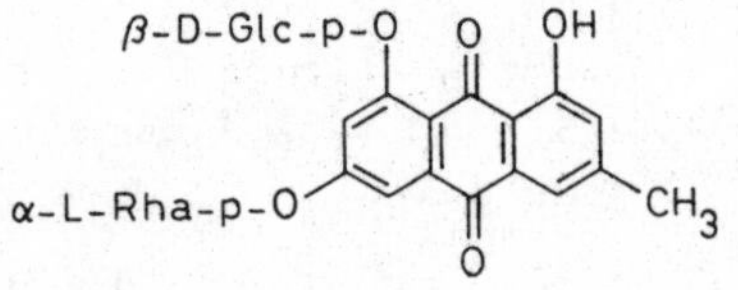   Glucofrangulin A 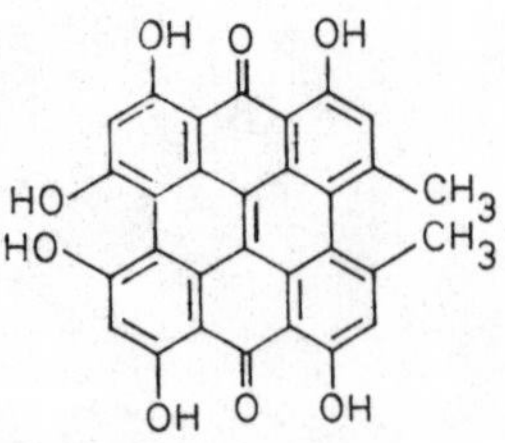   Hypericin

The ergochromes are light yellow dimeric xanthone derivatives resembling the anthraquinones in their pattern of substitution. Two xanthone moieties (Fig. 92) are interlinked at position 2. Ergochromes have been found in molds and lichens.

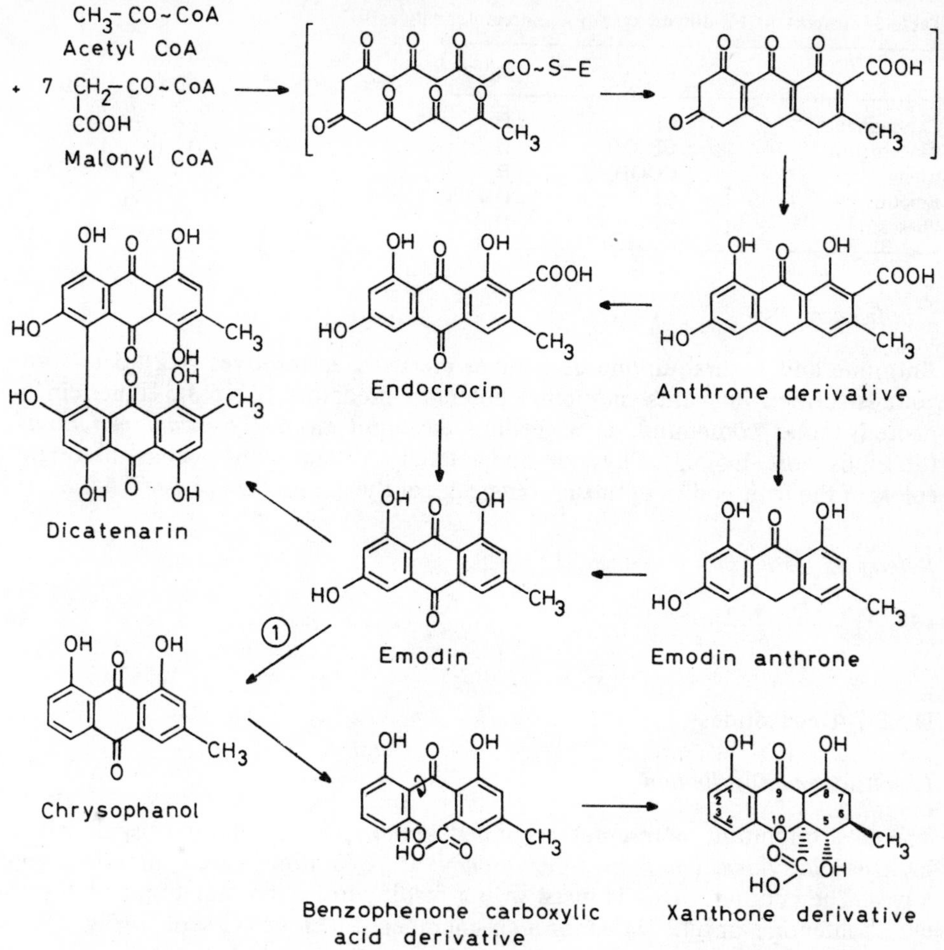

Fig. 92. Formation of anthracene and xanthone derivatives
1 Emodin dehydroxylase

Biosynthesis (Fig. 92)

Cyclization of the precursor octaoxo acid yields an anthrone derivative carrying a carboxy group in position 2 and a methyl group in position 3. The carboxy group may be subsequently eliminated and the methyl group oxidized to a CH_2OH or COOH-group (see Table 34). Anthrones easily undergo oxidation to anthraquinones. Dimerization proceeds via intermediate radicals formed by phenol oxidases (C 2.3.1) or peroxidases (C 2.4.1).

Oxidative cleavage of ring B of anthaquinones yields benzophenone derivatives which are transformed to xanthones. Both hydroxyl groups of the symmetrically substituted benzene nucleus of the benzophenone carboxylic acid intermediate can participate in the closure of the xanthone ring. From the xanthones (secalonic acids) dimeric products (ergochromes) are formed probably via phenol oxidases or peroxidase-derived radicals (C 2.3.1 and C 2.4.1).

Table 34. Important 1,8-dihydroxyanthraquinone derivatives

	R_1	R_2
Chrysophanol	CH_3	H
Aloeemodin	CH_2OH	H
Rhein	COOH	H
Emodin	CH_3	OH
Physcion	CH_3	OCH_3

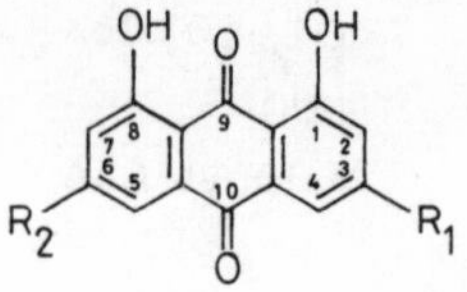

Significance

Anthrone and anthraquinone derivatives are used as laxatives (F 2). The compounds formed in plants may repel potential predators (E 5.5.3). Hypericin, a photodynamic compound, is a feeding deterrent from *Hypericum perforatum* (St. Johns wort, E 5.5.3). Physcion and related anthraquinones cause the bright colors of the fruit bodies of mushrooms, e.g., of the genus *Dermocybe* (E 5.6).

References for Further Reading

248, 457, 667, 708

D 5.3.7 Decaketides

Chemistry and Distribution

The most important representatives of decaketides are the aflatoxins synthesized by *Aspergillus flavus* and some other *Aspergillus sp.* Aflatoxins are coumarin derivatives. The coumarin ring is fused with a bisdihydrofurano moiety plus either a cyclopentenone ring (B series) or a six-membered lactone (G series) (Fig. 93).

Biosynthesis

The aflatoxins are derived from anthraquinone derivatives of the averufin type (Fig. 93). Of special interest is the conversion of the C_6-side-chain present in averantin to the C_4-chain found in versiconal A. A key intermediate is the xanthone sterigmatocystine. It is probably formed via intermediate I with rotation of the right-hand part of the molecule.

Significance

Aflatoxins are carcinogenic in animals and humans (F 4). Higher doses cause death due to liver damage.

References for Further Reading

53, 346, 708

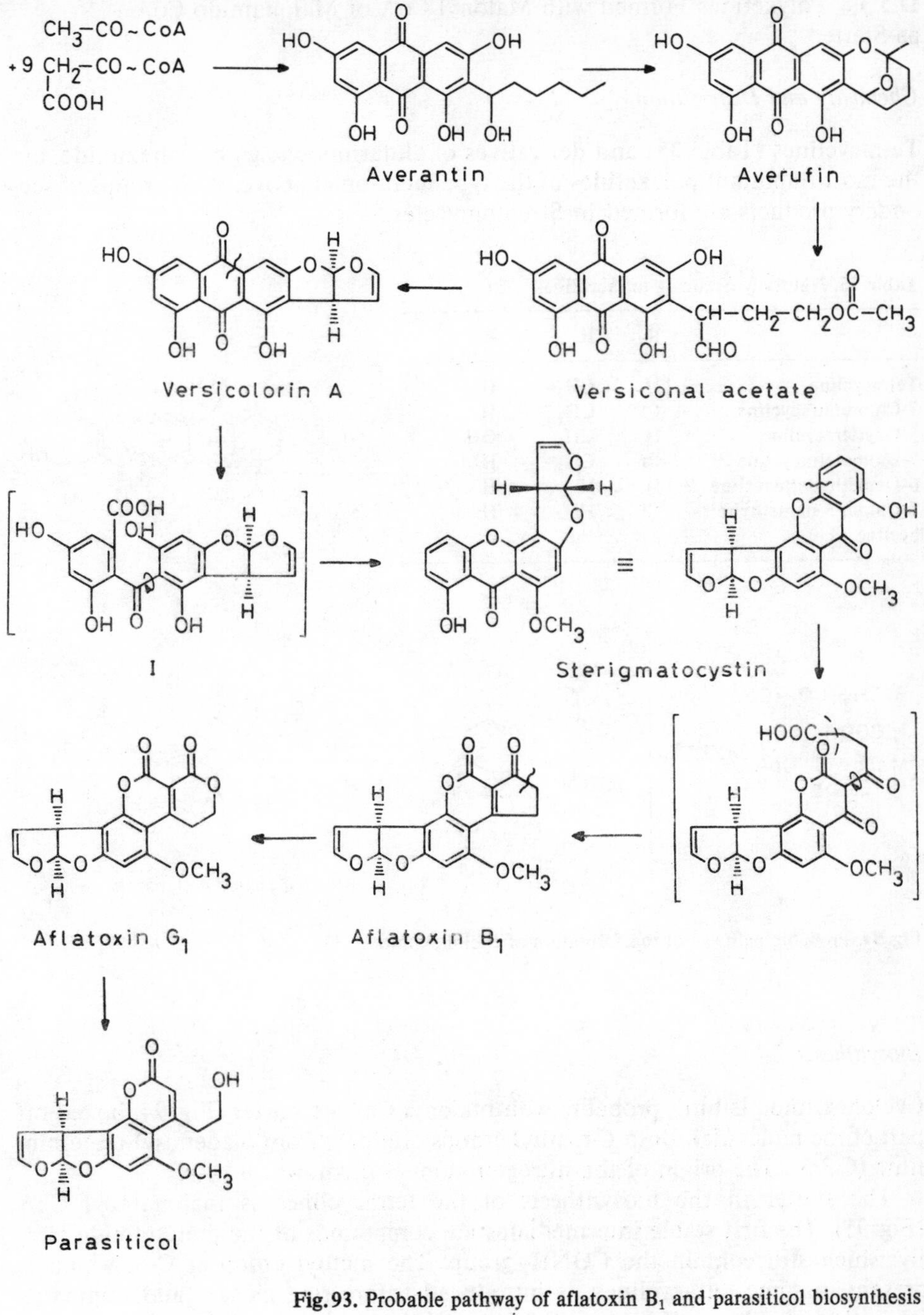
CH3-CO~CoA
+9 CH2-CO~CoA
COOH
Averantin
Averufin
Versicolorin A
Versiconal acetate
I
Sterigmatocystin
Aflatoxin G1
Aflatoxin B1
Parasiticol

Fig. 93. Probable pathway of aflatoxin B_1 and parasiticol biosynthesis

D 5.3.8 Polyketides Formed with Malonyl CoA or Malonamido CoA as Starter

Chemistry and Distribution

Tetracyclines (Table 35) and derivatives of glutarimide, e.g., cycloheximide, are the most important polyketides of the type mentioned above. Both groups of secondary products are formed in Streptomycetes.

Table 35. Naturally occuring tetracyclines

	R_1	R_2	R_3
Tetracycline	H	CH_3	H
7-Chlorotetracycline	Cl	CH_3	H
5-Oxytetracycline	H	CH_3	OH
7-Bromotetracycline	Br	CH_3	H
6-Demethyltetracycline	H	H	H
7-Chloro-6-demethyltetra-cycline	Cl	H	H

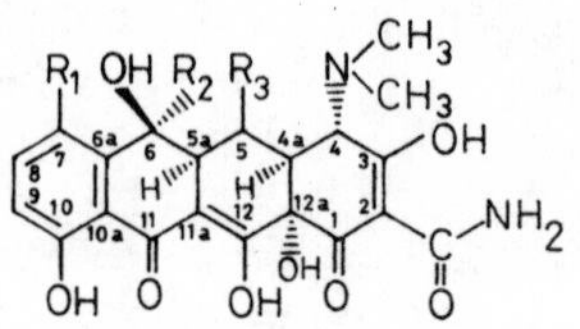

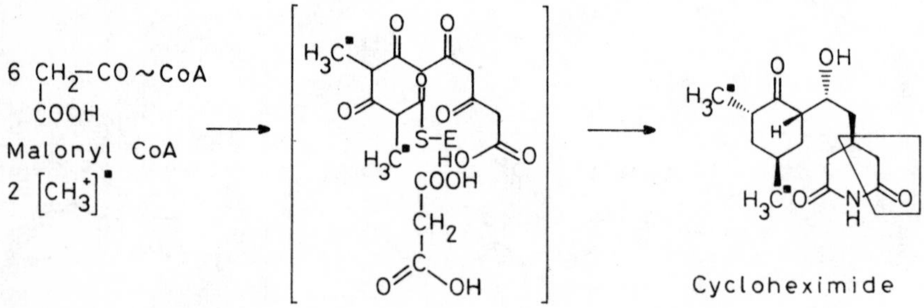

Fig. 94. Probable pathway of the formation of cycloheximide

Biosynthesis

Cycloheximide is built probably with malonyl CoA as starter (Fig. 94, boxed off part of the molecule). Both C-methyl groups originate from S-adenosyl-L-methionine (C 3.3). The origin of the nitrogen atom is unknown.

The starter in the biosynthesis of the tetracyclines is malonamoyl CoA (Fig. 95). The first stable intermediates are compounds of the pretetramide family which still contain the $CONH_2$-group. The methyl group at C-6, which is present in some tetracyclines, is introduced before ring closure and originates from S-adenosyl-L-methionine (C 3.3). L-Glutamine donates the amino group. The N-methyl groups come again from S-adenosyl-L-methionine. The chlorination proceeds by peroxidases (C 2.4.2). The hydroxyl group in position 6 is introduced by a monooxygenase (C 2.6.4).

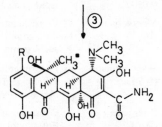

Fig. 95. Biosynthesis of tetracyclines
1 N-methyltransferase; 2 anhydrotetracycline oxygenase; 3 tetracycline (5a, 11a)-dehydrogenase

Significance

The tetracyclines and cycloheximide have antibiotic properties (E 5.2). Tetracyclines are used in medicine (F 2).

Reference for Further Reading

359

D 6 Biosynthesis of Isoprenoids

Chemistry and Distribution

Isoprenoids (terpenoids) are constructed from isoprene (2-methylbutadiene) units:

$$CH_2=\overset{\overset{\displaystyle CH_3}{|}}{C}-CH=CH_2$$

Isoprenoids are widespread in microorganisms, plants, and animals.

The terpenes still contain the original carbon skeleton of the isoprene units. In other groups of isoprenoids the skeleton is altered by addition or loss of carbon atoms. The isoprenoids are assigned to the groups given in Table 36 according to the number of isoprene units used as building blocks.

Isoprenoids may possess an *E-* or *Z*-configuration with respect to the double bonds. Most are derived from key products (prenyl diphosphates) with an all-*E*-configuration (geranyl diphosphate, farnesyl diphosphate, geranylgeranyl diphosphate, etc.). There are, however, also compounds with an all-*Z*-configuration such as rubber, or with a mixed configuration such as dolichols, betulaprenols, and ficaprenols.

The basic structure of isoprenoids may be modified by the introduction of a

Table 36. Secondary products derived from activated isoprene

Number of isoprene units	Group of compounds	Representatives
1	Hemiterpenes (D 6.1)	Isoprene, 3,3-dimethylallyl alcohol, isopentenol, isoamyl alcohol, tuliposides
2	Monoterpenes (D 6.2)	Constituents of essential oils, e. g., geraniol, menthol, and thymol, and of iridoids, e. g., loganin and secologanin
3	Sesquiterpenes (D 6.3)	Constituents of essential oils, e. g., farnesol and bisabolol, and of sesquiterpene lactones, e. g., matricin; abscisic acid, juvenile hormone, sirenin
4	Diterpenes (D 6.4)	Constituents of resins, e. g., abietic acid; gibberellins, phytol, vitamin A, crocetin
5	Sesterterpenes (D 6.5)	Constituents of unsaponifiable lipid extracts, waxes, etc.; ophiobilin, ircinin, variabilin
6	Triterpenes (D 6.6)	Squalene, steroids, pentacyclic triterpenes
7	Tetraterpenes (D 6.7)	Carotenes, xanthophylls
>8	Polyterpenes (D 6.8)	Rubber, gutta, solanesol, spadicol, dolichols, betulaprenols, ficaprenols

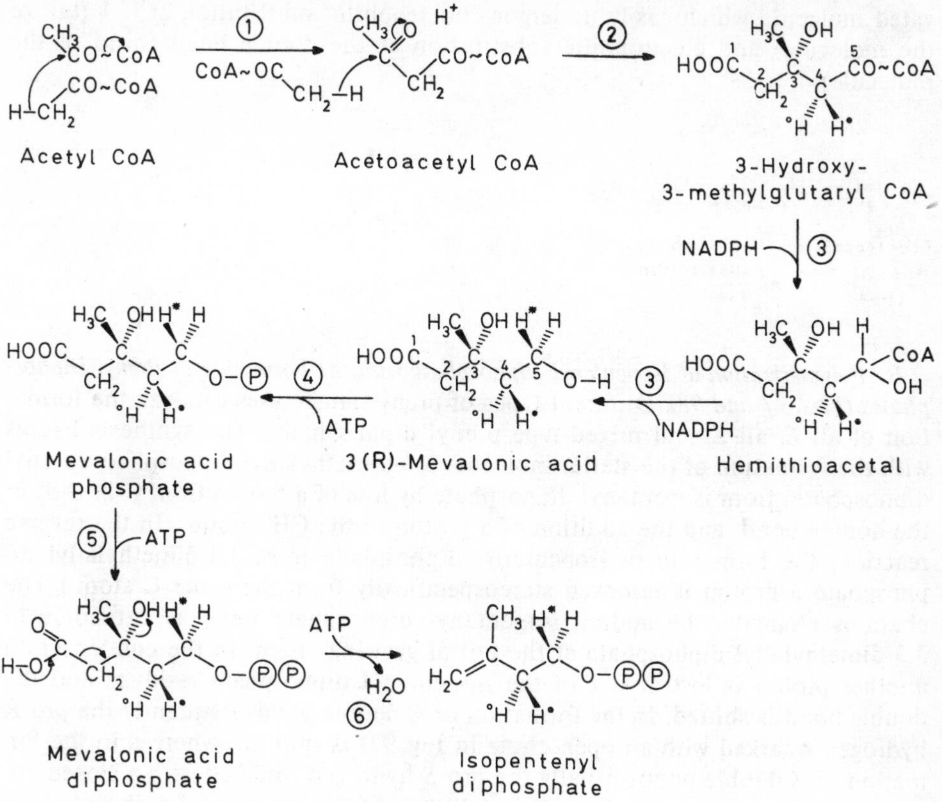

Fig. 96. Biosynthesis of isopentenyl diphosphate from acetyl CoA
1 Acetyl CoA acetyltransferase; *2* hydroxymethylglutaryl CoA synthase; *3* hydroxymethylglutaryl CoA reductase; *4* mevalonate kinase; *5* phosphomevalonate kinase; *6* diphosphomevalonate decarboxylase

wide variety of chemical groups, by isomerization, shift of double bonds, methyl groups, etc. Hence, a bewildering number of chemical structures arises. In addition, compounds derived from other biogenic pathways may contain isoprene residues (mixed terpenoids). For instance, the K vitamins (D 21.2), ubiquinones (D 21.3), chlorophylls (D 24.1), plastoquinones, and tocopherylquinones (D 23.4) have isoprenoid side chains with up to ten isoprene units. Polyketides (D 5.3), alkaloids (D 21.4.2), and coumarins (D 23.2.2) may be substituted by 3,3-dimethylallyl groups. The terpene residues are attached to nucleophilic sites, such as active methylene groups and phenolic oxygen atoms.

Biosynthesis

A. The Formation of Isopentenyl Diphosphate (Fig. 96). All isoprenoids originate from isopentenyl diphosphate which is formed from acetyl CoA (C 1.2). An important intermediate is mevalonic acid. Isopentenyl diphosphate is an activated molecule which easily undergoes nucleophilic substitution at C-1 (tail of the molecule) and electrophilic substitution at the double bond (head of the molecule):

Electrophilic	Nucleophilic
substitution	substitution
(Head)	(Tail)

B. Polymerization of Isopentenyl Diphosphate with the Formation of Prenyl Diphosphates (Figs. 97 and 98). Different types of prenyltransferases catalyze the formation of all-*E*, all-*Z*, and mixed-type prenyl diphosphates. The synthesis begins with the formation of the starter molecule 3,3-dimethylallyl diphosphate (prenyl diphosphate) from isopentenyl diphosphate by loss of a proton from C-2, shift of the double bond, and the addition of a proton to the CH_2-group. (In the reverse reaction, the formation of isopentenyl diphosphate from 3,3-dimethylallyl diphosphate a proton is removed stereospecifically from the same C-atom.) The chain is elongated by adding isopentenyl diphosphate units as extenders to 3,3-dimethylallyl diphosphate at the tail of growing chain. In the condensation another proton is lost at C-2 of the isopentenyl diphosphate residues and the double bond is shifted. In the formation of *E*-double bonds frequently the pro-R hydrogen (marked with an open circle in Fig. 97) is split off, whereas in the formation of *Z*-double bonds usually the pro-*S* hydrogen (marked with a closed circle in Fig. 98) is eliminated from C-2. Addition of the isopentenyl diphosphate to dimethylallyl diphosphate or the growing chain causes inversion of the configuration of the acceptor molecule at C-1.

C. The Tail-to-Tail Condensation of Prenyl Diphosphates (Fig. 99). In contrast to the head-to-tail condensations during polymerization of isopentenyl diphosphate (see above), the triterpene squalene and the tetraterpene phytoene are formed from two identical precursors by condensation.

Squalene is built from two farnesyl diphosphate molecules. The reaction is catalyzed by squalene synthase and includes the formation of presqualene diphosphate and the stereospecific reduction of one of the precursor farnesyl diphosphates by the 4_B-hydrogen of NADPH (C 2.1.1).

Squalene synthase produces lycopersene via prephytoene diphosphate in the presence of geranylgeranyl diphosphate. It is doubtful, however, whether this reaction is of significance in living cells.

The biosynthesis of phytoene proceeds without reduction of prephytoene diphosphate. The phytoene molecule therefore contains a central double bond. Phytoene synthases from different organisms form either *E*- or *Z*-phytoene. Formation of *Z*-phytoene requires the stereospecific loss of the H_B in the postulated intermediate, whereas removal of H_A yields *E*-phytoene (Fig. 99).

Certain bacteria are able to form C_{30}, C_{45}, and C_{50} isoprenoids of the phytoene type in addition to phytoene by tail-to-tail condensation of appropriate precursor molecules. In higher plants certain monoterpenes are probably synthesized in a similar manner. Chrysanthemyl alcohol, occurring in

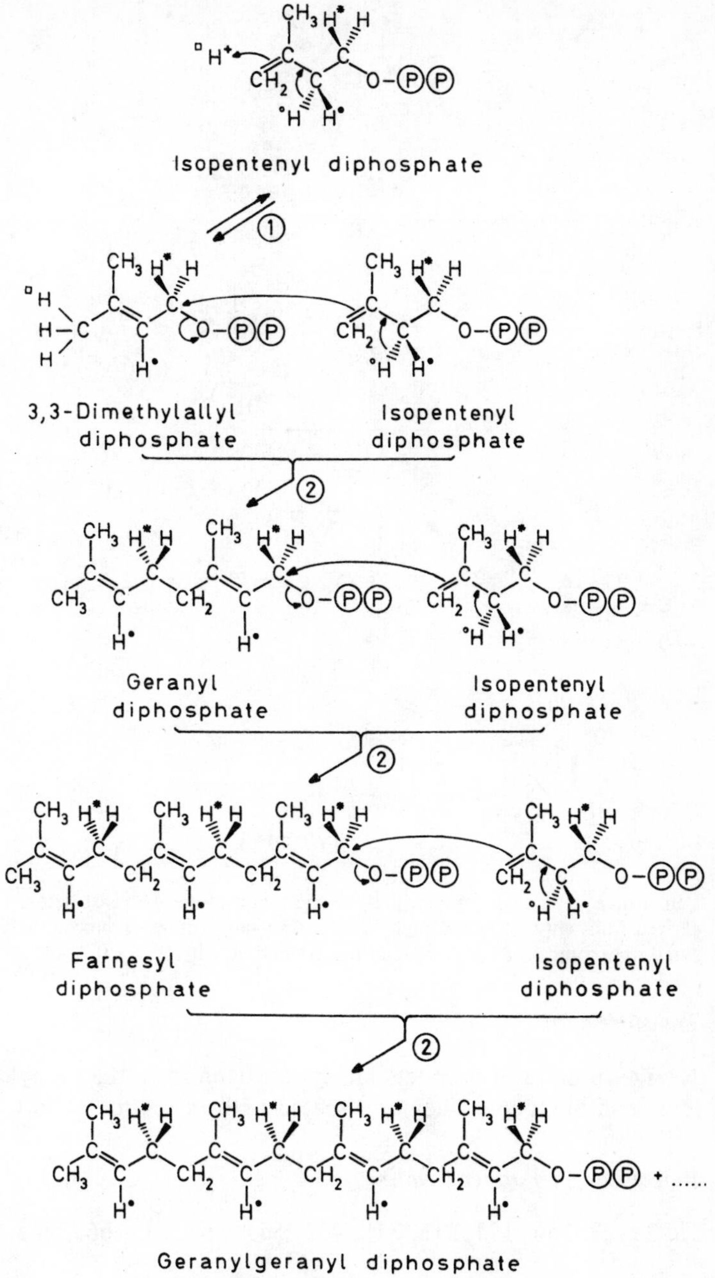

Isopentenyl diphosphate

3,3-Dimethylallyl
diphosphate

Isopentenyl
diphosphate

Geranyl
diphosphate

Isopentenyl
diphosphate

Farnesyl
diphosphate

Isopentenyl
diphosphate

Geranylgeranyl diphosphate

Fig. 97. Formation of all-*E*-prenyl diphosphates
1 Isopentenyl diphosphate *Δ*-isomerase; *2* prenyltransferase

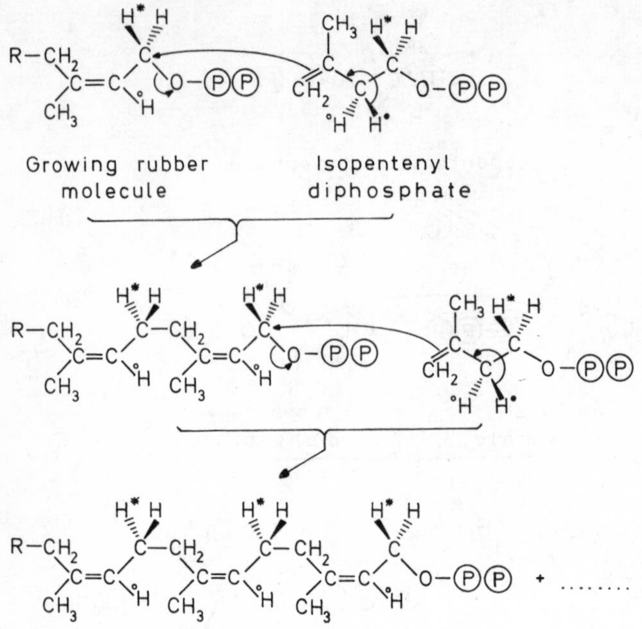

Fig. 98. Formation of all-Z-prenyl diphosphates during the biosynthesis of rubber

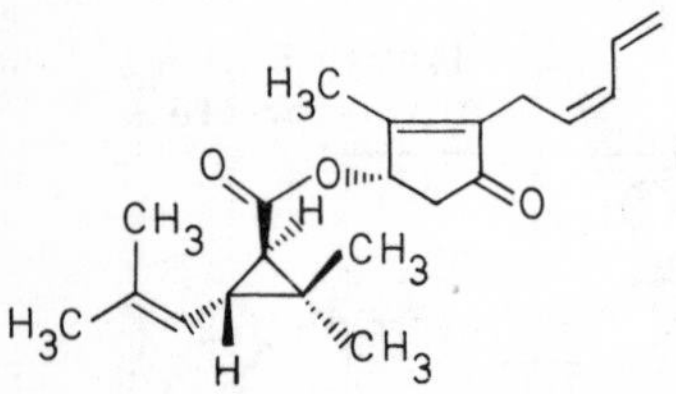

Pyrethrin I

Compositae, is an analog of presqualene and prephytoene. The C-skeletons shown in Fig. 100 may be derived from chrysanthemyl diphosphate. Chrysanthemum carboxylic acid is a constituent of the pyrethrins, compounds with marked insecticidal activity (E 5.5.3).

Transformation

A large number of products are synthesized from the prenyl diphosphates, squalene, and phytoene. They are treated with in sections D 6.1–D 6.8.

References for Further Reading

15, 26, 27, 156, 173, 215, 318, 472, 551, 563, 611, 662, 763

D 6.1 Hemiterpenes

Chemistry, Distribution, Biosynthesis

Hemiterpenes have the C-skeleton of isopentane. They are formed from isopentenyl diphosphate and dimethylallyl diphosphate. (Similar compounds may, how-

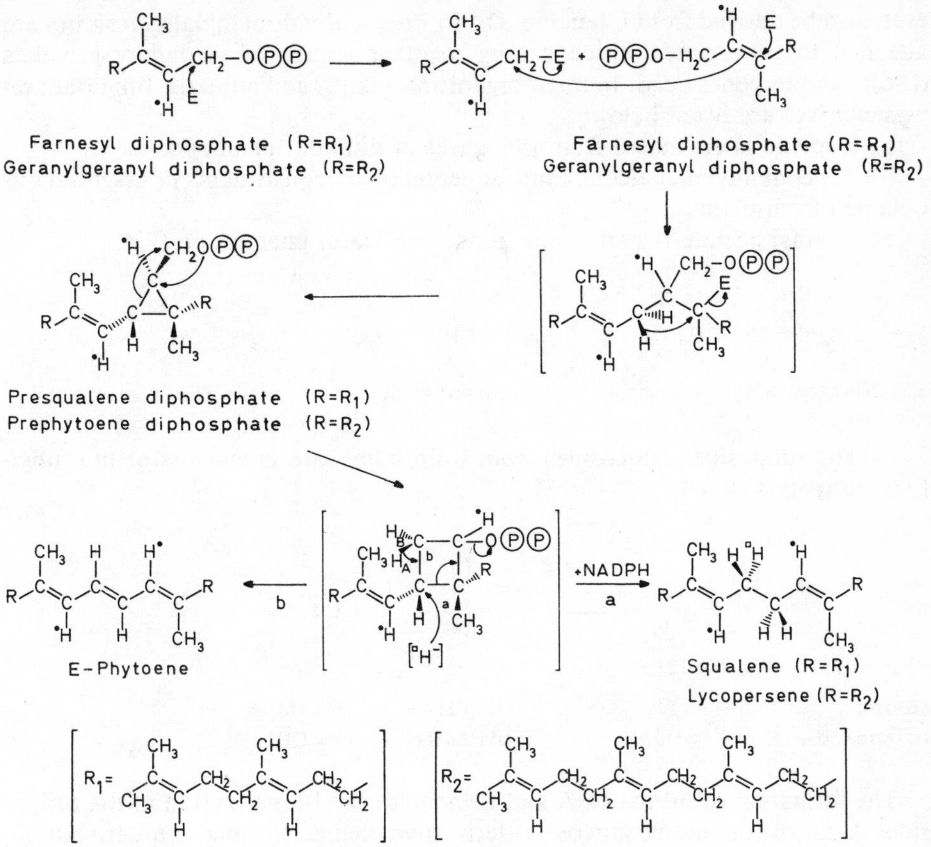

Fig. 99. Tail-to-tail condensation of prenyl diphosphates during the formation of squalene, lycopersene, and phytoene

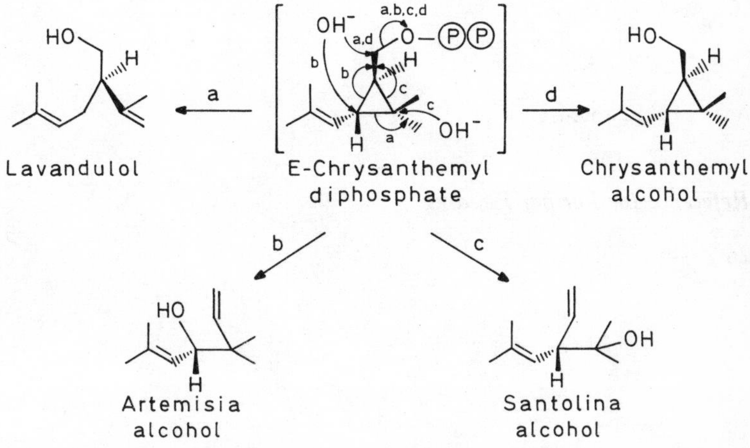

Fig. 100. Possible pathways leading to chrysanthemyl derivatives

ever, also be derived from L-leucine, D 14.) Frequently dimethylallyl residues are attached to compounds which belong to other groups of secondary products (D 6). Hemiterpenes occur in microorganisms, plants, and animals. Important representatives are listed below:

a) Isoprene is emanated from the leaves of different plant species.

b) 3,3-Dimethylallyl alcohol and isopentenol are constituents of essential oils obtained from plants.

c) Isoamyl acetate is part of the honey bee alarm pheromone (E 4).

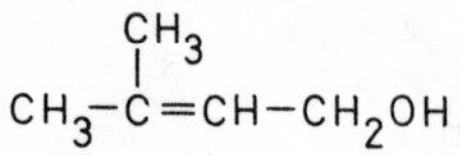

$$CH_3-\underset{\underset{CH_3}{|}}{C}=CH-CH_2OH \qquad CH_2=\underset{\underset{CH_3}{|}}{C}-CH_2-CH_2OH \qquad CH_3-\underset{\underset{CH_3}{|}}{CH}-CH_2-CH_2OH$$

3,3-Dimethylallyl alcohol 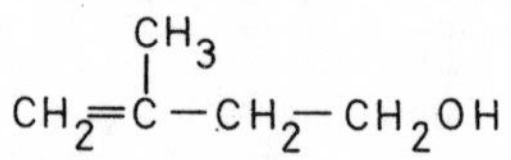   Isopentenol 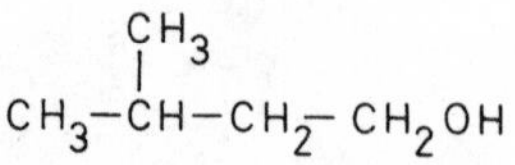   Isoamyl alcohol

d) The tuliposides, glucosides from tulip bulbs, are precursors of the fungitoxic tulipalins (E 5.4):

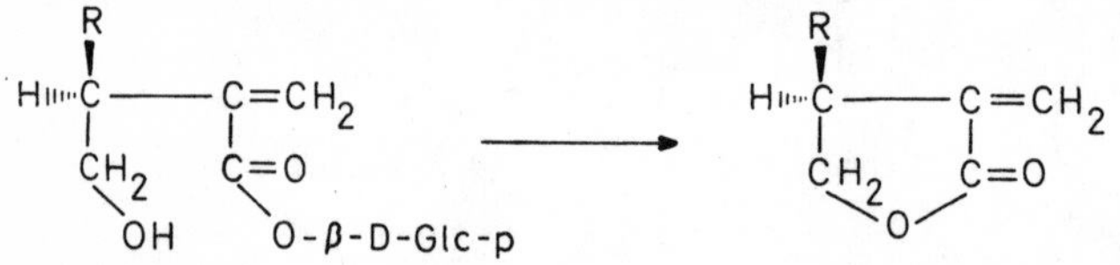

Tuliposide A (R=H)   Tulipalin A (R=H)
Tuliposide B (R=OH)   Tulipalin B (R=OH)

The formation of tulipalins A and B includes the isomerization of the tuliposides A and B (the native glucose-1-derivatives change to glucose-6-derivatives), hydrolysis of the glucose-6-derivatives and cyclization of the aglycones.

e) Ageratochromene is a compound with antijuvenile hormone activity in insects (E 5.5.3). It is a constituent of the plant Ageratum houstoniatum.

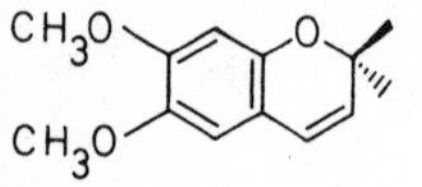

Ageratochromene

Reference for Further Reading

26

D 6.2 Monoterpenes

D 6.2.1 Aliphatic and Cyclohexanoid Monoterpenes

Chemistry and Distribution

Monoterpenes are aliphatic or cyclic isoprenoid C_{10}-compounds. They usually contain one or two isolated double bonds, but may also be saturated or aromatic. Often hydroxy, oxo, aldehyde, carboxylic acid, or ester groups are present (Fig. 101). Most compounds are volatile and lipohilic.

Monoterpenes are typical products of higher plants. They are synthesized, however, also in animals and in a few microorganisms.

Biosynthesis

The aliphatic or cyclohexanoid monoterpenes originate from geranyl diphosphate (D 6), linalyl, or neryl diphosphate. Of significance are the following types of reactions:

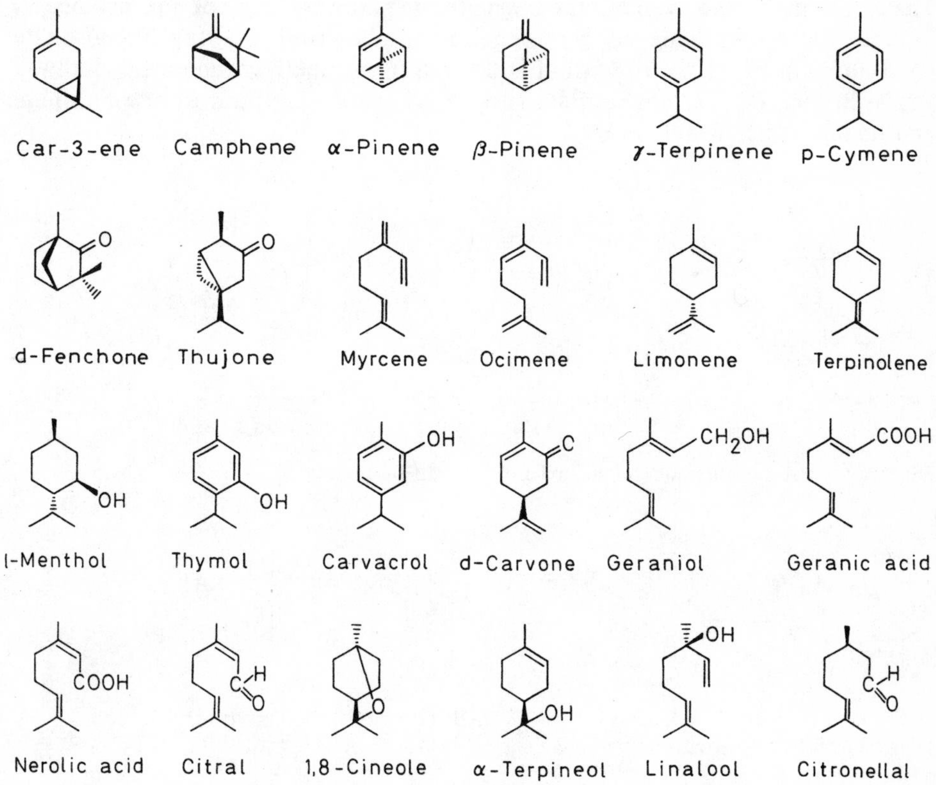

| Car-3-ene | Camphene | α-Pinene | β-Pinene | γ-Terpinene | p-Cymene |

| d-Fenchone | Thujone | Myrcene | Ocimene | Limonene | Terpinolene |

| l-Menthol | Thymol | Carvacrol | d-Carvone | Geraniol | Geranic acid |

| Nerolic acid | Citral | 1,8-Cineole | α-Terpineol | Linalool | Citronellal |

Fig. 101. Structural formulas of monoterpenes

a) *Hydrolytic cleavage of the diphosphate group:*

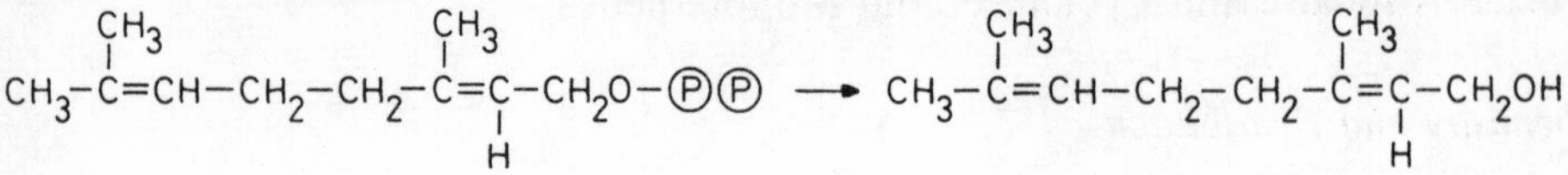

$$CH_3-\overset{\overset{\displaystyle CH_3}{|}}{C}=CH-CH_2-CH_2-\overset{\overset{\displaystyle CH_3}{|}}{C}=\underset{\underset{\displaystyle H}{|}}{C}-CH_2O-\textcircled{P}\textcircled{P} \longrightarrow CH_3-\overset{\overset{\displaystyle CH_3}{|}}{C}=CH-CH_2-CH_2-\overset{\overset{\displaystyle CH_3}{|}}{C}=\underset{\underset{\displaystyle H}{|}}{C}-CH_2OH$$

       Geranyl diphosphate   Geraniol

b) *E-Z-Isomerization*

An example is the transformation of geranyl and neryl diphosphate into each other, probably with linalyl diphosphate as intermediate (Fig. 102).

c) *Cyclization*

Cyclic monoterpenes are built from diphosphates of aliphatic monoterpenes, e. g., linalyl diphosphate, geranyl diphosphate, or neryl diphosphate by polyfunctional enzyme proteins, the so-called carbocyclases. No free diffusable intermediates occur.

d) *Hydroxylations, dehydrogenations, formation of carboxylic groups, ester groups, etc.*

The reactions are late steps in the biosynthetic pathway. Thymol, for instance, is formed from γ-terpinene via p-cymene, d-camphor from d-bornyl diphosphate via d-borneol (Fig. 103). l-Menthol is derived from piperitenone on the pathway shown in Fig. 104. Terpinyl acetate and linalyl acetate are built from α-terpineol and linalool with acetyl CoA.

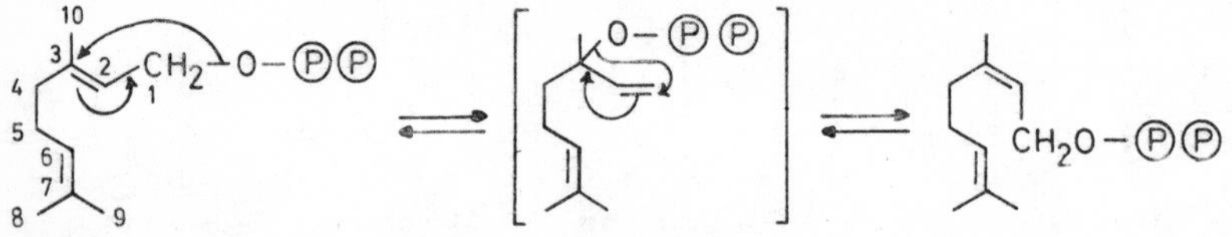

Geranyl   Linaloyl   Neryl
diphosphate   diphosphate   diphosphate

Fig. 102. Mutual transformation of geranyl and neryl diphosphates

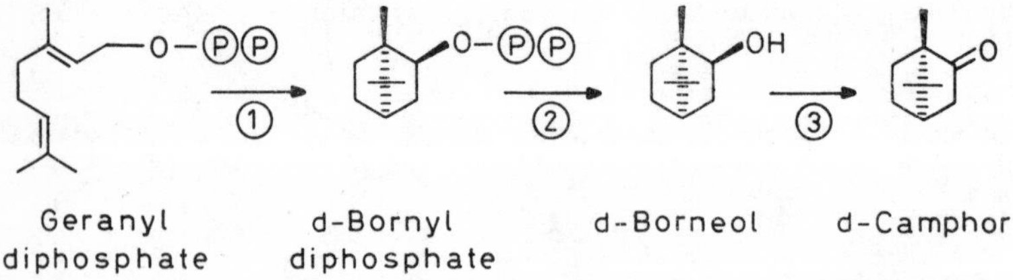

   Geranyl   d-Bornyl   d-Borneol   d-Camphor
diphosphate   diphosphate

Fig. 103. Formation of d-camphor
1 Bornyl diphosphate synthase; *2* phosphatase; *3* borneol dehydrogenase

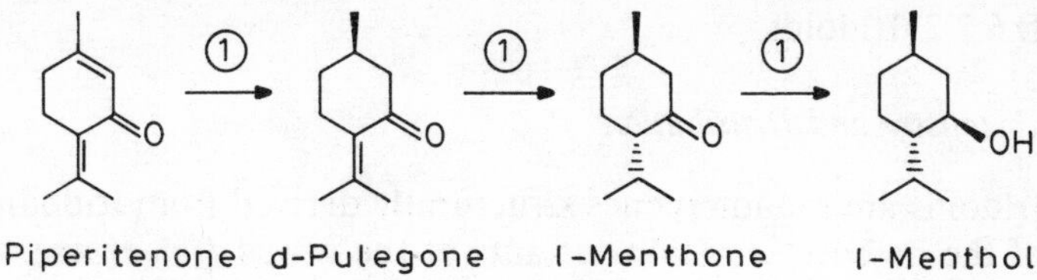

Piperitenone   d-Pulegone   l-Menthone   l-Menthol

Fig. 104. Transformation of piperitenone to l-menthol
1 Pulegone dehydrogenase; *2* menthone dehydrogenase; *3* menthol dehydrogenase

Degradation

In higher plants monoterpenes undergo a continuous turnover. As yet, however, monoterpene degradation is not well understood.

Significance

Monoterpenes are important constituents of essential oils. These complex mixtures of compounds, as well as individual monoterpenes are used in medicine (F 2) and industry (F 5). Commercially important monoterpenes are l-menthol, d-camphor, d-carvone, thymol, d-fenchone, and α-pinene. Monoterpenes flavor foodstuffs and beverages (F 1). Citral is an important constituent of the smell of lemon. Thymol is involved in the flavor of mandarin oranges. Other monoterpenes, e.g., limonene and geraniol, are constituents of flower scents and attract plant pollinators (E 5.5.1). High monoterpene concentrations in plants will repel most potential predators, but may attract some animals (E 5.5.3). Animals use citronellal, citral, as well as α- and β-pinene as feeding deterrents (E 5.1), and geraniol, geranyl esters, myrcene, and terpinolene as pheromones (E 4). Some monoterpenes, e.g., 1,8-cineole and d-camphor, are also involved in plant-plant interactions (E 5.3).

Transformation

The cannabinoids, e.g., $(-)-\Delta^1$-tetrahydrocannabinol, the major psychoactive constituent of *Cannabis sativa* (F 3), are in part of monoterpene origin (*C*-atoms 1–10). It is generally accepted that the rest of the molecule is derived from a polyketide (D 5.3). The detailed biosynthesis is, however, still unknown.

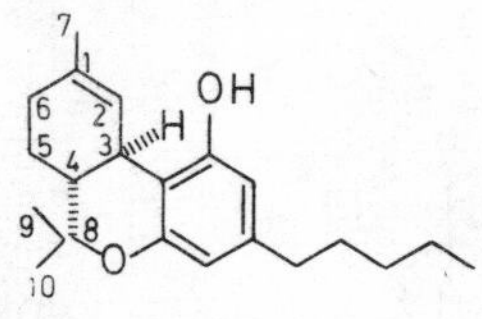

$(-)-\Delta^1$-Tetrahydrocannabinol

References for Further Reading

26, 158, 167, 189, 216, 284, 472, 551, 563, 611, 693, 856, 878

D 6.2.2 Iridoids

Chemistry and Distribution

Iridoids are monoterpenes structurally derived from iridodial, a constituent, e.g., of the defense secretions of ants of the genus *Iridomyrmex* (Fig. 105). They may be divided into two groups: the compounds of the loganin type (Fig. 105) and so-called secologanins. In the latter the ring system of loganin is cleaved between positions 7 and 8 (Fig. 106). One or two of the ten C-Atoms originally present in the iridoid basic skeleton may be lost during further transformation (cf. the formulas of aucubin, catalpol, and catalposide). Most iridoids occurring in plants are glucosides derived from the O-heterocyclic form of iridodial. Some may contain a heterocyclic nitrogen atom (iridoid alkaloids). Iridoids are formed de novo in plants and some animals. The iridoids found in larvae of the butterfly *Euphydryas cynthia* have been ingested with the diet.

Biosynthesis

Iridoids are formed from geranyl diphosphate (Fig. 106). 10-Hydroxygeraniol and/or 10-hydroxynerol are important intermediates. These compounds are transformed to loganin either via 10-oxogeranial/10-oxoneral (route a, common

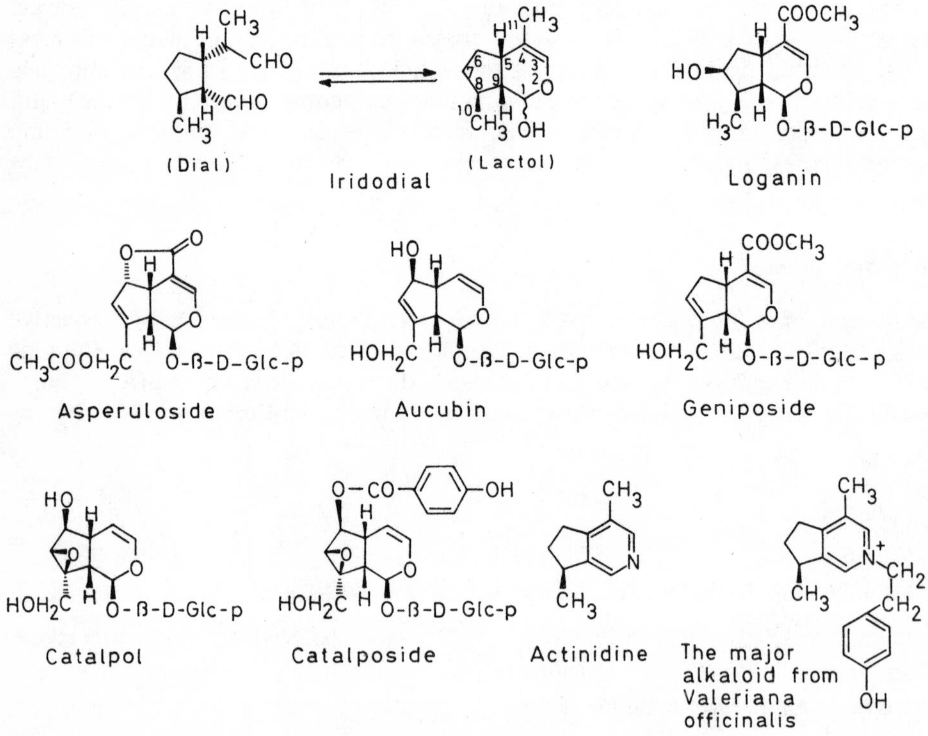

Fig. 105. Naturally occurring iridoids of the loganin type

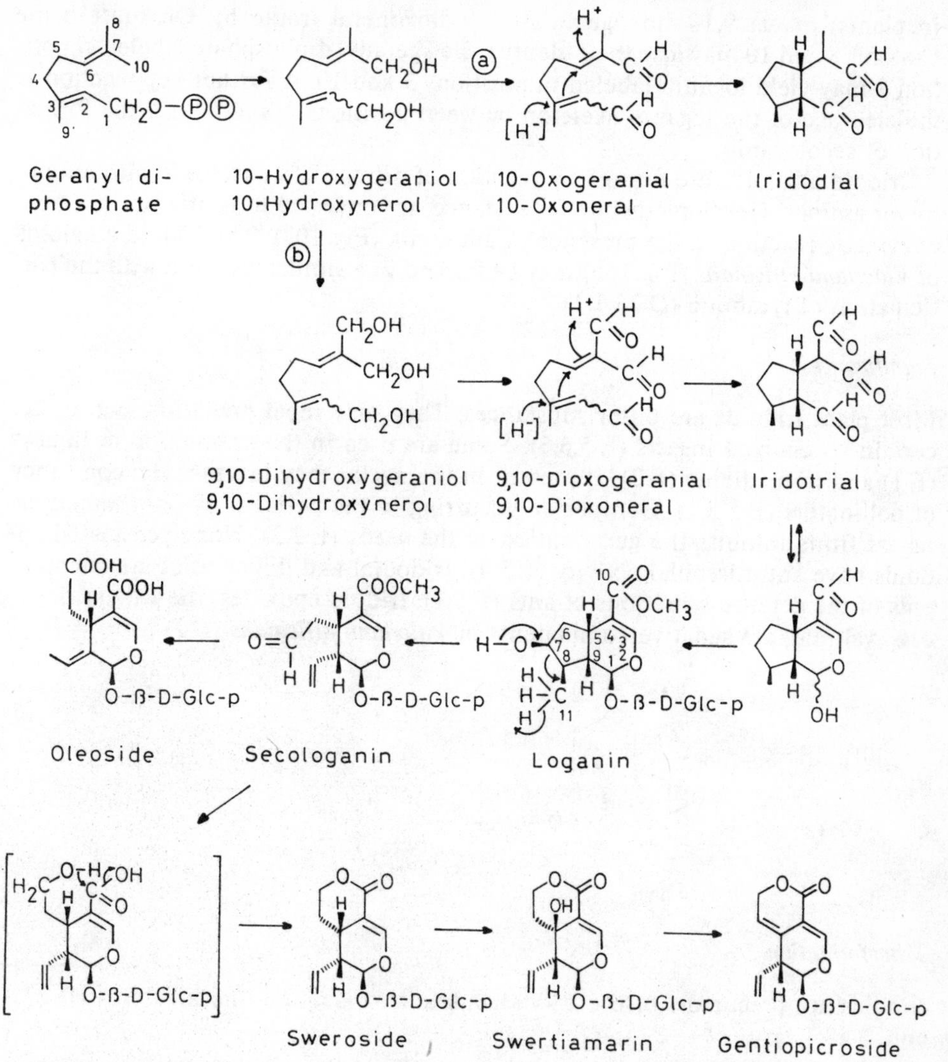

Fig. 106. Formation of loganin- and secologanin-type iridoids from geranyl diphosphate

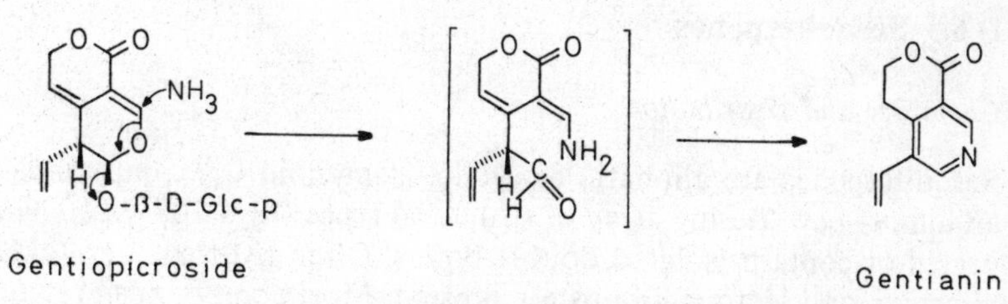

Fig. 107. Conversion of gentiopicroside to gentianin

in plants) or via 9,10-dioxogeranial/9,10-dioxoneral (route b). On route b the C-atoms 8 and 10 may lose their identity, i. e., geranyl diphosphate labeled at position 8 may yield loganin labeled in positions 3 and 10. A further key reaction is the cleavage of the loganin skeleton between positions 7 and 8 with the formation of secologanin.

Iridoid alkaloids are formed by reaction of nitrogen-free iridoids with ammonia or amines. Gentiopicroside, for instance, is converted to gentianin in a non-enzymatic reaction in the presence of ammonia (Fig. 107). The iridoid alkaloids of *Valeriana officinalis* (Fig. 105) may be formed in a similar reaction with the participation of tyramine (D 23.1.1).

Significance

Most plant iridoids are bitter substances. They may repel predators, but attract certain specialized insects (E 5.5.3). Some are used in the production of liquors (F 1) and in medicine (F 2). If present in the nectar they increase the constancy of pollinators (E 5.5.1). Geniposide occurring in the pericarp of *Gardenia jasmonoides* fruits inhibits the germination of the seeds (E 2.2). Nonglycosylated iridoids have antimicrobial activity (E 5.4). Iridodial and dolichodial are constituents of the defense secretions of ants (E 5.1). Iridoid epoxides, the valepotriates, e. g., valtrate, are sedative components of *Valeriana officinalis* (F 2).

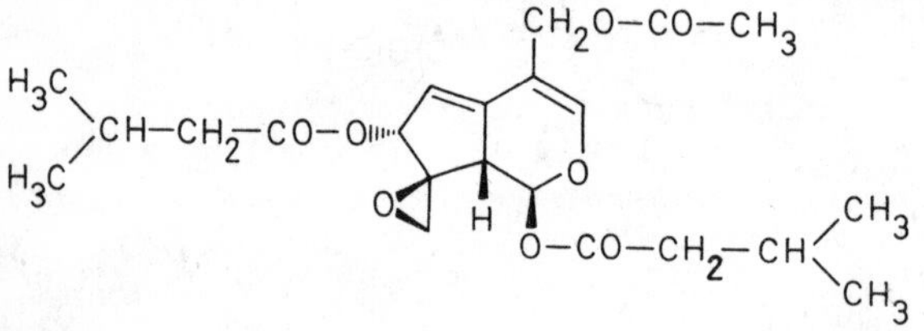

Valtrate

Transformation

Iridoids are precursors of the iridoid indole and isoquinoline alkaloids (D 22.3 and D 23.1.2).

References for Further Reading

26, 296, 382, 551, 648

D 6.3 Sesquiterpenes

Chemistry and Distribution

Sesquiterpenes are aliphatic or cyclic, isoprenoid C_{15}-compounds encompassing an almost bewildering array of structural types (Figs. 108–112). They may be saturated or contain isolated double bonds. Often hydroxy, oxo, aldehyde, carboxylic acid, and lactone groups are present. Most compounds are volatile and lipophilic. Sesquiterpenes are frequent plant constituents. They are built, however,

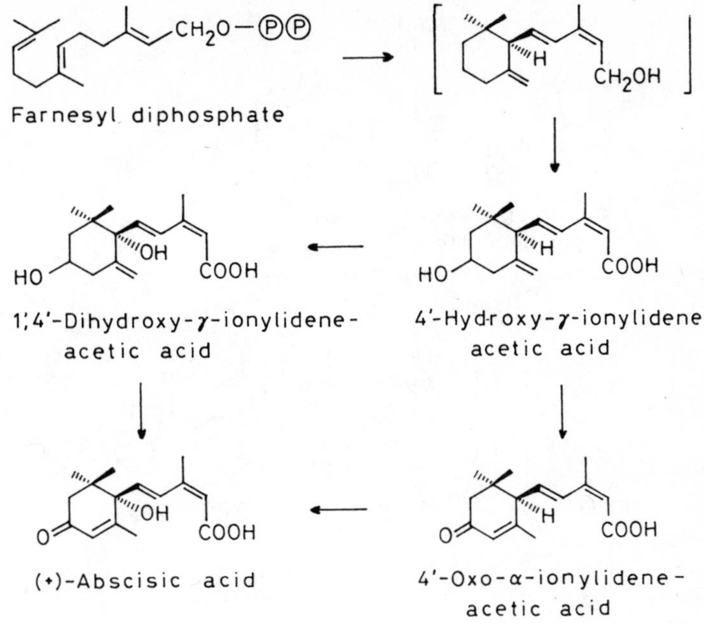

Farnesyl diphosphate

$CH_2O-\text{\textcircled{P}}\text{\textcircled{P}}$

CH_2OH

1,4'-Dihydroxy-γ-ionylidene-
acetic acid

4'-Hydroxy-γ-ionylidene-
acetic acid

(+)-Abscisic acid

4'-Oxo-α-ionylidene-
acetic acid

Fig. 108. Biosynthesis of (+)-abscisic acid

also in animals and microorganisms. Nitrogen-containing sesquiterpenes (sesquiterpenoid alkaloids) as yet have been found in some plant families only.

Biosynthesis

Sesquiterpenes originate from 2-*E*-6-*E*-farnesyl diphosphate (D 6). Of importance are the following types of reactions:

a) *Hydrolytic cleavage of the diphosphate group:*

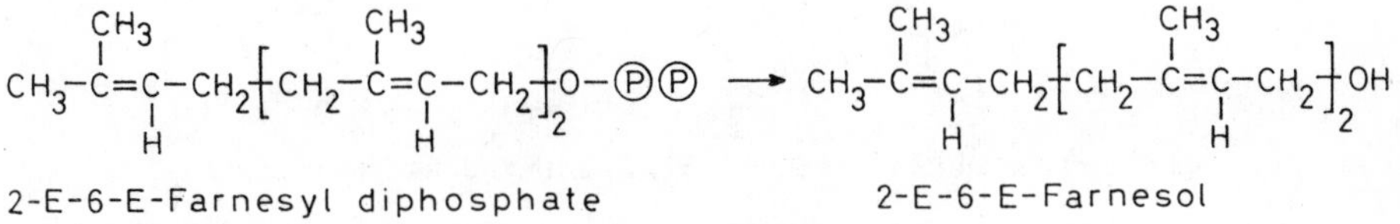

$$CH_3-\underset{H}{C}=C-CH_2\left[CH_2-\underset{H}{\overset{CH_3}{C}}=C-CH_2\right]_2 O-\text{\textcircled{P}}\text{\textcircled{P}} \longrightarrow CH_3-\underset{H}{\overset{CH_3}{C}}=C-CH_2\left[CH_2-\underset{H}{\overset{CH_3}{C}}=C-CH_2\right]_2 OH$$

2-E-6-E-Farnesyl diphosphate

2-E-6-E-Farnesol

b) *E-Z-Isomerization of the double bond at position 2:*

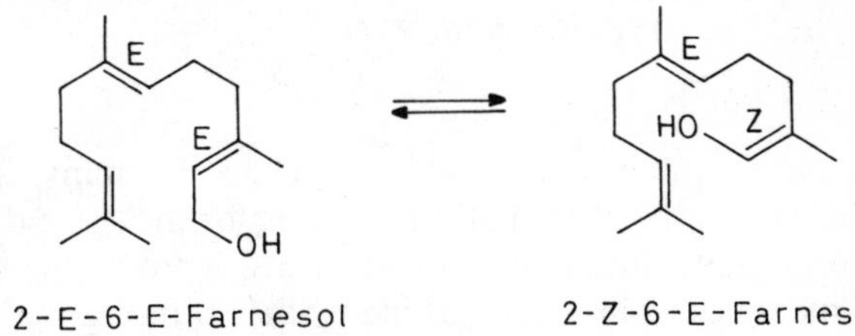

2-E-6-E-Farnesol

2-Z-6-E-Farnesol

13*

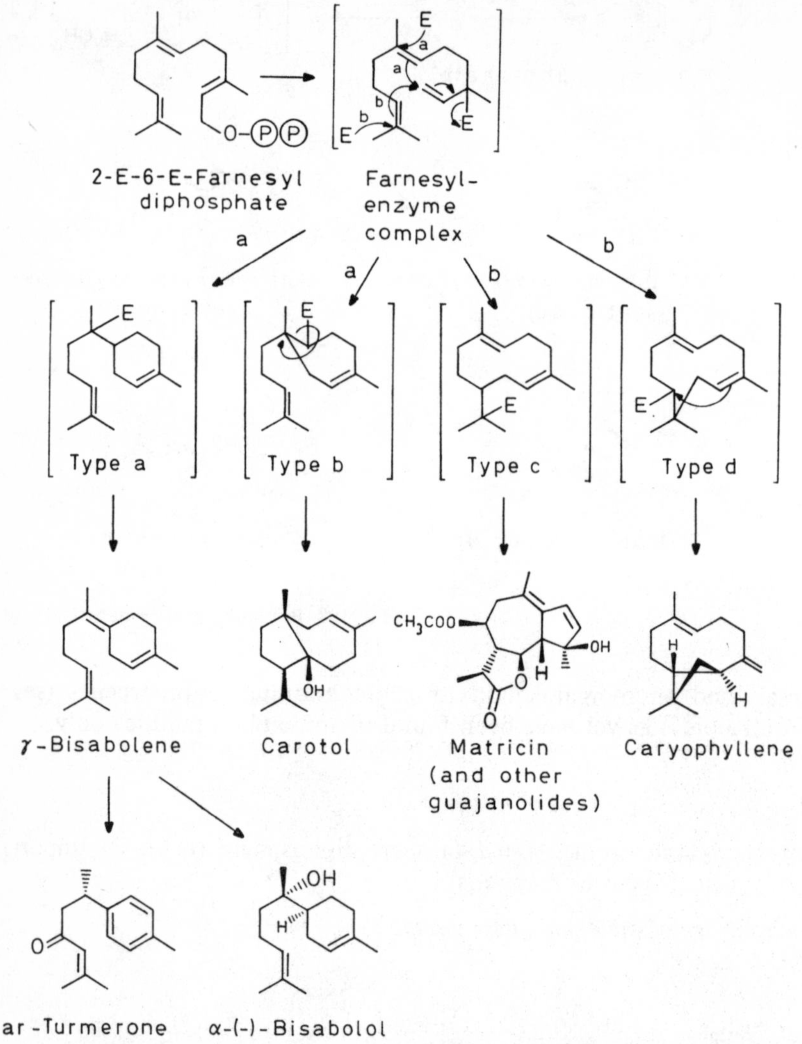

Fig. 109. Synthesis of cyclic sesquiterpenes from 2-*E*-6-*E*-farnesyl diphosphate, part I

Isomerization of farnesyl diphosphate is believed to involve reversible dephosphorylations with the formation of the corresponding alcohols and may proceed via 6-*E*-farnesyl aldehyde.

c) *Cyclization*

Farnesyl diphosphate is able to cyclize in many different ways (Figs. 108–110). This variability is one of the reasons for the great diversity of the chemical structures of sesquiterpenes. The cyclization probably proceeds via enzyme-bound intermediates. It is accompanied by the shift of double bonds, hydride ions, etc.

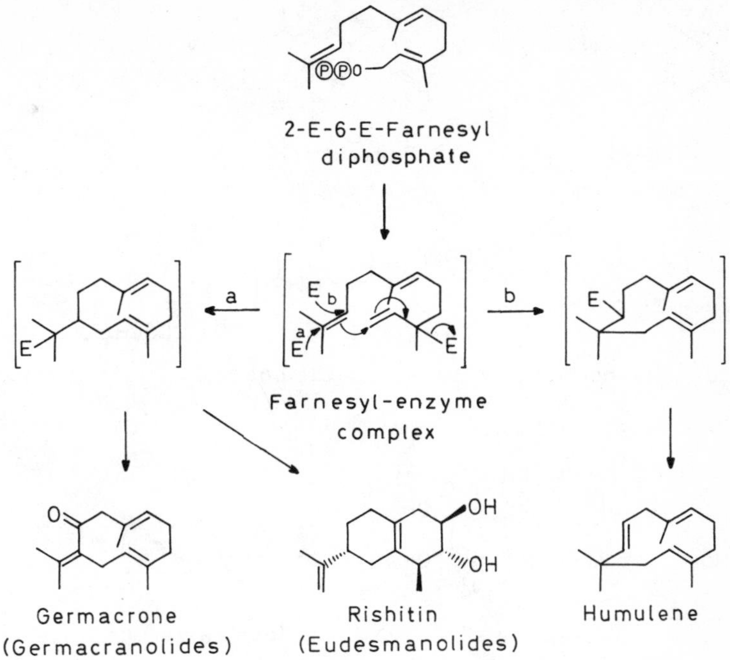

2-E-6-E-Farnesyl
diphosphate

Farnesyl-enzyme
complex

Germacrone   Rishitin   Humulene
(Germacranolides)   (Eudesmanolides)

Fig. 110. Synthesis of cyclic sesquiterpenes from 2-E-6-E-farnesyl diphosphate, part II

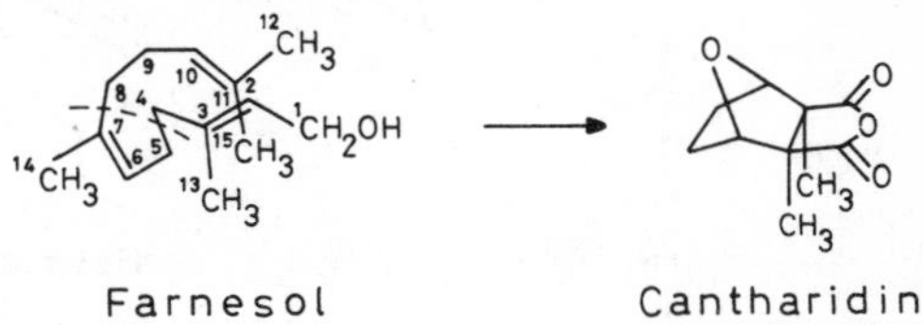

Fig. 111. Formation of cantharidin   Farnesol   Cantharidin

d) *Hydroxylations, dehydrogenations, formation of carboxylic groups, lactone groups, etc.*

These reactions are rather late steps in the biosynthesis pathways. Some typical products are represented in Figs. 109, 110, and 112.

e) *Loss of carbon atoms*

In the biosynthesis of cantharidin, for instance, C-1, C-5 to C-7, and C-14 of farnesol (or farnesyl diphosphate) are split off according to Fig. 111.

Significance

Lipophilic sesquiterpenes, e.g., farnesol, bisabolene, bisabolol, carotol, matricin, and caryophyllene are constituents of essential oils which are used in medicine (F 2) or are of commercial value (F 5). They also dominate the scents of certain flowers which attract pollinators, e.g., β-ionone, bisabolol, and (−)-δ-cadinene (E 5.5.1), and may contribute to the scents and tastes of spices, foodstuffs, and

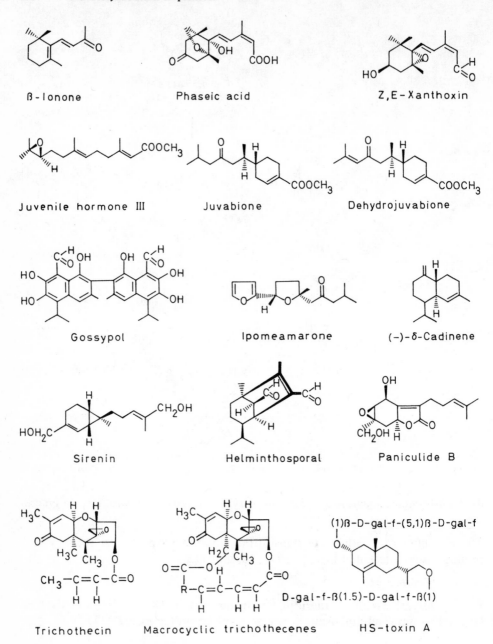

Fig. 112. Formulas of several sesquiterpenes

beverages. ar-Turmerone, for instance, is the pungent principle of *Curcuma* (F 1).

In higher plants and insects certain sesquiterpenes, e.g., (+)-abscisic acid, phaseic acid, xanthoxin, and juvenile hormone III act as hormones (E 3.1). Juvabione and dehydrojuvabione, compounds of plant origin, possess also hormone activity in insects. They are important as insecticides (E 5.5.3). Gossypol, a constituent of cotton seed oil, causes transient sterility of male human beings.

Some sesquiterpenes, e.g., rishitin and ipomeamarone, are phytoalexins formed in higher plants in response to microbial infections (E 5.4). Several sesquiterpenes are defense products. Sesquiterpene lactones of plant origin, e.g., matricin, have a bitter taste and may repel potential predators (E 5.5.3). Cantharidin is produced in beetles as a feeding deterrent (E 5.1).

Several sesquiterpenes act as pheromones (E 4). Sirenin is the sex pheromone of the water mold *Allomyces*, mediating the attraction of the male gametes by the female ones. 2-*E*-6-*E*-Farnesyl esters are sex pheromones of certain bees (E 4). 2-*E*-6-*E*-Farnesol regulates the contraction of stomata and in this way the respiration of plants (E 2.2).

Some sesquiterpenes are fungal toxins. Helminthosporal and HS-toxin A, cause plant diseases (E 5.4). The trichothecenes formed, e.g., in *Fusarium, Trichotecium*, and *Cephalosporium* sp. inhibit protein synthesis in most organisms including animals and man (F 4). They may be divided into nonmacrocyclic compounds, e.g., trichothecin, and macrocyclic derivatives. The macrocyclic trichothecanes found in the plant *Baccharis megapotamica* from Brazil (baccharinols) are probably taken up from microorganisms living in the soil and are transformed in the plant.

References for Further Reading

26, 229, 247, 273, 472, 551, 563, 596, 611, 648, 698, 776, 802

D 6.4 Diterpenes

Chemistry and Distribution

Diterpenes comprise four isoprene residues. They are either aliphatic substances or in most cases tri- or tetracyclic compounds, which may be substituted by different functional groups.

Diterpenes formed from geranylgeranyl diphosphate occur in microorganisms, plants, and animals. Diterpenes derived from tetraterpenes (D 6.7) are built in animals and plants. Nitrogen-containing diterpenes (diterpene alkaloids, cf. the formula of aconitine) have been found in higher plants.

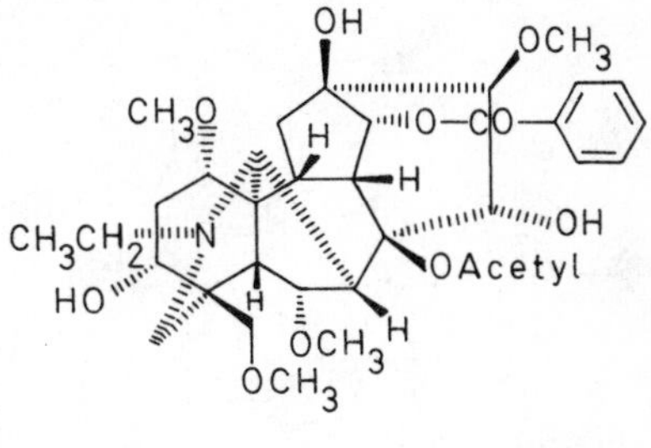

Aconitine

Biosynthesis from Geranylgeranyl Diphosphate

The key product in the formation of most diterpenes is geranylgeranyl diphosphate. Important reactions in its transformation are hydrolysis of the diphosphate group, reduction of double bonds, and cyclization.

Of significance is the transformation of geranylgeraniol to phytol during the biosynthesis of chlorophylls (D 24.1). Phytol is the precursor of phytane and pristane. Both compounds are widespread constituents of lake and river sediments and have been detected even in Precambrian fossils. They have the stereochemistry of phytol in recent sediments, but are more or less epimerized in older sediment layers.

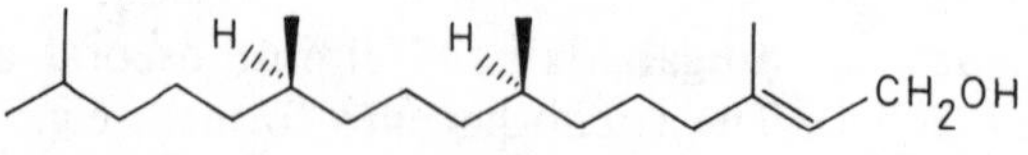

Phytol

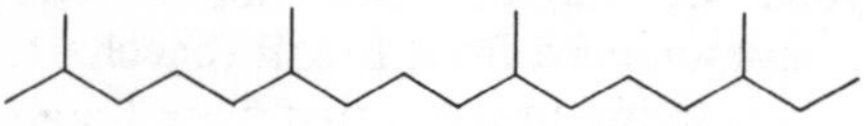

Phytane

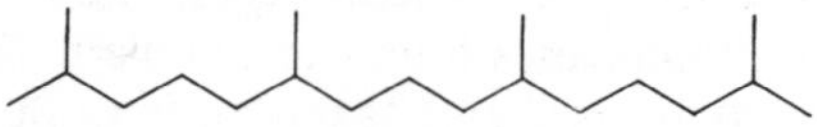

Pristane

Cyclization of geranylgeranyl diphosphate may proceed in different ways:
— In the biosynthesis of cembrene A the terminal double bond is joined to the tail of the molecule (Fig. 113). Cembrene derivatives occur in plants and animals.
— In the formation of labdanes cyclization is initiated by the addition of a proton to the terminal double bond (cf. the cyclization of squalene, D 6.6). Depending on the cyclase involved (+)- or (−)-labdadienyl diphosphates are formed.

(+)-Labdadienyl diphosphate is the precursor of the compounds shown in Fig. 114. (−)-Labdadienyl diphosphate (copalyl diphosphate) is the key product in gibberellin (GA) biosynthesis (Fig. 115). It cyclizes to ent-kaurene which is transformed to ent-7α-hydroxykaurenoic acid. The latter compound is converted to GA_{12}-aldehyde by ring contraction. GA_{12} aldehyde is the mother substance of many other GAs. Most important in nature is GA_1.

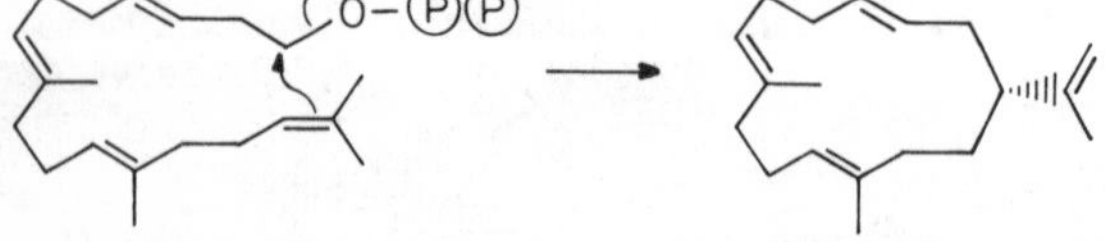

Geranylgeranyl
diphosphate

Cembrene A

Fig. 113. Biosynthesis of cembrene A

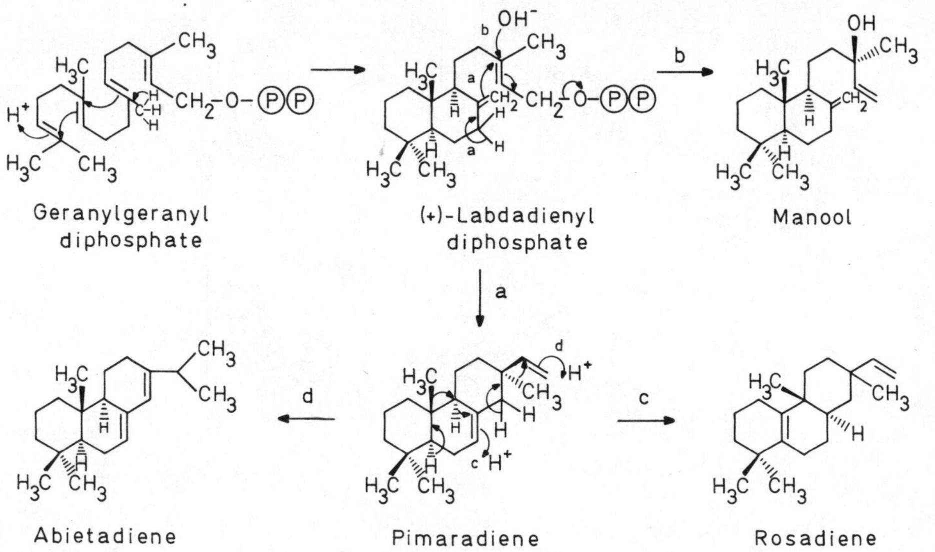

Fig. 114. Synthesis of cyclic diterpenoids from (+)-labdadienyl diphosphate

Significance

Diterpenes are frequent constituents of resins. Abietic acid (Fig. 116) is the main constituent of rosin (colophony, F 5). Larvae of certain insects, which feed on Pinaceae, ingest and sequester the lipophilic constituents, e.g., abietic acid, of these plants and use them to repel predators (E 5.1). Carnosolic acid causes the characteristic bitter taste of Labiatae. Geranylgeraniol acetate and cembrene A are constituents of the pheromones of bees, ants, and termites, respectively (E 4). The grayanotoxins, e.g., grayanotoxin I, are toxic compounds widespread in Ericaceae. They increase the constancy in pollination (E 5.5.1).

Gibberellins are active as hormones in higher plants (E 3.1). In the fungus *Fusarium moniliforme (Gibberella fujikuroi)* they are built as phytotoxins (E 5.4). The antheridiogens possess hormone activity in ferns. They are structurally related to the gibberellins (E 3.1).

Transformation

Diphytanyl glycerol ether and di-biphytanyl diglycerol tetraethers (Fig. 117) are important constituents of the membranes of archaebacteria. In the biphytanyl residues the phytane moieties are interlinkes "head-to-head", i.e., at positions 16 and 16'. The biphytanyl residues may contain one to four cyclopentane rings formed by desaturation.

References for Further Reading

26, 180, 188, 279, 280, 320, 563, 594, 611, 648

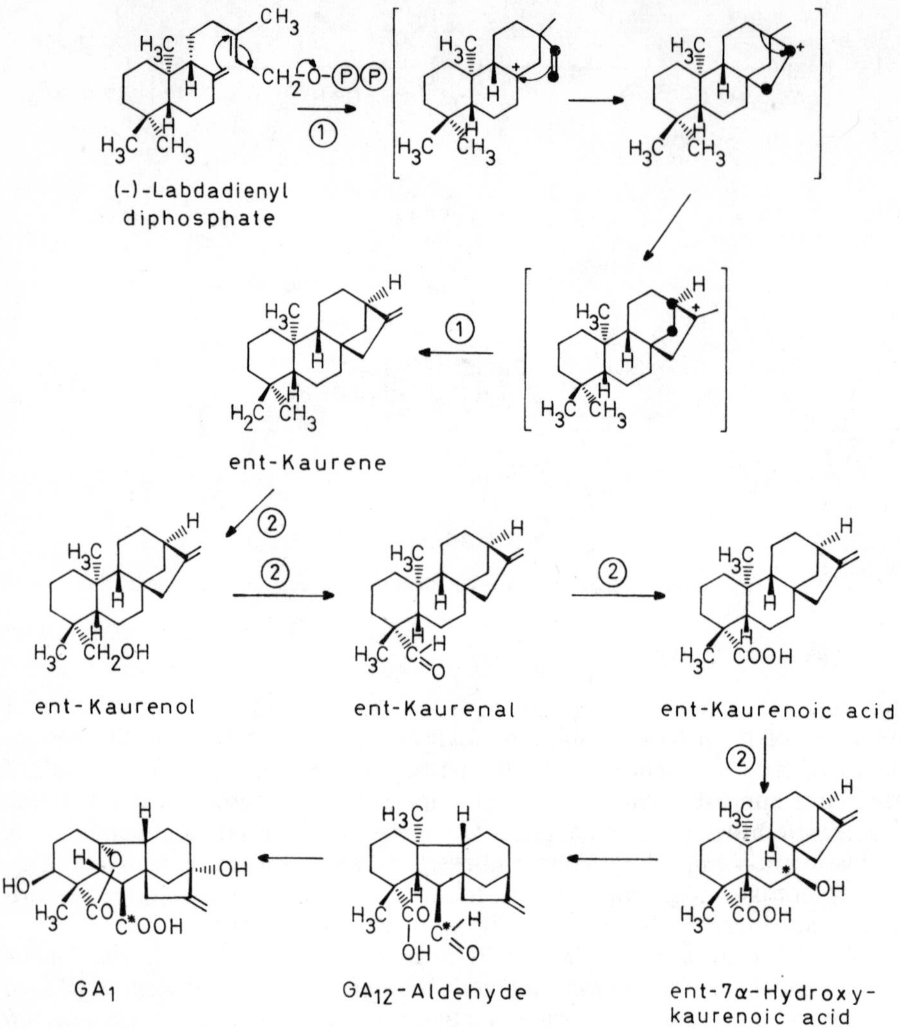

Fig. 115. Formation of gibberellins from (−)-labdadienyl diphosphate
1 Kaurene synthase; *2* monooxygenases (C 2.6)

D 6.5 Sesterterpenes

Chemistry, Distribution, Biosynthesis

Sesterterpenes probably arise from an all-*E*-pentaprenyl diphosphate (all-*E*-geranylfarnesyl diphosphate). Secondary products belonging to this group are relatively rare. Examples are the ophiobolins formed in fungi, higher plants, and animals, and compounds of the ircinin and variabilin group isolated from sponges (Fig. 118).

References for Further Reading

165, 319

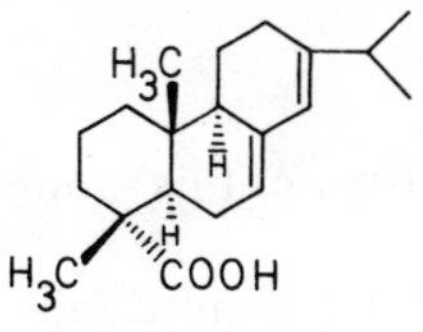

Abietic acid 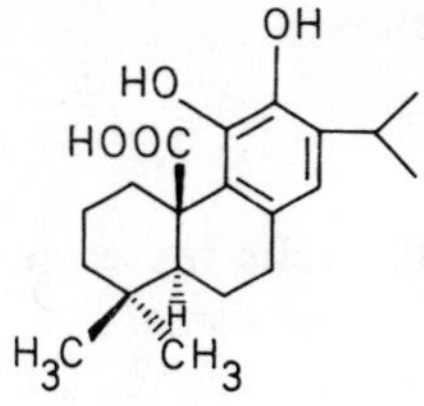   Carnosolic acid 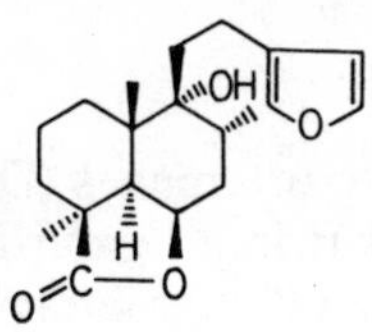   Marrubiin 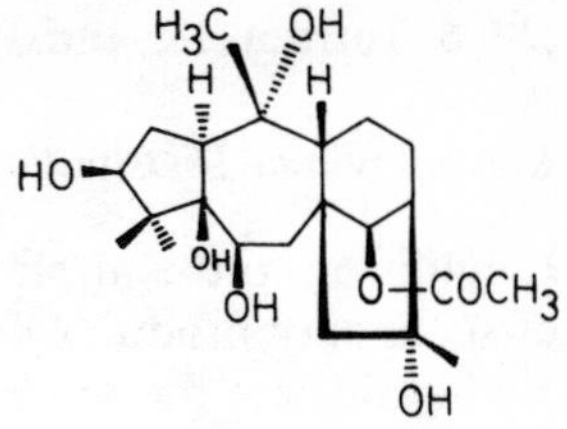   Grayanotoxin I

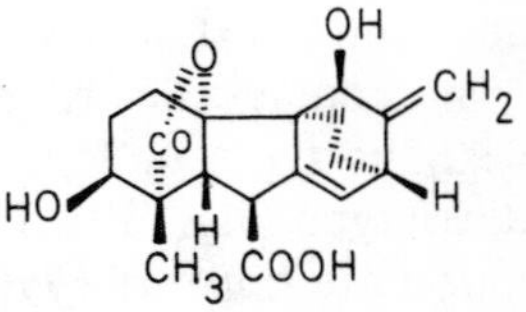

Fig. 116. Structural formulas of some diterpenes

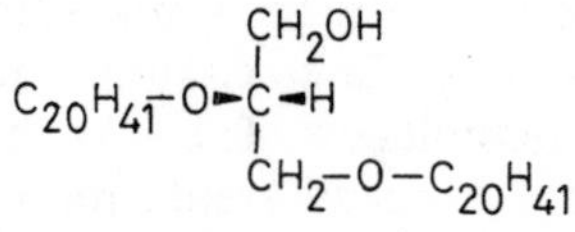

Antheridiogen

$$C_{20}H_{41}-O-\overset{CH_2OH}{\underset{CH_2-O-C_{20}H_{41}}{C-H}}$$

Diphytanyl
glycerol ether

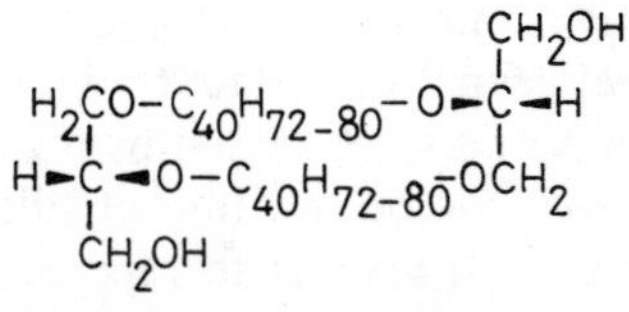

$$H_2CO-C_{40}H_{72-80}-O-\overset{CH_2OH}{\underset{CH_2OH}{C-H}}$$
$$H-\overset{}{C}-O-C_{40}H_{72-80}-OCH_2$$

Di-biphytanyl
diglycerol tetraether

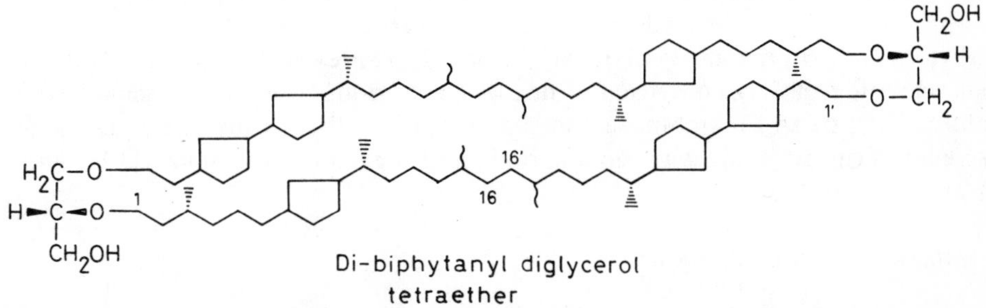

Di-biphytanyl diglycerol
tetraether

Fig. 117. Structural formulas of phytanyl glycerol ethers

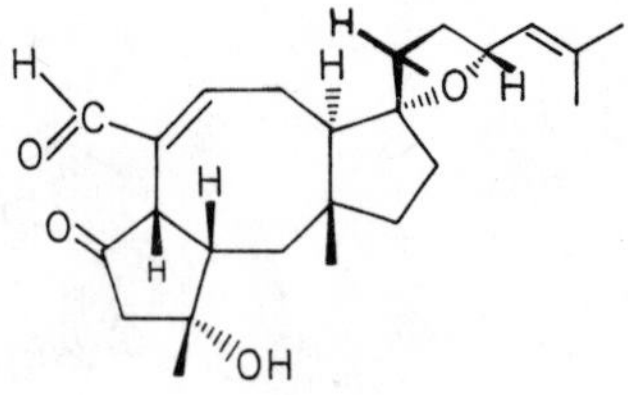

Ophiobolin A

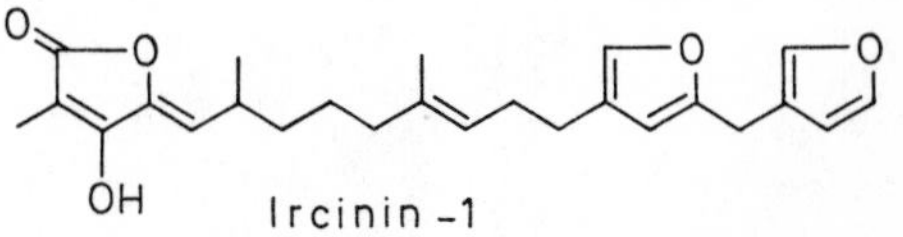

Ircinin -1

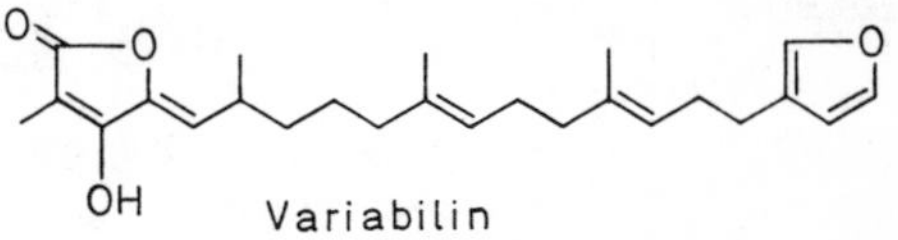

Variabilin

Fig. 118. Structural formulas of sesterterpenes

D 6.6 Tetracyclic and Pentacyclic Triterpenes

Chemistry and Distribution

In addition to the aliphatic triterpenes (D 6), cyclic representatives of this group of secondary products occur in all groups of organisms.

Steroids are the most significant tetracyclic triterpenes. They are derived from cyclopentanoperhydrophenanthrene (sterane, Fig. 119). Characteristic substituents are a hydroxy group at C-3, methyl groups at C-10 and C-13, and a side chain at C-17. Steroids are synthesized in most organisms. Exceptions are insects, crustaceans, mollusks, and protozoans. But even these animals are able to transform ingested steroids.

The pentacyclic triterpenes either comprise five anellated cyclohexane rings (oleanane type) or four cyclohexane rings connected with a cyclopentane ring (lupeol type, Fig. 119). These skeletons may carry a hydroxy group at C-3 (an exception are hopanoid-type compounds) and methyl groups at positions 4, 8, 10, 14, 17, and 20 and in the case of lupeol-type substances an isopropyl group at C-19. Pentacyclic triterpenes are widely distributed in higher plants (especially in dicotyledons) but occur also in echinoderms.

Connection between the anellated rings in the steroids and the pentacyclic triterpenes may be trans or cis, i.e., the substituents at the anellated C-atoms are either located at the same side or at different sides of the molecule. Table 37 shows the type of anellation with respect to representatives of the different groups of steroids. Substituents have the β-position if their spatial location relative to the plane of the ring system is the same as that of the methyl group at C-10, i.e., in front of the plane in the projection used in this book. They have the α-position in the opposite case, i.e., behind the plane of projection.

Most cyclic triterpenes of plant origin are glycosides containing a mono- or oligosaccharide moiety composed of hexoses, e.g., D-glucose and D-galactose, of pentoses, e.g., D- and L-arabinose and D-xylose, of methylpentoses, e.g., L-rhamnose, and of uronic acids, e.g., D-glucuronic and D-galacturonic acids (D 1.1 and D 1.2).

Biosynthesis of Cyclic Triterpenes

All types of cyclic triterpenes originate from squalene. This aliphatic compound may cyclize in different ways:

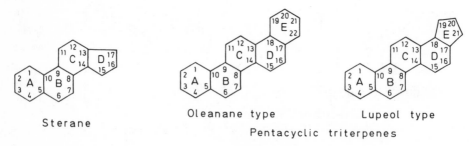

Sterane

Oleanane type

Lupeol type

Pentacyclic triterpenes

Fig. 119. Ring systems of steroids and pentacyclic triterpenes

Table 37. Ring annelation in different steroids

Group	Individual substance	Rings A/B	Substituents in position 5/10	Rings B/C	Substituents in position 8/9	Rings C/D	Substituents in position 13/14
Sterols (D 6.6 and D 6.6.1)	Lanosterol	trans	α/β	–	–	trans	β/α
	Cycloartenol	trans	α/β	cis	β/β	trans	β/α
	Euphol	trans	α/β	–	–	trans	α/β
	Cholesterol	–	–	trans	β/α	trans	β/α
Saponins (D 6.6.2)	Smilagenin	cis	β/β	trans	β/α	trans	β/α
	Tigogenin	trans	α/β	trans	β/α	trans	β/α
	Diosgenin	–	–	trans	β/α	trans	β/α
C_{27}-Steroid alkaloids (D 6.6.3)	α-Tomatine	trans	α/β	trans	β/α	trans	β/α
	Solasodine	–	–	trans	β/α	trans	β/α
Bile alcohols and bile acids (D 6.6.4)	Allocholic acid	trans	α/β	trans	β/α	trans	β/α
	Cholic acid	cis	β/β	trans	β/α	trans	β/α
Pregnanes and allopregnanes (D 6.6.5)	Urocortisol	cis	β/β	trans	β/α	trans	β/α
	Allocortolone	trans	α/β	trans	β/α	trans	β/α
	Progesterone	–	–	trans	β/α	trans	β/α
	Digipurpurogenin	–	–	trans	β/α	cis	β/β
Androstanes (D 6.6.6)	Testosterone	–	–	trans	β/α	trans	β/α
	5α-Androstane-17β-ol-3-one	trans	α/β	trans	β/α	trans	β/α
Estranes (D 6.6.7)	Estradiol	–	–	trans	β/α	trans	β/α
Cardiac glycosides (D 6.6.8)	Digitoxigenin	cis	β/β	trans	β/α	cis	β/β
	Uzarigenin	trans	α/β	trans	β/α	cis	β/β
	Scillarenin	–	–	trans	β/α	cis	β/β

A. Formation of Cyclic 3-Deoxytriterpenes. In some groups of prokaryotes squalene cyclization begins with the enzyme-mediated addition of a proton to one of the terminal double bonds (Fig. 120). In a concerted mechanism a C-21 or C-22 cation is formed, which may be stabilized by addition of a hydroxy ion (formation of tetrahymanol) or release of a proton [formation of 22(29)-hopene, diploptene].

Hopanoids are built in a similar way. The skeleton of these compounds is, however, substituted by a polyhydroxylated side chain. In contrast to the squalene-epoxide-derived cyclic triterpenes (see below), the hopanoid-type compounds and tetrahymanol lack a hydroxy group in position 3.

B. Formation of Tetracyclic and Pentacyclic Triterpenes from Squalene Epoxide. In most microorganisms as well as in plants and animals the cyclic triterpenes are derived from squalene-2,3-epoxide, which is formed from squalene by a monooxygenase (C 2.6.5). The action of different squalene epoxide cyclases results in the formation of different types of cyclic triterpenes.

The cyclization begins with opening of the epoxide ring by addition of a pro-

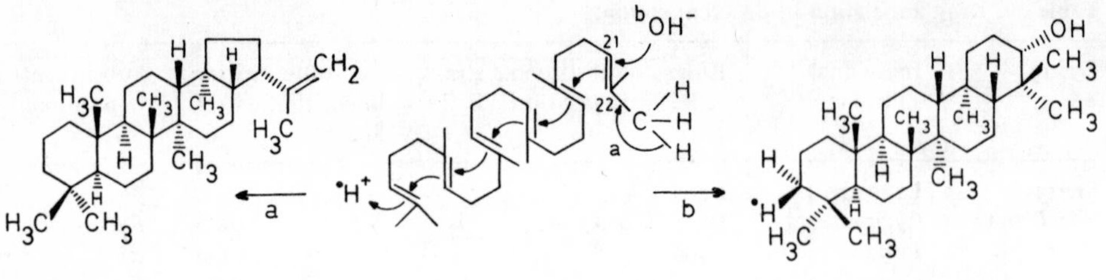

22(29)-Hopene   Squalene   Tetrahymanol

Fig. 120. Formation of hopene and tetrahymanol from squalene

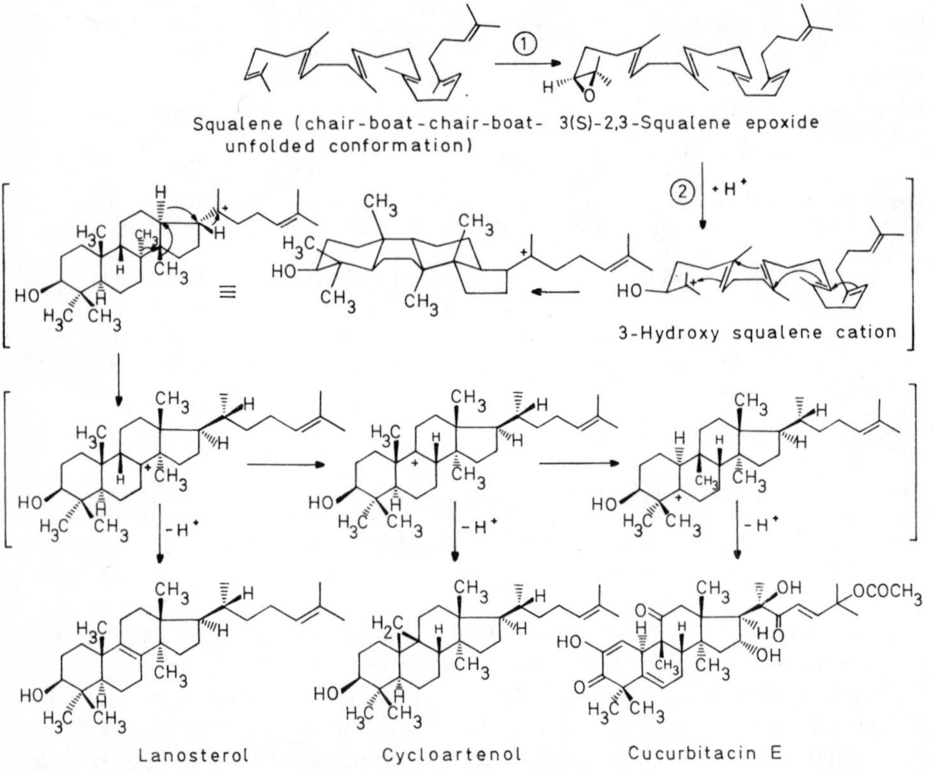

Squalene (chair-boat-chair-boat-   3(S)-2,3-Squalene epoxide
unfolded conformation)

3-Hydroxy squalene cation

Lanosterol   Cycloartenol   Cucurbitacin E

Fig. 121. Biosynthesis of lanosterol, cycloartenol, and cucurbitacin E
1 Squalene monooxygenase (2,3-epoxydizing); *2* 2,3-oxidosqualene cyclase (e.g., 2,3-oxidosqualene:
lanosterol cyclase and 2,3-oxidosqualene: cycloartenol cyclase)

ton (or an electron-accepting group of the enzyme) which causes the formation
of a hydroxy group at position 3. This leaves a positive charge at position 2. Ring
formation proceeds by addition of double bonds to this positively charged
C-atom. The number and configuration of the rings built depends on the folding of
the squalene chain, which is determined by the enzyme. The positive charge that
remains in the molecule after the cyclization is lost by elimination of a proton. It
may, however, migrate within the molecule prior to proton elimination by the

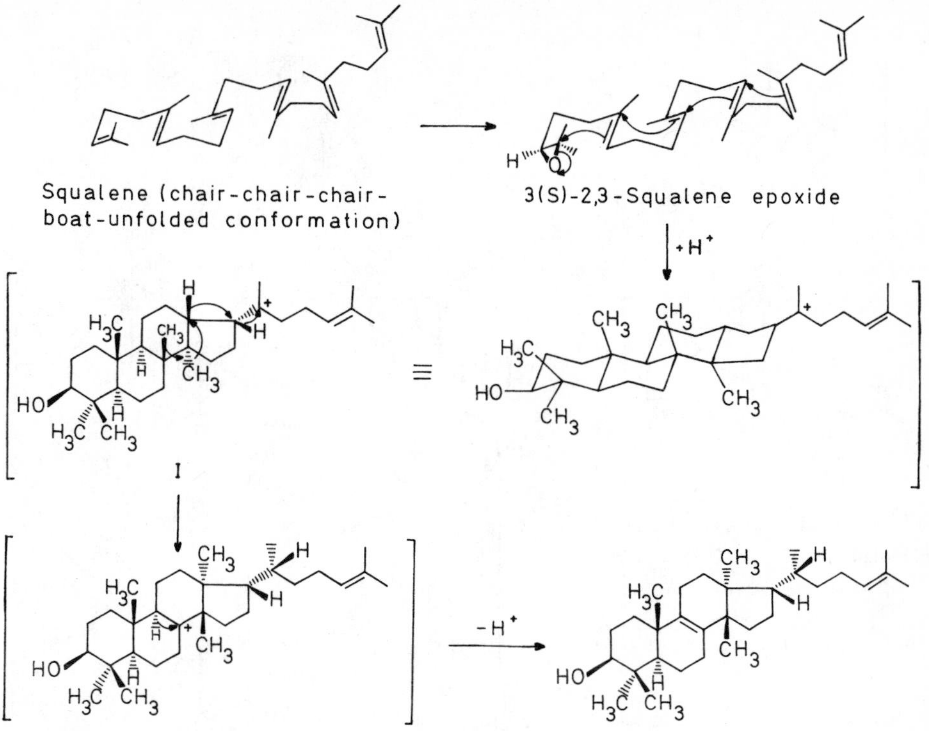

Fig. 122. Formation of euphol

shift of hydride ions and methyl anions while maintaining the spatial orientation of these groups in front of and behind the plane of the ring system (Z-anionotropy). Cyclization, Z-anionotropy, and elimination of the proton proceed in a concerted manner. (Intermediates are given for didactic reasons only).

Formation of steroids of the lanosterol-cycloartenol-cucurbitacin-type starts with squalene epoxide probably in a chair-boat-chair-boat-unfolded conformation (Fig. 121). It depends on the location of the proton that is split off, whether the cyclase forms lanosterol (animals, fungi), cycloartenol (algae, higher plants), or cucurbitacins (Cucurbitaceae, Cruciferae), e.g., cucurbitacin E (α-elaterin).

The steroid euphol as well as the pentacyclic triterpenes may be derived from 3(S)-2,3-squalene epoxide in chair-chair-boat-unfolded conformation (Figs. 122–124).

In pentacyclic triterpenes, ring D of the steroid cation I (Fig. 122) formed as an intermediate, may be enlarged by a Wagner-Meerwein rearrangement (Fig. 123), and an additional five-membered ring may be formed (biosynthesis of lupeol-type compounds). Enlargement of ring E by a further Wagner-Meerwein rearrangement leads to the cation II, from which a large number of different amyrin-type compounds are derived by Z-anionotropy and elimination of a proton (Fig. 124).

Compounds with the structure of α-onocerin (onocol), found in some higher

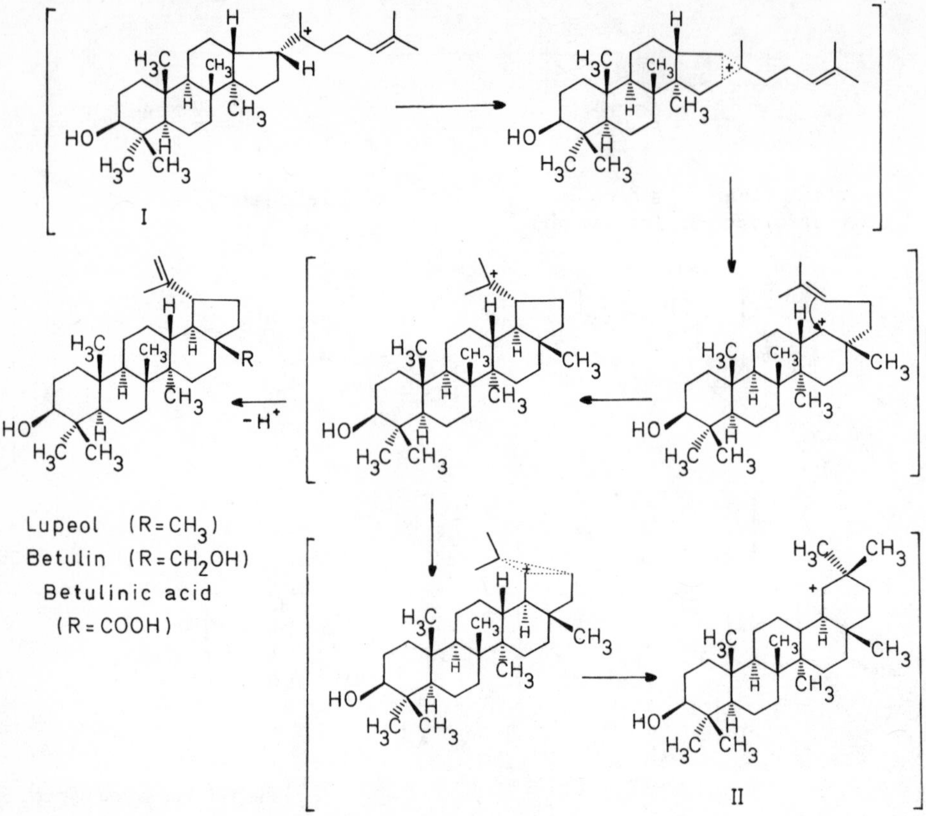

Fig. 123. Formation of pentacyclic triterpenes, part I

plants, e.g., in *Ononis spinosa*, are probably formed from 2,3;27,28-squalene di-epoxide (Fig. 125).

Significance

Hopanoids are membrane stabilizers in certain prokaryotes (E 2.2). Their hydrophilic, polyhydroxylated side chain is directed toward the aqueous face, whereas the lipophilic ring system is integrated in the lipophilic core of the membranes. Lanosterol is used as emulsifier in the pharmaceutical industry. Cucurbitacins, e.g., cucurbitacin E, due to their bitter taste may discourage animal predators, but attract certain beetles, which use Cucurbitaceae as food plants (E 5.5.3).

Transformation

a) Lanosterol in microorganisms and animals, and cycloartenol in plants are precursors of the steroids described in D 6.6.1–D 6.6.9.

b) Glycosides of pentacyclic triterpenes (triterpene saponins) are formed from the genins and activated sugar derivatives (C 6). They may contain one or two su-

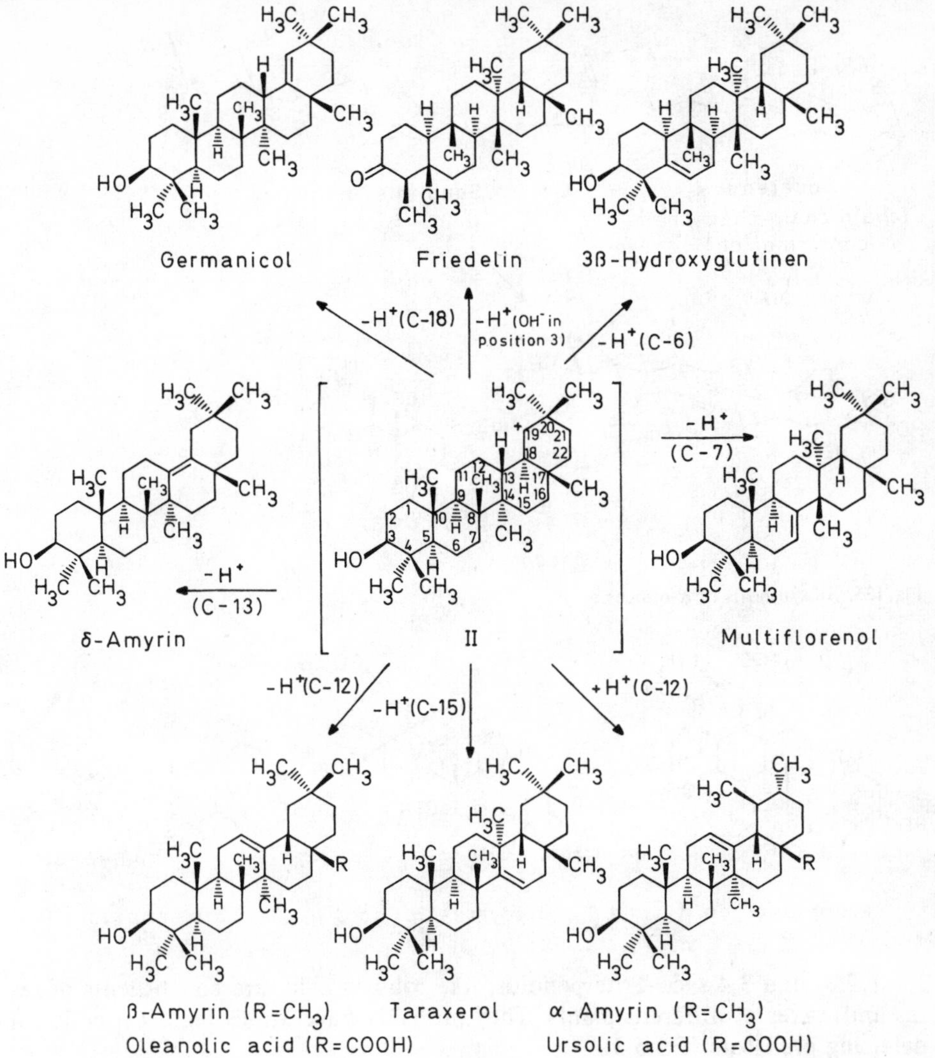

Fig. 124. Formation of pentacyclic triterpenes, part II

gar chains (monodesmosidic and bisdesmosidic triterpene saponins). Up to 11 sugar residues may be bound to one aglycone in straight or branched chains.

Triterpene saponins increase the permeability of membranes and therefore are toxic if ingested. They act as feeding deterrents for animals (E 5.5.3). A few triterpene saponins are active principles of drugs used in medicine (F 2). Of importance in this respect are protoaescigenin glycosides occurring in the seeds of the horse chestnut (*Aesculus hippocastanum*), and glycyrrhetic acid glycosides accumulating in the roots of *Glycyrrhiza glabra* (licorice roots).

c) The dammarane-type saponins found in *Panax ginseng*, e.g., ginsenoside Rg_1, probably act as immune stimulators in animals and human beings.

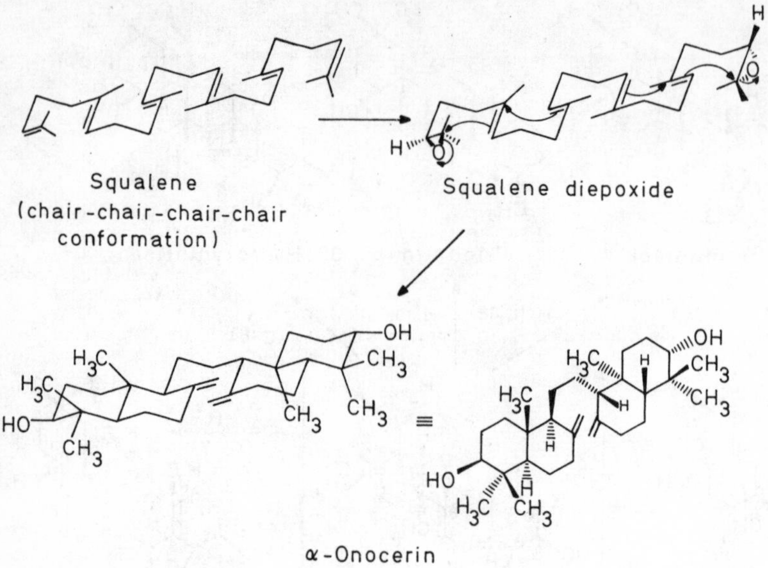

Squalene
(chair-chair-chair-chair
conformation)

Squalene diepoxide

α-Onocerin

Fig. 125. Biosynthesis of α-onocerin

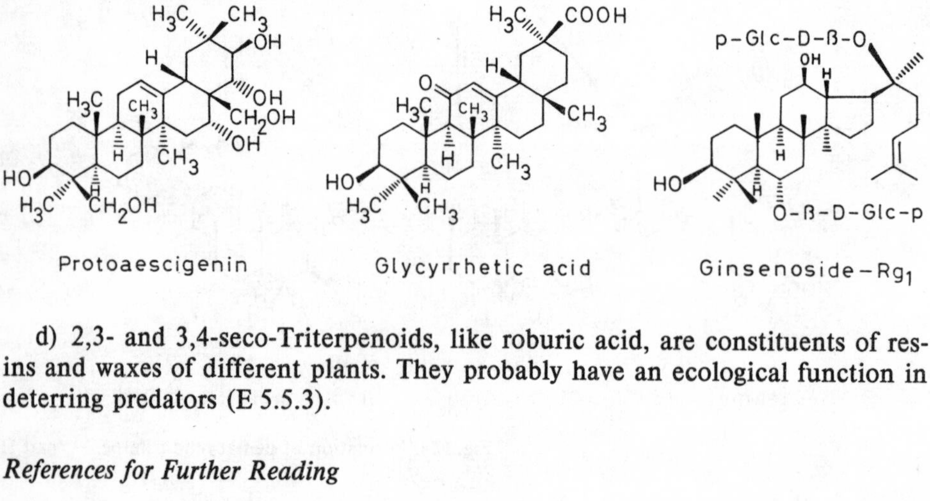

Protoaescigenin

Glycyrrhetic acid

Ginsenoside−Rg₁

d) 2,3- and 3,4-seco-Triterpenoids, like roburic acid, are constituents of resins and waxes of different plants. They probably have an ecological function in deterring predators (E 5.5.3).

References for Further Reading

6, 21, 23, 27, 132, 133, 135, 151, 173, 305, 333, 563, 582, 611, 621, 694, 756, 850

D 6.6.1 Sterols

Chemistry and Distribution

Sterols are steroids carrying an isoprenoid side chain at C-17 (Fig. 126). In plants they may be glycosylated (steryl glycosides) or acylated by fatty acids (steryl esters) at the C-3 hydroxy group. Acylated steryl glycosides, e.g., 6-*O*-palmitoyl-*β*-

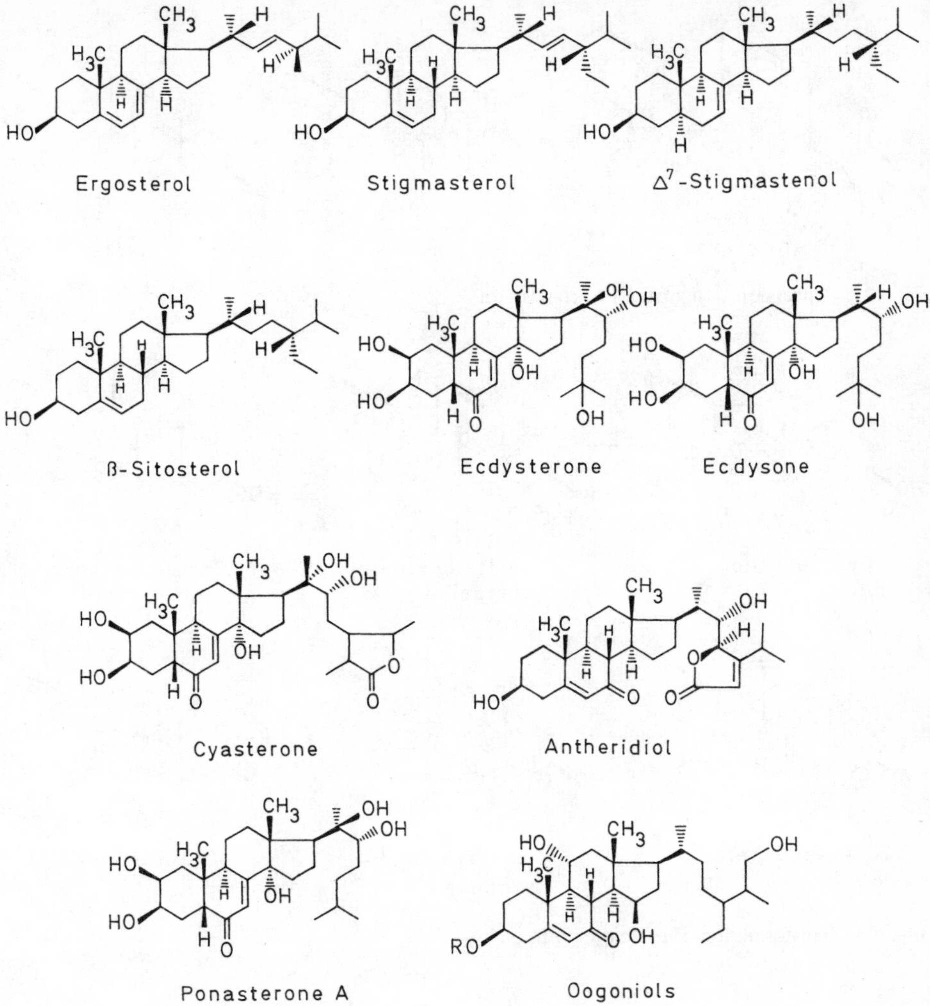

Ergosterol

Stigmasterol

Δ^7-Stigmastenol

ß-Sitosterol

Ecdysterone

Ecdysone

Cyasterone

Antheridiol

Ponasterone A

Oogoniols

Fig. 126. Some naturally occurring sterols

D-glucosylsitosterol, also occur in plants. Sterols are grouped according to the pattern of methylation into 4,4-dimethylsterols, e.g., lanosterol and cycloartenol, 4α-methylsterols, e.g., obtusifoliol, and 4-demethylsterols, e.g., cholesterol. Sterols are ubiquitous in fungi, higher plants, and animals. Most prokaryotes (bacteria, cyanobacteria, etc.) do not synthesize these secondary products.

Biosynthesis

Sterols are derived from cycloartenol (plants) or lanosterol (animals and microorganisms, D. 6.6). In the formation of most sterols the methyl groups in positions 4 and 14 are lost (Figs. 127 and 128).

In animals the methyl group in position 14 is removed before the two methyl

14*

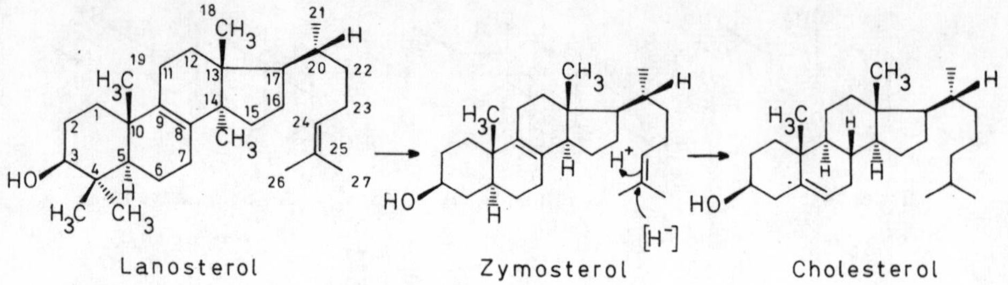

Fig. 127. Conversion of lanosterol to cholesterol

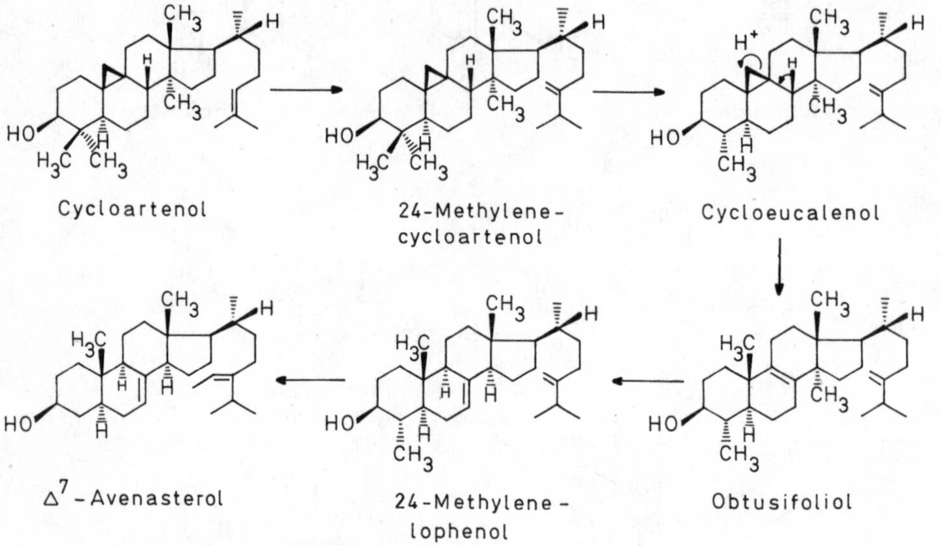

Fig. 128. Transformation of cycloartenol in plants

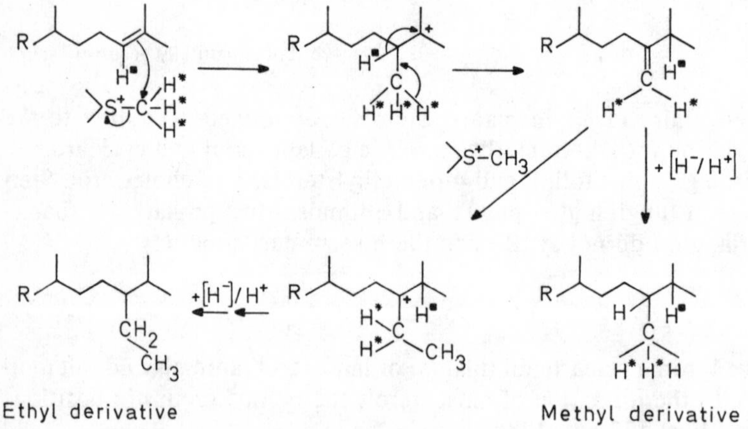

Fig. 129. C-Methylation of the sterol side chain

groups in position 4. On the contrary in plants the 4α-methyl group is eliminated first. During this reaction the 4β-methyl is epimerized to the 4α-position. In subsequent steps the methyl group in position 14 and the new 4α-methyl group are split off. All methyl groups are hydroxylated by monooxygenases (C 2.6.4) and are then oxidized to carboxy groups, which are decarboxylated.

Additional C-atoms in the side chain of certain sterols (cf. the formula of ergosterol, Fig. 126) are derived from the methyl group of S-adenosyl-L-methionine. Additional ethyl side chains as in stigmasterol and β-sitosterol (Fig. 126) are formed by two subsequent methylations. In most cases a hydrogen atom of the methyl group of S-adenosyl-L-methionine is lost during the methylation (Fig. 129). The mechanism in this respect resembles that of the synthesis of tuberculostearic acid from oleic acid (D 5.2.1).

Significance

Sterols are involved in the stabilization of cell membranes (E 2.2). Ergosterol is used in the technical preparation of vitamin D_2 (D 6.6.9). Ecdysone and ecdysterone are molting hormones in insects and crustaceans (E 3.1). Similar compounds (ecdysteroids), e.g., ponasterone A and cyasterone, which are widespread in higher plants, deter feeding insects (E 5.5.3). The hormone activity in insects of certain ecdysteroids is up to 20 times that of ecdysone. Certain *Drosophila sp.* are attracted by Δ^7-stigmastenol (schottenol) and use it as a precursor of ecdysone (E 5.5.3). Antheridiol and the oogoniols are sex pheromones of water molds of the genus *Achlea* (E 4). Fusidic acid and related compounds, produced in fungi, have antibiotic properties (E 5.2) and are used in medicine.

Brassinolide   Fusidic acid

Transformation

a) Cholesterol occupies a central position in the steroid metabolism of most organisms. Nearly all steroids described in the following sections (D 6.6.2–D 6.6.9) are directly or indirectly derived from cholesterol.

b) Brassinolide is the most important representative of the so-called brassinosteroids, i.e., sterol derivatives with a lactone group in ring B of the steroid nucleus. Brassinosteroids are plant hormones (E 3.1).

c) 15-aza-D-Homosterols, alkaloids with antifungal properties, are formed from sterols in microorganisms. They interfere with the biosynthesis of ergosterol, an important membrane constituent. Introduction of the N-atom may proceed as given in Fig. 130.

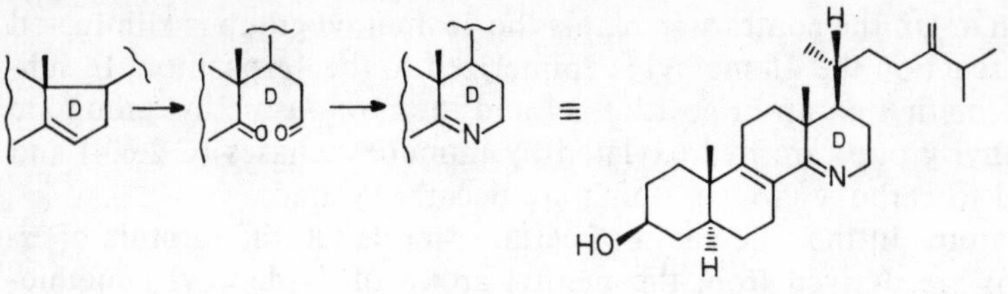

Azasterol B

Fig. 130. Possible pathway of azasterol biosynthesis

References for Further Reading

4, 57, 74, 123, 170, 173, 262, 273, 333, 409, 546, 563, 611, 688, 709

D 6.6.2 Steroid Saponins

Chemistry and Distribution

The steroid saponins are glycosides. The genins contain (one or) two *O*-hetero-cyclic rings, which are attached at C-16 and C-17 to the steroid ring system. These rings are connected by a spiroketal group. The methyl substituent at C-25 may be either equatorial (as in diosgenin) or axial (as in yamogenin). The con-nection of rings A and B may be cis (as in smilagenin) or trans (as in tigogenin Table 37). The genins are glycosidically linked to sugars, e.g., D-glucose, D-galac-tose, L-rhamnose, D-xylose, and L-arabinose, at the hydroxy group in position 3. Frequently several sugar moieties form a branched sugar chain (see the formula of digitonin).

Steroid saponins (Fig. 131) occur in some families of higher plants, e.g., Lilia-ceae, Scrophulariaceae, Solanaceae, and in some echinoderms, e.g. holothuroids and asteroids (sea cucumbers and starfish).

Biosynthesis

Steroid saponins are formed from cholesterol and in some cases from β-sitosterol (D 6.6.1). Hydroxylation in position 26 is part of the formation of diosgenin-type saponins, hydroxylation in position 27 is involved in the formation of yamoge-nintype saponins.

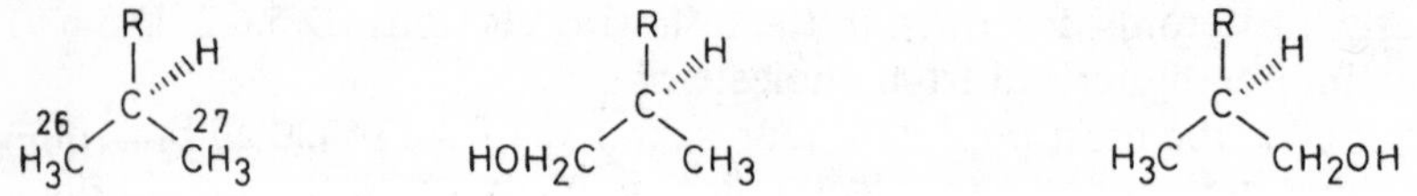

Cholesterol   C(26)-Hydroxyderivative   C(27)-Hydroxyderivative

In addition formation of the rings E and F needs functionalization at C-16 and C-22. Probably saponins devoid of the F-ring are built first. Protodioscin, for instance, is a precursor of dioscin (Fig. 132).

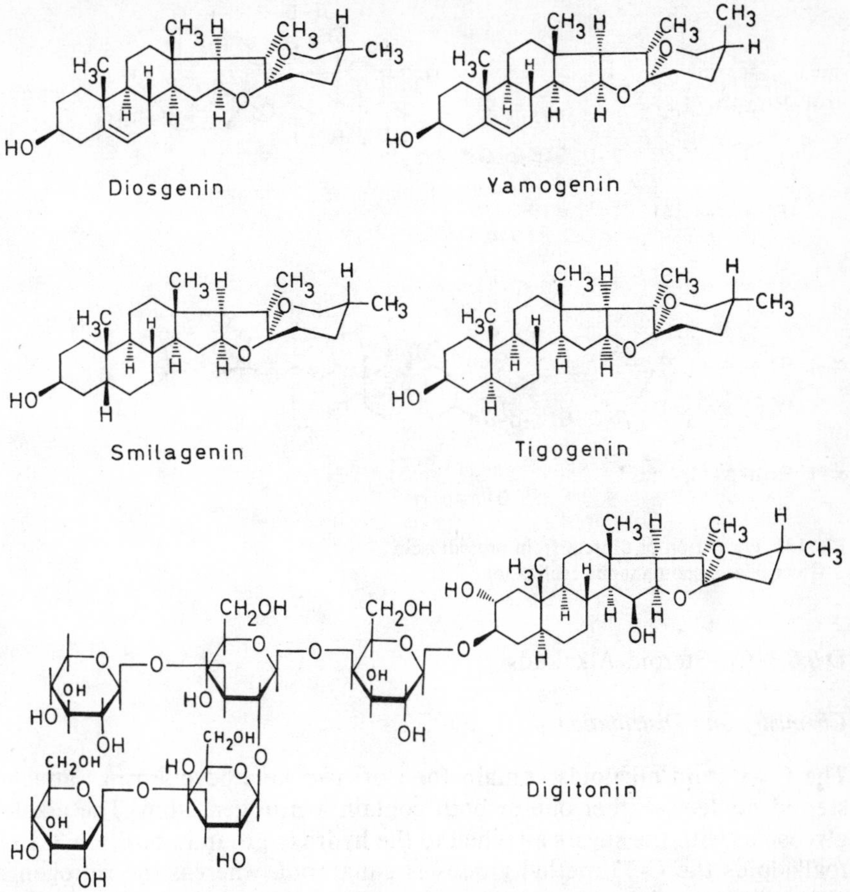

Fig. 131. Chemical structures of steroid sapogenins and saponins

Significance

Saponins (like soaps) lower the surface tension of aqueous solutions and cause the formation of stable foams. They increase the permeability of cell membranes and originate hemolysis of red blood cells. Hence they are poisonous to animals if ingested and due to their bitter taste may deter potential predators (E 5.5.3). The 3-O-glycosides of diosgenin, e.g., dioscin, occurring in *Dioscorea* sp., are important raw materials for the synthesis of steroid hormones used in medicine (F 2).

References for Further Reading

14, 305, 494, 563, 591, 793

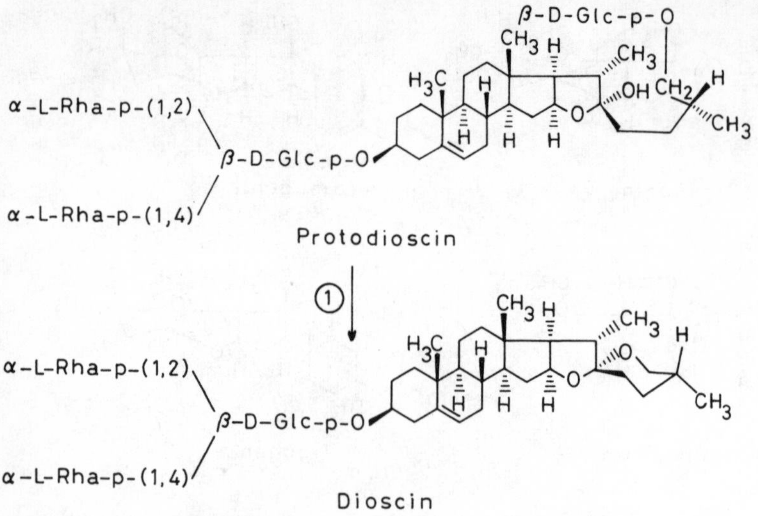

β-D-Glc-p-O
CH₃ H
H₃C
CH₃
OH CH₂ H
CH₃

α-L-Rha-p-(1,2)

β-D-Glc-p-O

α-L-Rha-p-(1,4)

H H

Protodioscin

① ↓

CH₃ H
H₃C
CH₃ H
O

α-L-Rha-p-(1,2)

β-D-Glc-p-O

α-L-Rha-p-(1,4)

CH₃

H H H

Dioscin

Fig. 132. Formation of dioscin from protodioscin
1 Glucosidase, spontaneous cyclization

D 6.6.3 C₂₇-Steroid Alkaloids

Chemistry and Distribution

The C₂₇-steroid alkaloids contain (one or) two heterocyclic rings attached to a steroid nucleus. Either one or both contain a nitrogen atom. The alkaloids are glycosides with the sugars attached to the hydroxy group in position 3. In the spiroalkaloids the C-25 methyl groups is equatorial, whereas the nitrogen may be either in front of the plane (as in α-tomatine) or behind the plane (as in solasodine). A well-known representative of the C₂₇-steroid alkaloids is α-tomatine, which contains a branched sugar chain frequently found also with steroid saponins (D 6.6.2).

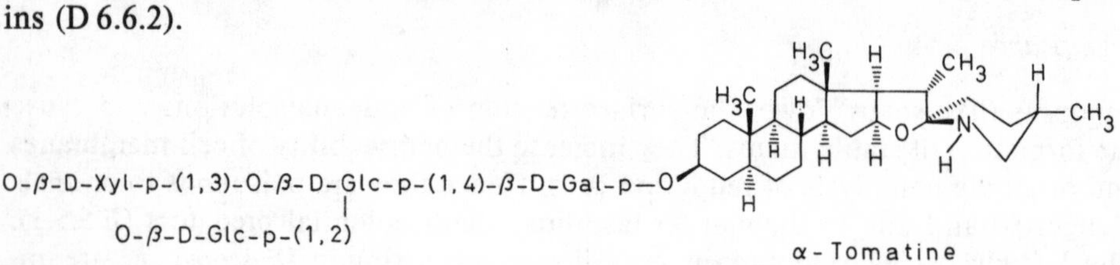

H₃C H CH₃ H
H₃C
CH₃
O N CH₃
H H H

O-β-D-Xyl-p-(1,3)-O-β-D-Glc-p-(1,4)-β-D-Gal-p—O
|
O-β-D-Glc-p-(1,2)

H

α-Tomatine

Biosynthesis

The alkaloids are formed from intermediates involved in the biosynthesis of steroid saponins (D 6.6.2). Hydroxylation at C-26 takes place in the biosynthesis of α-tomatine-type, hydroxylation at C-27 in the formation of solasodine-type alkaloids. Important is the replacement of these hydroxyl groups by an amino group. The amino group donors are amino acids, e.g., glycine, L-alanine, or L-arginine. Closure of ring F probably precedes formation of ring E (Fig. 133).

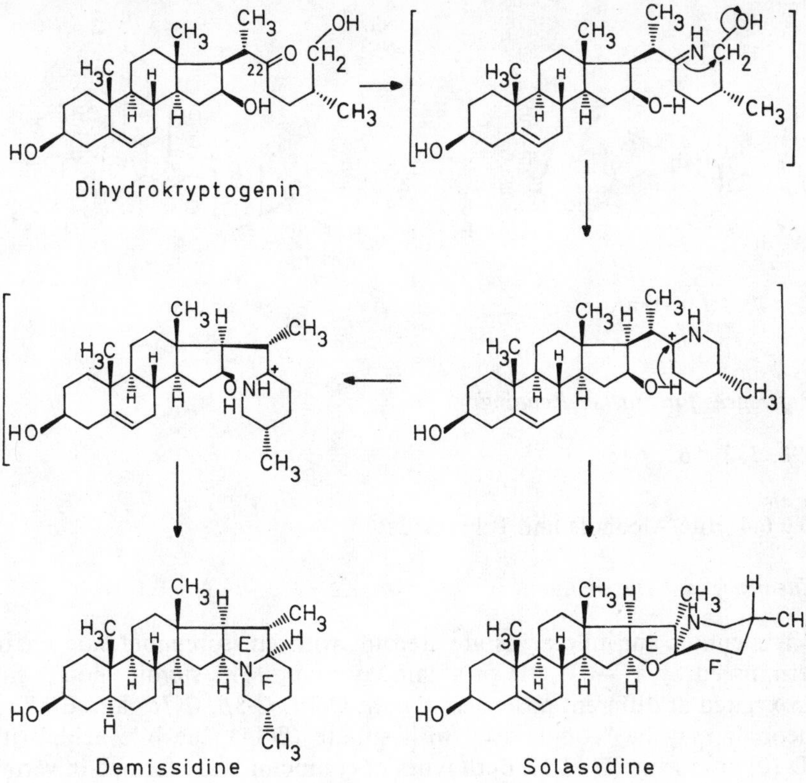

Fig. 133. Biosynthesis of demissidine and solasodine

Significance

Steroid alkaloids are poisonous to animals. Demissine, a demissidine glycoside, occurring in the green parts of the potato plant, for instance, acts as a feeding repellent, e.g., for the Colorado beetle (*Leptinotarsa decemlineata*, E 5.5.3). Solasodine is an important raw material in the production of steroid hormones (F 2).

Transformation

C-nor-D-Homosteroid alkaloids with 27 C-atoms have been found in *Veratrum* and *Fritillaria* sp. (Liliaceae). These substances are linked to sugars at the C-3 hydroxy group and many carry acyl residues attached to other OH-groups. Two typical representatives are veratramine and veracevine.

The C-nor-D-homosteroid alkaloids are formed from steroids by rearrangement of the ring system. The exact pathway is, however, still unknown. The alkaloids are poisonous to animals and act as feeding deterrents (E 5.5.3). They are effective insecticides.

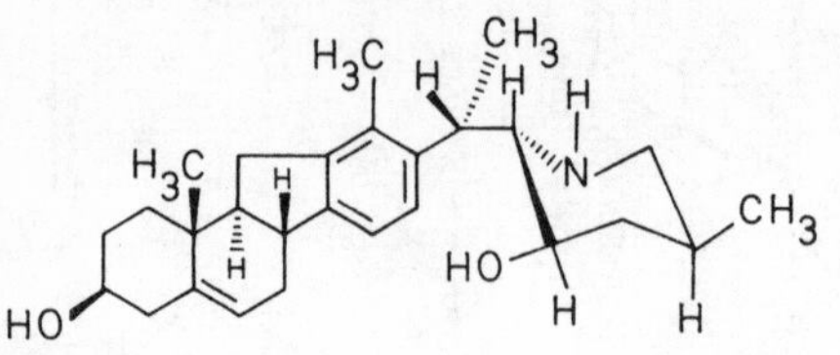

Veratramine 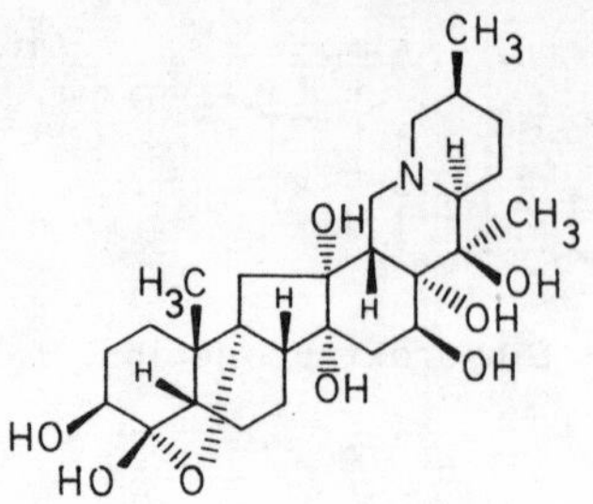   Veracevine

References for Further Reading

296, 332, 563, 648

D 6.6.4 Bile Alcohols and Bile Acids

Chemistry and Distribution

Bile alcohols and bile acids are steroids with an isoprenoid side chain at C-17 terminated by a —CH$_2$OH or a carboxy group. The steroid moiety may be hydroxylated at different positions, e.g., at C-3α, C-3β, C-7α, and C-12α. The bile alcohols may be "conjugated" with sulfate (D 11), the bile acids with glycine (D 10) or taurine (D 11.1). Both types of compounds are formed in vertebrates.

Biosynthesis (Fig. 134)

Bile alcohols and bile acids are derived from cholesterol. Several hydroxy groups are introduced in the steroid ring system and cholic acid-type compounds (anellation of rings A and B cis) or more seldom allocholic acid-type compounds (anellation of rings A and B trans, see Table 37) are formed by reduction of the Δ^5-double bond. The side chain is hydroxylated at one of the terminal C-atoms and from the CH$_2$OH-group a carboxy group may be formed by dehydrogenation.

In vertebrates higher on the phylogenetic scale than amphibians, the side chain is shortened to five C-atoms (Fig. 135). The reaction starts with the formation of CoA esters from bile acids and resembles the β-oxidation of fatty acids (D 5.2). Propionyl CoA is liberated and C$_{24}$-bile acids are formed, which may then be oxidized, reduced, dehydroxylated, and coupled with glycine or taurine. These reactions are carried out at least in part by the microorganisms of the intestine.

Significance

Bile alcohols and bile acids are emulsifiers produced in the liver and are excreted with the bile (E 1) into the intestine. They mediate the emulsification of

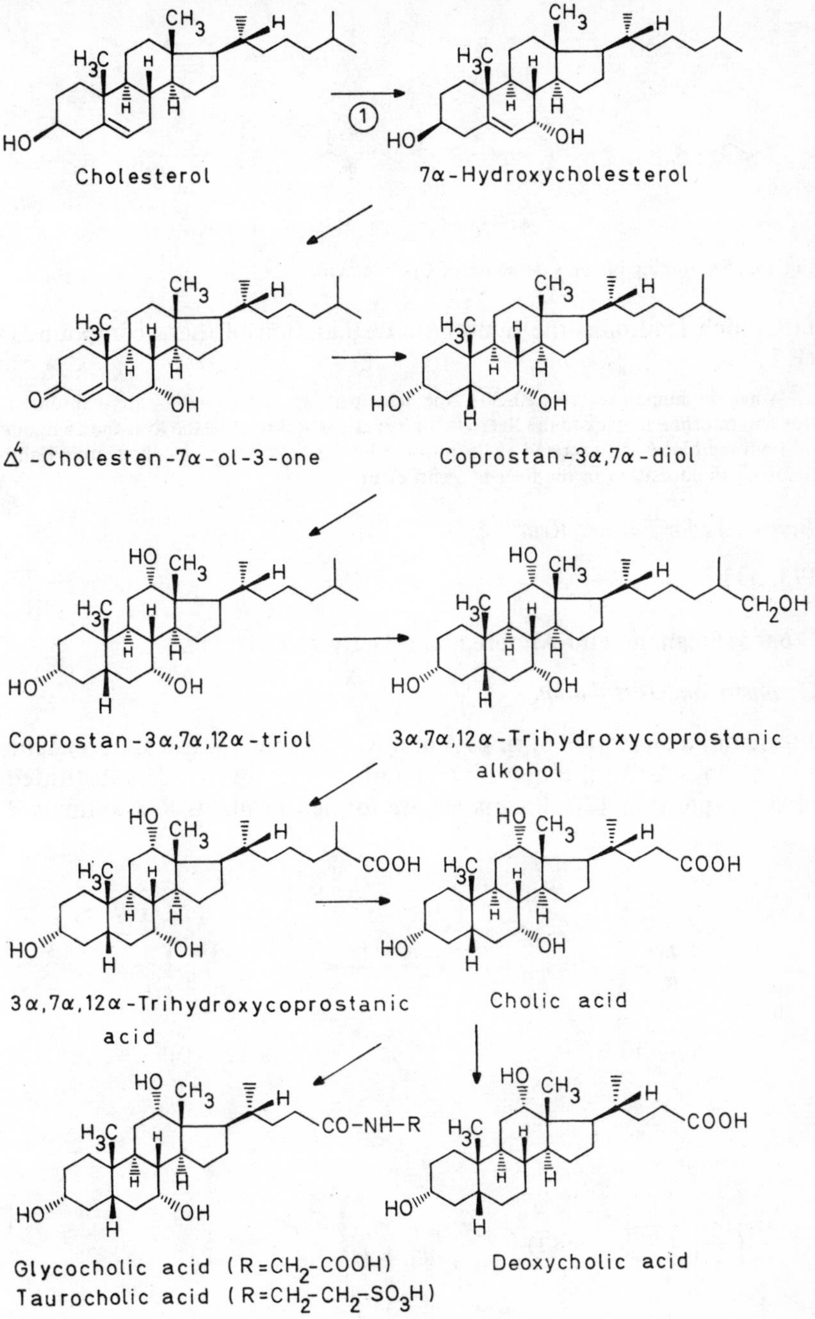

Cholesterol

7α-Hydroxycholesterol

Δ⁴-Cholesten-7α-ol-3-one

Coprostan-3α,7α-diol

Coprostan-3α,7α,12α-triol

3α,7α,12α-Trihydroxycoprostanic alkohol

3α,7α,12α-Trihydroxycoprostanic acid

Cholic acid

Glycocholic acid (R=CH$_2$-COOH)
Taurocholic acid (R=CH$_2$-CH$_2$-SO$_3$H)

Deoxycholic acid

Fig. 134. Formation of bile alcohols and bile acids
1 Cholesterol 7α-monooxygenase

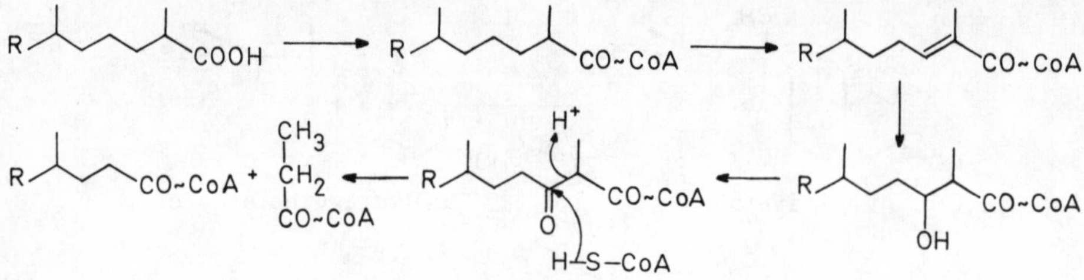

Fig. 135. Shortening of the side chain of C_{27}-bile acids

fats which facilitates the hydrolytic degradation of these compounds by lipases (E 2.2).

A healthy human secretes 20–30 g bile acids per day, of which 90 % are reabsorbed in the intestine and transported back to the liver via the hepatic portal rein. In the liver the compounds are taken up from the blood and excreted again into the bile. Bile acids and alcohols are administered to patients if bile formation in the liver is insufficient.

References for Further Reading

173, 333

D 6.6.5 Pregnane and Allopregnane Derivatives

Chemistry and Distribution

Pregnanes (connection of rings A and B cis, Table 37) and allopregnanes (connection of rings A and B trans) are C_{21}-steroids with an oxygen-substituted ethyl side chain in position 17β. Pregnanes are formed in plants and animals. Frequently

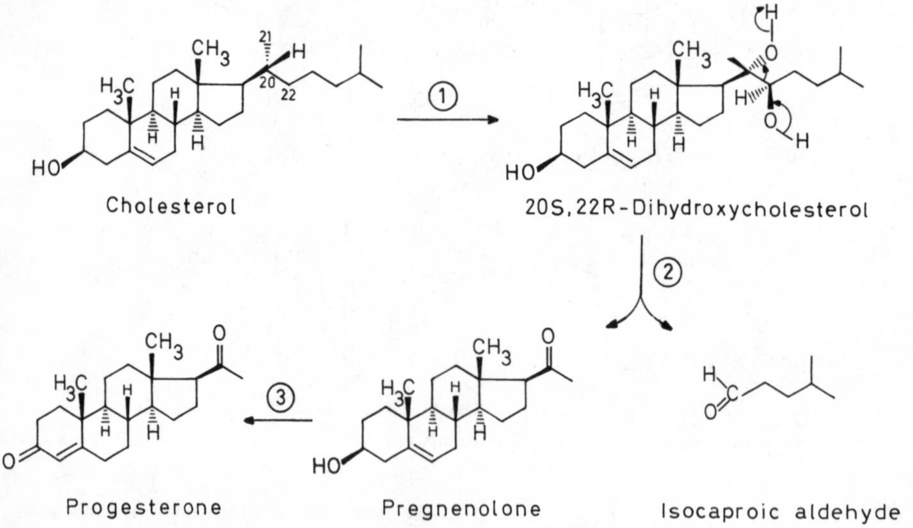

Fig. 136. Formation of progesterone from cholesterol
1 Monooxygenases; *2* 20S, 22R-dihydroxycholesterol aldolase; *3* 3-β-hydroxy-Δ^5-steroid dehydrogenase, steroid-Δ-isomerase

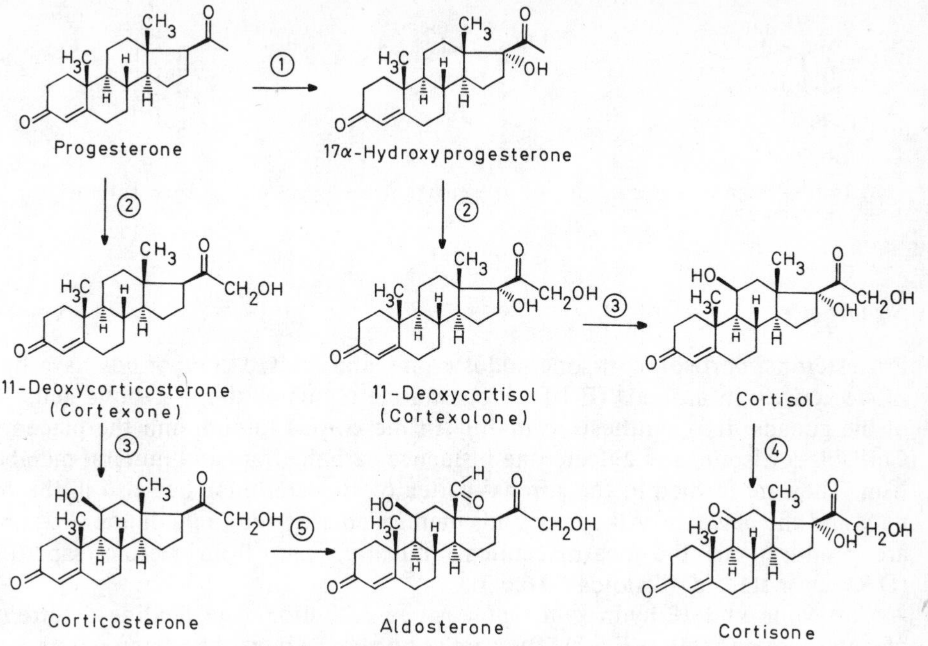

Fig. 137. Biosynthesis of corticosteroids from progesterone
1 Steroid 17α-monooxygenase; *2* steroid 21-monooxygenase; *3* steroid 11β-monooxygenase; *4* dehydrogenase; *5* corticosterone 18-monooxygenase, steroid 18-oxidase

$Δ^4$- and $Δ^5$-pregnene derivatives occur. Glycosides and nitrogen-containing pregnanes have been found in plants.

Biosynthesis

$Δ^5$-Pregnanes are derived from cholesterol by cleaving the side chain between positions 20 and 22 (Fig. 136). The reaction is initiated by introduction of two hydroxy groups. Then isocaproic aldehyde is liberated and pregnenolone is built. Pregnenolone may be transformed to progesterone. In the liver of animals progesterone is hydrogenated to pregnane and allopregnane derivatives. Also the oxo groups may be reduced.

Progesterone is the precursor of hydroxylated pregnane derivatives of the cortisone and aldosterone types (Fig. 137). In the transformation to these compounds monooxygenases (C 2.6.1) and dehydrogenases (C 2.1) are involved. In animals the hydroxylated pregnenes undergo a rapid turnover (A 5.2). Transformation takes place in the liver and includes formation of pregnane and allopregnane derivatives, e.g., 3α, 5β-tetrahydrocorticosterone, urocortisol, and 20β-allocortolone, as well as synthesis of 3α- and 3β-hydroxylated substances which may be conjugated with D-glucuronic acid (D 1.2) and are excreted with the urine (E 1).

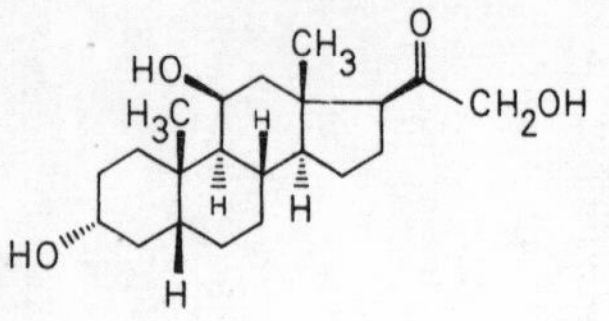

3α,5β-Tetrahydro-
corticosterone

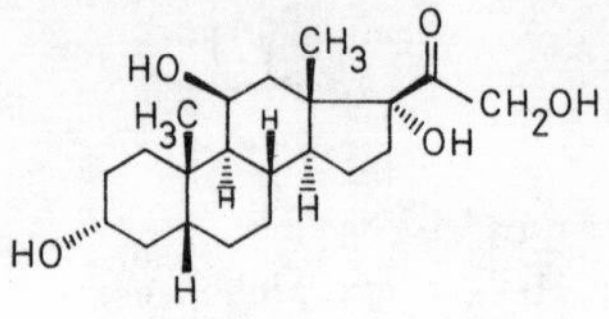

Urocortisol

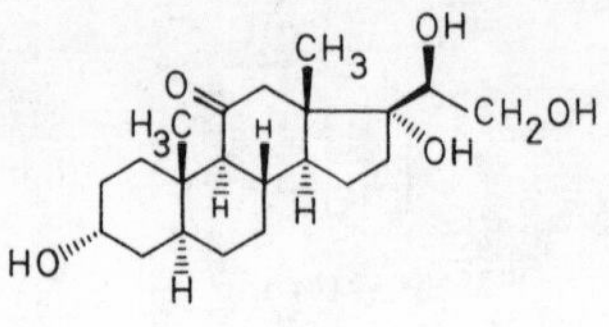

20β-Allocortolone

Significance

Progesterone, cortisol, cortisone, aldosterone, and related compounds have hormone activity in animals (E 3.1). Progesterone regulates the metabolic acitivity
of the gonads. It is synthesized mainly in the corpus luteum and the placenta.
Cortisol, cortisone, and aldosterone influence carbohydrate and mineral metabolism. They are formed in the adrenal cortex (corticosteroids), but also in the gonads and the placenta. All compounds mentioned are used in medicine (F 2) and
are produced in the pharmaceutical industry, e.g., from steroid saponins
(D 6.6.2) or steroid alkaloids (D 6.6.3).

Cortexone and 12-hydroxy-4,6-pregnadien-3,20-dione are feeding deterrents
of certain water beetles (E 5.1). They act as hormones in vertebrate predators.

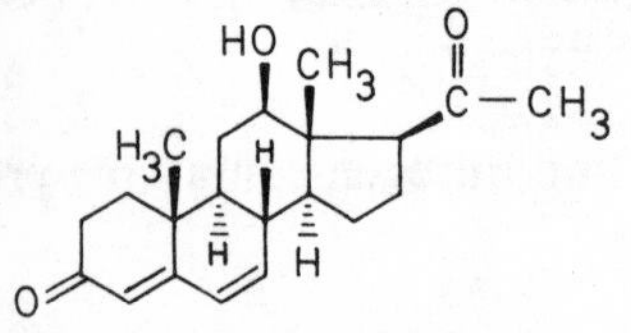

12-Hydroxy-4,6-pregnadien-
3,20-dione

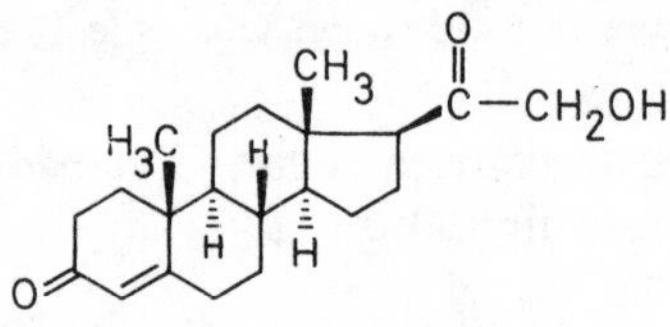

Cortexone

Transformation

a) Pregnenolone glycosides are formed in higher plants. *Digitalis* sp., for instance, synthesize digipurpurogenin and digifologenin glycosides. These compounds, like cardiac glycosides (D 6.6.8), have a cis-anellation of the C and D
rings and are linked to deoxy and methylated sugars.

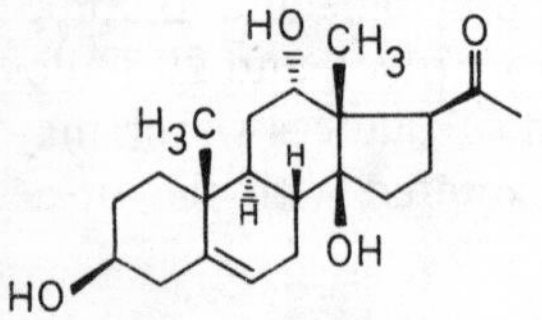

Digipurpurogenin

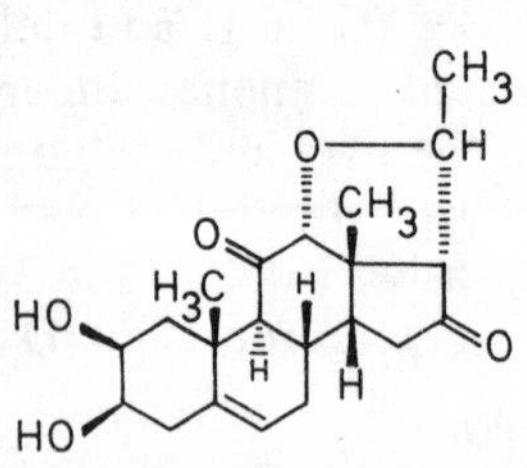

Digifologenin

b) Alkaloids of the holarrhimine and conarrhimine types are built from pregnenolone in Apocynaceae. Holarrhimine and related alkaloids are raw materials for the preparation of corticosteroids used in medicine.

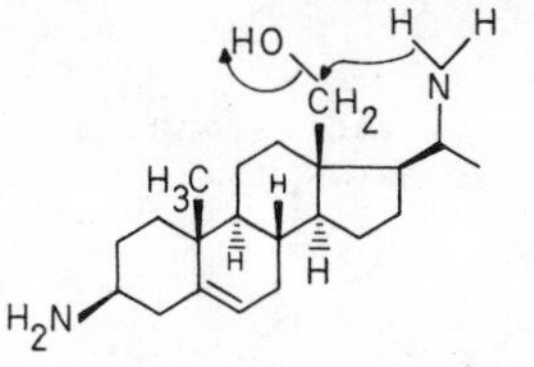

Holarrhimine 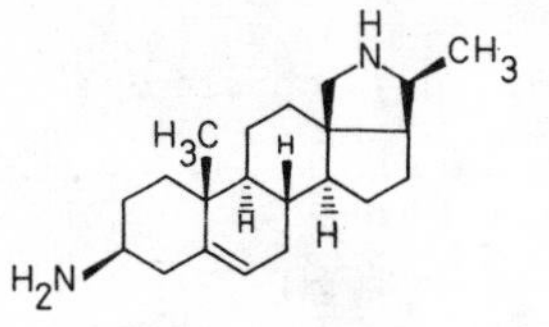   Conarrhimine 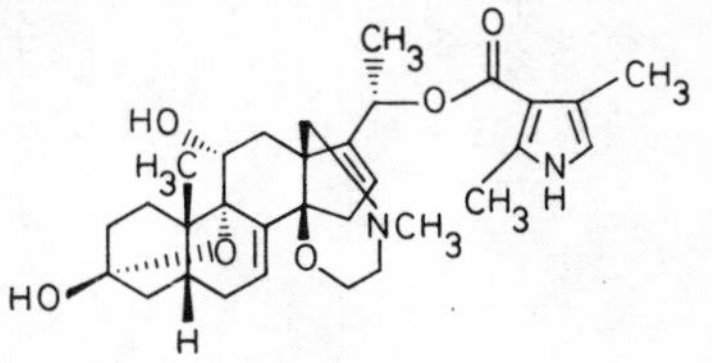   Batrachotoxin

c) Androstane (D 6.6.6) and estrane derivatives (D 6.6.7), as well as cardiac glycosides (D 6.6.8), are synthesized from progesterone.

d) The batrachotoxins are formed in the frog *Phyllobates aurotaenia* and accumulate in skin secretions as deterrents (E 5.1). The batrachotoxins possess a modified pregnane skeleton with cis-anellation of the C and D rings. They depolarize neurons and muscle cells by specific interaction with the sodium channels of the plasma membranes. Batrachotoxin-containing frog skin secretions have been used as poisons for arrows and blowdarts by the Indians of estern Colombia.

References for Further Reading

172, 296, 305, 332, 333, 648, 766

D 6.6.6 Androstanes

Chemistry and Distribution

Androstanes are steroids without a side chain at C-17 built in plants and animals. The steroid ring system carries hydroxy or carbonyl groups at C-3 and C-17.

Biosynthesis (Fig. 138)

Precursors of the androstane derivatives are 17α-hydroxylated pregnanes. The side chain at C-17 is cleaved oxidatively with formation of a 17-oxo steroid and acetic acid, or glycolic acid in the case of C-22 hydroxylated precursors. The carbonyl groups at C-3 and C-17 as well as the double bond at C-4 may be reduced and 5α- and 5β-androstane derivatives are formed, which may carry a 3α- or a 3β-hydroxy group.

Significance

Certain androstene derivatives are male sex hormones in higher animals (E 3.1). The most important ones in man are androstenedione and testosterone, which

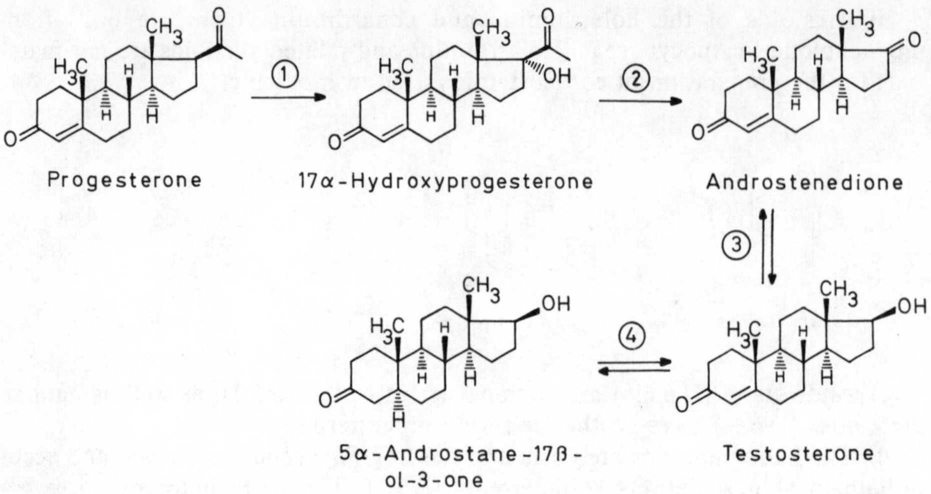

Fig. 138. Formation of androstane derivatives
1 Steroid 17α-monooxygenase; *2* steroid C-17, C-20 lyase; *3* 17β-hydroxysteroid dehydrogenase; *4* 3-oxo-5α-steroid Δ⁴-oxidoreductase

are produced in the testes. In the target tissues, e.g., the prostate gland, testosterone is reduced to 5α-androstane-17β-ol-3-one, which seems to be the real metabolic effector. The androstenedione and testosterone that accumulate in certain plant cells, e.g., in the pollen of *Pinus sylvestris*, probably have an ecological function (E 5.5.3). 5α-Androst-16-en-3-one and 5α-androst-16-en-3α-ol are sex pheromones of the boar (E 4). 5α-Androst-16-en-3-one is also a constituent of human urine (the urine of males containing three times the concentration of the urine of females) and 5α-androst-16-en-3α-ol was shown to occur in the sweat of male humans. It is unknown, however, whether both substances are human pheromones.

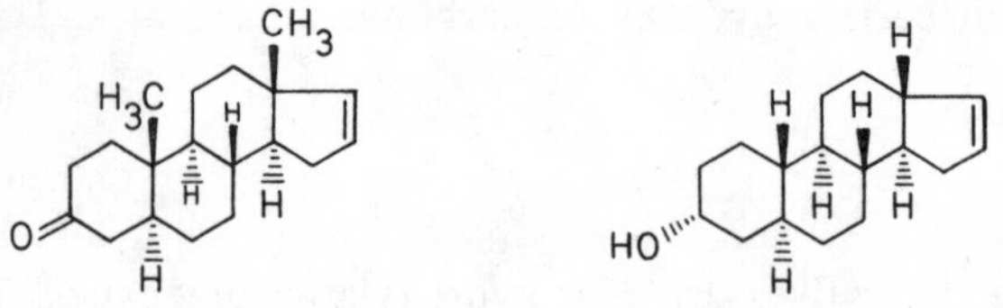

5α-Androst-16-en-3-one   5α-Androst-16-en-3α-ol

Transformation

Salamanders form alkaloids possessing an oxazolidin ring (see the formula of samandarin) and alkaloids whose nitrogen-containing ring includes the oxygenated C-19 of the steroid nucleus (see the formula of cycloneosamandione). Both types of compounds are derived from cholesterol (D 6.6.1) with loss of the side chain. Hence, androstane derivatives may be intermediates. The nitrogen atom is derived from glutamine. Detailed knowledge on the pathway of biosynthesis, however, is not yet available.

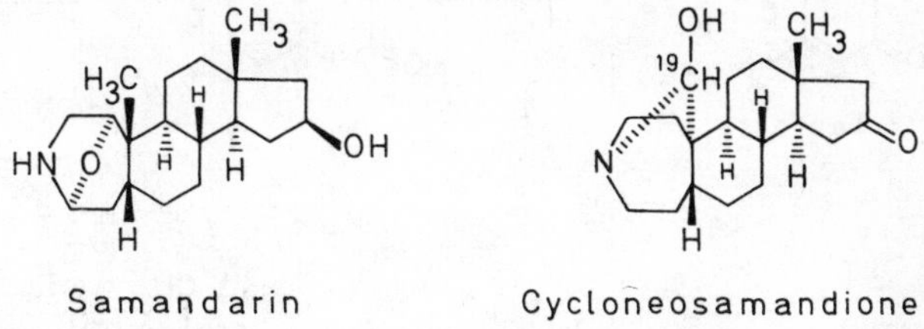

Samandarin   Cycloneosamandione

The alkaloids are produced in skin glands and may protect against microbial infections (E 5.1). In addition, they are poisonous to animals if ingested and may repel potential predators.

References for Further Reading

296, 333, 766

D 6.6.7 Estranes

Chemistry and Distribution

Estranes are steroid derivatives with an aromatic A ring formed in plants and animals. They are devoid of the angular methyl group at C-10. The ring system may be substituted by hydroxy or carbonyl groups.

Biosynthesis

Estrone and estradiol are derived from testosterone (Fig. 139). Aromatization of ring A is catalyzed by steroid aromatase, which mediates two subsequent hydroxylations (C 2.6.4) and cleaves off C-19 as formaldehyde. During this reaction the 1β and 2β H-atoms are lost (cis-elimination).

The most important estrane derivatives in animals are estradiol, estriol, and estrone. They are transformed to compounds with additional hydroxy and carbonyl groups as well as to sulfate esters (D 11) and glucuronides (D 1.2), e.g., estrone-3-O-β-D-glucuronide, which are excreted with the urine (E 1).

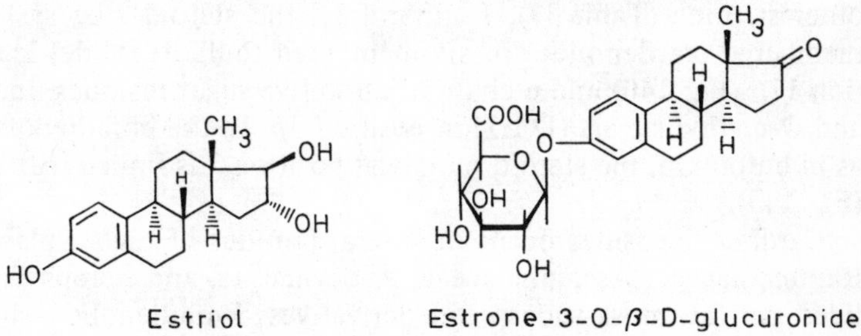

Estriol   Estrone-3-O-β-D-glucuronide

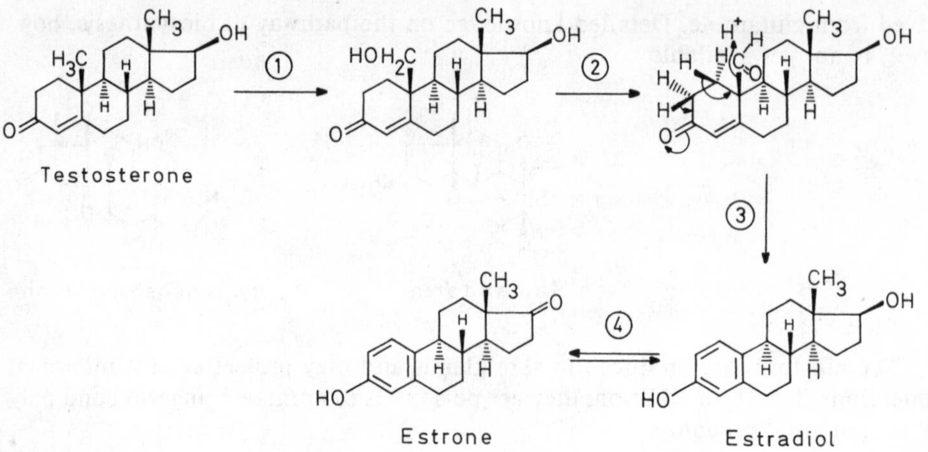

Fig. 139. Transformation of testosterone to estrane derivatives
1 Steroid 19-monooxygenase; *2* steroid 19-oxidase; *3* steroid C-10, C-19-lyase; *4* 17β-hydroxysteroid dehydrogenase

Significance

Estrane derivatives act as female sex hormones in higher animals including humans (estrogens, E 3.1). They are synthesized mainly in ovaries and placenta. Estrane derivatives formed in plants and accumulated, for instance, in seeds of pomegranate *(Punica granatum)* or flowers of willow *(Salix)* may have an ecological function (E 5.5.3). They probably interfere with the hormone metabolism of animal predators.

References for Further Reading

333, 453, 766, 808, 893

D 6.6.8 Cardiac Glycosides

Chemistry and Distribution

Cardiac glycosides are steroids substituted with a hydroxy group in position 14β, i.e., rings C and D of the steroid ring system are cis-connected, in contrast to most other steroids (Table 37). Furthermore, the steroid ring system carries a five-membered (cardenolide) or six-membered (bufadienolide) lactone ring in position 17β (Fig. 140) and a chain of up to five sugar residues, including rare deoxy and O-methyl sugars (D 1.1) in position 3β. In the bufadienolide esters of toads, as in bufotoxin, the steroid moiety at position 3 is linked with suberyl-L-arginine (Fig. 140).

Cardiac glycosides occur in several families of higher plants, e.g., Liliaceae, Ranunculaceae, Asclepiadaceae, Apocynaceae, and Scrophulariaceae. Cardenolides are the most widespread derivatives. Bufadienolides have been isolated

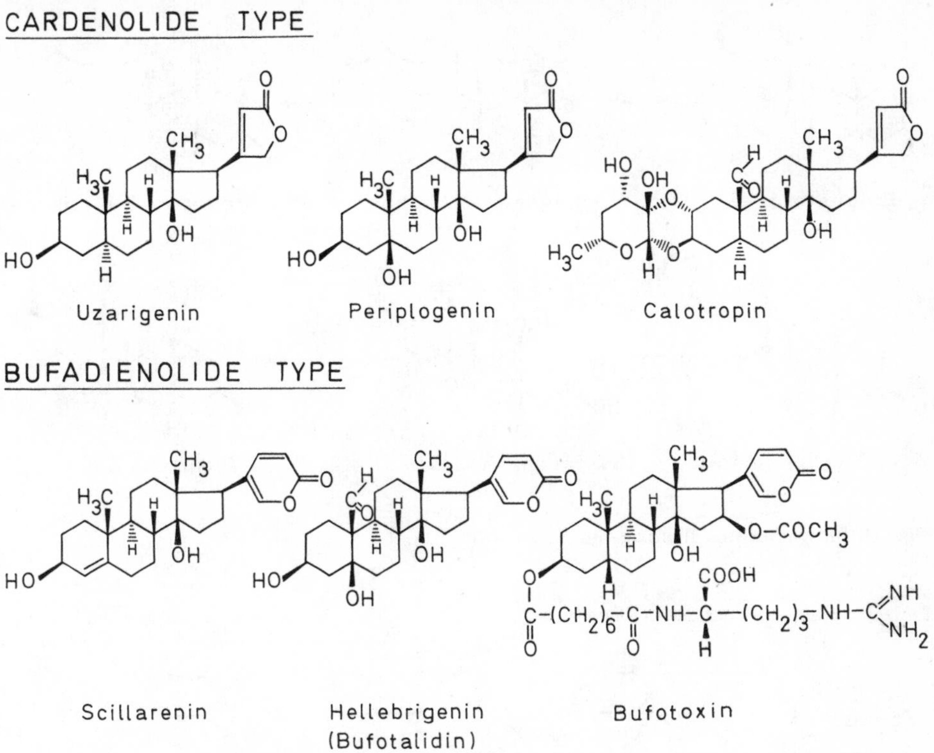

CARDENOLIDE   TYPE

Uzarigenin   Periplogenin   Calotropin

BUFADIENOLIDE   TYPE

Scillarenin   Hellebrigenin   Bufotoxin
                     (Bufotalidin)

Fig. 140. Chemical structures of some cardiac genins, glycosides, and esters

from only a few genera, e.g., *Scilla* (Liliaceae) and *Helleborus* (Ranunculaceae). In addition, cardenolides are synthesized in certain beetles and bufadienolides in toads (*Bufo* sp.). The cardenolides found in the locust *Poekilocerus bufonius*, the monarch butterfly *Danaus plexippus*, and several other animals are derived from cardiac glycosides taken up with the diet (E 5.1).

Biosynthesis (Fig. 141)

The immediate precursor of cardiac glycosides is progesterone (D 6.6.5). The two additional carbon atoms of the cardenolide ring originate from acetate (acetyl or malonyl CoA, D 5). The three corresponding *C*-atoms of the bufadienolide ring are derived from propionic acid (propionyl or methylmalonyl CoA, D 3). Progesterone is reduced or hydrated to give derivatives with either a 5α or 5β pregnane structure. The hydroxylation at C-14 precedes formation of the lactone ring.

Significance

The cardiac glycosides found in plants and animals and the bufadienolide esters, like bufotoxin, have a bitter taste and a high toxicity, properties which repel potential predators (E 5.1 and E 5.5.3). Several cardiac glycosides, e.g., digitoxin, di-

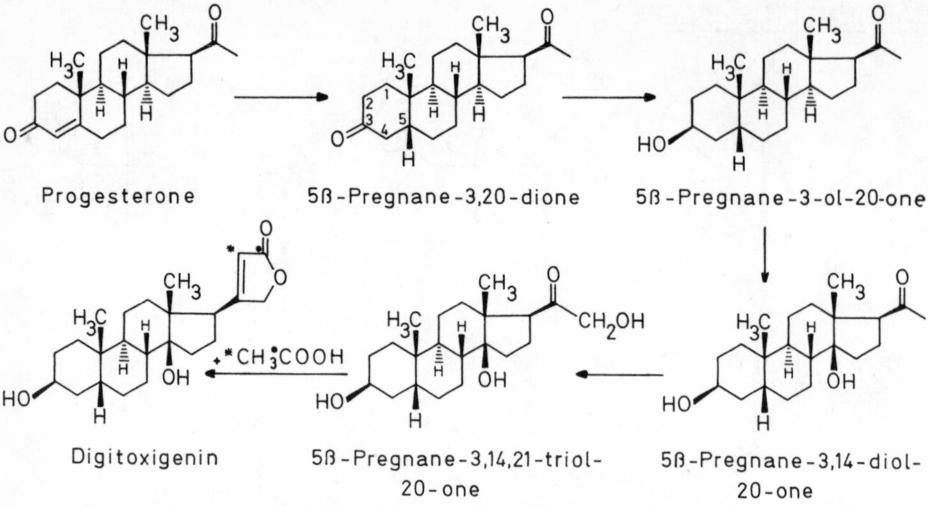

Fig. 141. Biosynthesis of digitoxigenin

Table 38. Main cardenolides of *Digitale sp.*

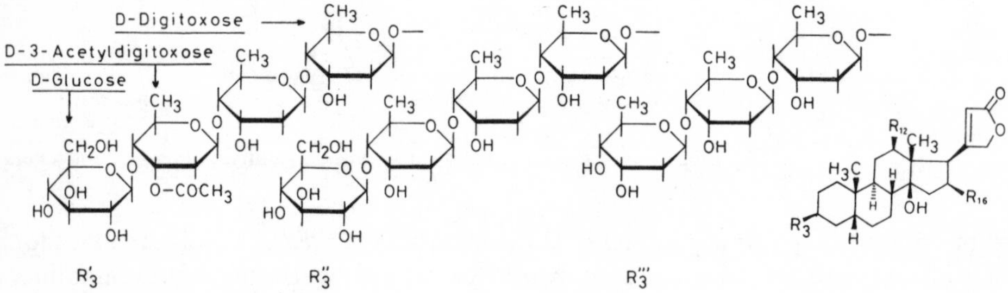

Cardenolides	R_3	R_{12}	R_{16}	Cardenolides	R_3	R_{12}	R_{16}
Digitoxigenin	—OH	—H	—H	*Gitoxigenin*	—OH	—H	—OH
Lanatoside A	—R_3'	—H	—H	Lanatoside B	—R_3'	—H	—OH
Purpurea-glycoside A	—R_3''	—H	—H	Purpurea-glycoside B	—R_3''	—H	—OH
Digitoxin	—R_3'''	—H	—H	Gitoxin	—R_3'''	—H	—OH
Digoxigenin	—OH	—OH	—H	*Diginatigenin*	—OH	—OH	—OH
Lanatoside C	—R_3'	—OH	—H	Lanatoside D	—R_3'	—OH	—OH
Digoxin	—R_3'''	—OH	—H	Diginatin	—R_3'''	—OH	—OH

goxin, and lanatoside C, synthesized in *Digitalis lanata* (Table 38), are used as cardiotonics in medicine (F 2).

References for Further Reading

39, 43, 305, 551, 563, 718, 780

D 6.6.9 Vitamin D Derivatives

Chemistry and Distribution

The D vitamins are methylated trans-hydrindane derivatives with an isoprenoid side chain. A system of two conjugated double bonds and a hydroxylated cyclohexane ring are attached to the hydrindane moiety. The latter is substituted by a methylene group. Most important are vitamin D_3 (cholecalciferol, calciol) and vitamin D_2.

Compounds of the vitamin D group are synthesized in animals, certain higher plants, mushrooms, and algae.

Biosynthesis (Fig. 142)

The D vitamins originate from $\Delta^{5,7}$-unsaturated sterols, e.g., 7-dehydrocholesterol or ergosterol (D 6.6.1), by opening of ring B. This nonenzymatic reaction is

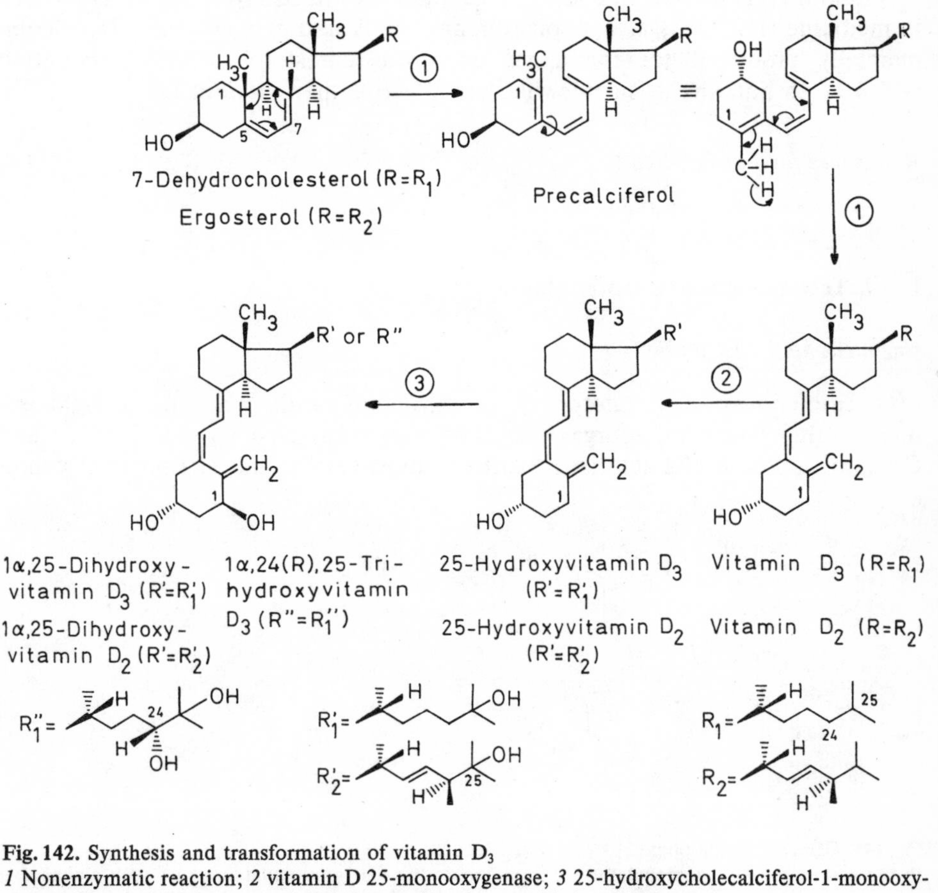

Fig. 142. Synthesis and transformation of vitamin D_3
1 Nonenzymatic reaction; *2* vitamin D 25-monooxygenase; *3* 25-hydroxycholecalciferol-1-monooxygenase. The designations α and β correspond to the steroid ring system

mediated by UV light (λ 260–285 nm). In animals D vitamins are hydroxylated by monooxygenases to 25-hydroxy, $1\alpha,25$-dihydroxy, $24(R),25$-dihydroxy, and $1\alpha,24(R),25$-trihydroxy compounds.

Significance

$1\alpha,25$-Dihydroxyvitamin D_3 (calcitriol) acts as a hormone in humans and higher animals, regulating the absorption of Ca^{2+} in the intestine and the Ca^{2+} metabolism in the bone (E 3.1). Reduced hormone levels result in calcium deficiency, which causes rickets. Surplus amounts cause calcification of soft tissues. Vitamin D_3 synthesized and stored in certain plants is poisonous to animals (E 5.5.3).

In human beings and animals vitamin D_3 is formed in the skin under the influence of UV radiation. The rate of synthesis depends on hairiness and skin color, i.e., on the transparency of the skin. The pigmented skin of human races living near the equator protects against surplus formation of the hormone. In contrast, the light-skinned races living in higher latitudes are able to synthesize vitamin D_3 with the relatively weak UV radiation falling in these areas.

Vitamin D_3 is present in considerable quantities in fish liver oil, which is used in medicine (F 2). It is also a constituent of milk and eggs. Vitamin D_2 is commercially produced for medicinal use from ergosterol by UV irradiation (D 6.4.1). In humans it is somewhat less effective than vitamin D_3.

References for Further Reading

401, 443, 473, 551, 571, 900

D 6.7 Tetraterpenes (Carotenoids)

Chemistry and Distribution

Most carotenoids are tetraterpenes, i.e., C_{40}-compounds, made up of eight isoprene units. Some microorganisms, however, form also C_{30}-, C_{35}-, C_{45}-, and C_{50}-carotenoids. A characteristic feature of most carotenoids is a series of conju-

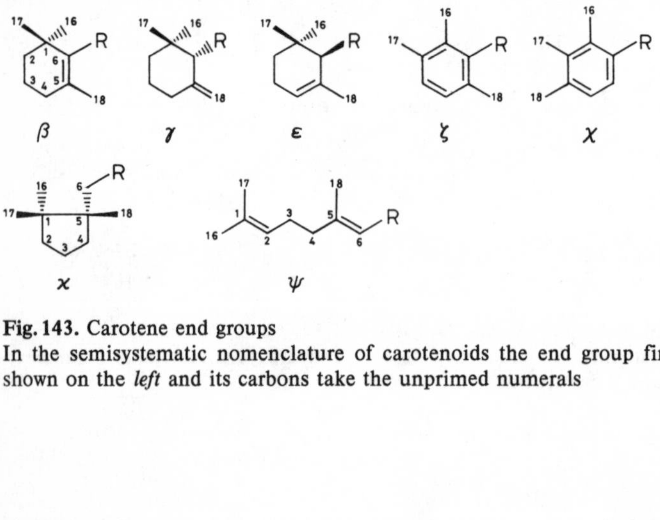

Fig. 143. Carotene end groups
In the semisystematic nomenclature of carotenoids the end group first in the Greek alphabet is shown on the *left* and its carbons take the unprimed numerals

gated *E*-double bonds. But several compounds contain also hydrogenated areas and/or acetylenic or allenic bonds. Carotenoids are either free of oxygen (carotenes) or contain —OH, —OR, —O—COR, —C=O, —CHO, —COOH, —COOR, or epoxide groups (xanthophylls). They are aliphatic or possess carbocyclic five- or six-membered rings at one or both extremities of the molecule (Fig. 143).

The yellow to red color of most carotenoids is caused by the conjugated *E*-double bonds. Carotenoid-protein complexes (carotenoproteins) may be green, yellow, red, or blue (E 5.6).

Carotenoids are formed in microorganisms and with a smaller structural variability in plants. Animals process carotenoids taken up with the diet (fish, birds, mammals, human beings) or from symbiotic microorganisms (e.g., ladybird beetles).

Biosynthesis

The C_{40}-carotenoids originate from the colorless tetraterpene all-*E*-phytoene (D 6). The carotenoids with shorter or longer carbon chains are formed from the corresponding isoprenoid hydrocarbons. all-*E*-Phytoene is dehydrogenated step

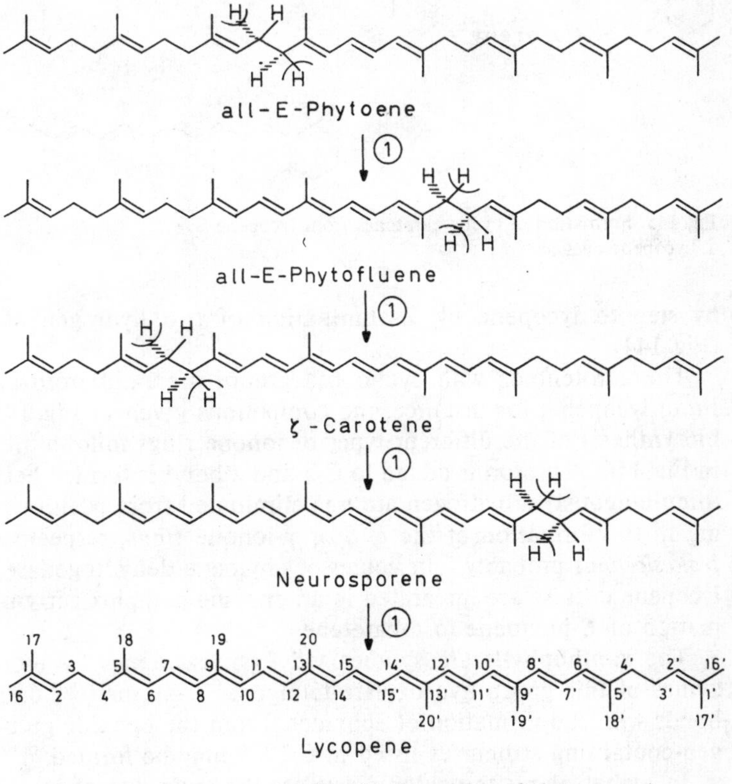

Fig. 144. Conversion of phytoene to lycopene
1 Phytoene dehydrogenase

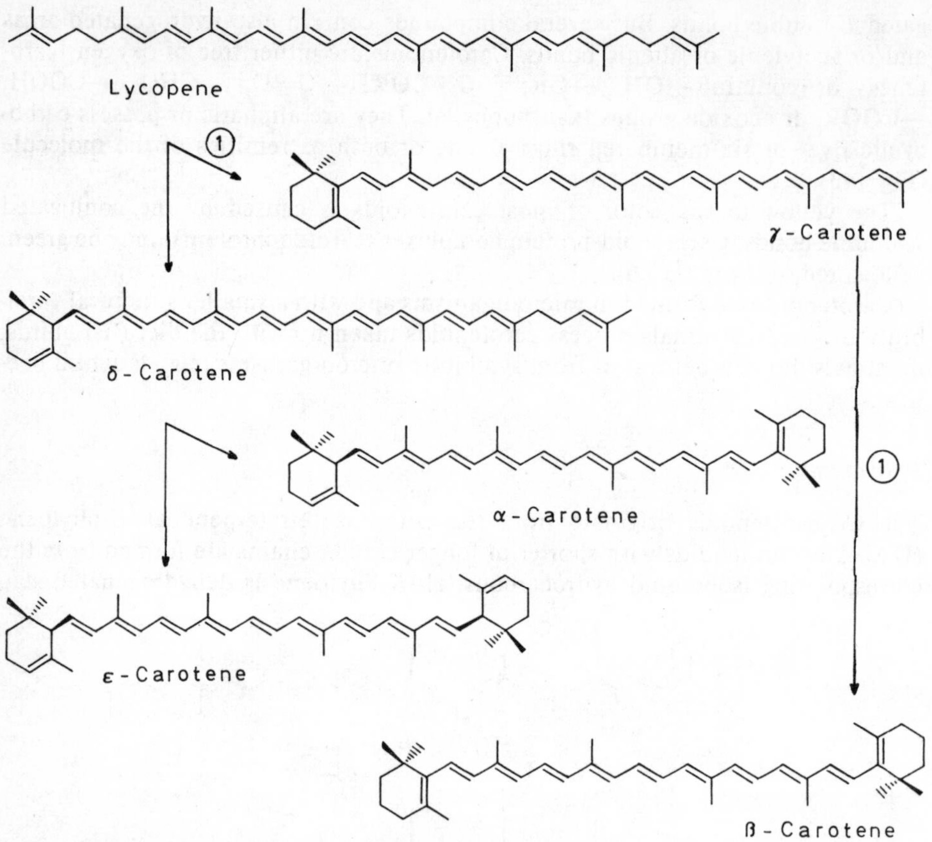

Fig. 145. Formation of cyclic carotenes from lycopene
1 Lycopene cyclase

by step to lycopene by Z-elimination of two hydrogen atoms at each step (Fig. 144).

The carotenoids with cyclic end groups are built from aliphatic carotenes. From lycopene, for instance, the compounds given in Fig. 145 are formed. The biosynthesis of the different types of ionone rings follows the pathway outlined in Fig. 146. A proton is added to C-2 and a bond is formed between C-1 and C-6. Simultaneously a hydrogen atom is eliminated from positions 4, 6, or 18, resulting in the formation of the ε, β or γ-ionone rings, respectively. In *Phycomyces blakesleeanus* probably four copies of phytoene dehydrogenase and two copies of lycopene cyclase are integrated in an enzyme complex carrying out the transformation of *E*-phytoene to β-carotene.

The xanthophylls are synthesized from carotenes by monooxygenases. Enzymes of this group hydroxylate tetragonal *C*-atoms (C 2.6.4) or attack double bonds with the formation of epoxides. From the epoxide group the various oxygen-containing structures listed in C 2.6.5 may be formed.

In higher plants the chloroplasts are the main site of carotenoid biosynthesis (A 3.1.3).

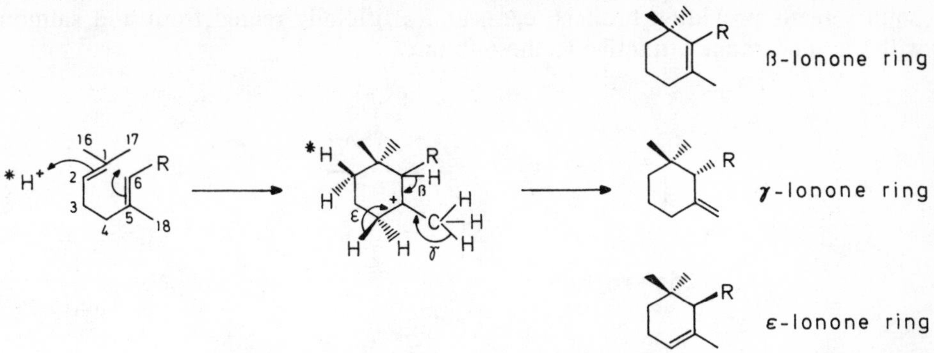

Fig. 146. Biosynthesis of the ionone rings

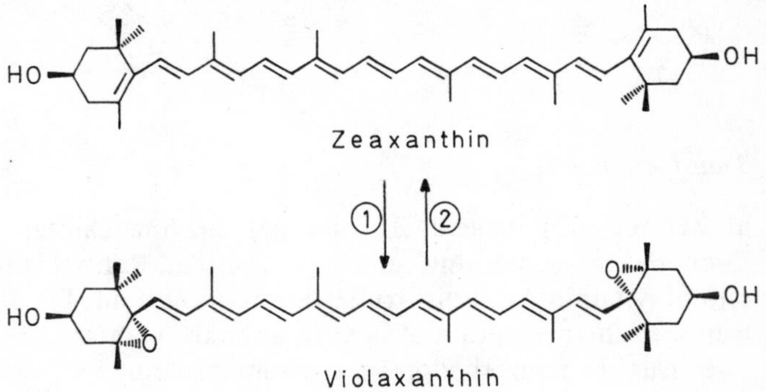

Zeaxanthin

Violaxanthin

Fig. 147. The violaxanthin cycle
1 Zeaxanthin monooxygenase; *2* violaxanthin deepoxidase (cosubstrate L-ascorbic acid, which is transformed to dehydroascorbic acid, D 1.2)

Significance

Carotenoids are copigments in photosynthesis and protect plant and microbial cells from UV and short wavelength visible radiation (E 2.2).

Zeaxanthin is transformed to violaxanthin by an epoxidase which is located at the stroma side of the thylakoid membranes. Violaxanthin is transformed to zeaxanthin at the locullus side by violaxanthin deepoxidase. Epoxidation and deepoxidation form the so-called violaxanthin cycle (Fig. 147) which seems to be involved in the regulation of photosynthesis at the membrane level.

Carotenoids mediate phototropic responses in higher plants and fungi and strengthen membrane rigidity in microorganisms and plants (E 2.2). They are pigments of flowers and fruits attracting pollinators (E 5.5.1) and animals mediating seed dispersal (E 5.5.2). Lycopene is the most important coloring matter of ripe tomatoes, capsanthin that of red pepper. Several carotenoids are accumulated in animals. Astaxanthin is one of the characteristic pigments of goldfish, trout, and salmon. Other carotenoids cause the yellow to orange color of egg yolk, the pink color of the plumage of flamingoes, and the red color of the wing cases of ladybird beetles (E 5.6). They are important constituents of fish and poultry feeds providing broilers, eggs, and artificially reared trout and salmon with an appearance attractive to the public.

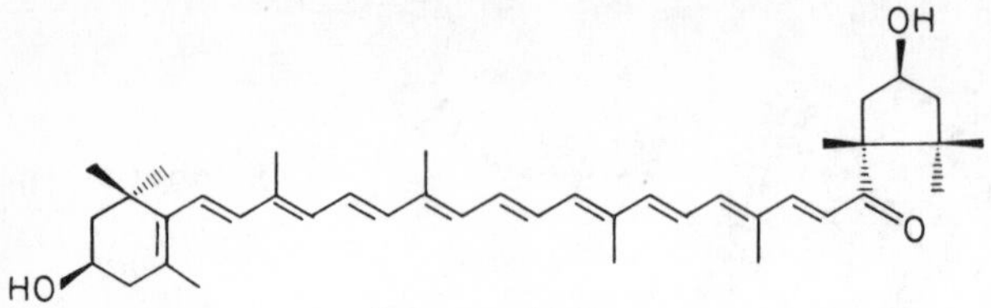

Capsanthin

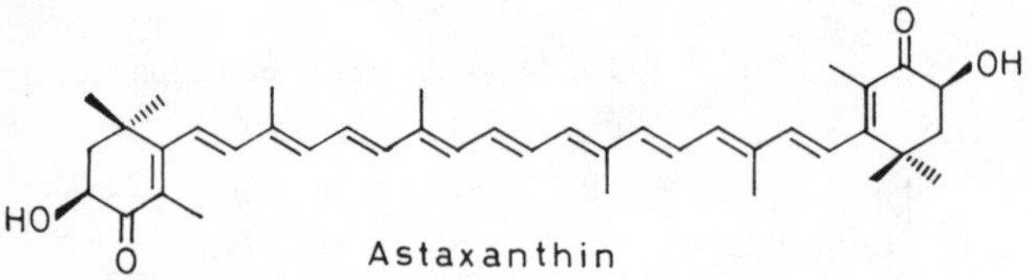

Astaxanthin

Transformation

a) Retinol/retinoic acid. The animal enzyme carotene dioxygenase converts β-ionone ring-containing carotenes to retinal. Retinal is subsequently reduced to retinol (vitamin A_1), or is oxidized to retinoic acid (Fig. 148). Retinal is of significance in the perception of light in animals, plants, and microorganisms (E 2.2). The visual pigment rhodopsin, a chromoprotein, e.g., of the retina of the human eyes, contains 11-Z-retinal. Rhodopsin is bleached during irradiation with visible light. One step in this process is photoisomerization of 11-Z-retinal to all-E-retinal, which is released from the protein. In the dark rhodopsin is regenerated. In halobacteria, prokaryotes that occur in natural brines, where the NaCl concentration is at or near saturation, bacteriorhodopsin, which also contains covalently-bound retinal, is a constituent of the cell membrane. When absorbing light, it ejects protons from the cell, thus generating an electrochemical gradient across the cell membrane, which directly drives ATP synthesis. In animals retinoic acid epoxide possesses hormone activity (E 3.1).

b) Crocetin esters. Crocetin mono- and diglycosyl esters, containing D-glucose and gentiobiose (D 1.4.1) as carbohydrate residues, are the main pigments of saffron (stigmas of *Crocus sativus*). Crocetin is probably formed from C_{40}-carotenoids by oxidative degradation.

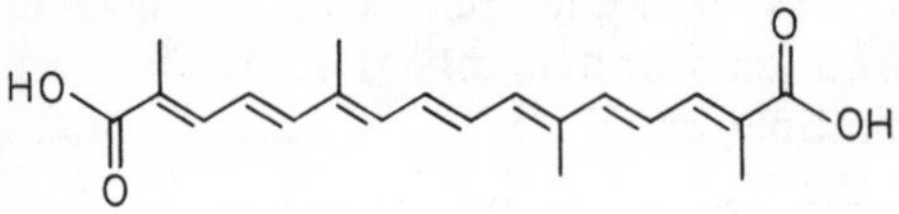

Crocetin

c) Trisporic acids are C_{18}-terpenoids (Fig. 149) active as sex pheromones in Mucorales (E 4). They are built from all-E-retinal by the loss of a C_2-fragment from the side chain and further modification.

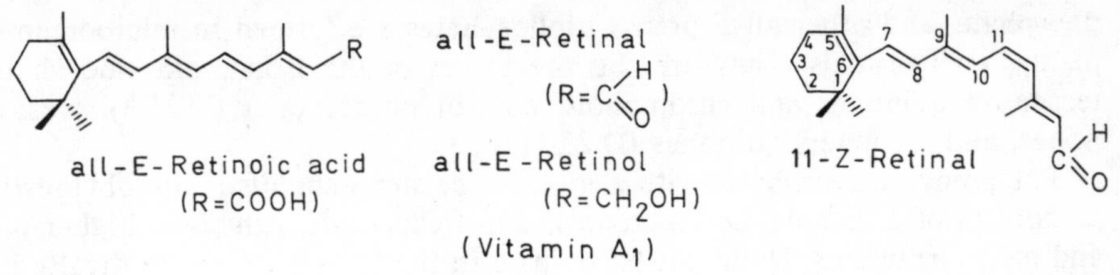

Fig. 148. Vitamin A₁ and some of its derivatives

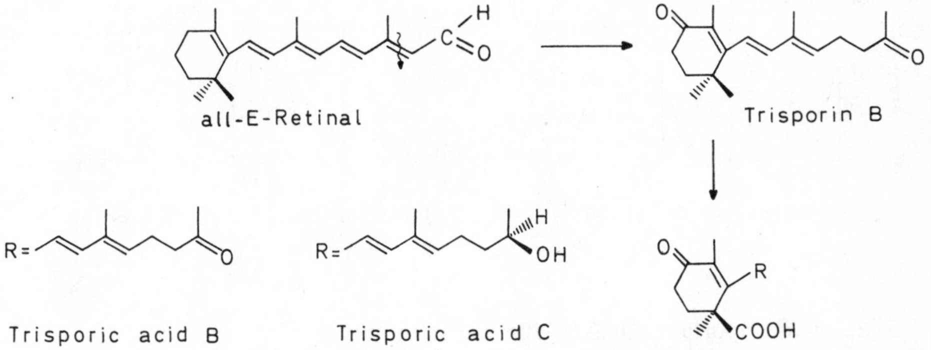

Fig. 149. Formation of trisporic acids

d) Sporopollenins are oxidative polymers of carotenoids and/or carotenoid esters present, e.g., in the wall of pollen grains and microbial or plant spores. They are extremely resistant to decay so that the walls of pollen grains and spores survive for millions of years. As yet the exact chemical structure of sporopollenins is unknown.

References for Further Reading

101, 103, 177, 178, 183, 199, 273, 274, 275, 313, 330, 334, 399, 600, 611, 659, 725, 735, 769, 893

D 6.8 High Molecular Terpenes

Chemistry and Distribution

Polyterpenes made up from more than eight isoprene units (high molecular terpenes) occur in microorganisms, plants, and animals. With respect to the chemical structure all-*E*-polyterpenes, e.g., gutta, all-*Z*-polyterpenes, e.g., rubber, and compounds with mixed stereochemistry may be distinguished.

Biosynthesis and Significance

By head-to-tail condensation of isoprene units (D 6) all-*E*-octaprenyl diphosphate, all-*E*-nonaprenyl (solanesyl) diphosphate, all-*E*-decaprenyl (spadicyl) diphosphate, and other all-E-prenyl diphosphates are formed in microorganisms, plants, and animals. They are the precursors of the isoprenoid side chain of terpenoid quinones and chromanols, e.g., of ubiquinones (D 21.3), plastoquinones, and tocopherylquinones (D 23.4).

Polyprenyl diphosphates with a mixed-type stereochemistry are obviously the precursors of dolichols, betulaprenols, and ficaprenols formed in higher plants and microorganisms. Dolichols participate in the transfer of glycosidically inter-linked sugar residues from nucleoside diphosphate sugars to proteins (biosynthesis of glycoproteins, D 1.4.2). In bacteria and plants other polyprenols may have a similar function.

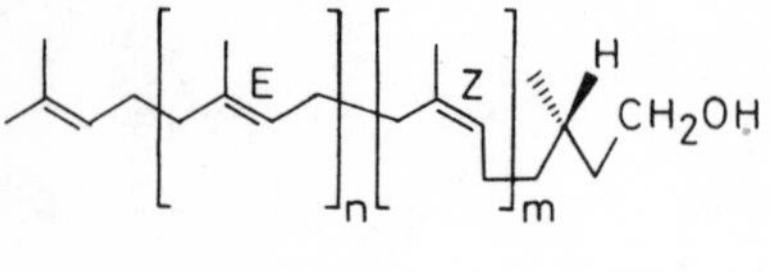

Dolichols (n = 2, m = 10 – 16)

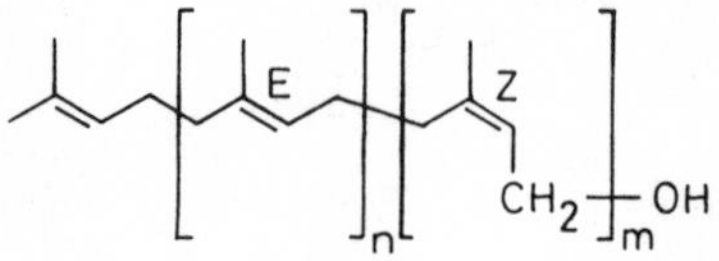

Betulaprenols (n = 2, m = 3 – 25)
Ficaprenols (n = 3, m = 5 – 9)

The most important all-Z-polyprenol is rubber (F 5).

Rubber is obtained from the latex of *Hevea brasiliensis* which occurs in a system of specialized, connected cells, the latex vessels. Latex contains rubber particles with a diameter of up to 3 µm. The rubber particles are suspended in the liquid cell plasm along with nuclei, mitochondria, ribosomes, soluble enzymes, etc. While formation of isopentenyl diphosphate occurs in the latex fluid, polymerization to rubber takes place on the surface of rubber particles. It is catalyzed by a polymerase located at the boundary between the particles and the aqueous phase. The rubber molecules formed vary in size from 500 to 5 000 isoprene units.

High molecular all-E-terpenes (gutta) have chemical and physical properties resembling those of rubber (F 5).

Reference for Further Reading

15

D 7 Derivatives of Homoisopentenyl Diphosphate

Chemistry, Distribution, Significance

The juvenile hormones I and II (Table 39) are isoprenoidlike compounds which contain one or two additional C-atoms if compared with juvenile hormone III, a true sesquiterpene (D 6.3). In the majority of insects they act as molting hormones (E 3.1)

Table 39. Biosynthesis of juvenile hormones.
The "additional" C-atoms derived from the methyl group of propionate are marked by "●".

PRECURSORS		HORMONES
Homoisopentenyl diphosphate	Isopentenyl diphosphate	
–	3	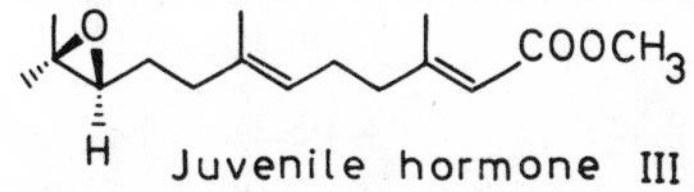 Juvenile hormone III
1	2	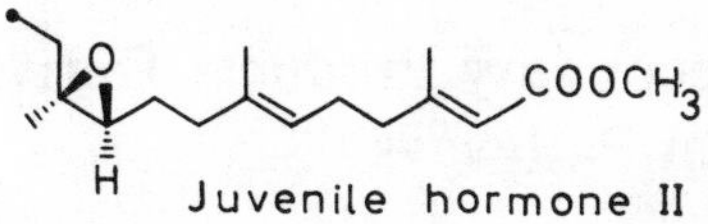 Juvenile hormone II
2	1	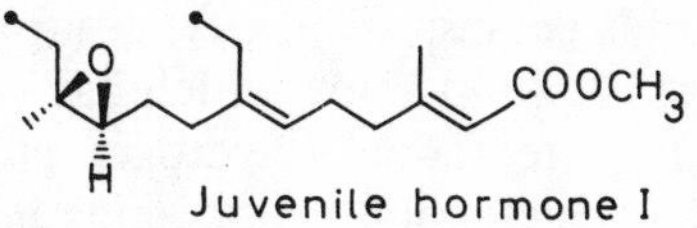 Juvenile hormone I

Biosynthesis and Degradation

Propionate is a precursor of the juvenile hormones I and II (in contrast to the juvenile hormone III). It is expected that with propionyl CoA (D 3) as the starter molecule, homoisopentenyl diphosphate is built in a pathway similar to that of isopentenyl diphosphate biosynthesis (Fig. 150). The different juvenile hormones

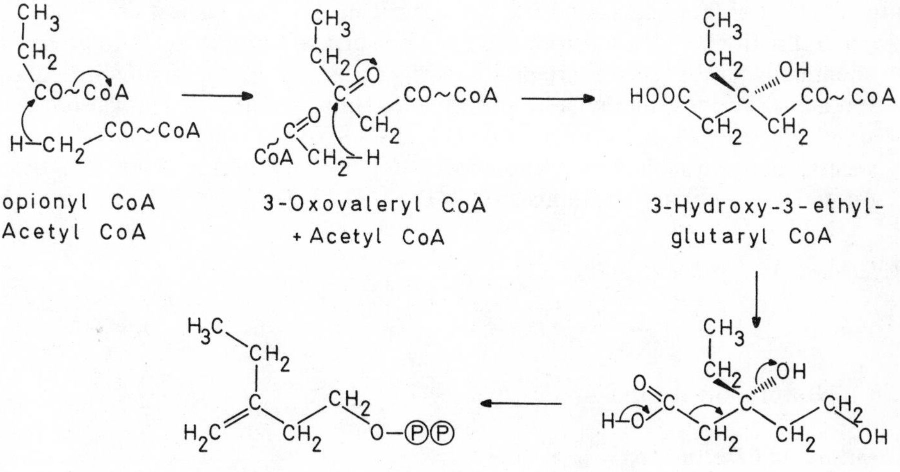

Propionyl CoA + Acetyl CoA   →   3-Oxovaleryl CoA + Acetyl CoA   →   3-Hydroxy-3-ethyl-glutaryl CoA

Homoisopentenyl diphosphate   ←   Homomevalonic acid

Fig. 150. Possible pathway of the biosynthesis of homoisopentenyl diphosphate

may be derived from homoisopentenyl diphosphate and isopentenyl diphosphate in the relations shown in Table 39.

Juvenile hormone degradation involves hydrolysis of the ester bond and hydratation of the epoxide group.

References for Further Reading

611, 638

D 8 Secondary Products Derived from Primary Amino Acids on General Pathways

Amino acids possess a carboxyl group and an amino or imino group. Most widespread are α-amino acids, which have the general formula R—CHNH$_2$—COOH. If R does not represent a hydrogen atom, α-amino acids have a chiralic center. The primary metabolic amino acids belong to the L-series [2(S)-amino acids].

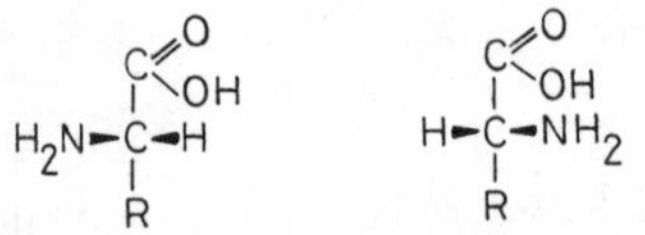

L-Amino acid   D-Amino acid

About 30 L-amino acids are regularly encountered in the biosynthesis of proteins or as intermediates in primary metabolism. They are the precursors of a large number of secondary products. Of significance in this respect are either

— general pathways leading from the different primary amino acids to groups of chemically similar secondary products, e.g., D-amino acids, N-substituted amino acids, amines, methylated amino acids and betanines, cyanogenic glycosides, glucosinolates, and alkaloids (D 8.1–D 8.5), or
— special pathways on which the individual primary amino acids are transformed to secondary substances (D 9–D 20, D 22, D 23).

References for Further Reading

31, 48

D 8.1 Nonprotein Amino Acids

Chemistry and Distribution

Amino acids not encountered in protein biosynthesis occur in all groups of organisms. They are heterogeneous in their chemical structures and may be derivatives of L-3-cyanoalanine (D 8.3), glycine (D 10), L-cysteine (D 11.1, D 11.2),

L-methionine (D 12), L-asparagine (D 16), L-glutamic acid, L-proline, L-ornithine (D 17), L-lysine (D 18), L-arginine (D 19), chorismic acid (D 21.1), L-tryptophan (D 22, 22.1, 22.5), L-phenylalanine, L-tyrosine (D 23), etc.

Nonprotein amino acids are usually found in the free state or as part of simple products such as γ-glutamyl, acetyl, or oxalyl derivatives. They also occur as constituents of small peptides, such as phalloin and α-amanitin (D 9.2). Several nonprotein amino acids are precursors of other groups of secondary products (cf. the formation of glucosinolates, D 8.4).

Biosynthesis

Formation of nonprotein amino acids proceeds either
— by biosynthesis on novel routes (cf. the formation of lathyrine, D 26).
— by modification of primary amino acids (see hydroxylation of L-tryptophan to 5-hydroxy-L-tryptophan, D 22.1), or
— by modification of the biosynthetic pathways leading to primary amino acids.

Examples are:
a) the formation of 3-carboxy-L-phenylalanine and 3-carboxy-L-tyrosine from chorismic acid, a normal intermediate in the formation of L-phenylalanine and L-tyrosine (D 23.1), and
b) the biosynthesis of selenium-containing amino acids, like L-selenocysteine, which is a constituent of glutathione peroxidase. L-Selenocysteine and the related selenoamino acids given below are accumulated in several plants, e.g., *Astragalus* sp. They are of ecological importance due to their toxicity to potential predators (E 5.5.3).

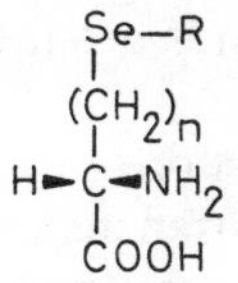

Se—R	L-Selenocysteine	(n=1, R=H)
(CH$_2$)$_n$	L-Selenohomocysteine	(n=2, R=H)
H—C—NH$_2$	L-Methylselenocysteine	(n=1, R=CH$_3$)
COOH	L-Selenomethionine	(n=2, R=CH$_3$)

The formation of L-selenocysteine, L-selenohomocysteine, and L-selenomethionine from selenate (SeO_4^{2-}) and selenite (SeO_3^{2-}) resembles the biosynthesis of the sulfur analogs L-cysteine, L-homocysteine, and L-methionine from sulfate (D 11 and D 12). L-Selenocysteine, for instance, is built by the reaction given in Fig. 151.

Modification of primary amino acids leads to the following groups of nonprotein amino acids:

a) D-Amino Acids. D-Amino acids have the configuration shown above. They occur in microorganisms, plants, and animals. D-Alanine, D-serine, D-cysteine, D-glutamic acid, and D-ornithine are especially widespread. D-Amino acids constitute a major portion of the free amino acids in the body fluid of certain insects and have also been found in human urine (E 1).

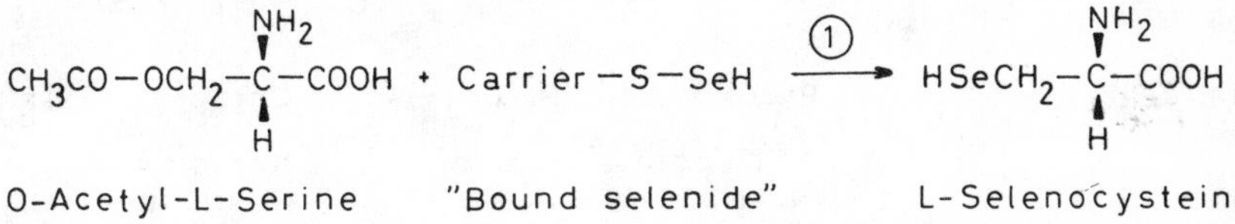

Fig. 151. Biosynthesis of L-selenocysteine
1 selenocysteine synthase

D-Amino acids are formed by the following reactions
— racemization of L-amino acids by pyridoxal-5′-phosphate-containing enzymes (C 5).

Probably the Schiff base between the amino group and pyridoxal phosphate is transformed to the R-configuration by the transient elimination of a proton (Fig. 29, intermediate I).

— racemization of L-amino acids by flavin-containing enzymes (C 2.2.2) via imino acids formed by reversible dehydrogenation.
— transamination by specific D-amino acid transaminases (C 5) according to the following equation:

D-amino acid$_1$ + 2-oxo acid$_2$ ⇌ D-amino acid$_2$ + 2-oxo acid$_1$

Bound L-amino acids may also be racemized (see formation of penicillin N and cephalosporin C, D 9.2.3).

The degradation of D-amino acids is catalyzed by specific D-amino acid oxydases (C 2.2.2).

b) *N*-Acylated Amino Acids. These compounds are amides usually formed by the reaction of activated acid derivatives, e.g., carboxylic acid CoA esters, with the amino group of L- or D-amino acids (exceptions are, however, the formyl derivatives which are built by formyltransferases, C 3.2). Most widespread are:
— formyl derivatives (*N*-formyl-L-methionine, for instance, is the starter in the biosynthesis of proteins in bacteria)
— acetyl derivatives (*N*-acetylated D-amino acids have been found in fungi and other organisms. *N*-Acetylated L-amino acids are involved in the synthesis of methionine, D 12)
— malonyl derivatives (*N*-malonyl derivatives of D-amino acids are synthesized in several plants by D-amino acid malonyltransferases, and
— oxalyl derivatives (4-*N*-oxalyl-L-2,4-diaminobutyric acid is formed in Leguminosae, D 8.3).

Free amino acids, as well as bound amino acid residues, may be acylated.

c) Opines. In crown gall tumors (A 2.1) and "hairy roots" 2-oxo acids condense with basic amino acids, e.g., L-arginine, L-histidine, and L-lysine, to the so-called opines (Fig. 152). The reaction takes place with participation of either pyruvate or 2-oxo glutarate. The opines serve as nutrients for the crown

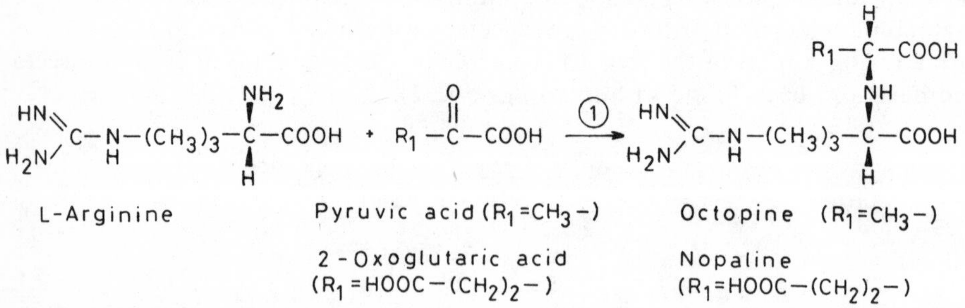

Fig. 152. Biosynthesis of opines
1 Octopine synthase or nopaline synthase

gall-inducing microorganisms (*Agrobacterium tumefaciens* and *Agrobacterium rhizogenes*).

d) *N*-Methylated Amino Acids and Betaines. The hydrogen atoms of the amino or imino groups of amino acids may be substituted by methyl groups (formation of methylated amino acids) and the nitrogen atom may be quarternized (for-

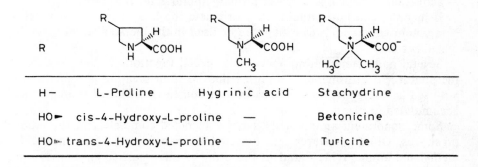

R	R–CH(NH₂)–COOH	R–CH(NHCH₃)–COOH	R–CH(N(CH₃)₂)–COOH	R–CH(N⁺(CH₃)₃)–COO⁻
H–	Glycine	Sarcosine	N,N-Dimethyl-glycine	Betaine
L-Histidine (imidazole CH₂)	L-Histidine	—	—	Hercynine
Indolyl-CH₂–	L-Tryptophan	Abrine	N,N-Dimethyl-L-Tryptophan	Hypaphorine
R₁—(phenyl)—CH₂–	L-Tyrosine (R₁=OH, R₂=H)	—	N,N-Dimethyl-L-tyrosine	L-Tyrosine betaine
	L-DOPA (R₁=R₂=OH)	—	—	L-DOPA betaine

R	pyrrolidine-COOH (NH)	pyrrolidine-COOH (NCH₃)	pyrrolidine-COO⁻ (N⁺(CH₃)₂)
H–	L-Proline	Hygrinic acid	Stachydrine
HO►	cis-4-Hydroxy-L-proline	—	Betonicine
HO˖˖	trans-4-Hydroxy-L-proline	—	Turicine

piperidine-COOH (NH)	piperidine-COOH (NCH₃)	piperidine-COO⁻ (N⁺(CH₃)₂)
L-Pipecolic acid	N-Methyl-L-pipecolic acid	Homostachydrine

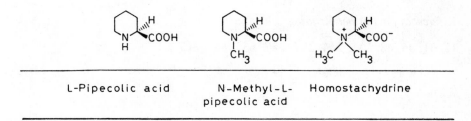

Fig. 153. Formulas of several methylated amino acids and betaines

mation of betaines). Some important methylated amino acids and betaines are shown in Fig. 153. The methylations are catalyzed by S-adenosyl-L-methionine-dependent methyltransferases (C 3.3) which may either react with free amino acids or with bound amino acid moieties.

Most methylated amino acids and betaines occur in the free state. N^6-Trimethyl-L-lysine (D 18), different L-arginine derivatives methylated in the guanidine group, and others, however, are constituents of various proteins and polypeptides. Methylation of these residues occurs at the polypeptide level.

The methylated amino acids are set free during the degradation of the methylated proteins.

e) 2,3-Dehydroamino Acids (R—CH=C(NH$_2$)—COOH). Amino acids of this type are constituents, e.g., of cyclic peptides, e.g., dehydrocyclopeptine (D 21.4.2). They are unstable in the free state.

Significance

Nonprotein amino acids may be used for storage of nitrogen.

Seeds and roots of plants, for instance, may contain these substances in high concentration. 5-Hydroxy-L-tryptophan (D 22.1) accounts for 14 % of the seed weight of *Griffonia simplicifolia*, L-DOPA (D 23) for more than 8 % of the seed weight of *Mucuna mutisiana*, and N^5-acetyl-L-ornithine (D 17) for 10 % of the roots of *Corydalis cava*.

In addition, most nonprotein amino acids are toxic and function as feeding deterrents (E 5.5.3). In nonadapted organisms they may interfere with
— uptake and transport of protein amino acids,
— biosynthesis of primary amino acids, and/or
— activation and incorporation of primary amino acids into proteins.

If incorporated into proteins they originate the loss of biological functions. Nonprotein amino acids can therefore be used in the metabolism of specialized organisms only.

Several nonprotein amino acids have special functions in the physiology of the producer organisms, cf. L-arginine phosphate or creatine phosphate, which are used as energy-rich substances in metabolism (E 2.2), or betaine, which is accumulated in plants as an osmotic (E 2.2).

Some nonprotein amino acids may be degraded completely in the producer organisms. Others, however, are excreted unchanged, e.g., D-amino acids or methylated amino acids found in the urine of mammals (E 1), or they are removed from the body after slight modification, e.g., creatinine (E 1).

References for Further Reading

12, 31, 45, 48, 131, 244, 287, 387, 655, 680, 807

D 8.2 Amines

Chemistry and Distribution

Amines are synthesized in all classes of organisms. They may be divided in primary amines ($R-NH_2$), secondary amines (R_1,R_2-NH), tertiary amines (R_1,R_2R_3N), and quarternary amines ($R_1,R_2,R_3,R_4N^+OH^-$). Secondary amines may react with nitrous acid to give nitrosamines, e.g., dimethylnitrosamine, $(CH_3)_2N-N \rightarrow O$, a carcinogenic compound isolated from *Solanum incanum*. Tertiary amines may form *N*-oxides, e.g., trimethylamine *N*-oxide, $(CH_3)_3N \rightarrow O$, found in algae. Monoamines, e.g., ethylamine and n-propylamine, contain one amino group. Diamines, e.g., putrescine and cadaverine, have two amino groups. Polyamines, e.g., spermidine and spermine (D 12.1), comprise three or more amino groups.

Biosynthesis

Amines are formed by the following reactions:

a) Decarboxylation of Amino Acids. Amino acid decarboxylases (C 5) occur in microorganisms, plants, and animals. Some amines formed by decarboxylation are collected in Table 40.

b) Transamination of Aldehydes. This reaction is the most common pathway of amine formation in plants. It is catalyzed by aminotransferases (C 5) according to the following equation:

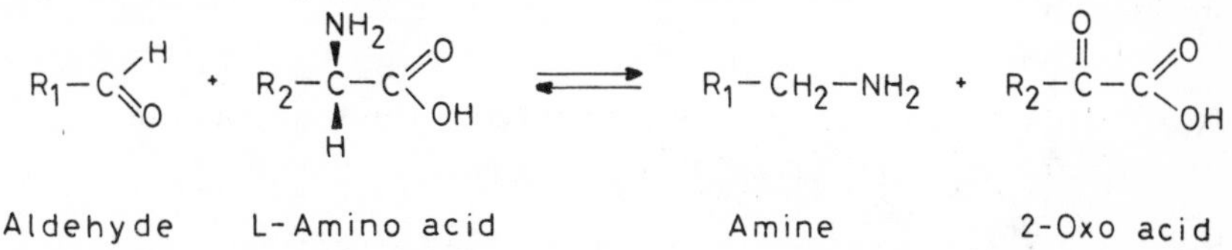

Aldehyde   L-Amino acid   Amine   2-Oxo acid

L-Alanine, L-glutamine, and other amino acids act as amino group donors. The amines given in Table 41 are synthesized by transamination in microorganisms or higher plants. Amines are degraded by amine oxidases (C 2.3.2) found in all classes of organisms. These enzymes usually possess a relatively low substrate specificity.

Significance

Volatile amines, in most instances monoamines, are constituents of flower scents attracting pollinators (E 5.5.1). They cause also the smell of the fetid morel *(Phallus impudicus)* which attracts carrionfeeding insects distributing the spores of this mushroom. 4-Aminobutyric acid, dopamine, serotonin, and histamine serve as hormones and/or neurotransmitters in animals (E 3.1 and E 3.2).

References for Further Reading

287, 690, 716, 751, 752, 817, 852

Table 40. Amines formed by decarboxylation of amino acids

Formula	Amino acid $R=COOH$	Amine $R=H$	Enzymes
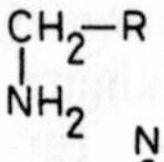 CH_2-R $\mid$ NH_2	Glycine (D 10)	Methylamine	
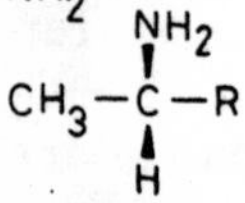 $CH_3-\overset{NH_2}{\underset{H}{C}}-R$	L-Alanine (D 10)	Ethylamine	
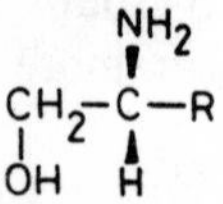 $\underset{OH}{CH_2}-\overset{NH_2}{\underset{H}{C}}-R$	L-Serine (D 10)	Ethanolamine (cholamine)	
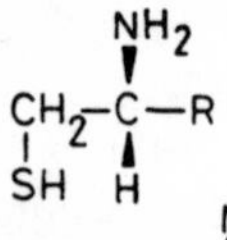 $\underset{SH}{CH_2}-\overset{NH_2}{\underset{H}{C}}-R$	L-Cysteine (D 11)	2-Thioethyl- amine (cysteamine)	
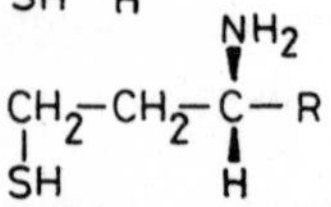 $\underset{SH}{CH_2}-CH_2-\overset{NH_2}{\underset{H}{C}}-R$	L-Homocysteine (D 12)	3-Thiopropyl- amine	
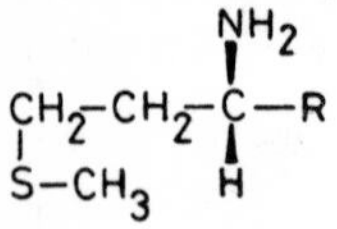 $\underset{S-CH_3}{CH_2}-CH_2-\overset{NH_2}{\underset{H}{C}}-R$	L-Methionine (D 12)	3-Methylthio- propyl amine	Methionine decarboxylase
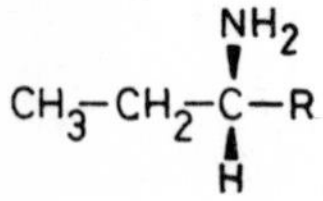 $CH_3-CH_2-\overset{NH_2}{\underset{H}{C}}-R$	L-α-Aminobutyric acid	n-Propylamine	
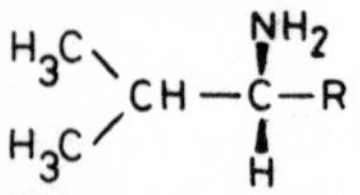 $\underset{H_3C}{\overset{H_3C}{>}}CH-\overset{NH_2}{\underset{H}{C}}-R$	L-Valine (D 13)	i-Butylamine	Valine decarboxylase
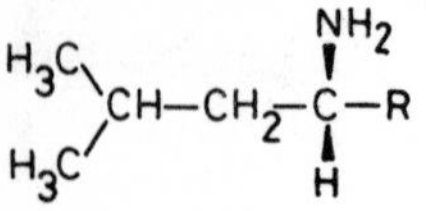 $\underset{H_3C}{\overset{H_3C}{>}}CH-CH_2-\overset{NH_2}{\underset{H}{C}}-R$	L-Leucine (D 14)	i-Amylamine	
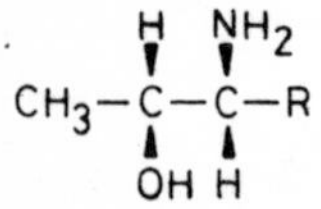 $CH_3-\overset{H}{\underset{OH}{C}}-\overset{NH_2}{\underset{H}{C}}-R$	L-Threonine (D 16)	1-Amino-2- propanol	
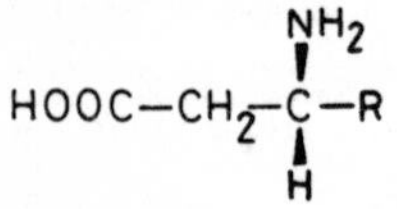 $HOOC-CH_2-\overset{NH_2}{\underset{H}{C}}-R$	L-Aspartic acid (D 16)	β-Alanine	Glutamate decarboxylase
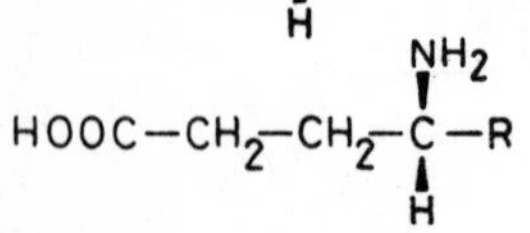 $HOOC-CH_2-CH_2-\overset{NH_2}{\underset{H}{C}}-R$	L-Glutamic acid (D 17)	4-Aminobutyric acid	Glutamate decarboxylase

Formula	Amino acid R = COOH	Amine R = H	Enzyme
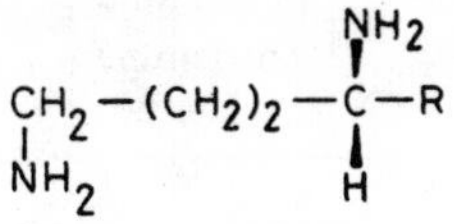 $CH_2-(CH_2)_2-\overset{\overset{NH_2}{\mid}}{\underset{\underset{H}{\mid}}{C}}-R$, NH_2	L-Ornithine (D 17)	Putrescine	Ornithine decarboxylase
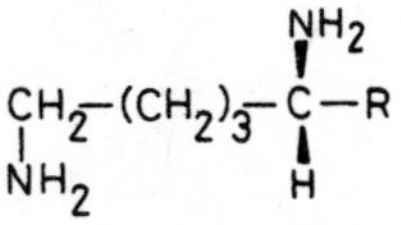 $CH_2-(CH_2)_3-\overset{\overset{NH_2}{\mid}}{\underset{\underset{H}{\mid}}{C}}-R$, NH_2	L-Lysine (D 18)	Cadaverine	Lysine decarboxylase
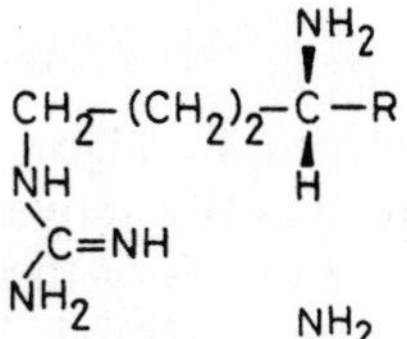 $CH_2-(CH_2)_2-\overset{\overset{NH_2}{\mid}}{\underset{\underset{H}{\mid}}{C}}-R$, NH, C=NH, NH_2	L-Arginine (D 19)	Agmatine	Arginine decarboxylase
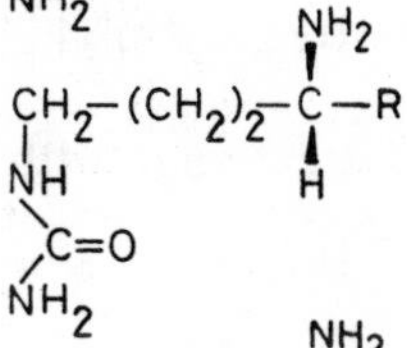 $CH_2-(CH_2)_2-\overset{\overset{NH_2}{\mid}}{\underset{\underset{H}{\mid}}{C}}-R$, NH, C=O, NH_2	L-Citrulline (D 19)	N-Carbamyl putrescine	
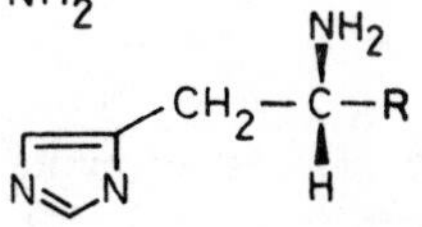 (imidazole ring)$-CH_2-\overset{\overset{NH_2}{\mid}}{\underset{\underset{H}{\mid}}{C}}-R$	L-Histidine (D 20)	Histamine	Histidine decarboxylase
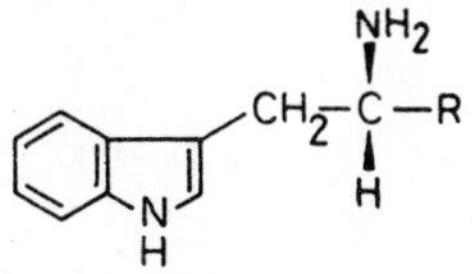 (indole ring)$-CH_2-\overset{\overset{NH_2}{\mid}}{\underset{\underset{H}{\mid}}{C}}-R$	L-Tryptophan (D 21)	Tryptamine	Aromatic-L-amino-acid decarboxylase
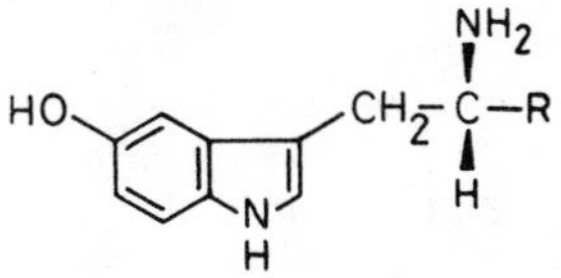 HO-(indole ring)$-CH_2-\overset{\overset{NH_2}{\mid}}{\underset{\underset{H}{\mid}}{C}}-R$	L-5-Hydroxy-tryptophan (D 21.1)	Serotonin	Aromatic-L-amino-acid decarboxylase
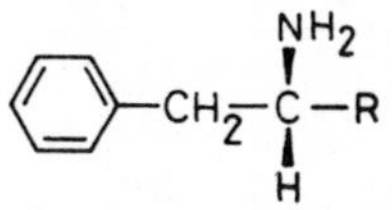 (phenyl ring)$-CH_2-\overset{\overset{NH_2}{\mid}}{\underset{\underset{H}{\mid}}{C}}-R$	L-Phenylalanine (D 22)	Phenylethyl-amine	Phenylalanine decarboxylase
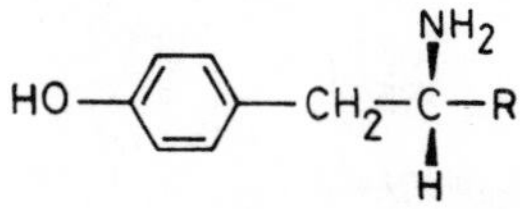 HO-(phenyl ring)$-CH_2-\overset{\overset{NH_2}{\mid}}{\underset{\underset{H}{\mid}}{C}}-R$	L-Tyrosine (D 22)	Tyramine	Tyrosine decarboxylase
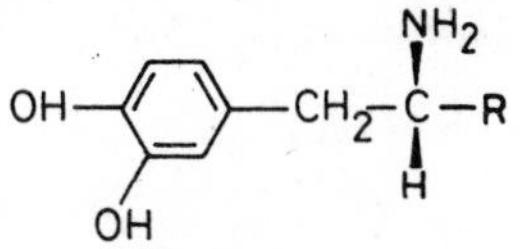 OH-(phenyl ring)-OH$-CH_2-\overset{\overset{NH_2}{\mid}}{\underset{\underset{H}{\mid}}{C}}-R$	L-3,4-Dihydroxy-phenylalanine (D 22)	Dopamine	Aromatic-L-amino-acid decarboxylase

Table 41. Primary amines originating from aldehydes by transamination

Aldehyde	Amine	Aldehyde	Amine
Formaldehyde	Methylamine	i-Butyraldehyde	i-Butylamine
Acetaldehyde	Ethylamine	i-Valeraldehyde	i-Amylamine
Propionaldehyde	n-Propylamine	Hexanal	Hexylamine

D 8.3 Cyanogenic Glycosides

Chemistry and Distribution

Cyanogenic glycosides are derivatives of 2-hydroxynitriles formed in higher plants and some insects, e.g., *Zygaena* and *Heliconius* sp. They are optically active due to the chirality of the hydroxylated C-atom 2 which has either the *S*- or the *R*-configuration (Table 42). The sugar moiety in most representatives is β-D-glucose.

Cyanogenic glycosides release hydrocyanic acid (HCN) on treatment with acids or on incubation with glycosidases.

Biosynthesis (Fig. 154)

Cyanogenic glycosides are formed from L-amino acids (Table 42). An enzyme complex comprising enzymes 1, 2, 4, and 5 yields a 2-hydroxynitrile (A 3.1.1),

Table 42. Amino acid precursors of cyanogenic glycosides

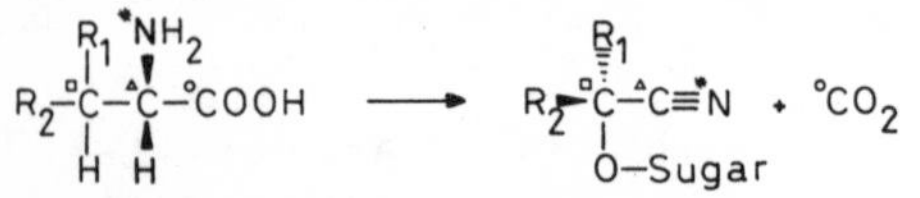

L-Amino acid	Cyanogenic Glycoside	R_1	R_2	Sugar	Configuration at C-2
L-Valine	Linamarin	$-CH_3$	$-CH_3$	Glucose	
L-Isoleucine	Lotaustralin	$-CH_2-CH_3$	$-CH_3$	Glucose	R
L-Leucine	Proacacipetalin	$-H$	$-C{\overset{CH_3}{\underset{CH_2}{}}}$	Glucose	S
L-Phenylalanine	Prunasin	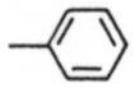 $-\langle\bigcirc\rangle$	$-H$	Glucose	R
	Sambunigrin	$-H$	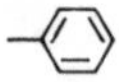 $-\langle\bigcirc\rangle$	Glucose	S
	Amygdalin	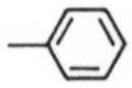 $-\langle\bigcirc\rangle$	$-H$	Gentiobiose	R
L-Tyrosine	Taxiphyllin	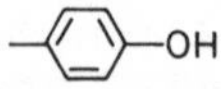 $-\langle\bigcirc\rangle-OH$	$-H$	Glucose	R
	Dhurrin	$-H$	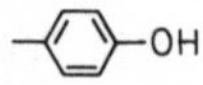 $-\langle\bigcirc\rangle-OH$	Glucose	S

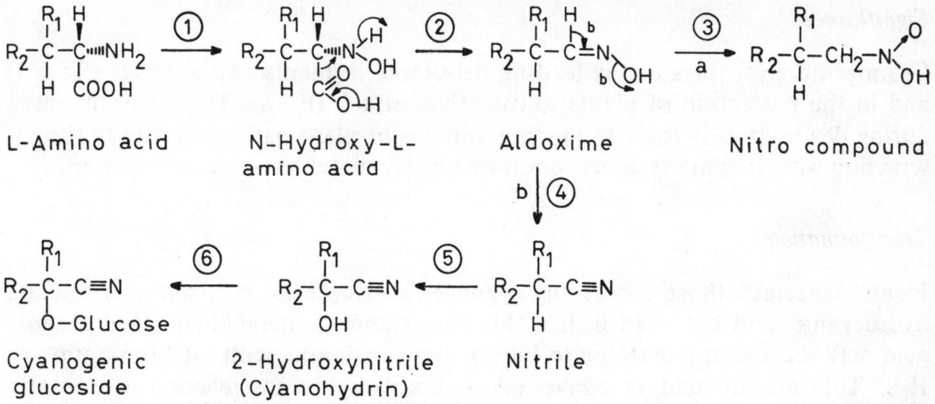

$$R_2-\overset{R_1}{\underset{H}{\underset{|}{C}}}-\overset{H}{\underset{COOH}{\underset{|}{C}}}-NH_2 \quad \overset{①}{\longrightarrow} \quad R_2-\overset{R_1}{\underset{H}{\underset{|}{C}}}-C-N\overset{H}{\underset{OH}{}} \quad \overset{②}{\longrightarrow} \quad R_2-\overset{R_1}{\underset{H}{\underset{|}{C}}}-C=N\overset{b}{\underset{OH}{}} \quad \overset{③}{\underset{a}{\longrightarrow}} \quad R_2-\overset{R_1}{\underset{H}{\underset{|}{C}}}-CH_2-N\overset{O}{\underset{OH}{}}$$

L-Amino acid   N-Hydroxy-L-amino acid   Aldoxime   Nitro compound

$$b \Big\downarrow ④$$

$$R_2-\overset{R_1}{\underset{O-Glucose}{\underset{|}{C}}}-C\equiv N \quad \overset{⑥}{\longleftarrow} \quad R_2-\overset{R_1}{\underset{OH}{\underset{|}{C}}}-C\equiv N \quad \overset{⑤}{\longleftarrow} \quad R_2-\overset{R_1}{\underset{H}{\underset{|}{C}}}-C\equiv N$$

Cyanogenic   2-Hydroxynitrile   Nitrile
glucoside   (Cyanohydrin)

Fig. 154. Formation of cyanogenic glycosides from amino acids
1 L-Amino acid *N*-monooxygenase; *2* *N*-hydroxy amino acid dehydrogenase (oxidative decarboxylating); *3* aldoxime monooxygenase; *4* aldoxime dehydratase; *5* nitrile monooxygenase; *6* glucosyltransferase

which is glucosylated in the last step. In a side reaction nitro compounds may be formed (Fig. 154). Of special significance in the biosynthesis of triglochinin is the oxidative splitting of the aromatic ring of L-tyrosine.

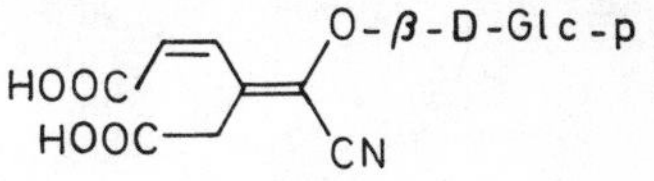

$$\begin{array}{c} HOOC \\ HOOC \end{array}\diagdown \cdots \diagup \overset{O-\beta-D-Glc-p}{\underset{CN}{}}$$

Triglochinin

Degradation of cyanogenic glycosides begins with hydrolysis and release of hydrocyanic acid (HCN) (Fig. 155). By means of pulse-chase experiments (B 3.3) it was demonstrated that continuous synthesis and degradation of cyanogenic glycosides takes place and that the balance of these reactions determines the concentration of cyanogenic glycosides in the tissues of cyanogenic plants.

The best-known *β*-glycosidase degrading cyanogenic glycosides is emulsin, which occurs in seeds of *Prunus amygdalus*. The natural substrate of emulsin is amygdalin. The substrate specificity of the enzyme, however, is low and in addition to amygdalin a large number of other *β*-glycosides are hydrolized.

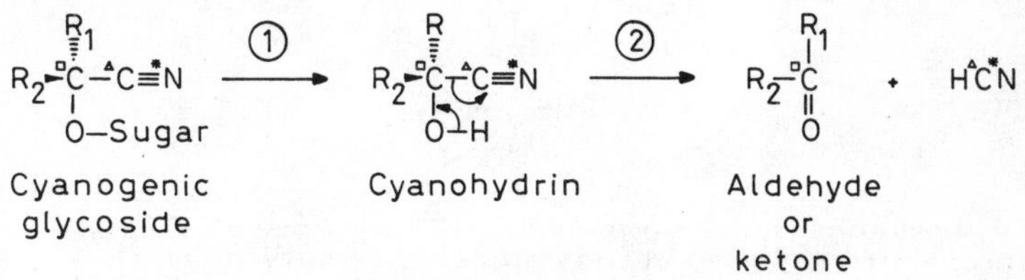

$$R_2-\overset{R_1}{\underset{O-Sugar}{\underset{|}{C}}}-C\equiv N \quad \overset{①}{\longrightarrow} \quad R_2-\overset{R}{\underset{O-H}{\underset{|}{C}}}-C\equiv N \quad \overset{②}{\longrightarrow} \quad R_2-\overset{R_1}{\underset{O}{\underset{\|}{C}}} \quad + \quad H\overset{\cdot\cdot}{C}N$$

Cyanogenic   Cyanohydrin   Aldehyde
glycoside   or
                                 ketone

Fig. 155. The degradation of cyanogenic glycosides
1 *β*-Glucosidase; *2* hydroxynitrile lyase (e.g., linamarase and emulsin)

Significance

Cyanogenic glycosides act as feeding deterrents in plants and animals (E 5.5.3) and in the protection of plants to microbial attack (E 5.4). The HCN liberated during destruction is toxic to grazing animals or plant pathogens due to the interaction with enzyme systems containing heavy metal ions, e.g., cytochromes.

Transformation

Plants, especially those containing cyanogenic glycosides, may not only tolerate hydrocyanic acid but even utilize this compound in metabolism. Hydrocyanic acid may react with L-cysteine to form L-3-cyanoalanine with the elimination of H_2S. This amino acid is converted to L-asparagine by reaction with water (Fig. 156). In this way hydrocyanic acid is converted to the —CO—NH$_2$ grouping, which can be further utilized in primary metabolism. Since the conversion to L-asparagine takes place very quickly, L-3-cyanoalanine is detectable only in minute quantities in most plants assimilating hydrocyanic acid. In *Vicia sativa* the predominant form of L-3-cyanoalanine is the γ-glutamylderivative.

In *Lathyrus sp.* L-3-cyanoalanine is a precursor of other secondary products (Fig. 156). On the one hand, 3-aminopropionitrile is formed by decarboxylation and, on the other hand, L-2,4-diaminobutyric acid, an ornithine homologue, is built by reduction. Oxalyl and glutamyl derivatives of these amino acids occur in Leguminosae.

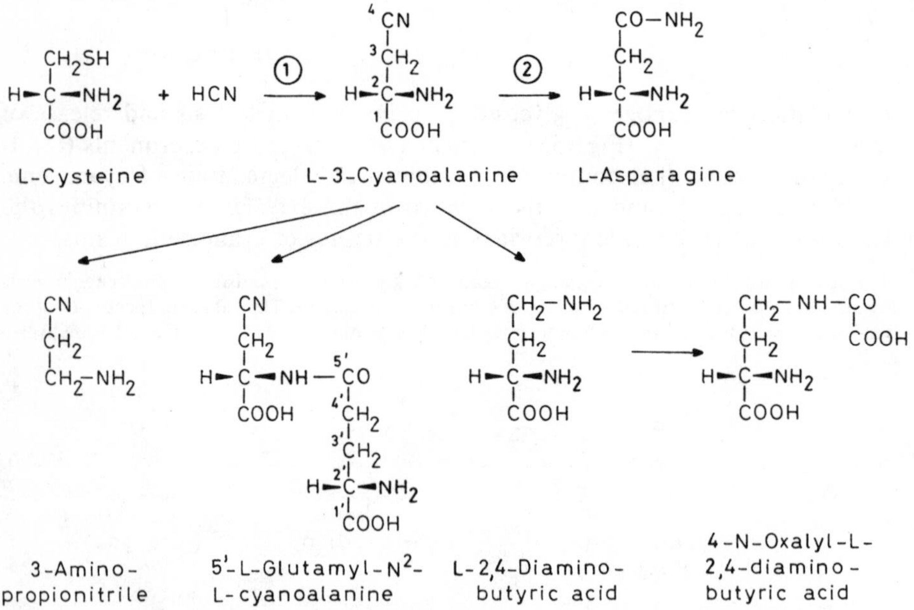

Fig. 156. Synthesis and transformation of L-3-cyanoalanine in higher plants
1 3-Cyanoalanine synthase (coenzyme: pyridoxal-5'-phosphate); *2* 3-cyanoalanine hydrolase

3-Aminopropionitrile, 2,4-diaminobutyric acid, and their derivatives cause the syndrome of lathyrism in animals or human beings which is characterized either by nervous disorders (neurolathyrism) or by anomalies of bone and mesenchymal tissue (osteolathyrism; cf. E 5.5.3). 3-Aminopropionitrile also inhibits growth and development of plants in the neighborhood of *Lathyrus odorata* seedlings (E 5.3).

References for Further Reading

148, 149, 191, 551, 656, 786, 848

D 8.4 Glucosinolates and Products of their Hydrolysis

Chemistry and Distribution

Glucosinolates are thioglucosides. They contain a C=N-group substituted with sulfate and a residue R in anti position, as well as a sulfur atom connected with a β-D-glucosyl moiety (Table 43). On hydrolysis they liberate D-glucose, sulfate, and an unstable aglycone. From the latter isothiocyanates (mustard oils), thiocyanates, nitriles, or cyanoepithioalkanes may be formed.

Glucosinolates have been found in a limited number of families of dicotyledonous plants (e.g., Cruciferae, Capparaceae, Resedaceae, and Tropaeolaceae).

Biosynthesis

Glucosinolates are formed from amino acids (Table 43). Either protein or nonprotein amino acids function as precursors. Relatively often the nonprotein amino acids are homologues of protein amino acids built by the pathway of chain elongation described for the biosynthesis of L-leucine from L-valine (D 14).

Aldoximes are the first intermediates of the biosynthetic pathway which are firmly established. They are probably formed by the reactions shown to proceed

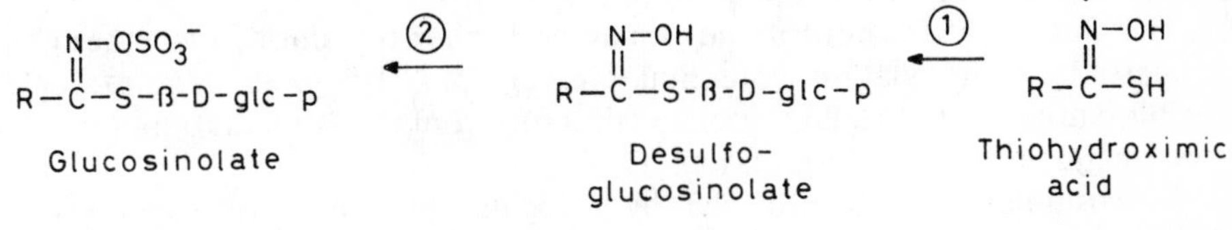

Fig. 157. Pathway of glucosinolate biosynthesis
1 UDP-glucose: thiohydroximate glucosyltransferase; *2* sulfotransferase (using PAPS, D 11)

Table 43. Precursors of glucosinolates and mustard oils

$$\underset{\substack{\uparrow\\H}}{\overset{\overset{\bullet}{N}H_2}{R-\overset{\bullet}{C}-\overset{\circ}{C}OOH}} \longrightarrow \underset{}{\overset{\overset{\bullet}{N}-SO_3^-}{R-\overset{\bullet}{C}-S-Glucose}} \longrightarrow R-\overset{\bullet}{N}\overset{\bullet}{=}\overset{\bullet}{C}=S$$

L-Amino acid	Glucosinolate	Mustard oil	R
Protein amino acids			
L-Leucine	Isopropylglucosinolate (Glucoconringiin)	Isopropyl mustard oil	$(CH_3)_2CH-$
L-Phenylalanine	Benzylglucosinolate (Glucotropaeolin)	Benzyl mustard oil	$C_6H_5-CH_2-$
L-Tyrosine	4-Hydroxybenzylglucosinolate	4-Hydroxybenzyl mustard oil	$HO-C_6H_4-CH_2-$
L-Tryptophan	3-Indolylmethyl glucosinolate (Glucobrassicin)	3-Indolylmethyl mustard oil	3-indolyl-CH_2-
Homologs of protein amino acids			
L-Homophenylalanine	Phenylethylglucosinolate (Gluconasturtiin)	Phenylethyl mustard oil	$C_6H_5-CH_2-CH_2-$
L-Homomethionine	ω-Methylthiopropyl-glucosinolate (Glucoibervirin)	ω-Methylthiopropyl mustard oil	$CH_3-S-CH_2-CH_2-CH_2-$
	Allylglucosinolate	Allyl mustard oil	$CH_2=CH-CH_2-$
L-Dihomomethionine	3-Butenylglucosinolate (Gluconapin)	3-Butenyl mustard oil	$CH_2=CH-CH_2-CH_2-$
L-Trihomomethionine	4-Pentenylglucosinolate (Glucobrassicanapin)	4-Pentenyl mustard oil	$CH_2=CH-CH_2-CH_2-CH_2-$

in the biosynthesis of cyanogenic glycosides (D 8.3). The first well-defined sulfur-containing intermediates are thiohydroximic acids. The most efficient source of the sulfur atom is L-cysteine. Probably nitro compounds function as the oxime-derived sulfur acceptors. No definite information exists about the configuration around the C=N-bond in the intermediates. It is assumed, however, that it is that found in the glucosinolates (cf. Fig. 157).

The glucosinolates built de novo may be further transformed to other glucosinolates by hydroxylation, oxidation, cleavage of CH_3S-, etc. One example is the biosynthesis of allylglucosinolate (sinigrin) from the ω-methyl thiopropyl derivative glucoibervirin (formulas cf. Table 43).

Glucosinolates are hydrolyzed by thioglucoside glucohydrolases (myrosinases) to D-glucose and an aglycone, which in dependence on the chemical structure of the glucosinolate, environmental conditions, additional factors, e.g., the

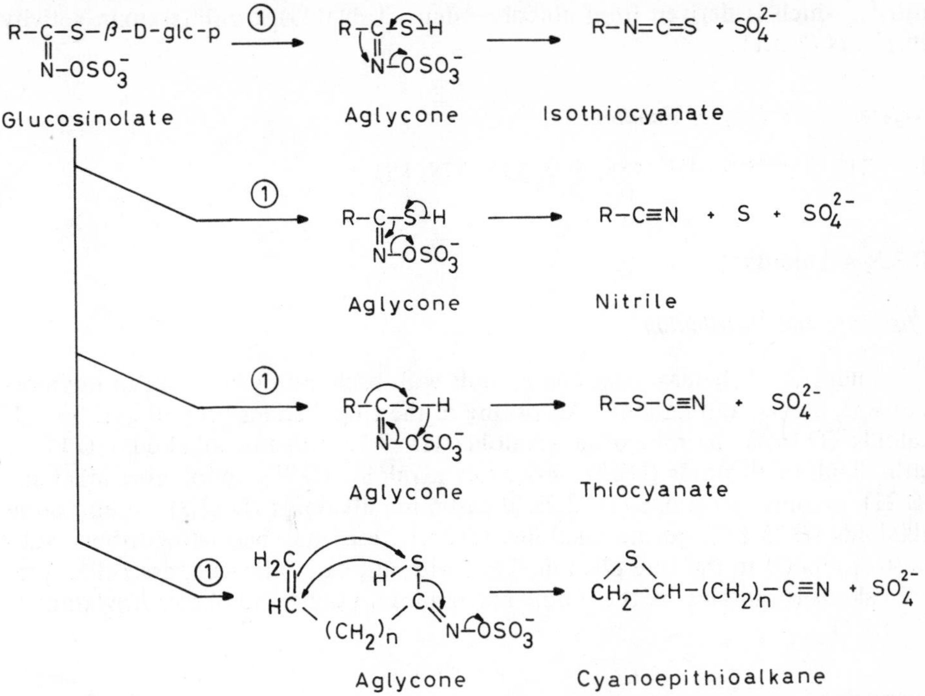

Fig. 158. Degradation of glucosinolates
1 Thioglucoside glucohydrolase (myrosinase)

epithiospecifier protein, etc., yields isothiocyanates, nitriles, thiocyanates, or cyanoepithioalkanes (Fig. 158). Isothiocyanates with a hydroxyl group in position 2 cyclize spontaneously to substituted oxazolidine-2-thiones (Fig. 159). Whereas myrosinase is localized in the cell wall and in membrane-surrounded vesicles, glucosinolates (and L-ascorbic acid, a myrosinase activator) are accumulated in the vacuoles. Cell disintegration or increase of membrane permeability combine myrosinase and its substrate and cause hydrolysis.

Significance

The isothiocyanates formed from glucosinolates after destruction of plant tissues are responsible for the pungent flavor and odor of the common species of Cruciferae used as vegetables and condiments (cabbage, radish, horseradish, mustard, F 1). They deter predators (E 5.5.3) and protect plants from microbial attack (E 5.4), but on the other hand attract specialized animals (E 5.5.3). Indoleaceto-

Fig. 159. Spontaneous formation of oxazolidine-2-thiones

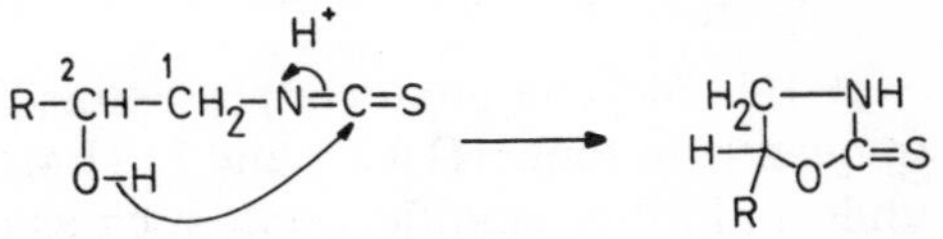

nitrile, which is derived from glucobrassicin (Table 43), exhibits auxin activity in plants (E 3.1).

References for Further Reading

149, 217, 336, 391, 450, 489, 509, 599, 779, 891

D 8.5 Alkaloids

Chemistry and Distribution

Alkaloids, i.e., N-heterocyclic compounds with basic nitrogen, occur in microorganisms, plants, and animals. According to the ring systems present pyridine alkaloids (D 16.1), pyrrolizidine alkaloids (D 17.1), tropane alkaloids (D 17.2), quinolizidine alkaloids (D 18), imidazole alkaloids (D 20), indolenine alkaloids (D 22), ergoline alkaloids (D 22.2), β-carboline alkaloids (D 22.3), isoquinoline alkaloids (D 23.1.2), purine alkaloids (D 25.1), etc. may be distinguished. Substances related to the true alkaloids, but with aliphatic nitrogen, are called protoalkaloids (examples are the indole alkylamines, D 22.1 and phenylethylamines, D 23.1.1).

Biosynthesis

Most alkaloids are derived from amino acids. Pyrrolizidines and tropanes, for instance, are formed from compounds of the L-glutamic acid-L-proline-L-ornithine group (D 17.1, D 17.2); quinolizidines are derived from L-lysine (D 18); imidazole alkaloids come from L-histidine (D 20); indole alkaloids from L-tryptophan (D 22); isoquinolines arise from L-tyrosine and L-DOPA (D 23.1.2), etc. Pseudoalkaloids are formed by the introduction of ammonia to nitrogen-free precursors such as isoprenoid substances (D 6.2.2, D 6.6.1, 6.6.3, etc.).

The biosynthesis of alkaloids from amino acids with aliphatic nitrogen requires the formation of heterocyclic rings, which is brought about by the formation of new C—N-bonds. Intramolecular as well as intermolecular reactions may be involved. Most important are the following:

a) Formation of Carboxylic Acid Amide Bonds. Carboxylic acid amides originate by reaction of an activated carboxylic acid with an amine. Normal carboxylic acids react as their CoA-esters (C 1.2), amino acids as adenylates or as thioesters with SH-groups of the participating enzymes (A 3.1.1). During formation of the carboxylic acid amide the activated group is eliminated:

$$R_1-C{\overset{O}{\underset{X}{\diagdown}}} \quad + \quad H_2N-R_2 \quad \longrightarrow \quad R_1-C{\overset{O}{\underset{NH-R_2}{\diagdown}}} \quad + \quad X-H$$

N-Heterocyclic products possessing carboxylic acid amide groups are found frequently in fungi (D 9.2.1 and D 21.4.2). The carboxylic acid amide group may undergo further modifications, such as oxygenation (D 9.2.2).

b) Synthesis of Azomethine Bonds. Azomethines (Schiff bases) may be formed either spontaneously or catalyzed by enzymes from compounds with amino and carbonyl groups:

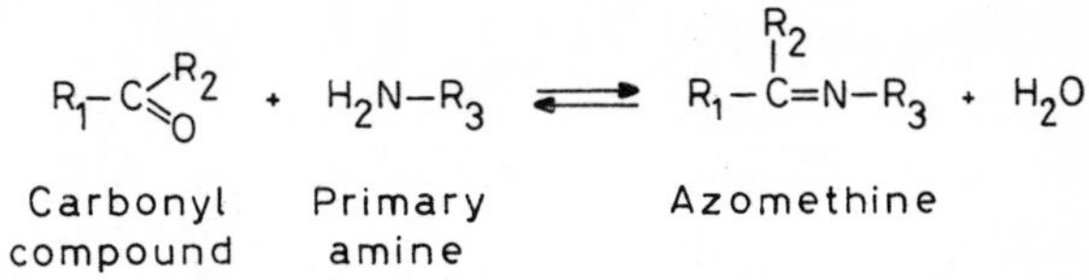

$$R_1-C\underset{O}{\overset{R_2}{\big<}} \;+\; H_2N-R_3 \;\rightleftharpoons\; R_1-C=N-R_3 \;+\; H_2O$$

Carbonyl compound	Primary amine	Azomethine

If secondary amines take part in the reaction, compounds with a quarternary nitrogen atom are built. These are particularly reactive due to the strong polarization of the molecule and easily undergo further modification.

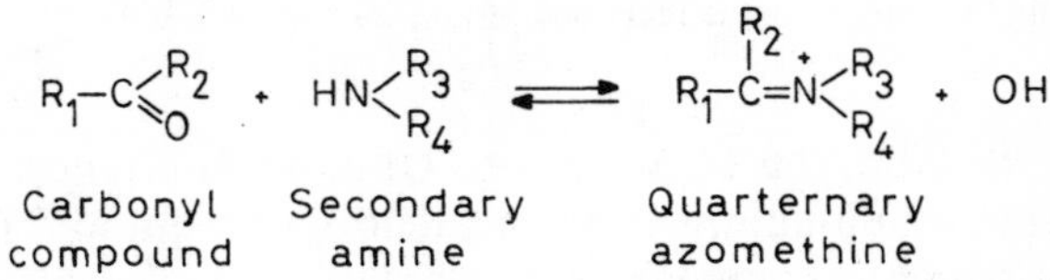

$$R_1-C\underset{O}{\overset{R_2}{\big<}} \;+\; HN\underset{R_4}{\overset{R_3}{\big<}} \;\rightleftharpoons\; R_1\overset{R_2}{\underset{}{-C}}=N\underset{R_4}{\overset{+}{\big<}}{}^{R_3} \;+\; OH^-$$

Carbonyl compound	Secondary amine	Quarternary azomethine

The amines taking part in the formation of the azomethines usually derive from amino acids (D 8.2). They are frequently methylated. The carbonyl compounds may be formed from amines by transamination (C 5), but may originate also from acetate/malonate (D 5.3.2, formation of coniine).

c) Mannich condensation. The reaction of a carbonyl compound with an amine and a substance with an acidic hydrogen atom is called Mannich condensation:

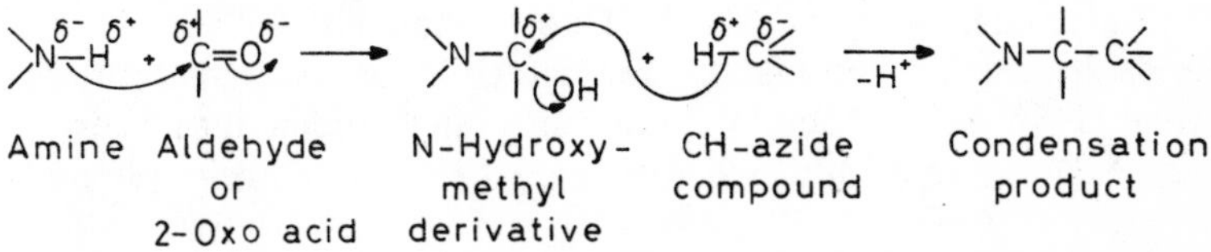

Amine	Aldehyde or 2-Oxo acid	N-Hydroxy-methyl derivative	CH-azide compound	Condensation product

Mannich condensations may proceed spontaneously even under physiological conditions, i.e., at room temperature, physiological pH, etc., but may also be enzyme catalyzed. In vivo aldehydes or 2-oxo acids act as carbonyl compounds. Typical Mannich condensations occur in the formation of the iridoid indole alkaloids (D 22.3) and the tetrahydroisoquinolines (D 23.1.2).

In the formation of certain imidazole-derived alkaloids (D 20) and indole alkaloids of the harman type (D 22.3), the condensation obviously takes place via a stable amide. This amide is formed as described above from the amine and acyl CoA, in case of harman, for instance, from tryptamine and acetyl CoA:

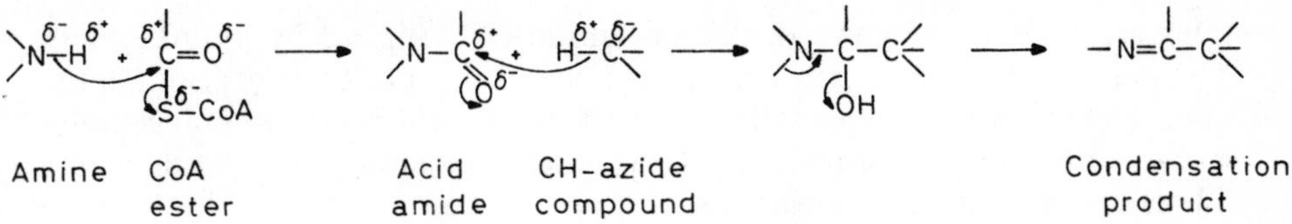

Amine	CoA ester	Acid amide	CH-azide compound		Condensation product

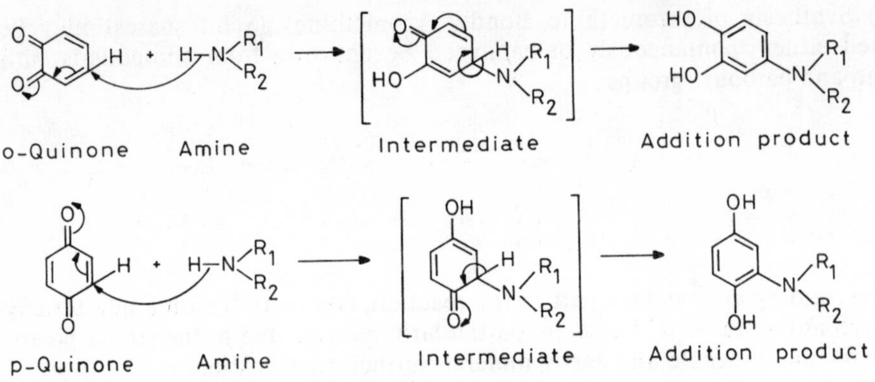

Fig. 160. Spontaneous reactions of amines with quinones

   d) Addition of Amines to Quinoid Substances (Fig. 160). Amines may react with o- or p-quinones to transient o- or p-quinoid intermediates, which rearrange to aromatic compounds. Examples are the formation of the heterocyclic rings of phenoxazines (D 21.4.1), ommochromes (D 22), and melanins (D 23.1.3).
   e) Formation of N-Glycosidic Bonds. Synthesis of N-glycosides (D 1.4.2) plays a role in the biosynthesis of the indole ring of L-tryptophan (D 22) and in the formation of pteridines (D 25.3).

Significance

Many alkaloids are toxic (F 4) and act as feeding deterrents (E 5.1, E 5.5.3) or as antimicrobial products (E 5.4). Due to their physiological activity, several alkaloids are used in the materia medica (F 2), as well as stimulants and hallucinogens (F 3). Some plant alkaloids lure specialized insects to their food source (E 5.5.3) and in a modified form may be used as insect pheromones (E 4).

References for Further Reading

104, 157, 544, 593, 647, 721

D 9 Peptides

Chemistry and Distribution

Peptides are formed from two or several amino acids linked by amide bonds. Coupling of two, three, four, etc. amino acids yields dipeptides, tripeptides, tetrapeptides, etc. Oligopeptides contain up to 10 amino acid residues, polypeptides 10 to about 100 amino acid residues. The sequence of amino acid residues in a peptide is called its primary structure.

The name of peptides is formed using the ending "yl" for those amino acids whose carboxyl group participates in the peptide bond:

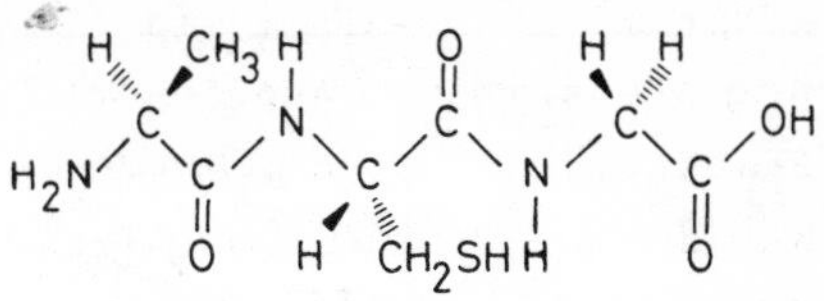

L-Alanyl-L-cysteinyl-glycine

With linear peptides the *N*-terminal amino acid is written on the left-hand-side. In more complicated peptides, e.g., cyclic compounds, the direction of the acid amide bond is indicated by an arrow (—CO→NH—) or the *N*-terminal side is marked by —H and the *C*-terminal side by —OH.

A terminal —NH$_2$ on the right-handside of the molecule means an acid amide group as in the formulas of thyreotropin-releasing hormone, luteinizing hormone/follicle-stimulating hormone (LH/FSH)-releasing hormone, oxytocin, vasopressin, and α-melanotropin (Table 44).

There are two groups of cyclic peptides:
a) compounds in which the amino acids are linked exclusively by peptide bonds (homodet cyclic substances, cf. the formula of gramicidin S, Fig. 166), and
b) compounds in which additional types of bondings, e.g., —S—S-groups, are present (heterodet cyclic substances, cf. the formula of oxytocin, Table 44).

Reference for Further Reading

390

D 9.1 Peptides Formed by the Degradation of Proteins

Chemistry and Distribution

Protein-derived peptides contain the common protein amino acids as building blocks (D 8). These may, however, be modified, e.g., by the addition of groups with other chemical structures (cf. the formulas of α-melanotropin, rhodotorucine A, and tremerogen A 10, Table 44). Peptides of this type are characteristic for animals and human beings. A few occur in microorganisms.

Biosynthesis

The peptides are split from the precursor proteins by endopeptidases. Examples are:
— the formation of angiotensin II from the protein angiotensinogen (Fig. 161),
— the biosynthesis of ACTH and several other peptides active as neurotransmitters from pre-proopiomelanocortin (Fig. 162), and
— the derivation of insulin from proinsulin (Fig. 163).

Table 44. Formulas of some linear or heterodet cyclic peptides

Name	Structure
Angiotensin II	Asp-Arg-Val-Tyr-Ile-His-Pro-Phe
Bradykinin	Arg-Pro-Pro-Gly-Phe-Ser-Pro-Phe-Arg
Kallidin	Lys-Arg-Pro-Pro-Gly-Phe-Ser-Pro-Phe-Arg
TRH	⌐Glu-His-Pro-NH₂
LH/FSH-RH	⌐Glu-His-Trp-Ser-Tyr-Gly-Leu-Arg-Pro-Gly-NH₂
GH-RH	Val-His-Leu-Ser-Ala-Glu-Glu-Lys-Glu-Ala
Oxytocin	$\underset{\text{Cys-Tyr-Ile-Gln-Asn-Cys-Pro-Leu-Gly-NH}_2}{\overset{\text{S}\underline{\hspace{3cm}}\text{S}}{}}$
Vasopressin	$\underset{\text{Cys-Tyr-Phe-Gln-Asn-Cys-Pro-Arg-Gly-NH}_2}{\overset{\text{S}\underline{\hspace{3cm}}\text{S}}{}}$
α-Melanotropin	Ser-Tyr-Ser-Met-Glu-His-Phe-Arg-Trp-Gly-Lys-Pro-Val-NH₂ (OCOCH₃ above Ser)
α-Factor	Trp-His-Trp-Leu-Gln-Leu-Lys-Pro-Gly-Gln-Pro-Met-Tyr
Rhodotorucine A	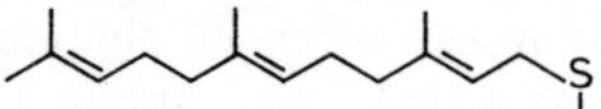 Tyr-Pro-Glu-Ile-Ser-Trp-Thr-Arg-Asn-Gly-Cys
Tremerogen A10	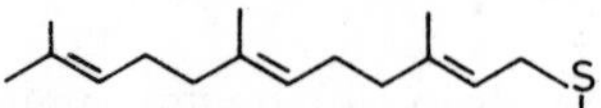 Glu-His-Asp-Pro-Ser-Ala-Pro-Gly-Asn-Gly-Tyr-Cys-OMe
Phytochelatins	(γ-Glu-Cys)ₙ-Gly n = 2 - 11

Asp-Arg-Val-Tyr-Ile-His-Pro-Phe-His-Leu⌇Leu-Val-Tyr-Ser-.......
 1   2   3   4   5   6   7   8   9   10 ↓ 11   12   13   14

Angiotensinogen

Asp-Arg-Val-Tyr-Ile-His-Pro-Phe⌇His-Leu
 1   2   3   4   5   6   7   8 ↓ 9   10

Angiotensin I

Asp-Arg-Val-Tyr-Ile-His-Pro-Phe
 1   2   3   4   5   6   7   8

Angiotensin II

Fig. 161. Formation of angiotensin II
1 Renin; *2* angiotensin converting enzyme

In the first step of insulin biosynthesis the *N*-terminal sequence of 24 amino acids (signal sequence) of pre-proinsulin is split off during the secretion into the cisternae of the endoplasmic reticulum. The formed proinsulin has a similar structure in all vertebrates. It consists of the later A- and

B-chains of insulin which are connected by the so-called C-chain. Proinsulin is transformed to insulin by cleaving the C-peptide by a specific endopeptidase. Insulin is stored in secretory granules as crystals of a zinc insulin hexamer.

The COOH-terminal amide group, e.g., of oxytocin and α-melanotropin, is derived from the amino group of a glycine residue, which is degraded by a monooxygenase (C 2.6) with the formation of glycolaldehyde.

Significance

Several peptides, e.g., angiotensin II, bradykinin, thyreotropin-releasing hormone, LH/FSH releasing hormone, and growth hormone releasing hormone, oxytocin, vasopressin, corticotropin, β-melanotropin, and insulin possess hormone activity in animals and humans (E 3.1). In addition, insulin acts as a neuronal growth factor. Met-enkephalin and β-endorphin, as well as a number of other peptides

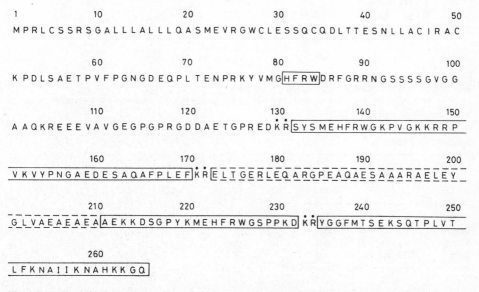

Fig. 162. Amino acid sequence of pre-proopiomelanocortin and derived peptide hormones
The amino acid residues 1–20 form a typical signal sequence, which is necessary for the transfer of the protein from plasm to the extraplasmic space (lumen of ER, cf. A 3.1.2). It is later on split off.
The proopiomelanocortin formed is the precursor of the following peptide hormones:

ACTH (corticotropin, amino acids 132–170)   *β-MSH* (β-melanotropin, amino acids 211–232)
α-MSH (α-melanotropin, amino acids   *Met-enkephalin* (amino acids 235–239) and
    132–144)
β-Lipotropin (amino acids 173–265)   *β-Endorphin* (amino acids 235–265)
Basic dipeptide sequences, indicated by *dots*, are the splitting sites of peptidases, which release the hormones from the precursor.
The amino acids are represented in the one letter code that corresponds to the three-letter notation as follows:

A	Ala	F	Phe	K	Lys	P	Pro	T	Thr
C	Cys	G	Gly	L	Leu	Q	Gln	V	Val
D	Asp	H	His	M	Met	R	Arg	W	Trp
E	Glu	I	Ile	N	Asn	S	Ser	Y	Tyr

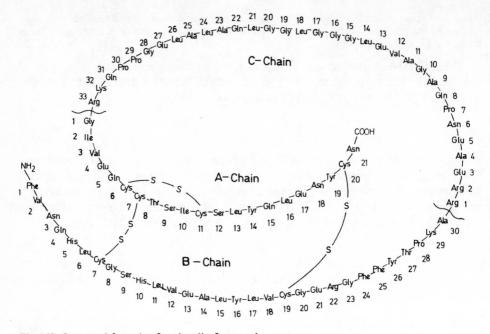

Fig. 163. Structural formula of proinsulin from pork

Insulins of human beings, bovine, and sheep show the following differences in the primary structure when compared with pork insulin:

Species	Positions			
	A-Chain			B-Chain
	8	9	10	30
Human	Thr	Ser	Ile	Thr
Bovine	Ala	Ser	Val	Ala
Sheep	Ala	Gly	Val	Ala
Pork	Thr	Ser	Ile	Ala

with similar structure, which occur in the brain of animals and human beings, reduce the perception of pain (E 3.2). The α-factor, rhodotorucine A, and tremerogen A 10 are mating pheromones of yeast (E 4).

References for Further Reading

12, 36, 130, 207, 271, 273, 353, 390, 427, 429, 439, 487, 578, 583, 633, 711, 774, 801, 807, 847, 854, 883, 905

D 9.2 Peptides Built from Activated Amino Acids

Chemistry and Distribution

Peptides of this type occur in all groups of organisms. They may contain protein amino acid residues as well as nonprotein amino acid residues and often have a homodet cyclic structure. In some compounds the amino acid moieties are inter-linked by unusual bonds (cf. the formulas of phalloin and α-amanitin, Fig. 165). Many of the peptides built directly from amino acids contain residues derived from other types of precursors as part of the molecule (cf. the formulas of edeine A, Fig. 165, of pteroylheptaglutamic acid, Fig. 166, as well as those of actinomy-cin D, D 21.4.1, and of ergotamine D 22.2). In the depsipeptides enniatin B, and valinomycin, amide and ester bonds alternate regularly (Fig. 165).

Biosynthesis

The formation of peptides from activated amino acids is brought about either by a set of monofunctional enzymes, or by one or a few polyfunctional enzymes or enzyme complexes.

Glutathione, for instance, a tripeptide containing a γ-glutamyl residue is built by two individual enzymes according to Fig. 164. Possibly amino acid phosphates are the activated intermediates.

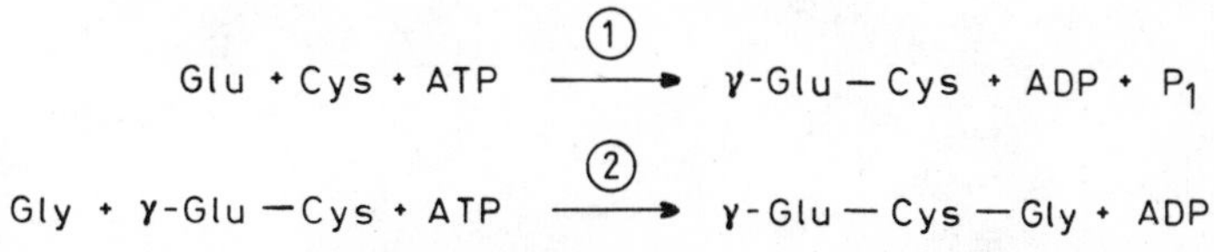

$$\text{Glu} + \text{Cys} + \text{ATP} \xrightarrow{\text{(1)}} \gamma\text{-Glu} - \text{Cys} + \text{ADP} + \text{P}_1$$

$$\text{Gly} + \gamma\text{-Glu} - \text{Cys} + \text{ATP} \xrightarrow{\text{(2)}} \gamma\text{-Glu} - \text{Cys} - \text{Gly} + \text{ADP} + \text{P}_1$$

Fig. 164. Biosynthesis of glutathione
1 γ-Glutamyl-cysteine synthetase; *2* glutathione synthetase

Glutathione is active as coenzyme (E 2.1) and, together with its dimeric disulfide, is an important redox system. In animals it forms "conjugates" with different secondary products (E 1).

More complicated peptides, e.g., edeine A (Fig. 165), gramicidin S, tyrocidine A, and bacitracin A (Fig. 166) are formed on polyfunctional proteins (A 3.1.1). One to three complementary enzymes are involved, each activating one or several (up to eight) amino acids. The amino acids are transformed to AMP de-rivatives which build enzyme-bound thioesters in a second reaction (thiotem-plate mechanism). Biosynthesis of the peptide bonds proceeds by transfer of the amino acids from the thiol groups to the terminal amino group of the growing peptide chain. All intermediates are covalently bound to SH-groups of the en-zyme. Some of the polyfunctional enzyme proteins contain pantetheine residues which allow the sequential transpeptidation by a shuttle mechanism (cf. D 5.2). In several instances the formed peptides are released from the enzyme protein by cyclization.

17*

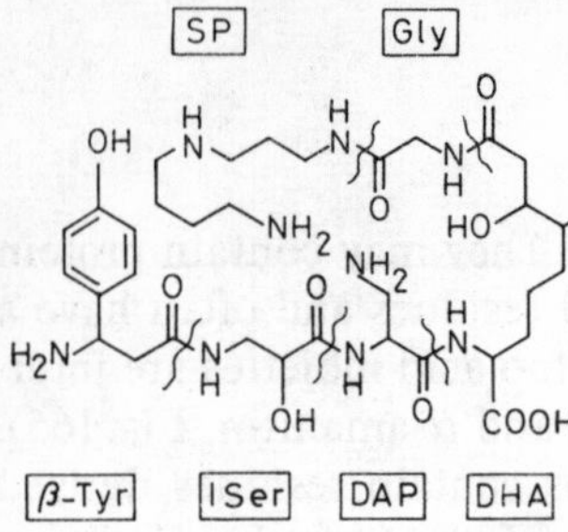

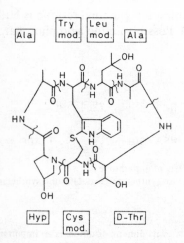

Edeine A

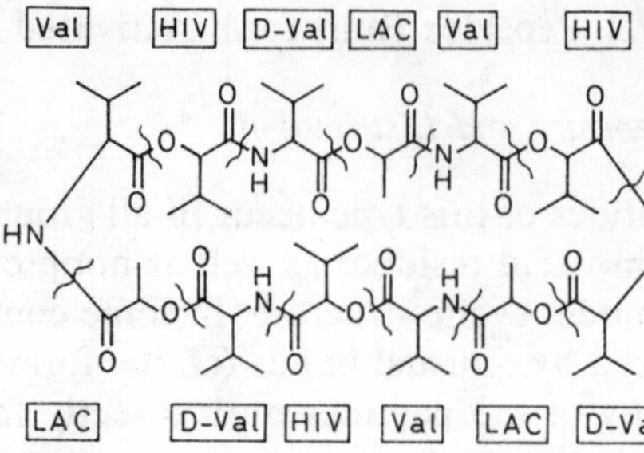

SP Gly
β-Tyr lSer DAP DHA

Valinomycin

Val HIV D-Val LAC Val HIV
LAC D-Val HIV Val LAC D-Val

Enniatin B

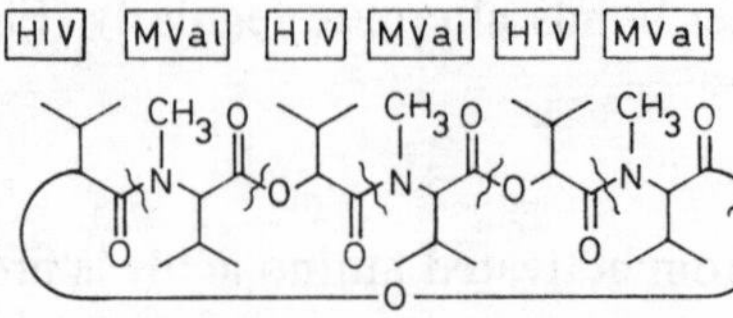

HIV MVal HIV MVal HIV MVal

Phalloin

Ala Try mod. Leu mod. Ala
Hyp Cys mod. D-Thr

α-Amanitin

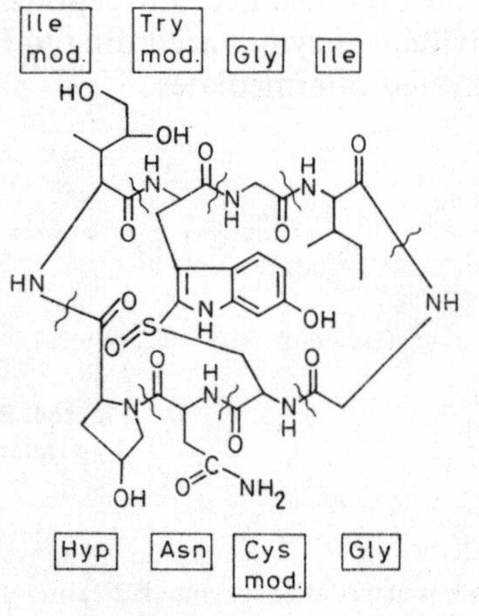

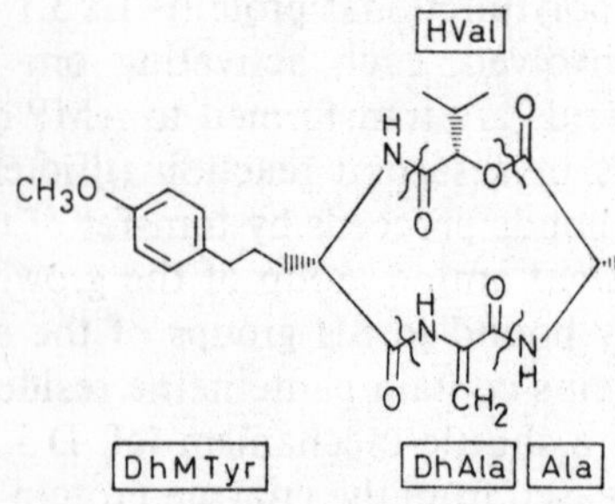

Ile mod. Try mod. Gly Ile
Hyp Asn Cys mod. Gly

AM-toxin I

HVal
DhMTyr DhAla Ala

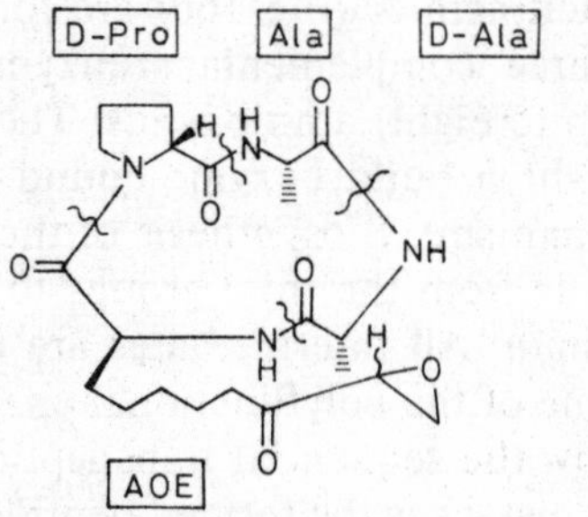

HC-toxin

D-Pro Ala D-Ala
AOE

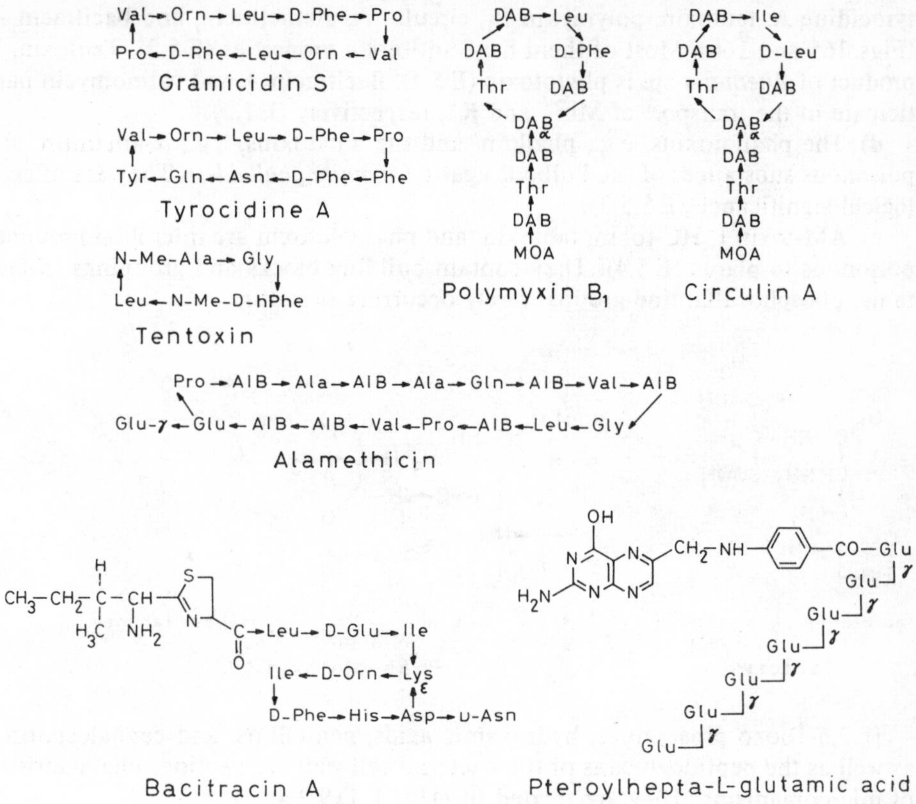

Val→Orn→Leu→D-Phe→Pro
Pro←D-Phe←Leu←Orn←Val
Gramicidin S

Val→Orn→Leu→D-Phe→Pro
Tyr←Gln←Asn←D-Phe←Phe
Tyrocidine A

N-Me-Ala→Gly
Leu←N-Me-D-hPhe
Tentoxin

DAB→Leu
DAB   D-Phe
Thr   DAB
DAB
↑α
DAB
Thr
DAB
MOA
Polymyxin B₁

DAB→Ile
DAB   D-Leu
Thr   DAB
DAB
↑α
DAB
Thr
DAB
MOA
Circulin A

Pro→AIB→Ala→AIB→Ala→Gln→AIB→Val→AIB
Glu-γ←Glu←AIB←AIB←Val←Pro←AIB←Leu←Gly
Alamethicin

CH₃-CH₂-C-CH
H₃C NH₂
S
N
C→Leu→D-Glu→Ile
Ile←D-Orn←Lys
D-Phe→His→Asp→D-Asn
Bacitracin A

OH
N
H₂N   N
CH₂-NH—   —CO-Glu
Glu⌋γ
Glu⌋γ
Glu⌋γ
Glu⌋γ
Glu⌋γ
Glu⌋γ
Pteroylhepta-L-glutamic acid

Fig. 166. Formulas of some cyclic polypeptides, part II
Abbreviations see Fig. 165

Groups of Secondary Products

a) The phytochelatins, which have the general formula $(\gamma$-Glu-Cys$)_n$Gly (n = 2
− 11), are probable glutathione derivatives. These compounds, occurring in
plants and some fungi, chelate heavy metal ions (E 2.2).

b) Pteroylpoly-L-glutamic acids are vitamins for animals (E 2.1). They are
used in the formation of tetrahydrofolic acid (D 25.3).

c) Cyclic peptides and depsipeptides are formed in several groups of microorganisms. Representatives are edeine A, valinomycin, enniatin B, gramicidin S,

◀

Fig. 165. Formulas of some cyclic polypeptides, part I
AIB aminoisobutyric acid; *AOE* L-2-amino-9,10-epoxy-8-oxodecanoic acid; *DAB* 2,4-diaminobutyric
acid; *DAP* diaminopropionic acid; *DHA* 2,6-diamino-7-hydroxyazelaic acid; *DhAla* dehydroalanine;
DhMTyr L-dihomomethyltyrosine; *HIV* D-2-hydroxyisovaleric acid; *HVal* 2*(S)*-hydroxyisovaleric acid;
ISer isoserine; *LAC* L-lactic acid; *MOA* 6-methyloctanoic acid (isopelargonic acid); *MVal* N-methyl-
L-valine; *N-Me-D-hPhe* N-methyl-Z-dehydrophenylalanine; *SP* spermidine; *β-Tyr* β-tyrosine; *mod.*
modified

tyrocidine A, tentoxin, polymyxin B_1, circulin A, alamethicin, and bacitracin A (Figs. 165 and 166). Most of them have antibiotic properties (E 5.2). Tentoxin, a product of *Alternaria* sp. is phytotoxic (E 5.4). Bacitracin A and valinomycin participate in the transport of Mn^{2+} and K^+, respectively (E 2.2).

d) The phallotoxins, e.g., phalloin, and the amatoxins, e.g., α-amanitin, are poisonous substances of the bulbous agaric *(Amanita phalloides)*. They are of ecological significance (E 5.5.3).

e) AM-toxin I, HC-toxin, tabtoxin, and phaseolotoxin are microbial products poisonous to plants (E 5.4). They contain building blocks and groupings (β-lactame, phosphosulfamine groups) rarely occurring in nature.

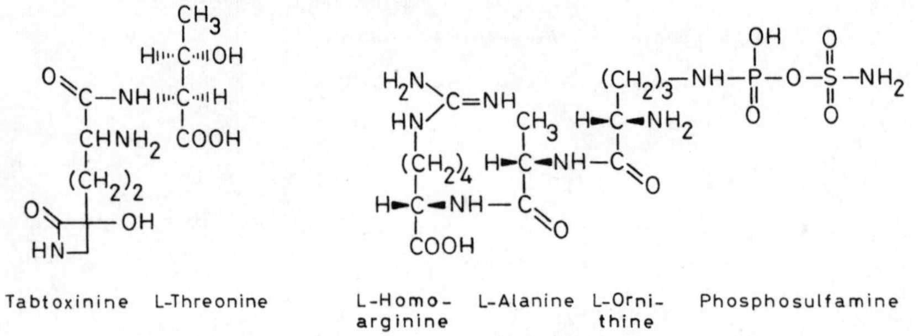

Tabtoxinine   L-Threonine   L-Homo-   L-Alanine L-Orni-   Phosphosulfamine
                                 arginine   thine

         Tabtoxin   Phaseolotoxin

f) 2,5-Dioxo piperazines, hydroxamic acids, penicillins, and cephalosporins, as well as the peptidoglucans of the bacterial cell wall are peptides characteristic of microorganisms. They are treated in D 9.2.1–D 9.2.4.

References for Further Reading

12, 428, 462, 807, 809, 814, 815, 894

D 9.2.1 2,5-Dioxo Piperazines

Chemistry and Distribution

2,5-Dioxo piperazines occur in microorganisms. They carry carbonyl groups in positions 2 and 5 and different substituents in positions 3 and 6. This fundamental structure may be modified by oxidation, reduction, etc.

Biosynthesis

2,5-Dioxo piperazines are cyclic dipeptides synthesized from two activated α-amino acids. The cyclic peptides built may be further transformed by monooxygenation (formation of *N*-oxides, C 2.6.2, cf. also D 9.2.2), reduction, and modification of the side chains (Fig. 167).

Gliotoxin is a representative of the 2,5-dioxo piperazine-derived fungal toxins. It is formed from L-tryptophan and L-serine. A key intermediate is cyclo-

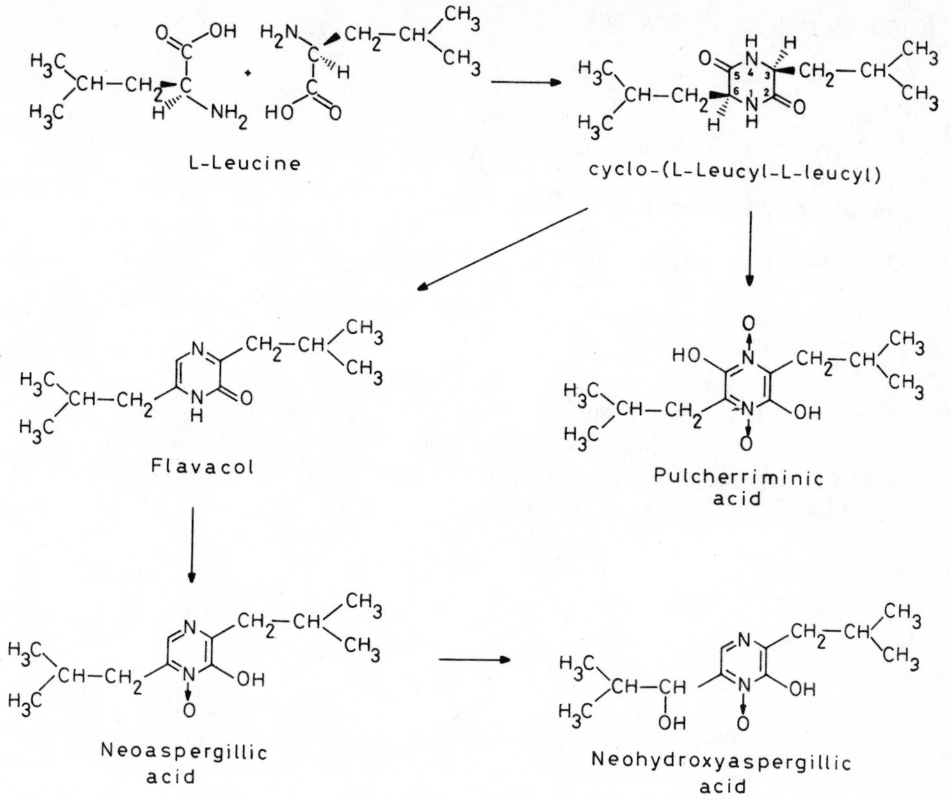

Fig. 167. Formation of dioxopiperazines from L-leucine

L-phenylalanyl-L-seryl) (Fig. 168). The biosynthesis of the cyclohexadienol ring probably involves an epoxy intermediate (C 2.6.5). The incorporated sulfur may come from L-cysteine.

Significance

N-Oxides, e.g., pulcherriminic acid, neoaspergillic acid, and neohydroxyaspergillic acid, chelate Fe^{3+} and participate in the transport of iron (E 2.2).

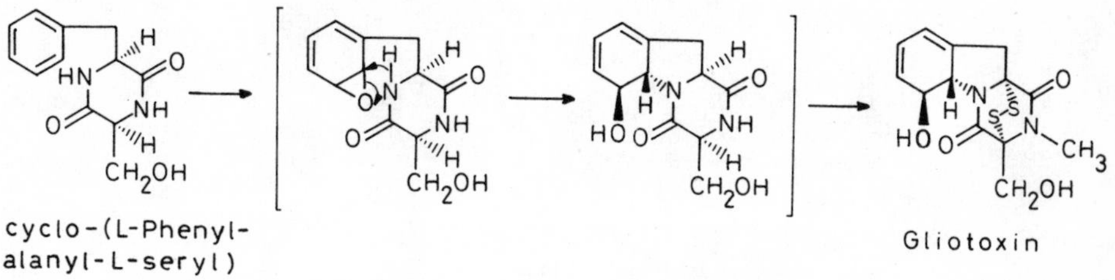

Fig. 168. Biosynthesis of gliotoxin

References for Further Reading

12, 426, 832

D 9.2.2 Hydroxamic Acids

Chemistry and Distribution

Hydroxamic acids possess one or several oxygenated peptide bonds:

$$-\overset{\underset{\|}{O}}{C}-\overset{\underset{|}{OH}}{N}- \rightleftharpoons -\overset{\overset{HO}{|}}{C}=\overset{\overset{O}{\uparrow}}{N}-$$

They are able to chelate metal ions, e.g., Fe^{3+}. Compounds with an iron-trihydroxamate center, e.g., ferrichrome, are called siderochromes (Fig. 169).

Hydroxamic acids are built in microorganisms. A few are formed also in higher plants (cf. DIMBOA, D 21.4).

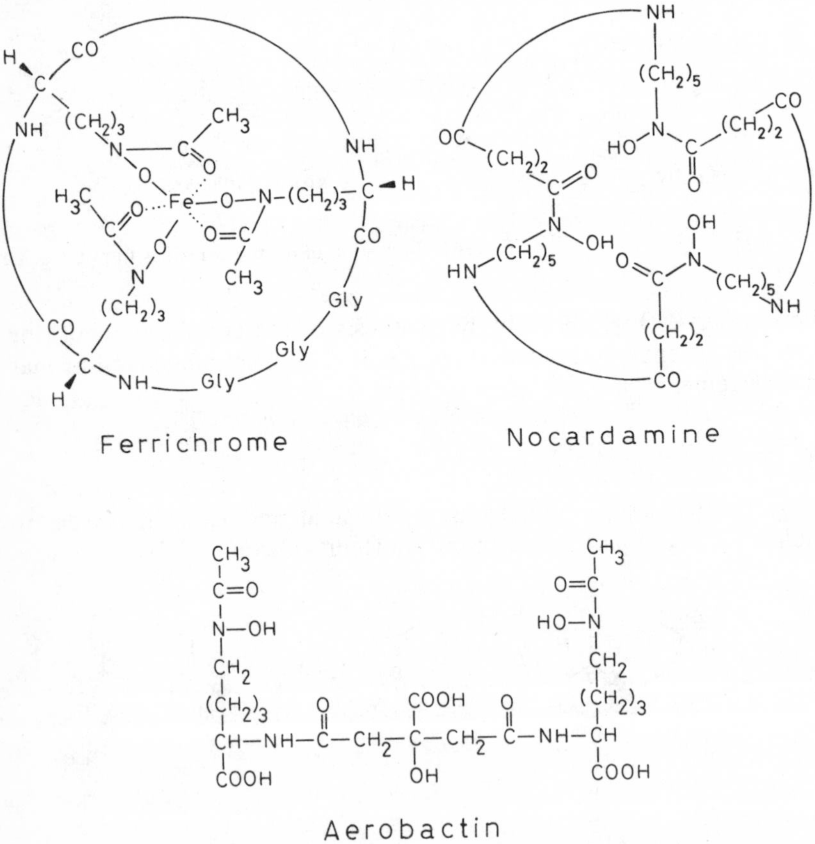

Ferrichrome

Nocardamine

Aerobactin

Fig. 169. Structural formulas of hydroxamic acids

Biosynthesis

The hydroxamic acid group originates by monooxygenation (C 2.6.2)
— either of an amide bond (cf. the formation of pulcherriminic acid and neoaspergillic acid, Fig. 167)
— or of a free amino group followed by amide formation:

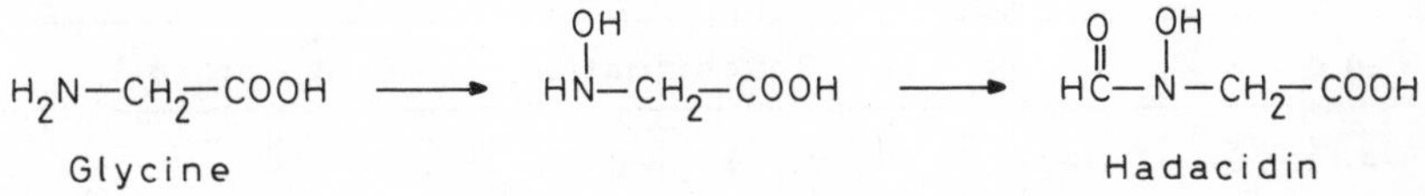

$$H_2N-CH_2-COOH \longrightarrow HN-CH_2-COOH \longrightarrow HC-N-CH_2-COOH$$

Glycine   Hadacidin

N-Hydroxylated amino acids, e.g., N^5-hydroxy-L-ornithine, serve in the biosynthesis of ferrichrome. They are converted to hydroxamic acids by acetylation.

Significance

Due to the chelating properties hydroxamic acids are of importance in the solubilization and usage of iron by many microorganisms. Aerobactin participates in
the uptake of iron in *Escherichia coli*. Sideramines, e.g., ferrichrome, are involved
in the Fe^{3+} uptake in fungi (E 2.2). Some iron-free sideraminelike compounds,
e.g., nocardamine, are antibiotics (E 5.2) and inhibit the activity of sideramines
by competition. If the Fe^{3+} ion is reduced to Fe^{2+}, the stability of the hydroxamic
acid complexes is reduced and iron may be transferred, for instance, from ferrichrome, to compounds with a porphyrin ring system (D 24).

References for Further Reading

560, 651, 707

D 9.2.3 Penicillins and Cephalosporins

Chemistry and Distribution

Penicillins and cephalosporins are the most important compounds which possess
a four-membered β-lactam ring. In the penicillins this ring is fused with a thiazolidine ring. In the cephalosporins it is anellated with the six-membered dihydrothiazine ring (Fig. 170). The penicillins are amides of 6-aminopenicillanic acid
(Table 45).
   Penicillins and cephalosporins are formed in prokariotic and eukariotic microorganisms, e.g., *Streptomyces, Penicillium, Aspergillus,* and *Cephalosporium* sp.

Biosynthesis

Penicillins and cephalosporins are built from L-2-aminoadipic acid, L-cysteine,
and L-valine (Fig. 170). An important intermediate is 6-(L-2-aminoadipyl)-L-cys-

Table 45. Some naturally occurring penicillins

$$R-NH-\underset{\underset{O}{\overset{H}{\underset{\|}{C}}}}{\overset{H}{\underset{\|}{C}}}-\underset{N}{\overset{H}{\underset{\|}{C}}}\cdots\overset{S}{\underset{C-H}{\underset{COOH}{}}}\overset{CH_3}{\underset{CH_3}{}}$$

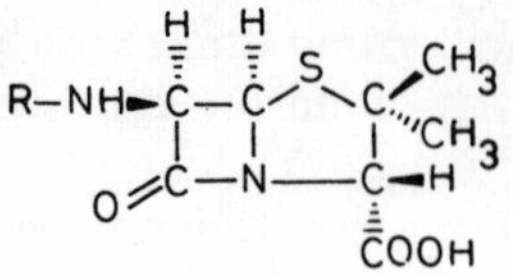

Name	R: Designation	R: Structure
6-Aminopenicillanic acid	Hydrogen	H –
p-Hydroxybenzylpenicillin (Penicillin X)	p-Hydroxyphenyl-acetyl-	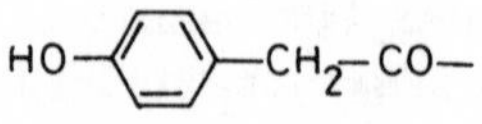 $HO-\langle\rangle-CH_2-CO-$
Benzylpenicillin (Penicillin G)	Phenylacetyl-	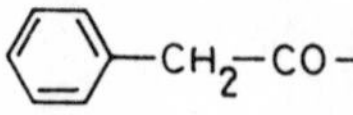 $\langle\rangle-CH_2-CO-$
n-Propylpenicillin	n-Butyryl-	$CH_3-(CH_2)_2-CO-$
n-Butylpenicillin	n-Valeryl-	$CH_3-(CH_2)_3-CO-$
n-Amylpenicillin	n-Capronyl-	$CH_3-(CH_2)_4-CO-$
Δ^2-Pentenylpenicillin (Penicillin F)	Δ^3-Dehydro-n-capronyl-	$CH_3-CH_2-\overset{H}{\underset{H}{C}}=C-CH_2-CO-$
n-Heptylpenicillin	n-Caprylyl-	$CH_3-(CH_2)_6-CO-$

teinyl-D-valine. The β-lactam ring of the 6-aminopenicillanic acid nucleus is formed by dehydrogenation. Synthesis of the thiazolidine ring proceeds with retention of the 2-H-atom of the D-valine unit. The dihydrothiazine ring of the cephalosporins is formed by enlargement of the thiazolidine ring of penicillin N by inclusion of one of the methyl groups of the valine-derived part. Replacement of the 2-aminoadipic acid group of isopenicillin N by other acids yields the penicillins given in Table 45. Most important in economy is benzylpenicillin (penicillin G).

Significance

Penicillins and cephalosporins are antibiotics (E 5.2) with great significance in medicine (F 2). They inhibit cell wall formation in bacteria. The naturally occurring penicillins and cephalosporins are inactivated, e.g., by bacterial β-lactamases and by the acid in the stomach. 6-Aminopenicillanic acid, obtained by enzymatic hydrolysis of penicillin G, is commercially converted to semisynthetic penicillins which are stable to acids and/or resistant to penicillin-β-lactamase and may possess a broader spectrum of antibiotic activity in comparison to the naturally occurring penicillins. Also semisynthetical cephalosporin C derivatives are used in medicine.

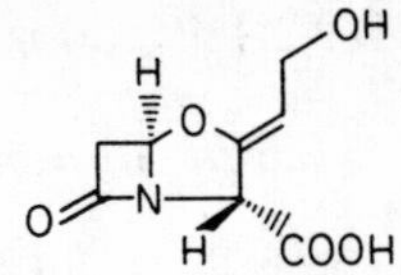

Clavulanic acid

Clavulanic acid, a compound with a bicyclic lactam structure resembling those of penicillins and cephalosporins, is a natural inhibitor of β-lactamases. It is produced in *Streptomyces clavuligerus* and is used together with lactam antibiotics in medicine.

References for Further Reading

2, 12, 13, 105, 206, 354, 535, 576, 588, 643, 720, 886

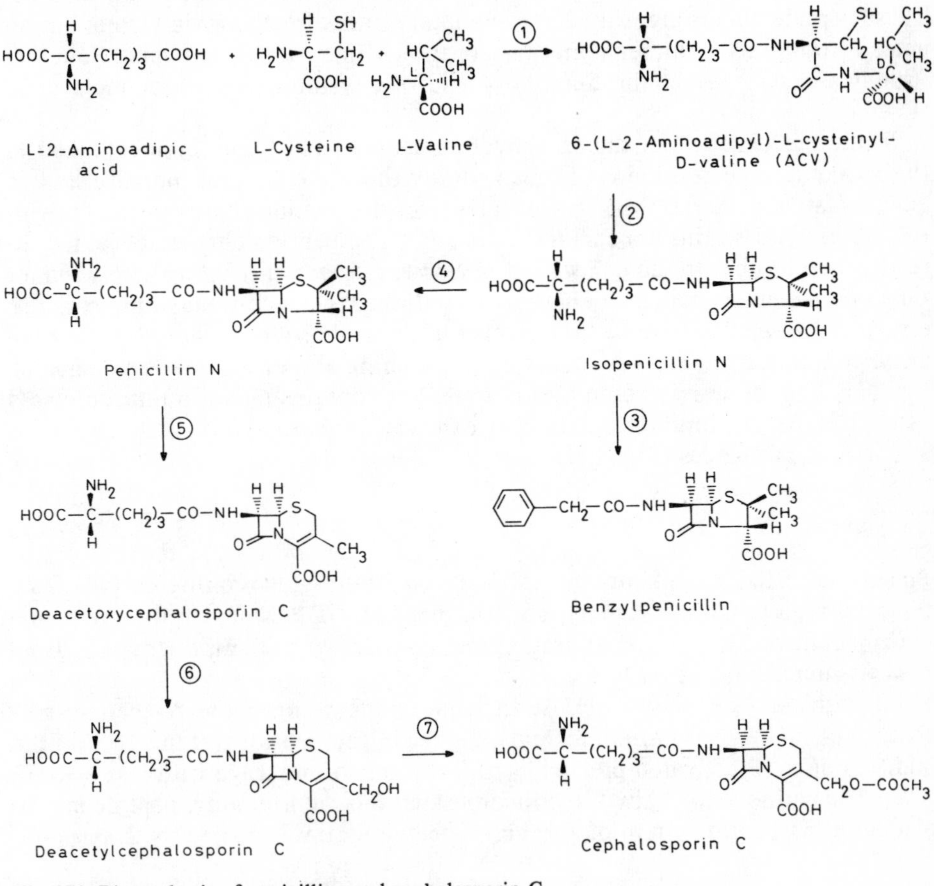

L-2-Aminoadipic acid   L-Cysteine   L-Valine   6-(L-2-Aminoadipyl)-L-cysteinyl-D-valine (ACV)

Penicillin N   Isopenicillin N

Deacetoxycephalosporin C   Benzylpenicillin

Deacetylcephalosporin C   Cephalosporin C

Fig. 170. Biosynthesis of penicillins and cephalosporin C
1 ACV synthetase; *2* isopenicillin N synthase (O$_2$-dependent); *3* penicillin acyltransferase; *4* isopenicillin N epimerase; *5* deacetoxycephalosporin C synthase (expandase); *6* deacetoxycephalosporin C monooxygenase; *7* deacetylcephalosporin C acetyltransferase

D 9.2.4 Peptidoglycans of the Bacterial Cell Wall

Chemistry, Distribution, Significance

The cell wall of bacteria has a complex structure. The matrix is formed by the peptidoglycans murein (eubacteria, e.g., *Escherichia coli*) or pseudomurein (Methanobacteriales) which are single macromolecules forming a sacculus which shapes the bacterial protoplast (E 2.2). They are built from polysaccharide chains interlinked by peptide bridges (Fig. 171). This network is covered by a layer of proteins. Additionally, the cell wall of gram negative bacteria, e.g., *Escherichia coli*, *Salmonella*, and *Shigella* sp. contains large quantities of lipids.

The carbohydrate chains of murein consist of up to 12 alternating N-acetyl-D-glucosamine and N-acetylmuramic acid units (D 1.1) linked to each other through β-1,4-bonds. All or several muramic acid residues are substituted by peptide chains which may consist of L-alanine, D-glutamic acid, a dibasic amino acid, e.g., L-lysine or meso-, D,D-, and L,L-diaminopimelic acid, and D-alanine. These peptide chains are linked to each other either directly or via further amino acids forming a two-dimensional (rarely three-dimensional) network. Coupling usually takes place via the 6-amino group of L-lysine or diaminopimelic acid (Fig. 171).

While the structure of the carbohydrate chains and of the peptides linked to the muramic acid residues is fundamentally the same in gram positive and in gram negative bacteria, the mode of linkage of the peptide chains in the individual species and strains appears to be specific. The peptide chains are joined directly to each other in the cell wall of *Micrococcus lysodeikticus*, *Escherichia coli*, or *Corynebacterium diphtheriae*. They are interlinked via additional amino acids, e.g., in *Staphylococcus aureus* (five glycine units) or *Micrococcus roseus* (three L-alanine residues, or in certain strains three L-alanine and one L-threonine residue, Fig. 171). Pseudomurein resembles murein, but contains L-talosaminuronic acid instead of N-acetylmuramic acid and exhibits different types of linkages and amino acid sequences (Fig. 171).

Biosynthesis

Precursors of the peptidoglycan are UDP-N-acetyl-D-glucosamine derived from N-acetyl-D-glucosamine (D 1.1), and the peptide UDP-N-acetylmuramyl-L-alanyl-D-glutamyl-L-lysyl- (or diaminopimelyl)-D-alanyl-D-alanine formed from N-acetylmuramic acid (D 1.1).

Glycosidases, e.g., lysozyme, destroy the bonds between the N-acetyl-D-glucosamine and N-acetylmuramic acid residues of the peptidoglycan allowing the addition of UDP-activated precursors to the points of breakage with loss of UDP (C 6). The amino acid chain of the incorporated N-acetylmuramyl peptide is coupled with the peptide chain of another N-acetylmuramine moiety by transpepti-

▶

Fig. 171. Structure of mureins and pseudomurein
A N-Acetyl-D-glucosamine residue; B N-acetylmuramic acid residue; C L-talosaminuronic acid residue

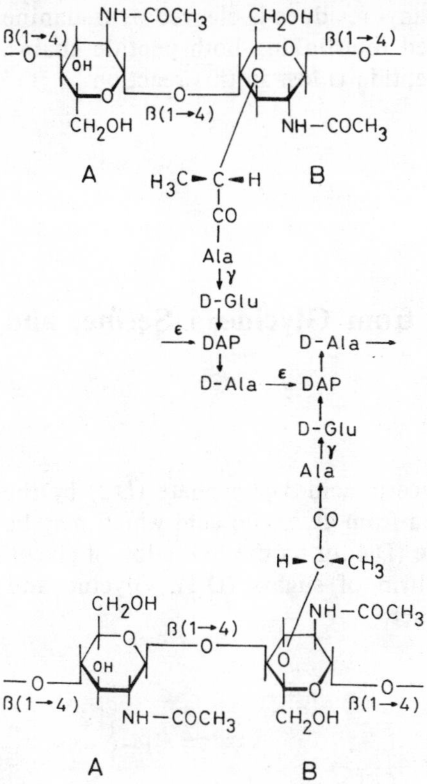

Section of murein of Escherichia coli

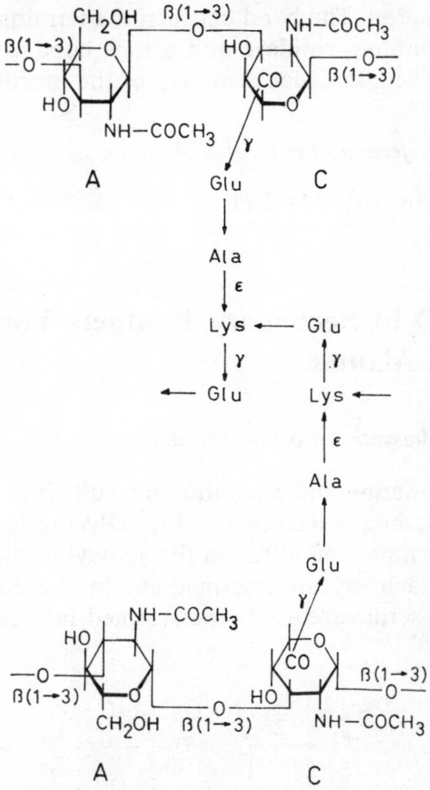

Section of pseudomurein of Methanobacteriales

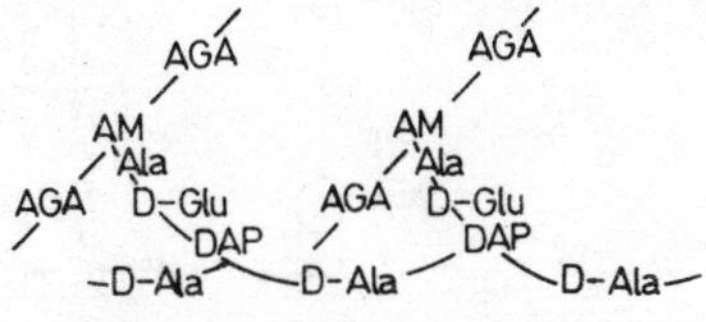

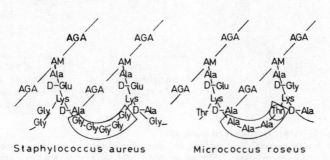

Escherichia coli or
Corynebacterium diphtheriae

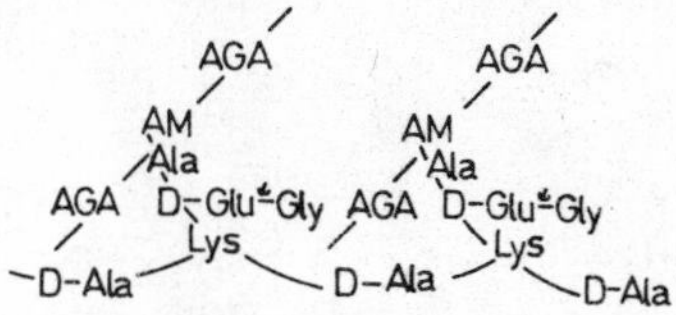

Micrococcus lysodeikticus

Staphylococcus aureus

Micrococcus roseus

Sections of mureins of several bacteria

dation. The bond between the terminal D-alanyl residues is cleaved by D-alanine carboxypeptidase and a new bond is formed interlinking both peptide chains. The last D-alanyl moiety of the precursor peptide is lost in this reaction.

References for Further Reading

316, 701, 705, 871

D 10 Secondary Products Formed from Glycine, L-Serine, and L-Alanine

Biosynthesis of the Precursors

L-Serine and L-alanine are built from D-glyceric acid 3-phosphate (D 2) by the pathway given in Fig. 172. Glycine is formed from glyoxylic acid which may be synthesized either in the glyoxylic acid cycle (D 4) or by the oxidation of glycol-aldehyde, an intermediate in the metabolism of sugars (D 1). Glycine and L-serine are easily transformed into each other.

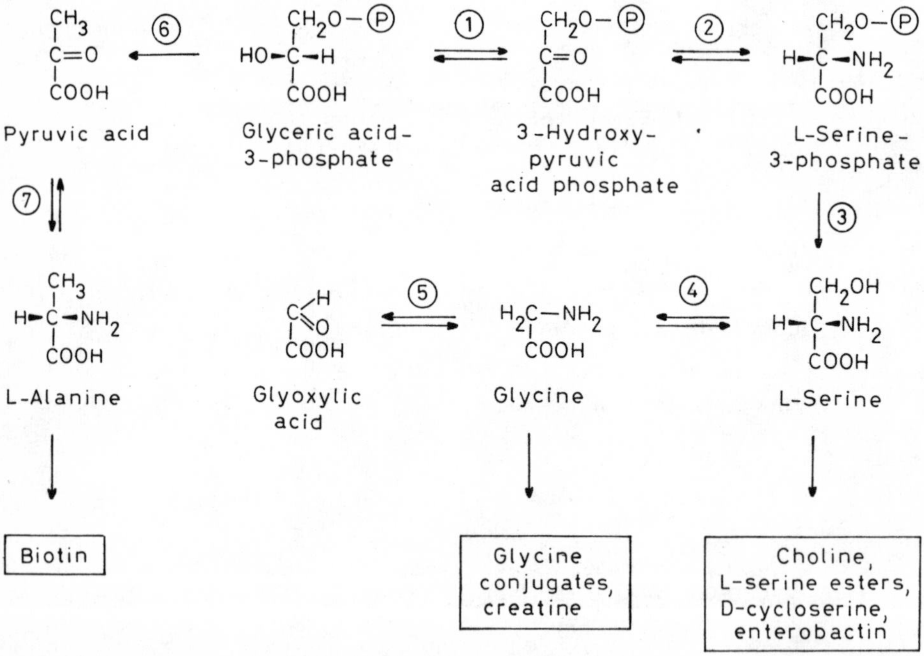

Fig. 172. Formation of glycine, L-serine, and L-alanine
1 Phosphoglycerate dehydrogenase; *2* phosphoserine aminotransferase; *3* phosphoserine phosphatase; *4* serine hydroxymethyltransferase; *5* glycine aminotransferase; *6* phosphoglyceromutase, enolase, pyruvate kinase (Fig. 49); *7* alanine: oxo acid aminotransferase

Chemistry and Distribution

Secondary products are derived from glycine, L-serine, and L-alanine in microorganisms, plants, and animals. The most important groups are listed in Fig. 172. They have either the skeleton of the precursor amino acids, like the glycine conjugates or serine esters, or show more complicated chemical structures (cf. the formula of biotin).

Groups of Secondary Products

a) Glycine conjugates (amides) are formed in animals, e.g., mammals, including man, with the CoA-esters of different acids, e.g., benzoic acid, phenylacetic acid, and nicotinic acid by glycine acyltransferase. The reaction takes place in liver and kidneys. The conjugates are excreted with the urine (E 1). The best-known conjugate is hippuric acid which is formed by the following reaction:

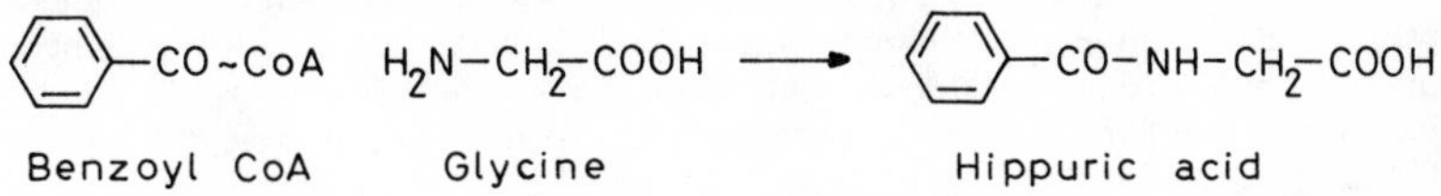

Benzoyl CoA   Glycine   Hippuric acid

b) Creatine and creatinine, derivatives of guanidino acetic acid, are derived from glycine and the imidine group of L-arginine (D 19, Fig. 173). Creatine phosphate is an energy-rich compound used in the regeneration of ATP in muscle cells of vertebrates (E 2.2). Creatinine is excreted with the urine. Its excretion removes excess methyl groups from the body of animals and human beings (E 1). The amount of creatinine in the urine corresponds to the amount of labile methyl groups in the diet.

c) Ethanolamine and choline originate from L-serine (Fig. 174). The cytidine diphosphate (CDP) derivatives of ethanolamine and choline are formed with cytidine triphosphate. They are precursors of membrane lipids (D 5.2.4). Choline

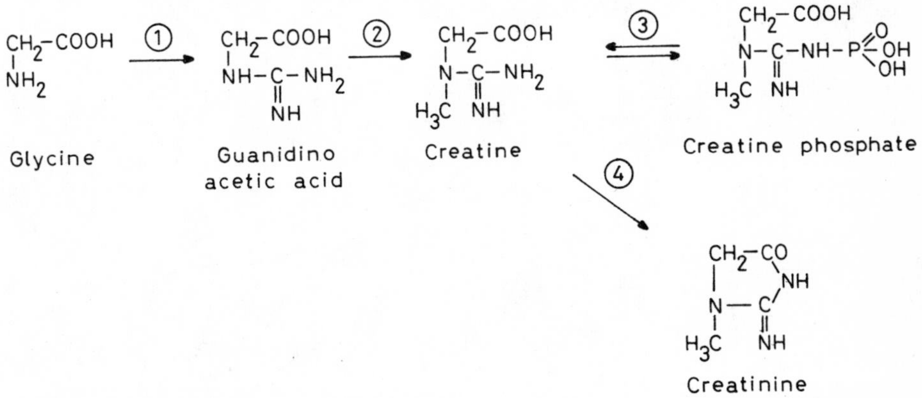

Fig. 173. Biosynthesis and degradation of creatine
1 Glycine amidinotransferase; 2 guanidinoacetate methyltransferase; 3 creatine kinase; 4 creatininase

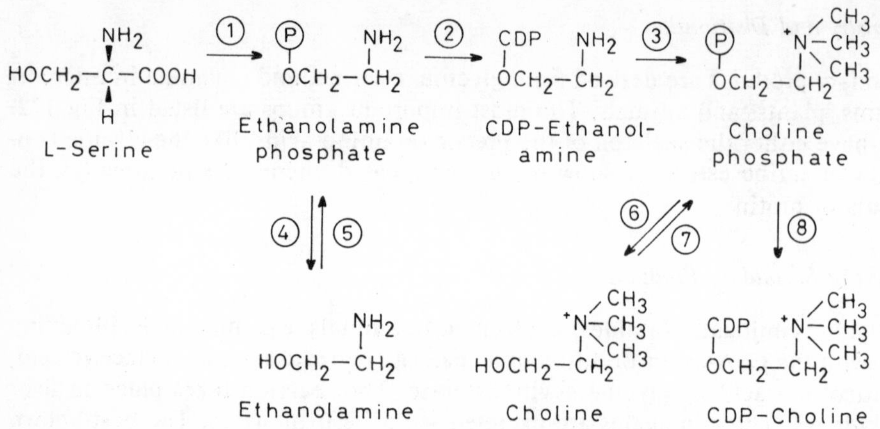

Fig. 174. Formation of ethanolamine and choline derivatives
1 L-Serine kinase, L-serine-3-phosphate decarboxylase; *2* ethanolamine phosphate cytidylyltransferase; *3* ethanolamine phosphotransferase (D 3.2.4), phosphatidyl ethanolamine methyltransferase, phospholipase D, choline kinase; *4* phosphoethanolamine phosphatase; *5* ethanolamine kinase; *6* phosphocholine phosphatase; *7* choline kinase; *8* choline phosphate cytidylyltransferase

occurs as an ester in several plants (D 23.2.1 and D 23.2.5). Acetylcholine, $CH_3CO—OCH_2CH_2N^+(CH_3)_3$, acts as neurotransmitter in animals (E 3.2). It is hydrolyzed to choline and acetic acid by acetylcholine esterase.

d) L-Serine esters, e.g., O-acetyl-L-serine, are formed in several groups of organisms. O-Acetyl-L-serine is an intermediate in the biosynthesis of L-cysteine (D 11) and participates in the biosynthesis of several secondary products, e.g., 3′-pyrazol-l-yl-L-alanine (D 12.1).

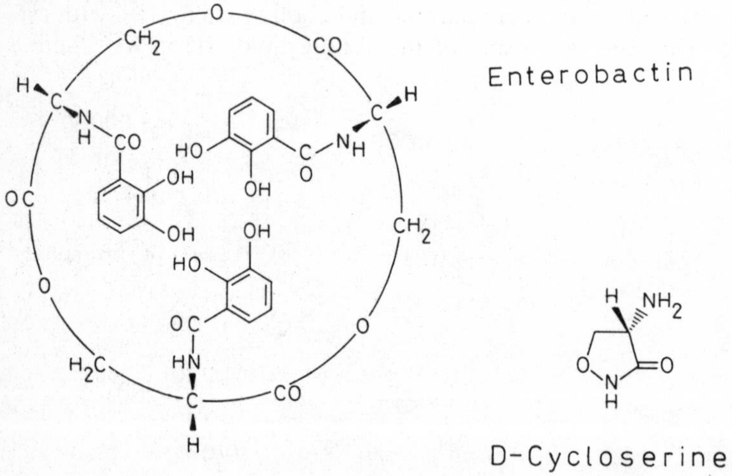

Enterobactin

D-Cycloserine

e) D-Cycloserine is a nonprotein amino acid formed in several Streptomycetes. It has antibiotic properties and is used in the treatment of tuberculosis (E 5.2).

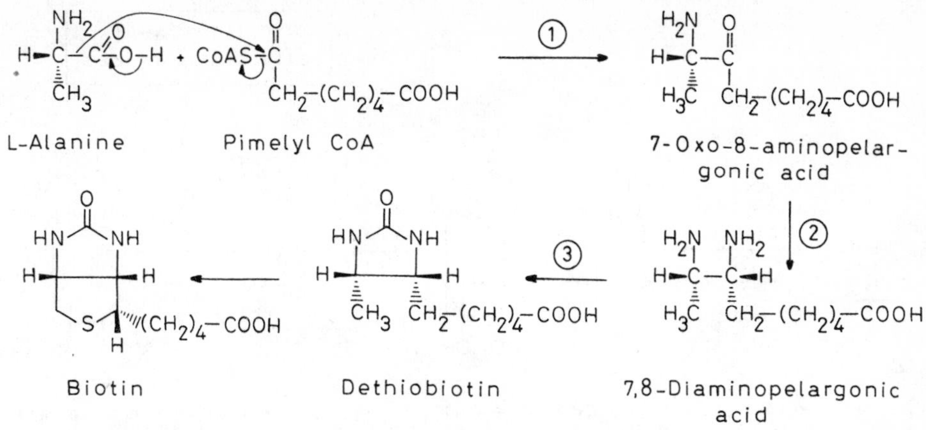

L-Alanine   Pimelyl CoA   7-Oxo-8-aminopelar-
gonic acid

Biotin   Dethiobiotin   7,8-Diaminopelargonic
acid

Fig. 175. Biosynthesis of biotin
1 7-Oxo-8-aminopelargonate synthase; *2* adenosylmethionine: 7-oxo-8-aminopelargonate aminotransferase; *3* dethiobiotin synthetase

f) Enterobactin is built from L-serine and 2,3-dihydroxybenzoic acid in bacteria. Three L-serine residues are joined by ester groups and bound to the 2,3-dihy-
droxybenzoic acid moieties by amide bonds. Enterobactin chelates Fe^{3+}-ions
and is involved in the uptake of iron (E 2.2).

g) Biotin originates from L-alanine, pimelyl CoA, CO_2, ATP, and the sulfur
atom of L-methionine (Fig. 175). Dethiobiotin is an important intermediate. Biotin is the coenzyme of carboxylases and carboxyltransferases (C 3.1). It is
bound covalently via the carboxyl group to the amino group in position 6 of an
L-lysine residue of the enzyme protein. Biotin is a vitamin for higher animals
and human beings (vitamin H, E 2.1).

References for Further Reading

31, 422, 566, 795, 820

D 11 Sulfuric Acid and L-Cysteine Derivatives

Biosynthesis and Degradation of the Precursors

Inorganic sulfate is transformed to adenosine-5'-sulfatophosphate (ASP) and ad-
enosine-3'-phosphate-5'-sulfatophosphate (3'-phosphoadenosine-5'-phosphosul-
fate, PAPS, "activated sulfate", Fig. 176) in microorganisms, plants, and animals.
The sulfate group of ASP and PAPS may be reduced to the level of sulfide in
microorganisms and plants. In most oxygen-evolving, photosynthesizing organisms APS, and in organisms lacking oxygen evolution, PAPS is used as the pre-

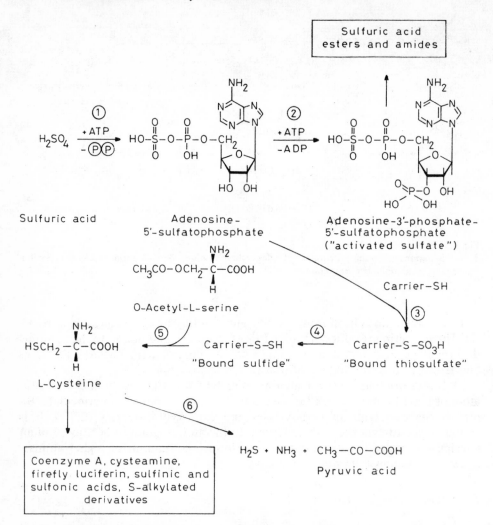

Fig. 176. Biosynthesis of L-cysteine
1 Sulfate adenylyltransferase (ATP sulfurylase); *2* adenosine 5'-sulfatophosphate kinase; *3* adenosine 5'-sulfatophosphate sulfotransferase; *4* ferredoxin: sulfoglutathione oxidoreductase; *5* O-acetyl serine sulfhydrase (cysteine synthase). The carrier probably is glutathione; *6* cysteine desulfhydrase

cursor. The carrier-bound sulfide (Fig. 176), or free sulfide formed as intermediate, reacts with O-acetyl-L-serine (D 10) to L-cysteine.

L- and D-Cysteine are degraded by cysteine desulfhydrase to hydrogen sulfide, ammonia, and pyruvic acid.

Chemistry and Distribution

L-Cysteine-derived secondary compounds occur in microorganisms, plants, and animals. The most important groups are listed in Fig. 176. They contain the sulfur in different states of oxidation. Of importance are sulfides (—S—), disulfides (—S—S—), sulfenic acids (—S—OH), sulfoxides (—SO—), sulfinic acids (—SO—OH), and sulfonic acids (—SO$_2$—OH).

Groups of Secondary Products

a) Sulfuric acid esters are synthesized in microorganisms, plants, and animals. Sulfuric acid amides occur in plants.

Sulfate esters   Sulfate amides

Both groups of compounds are formed by the transfer of the sulfate group from PAPS to acceptor molecules:

The reaction is catalyzed by sulfotransferases (cf. the synthesis of the sulfate esters of carbohydrates, D 1.1, Table 21, of indole derivatives, D 22, of flavonoids, D 23.3.3, and the synthesis of glucosinolates, D 8.4). In animals sulfate esters are of significance in the elimination of secondary products (E 1). They are excreted with the urine.

b) Coenzyme A is synthesized from pantoic acid (D 13), β-alanine (D 16), and L-cysteine in microorganisms and plants (Fig. 177). Pantothenic acid, which may be a precursor of 4'-phosphopantothenic acid, is a vitamin for animals (E 2.1).

c) Cysteamine is built by decarboxylation of L-cysteine (D 8.2) and by the degradation of coenzyme A.

d) The luciferin of the firefly, *Photinus pyralis*, is a compound structurally derived from D-cysteine (cf. the bold part of the formula in Fig. 178). Firefly luciferin reacts with ATP to form luciferyl-AMP which is transformed by molecular oxygen to a cyclic peroxide (a dioxethanone derivative). This compound is degraded to oxyluciferin (Fig. 178). Emission of visible radiation by enzyme-catalyzed reactions (bioluminescence) is of significance for *Photinus* (and other animals) in the attraction of sexual partners and of animals used as prey (E 2.2).

e) Further secondary substances formed from L-cysteine are
— the products of L-cysteine and cysteamine oxygenation (D 11.1) and
— the S-substituted cysteine derivatives and sulfoxides (D 11.2).

References for Further Reading

31, 134, 159, 324, 326, 387, 672, 768

18*

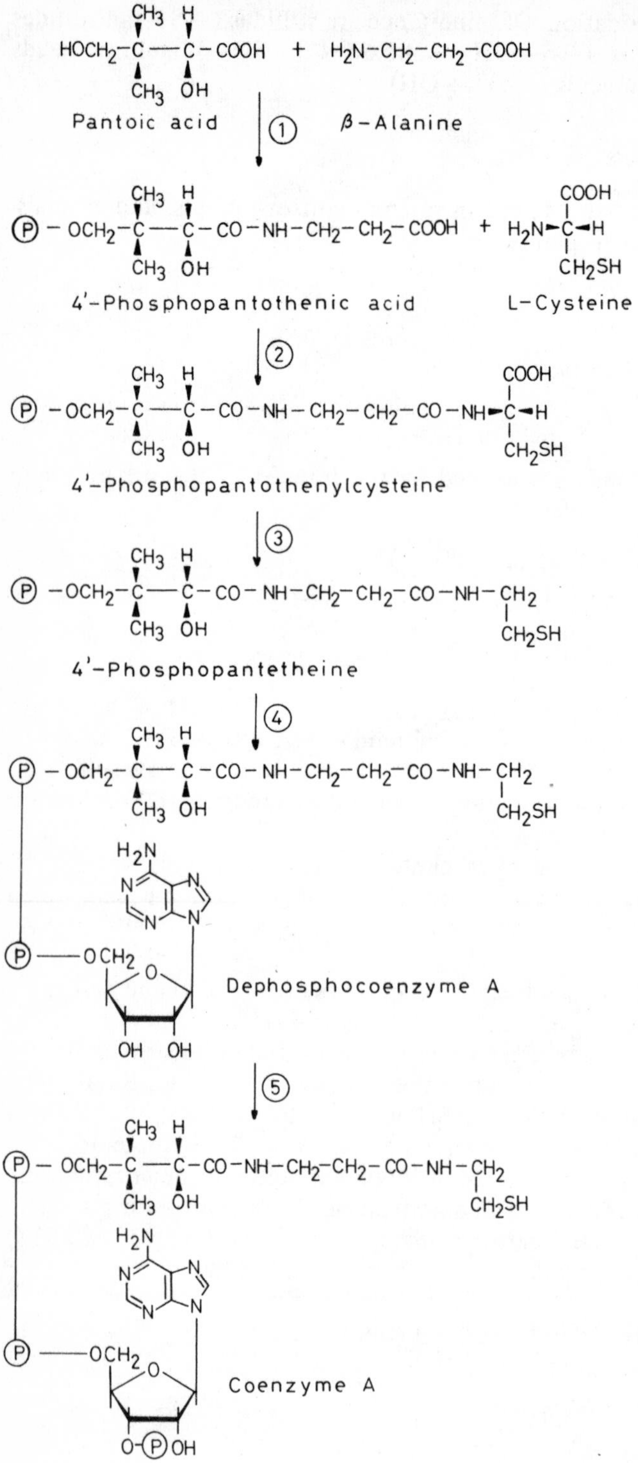

Fig. 177. Biosynthesis of coenzyme A
1 Pantothenate synthetase, pantothenate kinase; 2 phosphopantothenoylcysteine synthetase; 3 phosphopantothenoylcysteine decarboxylase; 4 dephospho-CoA synthetase; 5 dephospho-CoA kinase

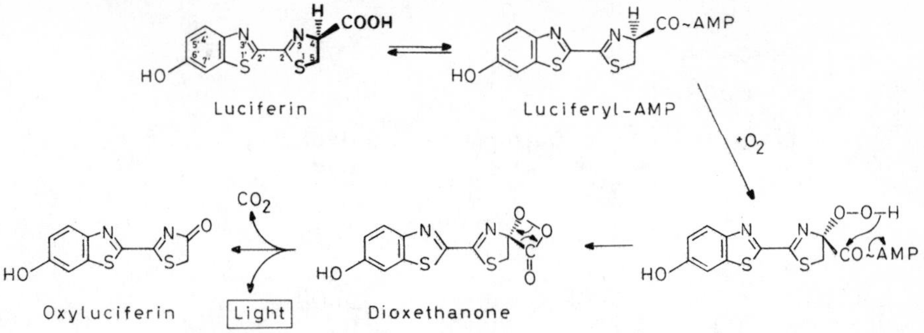

Fig. 178. Light emission by the action of luciferase (Photinus luciferin 4-monooxygenase) on firefly luciferin

D 11.1 Sulfinic and Sulfonic Acids

Chemistry and Distribution

Sulfinic acids, e.g., L-cysteine sulfinic acid and hypotaurine, and sulfonic acids, e.g., L-cysteic acid and taurine, are naturally occurring products of L-cysteine and cysteamine oxygenation. They occur in microorganisms, plants, and animals.

Biosynthesis

L-Cysteine is transformed to L-cysteine sulfinic acid and L-cysteic acid. Cysteamine (D 11) yields hypotaurine and taurine (Fig. 179). The latter compounds

Table 46. Naturally occuring *S*-substituted L-cysteine derivatives and sulfoxides

R	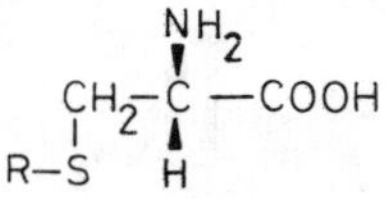 S-Substituted L-cysteine derivatives	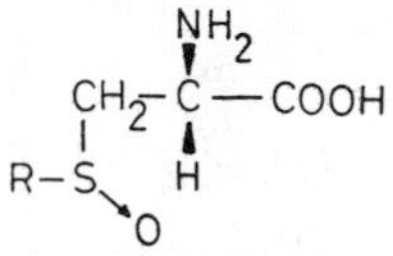 Alliins
CH_3-	S-Methyl-L-cysteine	Methylalliin
$CH_3-CH_2-CH_2-$	S-Propyl-L-cysteine	Propylalliin
$CH_2=CH-CH_2-$	S-Allyl-L-cysteine	Allylalliin (Alliin)
$CH_3-CH=CH-$	S-Δ^1-E-Propenyl-L-cysteine	Δ^1-Propenylalliin
$HOOC-CH_2-CH_2-$	S-2-Carboxyethyl-L-cysteine	—
$HOOC-CH(CH_3)-CH_2-$	S-2-Carboxy-n-propyl-L-cysteine	—

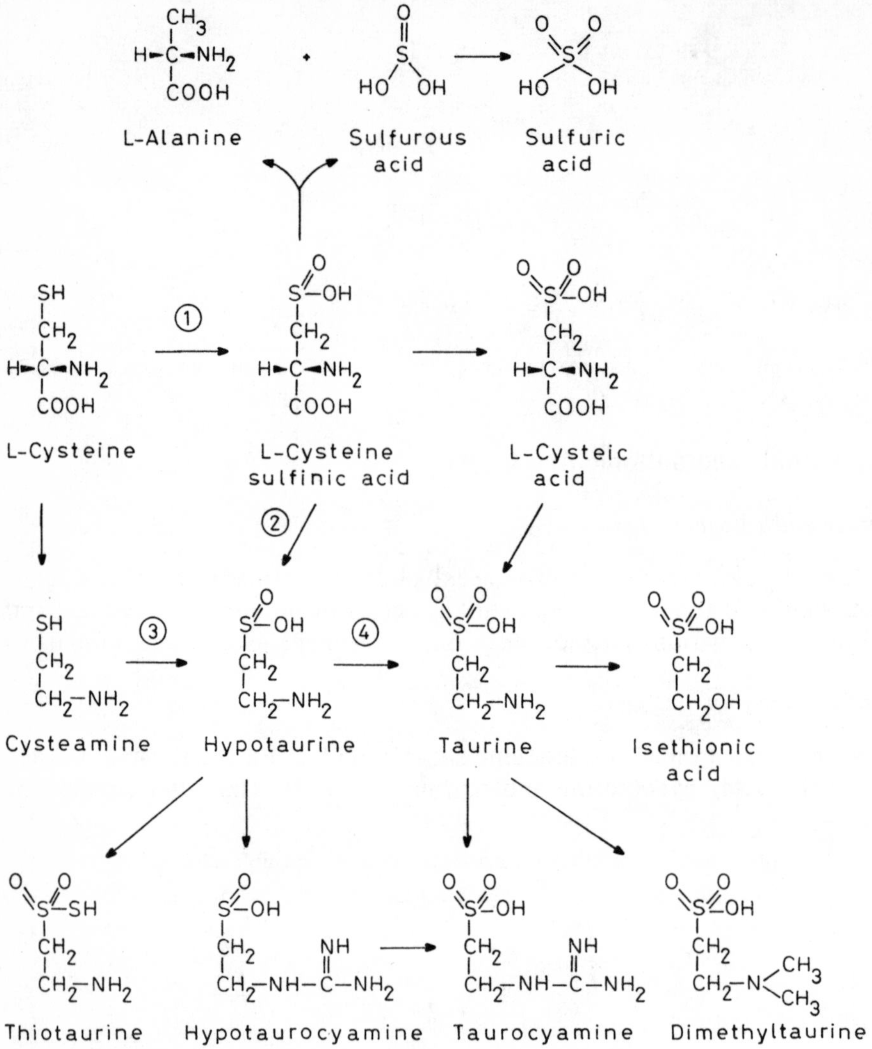

Fig. 179. Biosynthesis of L-cysteine derivatives with oxidized sulfur
1 Cysteine dioxygenase (C 2.5); *2* cysteine sulfinate decarboxylase; *3* cysteamine dioxygenase (C 2.5); *4* hypotaurine dehydrogenase

may be transformed to other secondary products by deamination, thiolation, guanylation, and methylation. L-Cysteine sulfinic acid may be degraded to L-alanine and sulfurous acid, which is oxidized sulfuric acid.

Significance

Hypotaurine and taurine occur in many tissues and in the urine of animals and human beings (E 1). Taurine is a constituent of several bile acids formed in animals (D 6.6.4) and is of significance as neurotransmitter (E 3.2).

References for Further Reading

134, 378, 387, 830

D 11.2 S-Substituted Cysteine Derivatives and Sulfoxides

Chemistry and Distribution

Several L-cysteine derivatives carry alkyl or related substituents at the S-atom. They occur at the oxidation level of sulfides or sulfoxides (alliins, Table 46). The substances are aliphatic or have a cyclic structure, like cycloalliin. Several products are dimers, cf. the formula of djenkolic acid.

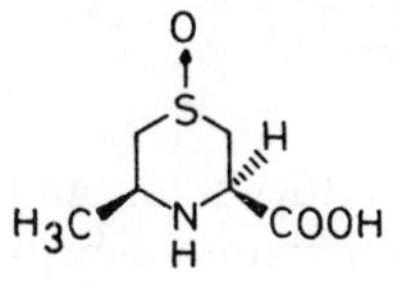

Cycloalliin 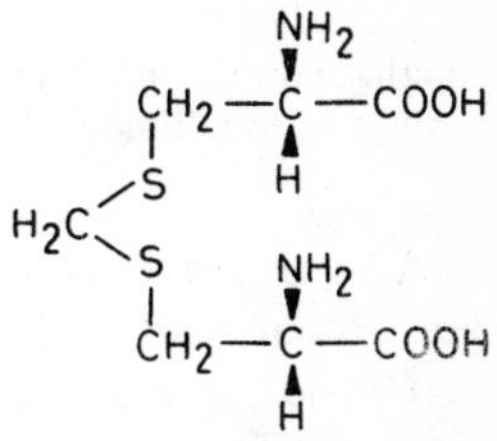   Djenkolic acid

S-Substituted cysteine derivatives and the corresponding sulfoxides are formed in plants, e. g., Liliaceae, Brassicaceae, and Mimosaceae.

Biosynthesis

The sulfhydryl group of cysteine is first substituted and then oxygenated to the alliins by a monooxygenase (C 2.6.2). Cycloalliin originates by addition of the amino group of Δ^1-propenylalliin to the double bond.

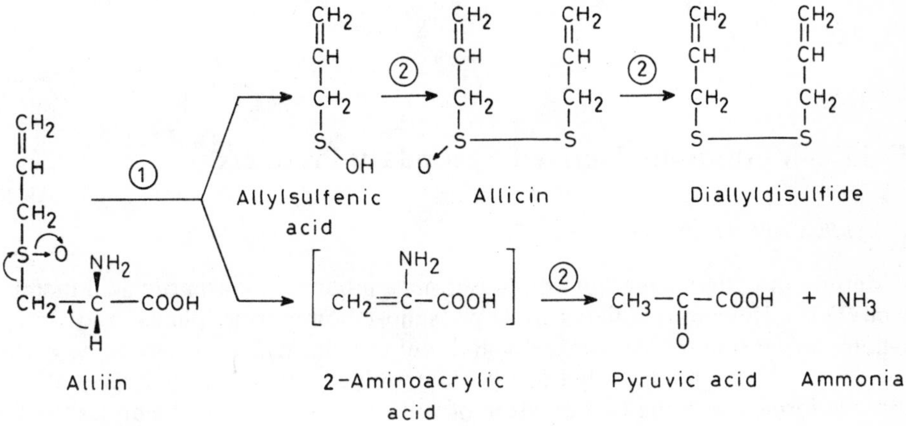

Alliin   Allylsulfenic   Allicin   Diallyldisulfide
                  acid

2-Aminoacrylic   Pyruvic acid   Ammonia
acid

Fig. 180. Degradation of alliin
1 Alliin lyase (allinase); *2* spontaneous reaction

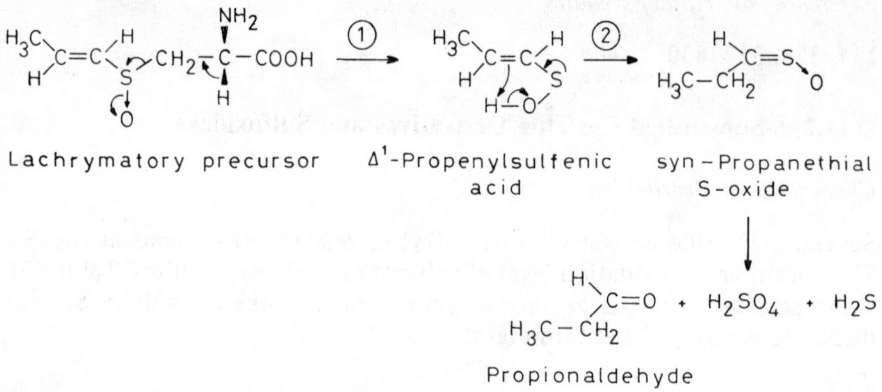

Fig. 181. Formation of syn-propanethial *S*-oxide
1 Alliin lyase (allinase); *2* spontaneous reaction

Alliins are degraded to sulfenic acids and 2-aminoacrylic acid (Fig. 180). 2-Aminoacrylic acid decomposes spontaneously to pyruvic acid and ammonia.

The sulfenic acids condense spontaneously to dimeric compounds of the allicin type. The allicins easily release oxygen and change to disulfides.

Significance

syn-Propanethial S-oxide derived from Δ^1-propenylsulfenic acid (Fig. 181) is the lachrymatory (tear-producing) substance liberated on crushing onion cells (F 1). In the presence of water (e. g., in the eye) it is hydrolyzed to propionaldehyde, sulfuric acid, and H_2S. It is a feeding deterrent to animals (E 5.5.3).

Compounds of the allicin type possess a considerable antimicrobial activity (E 5.4). Allicin and diallyldisulfide cause the smell of garlic (F 1).

References for Further Reading

75, 788

D 12 L-Methionine-Derived Secondary Products

Biosynthesis of the Precursor

L-Methionine originates from L-homoserine, a product of L-aspartic acid metabolism (D 16). Key intermediates are *O*-phospho-L-homoserine (plants) and *O*-acyl-L-homoserine (microorganisms). Cystathionine is formed with L-cysteine as sulfur donor. It may be degraded to L-homocysteine, which, however, may also be directly formed with the participation of sulfide. This reaction resembles the formation of L-cysteine from *O*-acetyl-L-serine (D 11). 5-Methyltetrahydrofolic acid acts as donor of the L-methionine methyl group (C 3.2).

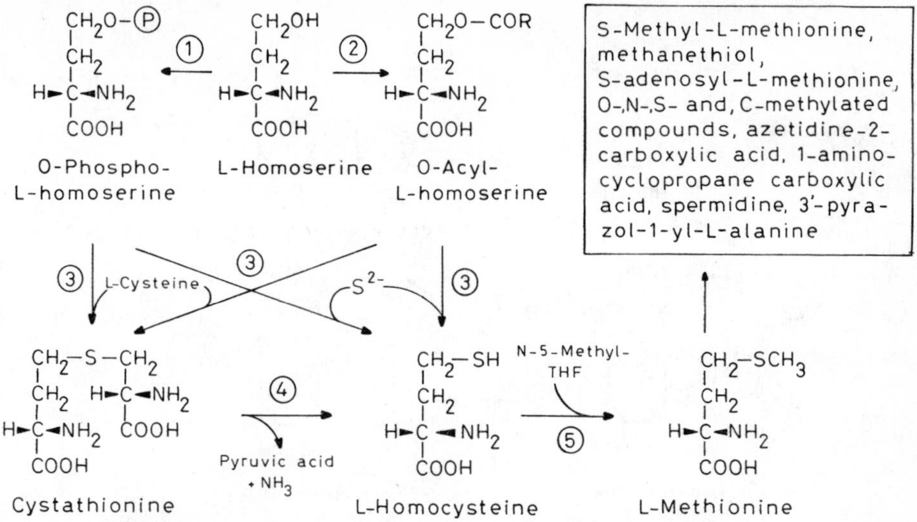

Fig. 182. Biosynthesis of L-methionine
1 Homoserine kinase; *2* homoserine acyltransferase; *3* O-succinylhomoserine (thiol)-lyase; *4* cystathionine β-lyase; *5* tetrahydropteroylglutamate methyltransferase

Chemistry and Distribution

Secondary products derived from L-methionine occur in microorganisms, plants, and animals. The most important compounds are listed in Fig. 182. They either contain still the more or less complete methionine skeleton (cf. the formulas of S-adenosyl-L-methionine, azetidine-2-carboxylic acid, 1-aminocyclopropane carboxylic acid, Fig. 183, spermidine and spermine, Fig. 184), or only the methyl group of L-methionine (most O-, N-, S-, and C-methylated secondary products). The sulfur-containing derivatives possess either a sulfide (—S—) or a sulfonium (=S+—) structure.

Groups of Secondary Products

a) S-Methyl-L-methionine is built in higher plants, dimethyl-β-propiothetin is synthesized in algae, bryophytes, and animals by special methyltransferases. Both substances may act as methyl group donors in several biochemical reactions. In algae and liverworts dimethyl-β-propiothetin is degraded enzymatically to dimethylsulfide (CH_3SCH_3) and acrylic acid.

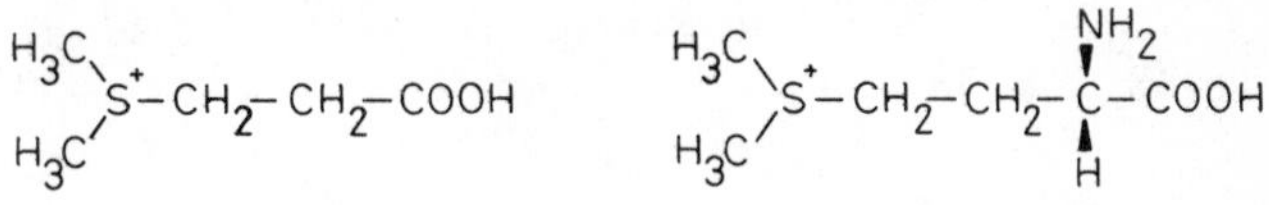

Dimethyl-β-propiothetin

S-Methylmethionine

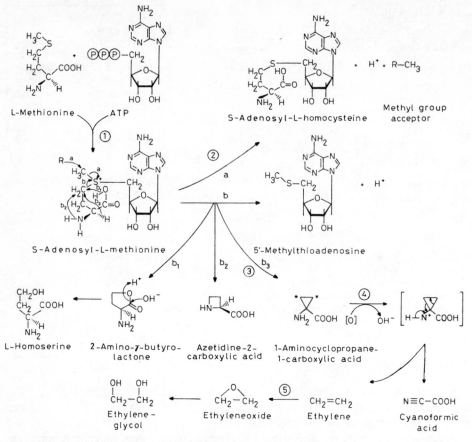

Fig. 183. Decomposition of *S*-adenosyl-L-methionine
1 Methionine adenosyltransferase; *2* *S*-adenosylmethionine-dependent methyltransferases; *3* 1-aminocyclopropanecarboxylic acid (ACC) synthase; *4* ACC oxidase; *5* ethylene monooxygenase

b) Methanethiol (methylmercaptane, CH_3SH) and 2-oxobutyric acid ($CH_3CH_2COCOOH$) are formed from L-methionine by methionine lyase as result of a 2,4-elimination. Methanethiol is easily oxidized to dimethyldisulfide (CH_3—SS—CH_3). This compound has an unpleasant smell (like other volatile sulfides) and is responsible for the *Foetor hepaticus* accompanying several liver diseases.

c) *S*-Adenosyl-L-methionine is built as described in D 12.1.

References for Further Reading

31, 48, 266, 387

D 12.1 Secondary Products Formed from S-Adenosyl-L-Methionine

Biosynthesis of the Precursor

S-Adenosyl-L-methionine ("activated methionine") is formed by the condensation of L-methionine with ATP in microorganisms, plants, and animals (Fig. 183). It is a sulfonium compound which possesses the reactive $=S^+-$ structure.

Groups of Secondary Products

a) The methyl groups of most O-, S-, N-, and C-methylated secondary products are derived from S-adenosyl-L-methionine by methyltransferases (C 3.3). During transmethylation, the methyl group is attacked nucleophilically and S-adenosyl-L-homocysteine is formed (Fig. 183, pathway a).

b) 5'-Methylthioadenosine is synthesized from S-adenosyl-L-methionine by splitting the bond between the methionine carbon skeleton and the sulfur atom (Fig. 183, pathway b). As additional products, the following may be formed:
— 2-amino-γ-butyrolactone, a compound yielding homoserine on hydrolysis
— azetidine-2-carboxylic acid, a nonprotein amino acid found in plants. Azetidine-2-carboxylic acid is toxic to animals and acts as feeding deterrent (E 5.5.3). It is activated by the prolyl-tRNA synthetase of most organisms and inhibits the biosynthesis of biologically active proteins. Azetidine-2-carboxylic acid shows structural relations to nicotianamine, a secondary plant product involved in Fe^{3+} transport (E 2.2)

Nicotianamine

— 1-aminocyclopropane-1-carboxylic acid (ACC). ACC synthase contains pyridoxal-5'-phosphate as coenzyme (C 5). Ethylene (ethene) and cyanoformic acid (which decarboxylates spontaneously to hydrocyanic acid and CO_2) are built from ACC in a oxygen-dependent reaction. Ethylene is transformed to ethyleneoxide and further on to ethyleneglycol. It is a plant hormone (E 3.1) which accelerates defoliation and the ripening of fruits and is used commercially for this purpose. In molds ethylene acts as "biochemical radar", allowing the hyphae to pass obstacles without touching (E 2.2).

c) S-Methyl-S-adenosylhomocysteamine (S-adenosyl-5'-methylmercaptopropylamine) is derived from S-adenosyl-L-methionine by decarboxylation. Together with putrescine (D 17) it is used in the biosynthesis of the polyamines spermidine and spermine (Fig. 184). In plants spermidine forms amides with activated cinnamic acid derivatives (D 23.2.1) as well as macrocyclic spermidine and spermine alkaloids.

The oxidative degradation of spermine yields 1,3-diaminopropane and 1,3-aminopropylpyrroline (Fig. 185). 1,3-Diaminopropane is further transformed

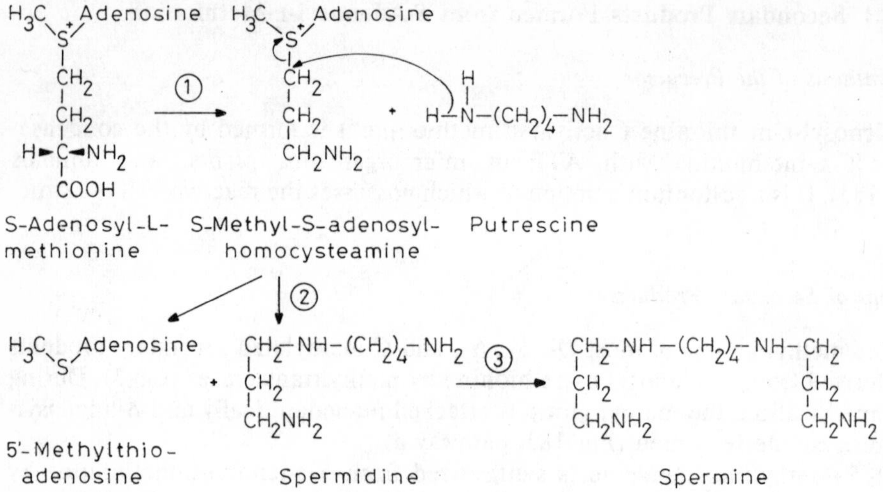

Fig. 184. Formation of spermidine and spermine
1 S-Adenosylmethionine decarboxylase; *2* spermidine synthase; *3* spermine synthase

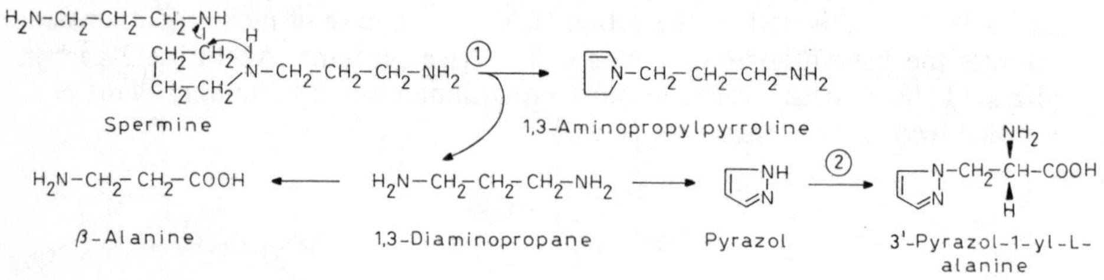

Fig. 185. Degradation of spermine
1 Amine oxidase; *2* 3'-pyrazol-1-yl-L-alanine synthase

to β-alanine and pyrazole, a precursor (together with *O*-acetyl-L-serine, D 10) of 3'-pyrazol-1-yl-L-alanine, a nonprotein amino acid found in Cucurbitaceae.

References for Further Reading

864, 865, 866

D 13 Secondary Products Originating from L-Valine

Biosynthesis and Degradation of the Precursor

L-Valine is built on a pathway resembling the formation of L-isoleucine (Fig. 186). It is degraded to succinyl CoA, as shown in Fig. 187.

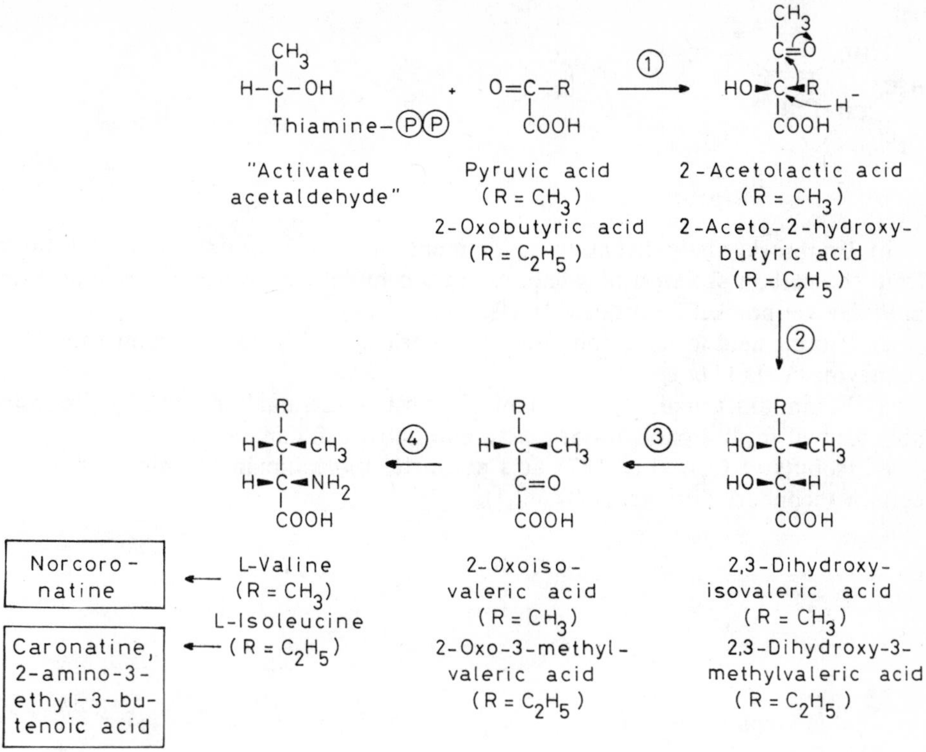

Fig. 186. Biosynthesis of L-valine and L-isoleucine
1 Acetolactate synthase; *2* acetolactate reducto-isomerase; *3* dihydroxyacid dehydratase; *4* aminotransferase

Chemistry and Distribution

Secondary compounds derived from L-valine and products of its degradation occur in microorganisms, plants, and animals. The most important substances are listed in Figs. 186 and 187. They still contain the isopentane skeleton of their precursors (cf. the formulas of pantoic acid and erythroskyrin).

Groups of Secondary Products

a) Norcoronatine is a phytotoxic compound produced in *Pseudomonas syringae* pvs. *atropurpurea* and *glycinea* (E 5.4). It is built from coronafacic acid and L-valine via *N*-coronafacoyl-L-valine.

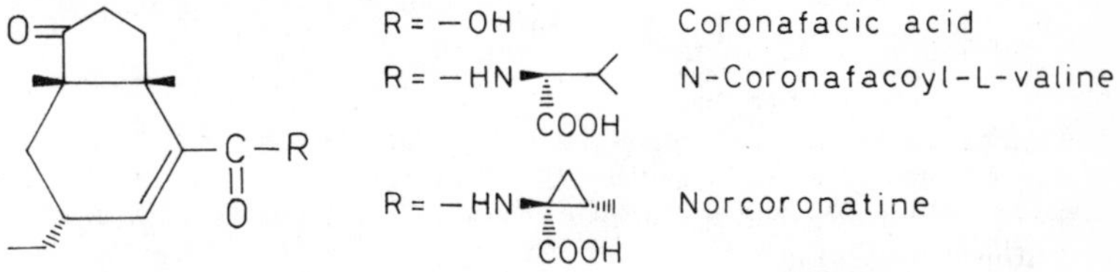

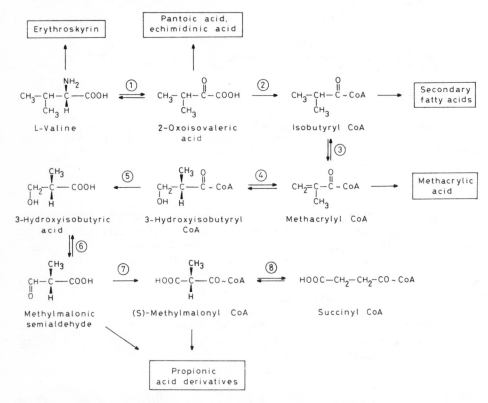

Erythroskyrin

b) Erythroskyrin, a hepatotoxic pigment of *Penicillium islandicum*, is built from the carbon skeleton of L-valine and ten molecules of acetate or malonate probably via polyketide intermediates.

c) Pantoic acid formed from 2-oxoisovaleric acid (Fig. 187) is a constituent of coenzyme A (D 11).

d) Echimidinic acid, a constituent of pyrrolizidine alkaloids (D 17.1), is probably formed from 2-oxoisovaleric acid and activated acetaldehyde (Fig. 188).

e) Isobutyryl CoA (Fig. 187) acts as starter molecule in the biosynthesis of certain secondary fatty acids (D 5.2.1).

Fig. 187. Degradation of L-valine
1 Branched-chain amino acid aminotransferase; *2* branched-chain oxo acid dehydrogenase complex (C 2.7); *3* 2-methyl-branched-chain acyl CoA dehydrogenase; *4* crotonase; *5* hydroxyisobutyryl-CoA hydrolase; *6* 3-hydroxybutyrate dehydrogenase; *7* methylmalonic semialdehyde dehydrogenase; *8* methylmalonyl-CoA mutase

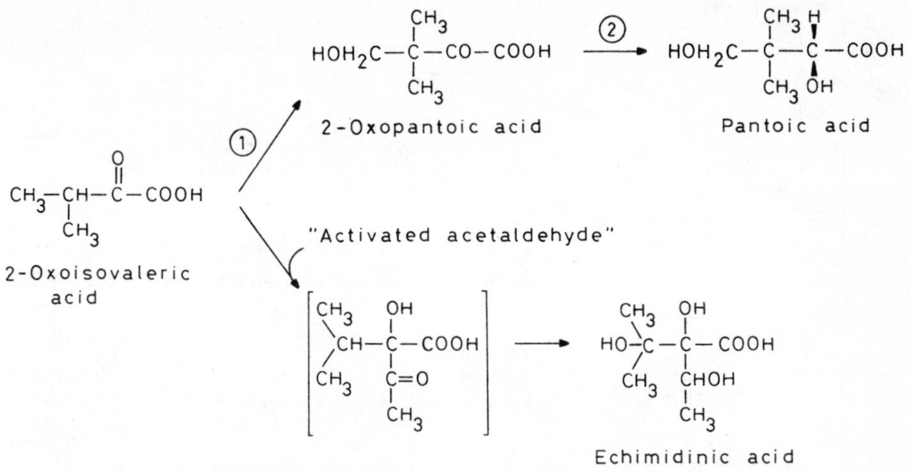

$$CH_3$$
$$HOH_2C-\overset{\overset{\displaystyle CH_3}{|}}{\underset{\underset{\displaystyle CH_3}{|}}{C}}-CO-COOH \quad \xrightarrow{\text{②}} \quad HOH_2C-\overset{\overset{\displaystyle CH_3}{|}}{\underset{\underset{\displaystyle CH_3}{|}}{C}}-\overset{\overset{\displaystyle H}{|}}{\underset{\underset{\displaystyle OH}{|}}{C}}-COOH$$

2-Oxopantoic acid

Pantoic acid

$$CH_3-\overset{\overset{\displaystyle O}{\|}}{\underset{\underset{\displaystyle CH_3}{|}}{CH}}-C-COOH$$

2-Oxoisovaleric acid

"Activated acetaldehyde"

$$\begin{bmatrix} CH_3 & OH \\ \diagdown CH-\overset{|}{\underset{\underset{\displaystyle CH_3}{|}}{C}}-COOH \\ CH_3 & C=O \end{bmatrix} \longrightarrow HO-\overset{\overset{\displaystyle CH_3}{\diagup}}{\underset{\underset{\displaystyle CH_3}{\diagdown}}{C}}-\overset{\overset{\displaystyle OH}{|}}{\underset{\underset{\displaystyle CH_3}{|}}{C}}-COOH$$

Echimidinic acid

Fig. 188. Formation of pantoic and echimidinic acids
1 Oxopantoate hydroxymethyltransferase; *2* oxopantoate oxidoreductase

f) Methacrylic acid, derived from methacrylyl CoA (Fig. 187), is found in high concentration in the defense secretions of insects (E 5.1).

It is easily substituted by functional groups of proteins, a reaction originating the polymerization to a plexiglaslike plastic which may cover (part of) the aggressor.

g) Methylmalonyl CoA is formed from L-valine, as well as from succinyl CoA, an intermediate of the tricarboxylic acid cycle (D 4) (Fig. 187). It is a precursor of polyketidelike compounds (D 3).

References for Further Reading

31, 107, 113, 369

D 14 Secondary Products derived from L-Leucine

Biosynthesis and Degradation of the Precursor

L-Leucine is an α-amino acid with six C-atoms and a branched carbon chain. It is derived from L-valine by the 2-oxo acid elongation system, i. e., via 2-oxoisovaleric acid and 3-carboxy-3-hydroxyisocaproic acid as outlined in Fig. 189. The transformation of 2-oxo-isovaleric acid closely resembles the conversion of oxaloacetic acid to 2-oxoglutaric acid in the tricarboxylic acid cycle (D 4). 2-Oxo acid elongation systems also participate in the formation of L-lysine via the 2-aminoadipic acid pathway (D 18) and in the biosynthesis of a number of secondary amino acids which are precursors of glucosinolates (D 8.4).

L-Leucine is degraded to acetoacetate and acetyl CoA (Fig. 190).

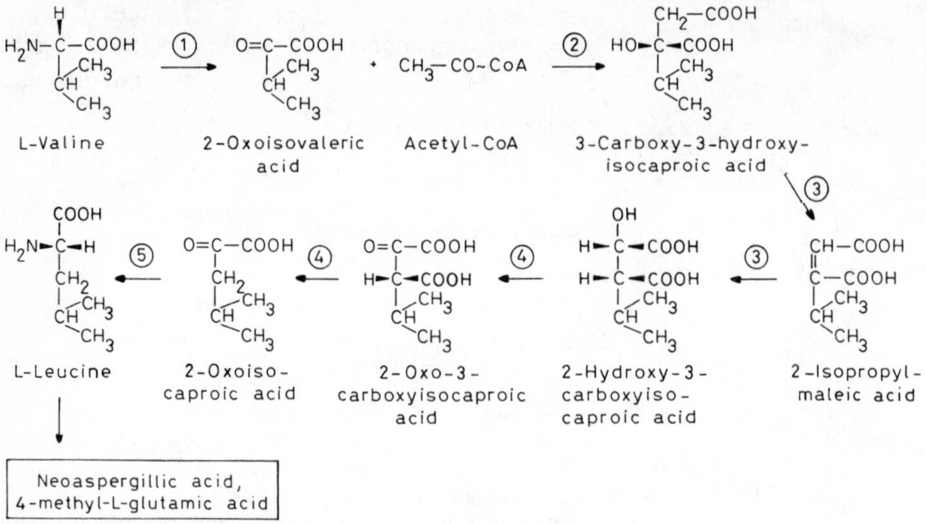

L-Valine   2-Oxoisovaleric acid   Acetyl-CoA   3-Carboxy-3-hydroxy-isocaproic acid

L-Leucine   2-Oxoiso-caproic acid   2-Oxo-3-carboxyisocaproic acid   2-Hydroxy-3-carboxyiso-caproic acid   2-Isopropyl-maleic acid

Neoaspergillic acid, 4-methyl-L-glutamic acid

Fig. 189. Biosynthesis of L-leucine
1 Branched-chain amino acid aminotransferase; *2* isopropylmalate synthase; *3* isopropylmalate isomerase; *4* carboxyhydroxyisocaproate decarboxylase; *5* leucine aminotransferase

L-Leucine   2-Oxoisocaproic acid   Isovaleryl CoA   Secondary fatty acids

3-Hydroxy-3-methyl-glutaryl CoA   E-3-Methylglutaconyl CoA   3-Methylcrotonyl CoA   Senecio alkaloids

Isopentenyl diphosphate   Acetoacetic acid   Acetyl CoA

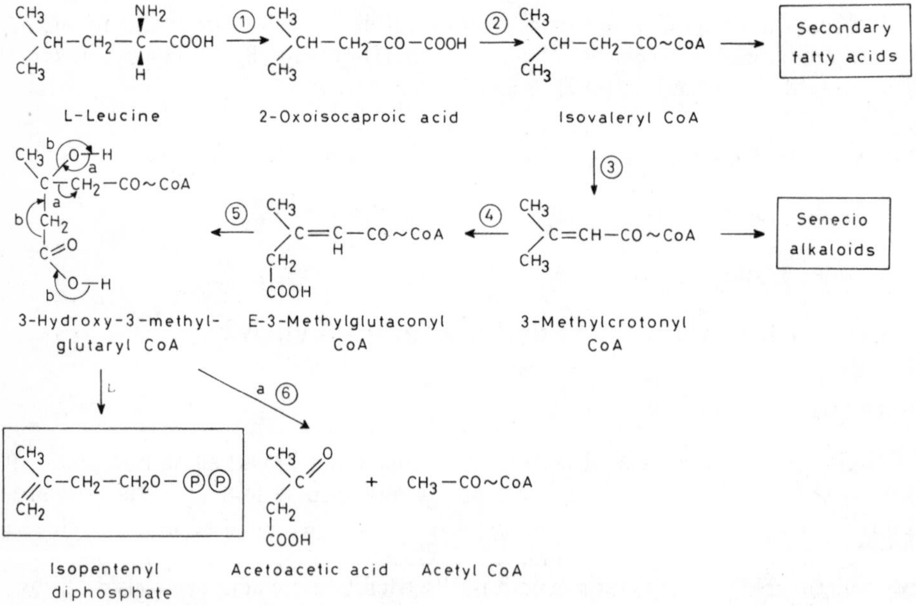

Fig. 190. Degradation of L-leucine
1 Branched-chain amino acid aminotransferase or L-leucine aminotransferase; *2* branched-chain oxo acid dehydrogenase complex (C 2.7); *3* 2-methyl-branched-chain acyl-CoA dehydrogenase; *4* methylcrotonyl-CoA carboxylase; *5* methylglutaconyl-CoA hydratase; *6* hydroxymethylglutaryl-CoA lyase

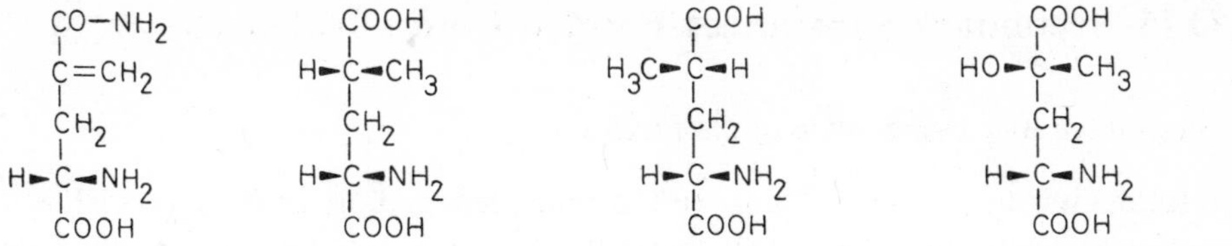

$$\begin{array}{cccc}
\text{CO}-\text{NH}_2 & \text{COOH} & \text{COOH} & \text{COOH} \\
| & | & | & | \\
\text{C}{=}\text{CH}_2 & \text{H}\!\!\blacktriangleright\!\!\text{C}\!\!\blacktriangleleft\!\!\text{CH}_3 & \text{H}_3\text{C}\!\!\blacktriangleright\!\!\text{C}\!\!\blacktriangleleft\!\!\text{H} & \text{HO}\!\!\blacktriangleright\!\!\text{C}\!\!\blacktriangleleft\!\!\text{CH}_3 \\
| & | & | & | \\
\text{CH}_2 & \text{CH}_2 & \text{CH}_2 & \text{CH}_2 \\
| & | & | & | \\
\text{H}\!\!\blacktriangleright\!\!\text{C}\!\!\blacktriangleleft\!\!\text{NH}_2 & \text{H}\!\!\blacktriangleright\!\!\text{C}\!\!\blacktriangleleft\!\!\text{NH}_2 & \text{H}\!\!\blacktriangleright\!\!\text{C}\!\!\blacktriangleleft\!\!\text{NH}_2 & \text{H}\!\!\blacktriangleright\!\!\text{C}\!\!\blacktriangleleft\!\!\text{NH}_2 \\
| & | & | & | \\
\text{COOH} & \text{COOH} & \text{COOH} & \text{COOH}
\end{array}$$

4-Methylene-   erythro-4-Methyl-   threo-4-Methyl-   4-Hydroxy-4-methyl-
L-glutamine   L-glutamic acid   L-glutamic acid   L-glutamic acid

Fig. 191. Structural formulas of 4-methylglutamic acid derivatives

Chemistry and Distribution

Secondary products are built from L-leucine and intermediates of L-leucine degradation in microorganisms and plants. The most important groups are listed in Figs. 189 and 190. Their carbon skeleton shows either an isohexane or isopentane structure.

Groups of Secondary Products

a) Cyanolipids synthesized in certain plants and neoaspergillic acid and related compounds found in microorganisms are derived from L-leucine. They are treated in D 5.2.4 and D 9.2.1, respectively.

b) 4-Substituted glutamic acid derivatives, e.g., 4-methylene-L-glutamine, erythro-4-methyl-L-glutamic acid, threo-4-methyl-L-glutamic acid, and 4-hydroxy-4-methyl-L-glutamic acid (Fig. 191), accumulating in certain plant species, are derived from L-leucine by the oxidation of one of the terminal methyl groups to a carboxyl group. 4-Methylene-L-glutamine is the major nitrogen transport compound from the roots to the shoots in peanuts. It is hydrolyzed by 4-methyleneglutaminase to 4-methylene-L-glutamic acid and ammonia (E 2.2). The 4-methyl-L-glutamic acid derivatives are toxic to animals and act as feeding deterrents (E 5.5.3).

c) Isovaleryl CoA is used as a starter in the biosynthesis of certain fatty acids (D 5.2.1).

d) 3-Methylcrotonyl CoA (senecioyl CoA) is a precursor of pyrrolizidine alkaloids (D 17.1).

e) 3-Hydroxy-3-methylglutaryl CoA (Fig. 190) is an intermediate in isopentenyl diphosphate formation (D 6). Synthesis of isopentenyl diphosphate from L-leucine does, however, not occur in nature.

References for Further Reading

31, 113, 612

D 15 Secondary Substances Formed from L-Isoleucine

Biosynthesis and Degradation of the Precursor

L-Isoleucine is an amino acid with a branched carbon chain comprising six C-atoms. It originates from 2-oxobutyric acid, a threonine derivative, and "activated acetaldehyde" (C 4) as outlined in Fig. 186. Both compounds condense to form 2-aceto-2-hydroxybutyric acid from which 2,3-dihydroxy-3-methylvaleric acid is formed by the shift of the ethyl side chain from C-2 to C-3. The latter compound is converted to L-isoleucine via 2-oxo-3-methylvaleric acid.

L-Isoleucine is degraded to acetyl CoA and propionyl CoA (Fig. 192).

Chemistry and Distribution

Secondary products formed from L-isoleucine and compounds built during its degradation have been found in plants and animals. The most important are given in Figs. 186 and 192. They show a characteristic branched carbon skeleton with either six or five C-atoms.

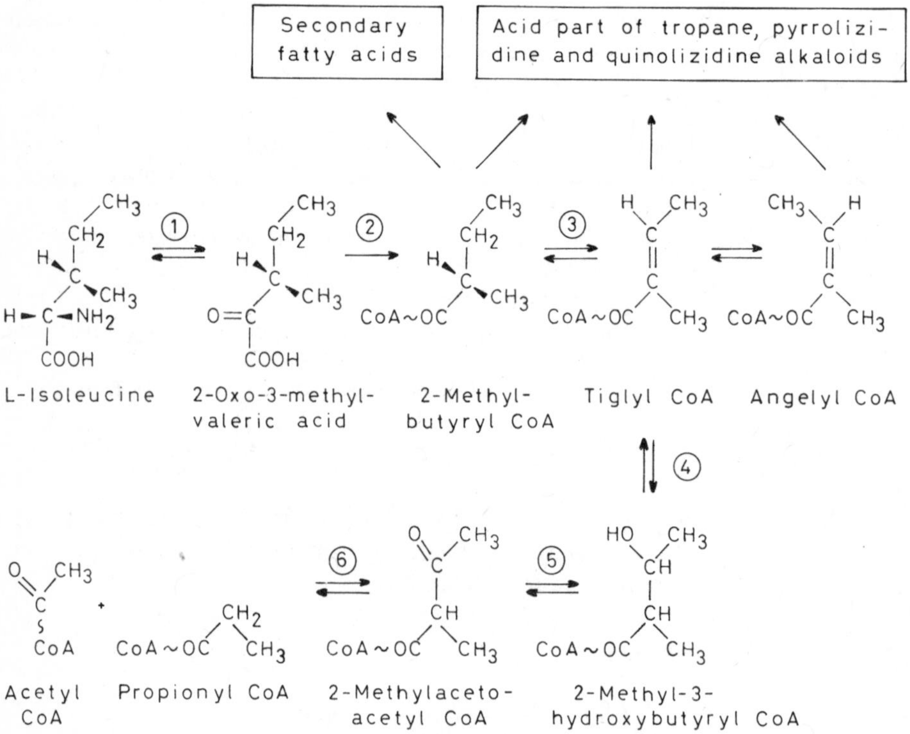

Fig. 192. Degradation of L-isoleucine
1 Branched-chain amino acid aminotransferase; *2* branched-chain oxo acid dehydrogenase complex (C 2.7); *3* 2-methyl-branched-chain acyl-CoA dehydrogenase; *4* crotonase; *5* methyl hydroxybutyryl-CoA dehydrogenase; *6* 2-methylacetoacetyl-CoA lyase

Groups of Secondary Products

a) Coronatine is a phytotoxic compound produced in *Pseudomonas syringae* pvs. *atropurpurea* and *glycinea* (E 5.4). It is formed from coronafacic acid and L-isoleucine probably via *N*-coronafacoyl-L-isoleucine (D 13).

b) *L*-2-amino-3-ethyl-3-butenoic acid and structurally related compounds are nonprotein amino acids synthesized in plants.

c) The CoA esters of 2-methylbutyric acid, tiglic acid, and angelic acid (Fig. 192) are precursors of pyrrolizidine (D 17.1), tropane (D 17.2), and quinolizidine alkaloids (D 18), which are formed in plants. Since all reactions involved in the degradation of the esters are reversible, they may also be built de novo from acetyl CoA and propioyl CoA.

d) Seneciphyllic acid and senecic acid are constituents of certain pyrrolizidine alkaloids (D 17.1). They are derived from the acids formed during L-isoleucine degradation.

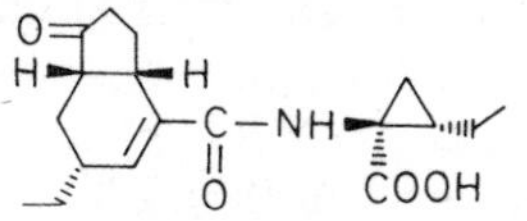

Coronatine

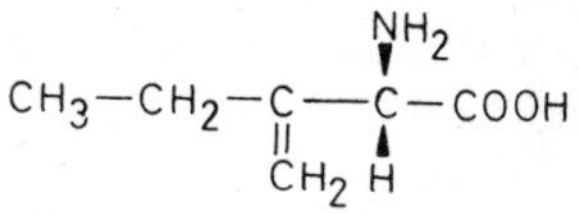

L-2-Amino-3-ethyl-
3-butenoic acid

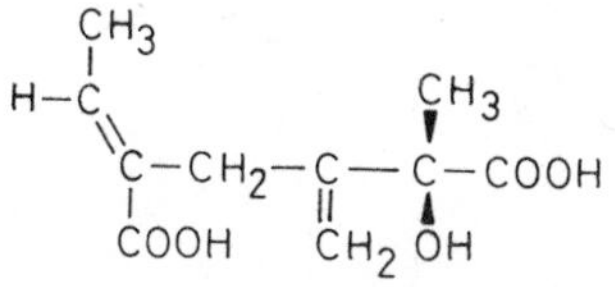

Seneciphyllic acid

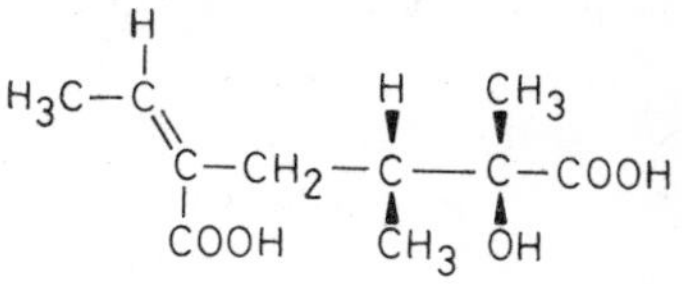

Senecic acid

e) 2-Methylbutyryl CoA acts as starter in the biosynthesis of some secondary fatty acids with a branched carbon chain (D 5.2.1).

f) Tiglic acid, which may be derived from tiglyl CoA, is a constituent of the defense secretions of certain insects (E 5.1).

References for Further Reading

31, 113, 240

D 16 Secondary Products Built from L-Aspartic Acid and L-Threonine

Biosynthesis of the Precursor

L-Aspartic acid is formed from oxaloacetic acid which originates in the tricarboxylic acid cycle (D 4). In addition, it may be derived from fumaric acid in a reversible reaction catalyzed by aspartase:

HOOC–CH   HOOC–CH₂
    ‖   ⇌   H
   HC–COOH   H₂N–C COOH

Fumaric acid   L-Aspartic acid

L-Threonine is built from L-aspartic acid via L-aspartic acid-4-phosphate and L-aspartic-4-semialdehyde (Fig. 193). The latter is a precursor of L-lysine (D 18) as well as of L-homoserine (D 12). L-Threonine is formed from L-homoserine by synthesis of the 4-phosphate and shift of the hydroxyl group by a pyridoxal-5'-phosphate-dependent enzyme.

Chemistry and Distribution

Secondary products are derived from L-aspartic acid or L-threonine in microorganisms, plants, and animals. The most important groups are given in Fig. 193. They
— either have still the 4-carbon skeleton of the precursor amino acids (cf. the formulas of the substituted L-asparagine derivatives and L-aspartic acid amides)
— are devoid of one carbon atom, like miserotoxin (Fig. 196),
— or possess additional carbon atoms, like hypoglycine A (Fig. 198).

Furthermore, the skeleton of aspartic acid is incorporated in cyclic compounds with pyridine (D 16.1) and pyrimidine rings (D 26).

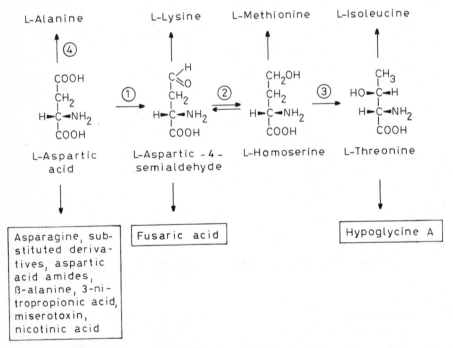

Fig. 193. Transformation of L-aspartic acid
1 Aspartate kinase, aspartate semialdehyde dehydrogenase; *2* homoserine dehydrogenase; *3* homoserine kinase, threonine synthase; *4* aspartate 4-decarboxylase

Groups of Secondary Products

a) L-Asparagine is synthesized from L-aspartic acid, ATP, and ammonia in microorganisms and plants. In most animals, however, the amide group of glutamine is transferred to aspartic acid. L-Asparagine is a nitrogen store in certain plants (E 2.2).

b) N^4-Substituted L-asparagine derivatives, e.g., N^4-methyl-, N^4-ethyl, and N^4-(2-hydroxyethyl)-L-asparagine are derived from L-aspartic acid, ATP, and the respective amines (Fig. 194). Compounds of this type are formed in Cucurbitaceae. They are accompanied by the respective L-glutamine derivatives (D 17).

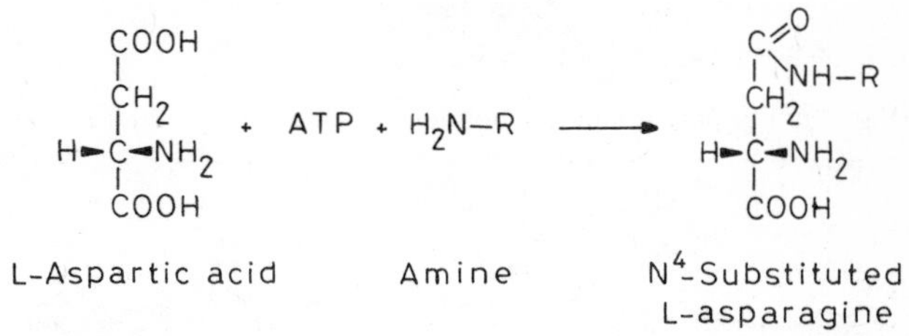

Fig. 194. Formation of N^4-substituted L-asparagine derivatives

L-Aspartic acid   Amine N^4-Substituted L-asparagine

c) L-Aspartic amides are synthesized in animals and plants from aspartic acid and activated acid derivatives according to Fig. 195. They are of significance in the detoxification of secondary products (E 1).

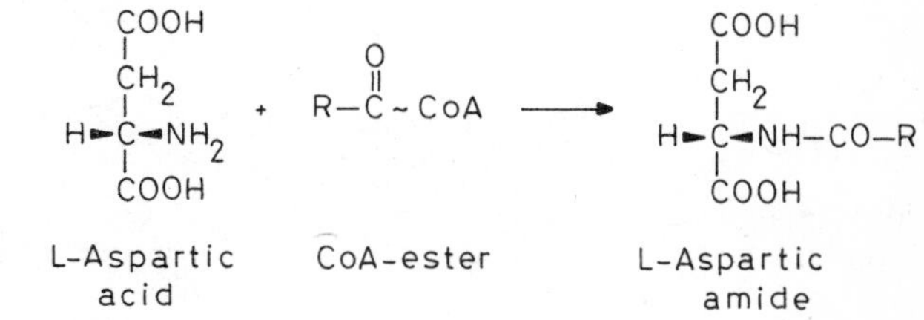

Fig. 195. Biosynthesis of L-aspartic acid amides

L-Aspartic acid   CoA-ester   L-Aspartic amide

d) β-Alanine is formed, e.g., by the decarboxylation of L-aspartic acid (D 8.2). It is a constituent of coenzyme A (D 11).

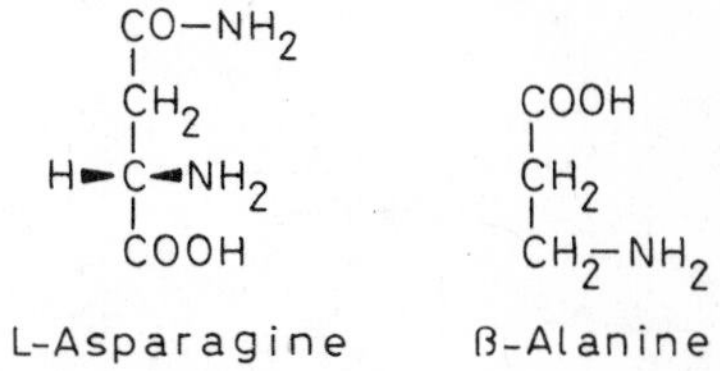

L-Asparagine   ß-Alanine

e) 3-Nitropropionic acid is derived from β-alanine. Esters of this acid with D-glucose are toxic compounds found in some plants. 3-Nitropropionic acid is probably also the precursor of miserotoxin, a substance found in *Astragalus* sp. (Fig. 196). Miserotoxin releases nitrite and is therefore highly toxic to animal predators (E 5.5.3).

f) Nicotinic acid derivatives are formed from L-aspartic acid and a C_3-unit (D 16.1).

g) Fusaric acid is derived from L-aspartic acid (most probably via L-aspartic semialdehyde) and acetate (Fig. 197). Fusaric acid is formed in various *Fusarium* sp. and causes, for instance, the wilting of tomato plants (E 5.4).

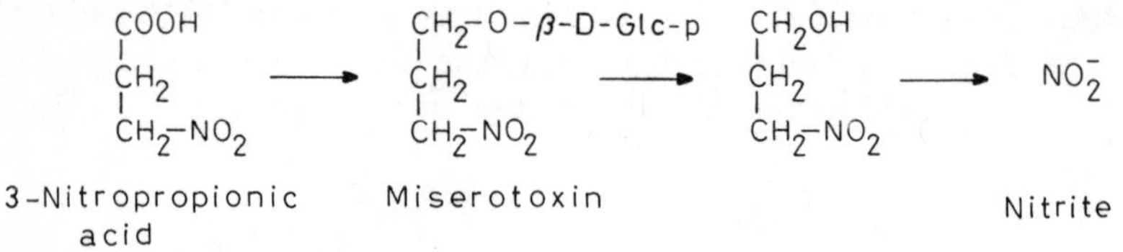

Fig. 196. Formation and degradation of miserotoxin

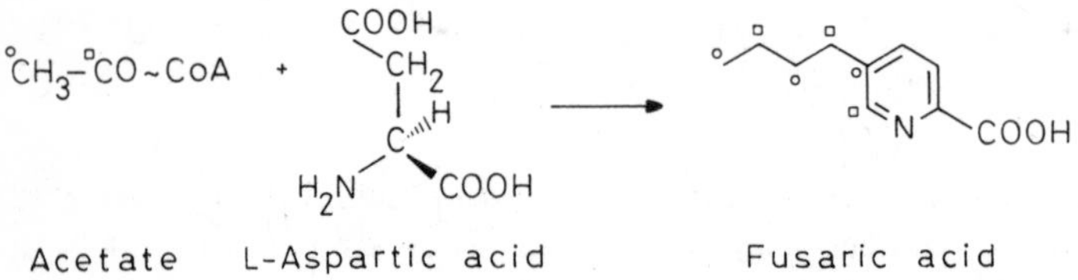

Fig. 197. Biosynthesis of fusaric acid

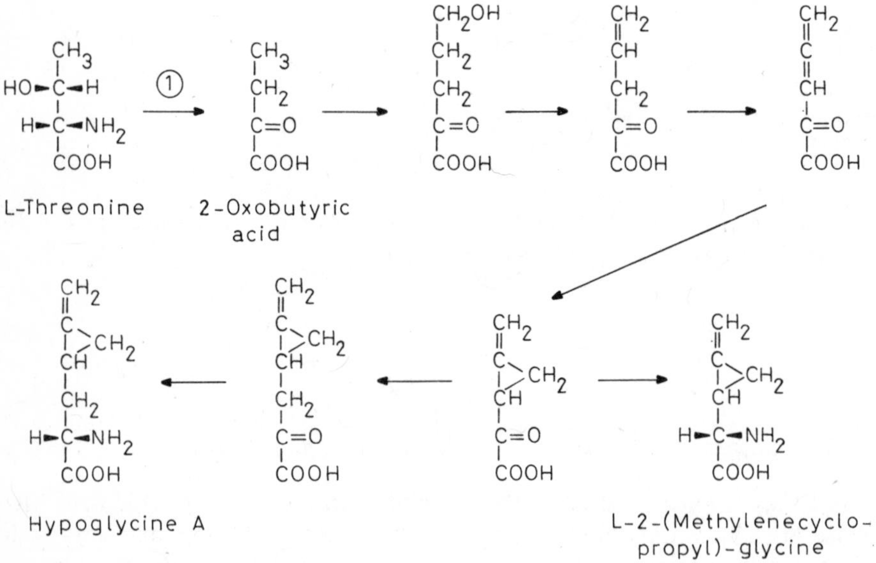

Fig. 198. Proposed pathway for the formation of cyclopropane amino acids from threonine
1 Threonine dehydratase

h) Hypoglycine A and L-2-(methylenecyclopropyl)-glycine, nonprotein amino acids found in plants, are derived from L-threonine (Fig. 198). The carbon skeleton of L-2-(methylenecyclopropyl)-glycine is completed by two additional C_1-units. In the formation of hypoglycine A the carbon chain may be extended by a further C-atom by the 2-oxo acid elongation system (D 14).

L-2-(Methylenecyclopropyl)-glycine occurs, e.g., in the seeds of *Litchi chinensis*. Hypoglycine A is the toxic principle of the unripe ackee fruit (fruits of the tree *Blighia sapida*, E 5.5.3). In animals it undergoes oxidative deamination and decarboxylation to 2-(methylenecyclopropyl)-acetic acid, which strongly inhibits the β-oxidation of fatty acids.

References for Further Reading

31, 77, 113, 416

D 16.1 Nicotinic Acid and Derivatives

Chemistry and Distribution

Nicotinic acid (pyridine 3-carboxylic acid) is the mother substance of several pyridine derivatives. Most of these substances contain a carboxy group or another substituent, e.g., an *N*-heterocyclic ring in position 3. Oxygen functions, especially in positions 2, 4, and 6 and *N*-methyl groups may be present.
Nicotinic acid and structurally related compounds are synthesized in microorganisms, plants, and animals.

Biosynthesis (Fig. 199)

In most bacteria and in higher plants nicotinic acid is formed from L-aspartic acid and a three-carbon unit derived from glycerol, probably D-glyceraldehyde-3-phosphate (D 2). A key intermediate is quinolinic acid, which in animals, however, is derived from L-tryptophan (D 22). Nicotinic acid originates from quinolinic acid via nicotinic acid mononucleotide formed with the participation of 5-phosphoribosyl-1-diphosphate. It changes either directly to nicotinic acid or is formed via nicotinamide adenine dinucleotide (NAD^+) in the nicotinic acid nucleotide cycle.

In bacteria nicotinic acid is degraded by the pathway outlined in Fig. 200. CO_2, NH_3, maleic, formic, acetic, and propionic acids are formed as final products.

Significance

Nicotinamide adenine dinucleotide (NAD^+) and its derivative nicotinamide adenine dinucleotide phosphate ($NADP^+$) are coenzymes or cosubstrates of oxidoreductases. $NADP^+$ differs from NAD^+ by an additional phosphate group bound to the C-2 hydroxyl of the ribose moiety of the adenosine portion of the molecule. Though nicotinic acid and nicotinamide can be synthesized in animals and hu-

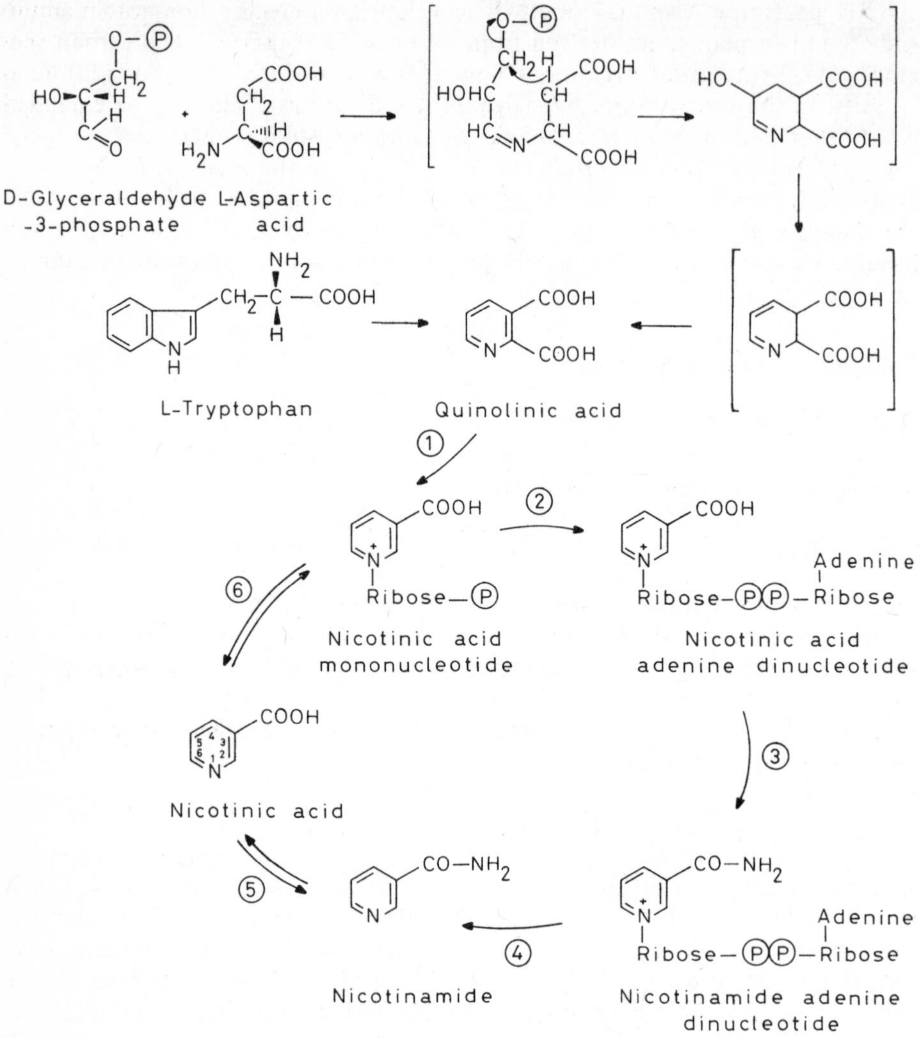

Fig. 199. Biosynthesis of nicotinic acid

1 Quinolinate phosphoribosyltransferase (decarboxylating); *2* nicotinate mononucleotide adenylyl-transferase; *3* nicotinamide adenine dinucleotide synthetase; *4* NAD(P)$^+$ nucleosidase; *5* nicotinamidase; *6* nicotinate phosphoribosyltransferase

man beings, they are vitamins for men living on a special diet, e.g., on corn as the main nutrient (E 2.1).

Transformation

a) Trigonelline, a betaine (D 8.1), is formed from nicotinic acid in plants. It influences cell proliferation and promotes arrest of cells in the mitotic cycle after DNA transcription during the G 2-phase (E 2.2).

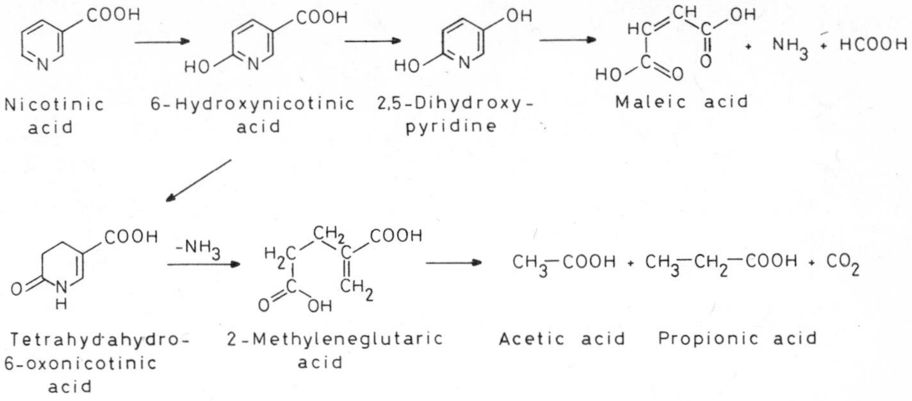

Fig. 200. Degradation of nicotinic acid

b) 1-Methylnicotinamide and 1-methyl-5-carboxamidopyridone-(2) are built in animals from nicotinamide which is derived from NAD$^+$ and NADP$^+$. Both compounds are excreted with the urine (E 1).

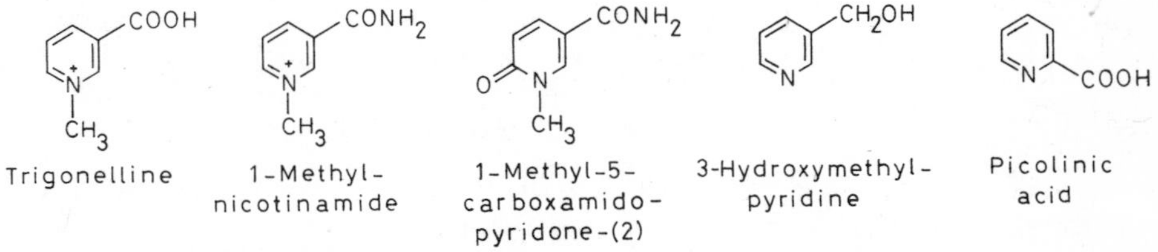

c) 3-Hydroxymethylpyridine is formed from nicotinic acid in Mycobacteria.

d) Picolinic acid, probably derived from quinolinic acid by decarboxylation (cf., however, D 22), is a microbial product which causes rice blast disease (E 5.4).

e) Ricinine, an alkaloid occurring in the plant *Ricinus communis*, is synthesized from nicotinic acid via 1-methyl-3-cyanopyridine (Fig. 201). The biosynthesis proceeds probably at the level of the nucleotides and displacement of the ribotide side chain by the methyl group is a late step. Ricinine undergoes a rapid turnover (A 5.2). It is poisonous to animals and may act as a feeding deterrent (E 5.5.3).

f) Acalyphin, a constituent of the plant *Acalypha indica*, is probably formed on a pathway resembling that leading to ricinine. Acalyphin releases hydrocyanic acid and D-glucose when treated with β-glucosidases (D 8.3).

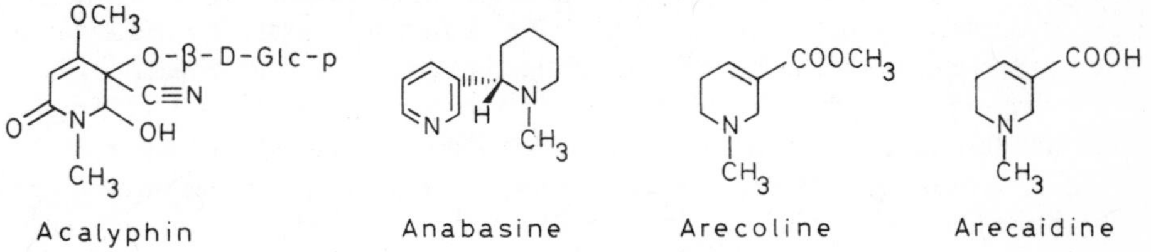

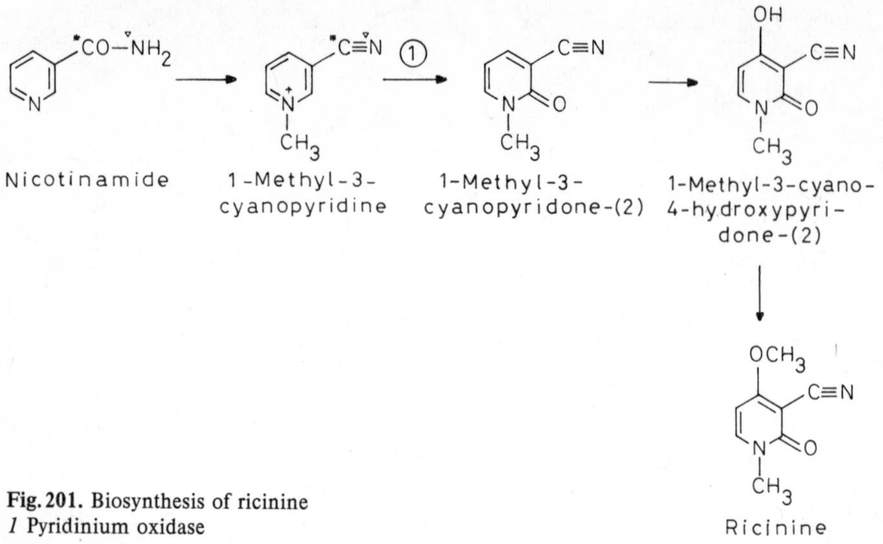

Fig. 201. Biosynthesis of ricinine
1 Pyridinium oxidase

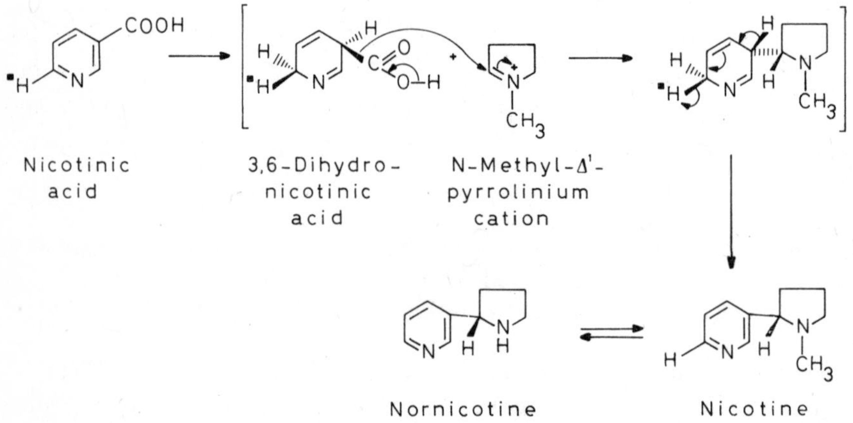

Nicotinamide

1-Methyl-3-cyanopyridine

1-Methyl-3-cyanopyridone-(2)

1-Methyl-3-cyano-4-hydroxypyridone-(2)

Ricinine

Fig. 202. Formation of nicotine and nornicotine

Nicotinic acid

3,6-Dihydronicotinic acid

N-Methyl-Δ'-pyrrolinium cation

Nornicotine

Nicotine

g) Nicotine- and anabasine-type alkaloids are formed from nicotinic acid in a number of plant families. In *Nicotiana* sp. dihydronicotinic acid condenses with the *N*-methyl-Δ^1-pyrrolinium cation (D 17) to form nicotine (Fig. 202). Condensation with Δ^1-piperideine (D 18) results in the biosynthesis of anabasine. Nicotine is oxidatively degraded to nornicotine, which may again be methylated to nicotine. Nicotine is the physiologically active compound of tobacco. It acts as an insecticide and is a feeding deterrent to most animals (E 5.5.3). Human beings use fermented, nicotine-containing tobacco leaves as stimulant (F 3).

h) Arecoline, a constituent of the fruits of *Areca catechu* (betel nuts), is probably built from nicotinic acid. It is toxic to potential predators and acts as feed-

ing deterrent (E 5.5.3). Pieces of betel nuts mixed with lime and further ingredients and enveloped in leaves of *Piper betle* are chewed in Southeast Asia as
stimulant. The physiologically active component is arecaidine, formed from
arecoline up on hydrolysis (F 3).

References for Further Reading

203, 220, 294, 354, 455, 605, 739, 806

D 17 Secondary Products Derived from L-Glutamic Acid, L-Proline, and L-Ornithine

Biosynthesis of the Precursor

L-Glutamic acid, L-proline, and L-ornithine are amino acids with five C-atoms.
L-Glutamic acid is formed by glutamate dehydrogenase from 2-oxoglutaric acid,
an intermediate in the tricarboxylic acid cycle (D 4). It is transformed reversibly
to L-proline and L-ornithine (Fig. 203).

Chemistry and Distribution

Secondary products are built from amino acids of the L-glutamic acid-L-proline-
L-ornithine family in microorganisms, plants, and animals. The most significant
are listed in Fig. 203. They
— possess either the aliphatic carbon chain of the amino acids L-glutamic acid
  and L-ornithine, e.g., N^5-ethyl-L-glutamine, ornithuric acid, and phenylacetyl-
  L-glutamine,
— are pyrrolidine derivatives like L-proline, cf. the formulas of N-methyl-Δ^1-pyr-
  rolinium cation and cuscohygrine, or
— possess the more complicated quinolizidine and tropane ring systems.

Secondary Products Formed from L-Glutamic Acid

a) L-Glutamine (Fig. 203) is derived from L-glutamic acid, ATP, and ammonia
by glutamine synthetase. It is a nitrogen storage in plants, animals, and microorganisms (E 2.2).

b) N^5-Substituted L-glutamine derivatives ($=\gamma$-glutamyl derivatives of primary amines) are formed in microorganisms and plants according to Fig. 204 (cf.
also the biosynthesis of N^4-substituted L-asparagine derivatives, D 16). Best
known is N^5-ethyl L-glutamine (theanine) isolated, e.g., from *Camellia thea*. However, also N^5-methyl, N^5-isopropyl-, N^5-(2'-hydroxyethyl)- and N^5-(4'-hydroxy-
phenyl)-L-glutamine have been shown to occur.

c) γ-Glutamyl derivatives of amino acids, like γ-glutamyl-2-phenylalanine,
are synthesized in microorganisms and plants by γ-glutamyltranspeptidases.
They include derivatives of most protein amino acids as well as derivatives of

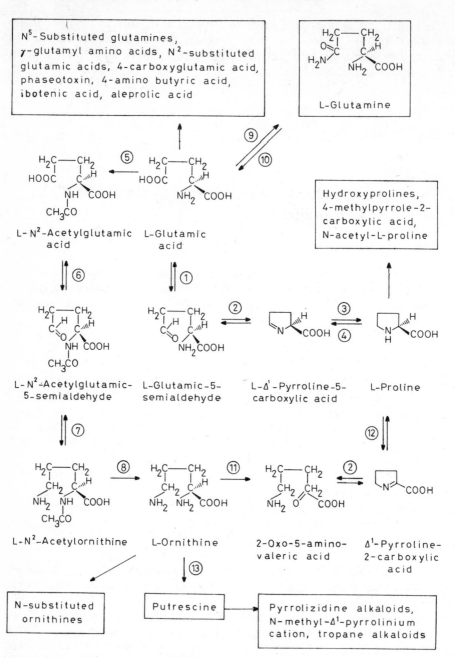

Fig. 203. Biosynthesis and transformation of the amino acids of the L-glutamic acid-L-proline-L-orni-thine family

1 Glutamyl kinase, glutamate semialdehyde dehydrogenase; *2* spontaneous cyclization; *3* pyrroline-5-carboxylate reductase; *4* proline oxidase; *5* acetylglutamate synthase; *6* acetylglutamate kinase, acetyl-γ-glutamylphosphate reductase; *7* acetylornithine aminotransferase; *8* acetylornithine deacet-ylase; *9* glutamine synthetase; *10* glutaminase; *11* ornithine aminotransferase; *12* pyrroline-2-carb-oxylate reductase; *13* ornithine decarboxylase

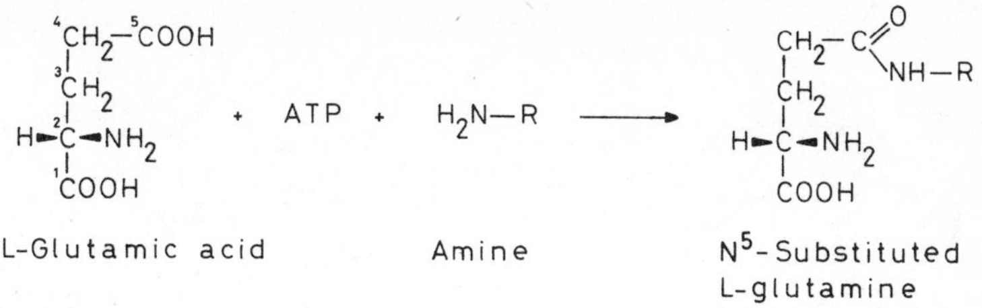

$$\begin{array}{c}
{}^4CH_2-{}^5COOH \\
| \\
{}^3CH_2 \\
| \\
H\!\blacktriangleright\!{}^2\!C\!\blacktriangleleft\!NH_2 \\
| \\
{}^1COOH
\end{array}
\quad + \quad ATP \quad + \quad H_2N-R \quad \longrightarrow \quad
\begin{array}{c}
CH_2-C{\diagup}{\diagdown}^{O}_{NH-R} \\
| \\
CH_2 \\
| \\
H\!\blacktriangleright\!C\!\blacktriangleleft\!NH_2 \\
| \\
COOH
\end{array}$$

L-Glutamic acid   Amine   N⁵-Substituted
                                                    L-glutamine

Fig. 204. Formation of N^5-substituted L-glutamine derivatives by glutamine synthetase
The enzyme has a relatively low substrate specificity

many nonprotein amino acids. Obviously high concentrations of nonprotein amino acids give rise to the formation of γ-glutamyl derivatives.

d) N^2-Substituted L-glutamyl and L-glutaminyl derivatives are formed in liver and kidney of human beings and of some anthropoid monkeys from L-glutamic acid or L-glutamine and acid CoA esters, e.g., phenylacetic acid CoA ester, D 23.1.1, or indole-3-acetic acid CoA ester, D 22 (E 1; cf. the formula of phenylacetyl-L-glutamine).

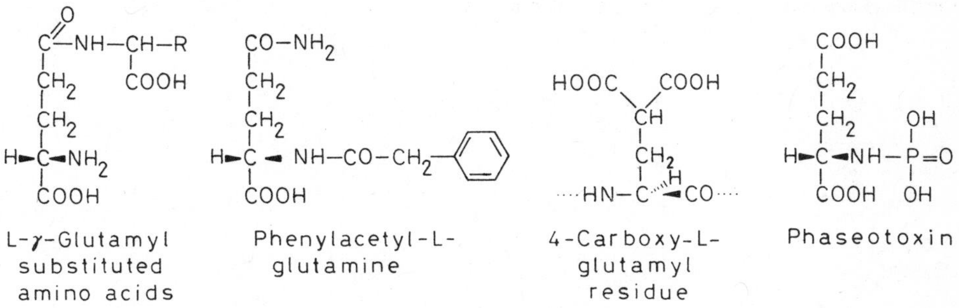

$$\begin{array}{cc}
C{\diagup}^{O}-NH-CH-R & \\
| \qquad\qquad | & \\
CH_2 \qquad COOH & \\
| & \\
CH_2 & \\
| & \\
H\!\blacktriangleright\!C\!\blacktriangleleft\!NH_2 & \\
| & \\
COOH &
\end{array}$$

L-γ-Glutamyl   Phenylacetyl-L-   4-Carboxy-L-   Phaseotoxin
substituted   glutamine   glutamyl
amino acids   residue

e) 4-Carboxy-L-glutamic acid is derived from L-glutamic acid residues bound in proteins, e.g., in the hepatic precursor protein of prothrombin. Prothrombin binds Ca^{2+} and is involved in blood coagulation. Glutamic acid carboxylase is vitamin K-dependent (E 2.1).

f) Phaseotoxin (phosphoglutamic acid) is a phytotoxin formed in *Pseudomonas syringae* pv. *phaseolicola* (E 5.4). It contains a phosphamine group, which is scarcely found in nature.

g) 4-Aminobutyric acid (γ-aminobutyric acid, GABA) is a neurotransmitter in vertebrates and invertebrates (E 3.2). It is formed from L-glutamic acid by decarboxylation (D 8.2). Formation and degradation of GABA seems to be the main route of citrate degradation in brain cells (GABA shunt). This metabolic route starts from 2-oxoglutaric acid and proceeds via glutamic acid, GABA, and succinic semialdehyde to succinic acid. During formation of succinic semialdehyde, a hydrogen atom is removed stereospecifically from the amino C-atom (Fig. 205).

h) Ibotenic acid, a constituent of *Amanita muscaria*, the fly agaric, and pantherine (muscimol), an artifact obviously formed during drying of the mush-

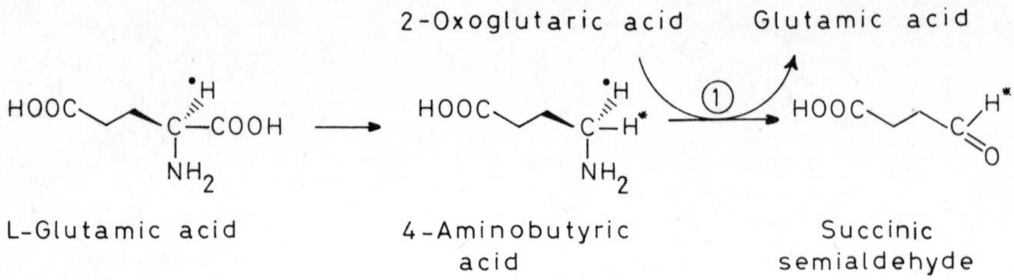

Fig. 205. Formation and degradation of γ-aminobutyric acid
1 2-Oxoglutarate aminotransferase

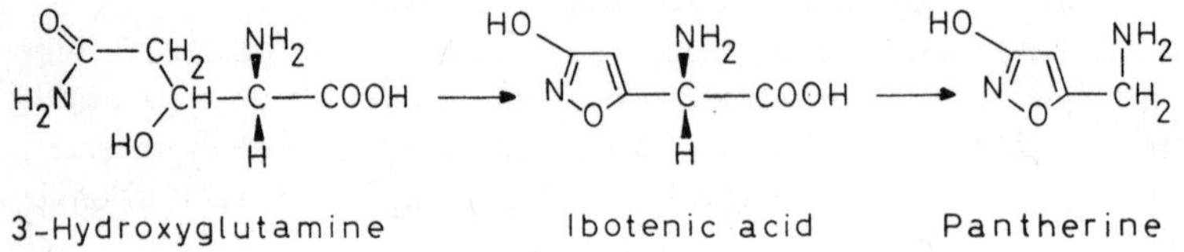

Fig. 206. Formation of ibotenic acid and pantherine

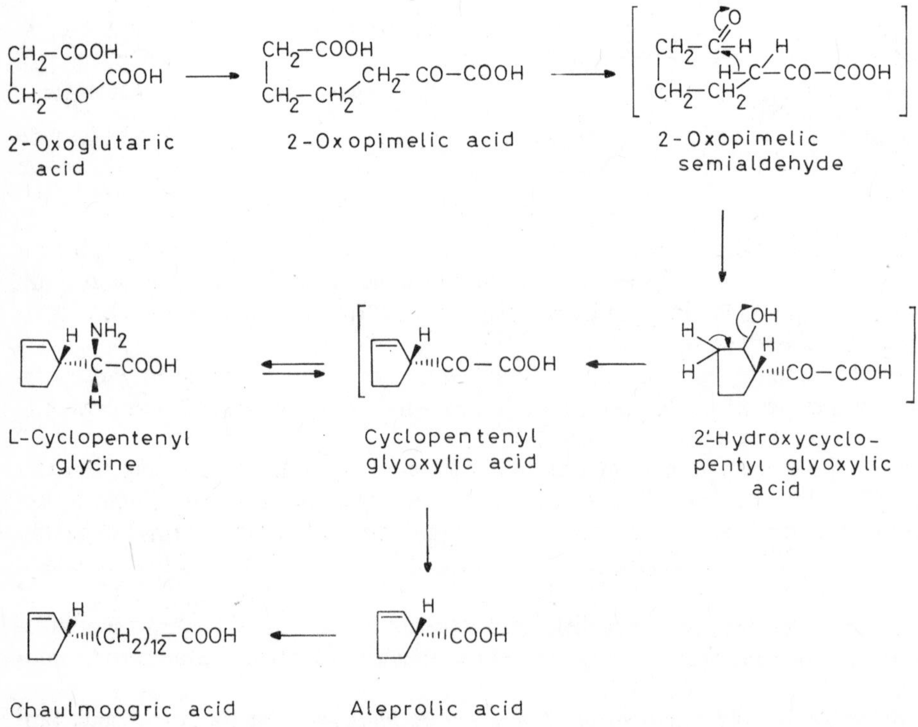

Fig. 207. Biosynthesis of L-cyclopentenyl glycine and chaulmoogric acid

rooms, are probably derived from 2-hydroxyglutamine (Fig. 206). Pantherine (and ibotenic acid?) are toxic and repel potential predators (E 5.5.3). They produce hallucinations in human beings (F 3).

i) Aleprolic acid and the nonprotein amino acid L-cyclopentenyl glycine are derived from 2-oxoglutaric acid in plants (Fig. 207). In a first step 2-oxoglutaric acid is transformed to 2-oxopimelic acid by the 2-oxo acid elongation system (D 14). 2-Oxopimelic acid probably cyclizes after reduction to the semialdehyde. Aleprolic acid is the starter molecule in the formation of cyclopentenyl fatty acids in Flacourtiaceae (D 5.2.1). L-Cyclopentenyl glycine is the precursor of cyanogenic glycosides, e.g., deidaclin, in the order Violales.

Secondary Products Derived from L-Proline

a) cis-4-Hydroxy-L-proline, trans-3-hydroxy-L-proline, trans-4-methyl-L-proline, and cis-4-hydroxymethyl-L-proline are nonprotein amino acids formed in plants. trans-4-Hydroxy-L-proline and trans-3-hydroxy-L-proline are also constituents of the connective tissue proteins collagen and elastin found in multicellular animals. In animals hydroxylation proceeds after formation of the protein molecule by intermolecular dioxygenases (C 2.5).

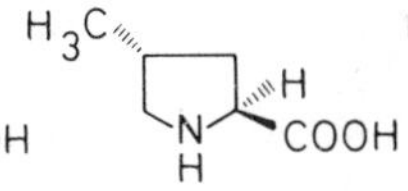

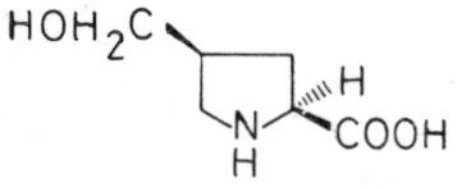

b) 4-Methylpyrrole-2-carboxylic acid methyl ester acts as an insect pheromone (E 4).

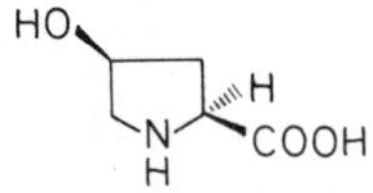

cis-4-Hydroxy-   trans-3-Hydroxy-   trans-4-Methyl-   cis-4-Hydroxy-
L-proline   L-proline   L-proline   methyl-L-proline

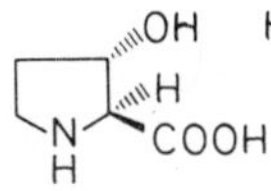

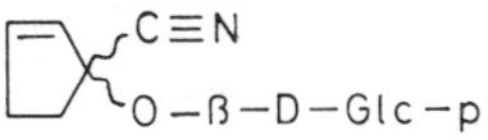

Deidaclin

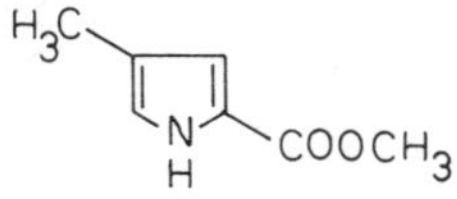

4-Methylpyrrole-2-
carboxylic acid
methyl ester

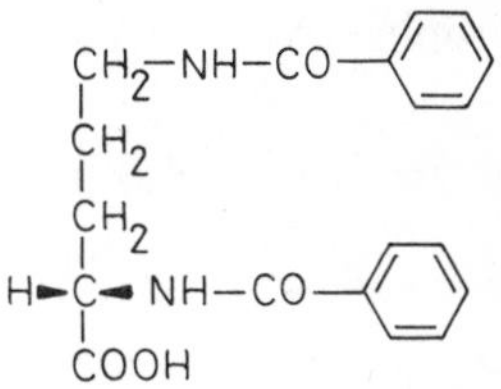

Ornithuric acid

Secondary Products Formed from L-Ornithine

a) N-Substituted L-ornithine derivatives, e.g., ornithuric acid (N, N'-dibenzoyl-L-ornithine), are built in birds and reptiles from L-ornithine and the CoA-esters of several carbonic acids, e.g., nicotinic acid (D 16.1), phenylacetic acid (D 23.1.1), and benzoic acid (D 23.2.5).

b) N^5-Acetyl-L-ornithine and the acetyl derivative of L-proline, act as storage products for nitrogen in certain plants (E 2.2).

c) Putrescine and N-methylputrescine are formed by the decarboxylation of L-ornithine and N-methyl-L-ornithine, respectively (Fig. 208). Putrescine is a

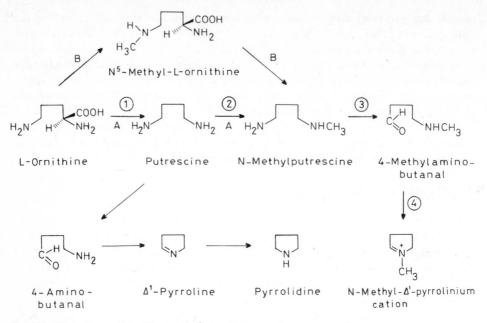

Fig. 208. Biosynthesis of the *N*-methyl-*Δ*¹-pyrrolinium cation
1 Ornithine decarboxylase; *2* putrescine methyltransferase; *3* methylputrescine oxidase; *4* spontaneous reaction

precursor of the pyrrolizidine alkaloids (D 17.1). *N*-Methylputrescine is transformed to the *N*-methyl-*Δ*¹-pyrrolinium cation which is a building block of several other types of alkaloids (D 16.1, D 17.2). In the biosynthesis of nicotine, hygrine, and ecgonine it is formed on route A, in contrast to the biosynthesis of tropine, where it is built on route B (Fig. 208).

References for Further Reading

31, 414, 455, 508, 605, 716, 732, 760

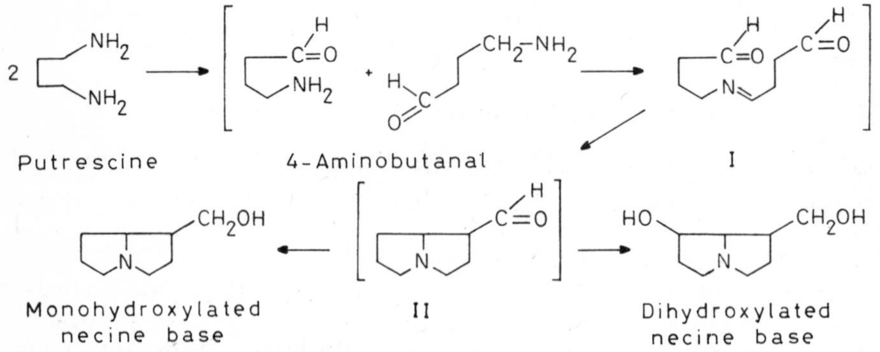

Fig. 209. Formation of necine bases

D 17.1 Pyrrolizidine Alkaloids

Chemistry and Distribution

Most pyrrolizidine alkaloids are esters consisting of a basic alcohol with the pyrrolizidine nucleus (necine base), and one or two characteristic monocarboxylic acids or one dicarboxylic acid (necic acids). Typical pyrrolizidine alkaloids are lindelofine, heliosupine, and senecionine (cf. formulas given below; in parentheses the necine base and the necic acid constituents).

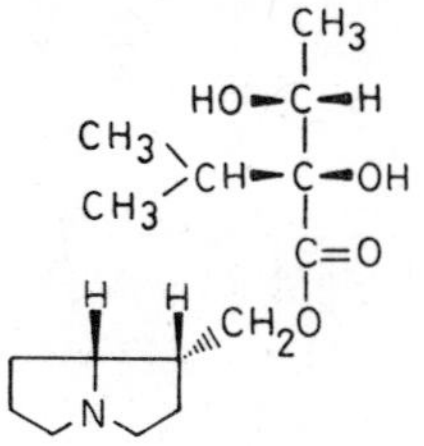

Lindelofine
((+)-isoretronecanol,
trachelanthic acid)

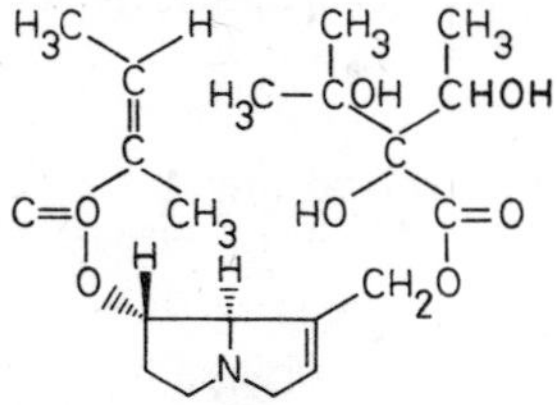

Heliosupine
(heliotridine, angelic and
echimidinic acids)

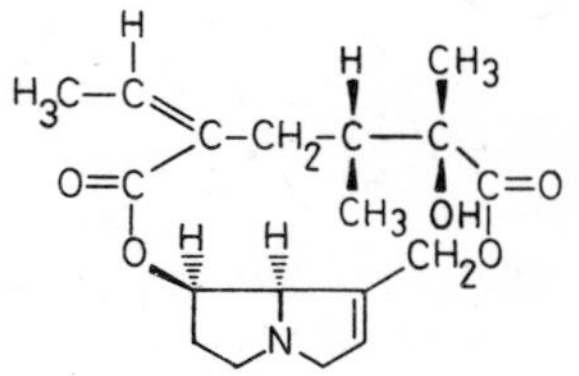

Senecionine
(retronecine, senecic
acid)

Pyrrolizidine alkaloids are synthesized in plants, e.g. in Compositae, Papilionaceae, Boraginaceae, and Leguminosae.

Biosynthesis

a) Necine bases. The necine bases originate from L-ornithine via the symmetrical intermediate putrescine (Fig. 209). Putrescine is transferred by oxidative deamination or transamination to 4-aminobutanal, two molecules of which (after further oxidative deamination or transamination) combine to give the Schiff base I. The corresponding compound lacking the double bond is assumed to cyclize by an intramolecular Mannich reaction to intermediate II which after reduction and further hydroxylation may give the monohydroxylated and dihydroxylated necine bases.

According to the configuration, the monohydroxylated necine base given in Fig. 209 is known as lindelofidine, isoretronecanol, trachelanthamidine, or laburnine. A representative of the dihydroxylated bases is retronecine, the basic part of senecionine (see above).

b) Necic acids. Most necic acids are formed from branched-chain amino acids. L-Valine, for instance, is the precursor of echimidinic acid (D 13), L-isoleucine that of tiglic, angelic, seneciphyllic, and senecic acids (D 15).

c) Esterification. Esterification of necine bases and (activated) necic acids is the last step in biosynthesis. Cyclic diesters probably originate by intramolecular coupling of two preexisting half-esters.

Significance

Pyrrolizidine alkaloids are toxic for animals and possess importance as feeding deterrents (E 5.5.3). Certain insects, however, are able to store pyrrolizidine alka-

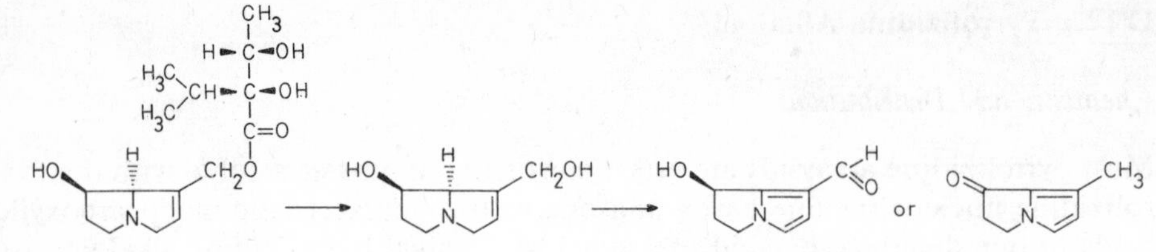

Fig. 210. Transformation of pyrrolizidine alkaloids in Danaid butterflies

loids, e.g., senecionine, ingested with the diet for repellence of potential predators (E 5.1). Some kinds of butterflies hydrolyze pyrrolizidine alkaloids taken up from plant sources and transform the necine bases to pyrrolizidine aldehydes and ketones (Fig. 210) used as sex pheromones (E 4).

References for Further Reading

82, 301, 354, 464, 511, 641, 642, 713, 723, 831

D 17.2 Tropane Alkaloids

Chemistry and Distribution

The basic structure of the tropane alkaloids is the bicyclic tropane ring system:

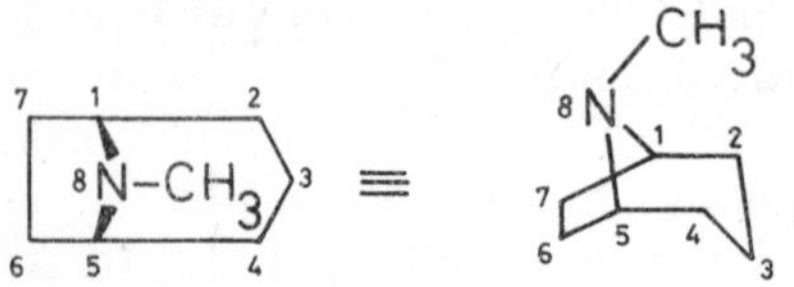

Tropane

Of significance are the following hydroxylated and carboxylated tropane derivatives:

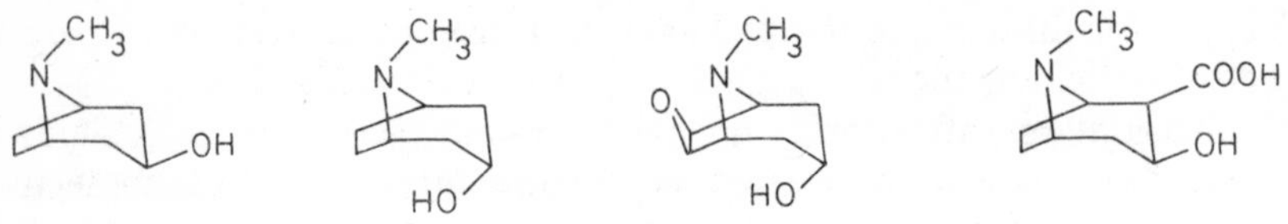

Pseudotropine   Tropine   Scopine   Ecgonine

In most of the alkaloids the tropane moiety at position 3 is ester-bound to an acid. Some important tropane alkaloids are listed in Table 47.
Tropane alkaloids are constituents of higher plants.

Table 47. Important alkaloids with tropane moiety

Alkaloid	Basic part	Acid part
L-Hyoscyamine	Tropine	L-Tropic acid
Atropine	Tropine	D, L-Tropic acid
Apoatropine	Tropine	Atropic acid
L-Scopolamine	Scopine	L-Tropic acid
Cocaine	Ecgoninemethylester	Benzoic acid

Biosynthesis

a) Tropine, ecgonine, hygrine, and cuscohygrine. The skeleton of these compounds originates from the N-methyl-Δ^1-pyrrolinium cation (D 17) and two molecules acetyl CoA or malonyl CoA (Fig. 211).

b) The acids of tropane alkaloids. The most frequently occurring acid components are L-tropic and atropic (D 23), benzoic (D 23.2.5), tiglic, and 2-methylbutyric acid (D 15).

c) Esterification. Linkage of the tropane alcohols with the (activated) acids is catalyzed by tropinesterases (Fig. 212). Hydrolytic enzymes of this group split the ester alkaloids to tropane alcohols and the free acids. They are important for the tolerance of certain animals, e.g., rabbits, to tropane alkaloids.

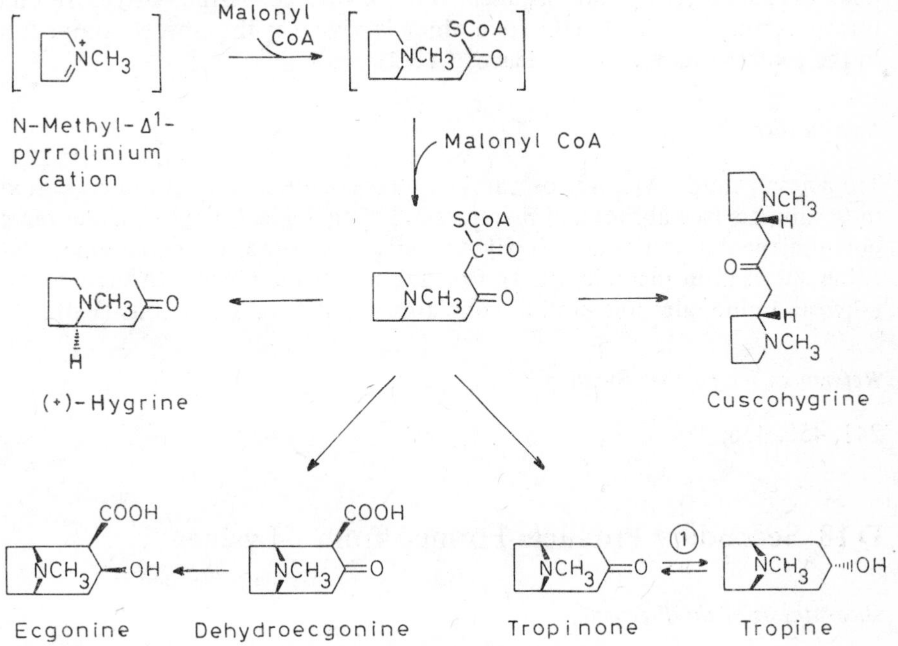

Fig. 211. Biosynthesis of tropine, ecgonine, and cuscohygrine
1 Tropine dehydrogenase

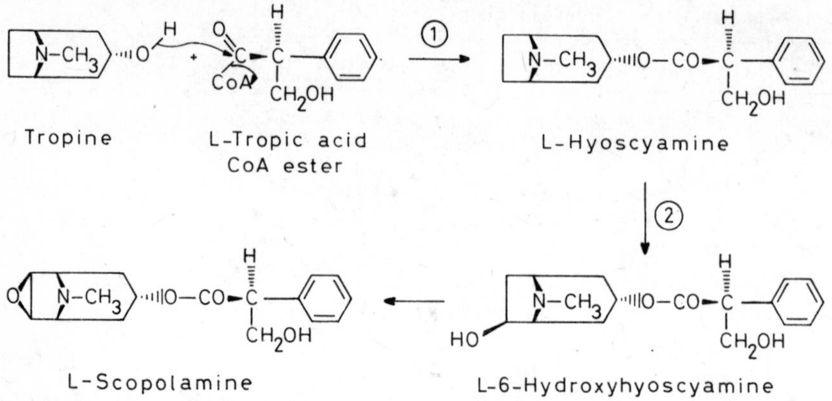

Fig. 212. Formation of L-hyoscyamine and L-scopolamine
1 Tropinesterase; *2* hyoscyamine 6β-hydroxylase (intermolecular dioxygenase, C 2.5)

d) Further transformation. The ester alkaloids of the tropine type may be epoxidized to scopine-derived compounds (Fig. 212). During formation of the epoxide ring, the β-hydrogen atoms at C-6 and C-7 of tropine are lost.

Reciprocal graftings using alkaloid-abundant and alkaloid-free Solanaceae, e.g., *Datura* and *Lycopersicon* sp., have shown that the major portion of the tropane alkaloids is synthesized in the roots, but accumulates in the leaves. During the transport to the leaves the chemical structure of the alkaloids may be modified. In *Datura ferrox*, for instance, the enzyme converting L-hyoscyamine to L-scopolamine is located in the stem which is passed by the L-hyoscyamine formed in the roots on its way to the leaves (A 3.2).

Significance

Tropane alkaloids, e.g., L-hyoscyamine, L-scopolamine, and cocaine are toxic to most animals (see above) and therefore are of ecological significance in repelling potential predators (E 5.5.3). L-Hyoscyamine, atropine, L-scopolamine, and cocaine are used in medicine (F 2). Cocaine is a hallucinogen. In high doses also L-hyoscyamine, atropine, and L-scopolamine cause hallucinations (F 3).

References for Further Reading

241, 455, 456, 464, 639

D 18 Secondary Products Formed from L-Lysine

Biosynthesis of the Precursor

L-Lysine is a basic amino acid with six C-atoms. It is formed on two independent sets of reactions:

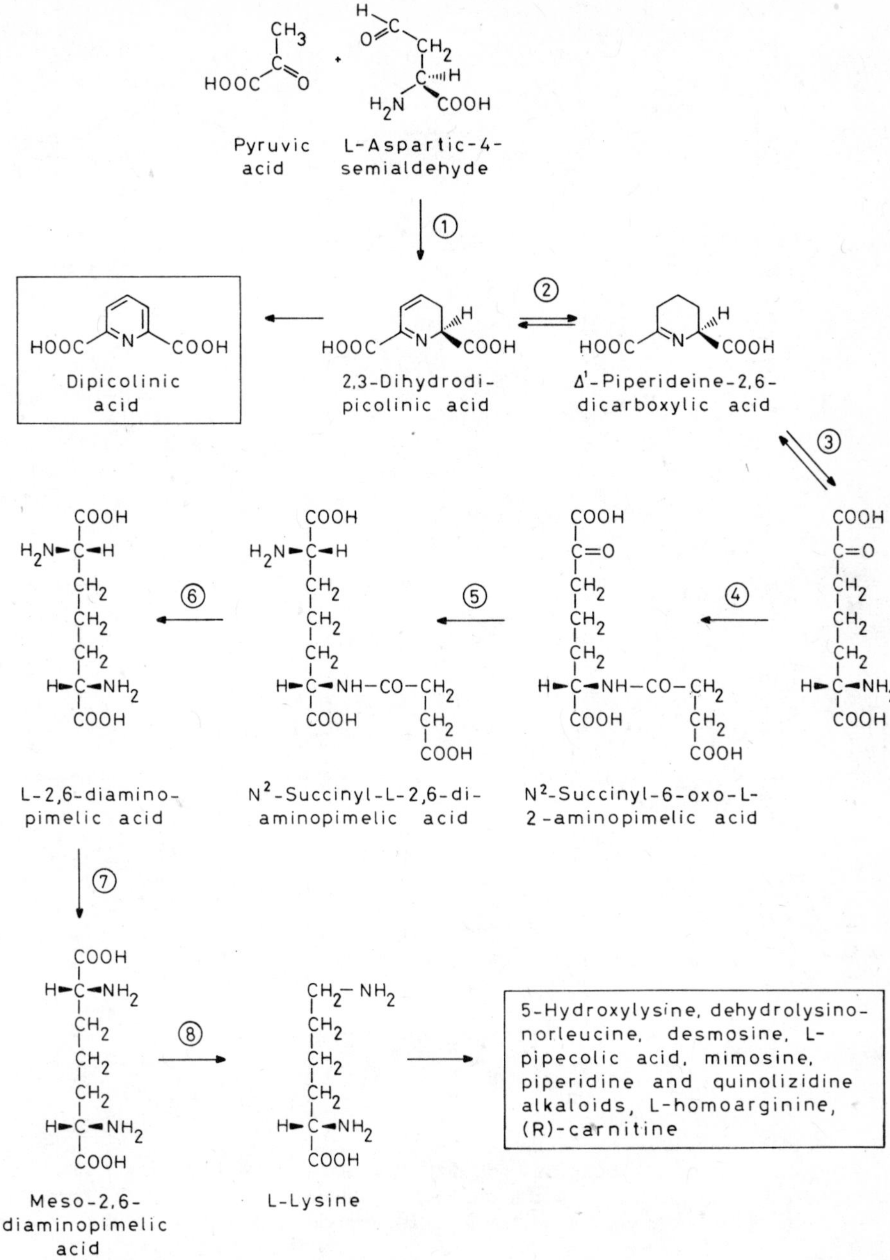

Fig. 213. Formation of L-lysine on the diaminopimelic acid pathway
1 Dihydrodipicolinate synthase; *2* dihydrodipicolinate hydrogenase; *3* spontaneous reaction; *4* succinylase; *5* succinyldiaminopimelate aminotransferase; *6* succinyldiaminopimelate desuccinylase; *7* diaminopimelate epimerase; *8* diaminopimelate decarboxylase (the decarboxylation proceeds with net inversion at C-6, in contrast to the steric course of amino acid decarboxylation at C-2; C 5)

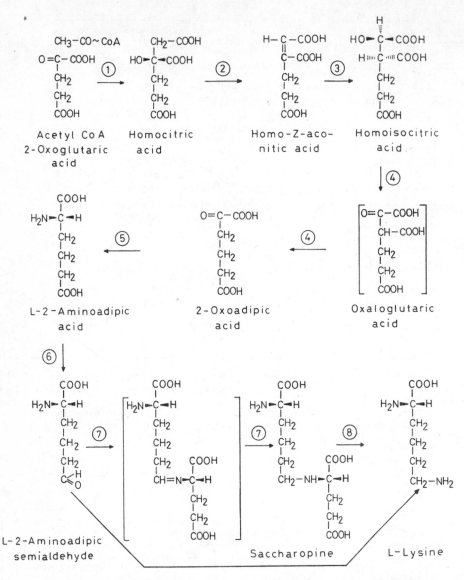

Fig. 214. Biosynthesis of L-lysine on the 2-aminoadipic acid pathway
1 Homocitrate synthase; *2* homocitrate dehydratase; *3* homoaconitate hydratase; *4* homoisocitrate dehydrogenase; *5* 2-aminoadipate aminotransferase; *6* aminoadipate semialdehyde dehydrogenase; *7* saccharopine dehydrogenase (L-glutamate forming); *8* saccharopine dehydrogenase (L-lysine forming)

a) The diaminopimelic acid pathway. In certain bacteria and in plants L-lysine originates from L-aspartic-4-semialdehyde and pyruvic acid (Fig. 213). Important intermediates are 2,3-dihydrodipicolinic acid and meso-2,6-diamino-pimelic acid.

b) The 2-aminoadipic acid pathway. In fungi, algae, and plants L-lysine is built from 2-oxoglutaric acid (Fig. 214). This precursor is transformed by the 2-oxo acid elongation system (D 14) to 2-oxoadipic acid. The corresponding amino acid is reduced to the semialdehyde which is transformed to L-lysine either directly or via saccharopine.

Chemistry and Distribution

Secondary products derived from L-lysine or its precursors are formed in micro-organisms, plants, and animals. Important groups are shown in Fig. 213. In most cases they are piperidine or pyridine derivatives (cf. the formulas of dipicolinic acid, desmosine, pipecolinic acid, lobeline, etc.). Also quinolizidines, e.g., the alkaloids lupinine and sparteine, are synthesized from L-lysine.

Transformation of L-Lysine and its Precursors to Secondary Products

a) Dipicolinic acid is derived from 2,3-dihydrodipicolinic acid (Fig. 213). Dipicolinic acid occurs, e.g., in the spores of *Bacillus cereus* (A 4.4).

b) 5(R)-Hydroxy-L-lysine is a constituent of the connective tissue proteins collagen and elastin. It is formed from L-lysine already incorporated in the proteins by an intermolecular dioxygenase (C 2.5).

5(R)-Hydroxy-L-lysine

N^6-Trimethyl-L-lysine

c) $\Delta^{6,7}$-Dehydrolysinonorleucine, desmosine, and isodesmosine interlink the chains of collagen and elastin, characteristic proteins of the connective tissues of animals. They are formed by oxidative deamination of the amino group in position 6 of L-lysine and condensation of the allysine residue formed with lysyl groups (Fig. 215). In the biosynthesis of desmosine and isodesmosine four L-lysyl moieties are involved. Key intermediates are probably dehydromerodesmosine residues.

d) L-Pipecolic acid is derived from D-lysine. It may be transformed into L-lysine (Fig. 216). Though most of the individual reactions of this pathway seem to be reversible, the transformation of L-lysine into L-pipecolic acid does not occur in nature. L-Pipecolic acid is widespread in microorganisms, plants, and animals. It is frequently accompanied by derivatives substituted in positions 4 and 5. In animals L-pipecolic acid is excreted with the urine (E 1). It may be, however, also further degraded via L-aminoadipic acid to 2-oxoadipic acid and glutaryl CoA. In

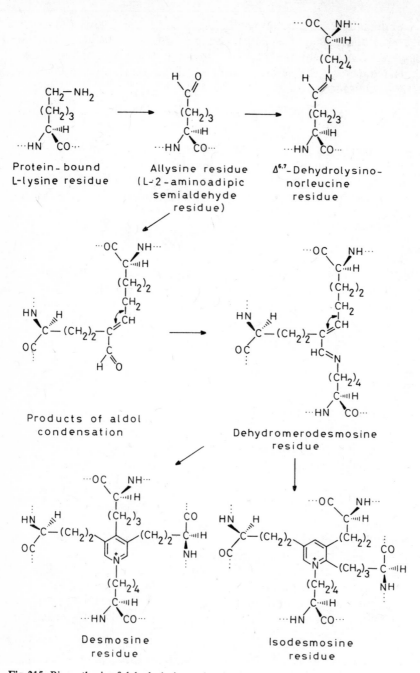

Protein-bound
L-lysine residue

Allysine residue
(L-2-aminoadipic
semialdehyde
residue)

$\Delta^{6,7}$-Dehydrolysino-
norleucine
residue

Products of aldol
condensation

Dehydromerodesmosine
residue

Desmosine
residue

Isodesmosine
residue

Fig. 215. Biosynthesis of dehydrolysinonorleucine and desmosines

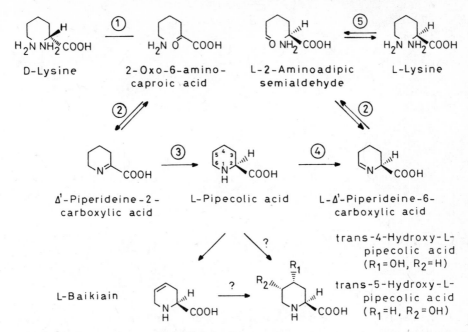

Fig. 216. Formation of L-pipecolic acid and some of its derivatives
1 Transaminase or oxidative deaminase; *2* spontaneous cyclization; *3* pyrroline 2-carboxylate reductase; *4* L-pipecolate dehydrogenase; *5* L-lysine 6-aminotransferase

plants L-pipecolic acid derivatives, such as baikiain, trans-4-hydroxypipecolic acid, and trans-5-hydroxypipecolic acid, are formed.

e) Mimosine, a nonprotein amino acid, is derived from 3,4-dihydroxypyridine, a probable product of L-lysine transformation, and *O*-acetyl-L-serine (Fig. 217). It is a constituent of plants, e.g., of *Mimosa* and *Leucaena* sp. Mimosine and its glucoside, mimoside, are toxic to animals by interaction with pyridoxal-5'-phosphate-containing enzymes. They act as feeding deterrents (E 5.5.3).

f) Piperidine alkaloids, found, for instance, in *Punica granatum*, are derived from L-lysine via asymmetrical intermediates, e.g., 5-aminopentanal (Fig. 218). It is suggested that L-lysine forms a Schiff base with a pyridoxal-5'-phosphate-de-

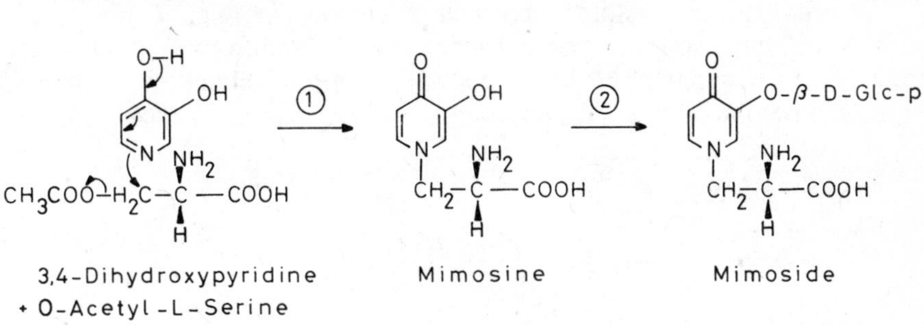

Fig. 217. Biosynthesis of mimosine
1 Mimosine synthase; *2* mimosine glucosyltransferase

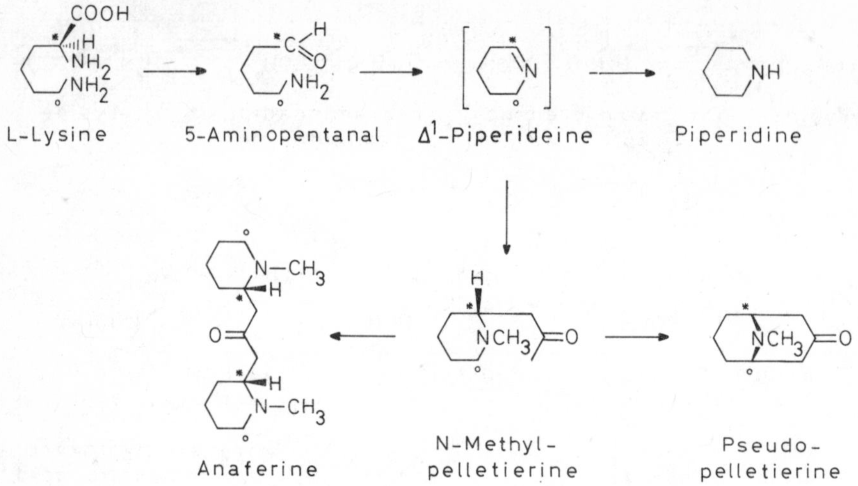

Fig. 218. Formation of Punica alkaloids

pendent enzyme (C 5) and in a concerted mechanism undergoes decarboxylation and deamination while attached to the enzyme protein. Δ^1-Piperideine formed spontaneously from 5-aminopentanal may condense with 2 mol acetyl CoA to alkaloids of the anaferine and pelletierine types in a reaction sequence resembling that of tropane alkaloid formation (D 17.2). Pseudopelletierine may be synthesized from N-methylpelletierine by a second cyclization.

g) Sedamine and lobeline, alkaloids found in plants of the genera *Sedum* and *Lobelia*, are derived from L-lysine and phenylpropanoids. The actual precursor is probably phenylacetic acid (D 23.1.1). In lobeline L-lysine is incorporated via symmetrical intermediates like putrescine, in sedamine via asymmetrical derivatives (Fig. 219). Sedamine- and lobeline-type alkaloids are toxic to animals and may be of ecological significance (E 5.5.3). Lobeline is used in medicine (F 2).

h) The piperidine ring of the alkaloid anabasine, which occurs, e.g., in *Anabasia* and *Nicotiana* sp. is derived from L-lysine via an asymmetrical intermediate, probably 5-aminopentanal. Δ^1-Piperideine formed by cyclization of this compound reacts with nicotinic acid (D 16.1).

i) Quinolizidine alkaloids found in Papilionaceae, e.g., *Lupinus* and *Sarothamnus sp.*, but also in many other plants, are derived from L-lysine via cadaverine (D 8.2) and probably its semialdehyde (5-aminopentanal which may cyclize to Δ^1-piperideine). Lupinine the simplest of the quinolizidine alkaloids is de-

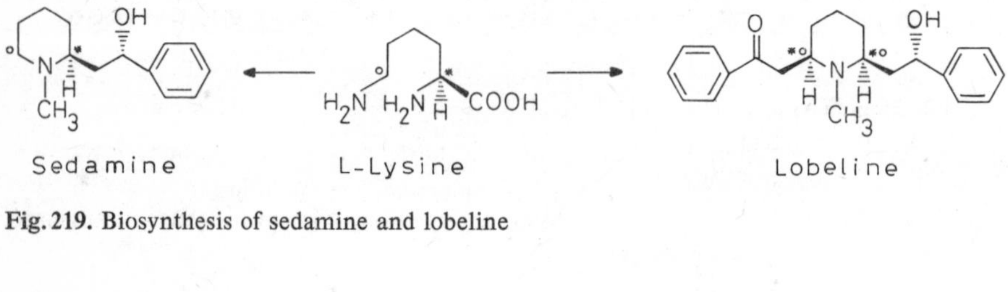

Fig. 219. Biosynthesis of sedamine and lobeline

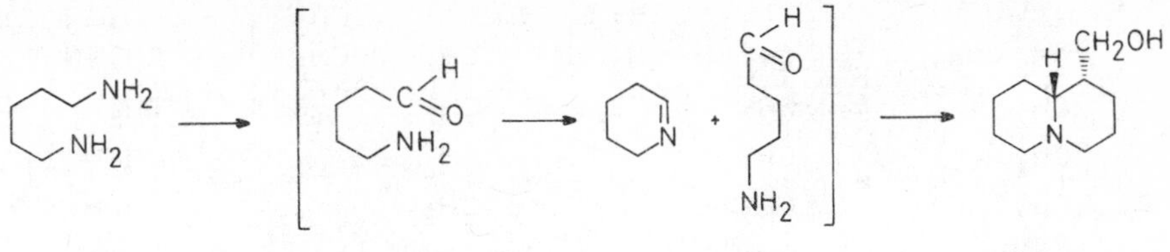

Cadaverine   5-Amino-   Δ¹-   Lupinine
                 pentanal   Piperideine

Fig. 220. Formation of lupinine

rived from two molecules of cadaverine (Fig. 220). Sparteine is formed from three molecules of cadaverine. From *Lupinus* cell cultures an enzyme preparation has been obtained which synthesizes 17-oxosparteine from three molecules cadaverine and four molecules pyruvate as acceptor of the excess amino groups. This 17-oxosparteine synthase contains pyridoxal-5′-phosphate as coenzyme (C 5). Intermediates are probably covalently bound as Schiff bases. Esters formed from quinolizidine alkaloids, which contain hydroxy groups, and cinnamic acid derivatives (D 23.2.1) or tiglic acid (D 15) occur frequently in nature.

Quinolizidine alkaloids, e.g., lupinine, sparteine, and cytisine are toxic to grazing animals and repel mollusks and insects (E 5.5.3). Only a few specialized animals are attracted by certain alkaloids e.g., by sparteine. Quinolizidine alkaloids inhibit seed germination as well as the growth of plants (E 5.3) and microorganisms (E 5.4). Sparteine is used in therapy (F 2).

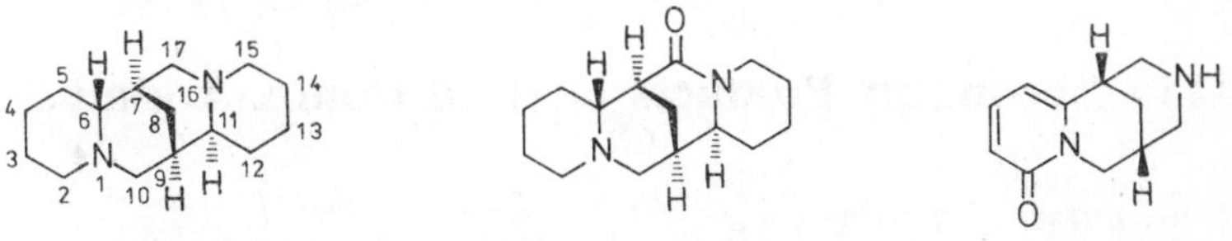

Sparteine   17-Oxosparteine   Cytisine

j) L-Homoarginine, a constituent of the microbial toxin phaseolotoxin (D 9.1), is formed by transfer of the imidino group of L-arginine to L-lysine.

k) (R)-Carnitine (L-carnitine), a betaine formed in plants and animals, is derived from protein-bound L-lysine (Fig. 221). Carnitine-O-acetylester acts as a storage and transport form of acetyl groups in animals and human beings. Fatty acid esters of carnitine are transported by carnitine-acylcarnitine translocase from cytoplasm into the lumen of mitochondria. After reaction with CoA to acyl CoA and free (R)-carnitine the fatty acids are degraded in the mitochondria by β-oxidation (D 5.2).

References for Further Reading

31, 48, 222, 224, 250, 272, 302, 354, 465, 605, 723, 739, 822, 823, 824

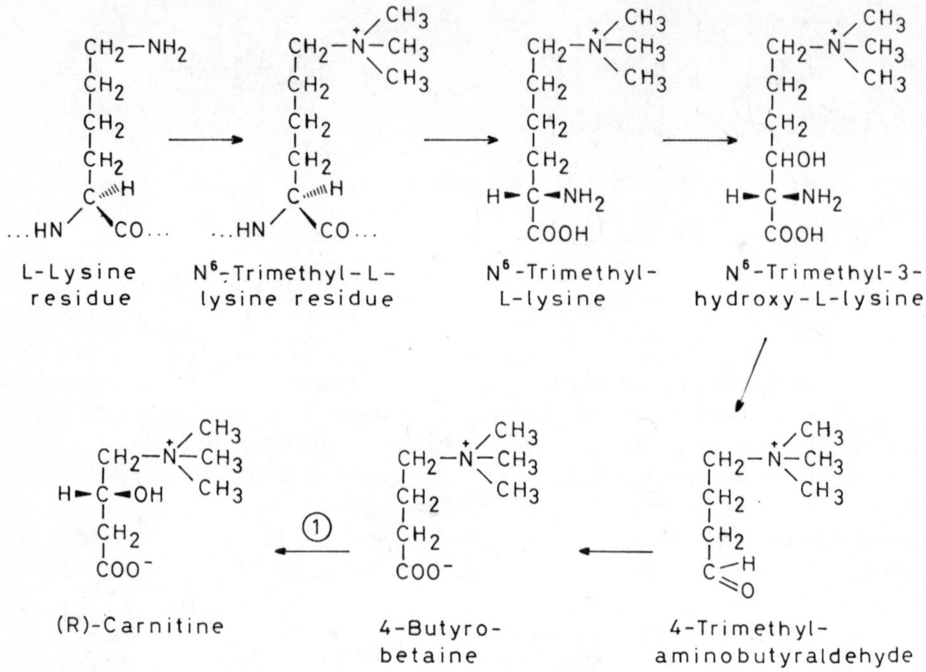

Fig. 221. Biosynthesis of *(R)*-carnitine
1 4-Butyrobetain dioxygenase (intermolecular oxygenase, C 2.5)

D 19 Secondary Products Derived from L-Arginine

Biosynthesis of the Precursor

L-Arginine is the most important representative of the guanidino amino acids. It originates from L-ornithine, carbamyl phosphate, and the amino group of L-aspartic acid (Fig. 222).

Chemistry and Distribution

Secondary products are synthesized from L-arginine in microorganisms, plants, and animals. The most remarkable compounds are listed in Fig. 222. A characteristic feature is the presence of a guanidino group (cf. the formulas of octopine, L-canavanine, galegine, and sphaerophysine). In some compounds this grouping is incorporated in a pyrimidine ring (cf. the formulas of lathyrine, D 26, and of tetrodotoxin). In plants a few secondary compounds are also derived from L-citrulline.

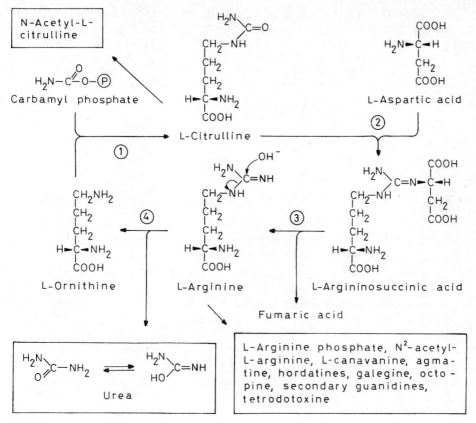

Fig. 222. The urea cycle
1 Ornithine carbamyltransferase; *2* argininosuccinate synthetase; *3* argininosuccinate lyase;
4 arginase

Groups of Secondary Products ·

a) L-Arginine phosphate is formed from L-arginine and ATP as an energy store
in invertebrates (E 2.2). It is used in the regeneration of ATP according to
Fig. 223.

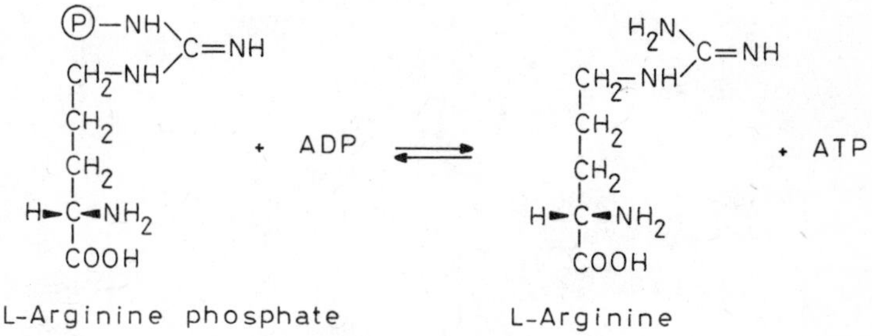

Fig. 223. Synthesis and breakdown of L-arginine phosphate

b) N^2-Acetyl-L-arginine and N^2-acetyl-L-citrulline are formed from the corresponding amino acids and acetyl CoA. They are used as nitrogen stores in certain plants (E 2.2).

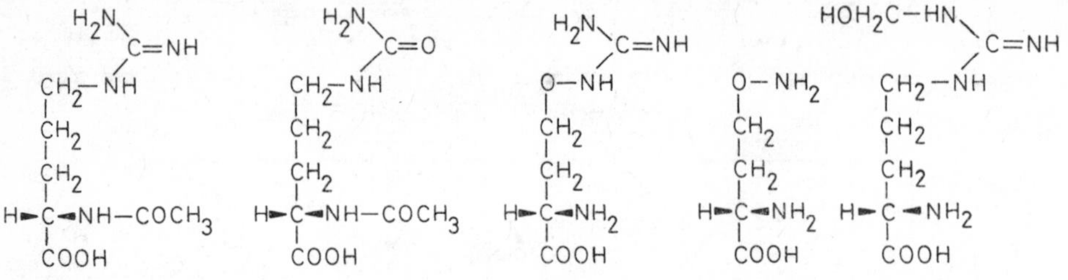

N²-Acetyl-L-arginine   N²-Acetyl-L-citrulline   L-Canavanine   L-Canaline   Hydroxymethyl-L-arginine

c) L-Canavanine and L-canaline are nonprotein amino acids built in plants. They are accumulated, e.g., in the jack bean, i.e., the seeds of *Canavalia ensiformis*. The amino acids are guanidinooxy structural analogs of L-arginine and ornithine, respectively. L-Canavanine is incorporated in proteins instead of L-arginine. L-Canaline is an inhibitor of pyridoxal-5′-phosphate-dependent enzymes. Both amino acids are toxic to most animals and are of significance as feeding deterrents (E 5.5.3). In addition, L-canavanine acts as nitrogen store (E 2.2). It is degraded during germination of the jack beans. L-Canavanine released from germinating seeds inhibits plant growth in the neighborhood (E 5.3).

d) Agmatine is an amine found in plants. It is the product of L-arginine decarboxylation (D 8.2). The hordatines, fungitoxins of *Hordeum vulgare* (E 5.4), obviously originate from cinnamic acids and agmatine. Agmatine is also a precursor of sphaerophysine found in Leguminosae. The isoprenoid portion of the molecule as well as that of the related alkaloid galegine is probably derived from 3,3-dimethylallyl diphosphate (D 6).

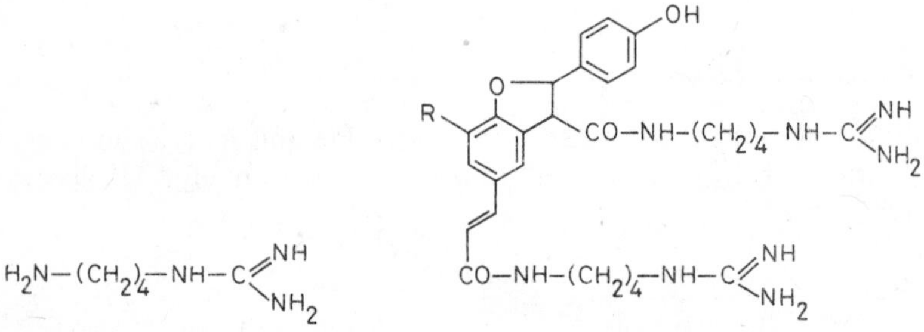

Agmatine

Hordatine A (R = H)
Hordatine B (R = OCH₃)

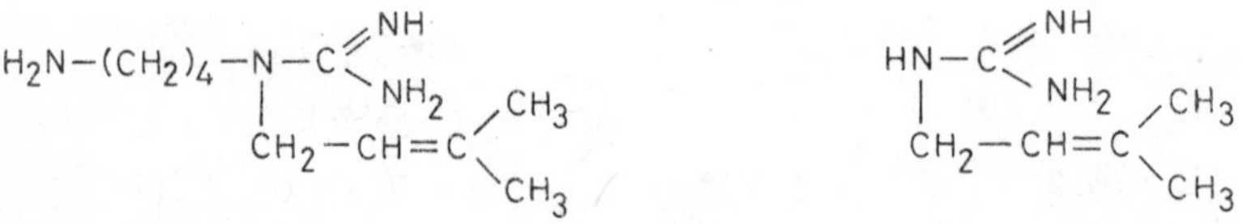

Sphaerophysine

Galegine

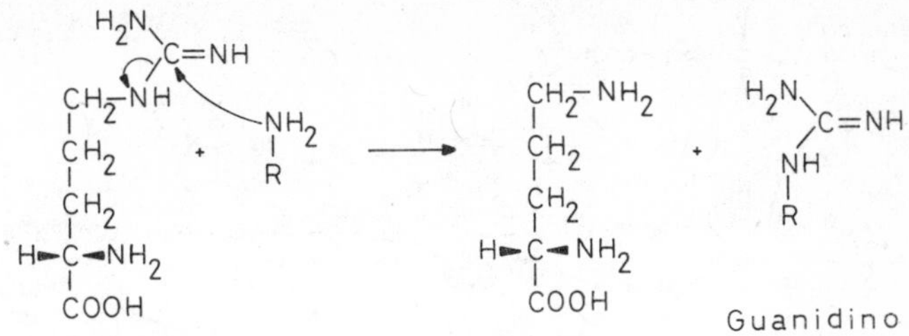

Fig. 224. Biosynthesis of guanidino compounds from L-arginine

e) Urea is formed from L-arginine in many animal species (Fig. 222). It is a detoxication product of excess ammonia and is excreted with the urine (E 1). Urea is formed in a cyclic set of reactions, the so-called urea cycle. The enzymes involved have been found also in plants and microorganisms, though in these organisms urea formation is without significance.

f) Octopine, accumulating in the muscles of *Octopus sp.* and in the crown gall tumors of plants, originates from L-arginine and pyruvic acid (D 8.1).

g) Guanidino derivatives, e.g., streptidine (D 1.3) and creatine (D 10), are formed by transfer of the amidine group of L-arginine to amines according to Fig. 224.

h) Tetrodotoxin is synthesized in the bowl fish *(Sphaeroides rubipes)*, in some amphibia, e.g., the frog *Atelopus chiriquensis*, the salamander *Taricha torose*, and Cephalopods (Hapalachaena). It inhibits Na^+-transport through membranes and acts as feeding deterrent (E 5.1).

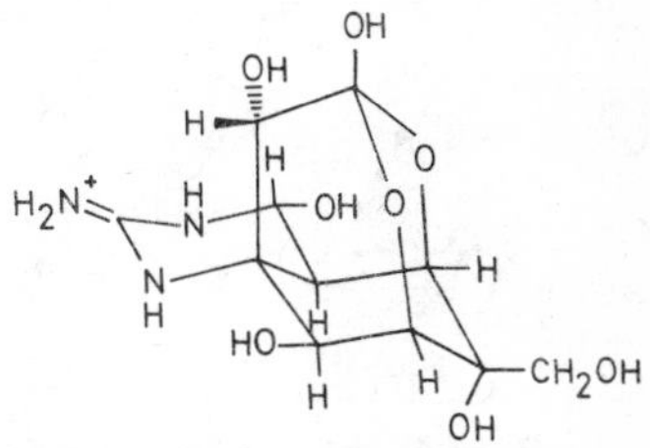

Tetrodotoxin

References for Further Reading

31, 533, 631, 656, 760

D 20 Secondary Products Synthesized from L-Histidine

Biosynthesis and Degradation of the Precursor

L-Histidine (imidazolyl-L-alanine) is one of the rare natural substances with an imidazole ring. Its formation is closely related to purine biosynthesis (D 25,

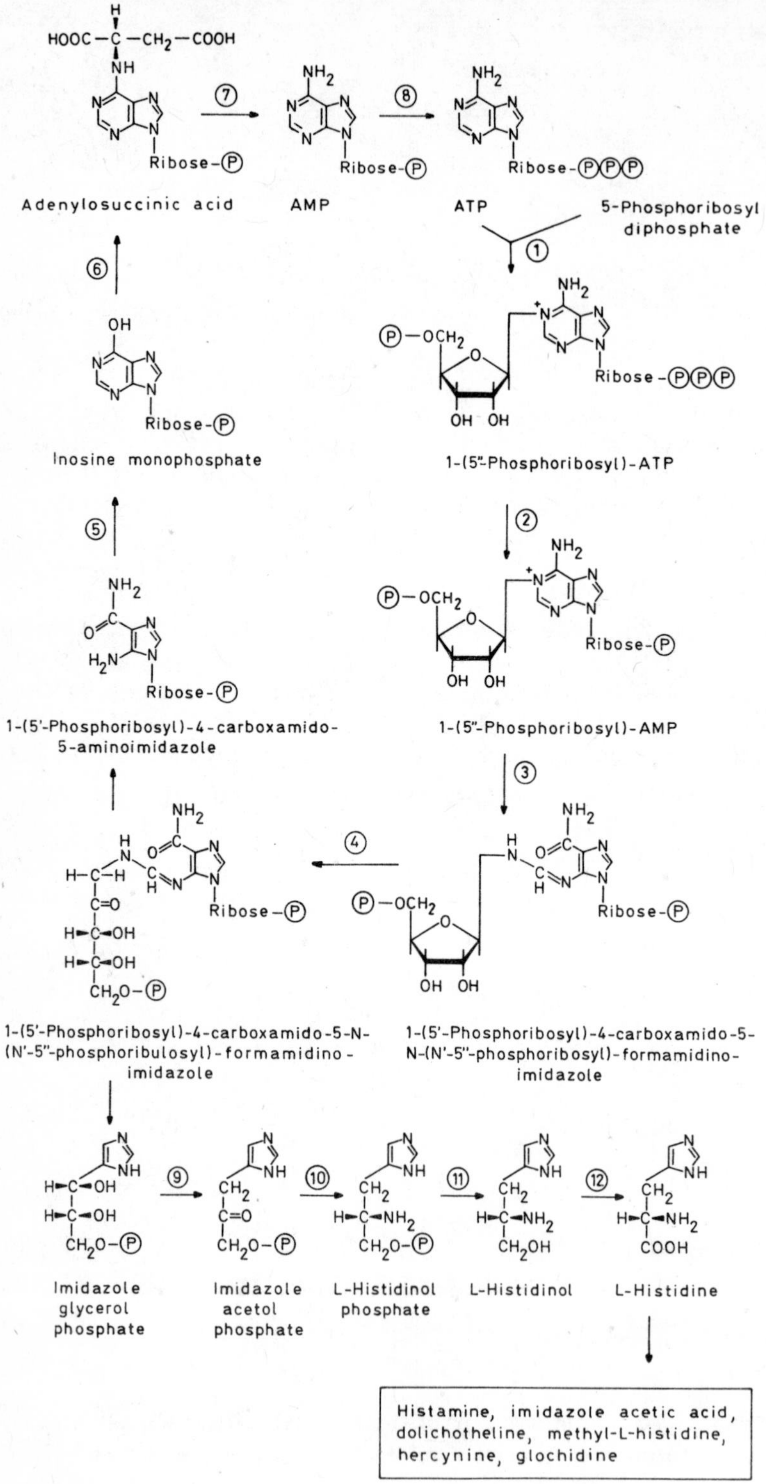

Adenylosuccinic acid

AMP

ATP

5-Phosphoribosyl diphosphate

Inosine monophosphate

1-(5''-Phosphoribosyl)-ATP

1-(5'-Phosphoribosyl)-4-carboxamido-5-aminoimidazole

1-(5''-Phosphoribosyl)-AMP

1-(5'-Phosphoribosyl)-4-carboxamido-5-N-(N'-5''-phosphoribulosyl)-formamidino-imidazole

1-(5'-Phosphoribosyl)-4-carboxamido-5-N-(N'-5''-phosphoribosyl)-formamidino-imidazole

Imidazole glycerol phosphate

Imidazole acetol phosphate

L-Histidinol phosphate

L-Histidinol

L-Histidine

Histamine, imidazole acetic acid, dolichotheline, methyl-L-histidine, hercynine, glochidine

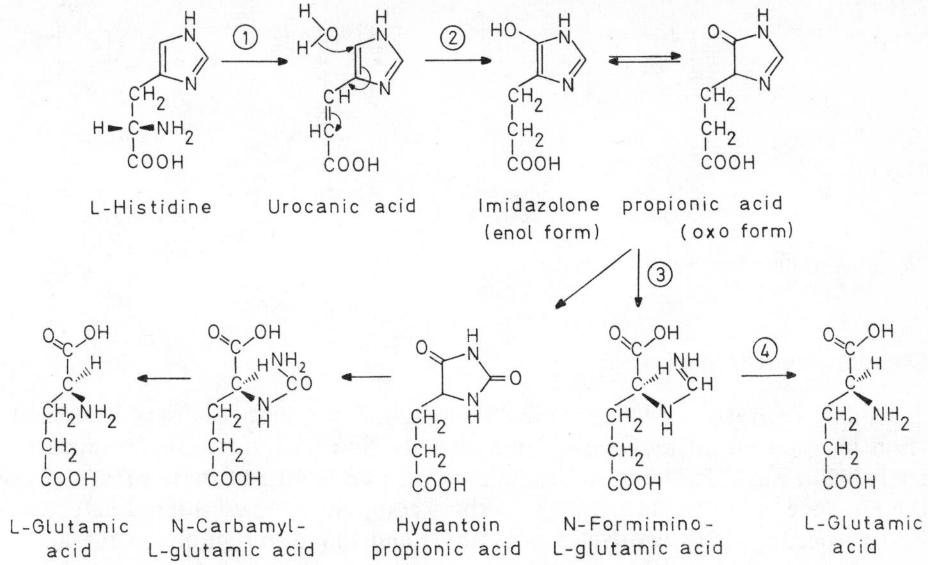

Fig. 226. Degradation of L-histidine
1 Histidine ammonia-lyase; *2* urocanate hydratase; *3* imidazolone propionase; *4* glutamate form-iminotransferase (the enzyme occurs in vertebrates and yields 5-formimino tetrahydrofolic acid)

Fig. 225). ATP is the actual precursor. It condenses with phosphoribosyl diphosphate at position 1. Then the purine ring is split between positions 1 and 6. An Amadori rearrangement in the two-prime-ribose unit yields the ribulose derivative 1-(5'-phosphoribosyl)-4-carboxamido-5-N-(N'-5''-phosphoribosyl)-formamidinoimidazole, which reacts with glutamine to D-erythroimidazole glycerol phosphate (the precursor of L-histidine) and 1-(5'-phosphoribosyl)-4-carboxamido-5-aminoimidazole, which may be regenerated to ATP (D 25).

The main pathway of L-histidine degradation includes the formation of urocanic acid and leads to glutamic acid (Fig. 226). Histidine ammonia-lyase (like phenylalanine ammonia-lyase, D 23.2.1) catalyzes the transelimination of the NH_2-group and a hydrogen atom at position 3. Of minor importance in L-histidine degradation is the formation and further degradation of imidazole pyruvic acid.

◀

Fig. 225. Formation of L-histidine
1 ATP phosphoribosyltransferase; *2* phosphoribosyl-ATP diphosphatase; *3* phosphoribosyl-AMP cyclohydrolase; *4* isomerase; *5* formyltransferase, inosine monophosphate cyclohydrolase; *6* adenylosuccinate synthetase; *7* adenylosuccinate lyase; *8* adenylate kinase; *9* imidazole glycerol phosphate dehydratase; *10* histidinol phosphate aminotransferase; *11* histidinol phosphatase; *12* histidinol dehydrogenase

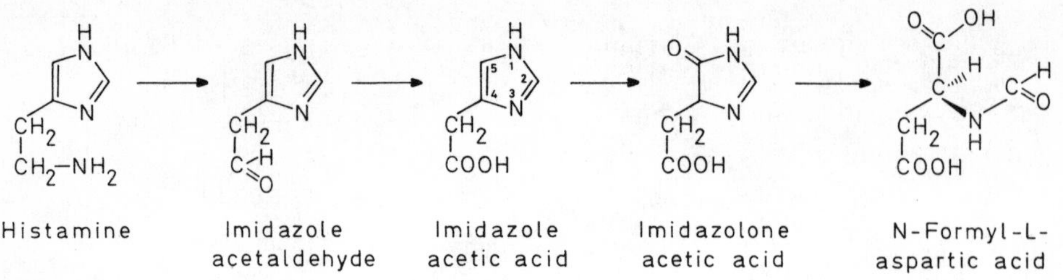

Fig. 227. Degradation of histamine

Chemistry and Distribution

Secondary derivatives are formed from L-histidine or intermediates of its degradation in microorganisms, plants, and animals. Some characteristic compounds are listed in Fig. 225. The most prominent chemical feature of these substances is the presence of the imidazole ring. In some compounds an additional heterocyclic ring occurs which involves the C-atoms and the amino group of the alanyl side chain.

Groups of Secondary Products

a) Histamine is formed by decarboxylation of L-histidine in microorganisms, plants, and animals (D 8.2). It is degraded to imidazole acetic acid, which either yields aspartic acid derivatives (Fig. 227) or is transformed to the riboside. Acylated histamine derivatives, e.g., dolichotheline and *N*-4'-oxodecanoyl histamine, are built in plants (Fig. 228).

b) Methylated derivatives, like 1-methyl-L-histidine, 1-methyl-histamine, and 1-methylimidazole acetic acid are synthesized in animals. Pilocarpine is formed in plants. The betaine hercynine (D 8.1) and its sulfur-containing derivative ergothioneine (Fig. 228) are built in microorganisms and plants.

c) Imidazole derivatives with a second heterocyclic ring have been found in plants. These compounds may be synthesized
— either by Mannich condensation (D 8.5); an example is glochicidine, which occurs together with its possible precursor *N*-4'-oxodecanoyl histamine
— or by closure of the new ring with participation of one of the imidazole ring nitrogen atoms (cf. the formula of glochidine).

Significance

Histamine has hormone activity in animals (E 3.1) and acts as neurotransmitter (E 3.2). It is a plant-feeding deterrent present, for instance, in nettle toxin (E 5.5.3). Also the alkaloid pilocarpine has strong physiological activity in vertebrates and deters potential predators (E 5.5.3). It is used as a cholinergic drug in medicine (F 2). Urocanic acid, hydantoin propionic acid, imidazole acetic acid riboside, 1-methylhistidine, 1-methylimidazole acetic acid, and other imidazole derivatives are involved in the removal of L-histidine and histamine residues

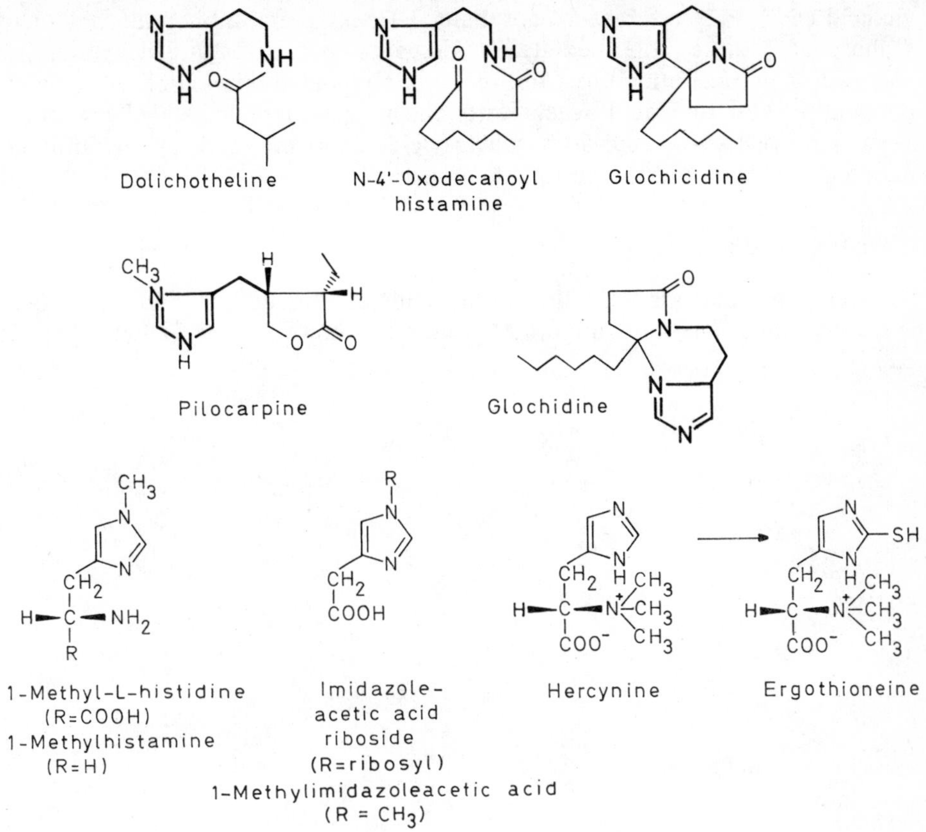

Dolichotheline   N-4'-Oxodecanoyl   Glochicidine
                      histamine

Pilocarpine   Glochidine

1-Methyl-L-histidine   Imidazole-   Hercynine   Ergothioneine
  (R=COOH)   acetic acid
1-Methylhistamine   riboside
  (R=H)   (R=ribosyl)
              1-Methylimidazoleacetic acid
                      (R = CH₃)

Fig. 228. Several secondary products with the imidazole ring

from the organism of animals. They are found in the urine (E 1). Urocanic acid is also a constituent of sweat (E 1). It is involved in the protection of skin to UV radiation (E 2.2).

References for Further Reading

31, 240, 462, 476, 522, 894

D 21 Derivatives of Dehydroquinic Acid, Dehydroshikimic Acid, Shikimic Acid, and Chorismic Acid

Biosynthesis of the Precursors

5-Dehydroquinic acid, 5-dehydroshikimic acid, shikimic acid, and chorismic acid are carboxylated compounds containing a six-membered carbocyclic ring with one or two double bonds (Fig. 229). They are formed from phosphoenolpyruvic acid (D 2) and erythrose-4-phosphate. 3-Deoxy-D-arabinoheptulosonic acid-7-phosphate is a key intermediate. This compound cyclizes to 5-dehydroquinic acid, which is transformed to 5-dehydroshikimic and shikimic acid. After phosphorylation shikimic acid reacts with phosphoenolpyruvic acid. The formed 3-enolpyruvylshikimic acid-5-phosphate yields chorismic acid by an antielimination of a proton and the phosphate group.

Chemistry and Distribution

Secondary products are formed from the acids mentioned above in microorganisms and plants. The most important groups are shown in Figs. 229 and 231. On

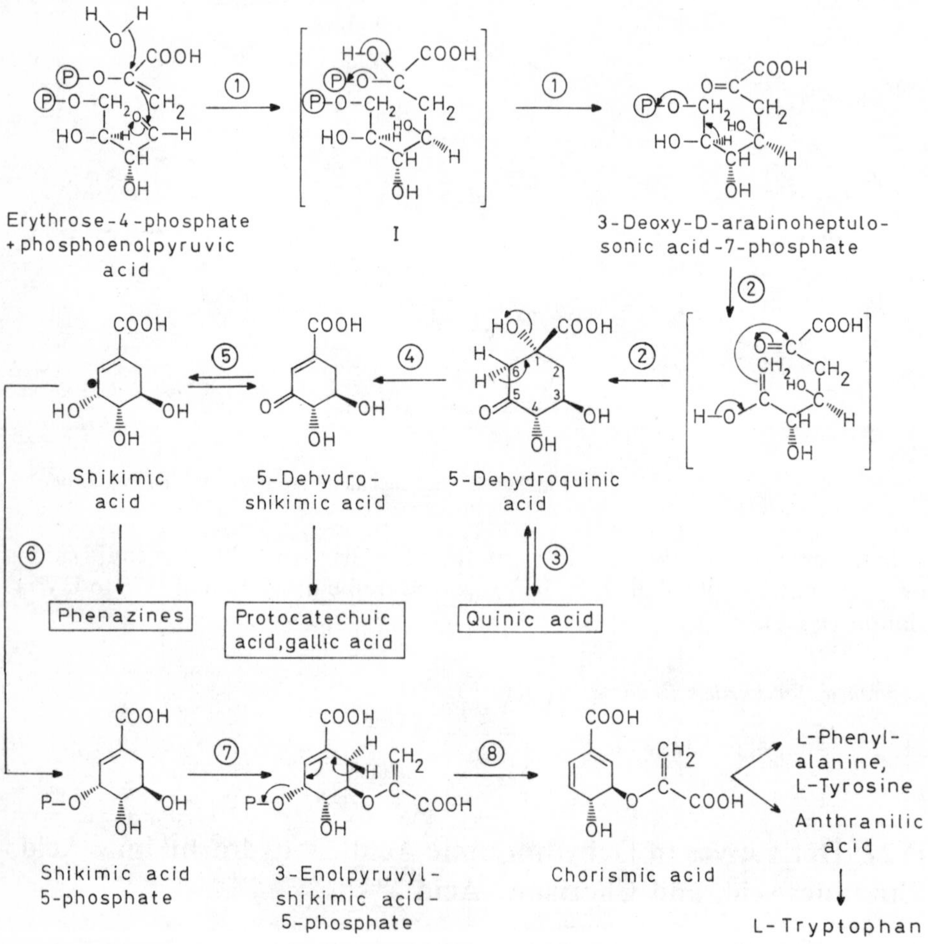

Fig. 229. Formation of dehydroquinic, dehydroshikimic, shikimic, and chorismic acids and secondary products derived from these compounds
1 3-Deoxy-D-arabinoheptulosonic acid-7-phosphate (DAHP) synthase; *2* 5-dehydroquinate synthase; *3* quinate dehydrogenase; *4* 5-dehydroquinate dehydratase; *5* shikimate dehydrogenase; *6* shikimate kinase; *7* 3-enolpyruvylshikimate-5-phosphate synthase; *8* chorismate synthase

the one hand, they may still contain the ring and the C_1-side chain of the acids (see the structure of the benzoic acid derivatives below, as well as the formulas of the anthranilic acid and 3-hydroxyanthranilic acid esters, D 21.1, D 21.4, D 21.4.1), or they have additional rings (see the formulas of naphthoquinones and anthraquinones, D 21.2, of quinoline, acridine, and benzodiazepine alkaloids, D 21.4.2). The carbon skeletons may be substituted by isoprenoid side chains (see the structure of ubiquinones, D 21.3) and may carry different functional groups, e.g., hydroxy, carboxy, methoxy, and amino groupings.

Groups of Secondary Products

Dehydroquinic Acid Derivatives

Quinic acid, a compound widespread in higher plants is derived from dehydroquinic acid by quinate dehydrogenase according to the following equation:

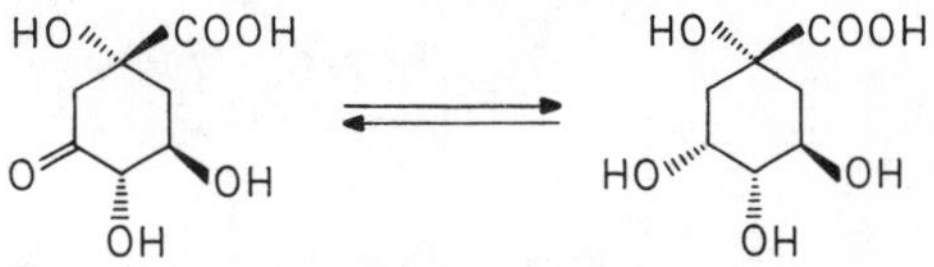

Dehydroquinic acid   Quinic acid

Quinic acid is a constituent of cinnamic acid esters (D 23.2.1).

Dehydroshikimic Acid and Shikimic Acid Derivatives

a) Protocatechuic and gallic acids are formed from dehydroshikimic acid, e.g., in *Phycomyces blakesleeanus* and in higher plants, e.g., *Rhus typhina*, *Acer saccharinum*, *Camellia sinensis*, and *Vaccinium vitis-idea*. (In most other producer organisms both acids are derived from cinnamic acids, D 23.2.5.) The formation of protocatechuic and gallic acids from dehydroshikimic acid is shown in Fig. 230. Gallic acid is a constituent of hydrolyzable tannins (D 23.2.5).

b) Phenazines, e.g., iodinin, are pigments formed in bacteria from shikimic acid according to the following equation:

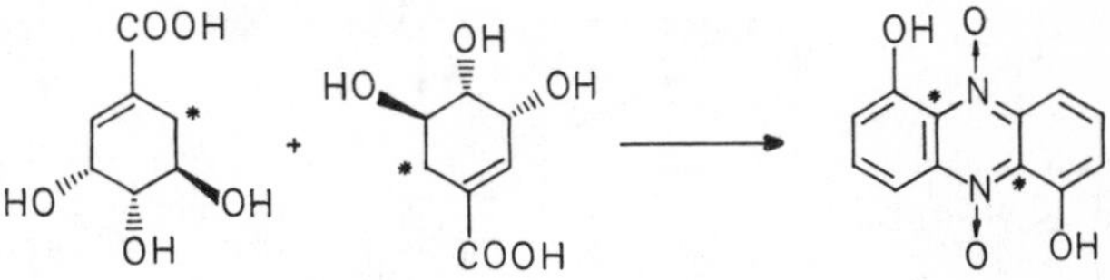

Shikimic acid   Iodinin

Chorismic Acid Derivatives

Secondary products derived from chorismic acid are treated in Sections D 21.1-D 21.4.

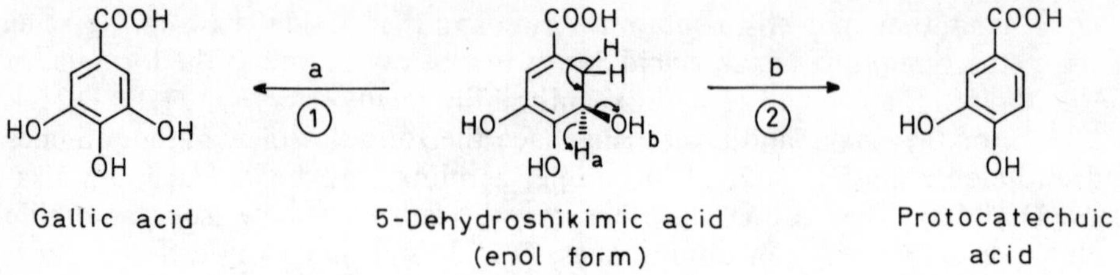

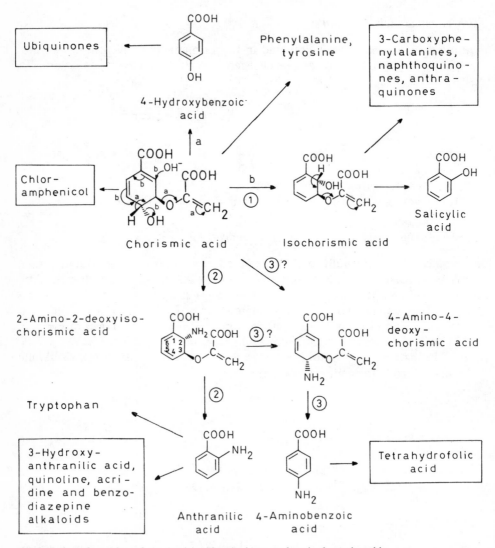

Gallic acid   5-Dehydroshikimic acid   Protocatechuic
                      (enol form)   acid

Fig. 230. Biosynthesis of hydroxylated benzoic acids from 5-dehydroshikimic acid
1 Dehydroshikimate dehydrogenase; *2* dehydroshikimate dehydratase

Ubiquinones

4-Hydroxybenzoic
acid

Phenylalanine,
tyrosine

3-Carboxyphe-
nylalanines,
naphthoquino-
nes, anthra-
quinones

Chlor-
amphenicol

Chorismic acid   Isochorismic acid   Salicylic
                                             acid

2-Amino-2-deoxyiso-
chorismic acid

4-Amino-4-
deoxy-
chorismic acid

Tryptophan

3-Hydroxy-
anthranilic acid,
quinoline, acri-
dine and benzo-
diazepine
alkaloids

Anthranilic   4-Aminobenzoic
acid   acid

Tetrahydrofolic
acid

Fig. 231. Transformation of chorismic acid to hydroxy and aminobenzoic acids
1 Isochorismate synthase; *2* anthranilate synthase; *3* 4-aminobenzoate synthase

References for Further Reading

48, 150, 190, 264, 367, 643, 808

D 21.1 Hydroxy and Aminobenzoic Acids Formed from Chorismic and Isochorismic Acid

Chemistry and Distribution

Hydroxy and aminobenzoic acids are aromatic compounds, derived from benzoic acid by an additional hydroxy or amino group in position 3 or 4. They are formed in microorganisms and plants.

Biosynthesis

Chorismic Acid-Derived Compounds

a) 4-Hydroxybenzoic, 4-aminobenzoic, and anthranilic acid are formed by the reactions given in Fig. 231. The amino groups of anthranilic acid and 4-aminobenzoic acid originate from L-glutamine. Intermediates in the transformation of chorismic acid to both acids are unknown and probably do not occur in the free state. 4-Hydroxybenzoic acid is a precursor of ubiquinones (D 21.3). 4-aminobenzoic acid is involved in the formation of tetrahydrofolic acid (D 25.3), a coenzyme in one-carbon metabolism (C 3.2). Anthranilic acid is the mother substance of the amino acid L-tryptophan (D 22) and the secondary products treated in D 21.4.

b) Chloramphenicol, an antibiotic built in *Streptomyces venezuelae* (E 5.2), is probably formed on the pathway given in Fig. 232.

Synthetically prepared chloramphenicol is widely used in medicine (F 2).

Isochorismic Acid-Derived Compounds

a) Salicylic and 2,3-dihydroxybenzoic acid occur in microorganisms. Salicylic acid is synthesized by the reactions shown in Fig. 231.

b) L-3-Carboxyphenylalanine and its derivatives are probably formed on the pathway given in Fig. 233. L-3-Hydroxymethylphenylalanine and L-4-hydroxy-3-hydroxymethylphenylalanine are toxic to animals by inhibition of L-phenylalanine uptake. They are of significance as feeding deterrents (E 5.5.3). L-3-Carboxyphenylalanine may be an intermediate in the biosynthesis of 3-acetyl-6-methoxybenzaldehyde, a compound involved in the allelopathic interactions between higher plants (E 5.3).

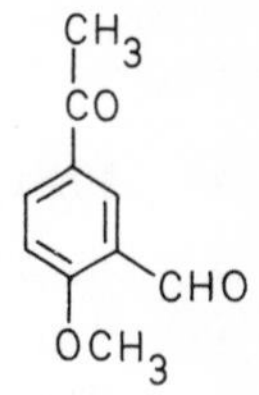

3-Acetyl-6-methoxybenzaldehyde

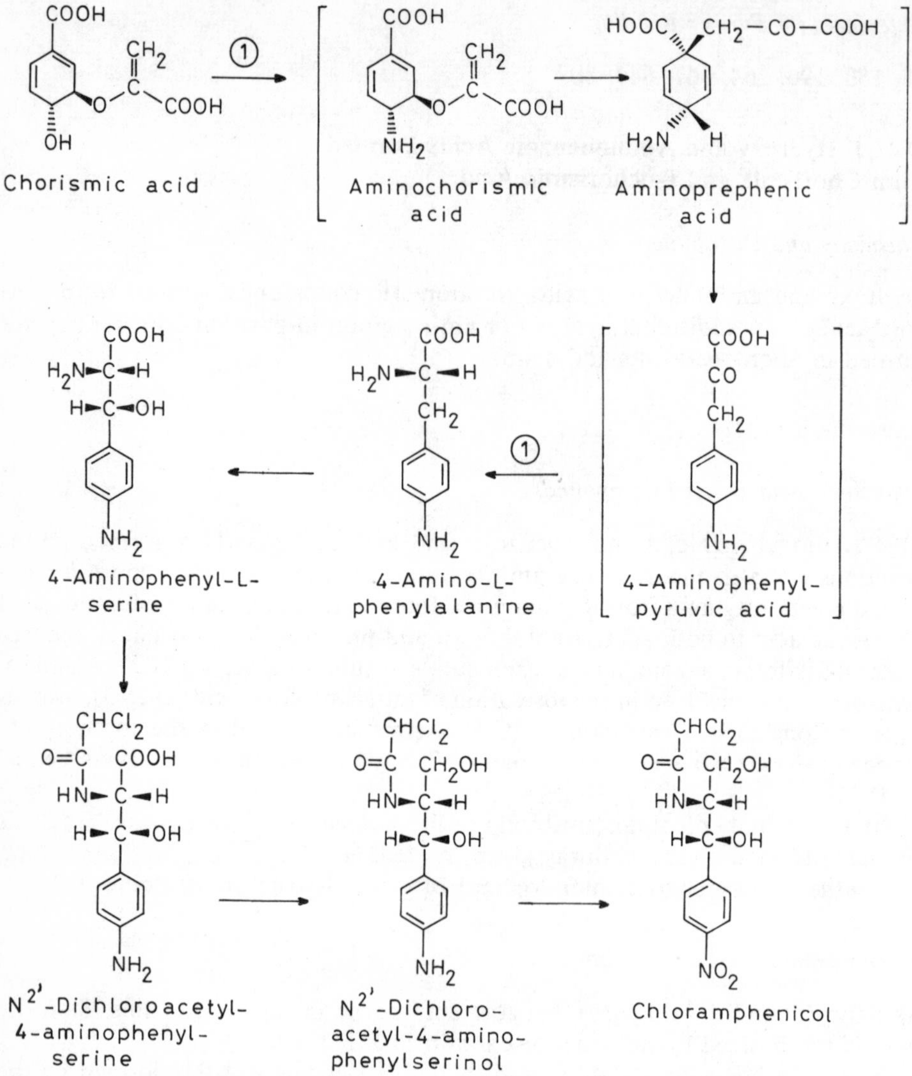

Fig. 232. Biosynthesis of chloramphenicol
1 Arylamine synthetase complex

c) Naphthoquinones and anthraquinones are formed by the reactions described in D 21.2.

References for Further Reading

48, 150, 190, 451, 808

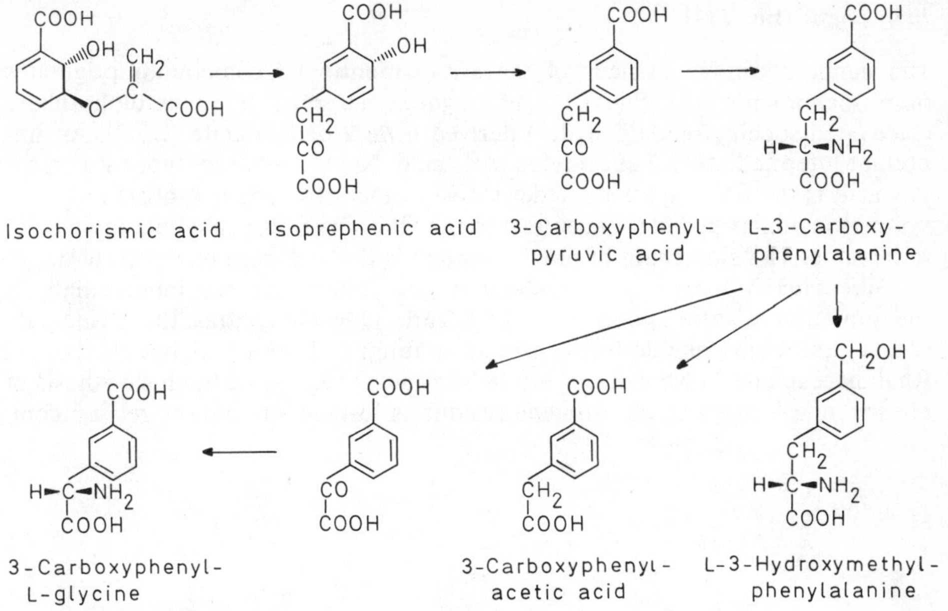

Isochorismic acid | Isoprephenic acid | 3-Carboxyphenyl-pyruvic acid | L-3-Carboxy-phenylalanine

3-Carboxyphenyl-L-glycine | | 3-Carboxyphenyl-acetic acid | L-3-Hydroxymethyl-phenylalanine

Fig. 233. Transformation of isochorismic acid to L-3-carboxyphenylalanine and derivatives

D 21.2 Naphthoquinones and Anthraquinones Derived from Isochorismic Acid

Chemistry and Distribution

Naphthoquinones and anthraquinones are derivatives of the bi- and tricyclic aromatic compounds, naphthene and anthracene, respectively. The naphthoquinones of the vitamin K type have a long isoprenoid side chain. Other naphthoquinones are devoid of a side chain, e.g., juglone (Fig. 234), or are substituted only by an isoprene residue, e.g., deoxylapachol. A typical representative of the anthraquinone derivatives is alizarin (Fig. 234).

Naphthoquinones with a long isoprenoid side chain are formed in microorganisms (menaquinones, vitamins K_2) and in higher plants (phylloquinone, vitamin K_1). Naphthoquinone derivatives with short or absent side chains, as well as anthraquinones of the alizarin type, are produced in a few families of higher plants, e.g., Balsaminaceae, Juglandaceae, Rubiaceae.

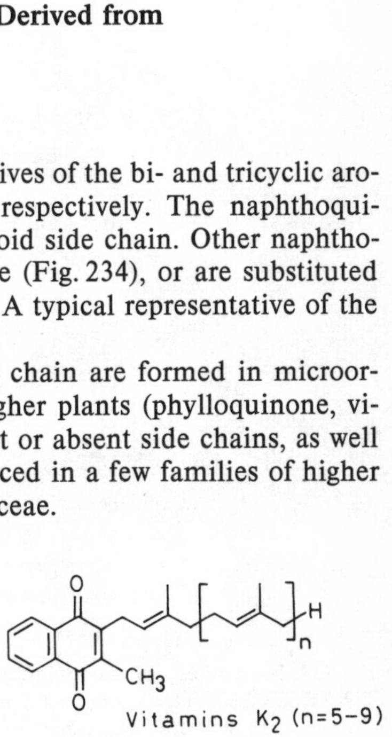

Vitamins K_2 (n=5–9)

Biosynthesis (Fig. 234)

The naphthoquinone nucleus of the above mentioned compounds originates from isochorismic acid (Fig. 231) and 4-hydroxybutyryl-2'-thiamine diphosphate ("activated succinylsemialdehyde") derived from 2-oxoglutarate (C 2.7). An important intermediate is 2-succinylbenzoic acid. Naphthohydroquinone-2-carboxylic acid is the first naphthalene derivative formed. Juglone is synthesized via a symmetrical intermediate, in contrast to the vitamins K, e.g., phylloquinone, and to lawsone. The side chain of phylloquinone is derived from phytol (D 6.4).

3-Prenylnaphthohydroquinone-2-carboxylic acid is the key intermediate in the formation of anthraquinones of the alizarin type. (In contrast the A-ring-substituted anthraquinone derivatives found in fungi and certain higher plants, e.g., Rhamnaceae and Polygonaceae, are polyketides, D 5.3.6.) In the biosynthesis of alizarin one C-atom of the isoprene residue is lost. In structurally related com-

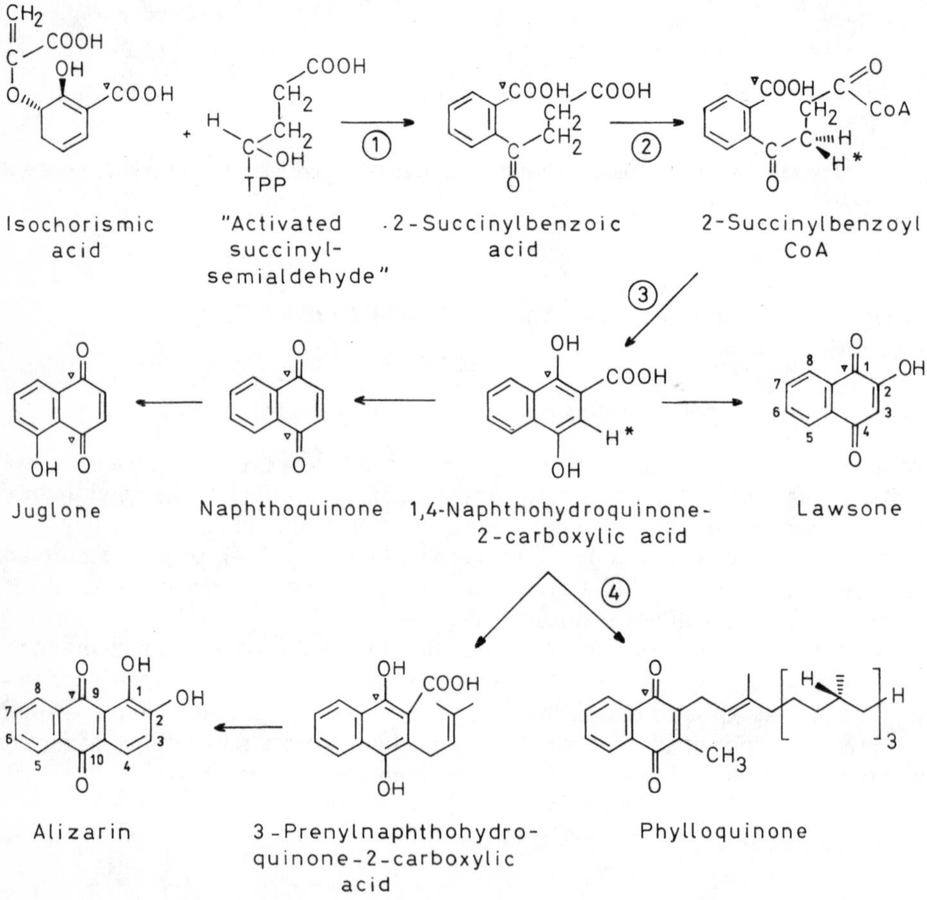

Fig. 234. Biosynthesis of naphthoquinone and anthraquinone derivatives from chorismic acid
1 2-Succinylbenzoate synthase; *2* 2-succinylbenzoate: CoA ligase; *3* naphthoate synthase; *4* 1,4-dihydroxy-2-naphthoate: polyprenyltransferase

pounds, however, this C-atom may still be present in the form of a methyl, hydroxymethyl, aldehyde, or carboxy group.

Significance

Vitamins K are coenzymes of glutamic acid carboxylase (D 17). In mammals and man naphthoquinones of the vitamin K_2 group are formed by bacteria in the intestine. Phylloquinone is taken up with the diet (vitamin K_1, E 2.1).

Juglone is involved in the allelopathic interactions between higher plants (E 5.3) and is a feeding repellent for insects (E 5.5.3). Lawsone is the main constituent of henna (leaves and flowers of *Lawsonia inermis*) used as dye in cosmetics in India and the Arab countries.

References for Further Reading

55, 56, 150, 199, 349, 457, 458, 745, 746, 808, 818

D 21.3 Ubiquinones

Chemistry and Distribution

Ubiquinones are prenylated benzoquinone derivatives carrying a side chain of five to ten isoprene residues. Ubiquinone 50 (coenzyme Q_{10}), for instance, has a side chain of 50 carbon atoms ($\triangleq$ 10 isoprene units). Ubiquinones are formed in microorganisms, plants, and animals.

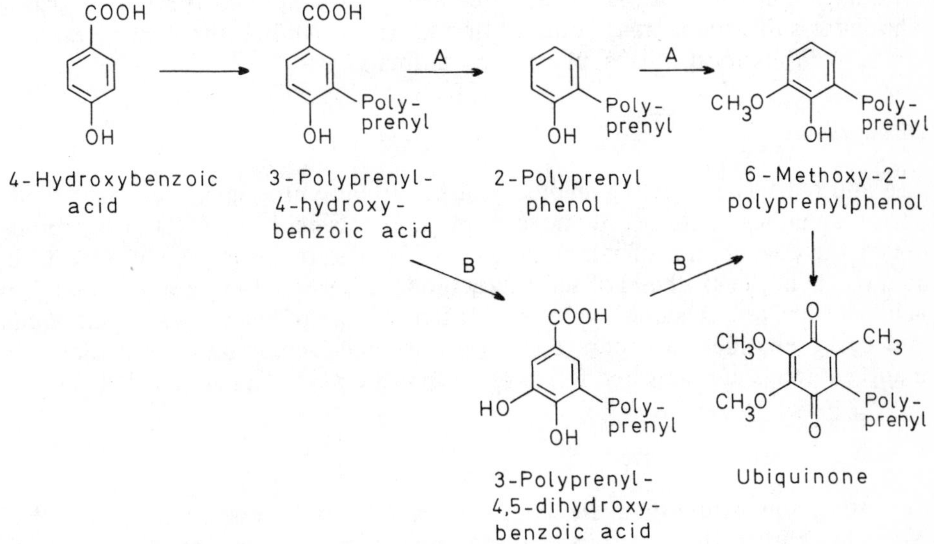

Fig. 235. Formation of ubiquinones from 4-hydroxybenzoic acid

Biosynthesis (Fig. 235)

4-Hydroxybenzoic acid derived from chorismic acid (bacteria, D 21.1) or p-coumaric acid (higher plants and animals, D 23.2.5) reacts with prenyl diphosphates
(D 6) of different chain length. In bacteria decarboxylation (pathway A), in
higher plants and animals hydroxylation (pathway B) is the first step in the further transformation. The *O*-methyl and *C*-methyl groups originate from L-methionine (C 3.3; see also the biosynthesis of plastoquinones and tocopherylquinones, D 23.4).

Significance

The ubiquinones participate in cell respiration (E 2.1). They are attached with
the hydrophobic side chain to lipid components of the membranes, whereas the
quinoid ring undergoes oxidoreductions.

References for Further Reading

55, 611, 763

D 21.4 Anthranilic Acid Derivatives

Chemistry and Distribution

The secondary products derived from anthranilic acid (2-amino benzoic acid)
may carry additional substituents at the aromatic ring, e.g., hydroxy groups (see
formula for hydroxyanthranilic acid) or at the carboxyl and the amino group.
The latter substituents may form additional rings. Anthranilic acid derivatives
are built in microorganisms, plants, and animals.

Biosynthesis

The anthranilic acid, acting as precursor in the biosynthesis of secondary products, may either be formed de novo from chorismis acid (Fig. 231) or is a product of the degradation of L-tryptophan (D 22). The mode in which anthranilic
acid reacts in the synthesis of secondary products resembles that of the α-amino
acids. Hence, protoalkaloids as well as different types of heterocyclic compounds
and cyclic peptides (benzodiazepines) are formed. Important intermediates in
many pathways are activated forms of anthranilic acid, e.g., anthranilyl CoA.

Groups of Secondary Products

a) Anthranilic acid esters are constituents of certain essential oils. Methyl
N-methyl anthranilate, for instance, dominates the smell of mandarin oranges
(F 1).

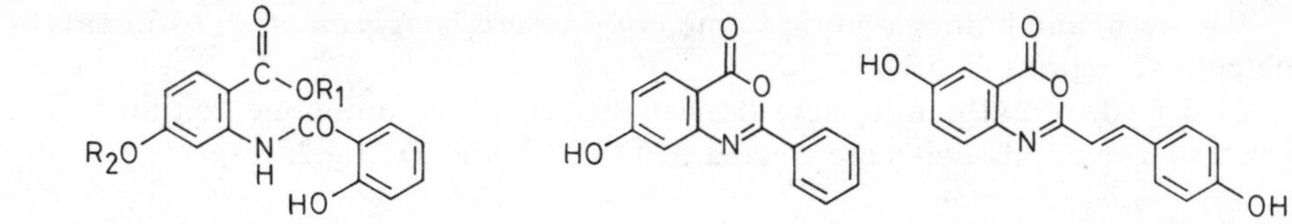

Dianthramide A (R_1=H, R_2=CH_3)
Dianthramide B (R_1=CH_3, R_2=H)

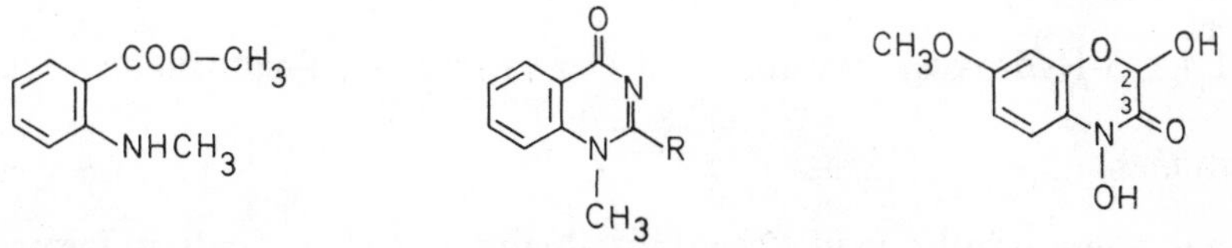

Dianthalexin

Avenalumin I

Methyl N-methyl
anthranilate

Glomerin (R=CH_3)
Homoglomerin (R=CH_2-CH_3)

DIMBOA

b) The dianthramides A and B as well as dianthalexin and avenalumin I are phytoalexins (E 5.4).

c) 2,4-Dihydroxy-7-methoxy-2H-1,4-benzoxazin-3-one (DIMBOA) is a constituent of higher plants. It is derived from anthranilic acid by loss of the carboxy group and incorporation of C-atoms 1 and 2 of ribose into C-atoms 3 and 2, respectively.

d) Quinazolines are alkaloids formed in microorganisms, e.g. *Pseudomonas* sp., in several families of higher plants, and in animals, e.g., in the millipede *Glomeris marginata*. The aromatic ring, the adjoining carbon atom, and one of the nitrogen atoms originate from anthranilic acid. In the biosynthesis of most quinazolines a second amino acid and an unidentified third precursor are involved (Fig. 236).

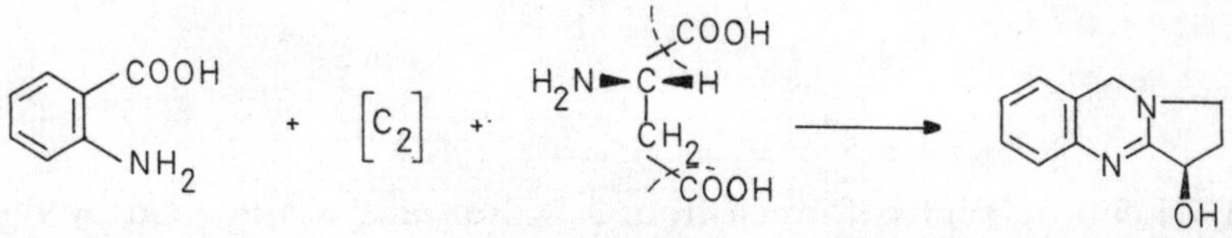

Anthranilic acid C_2-unit   L-Aspartic acid   (-)-Peganine

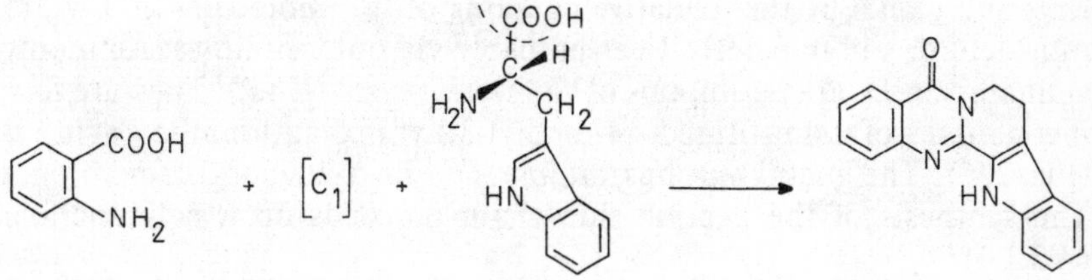

Anthranilic acid C_1-unit   L-Tryptophan   Rutaecarpine

Fig. 236. Biosynthesis of quinazoline alkaloids

Glomerin and homoglomerin formed in *Glomeris marginata* are constituents of defense secretions (E 5.1).

e) 3-Hydroxyanthranilic acid derivatives as well as quinoline, acridine, and benzodiazepine alkaloids are treated in D 21.4.1 and D 21.4.2, respectively.

References for Further Reading

150, 292, 303, 396, 483, 609

D 21.4.1 3-Hydroxyanthranilic Acid Derivatives and Phenoxazines

Biosynthesis

3-Hydroxyanthranilic acid (2-amino-3-hydroxybenzoic acid) is formed on two independent pathways: (a) from anthranilic acid by monooxygenation (C 2.6.5), a reaction occurring in microorganisms and plants and (b) by cleavage of 3-hydroxy-L-kynurenine (D 22) in plants and animals. In the formation of secondary products which proceeds in microorganisms, plants, and animals 3-hydroxy-anthranilic acid may be substituted at the functional groups, e.g., by methylation, or it may react with the formation of additional rings.

Groups of Secondary Products

a) Damascenine is a three-methylated product of 3-hydroxyanthranilic acid. It is built in the plant *Nigella damascena*.

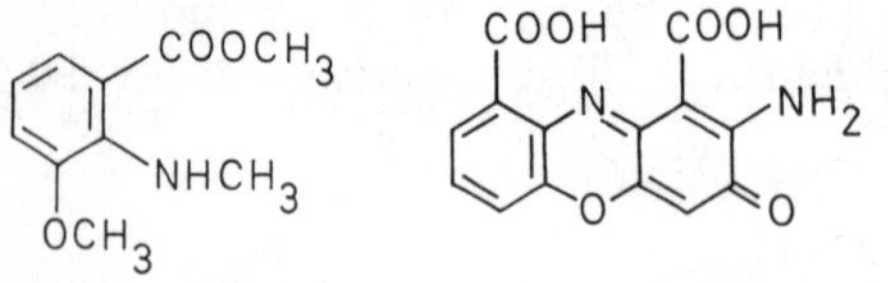

Damascenine

Cinnabarinic
acid

b) Nicotinic acid is formed from 3-hydroxyanthranilic acid by the pathway given in D 22.

c) Phenoxazinones, e.g., cinnabarinic acid, are synthesized in microorganisms and plants by the oxidative coupling of two molecules of 3-hydroxyanthranilic acid or suitable derivatives probably via quinonimines. Actinomycins, e.g., actinomycin D, are antibiotics of *Streptomyces* sp. (E 5.2). They are formed by dimerization of substituted 4-methyl-3-hydroxyanthranilic acid derivatives (Fig. 237). The methyl groups originate from *S*-adenosyl-L-methionine (C 3.3). The synthesis of the peptide side chain proceeds on a polyfunctional protein (A 3.1.1).

d) Polymeric products are formed from 3-hydroxyanthranilic acid via o-quinoid intermediates under the influence of phenol oxidases (C 2.3.1). They are constituents of the sclerotins of insects (D 23.2.5).

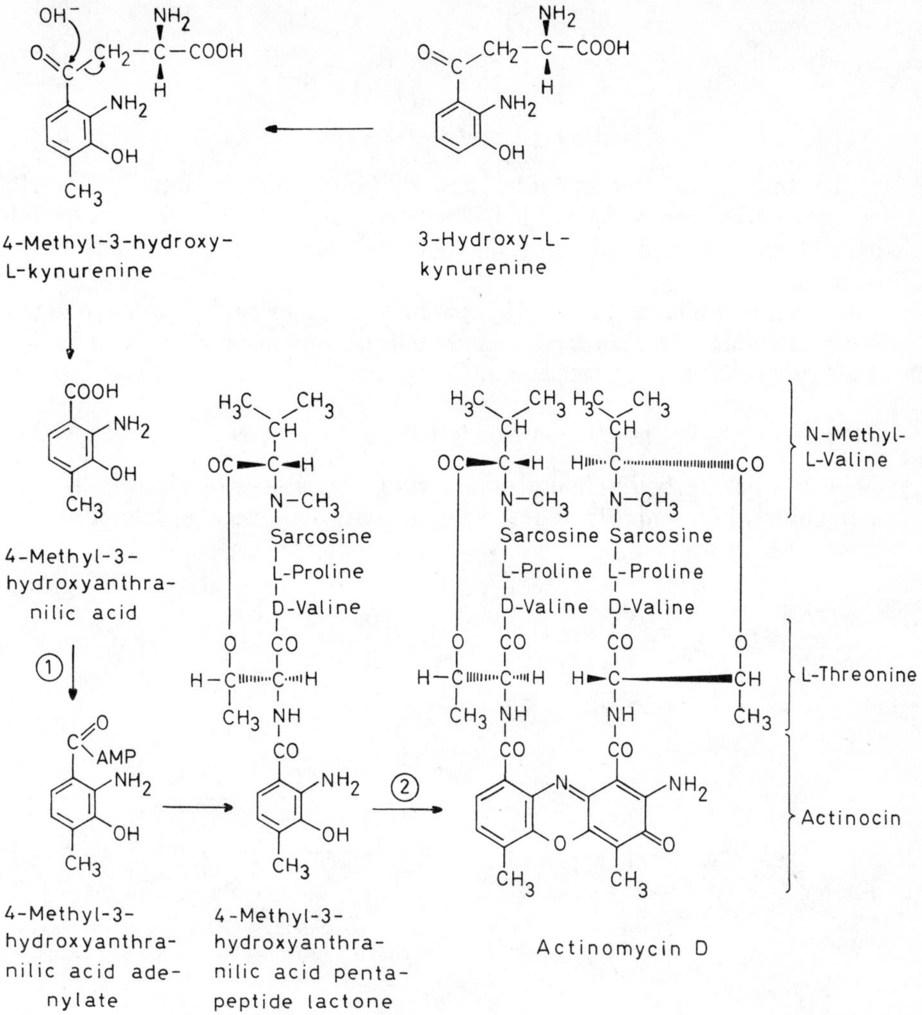

4-Methyl-3-hydroxy-
L-kynurenine

3-Hydroxy-L-
kynurenine

4-Methyl-3-
hydroxyanthra-
nilic acid

4-Methyl-3-
hydroxyanthra-
nilic acid ade-
nylate

4-Methyl-3-
hydroxyanthra-
nilic acid penta-
peptide lactone

Actinomycin D

N-Methyl-
L-Valine

L-Threonine

Actinocin

Fig. 237. Biosynthesis of actinomycin D
1 4-Methyl-3-hydroxyanthranilate activating enzyme; *2* phenoxazinone synthase

References for Further Reading

292, 354, 483

D 21.4.2 Quinoline, Acridine, and Benzodiazepine Alkaloids

Chemistry and Distribution

The following ring systems are the fundamental structures of the above mentioned groups of alkaloids:

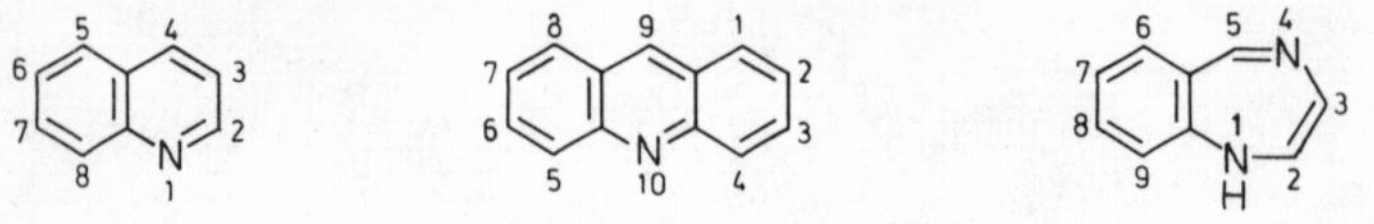

Quinoline   Acridine   1H-1,4-Benzodiazepine

In the compounds found in nature these ring systems are hydroxylated, methylated, prenylated, and *N*-methylated. They may carry aliphatic or aromatic side chains and in the case of quinoline alkaloids may be condensed with a furane or dihydrofurane ring.

Quinoline, acridine, and benzodiazepine alkaloids are formed in microorganisms, e.g., pseudanes in *Pseudomonas* and benzodiazepines in *Penicillium*, as well as in higher plants, e.g., Rutaceae and Compositae.

Biosynthesis

a) Quinoline and furoquinoline alkaloids (Fig. 238) are synthesized from (activated?) anthranilic acid which reacts with (activated?) acetic acid. Probably

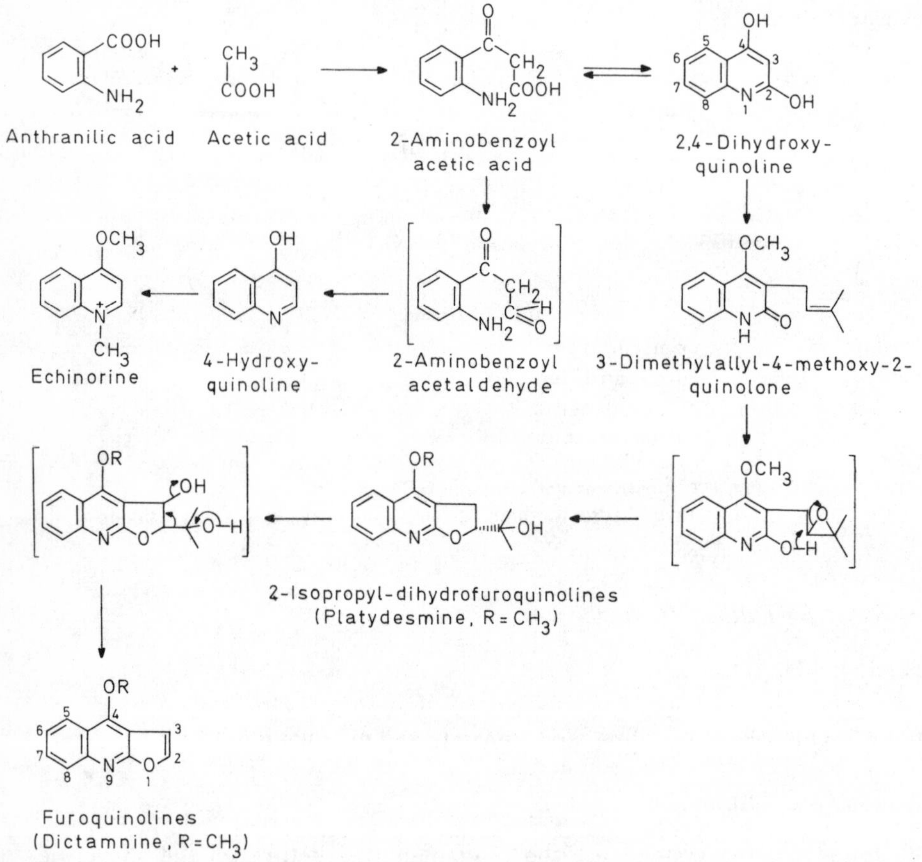

Fig. 238. Biosynthesis of quinoline and furoquinoline alkaloids

2-aminobenzoyl acetic acid, which is in pH-dependent equilibrium with 2,4-di-hydroxyquinoline, or an activated derivative of this compound are intermediates. On the one hand, 2,4-dihydroxyquinoline is transformed to echinorine. On the other hand, it is a precursor of 3-prenylated chinoline alkaloids, isopropyldi-hydrofurochinolines, and furoquinolines.

b) 2-n-Alkylquinolones (pseudanes) and 2-phenylquinolones (Fig. 239) are probably derived from anthranilyl CoA and aliphatic 2-oxo acids or benzoyl acetic acid. Intermediates may be 3-carboxylated quinolones.

c) Acridones (benzylquinolones, Fig. 240), found in plants, e.g., Rutaceae, are of polyketide origin (D 5.3). Probably anthranilyl CoA serves as starter molecule. Benzophenone derivatives occur as intermediates. They allow rotation of ring B before formation of the heterocyclic ring.

d) Benzodiazepine alkaloids and 4-phenylquinolines (Fig. 241) are formed in molds of the genus *Penicillium*. The benzodiazepines are cyclic peptides derived from anthranilic acid and L-phenylalanine. The key intermediate cyclopeptine is synthesized by a polyfunctional enzyme resembling those involved in the biosynthesis of peptide antibiotics (A 3.1.1). It is stereospecifically dehydrogenated

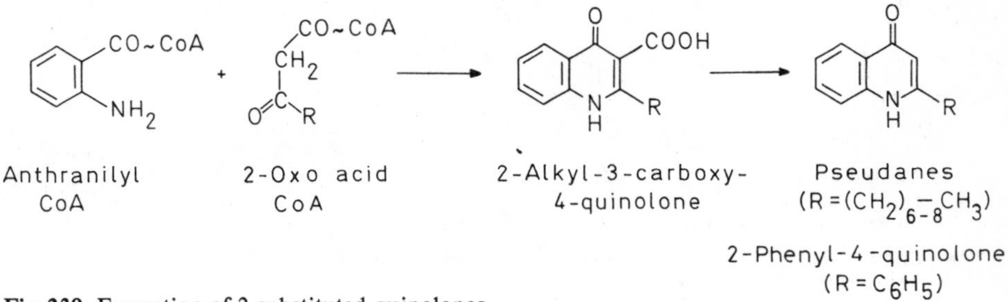

Fig. 239. Formation of 2-substituted quinolones

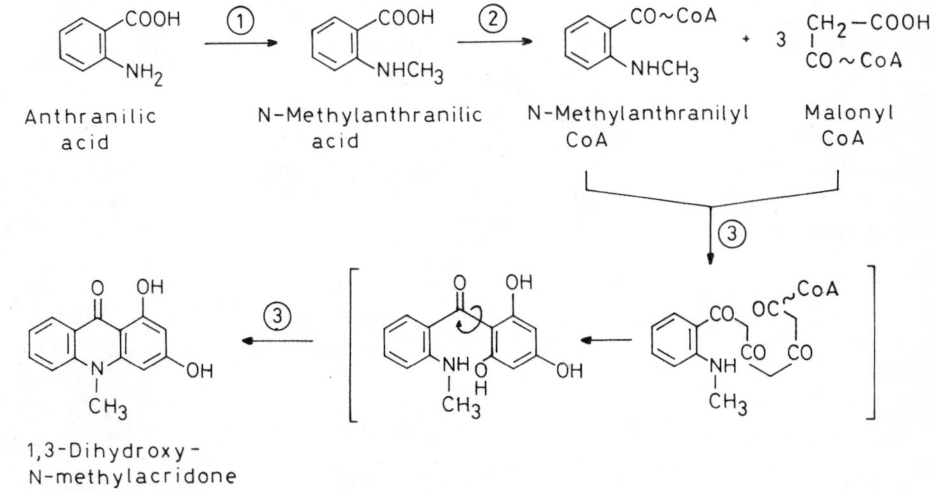

Fig. 240. Biosynthesis of acridone alkaloids
1 Anthranilate N-methyltransferase; *2* methylanthranilic acid-thiol ligase; *3* acridone synthase

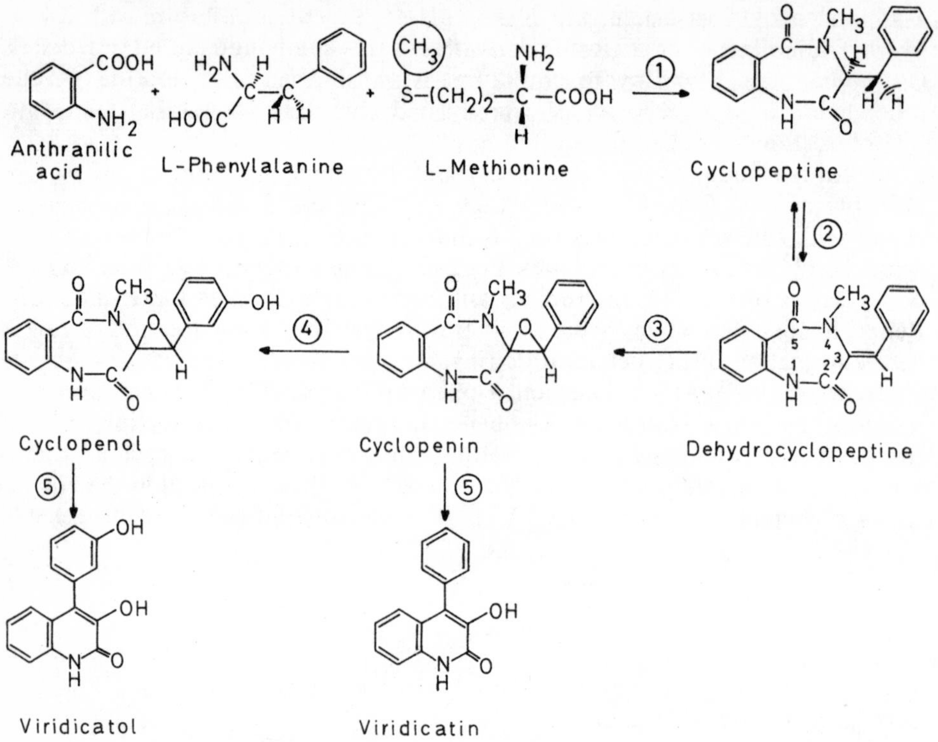

Fig. 241. Formation of the alkaloids of the cyclopenin-viridicatin group
1 Cyclopeptine synthetase; *2* cyclopeptine dehydrogenase; *3* dehydrocyclopeptine epoxidase;
4 cyclopenin m-hydroxylase; *5* cyclopenase

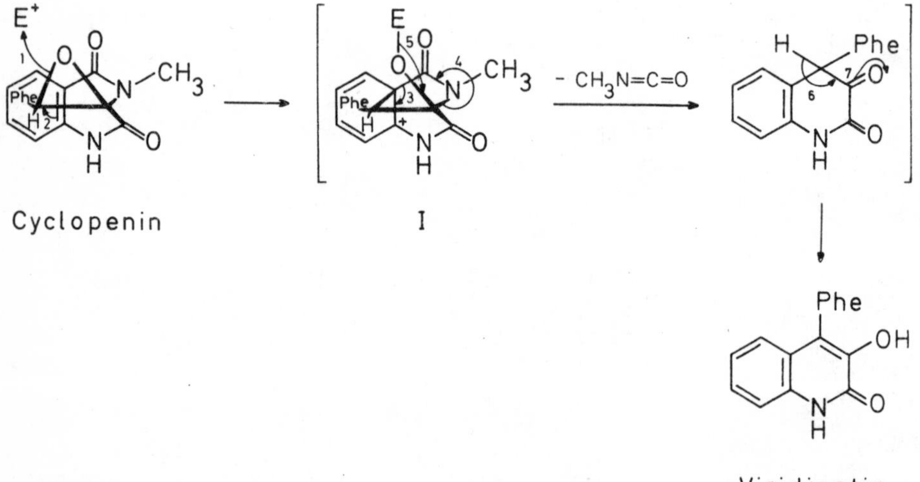

Fig. 242. Possible sequence of reactions for the conversion of cyclopenin to viridicatin

(C 2.1.1). The double bond introduced is epoxidized and the alkaloid cyclopenin formed may be hydroxylated in the unique m-position. The hydroxyl group is introduced by a monooxygenase. The reaction shows the typical NIH-shift (C 2.6.5).

Cyclopenin and cyclopenol may be rearranged to viridicatin and viridicatol, respectively. In this reaction the carbonyl group in position 5 of the benzodiazepine ring system and the $>$N—CH$_3$ group are lost (Fig. 242).

The rearrangement is probably initiated by the addition of an electron-attracting group to the epoxide oxygen. In a concerted mechanism the tricyclic compound I may be formed, which yields methylisocyanate and viridicatin during decomposition. Methylisocyanate may be hydrolyzed to methylamine and CO_2, the products observed on the action of cyclopenase.

Another group of benzodiazepines is formed in Actinomycetes (antibiotics of the tomaymycin-anthramycin type, Fig. 243). These compounds are built from anthranilic acid and L-DOPA. The O- and C-methyl groups as well as the carboxamido C-atom are derived from the CH$_3$-group of L-methionine. Probably the ring of L-DOPA is split by a dioxygenase and two carbon atoms are lost. It is not yet known, however, whether degradation of the ring occurs before or after benzodiazepine formation, or whether the ring is first split between C-atoms 2 and 3 or 4 and 5.

References for Further Reading

292, 303, 354, 481, 483, 666

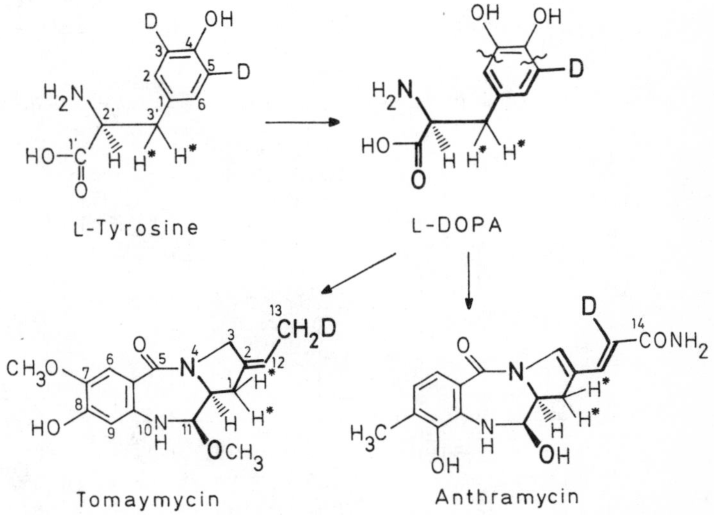

Fig. 243. Biosynthesis of tomaymycin and anthramycin from D-labelled L-tyrosine

22*

D 22 Secondary Products Built from L-Tryptophan

Biosynthesis and Degradation of the Precursor

L-Tryptophan originates from anthranilic acid, 5-phosphoribosyl-1-diphosphate, and L-serine (Fig. 244). An anthranilic acid riboside is built, which is transformed by an Amadori rearrangement to the corresponding ribuloside. This com-

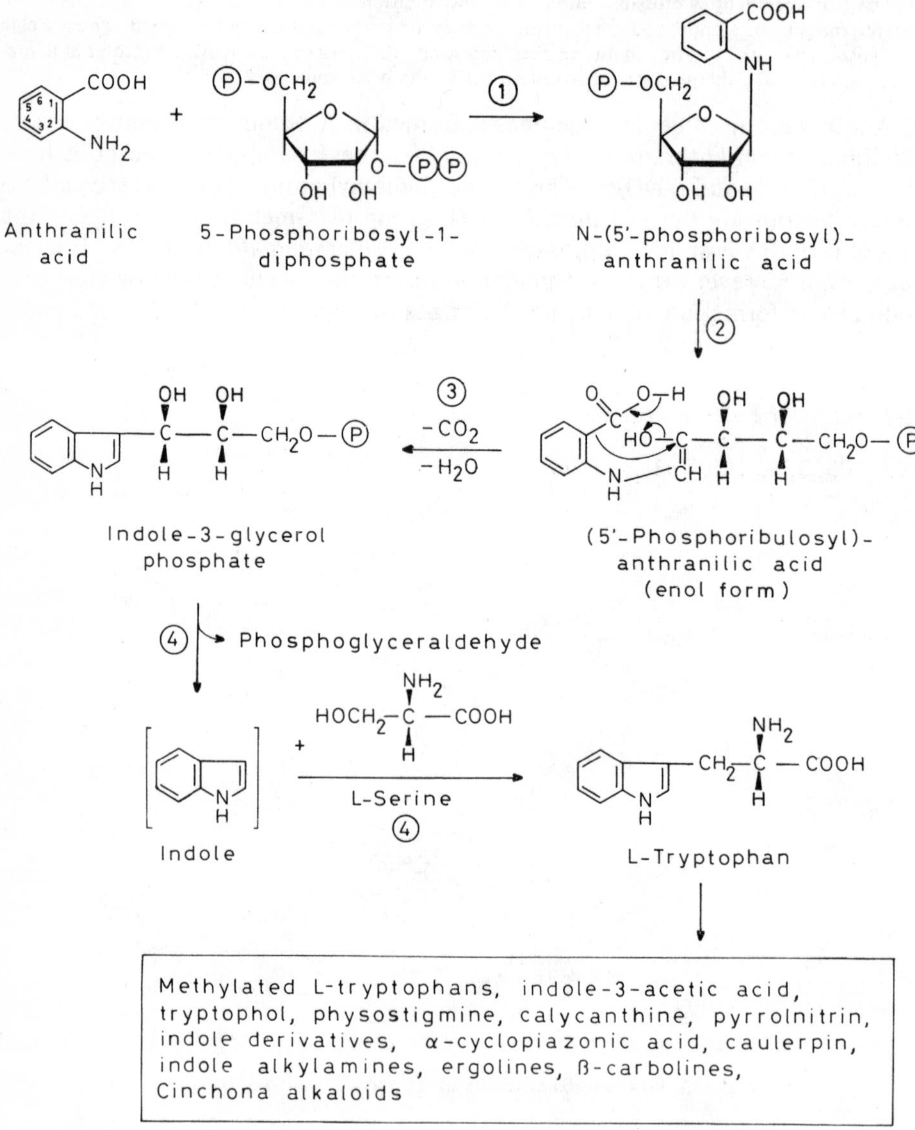

Anthranilic acid   5-Phosphoribosyl-1-diphosphate   N-(5'-phosphoribosyl)-anthranilic acid

Indole-3-glycerol phosphate

(5'-Phosphoribulosyl)-anthranilic acid (enol form)

Phosphoglyceraldehyde

L-Serine

Indole

L-Tryptophan

Methylated L-tryptophans, indole-3-acetic acid, tryptophol, physostigmine, calycanthine, pyrrolnitrin, indole derivatives, α-cyclopiazonic acid, caulerpin, indole alkylamines, ergolines, ß-carbolines, Cinchona alkaloids

Fig. 244. Biosynthesis of L-tryptophan from anthranilic acid
1 Anthranilate phosphoribosyltransferase; *2* phosphoribosyl anthranilate isomerase; *3* indole-3-glycerol phosphate synthase; *4* tryptophan synthase

pound cyclizes to indole-3-glycerol phosphate which, by exchange of the side chain, yields L-tryptophan.

Degradation of L-tryptophan in most organisms proceeds via L-kynurenine, 3-hydroxy-L-kynurenine and 3-hydroxyanthranilic acid, to acetyl CoA und CO_2 (Fig. 245). Anthranilic acid formed from L-kynurenine may be recycled to L-tryptophan (see above). The ring of 3-hydroxyanthranilic acid is cleaved by a dioxygenase (C 2.5). The formed 2-amino-3-carboxymuconic-6-semialdehyde, on the one hand, undergoes an E-Z-isomerization of the Δ^2-double bond and cyclization to quinolinic acid which subsequently is decarboxylated to nicotinic acid. (In microorganisms and plants, however, nicotinic acid is synthesized from aspartic acid and D-glyceraldehyde-3-phosphate, D 16.1.) On the other hand, 2-amino-3-carboxymuconic acid-6-aldehyde may be decarboxylated and is then the immediate precursor of NH_3, acetic acid, and CO_2.

In a group of bacteria tryptophan degradation follows another pathway and yields indole, pyruvic acid, and ammonia (Fig. 246).

Chemistry and Distribution

Secondary products derived from L-tryptophan and products of its degradation occur in microorganisms, plants, and animals. Important groups are listed in Figs. 244 and 245. Most of them still possess the indole ring system. Some compounds, however, are quinoline, pyrrole, or benzene derivatives. Additional rings may be present yielding complicated structures, like that of ergoline and β-carboline alkaloids (cf. the formulas of ergotamine, Corynanthe, Strychnos, Iboga, and Aspidosperma-type alkaloids).

Groups of Compounds Derived from L-Tryptophan

a) N-Methyl-L-tryptophan (abrine) and N, N-dimethyl-L-tryptophan (D 8.1) are formed from L-tryptophan and S-adenosyl-L-methionine by the action of methyltransferases (C 3.3). They are constituents of the seeds of *Abrus precatorius*. N, N-Dimethyl-L-tryptophan inhibits the germination of plant seeds (E 2.2). Hypaphorine (D 8.1) found in the seeds of *Pterocarpus officinalis* acts as a feeding deterrent (E 5.5.3).

b) Indole-3-acetic acid (β-indoleacetic acid) is formed either via indole-3-acetamide (C 2.6.3), via tryptamine (D 22.1), or via indolepyruvic acid (Fig. 247). It is a growth hormone of higher plants formed in the shoot tips (E 3.1), but is built also in certain microorganisms and in animals where it is without physiological activity. Mammals excrete indole-3-acetic acid with the urine (E 1). Leaf-cutting ants, but also plant gall-producing microorganisms and insects synthesize indole-3-acetic acid as an ecological factor. The transformation of plant cells with the Ri-plasmid of *Agrobacterium rhizogenes*, which causes the biosynthesis of indole-3-acetic acid, leads to the formation of "hairy roots". The excess indole-3-acetic acid (together with extra N^6-dimethylallyladenine, D 25.1), built in plant cells transformed with the Ti-plasmid of *Agrobacterium tumefaciens*, causes uncontrolled cell division and the formation of tumors

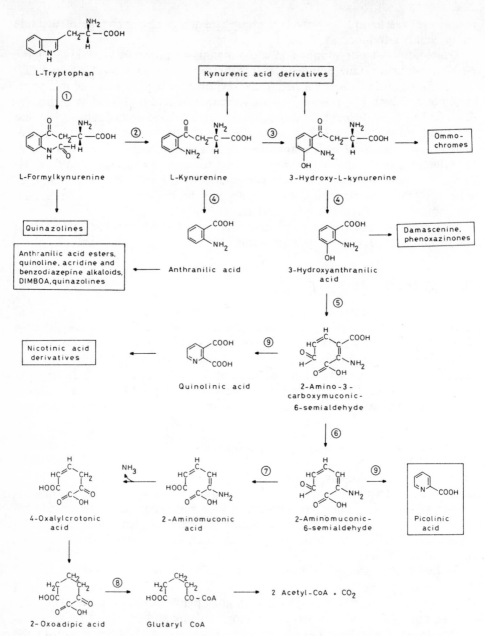

Fig. 245. Degradation of L-tryptophan via L-kynurenine and 3-hydroxyanthranilic acid
1 Tryptophan 2,3-dioxygenase (tryptophan pyrrolase, C 2.5); *2* formamidase; *3* kynurenine 3-mono-oxygenase; *4* kynureninase; *5* 3-hydroxyanthranilate 3,4-dioxygenase; *6* aminocarboxymuconate semialdehyde decarboxylase; *7* aminomuconate semialdehyde dehydrogenase; *8* oxoglutarate dehydrogenase system; *9* spontaneous cyclization

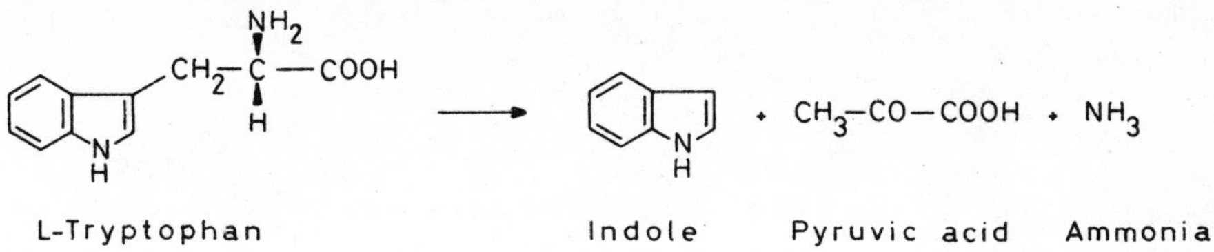

Fig. 246. Formation of indole by the action of tryptophanase on tryptophan

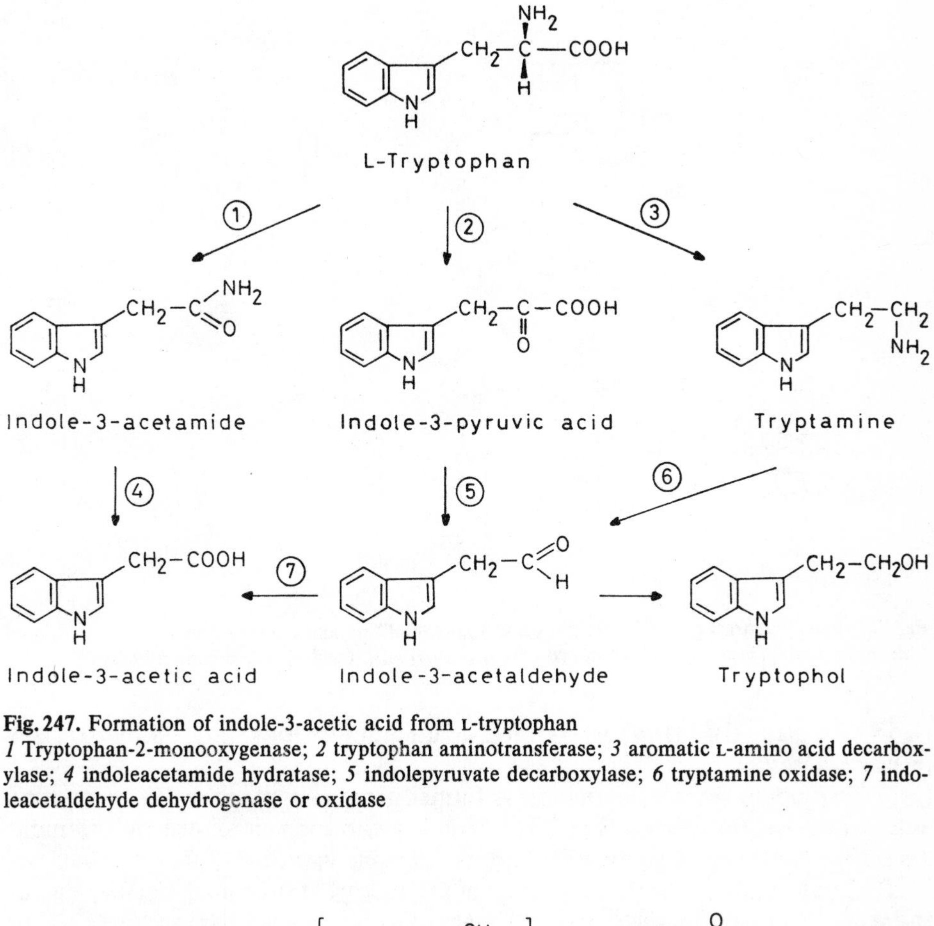

Fig. 247. Formation of indole-3-acetic acid from L-tryptophan
1 Tryptophan-2-monooxygenase; *2* tryptophan aminotransferase; *3* aromatic L-amino acid decarboxylase; *4* indoleacetamide hydratase; *5* indolepyruvate decarboxylase; *6* tryptamine oxidase; *7* indoleacetaldehyde dehydrogenase or oxidase

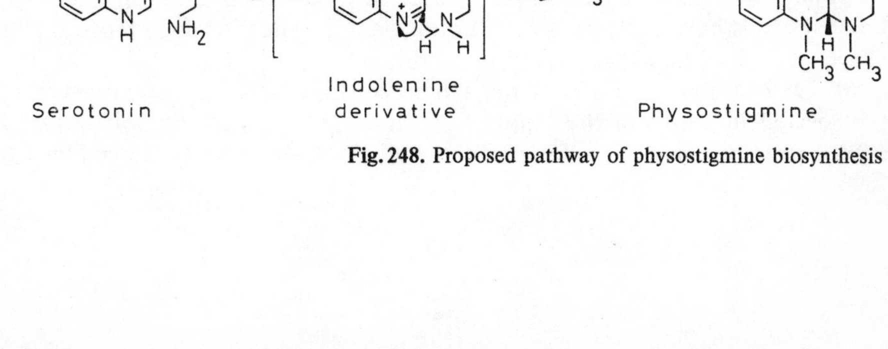

Fig. 248. Proposed pathway of physostigmine biosynthesis

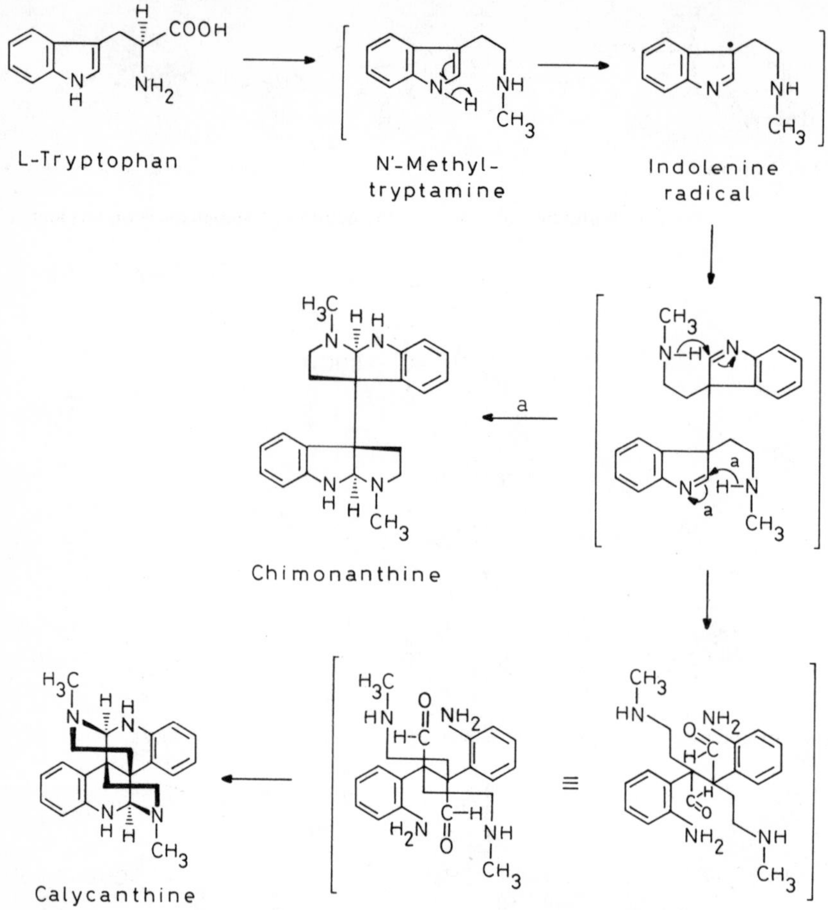

Fig. 249. Possible pathway for the formation of chimonanthine and calycanthine
The configuration of the alkaloids shown is that occurring in the frog *Phyllobates terribilis*

(E 5.4). In plant tissues indole-3-acetic acid is rapidly broken down by oxidases or peroxidases.

c) Tryptophol (indole-3-ethanol) is formed from L-tryptophan via tryptamine and indole-3-acetaldehyde (Fig. 247). It is a phytotoxin produced by the mold *Drechslera nodulosum* infecting goosegrass *(Eleusine indica)* (E 5.4).

d) Physostigmine (eserine), an ester of methylcarbaminic acid, occurs, e.g., in the seeds of *Physostigma venenosum* (Calabar beans). It is probably formed via an indolenine derivative from serotonin (Fig. 248). Physostigmine repels most potential animal predators (E 5.5.3). It is used in medicine (F 2). In Western Africa Calabar beans have been (or are?) administered in ordeal trials to prove the innocence or guilt of persons accused of a crime.

e) Calycanthus alkaloids of the chimonanthine and calycanthine types are synthesized in plants of the family Calycanthaceae and in the poison-dart frog *Phyllobates terribilis*. The skeleton of these alkaloids is probably formed by the di-

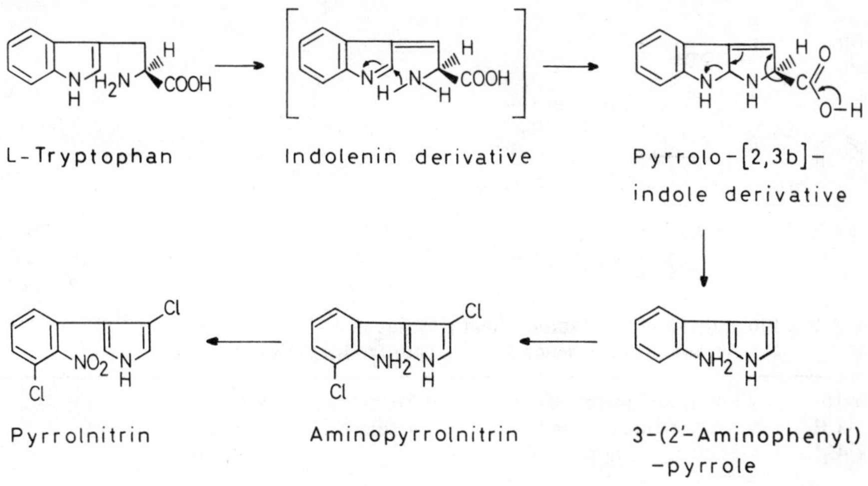

L-Tryptophan   Indolenin derivative   Pyrrolo-[2,3b]-
indole derivative

Pyrrolnitrin   Aminopyrrolnitrin   3-(2'-Aminophenyl)
-pyrrole

Fig. 250. Possible pathway of pyrrolnitrin biosynthesis

merization of indolenine radicals (Fig. 249). The alkaloids present in the skin secretions of the frogs are the optical antipodes of the corresponding compounds
from the plant kingdom. They act as feeding deterrents (E 5.1).

f) Pyrrolnitrin, an antifungal antibiotic (E 5.2), is produced in various species of *Pseudomonas*. It is built from L-tryptophan probably by the pathway shown
in Fig. 250.

g) α-Cyclopiazonic acid and structurally related compounds are toxins produced, e.g., in certain strains of *Aspergillus* and *Penicillium*. They are derived
from L-tryptophan, a prenyl residue, and two molecules of acetate. The compounds resemble structurally the ergoline alkaloids (D 22.2).

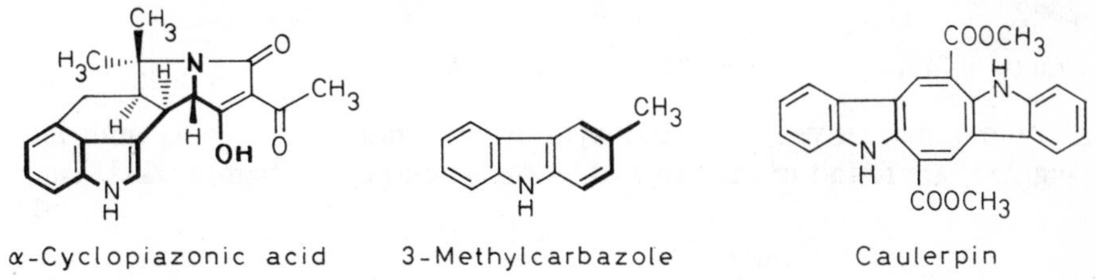

α-Cyclopiazonic acid   3-Methylcarbazole   Caulerpin

h) Caulerpin, a dimeric tryptophan derivative has been isolated from green algae.

i) Indole alkylamines, ergolines, β-carbolines, and *Cinchona* alkaloids are
treated in D 22.1–D 22.4.

Groups of Compounds Formed from L-Kynurenine

a) 4-Methylquinazoline, 2,4-dimethylquinazoline, and 2-ethyl-4-methylquinaz-
oline are formed in *Pseudomonas*. Probably L-kynurenine is an intermediate. It

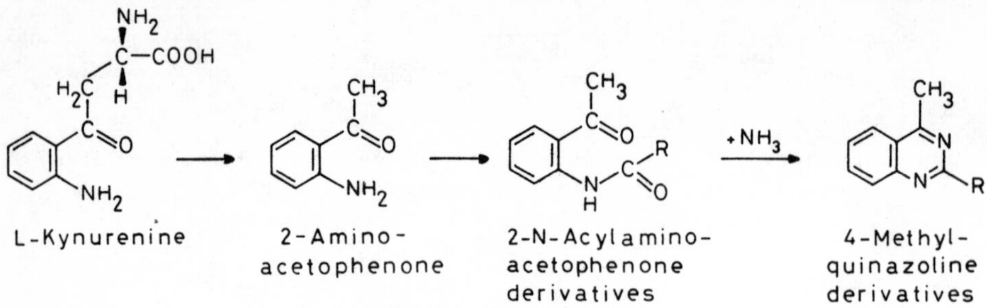

Fig. 251. Biosynthesis of quinazolines from L-kynurenine

R	2-N-Acylaminoacetophenones	Quinazoline derivatives
—H	2-N-Formylaminoacetophenone	4-Methylquinazoline
—CH₃	2-N-Acetylaminoacetophenone	2,4-Dimethylquinazoline
—C₂H₅	2-N-Propionylaminoacetophenone	2-Ethyl-4-methylquinazoline

may be transformed to 2-aminoacetophenone and *N*-acylated 2-amino acetophenones which are able to cyclize in the presence of ammonia (Fig. 251).

b) Ommochromes, a group of protein-bound pigments with a phenoxazinone ring system are derived from 3-hydroxy-L-kynurenine. They are yellow-brown in the oxidized form and red-violet in the reduced form. Ommochromes are abundant in arthropods, e.g., worms, crabs, and insects and may be located, for instance, in the eyes, the integuments, and the wings. Xanthommatin, a representative found frequently in insects, is formed from 3-hydroxy-L-kynurenine according to Fig. 252 (cf. the biosynthesis of phenoxazinones from 3-hydroxyanthranilic acid, D 21.4.1, which, like cinnabaric acid, are also found in certain insects). Ommochromes form a screen to damaging radiation (E 2.2). Xanthommatin acts as a cofactor of dihydrolipoamide reductase (C 2.7).

c) Kynurenic acid derivatives are described in D 22.5.

Groups of Secondary Products Built from Indole

a) Indole derivatives, like indoxyl, urinary indican, indigo, and indirubin (Fig. 253) are found in the urine of animals and human beings (E 1). They origi-

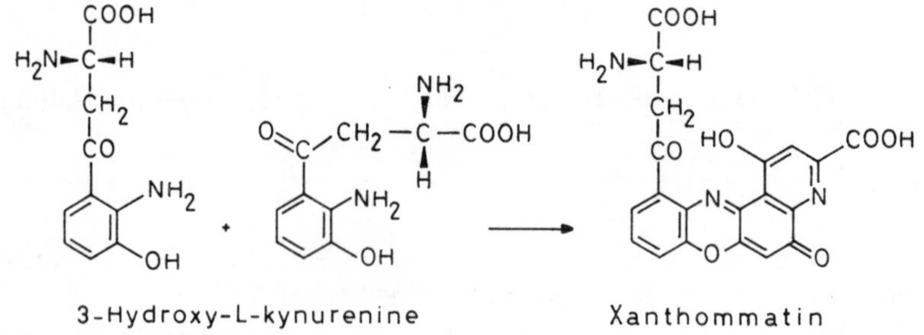

Fig. 252. Formation of xanthommatin from 3-hydroxy-L-kynurenine by monophenol monooxygenase in the presence of small amounts of L-DOPA

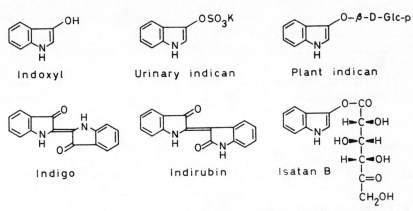

Fig. 253. Structural formulas of indoxyl derivatives

nate from indole formed by the enzyme tryptophanase of intestinal bacteria. Indoxyl is built from indole in liver mitochondria. It may be either transformed to urinary indican or oxidatively to the dimeric products indigo and indirubin.

b) Plant indican, a characteristic constituent of *Indigo* sp. and isatan B (indoxyl-5-oxogluconic acid, a compound from *Isatis tinctoria*, liberates indoxyl upon hydrolysis. Indoxyl is oxidized by atmospheric oxygen to indigo. Natural indigo has been of great importance as a blue dye.

c) Indoxyl sulfates, which are in part brominated, are produced in a special gland of marine mollusks, e.g., of the genus *Murex* (Fig. 254). Hydrolysis of these compounds yields indoxyl derivatives which may be dehydrogenated to indoleninones. Radiation of the built quinhydrone complex cleaves the substituents in

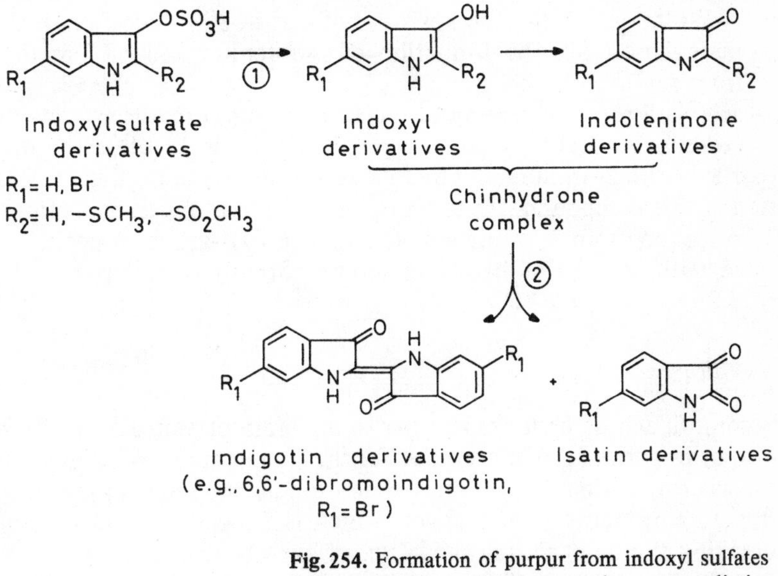

Indoxylsulfate derivatives

$R_1 = H, Br$
$R_2 = H, -SCH_3, -SO_2CH_3$

Indoxyl derivatives

Indoleninone derivatives

Chinhydrone complex

Indigotin derivatives
(e.g., 6,6'-dibromoindigotin, $R_1 = Br$)

Isatin derivatives

Fig. 254. Formation of purpur from indoxyl sulfates
1 Purpurase; *2* spontaneous reaction upon radiation

position 2 of the indole nucleus and causes dimerization. The mixture of indigotin and isatin derivatives formed has a red-violet tint. It is designated as purpur and has been the most appreciated dye in antiquity and the Middle Ages in Europe.

d) Carbazole alkaloids, e.g., 3-methylcarbazole, occur in Rutaceae. They are probably derived from indole and an isoprenoid C_5-unit (cf. the bold part in the formula of 3-methylcarbazole at p. 345).

References for Further Reading

31, 44, 48, 67, 150, 172, 190, 203, 264, 369, 395, 415, 436, 743, 806, 882

D 22.1 Indole Alkylamines

Chemistry and Distribution

Most indole alkylamines are derived from indole ethylamine (tryptamine). In addition, however, a few indole methylamines, e.g., gramine, occur in plants. Indole alkylamine derivatives are synthesized in microorganisms, plants, and animals.

Biosynthesis

Decarboxylation of L-tryptophan and 5-hydroxy-L-tryptophan yields tryptamine and serotonin, respectively (Fig. 255). These amines may be further modified by
— hydroxylation (cf. the biosynthesis of serotonin or 6-hydroxytryptamine from tryptamine),
— *N*-oxidation (cf. the synthesis of bufotenine-N^1-oxide from bufotenine),
— methylation (cf. the formation of bufotenine and of 5-methoxytryptamine from serotonin), and
— acetylation (cf. formation of melatonin from 5-methoxytryptamine).

Gramine is derived from L-tryptophan, e.g., in seedlings of *Hordeum vulgare* (barley), with 3-aminomethylindole as an intermediate (Fig. 256). During formation of this compound, the amino group of L-tryptophan shifts from carbon atom 2′ to carbon atom 3′ of the side chain. The CH_2-group in position 3′ is incorporated without loss of hydrogen atoms. Gramine undergoes a rapid turnover (A 5.2).

Significance

Serotonin acts as neurotransmitter in the brain of animals (E 3.2). It is, however, also of importance as a hormone for the maintenance of normal gut mobility in the gastrointestinal tract (E 3.1). Serotonin accumulates in the trichomal fluid of the stinging nettle *(Urtica dioica)* which is "injected" into the body of animals breaking the hairs during predation of even touching. In addition to the other constituents of the nettle toxin, e.g., acetylcholine (D 10) and histamine (D 20),

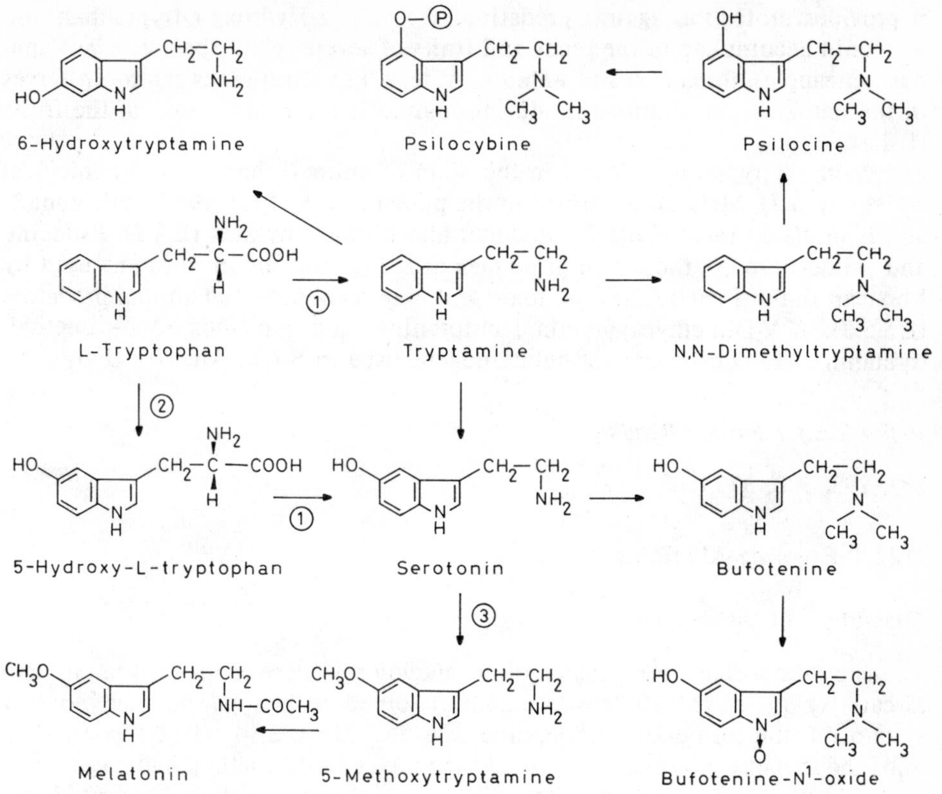

Fig. 255. Formation of indole ethyl amines
1 Aromatic-L-amino acid decarboxylase; *2* tryptophan 5-monooxygenase (C 2.6.5), 5-methyltetrahydrofolate-dependent methyltransferase (C 3.2)

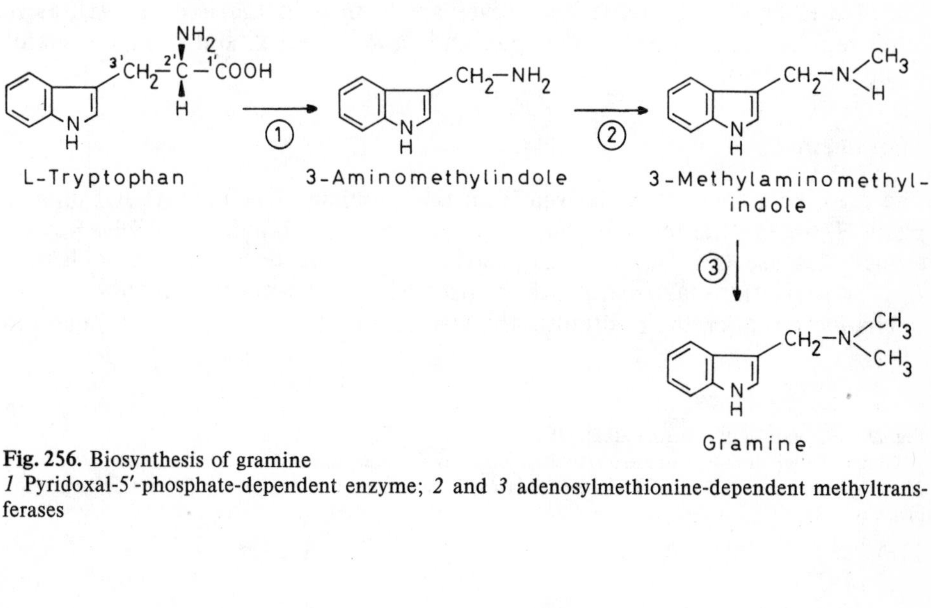

Fig. 256. Biosynthesis of gramine
1 Pyridoxal-5'-phosphate-dependent enzyme; *2* and *3* adenosylmethionine-dependent methyltransferases

it provides protection against predation (E 5.5.3). 5-Hydroxy-L-tryptophan and serotonin accumulate in the seeds and fruits of several plant species, e.g., bananas, pineapples, tomatoes, and walnuts. At least in walnut seeds serotonin serves in the detoxication of ammonia during maturation and abscission of the fruits (E 2.2).

5-Methoxytryptamine, found in the brain of animals, has neurophysiological activity (E 3.2). Melatonin, formed in the pineal gland, effects the female gonads and is probably responsible for seasonal and diurnal rhythms (E 3.1). Psilocine and psilocybine are the active principles of hallucinogenic mushrooms used by Mexican Indians (F 3). They are toxic and may repel potential animal predators (E 5.5.3). *N,N*-Dimethyltryptamine, bufotenine, and 5-methoxy-*N,N*-dimethyltryptamine are constituents of hallucinogens used in South America (F 3).

References for Further Reading

292, 293, 300, 375, 668

D 22.2 Ergoline Alkaloids

Chemistry and Distribution

The most important group of ergoline alkaloids is derived from lysergic acid (8-carboxy-6-methyl-9,10-dehydroergoline) joined with a peptide-derived ring system (cf. the formula of ergotamine, Fig. 257). In the so-called clavine alkaloids the carboxy group of lysergic acid is replaced by a methyl, hydroxymethyl, or aldehyde grouping (cf. the formulas of agroclavine or elmyoclavine). Some clavine alkaloids contain an incomplete ergoline ring system (cf. the formula of chanoclavine I).

The most important source of ergoline alkaloids are sclerotia of fungi of the genus *Claviceps. Claviceps* sp. are parasites living on the flowers of rye and a number of wild grasses. Ergoline derivatives are formed in *Claviceps* as well as in other fungi, e.g., *Aspergillus, Rhizopus,* and *Penicillium* sp., and in higher plants (Convulvulaceae).

Biosynthesis

The ergoline alkaloids are derived from L-tryptophan, 3,3-dimethylallyl diphosphate (D 6), and the methyl group of *S*-adenosyl-L-methionine (D 12.1). Key intermediates are the tricyclic chanoclavine I and the tetracyclic agraoclavine. *E-Z*-Isomerizations take place in the transformation of 4-dimethylallyltryptophan to chanoclavine (cf. the position of the ■-labeled methyl group, Fig. 257) and in

▶

Fig. 257. Formation of ergoline alkaloids
1 Dimethylallyl diphosphate: L-tryptophan dimethylallyltransferase (DMAT synthase); *2* DMAT *N*-methyltransferase; *3* chanoclavine-I-cyclase; *4* agroclavine hydroxylase; *5* lysergic acid-thiol ligase (C 1.2)

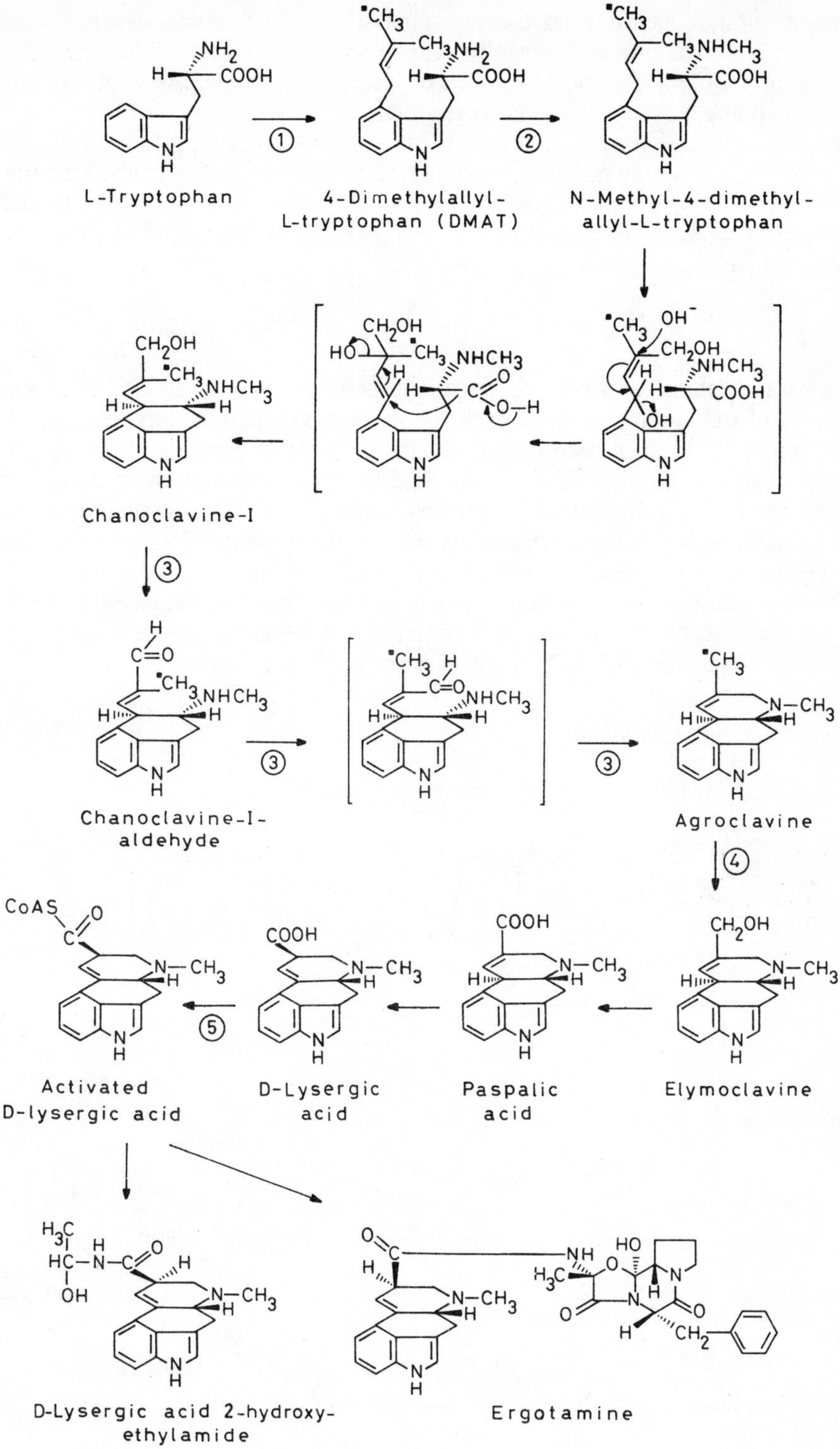

L-Tryptophan

4-Dimethylallyl-
L-tryptophan (DMAT)

N-Methyl-4-dimethyl-
allyl-L-tryptophan

Chanoclavine-I

Chanoclavine-I-
aldehyde

Agroclavine

Activated
D-lysergic acid

D-Lysergic
acid

Paspalic
acid

Elymoclavine

D-Lysergic acid 2-hydroxy-
ethylamide

Ergotamine

the synthesis of agroclavine from chanoclavine. The formation of D-lysergic acid proceeds via elymoclavine and paspalic acid to activated lysergic acid. The activation is a prerequisite for the formation of lysergic acid peptides, e. g., ergotamine, and D-lysergic acid 2-hydroxyethylamide.

Formation of the peptide part proceeds on a polyfunctional protein (A 3.1.1). The key intermediate may be a linear tripeptide bound covalently to the enzyme proteins (Fig. 258). Cyclization probably starts with the elimination of the enzyme. Hydroxylation of the alanyl residue precedes the formation of the characteristic cyclol ring.

Significance

Ergotamine and related alkaloids, as well as derivatives obtained by chemical hydrogenation of the isolated double bond in the lysergic acid moiety, e.g., dihydroergotamine, are used in medicine (F 2). Large doses of ergoline alkaloids are toxic to animals and man (E 5.5.3). In the Middle Ages in Europe intoxication of human beings with ergoline alkaloids was widespread and in those years rye was heavily infected with *Claviceps purpurea*, it had the dimensions of a plague. Seeds of *Ipomoea* sp. (Convulvulaceae), which contain lysergic acid amide and lysergic acid α-hydroxyethylamide, are used as an hallucinogen by Mexican Indians (F 3). Lysergic acid diethylamide (LSD), derived synthetically from lysergic acid, is one of the most powerful hallucinogens known (F 3).

References for Further Reading

233, 236, 292, 293, 668

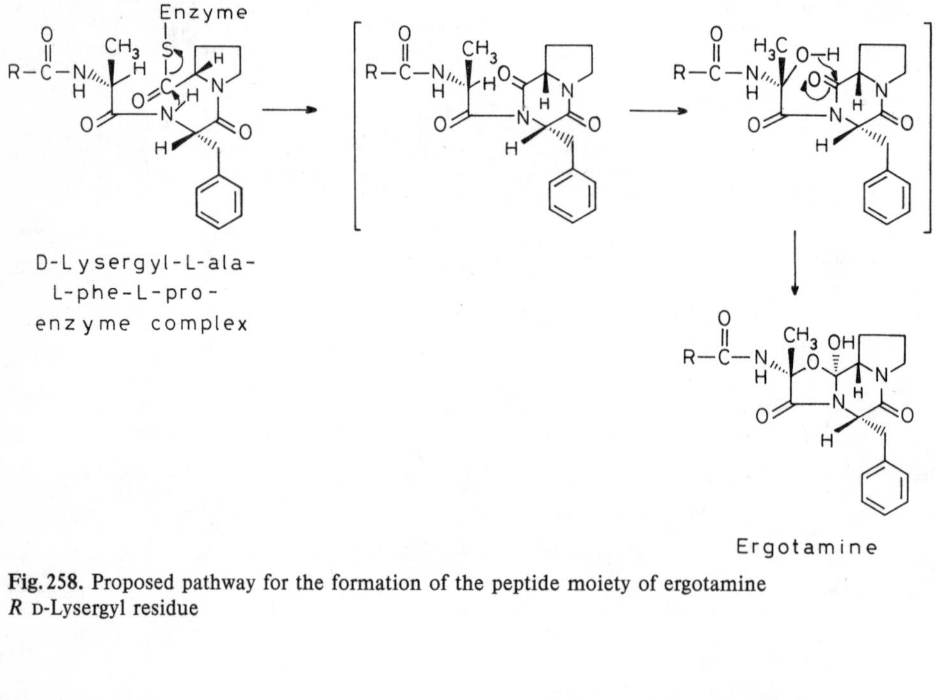

D-Lysergyl-L-ala-
L-phe-L-pro-
enzyme complex

Ergotamine

Fig. 258. Proposed pathway for the formation of the peptide moiety of ergotamine
R D-Lysergyl residue

D 22.3 β-Carbolines and Related Alkaloids

Chemistry and Distribution

β-Carboline alkaloids possess the tricyclic ring system shown in the formula below. Often the six-membered nitrogen-containing ring is hydrogenated (1,2,3,4-tetrahydro-β-carbolines). Additional rings may be present (cf. the formulas of the iridoid carbolines in Fig. 260).

β-Carboline   1,2,3,4-Tetrahydro-
                     β-carboline

Many β-carboline alkaloids have been isolated from plants, e.g., Apocynaceae, Loganiaceae, and Rubaceae. More than 1200 representatives of this group are known today. A few tetrahydro-β-carbolines have been found in mammalian tissues. Harman has been identified in cultures of a mushroom.

Biosynthesis

The β-carboline ring system is formed by a Mannich condensation (D 8.5) with the participation of tryptamine. According to the carbonyl component involved, two types of alkaloids may be distinguished:

A. Harman Alkaloids. Harman is formed in plants by the reaction of tryptamine with pyruvic acid on the pathway shown in Fig. 259. 1,2,3,4-Tetrahydro-β-carboline has been found in the urine of humans (E 1). It is probably formed from tryptamine and formaldehyde.

B. Iridoid Indole Alkaloids. Most β-carbolines are derived from tryptamine and the iridoid secologanin (D 6.2.2). In dependence on the structure of the iridoid part alkaloids of the Corynanthe-Strychnos type as well as of the Aspidosperma-Hunteria and Iboga types may be distinguished. The latter are formed by rearrangement of the iridoid part which is shown schematically in Fig. 260 and in detail in Fig. 262.

A key intermediate in the biosynthesis of all iridoid indole alkaloids is the glucoside strictosidine (isovincoside, Fig. 261). Strictosidine is the precursor of

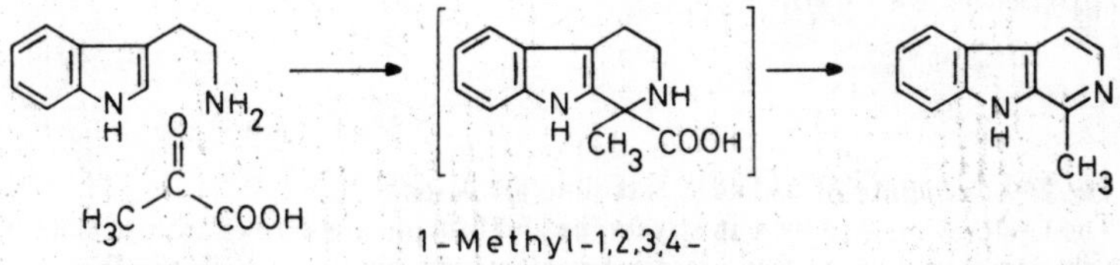

Tryptamine +   1-Methyl-1,2,3,4-
Pyruvic acid   tetrahydro-β-carboline-   Harman
                      1-carboxylic acid

Fig. 259. Biosynthesis of harman

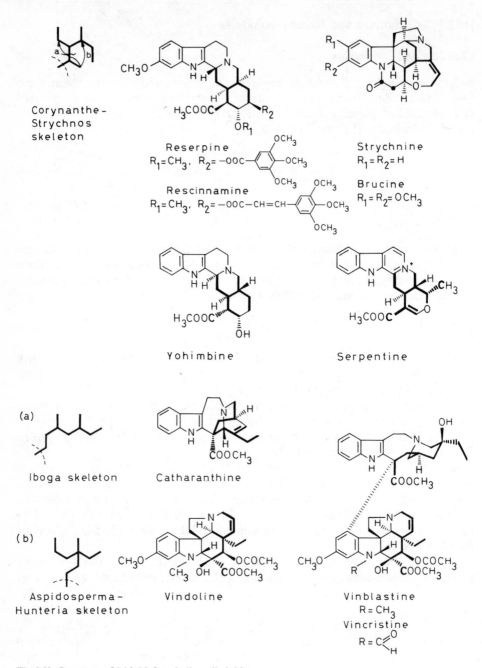

Fig. 260. Structure of iridoid β-carboline alkaloids

The C-atoms marked by ● in strychnine and brucine are derived from acetate. Vinblastine (vincaleucoblastine) and vincristine are dimeric alkaloids containing both the Iboga and the Aspidosperma-Hunteria skeletons

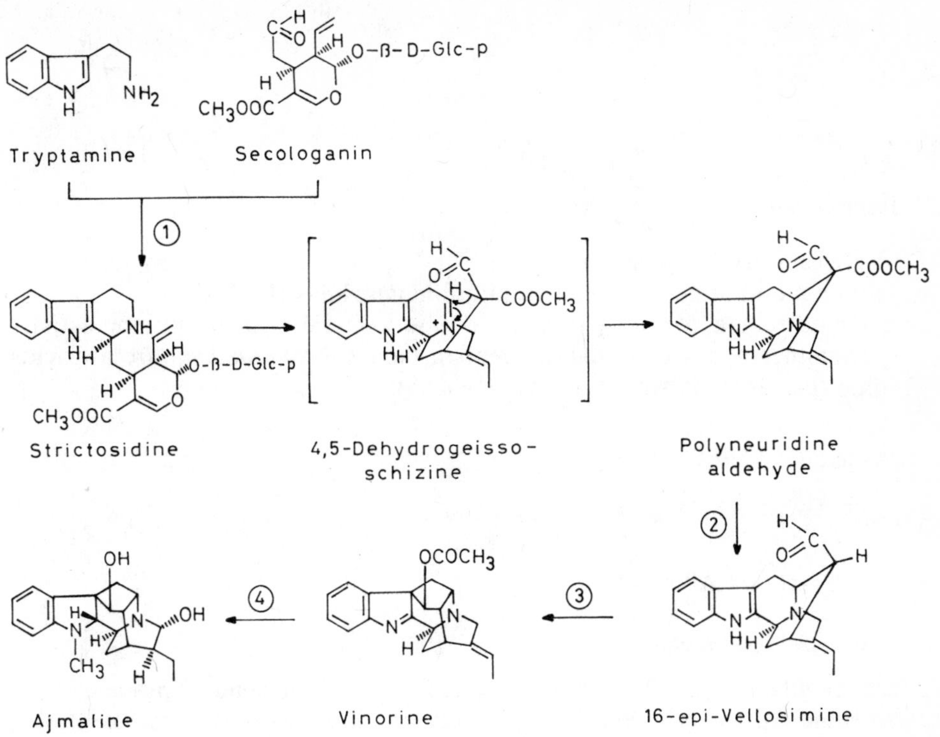

Fig. 261. Biosynthesis of ajmaline
1 Strictosidine synthase; *2* polyneuridine aldehyde esterase, spontaneous decarboxylation; *3* vinorine synthase; *4* monooxygenase, esterase, reductase(s), methyltransferase

ajmaline as well as of ajmalicine. The latter alkaloid may be further transformed to stemmadenine, tabersonine, vindoline, and catharanthine (Fig. 262). The dimeric alkaloids, like vinblastine (vincaleucoblastine) and vincristine (Fig. 260), are derived from monomeric precursors, e. g., catharanthine and vindoline.

Significance

Most β-carboline alkaloids are toxic to animals and may repel potential predators (E 5.5.3). Reserpine, rescinnamine, yohimbine, vinblastine, vincristine, and strychnine are used in the materia medica (F 2). 2-Methyl- and 1,2-dimethyl-6-methoxytetrahydro-β-carboline, harmine, harmaline, and tetrahydroharmine as well as ibogaine are constituents of hallucinogenic preparations used by American and African natives (F 3).

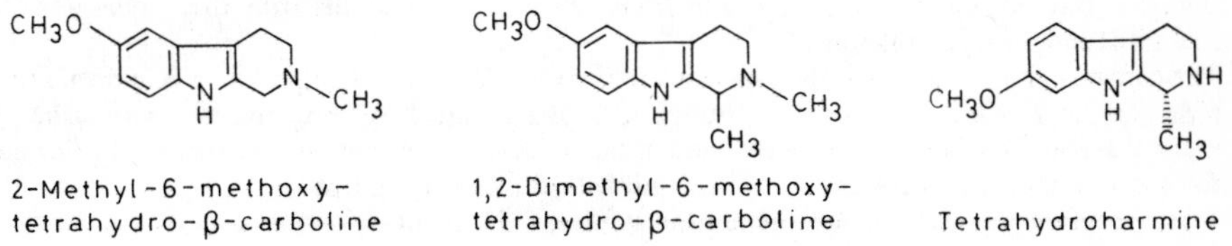

2-Methyl-6-methoxy-
tetrahydro-β-carboline

1,2-Dimethyl-6-methoxy-
tetrahydro-β-carboline

Tetrahydroharmine

23*

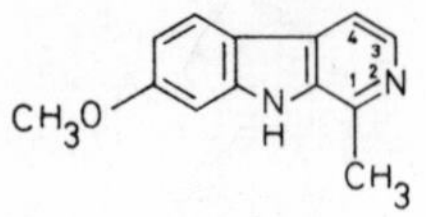

Harmine 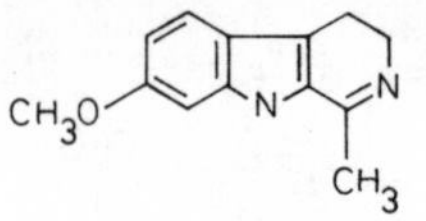   Harmaline 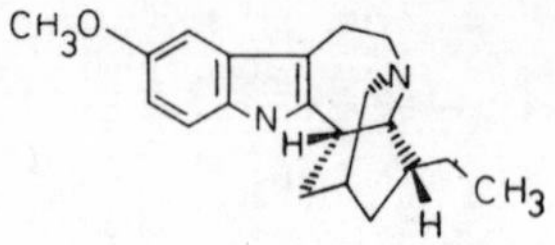   Ibogaine

Transformation

Derivatives of iridoid β-carbolines are:

a) Quinine, quinidine, and related quinoline alkaloids (D 22.4) and

b) camptothecin a pyrrolo [3,4-b] quinoline alkaloid with antitumor activity. Camptothecin is synthesized in the plant *Camptotheca acuminata* from strictosidine (Fig. 261) via strictosamide (Fig. 263).

References for Further Reading

19, 293, 354, 375, 377, 444, 668, 691, 736

D 22.4 Cinchona Alkaloids

Chemistry and Distribution

Cinchona alkaloids are formed in Rubiaceae, e.g., in the genus *Cinchona*, as well as in some Annonaceae and Loganiaceae. The most important representatives are derivatives of quinoline. In these compounds the quinoline ring is linked at position 4 via a secondary alcoholic group to the so-called quinuclidine nucleus, a system of two piperidine rings sharing a nitrogen atom and three C-atoms. A number of differently substituted and isomeric alkaloids exist. Of special significance are quinine and quinidine.

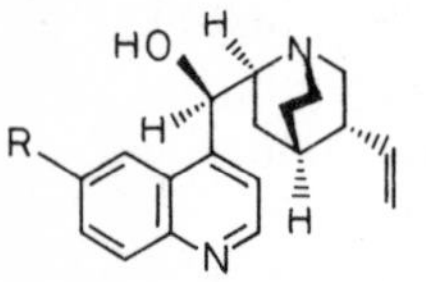

Cinchonidine (R = H)
Quinine   (R = OCH₃)

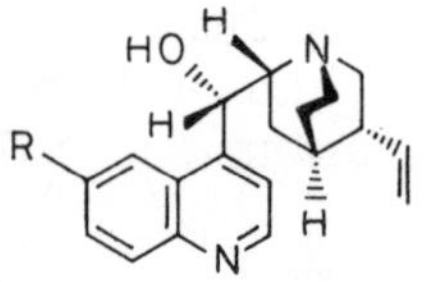

Cinchonine (R = H)
Quinidine (R = OCH₃)

Indole alkaloids occur in addition to the quinoline derivatives. They carry the quinuclidine ring system in position 2 (cf. the formula of cinchonamine, Fig. 264).

▶

Fig. 262. Biosynthesis of ajmalicine and transformation to alkaloids with Iboga-type and Aspido-sperma-Hunteria-type skeletons

The conversion of the Corynanthe-type skeleton present in stemmadenine to the Aspidosperma-Hunteria- or the Iboga-type skeletons requires cleavage of the bond between the two carbon atoms marked ● and the formation of a new bond at the o-position. In the first case the bond is formed with the carbon atom marked ▼ and in the second with the one marked ■

1 Strictosidine glucosidase; *2* cathenamine synthase; *3* ajmalicine synthase

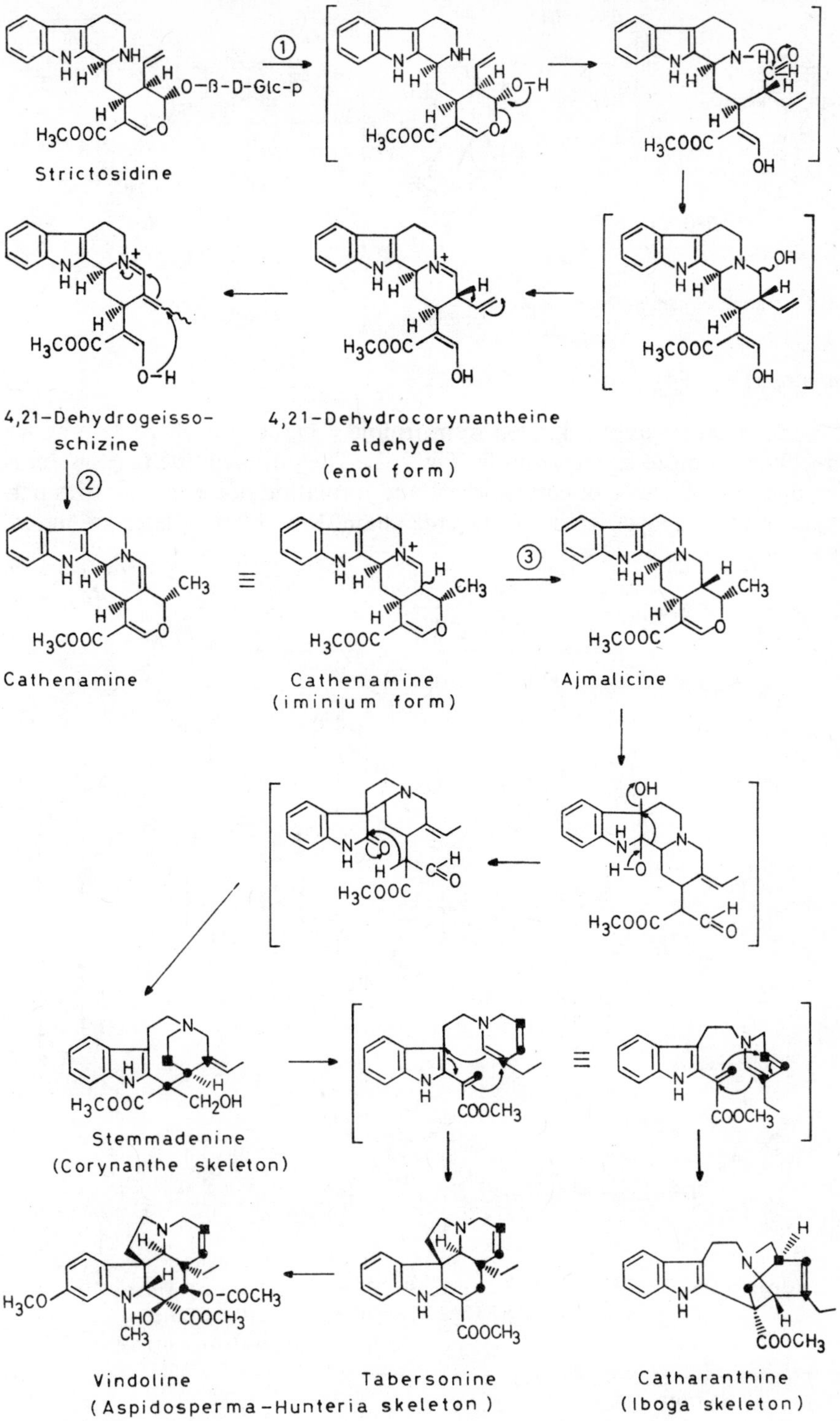

Strictosidine

4,21-Dehydrogeisso-
schizine

4,21-Dehydrocorynantheine
aldehyde
(enol form)

Cathenamine

Cathenamine
(iminium form)

Ajmalicine

Stemmadenine
(Corynanthe skeleton)

Vindoline
(Aspidosperma-Hunteria skeleton)

Tabersonine

Catharanthine
(Iboga skeleton)

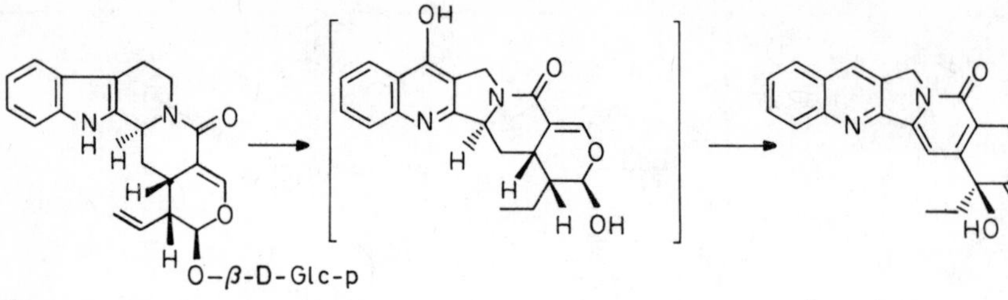

Fig. 263. Formation of camptothecin

Biosynthesis (Fig. 264)

The Cinchona alkaloids are derived from iridoid β-carbolines (D 22.3). The actual precursor seems to be stictosidine (Fig. 261). The quinuclidine ring system is built by opening of ring C of corynantheal and formation of a new ring with participation of the aldehyde group of the side chain. Key intermediates are indole

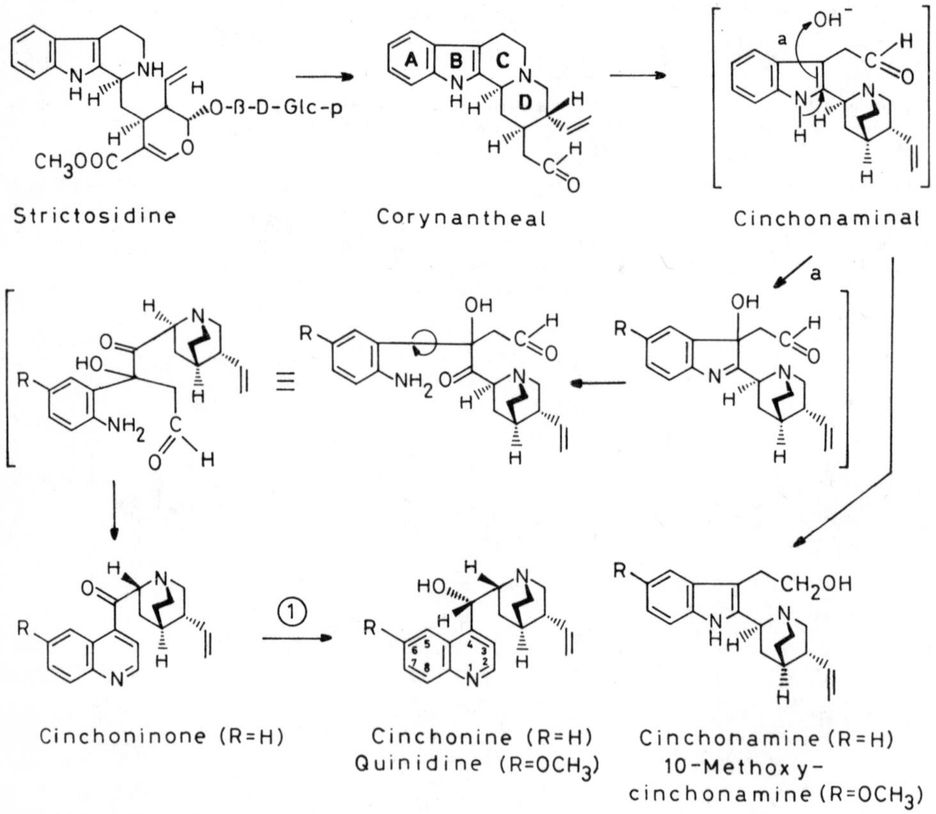

Fig. 264. Biosynthesis of *Cinchona* alkaloids
1 Cinchoninone: NADPH oxidoreductase

alkaloids of the cinchonaminal type. The indole ring of these compounds may be cleaved and the quinoline ring system may be formed after turning the right-hand part of the molecule. From cinchoninone, compounds of the cinchonine as well as of the cinchonidine series are derived.

Significance

The alkaloids have a bitter taste and cause intoxication in animals. Their formation is of ecological significance (E 5.5.3). Quinine and quinidine are used in the materia medica (F 2).

References for Further Reading

157, 293, 668

D 22.5 Kynurenic Acid Derivatives

Chemistry and Distribution

Kynurenic acid is the mother substance of a group of quinolines formed in microorganisms, plants, and animals. Most of these substances carry a carboxy group in position 2 and a hydroxy group in position 4. Additional hydroxy groups may be present.

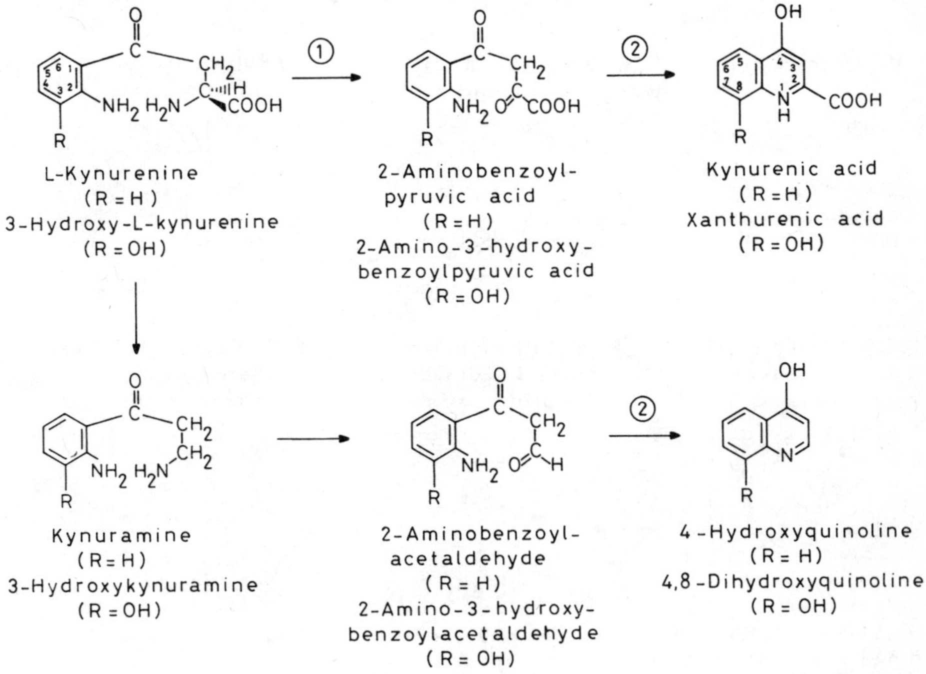

Fig. 265. Formation of quinoline derivatives from L-kynurenine
1 L-kynurenine aminotransferase; *2* spontaneous cyclization

Biosynthesis (Fig. 265)

The precursors of kynurenic acid and its derivatives are L-kynurenine and hydroxylated L-kynurenines, like 3-hydroxy-L-kynurenine (D 22), 5-hydroxy-L-kynurenine (derived from 5-hydroxy-L-tryptophan), 3,4-dihydroxy-L-kynurenine, etc. Quinaldic acid and related substances not hydroxylated in position 4 are derived from the hydroxylated compounds in animals. The reduction is probably performed by the microorganisms living in the intestinal tract. 4-Hydroxyquinoline and derivatives of this compound, which are devoid of the carboxy group in position 2 are derived from kynuramine.

Certain microorganisms degrade L-tryptophan (in some cases also D-tryptophan) via kynurenic acid to compounds of primary metabolism (Fig. 266, the quinolinic pathway of tryptophan degradation, in contrast to the aromatic pathway via 3-hydroxyanthranilic acid shown in Fig. 245).

Significance

Compounds of the kynurenic acid group are end products of L-tryptophan metabolism in mammals, birds, insects, and some plants. They are either excreted with

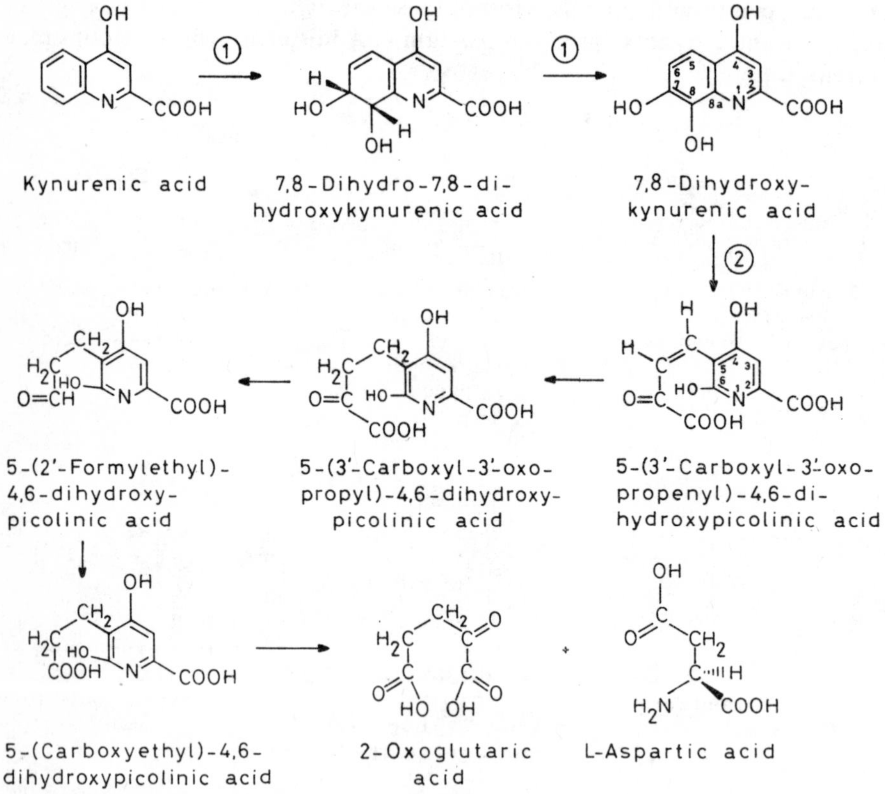

Fig. 266. Degradation of kynurenic acid in microorganisms
1 Kynurenate 7,8-hydroxylase (C 2.5); *2* 7,8-dihydroxykynurenate 8,8a-dioxygenase

the urine (E 1) or are accumulated in the organism. 8-Hydroxyquinaldic acid methyl ester is a feeding deterrent and aseptic of certain water beetles (E 5.1). Xanthurenic acid inhibits the synthesis of ecdysone in crustaceans and in this way participates in the regulation of molting (E 3.1).

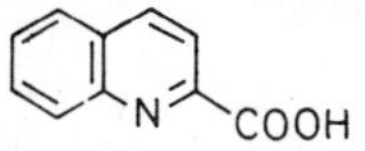

Quinaldic
acid

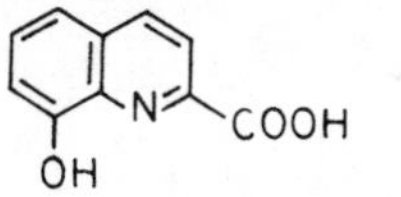

8-Hydroxyquinaldic
acid

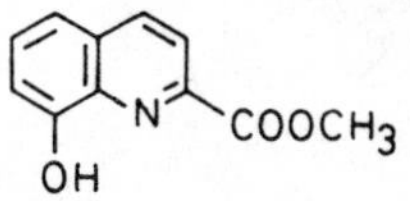

8-Hydroxyquinaldic acid
methyl ester

Reference for Further Reading

475

D 23 Secondary Products Derived from L-Phenylalanine and L-Tyrosine (Phenylpropanoids)

Biosynthesis and Degradation of the Precursors

L-Phenylalanine and L-tyrosine are formed from chorismic acid (D 21) (Fig. 267). Two pathways exist for the biosynthesis of L-tyrosine: the 4-hydroxyphenylpyruvic acid and the L-pretyrosine (arogenic acid) route. Both pathways occur in microorganisms and plants. Higher animals are unable to synthesize L-phenylalanine and L-tyrosine de novo, but hydroxylate L-phenylalanine to L-tyrosine. Certain insects contain colonies of bacteria in the fat body synthesizing L-phenylalanine and L-tyrosine, which may be used by their hosts.

   L-Tyrosine is degraded via 4-hydroxyphenylpyruvic acid and homogentisic acid to acetoacetic acid and fumaric acid (Fig. 268).

Chemistry and Distribution

Secondary substances are derived from L-phenylalanine and L-tyrosine as well as from products of their biosynthesis and degradation in microorganisms, plants, and animals. The most important groups are given in Figs. 267 and 268. They possess the typical carbon skeleton of the so-called phenylpropanoids consisting of a phenyl ring linked to an n-propyl side chain.

Groups of Secondary Products

Derivatives of Prephenic Acid

a) L-2,5-Dihydrophenylalanine, a nonprotein amino acid, is derived from prephenic acid by the pathway outlined in Fig. 269. It is a rather ubiquitous constituent of Actinomycetes and exerts antibiotics activity as an antagonist of L-phenylalanine (E 5.2).

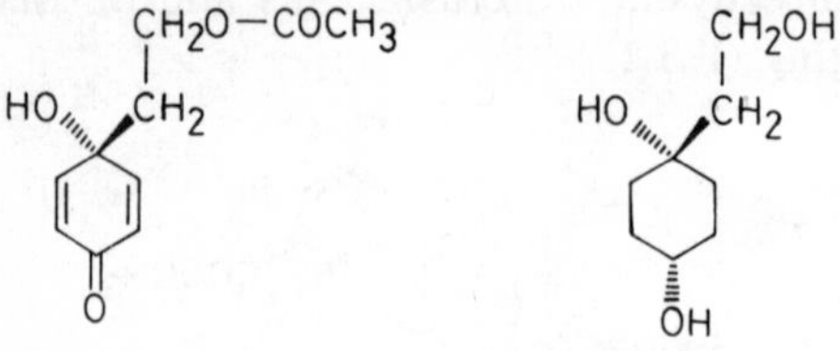

Hallerone   Rengyol

b) Ethylcyclohexanols, e. g., hallerone and rengyol, are constituents of the plant *Halleria lucida*. Structural similarity indicates their derivation from prephenic acid.

Derivative of L-Phenylalanine and L-Tyrosine

a) L-Tropic and atropic acids are phenylpropane derivatives with a branched carbon chain. They are formed in certain plants and are constituents of tropane alkaloids (D 17.2). L-Tropic acid is derived from L-phenylalanine by an intramolecular 1,2-migration of the carboxyl group from C-2′ to C-3′ of the side chain. One of the hydrogen atoms at C-3′ is shifted concomitantly to C-2′ (Fig. 270). Hence, the reaction resembles the formation of methylmalonyl CoA from succinyl CoA by methylmalonyl CoA mutase (D 24.3). Atropic acid originates from tropic acid. It is, however, not known whether its formation takes place before or after esterification with tropane alcohols (D 17.2).

c) L-Di- and L-trityrosine units are constituents of resilin, an elastic protein from insects participating in flight and jumping. Isodityrosine residues occur in the protein extensin present in the primary cell wall of higher plants. The peptide chains of resilin and extensin are cross-linked by these units, which are formed by oxidative coupling of L-tyrosine moieties.

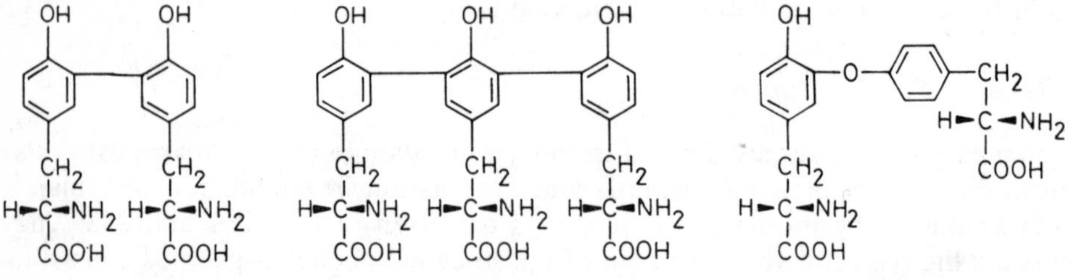

L-Dityrosine   L-Trityrosine   L-Isodityrosine

d) L-3,4-Dihydroxyphenylalanine (L-DOPA) is a nonprotein amino acid formed from L-tyrosine in microorganisms, plants, and animals (Fig. 267). In plants it occurs in the free state as well as bound to D-glucose. It is toxic to animals if ingested and functions as feeding deterrent (E 5.5.3).

e) For further derivatives of L-phenylalanine, L-tyrosine, and L-DOPA see D 23.1—D 23.4.

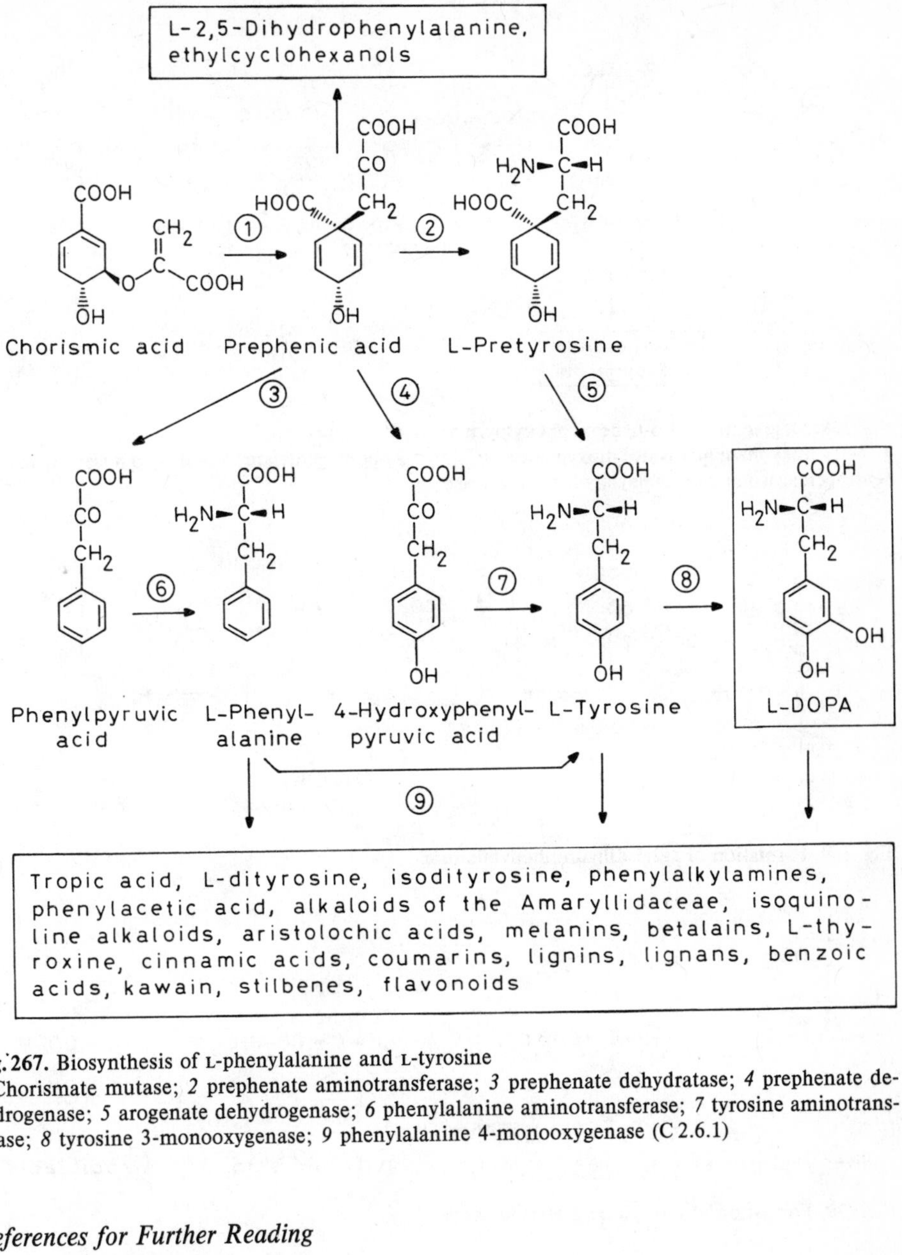

Fig. 267. Biosynthesis of L-phenylalanine and L-tyrosine
1 Chorismate mutase; *2* prephenate aminotransferase; *3* prephenate dehydratase; *4* prephenate dehydrogenase; *5* arogenate dehydrogenase; *6* phenylalanine aminotransferase; *7* tyrosine aminotransferase; *8* tyrosine 3-monooxygenase; *9* phenylalanine 4-monooxygenase (C 2.6.1)

References for Further Reading

31, 48, 150, 190, 264, 415, 632

D 23.1 Secondary Products Retaining the Amino Nitrogen of L-Phenylalanine, L-Tyrosine, and L-DOPA

D 23.1.1 Phenylethylamines and Products of Further Transformation

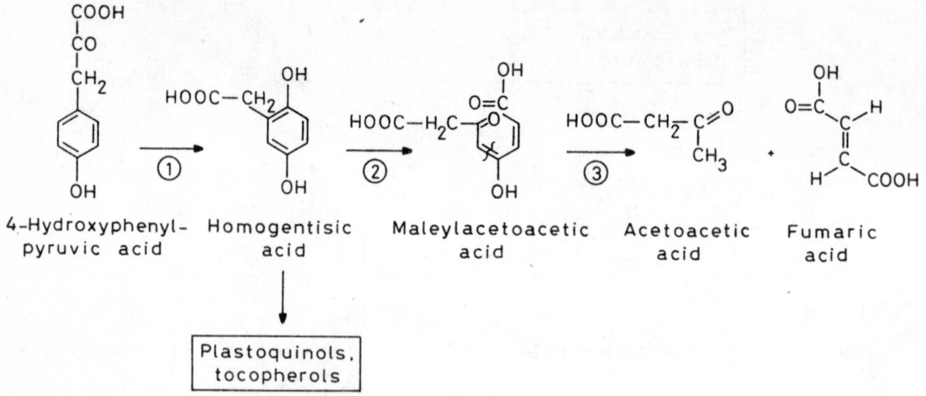

Fig. 268. Degradation of p-hydroxyphenylpyruvic acid
1 4-Hydroxyphenylpyruvate dioxygenase (C 2.5); *2* homogentisate 1,2-dioxygenase; *3* maleyl-acetoacetate isomerase, maleylacetoacetate lyase

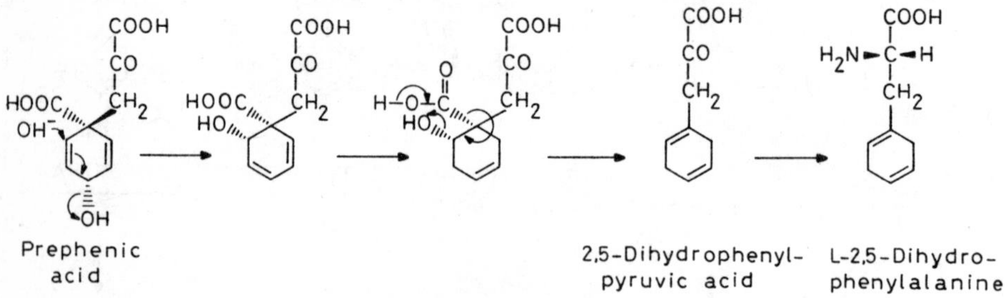

4-Hydroxyphenyl-pyruvic acid · Homogentisic acid · Maleylacetoacetic acid · Acetoacetic acid · Fumaric acid · Plastoquinols, tocopherols

Fig. 269. Formation of L-2,5-dihydrophenylalanine

Prephenic acid · 2,5-Dihydrophenyl-pyruvic acid · L-2,5-Dihydro-phenylalanine

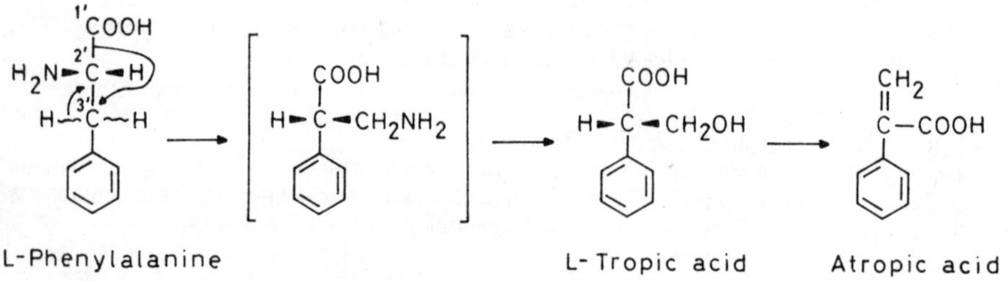

Fig. 270. Formation of L-tropic and atropic acids

L-Phenylalanine · L-Tropic acid · Atropic acid

Chemistry and Distribution

Phenylethylamines (phenethylamines) are formed in microorganisms, plants, and animals. They contain a phenyl ring adjacent to an ethylamine side chain. The phenyl ring is frequently hydroxylated and methoxylated. The amino group may be methylated.

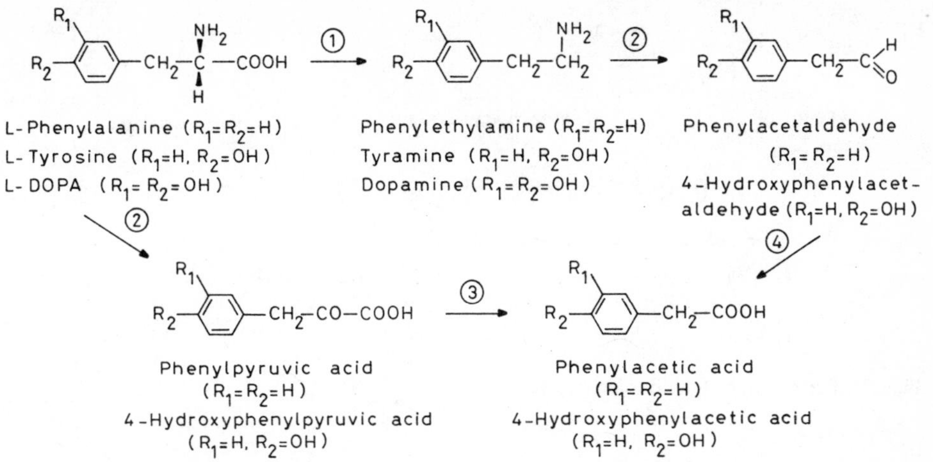

L-Phenylalanine ($R_1=R_2=H$)
L-Tyrosine ($R_1=H, R_2=OH$)
L-DOPA ($R_1=R_2=OH$)

Phenylethylamine ($R_1=R_2=H$)
Tyramine ($R_1=H, R_2=OH$)
Dopamine ($R_1=R_2=OH$)

Phenylacetaldehyde
($R_1=R_2=H$)
4-Hydroxyphenylacet-
aldehyde ($R_1=H, R_2=OH$)

Phenylpyruvic acid
($R_1=R_2=H$)
4-Hydroxyphenylpyruvic acid
($R_1=H, R_2=OH$)

Phenylacetic acid
($R_1=R_2=H$)
4-Hydroxyphenylacetic acid
($R_1=H, R_2=OH$)

Fig. 271. Biosynthesis of phenylacetic acid and p-hydroxyphenylacetic acid
1 Amino acid decarboxylases; *2* aminotransferases; *3* 2-oxo acid dehydrogenase (C 2.7); *4* dehydrogenase

Biosynthesis

The phenylethylamines originate by decarboxylation of the amino acids L-phenylalanine, L-tyrosine, and L-DOPA (Fig. 271).

Significance

Dopamine is a neurotransmitter in animals and human beings (E 3.2). Also phenylethylamine has been identified in the human brain and is suspected to play a role in the central nervous system. *N*-Acetyldopamine may be transformed by phenol oxidases (C 2.3.1) to o-quinones which add amino groups of proteins (quinone tanning, e. g., in the sklerotinization of insects, D 23.2.5).

Transformation

a) Phenylacetic and 4-hydroxyphenylacetic acids are formed by the alternative pathways given in Fig. 271. Phenylacetic acid is used as a territorial marker by certain animals (E 4) and as an antibacterial agent by leaf-cutting ants (E 5.1). 4-Hydroxyphenylacetic acid is a growth regulator in certain higher plants and algae (E 2.2).

b) Hordenine, a protoalkaloid of the plant *Hordeum vulgare*, is derived from tyramine (Fig. 272). Hordenine undergoes a rapid turnover (A 5.2). Its degradation starts with demethylation and the formation of 4-hydroxyphenylacetic acid. The letter compound is hydroxylated to 3,4-dihydroxyphenylacetic, which undergoes ring fission and degradation to CO_2. In addition, hordenine and some of its degradation products are polymerized to insoluble materials probably via o-quinoid intermediates.

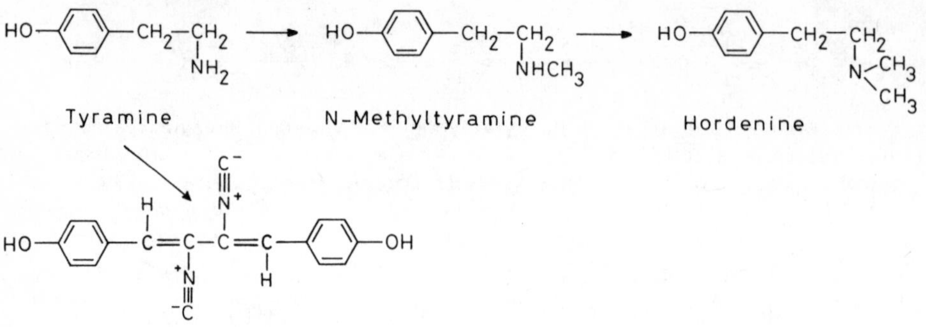

Tyramine   N–Methyltyramine   Hordenine

Xanthocillin X

Fig. 272. Formation of tyramine derivatives

c) Xanthocillin X ist an antibiotic (E 5.2) formed in *Penicillium notatum*. Probably one molecule of tyramine, including its nitrogen atom, and one molecule of 4-hydroxyphenylpyruvic acid are incorporated in the two halves of the molecule.

d) The alkaloids found in Amaryllidaceae are derived from tyramine and protocatechuic aldehyde (D 23.2.5) via formation of a Schiff base, which is stabilized by reduction (Fig. 273). The alkaloids of the norbelladine type are precursors of the more complicated representatives of the galanthamine, haemanthamine, haemanthidine, and tazettine types. An important reaction in the formation of the new rings is oxidative coupling (C 2.3.1).

e) Epinephrine (adrenaline) and norepinephrine (noradrenaline), which play a role as hormones and neurotransmitters in animals and humans (E 3.1 and E 3.2), are formed from dopamine (Fig. 274). The methyl group of epinephrine is directly derived from 5-methyltetrahydrofolic acid (C 3.2). In animals and humans epinephrine and norepinephrine are degraded to vanillic acid (Fig. 275). Vanillic acid is excreted with the urine (E 1).

f) Mescaline, a protoalkaloid of the cactus *Lophophora williamsii*, growing in Mexico and the southwestern United States, is derived from dopamine (Fig. 274). Mescaline is toxic to potential predators and acts as feeding deterrent (E 5.5.3). It is the hallucinogenic constituent of peyote (F 3).

g) Capsaicin, the pungent principle of the fruits of *Capsicum* sp. (red pepper, F 1), contains a substituted benzylamine moiety bound to an acyl chain. The amino nitrogen of L-phenylalanine is retained in capsaicin by a shift of the amino group from C-atom 2' to C-atom 3' of the side chain (cf. also the formation of gramine, D 22.1). Immediate precursors of capsaicin are vanillylamine and isocapric acid. The latter compound is probably built with isobutyryl CoA as starter (D 5.2.1, Fig. 276).

▶

Fig. 273. Possible pathways in the biosynthesis of different types of alkaloids formed in Amaryllidaceae

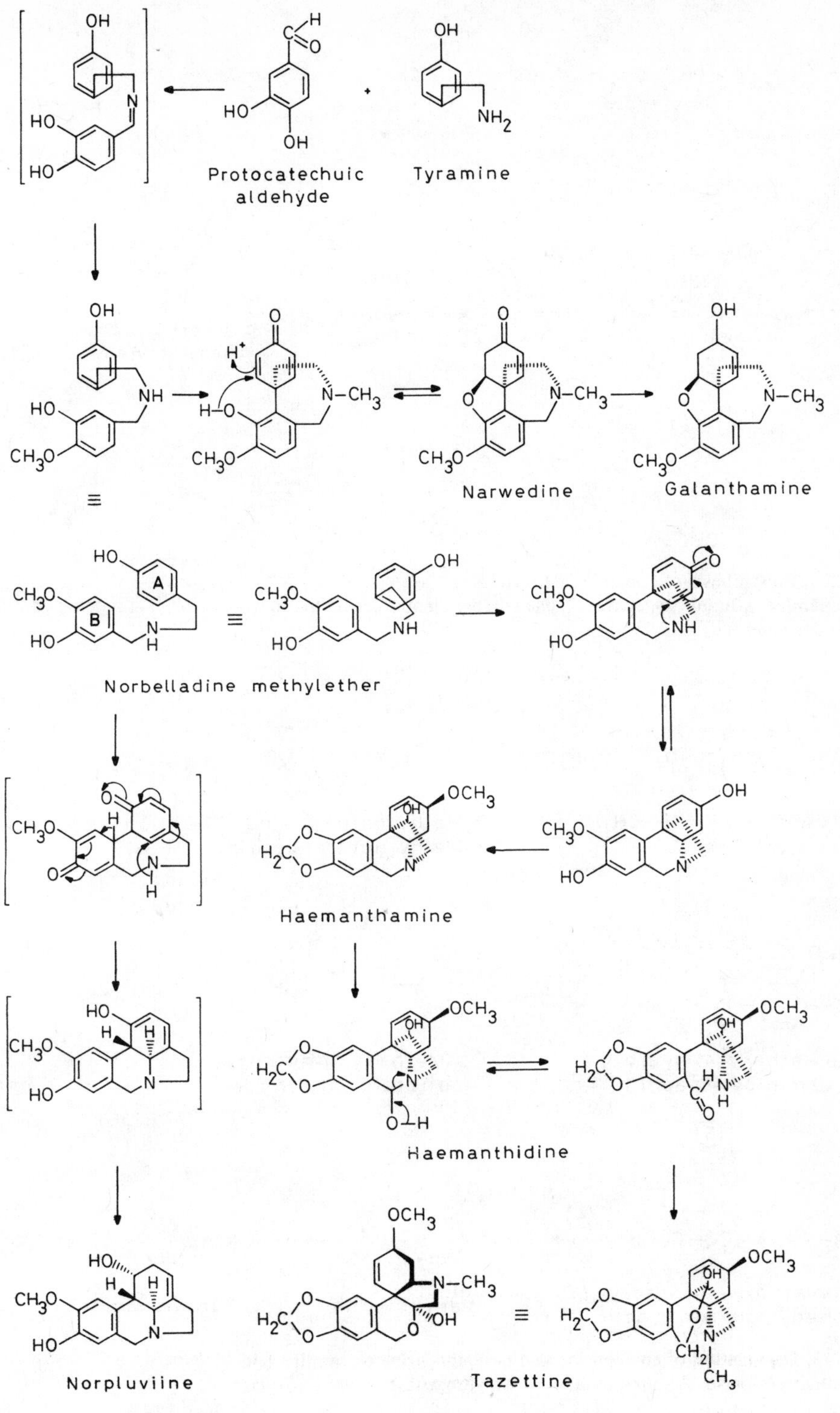

Protocatechuic
aldehyde   Tyramine

≡   Narwedine   Galanthamine

Norbelladine methylether

Haemanthamine

Haemanthidine

Norpluviine   Tazettine

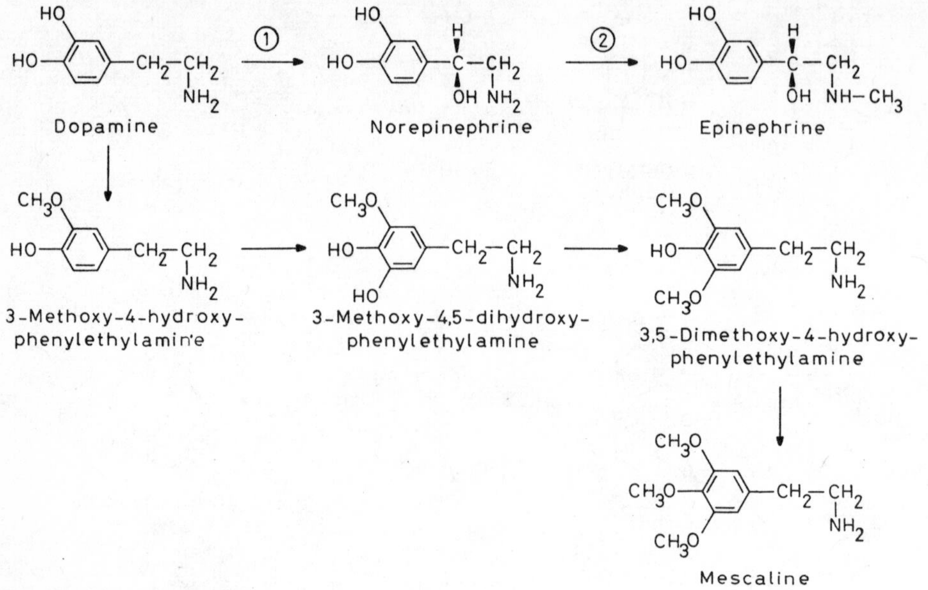

Fig. 274. Biosynthesis of dopamine derivatives
1 Dopamine β-monooxygenase; *2* phenylethanolamine-*N*-methyltransferase (5-methyltetrahydro-folate-dependent)

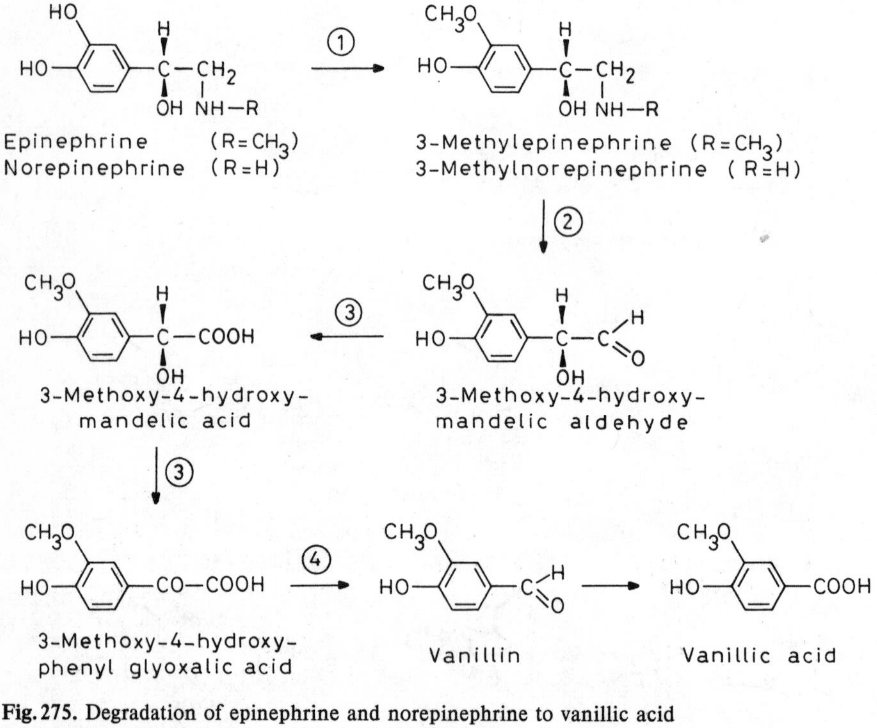

Fig. 275. Degradation of epinephrine and norepinephrine to vanillic acid
1 Methyltransferase; *2* amine oxidase; *3* dehydrogenase; *4* decarboxylase

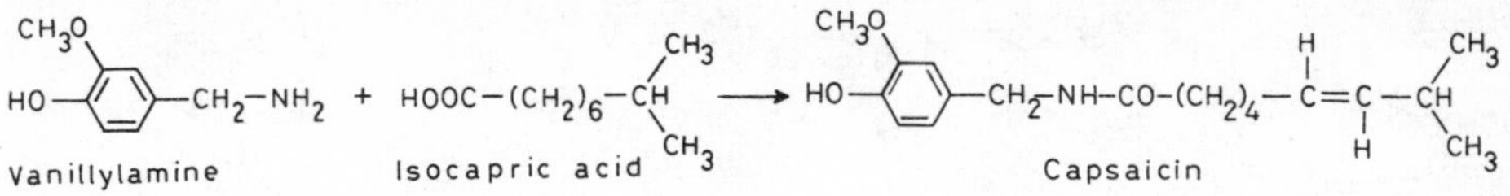

The chemical scheme for the biosynthesis of capsaicin:

Vanillylamine + Isocapric acid → Capsaicin

Fig. 276. Biosynthesis of capsaicin

References for Further Reading

3, 48, 54, 200, 239, 304, 507, 695, 790

D 23.1.2. Tetrahydroisoquinolines

Chemistry and Distribution

Tetrahydroisoquinoline alkaloids are formed in several families of higher plants. A few are synthesized also in animals and human beings. They are derived from the following ring system:

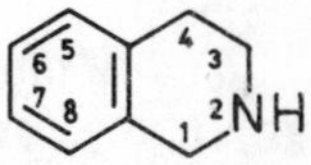

Most tetrahydroisoquinolines are substituted at position 1 by various substituents and carry hydroxy, methoxy, and methylenedioxy groups attached to the aromatic ring. In many cases additional rings may be present.

Biosynthesis

The tetrahydroisoquinoline ring system is built by a Mannich condensation (D 8.5) from a phenylethylamine derivative, e. g., tyramine or dopamine (D 23.1.1), and an aldehyde or a 2-oxo acid (Fig. 277). With regard to the carbonyl compounds involved, the following types of alkaloids may be distinguished:

a) Tetrahydroisoquinolines, unsubstituted in position 1. Compounds of this type, e. g., anhalamine (Fig. 278) are synthesized, with formaldehyde as second building block (pathway A in Fig. 277).

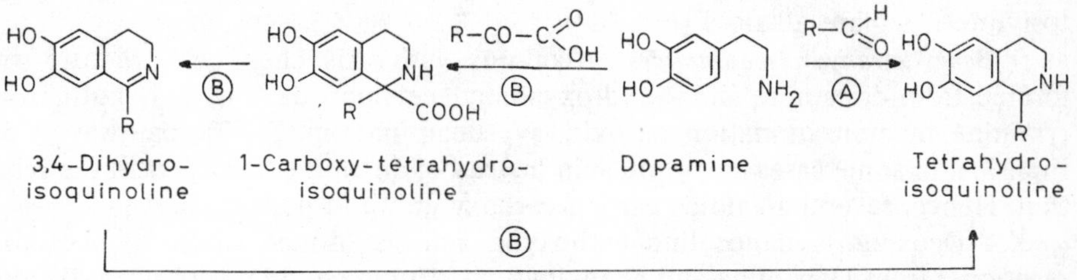

The reaction scheme shows:

3,4-Dihydro-isoquinoline ←(B)← 1-Carboxy-tetrahydro-isoquinoline ←(B)← $R-CO-C\diagdown\substack{O\\OH}$ → Dopamine →(A)→ $R-C\diagdown\substack{H\\O}$ → Tetrahydro-isoquinoline, with a (B) pathway connecting 1-Carboxy-tetrahydroisoquinoline and Tetrahydroisoquinoline at the bottom.

Fig. 277. Alternative routes for the biosynthesis of tetrahydroisoquinoline alkaloids

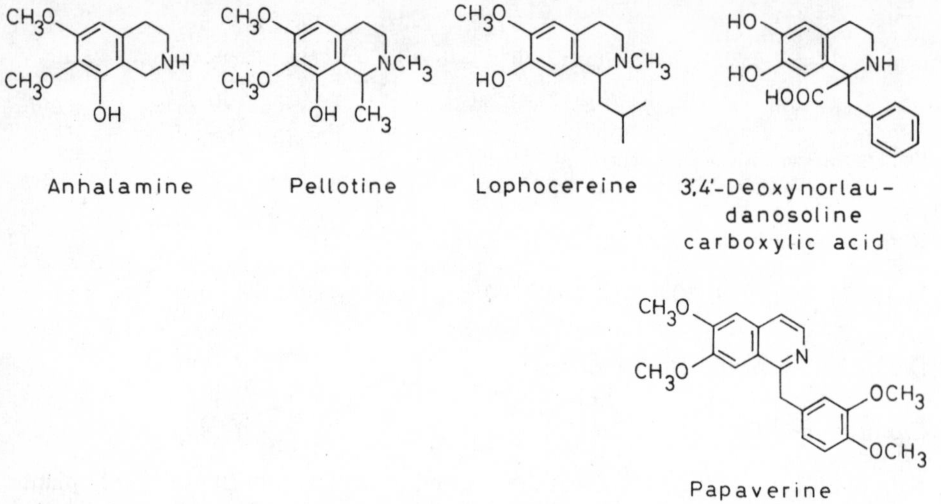

Anhalamine   Pellotine   Lophocereine   3',4'-Deoxynorlau-
                                                                  danosoline
                                                               carboxylic acid

Papaverine

Fig. 278. Structural formulas of simple tetrahydroisoquinolines

b) 1-Methyltetrahydroisoquinolines. In higher plants alkaloids belonging to this group, e.g., pellotine, are formed with participation of pyruvic acid (pathway B in Fig. 277). In mammalian tissues similar compounds are synthesized from tyramine, dopamine, epinephrine or norepinephrine, and acetaldehyde via pathway A. Acetaldehyde formed from ethanol by dehydrogenation (Fig. 51) causes an increased level of these isoquinolines (and related β-carbolines, D 22.3). after the consumption of alcoholic beverages.

c) Isobutyltetrahydroisoquinolines. Alkaloids of this type, e. g., lophocereine, have been found in some plants. They may be formed with mevalonic acid or a biological equivalent, e. g., isoamylaldehyde, as second precursor.

d) Iridoid tetrahydroisoquinolines. In the formation of the Ipecacuanha alkaloids one or two molecules of a phenylethylamine are linked with an iridoid of the secologanin type (D 6.2.2) (pathway A of Fig. 277). In the first step deacetylipecoside and deacetylisoipecoside are formed (Fig. 279). The first epimer is stabilized either by acethylation (formation of ipecoside) or by lactame formation (synthesis of alangiside). The iso-epimer is an intermediate in the building of the aldehyde protoemetine, which by a second Mannich reaction yields the alkaloids cephaeline and emetine.

e) Phenyltetrahydroisoquinolines. L-DOPA and vanillin are precursors of cryptostyline I, a plant alkaloid (Fig. 280).

f) Benzyltetrahydroisoquinolines. Alkaloids with this chemical structure are formed from dopamine and 4-hydroxyphenylacetaldehyde (Fig. 281) built from tyramine by transamination or oxidative deamination (D 8.2) (pathway A in Fig. 277). In some cases an amino acid instead of an amine takes part in the reaction. Hence, several alkaloids carry a carboxy group in position 3.

3',4'-Deoxynorlaudanosoline carboxylic acid is formed (in a spontaneous reaction?) from Dopamine and phenylpyruvic acid in children with phenylketonuria (pathway B in Fig. 277). It is excreted with the urine (E 1). Benzylisochinolines, e. g., papaverine, are formed from tetrahydrobenzylisoquinolines by dehydrogenation. The more complicated alkaloids are derived from *(S)*-coclaurine. Examples are the alkaloids of the berberine group (Fig. 281) and the morphinane type alkaloids (Fig. 282).

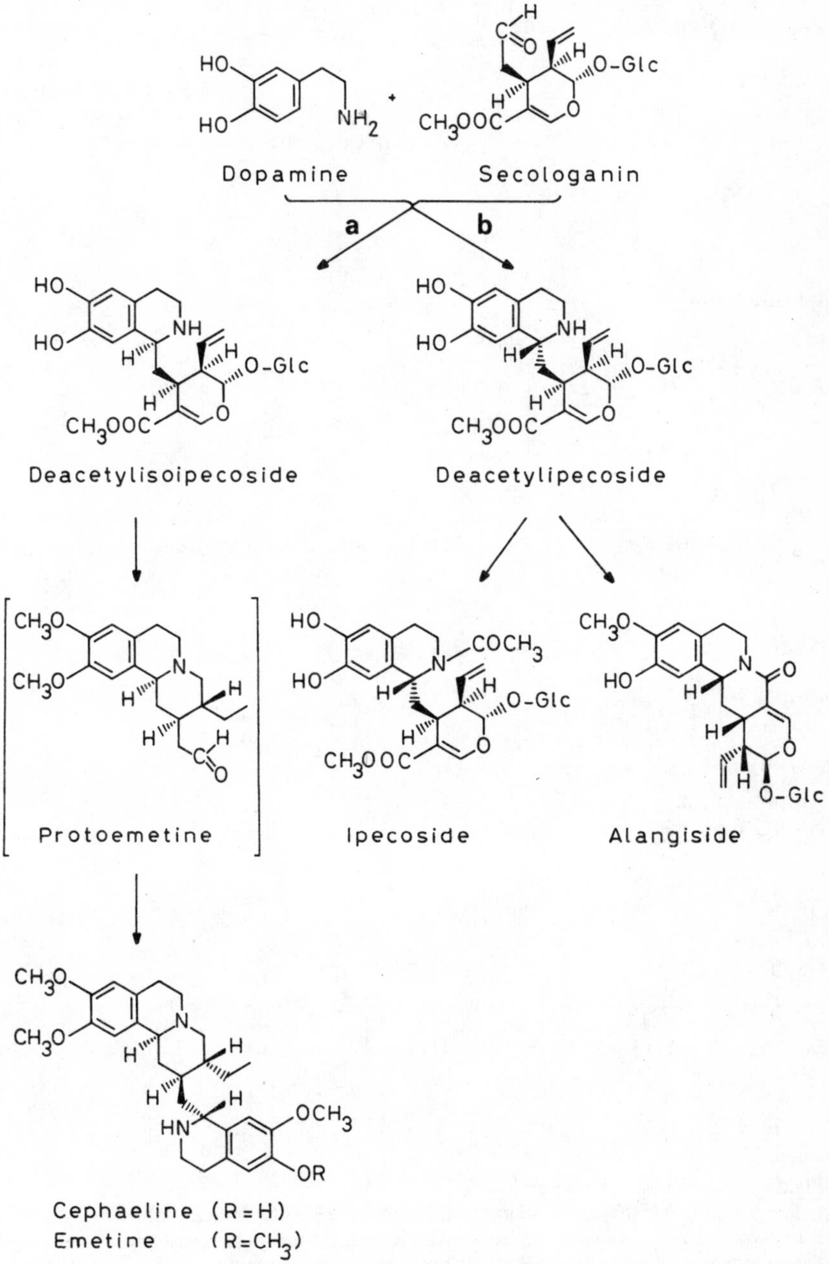

Fig. 279. Formation of Ipecacuanha alkaloids

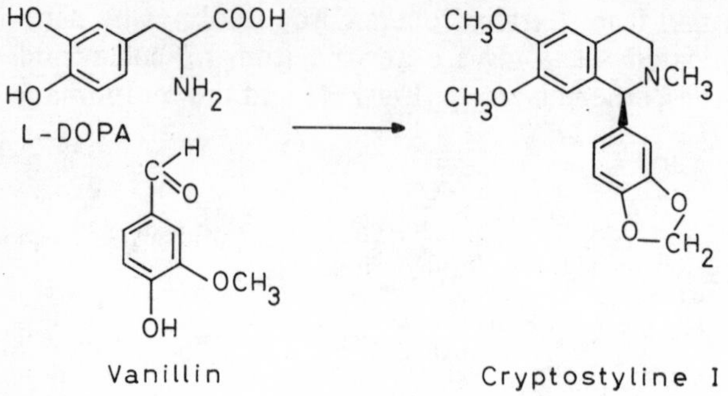

Fig. 280. Biosynthesis of the phenyltetrahydroisoquinoline cryptostyline I

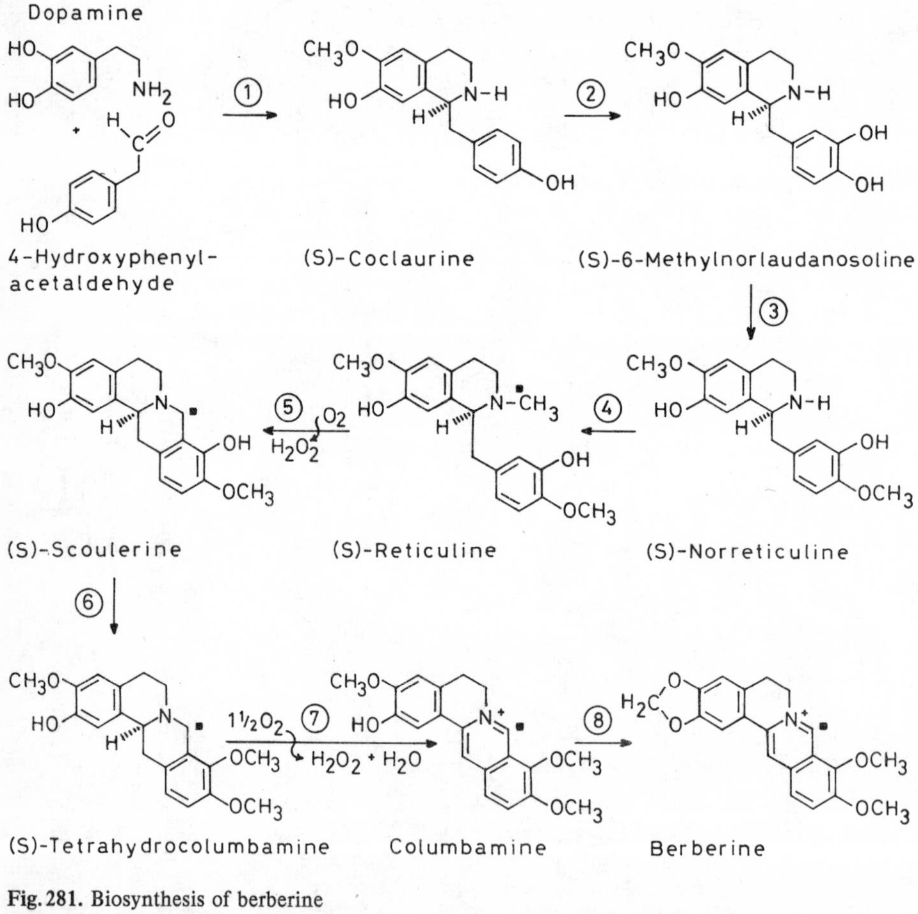

Fig. 281. Biosynthesis of berberine

1 Norcoclaurine synthase, *S*-adenosyl-ʟ-methionine: *(R,S)*-norlaudanosoline-6-*O*-methyltransferase; *2* unknown monooxygenase; *3 S*-adenosyl-ʟ-methionine: norlaudanosoline-4′-*O*-methyltransferase; *4* norreticuline-*N*-methyltransferase; *5* berberine bridge-forming enzyme; *6 S*-adenosyl-ʟ-methionine: (S)-scoulerine-9-*O*-methyltransferase; *7 (S)*-tetrahydroberberine oxidase; *8* berberine synthase

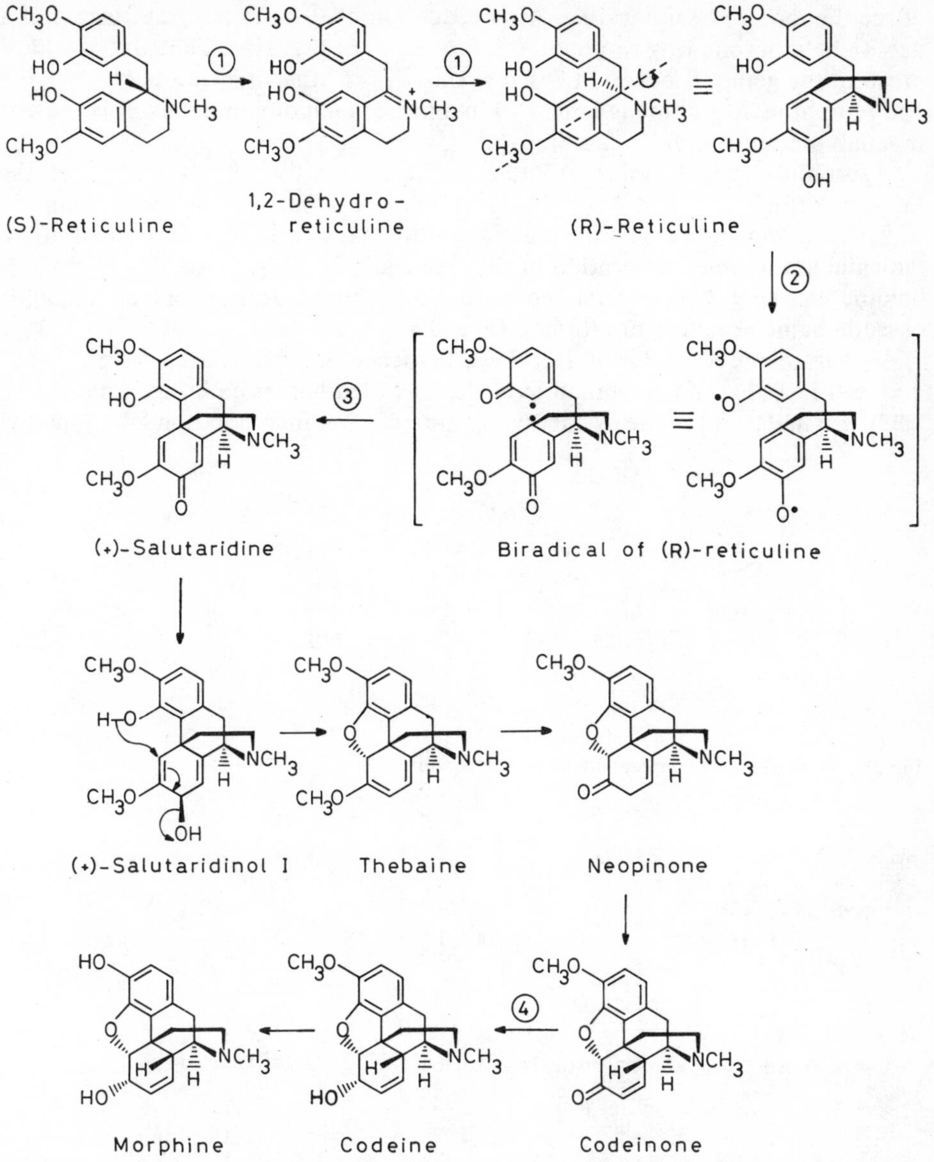

Fig. 282. Biosynthesis of codeine and morphine
1 Reticuline dehydrogenase; *2* phenol oxidase; *3* spontaneous reaction; *4* codeinone reductase

Alkaloids of the morphinane group. If the tetrahydroisoquinoline alkaloid *(R)*-reticuline is written in such a way that part of the molecule is rotated around the dotted line (Fig. 282), the relationship to the morphinane-type alkaloids becomes obvious. The actual precursor of these compounds is *(R)*-reticuline which is derived from *(S)*-reticuline (Fig. 282). *(R)*-Reticuline is probably attacked by a phenol oxidase (C 2.3.1) yielding a biradical which is stabilized by the formation of the dienone (+)-salutaridine. After reduction of (+)-salutaridine closure of a new O-heterocyclic ring results in the formation of thebaine. Thebaine is widespread in the genus *Papaver*. In *P. somniferum* it is further transformed to codeine and morphine. Morphine is built also in animals including humans. It is present in small amounts, e.g., in the brain.

Aporphine-type alkaloids. Aporphines belong to the *(R)* as well as to the *(S)*-series (Figs. 283 and 284). The ring system is built by radicalic coupling (C 2.3.1). In dependence on the position of the hydroxy group transformed into a radicalic group and the location of the free radicals, either alkaloids of the isoboldine type (Fig. 283) or via dienones and their further rearrangement, alkaloids of isothebaine structure are formed (Fig. 284).

Berberine and alkaloids derived from berberine are synthesized from *(S)*-reticuline (Fig. 281). The carbon atom of the so-called berberine bridge (marked by "■") originates from the *N*-methyl group (cf. the incorporation of *O*-methyl

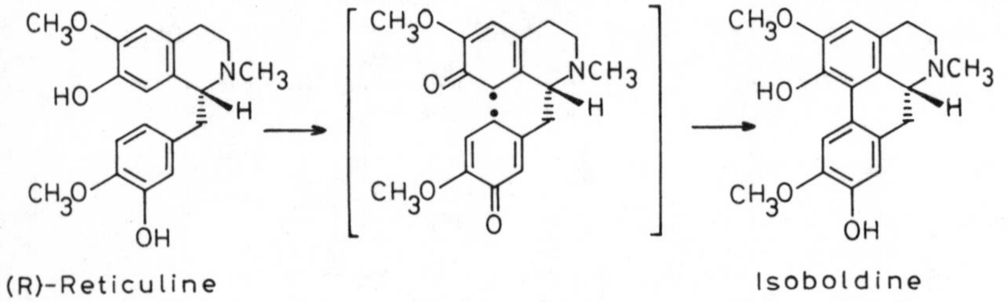

Fig. 283. Formation of the aporphine alkaloid isoboldine

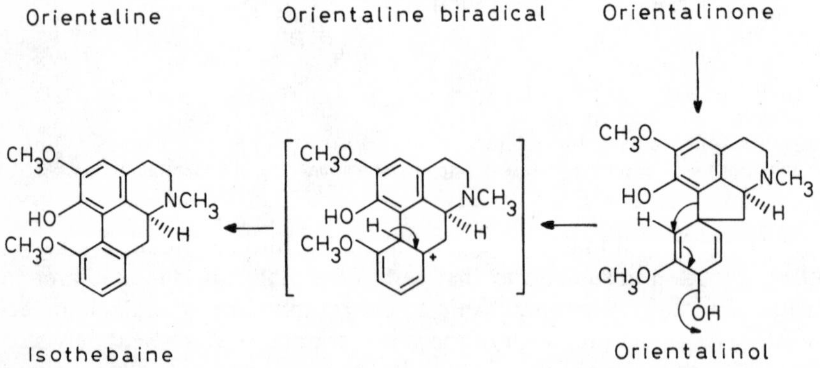

Fig. 284. Biosynthesis of isothebaine

C-atoms in the ring systems of flavonoids, D 23.3.3). *(S)*-Scoulerine is the branch point in the biosynthesis of protopine- and benzophenanthridine-type alkaloids (e. g., chelidonine and sanguinarine) and of phthalidisoquinolines (e. g., narcotine). These substances are derived by splitting different bonds of the ring system of *(S)*-scoulerine and further transformation (Fig. 285).

Bisbenzyltetrahydroisoquinolines. Biradicalic coupling (C 2.3.1) of two molecules of tetrahydroisoquinolines yields different types of dimeric alkaloids (cf. the formulas of tubocurarine and thalicarpine, Fig. 286).

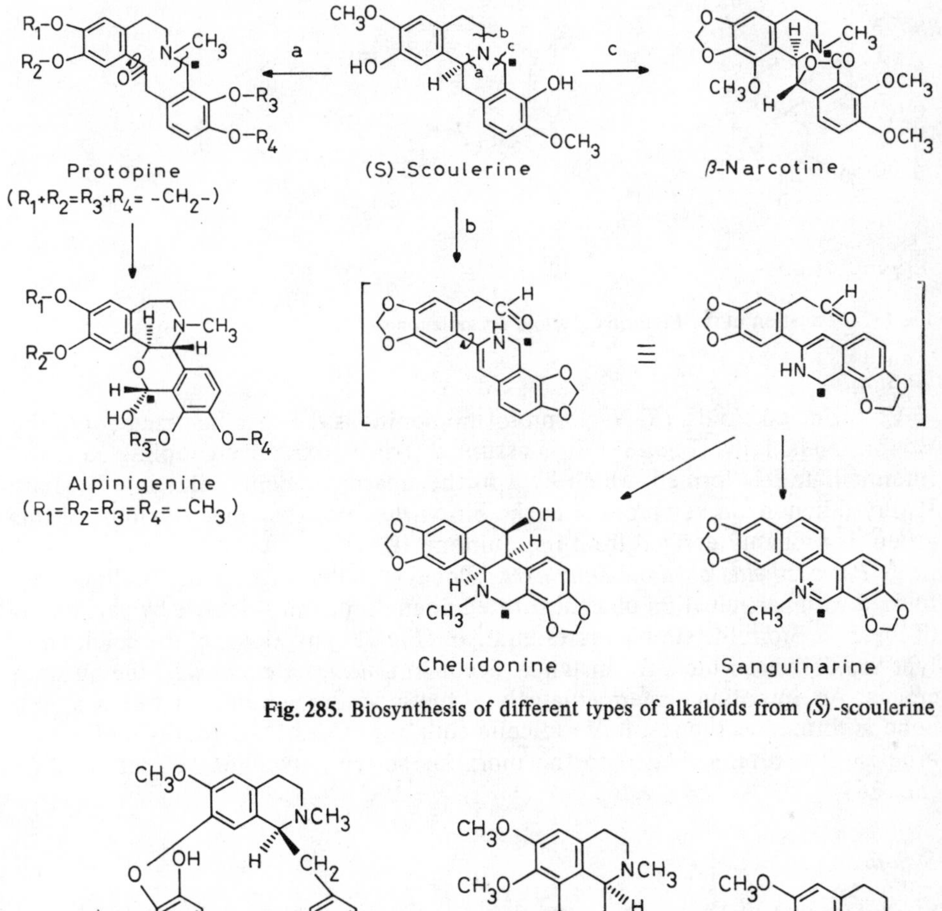

Fig. 285. Biosynthesis of different types of alkaloids from *(S)*-scoulerine

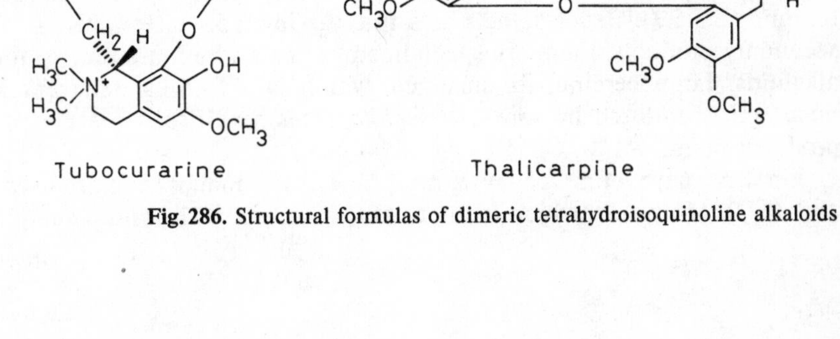

Fig. 286. Structural formulas of dimeric tetrahydroisoquinoline alkaloids

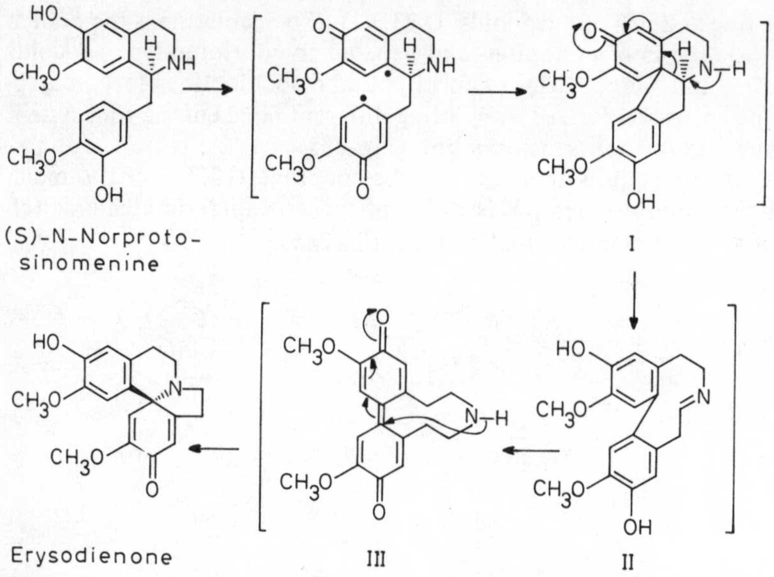

Fig. 287. Formation of the Erythrina alkaloid erysodienone

Erythrina alkaloids. *(S)-N*-Norprotosinomenine is the specific precursor of the Erythrina alkaloids (Fig. 287). It is assumed that by oxidative coupling (C 2.3.1) intermediate I is formed, which by a further rearrangement yields intermediate II. Erysodienone, a key product in the biosynthesis of the other alkaloids of this group, is probably derived from the quinone III.

g) *Phenylethyltetrahydroisoquinolines.* Phenylethyltetrahydroisoquinolines are formed with participation of a substituted phenylpropane aldehyde by pathway A of Fig. 277. From the simple representatives (Fig. 288) alkaloids of the colchicine-type are derived, which, for instance, occur in *Colchicum autumnale*, the autumn crocus. An important intermediate is *O*-methylandrocymbine in which a new bond is formed as the result of radicalic coupling (C 2.3.1). *O*-Methylandrocymbine ist structurally related to the morphinane-type alkaloids (+)-salutaridine (Fig. 289).

Significance

Many tetrahydroisoquinoline alkaloids are highly toxic. Berberine, for instance, interacts with DNA, colchicine causes polyploidy. They are therefore feeding deterrents (E 5.5.3). Beberine acts as a fungitoxin (E 5.4). However, as with other secondary products, there are specialized animals which are able to tolerate the alkaloids. Lophocereine, for instance, which is a feeding deterrent for *Drosophila* sp., is endured by some *Drosophila* races specifically associated with the producer plant.

Emetine, papaverine, codeine, morphine, and tubocurarine are used in medicine (F 2). Morphine, which is important as an analgesic, and diacetylmorphine

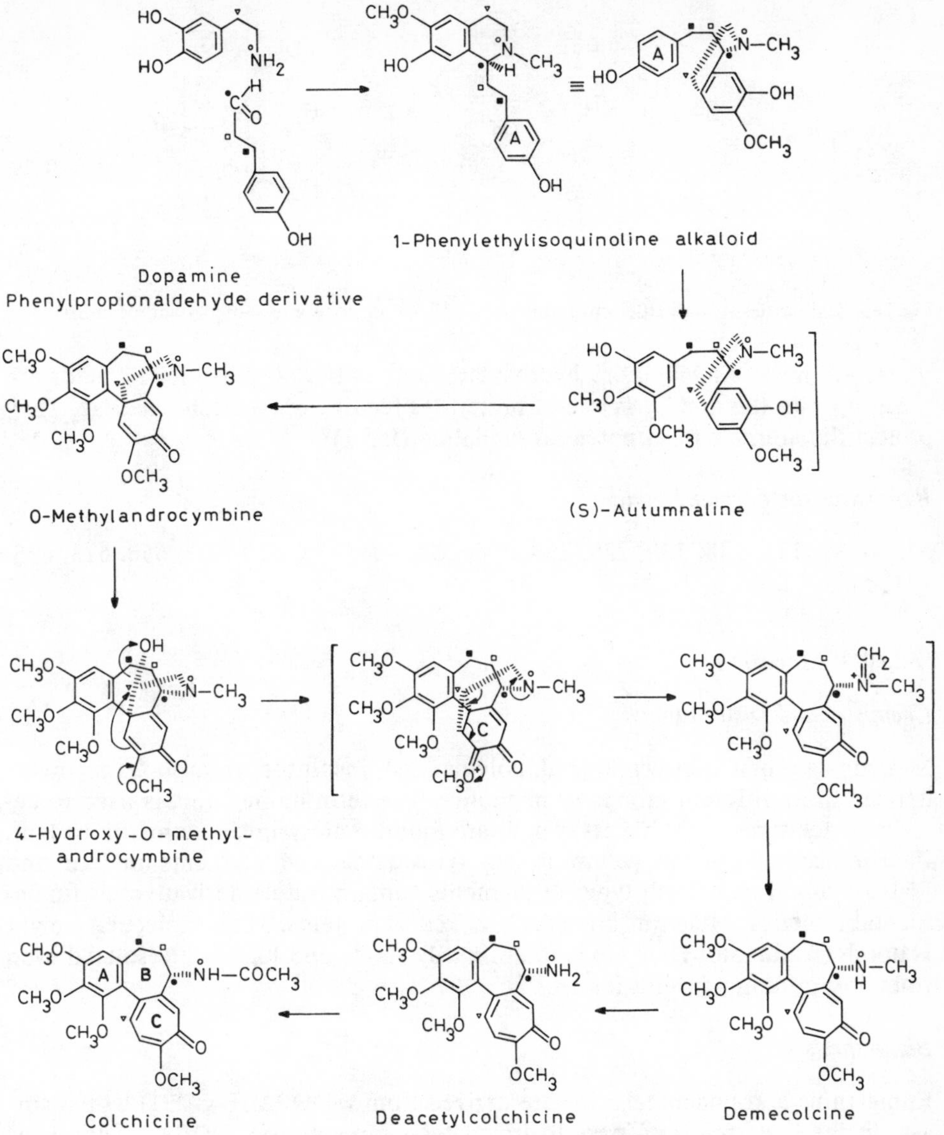

Fig. 288. Biosynthesis of colchicine

(heroine) prepared synthetically from morphine, are powerful hallucinogens (F 3). Both substances interact with the receptors of the endorphins (D 9.1), endogenic peptides mediating the perception of pain.

Transformation

Aristolochic acids, nitrocompounds found in *Aristolochia* sp., are derived from aporphine-type alkaloids by oxidative degradation (Fig. 290). Probably interme-

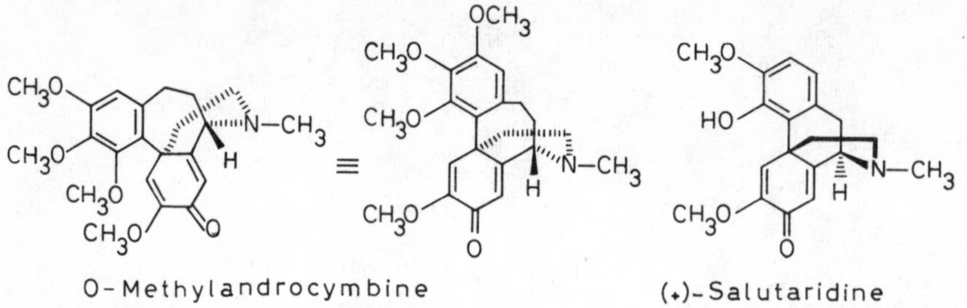

O-Methylandrocymbine   (+)-Salutaridine

Fig. 289. Similarities in the structural formulas of (+)-salutaridine and o-methylandrocymbine

diates are involved which are hydroxylated at position 4. Aristolochic acids are taken up with the diet by larvae of butterflies feeding on Aristolochiaceae. They protect the animals from potential predators (E 5.1).

References for Further Reading

54, 66, 99, 115, 138, 139, 239, 258, 306, 354, 405, 435, 518, 603, 650, 673, 695, 704, 845, 901

D 23.1.3 Melanins

Chemistry and Distribution

Melanins are black, brown, or red, polymerized, insoluble, amorphous pigments derived from different groups of phenolics. The term eumelanins is used to describe black tyrosine polymers, which are found mainly in the animal kingdom. Phaeomelanins are the yellow to red tyrosine-derived pigments in hair and feathers of animals. Both types of pigments contain indole derivatives as fundamental moieties. Melanins are closely linked to proteins. Their molecular weight is not determinable. They are extraordinarily stable and have been isolated even from 150-million-year-old fossils.

Biosynthesis

Eumelanins and phaeomelanins are derived from L-DOPA (Fig. 291) by tyrosinase. In the first step tyrosinase hydroxylates L-tyrosine to L-DOPA. In the later

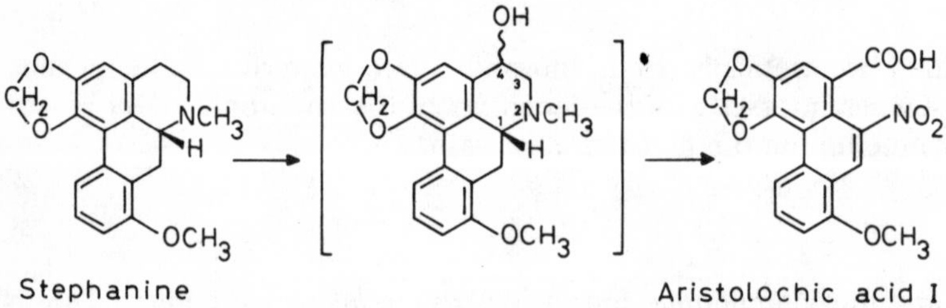

Stephanine   Aristolochic acid I

Fig. 290. Formation of aristolochic acid I

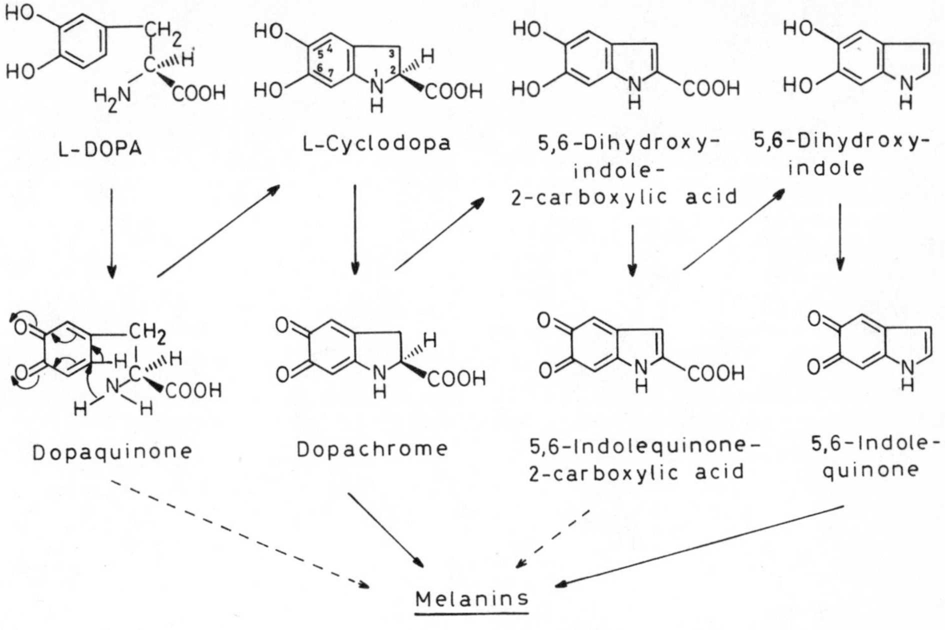

Fig. 291. Biosynthesis of melanin by tyrosinase

steps L-DOPA and other intermediates are dehydrogenated to compounds with o-quinoid structures (C 2.3.1). The indole ring system is formed by the intramolecular addition of the amino group to the o-quinoid system of dopaquinone. All quinoid intermediates are able to undergo additions, especially at positions 4 and 7, but also positions 2 and 3, and tend to polymerize. Due to polymerization highly interlaced three-dimensional macromolecules are formed (Fig. 292). These are red if the molecular weight is relatively low, and black if the molecular weight is high.

Addition to the qinoid structures of thiol or amino groups of proteins, e.g., HS-groups of L-cysteine or 6-amino groups of L-lysine residues, yields covalent bindings between melanins and proteins. Oxidative cleavage of the quinoid rings causes formation of pyrrol moieties, which are frequently part of the melanin molecule (Fig. 292).

Significance

L-DOPA-derived melanins are synthesized in melanocytes and are responsible for the dark pigmentation of skin and hair of animals and humans. They protect the body against excessive irradiation by UV-light, since they build free radicals which interact with the radicals formed during irradiation of skin cells. An eumelanin is also the pigment of the "ink" of Cephalopods (cuttlefish), which serves in defense (E 5.1). Cuttlefish melanin is the coloring constituent of the dye sepia used by artists.

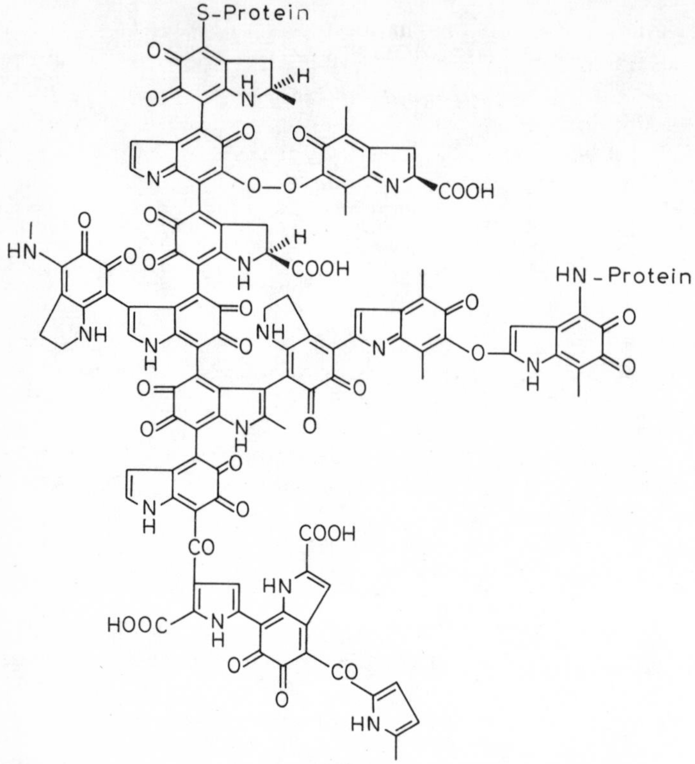

Fig. 292. Part of the molecule of the ink melanin of *Sepia officinalis*

References for Further Reading

884, 885

D 23.1.4. Compounds Formed by Cleavage of the Aromatic Ring of L-DOPA

Chemistry and Distribution

The O- and N-heterocyclic compounds muscaflavin, stizolobinic acid, stizolobic acid, and (S)-betalamic acid (Fig. 293) occur in mushrooms and higher plants. They are derived from L-DOPA. The O-heterocyclic derivatives still contain the L-alanyl side chain.

Biosynthesis

The aromatic ring of L-DOPA is cleaved by dioxygenases (C 2.5) adjacent to the both phenolic groups (Fig. 293). Cleavage of the ring proceeds
— either between the positions 4 and 5 in the biosynthesis of (S)-betalamic acid and probably stizolobic acid

— or between the positions 2 and 3 in the formation of muscaflavin and stizolobinic acid.

(S)-Betalamic acid is the precursor of betalains, a group of red (betacyanins) or yellow pigments (betaxanthins) formed in Centrospermae and the pili of some mushrooms (Fig. 294). Examples are betanin (betanidin-5-glucoside), the pigment of red beets, and indicaxanthin a pigment of *Opuntia ficus-indica* flowers and of the fly agaric *(Amanita muscaria)*. In the formation of betacyanins

Muscaflavin (Amanita, Hygrocybe)	Stizolobinic acid (Mucona, Leguminosae)	Stizolobic acid (Mucona, Leguminosae)	(S)–Betalamic acid (Centrospermae, Amanita)

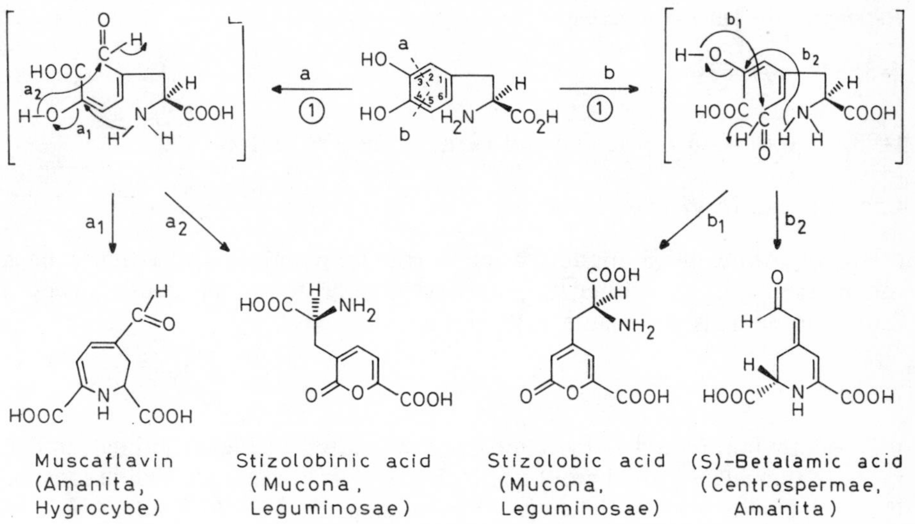

Fig. 293. Recyclization of L-DOPA after oxidative ring cleavage
1 Dioxygenases, e. g., stizolobate synthase and stizolobinate synthase

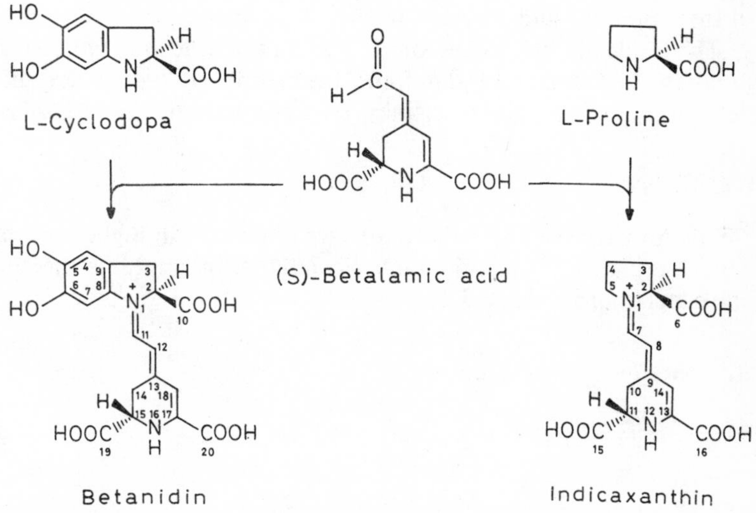

Fig. 294. Formation of betacyanins and betaxanthins from *(S)*-betalamic acid

(S)-betalamic acid condenses with L-cyclodopa (D 23.1.3), in case of betaxanthins with some other amines or amino acids, e.g., L-proline (D 17). Betacyanins and betaxanthins attract pollinators if present in plants flowers (E 5.5.1), and animals mediating seed dispersal, if they are constituents of fruits (E 5.6). It is of interest that betacyanins and anthocyanins (another type of red flower pigment, D 23.3.3) do not occur together in the same species, in contrast to the yellow betaxanthins and flavonols, which may be found side by side in certain plants.

References for Further Reading

488, 549, 604, 634

D 23.1.5 Iodinated L-Tyrosine and L-Thyronine Derivatives

Chemistry and Distribution

L-3-Iodotyrosine, L-3,5-diiodotyrosine, L-triiodothyronine, and L-thyroxine are the most important iodinated phenylpropanoids found in nature. They are formed in animals and plants, e.g., certain seaweeds.

Biosynthesis

L-Triiodothyronine and L-thyroxine are synthesized in higher animals and humans in the cells of the thyroid gland. These cells accumulate iodide ions from the blood with great efficiency. A halogenoperoxidase (C 2.4.2) reacts with L-tyrosine residues present in thyroglobulin, one of the thyroid proteins, to form L-monoiodo- and L-diiodotyrosine residues (Fig. 295). These residues, while still incorporated into thyroglobulin, condense in a radicalic reaction with each other to L-triiodothyronine and L-thyroxine residues, which are released by hydrolysis of thyroglobulin and secreted in the blood stream.

The levels of L-triiodothyronine and L-thyroxine are regulated mainly by deiodination. In the thyroid gland and the peripheric tissues deiodinating enzymes dehalogenate free L-iodotyrosines, L-triiodothyronine, and L-thyroxine.

Significance

L-Triiodothyronine and L-thyroxine are hormones in higher animals and humans which regulate growth and metabolic activity. In amphibia the compounds accelerate metamorphosis (E 3.1).

Reference for Further Reading

48

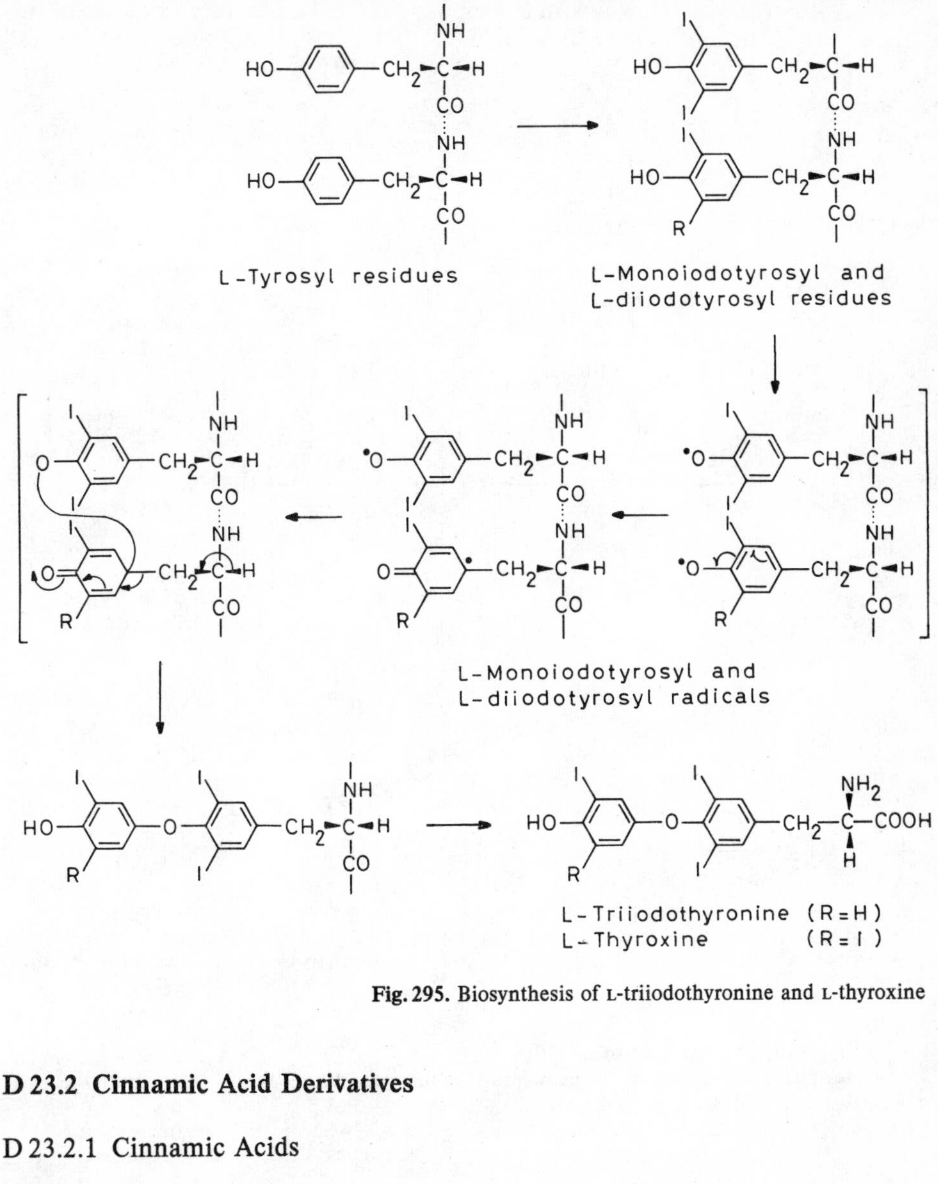

Fig. 295. Biosynthesis of L-triiodothyronine and L-thyroxine

D 23.2 Cinnamic Acid Derivatives

D 23.2.1 Cinnamic Acids

Chemistry and Distribution

Cinnamic acids are formed in higher plants and some microorganisms, e. g., several fungi and *Streptomyces verticillatus*, as well as in animals. They are phenyl-propenic acids. The phenyl ring may be substituted by hydroxy and methoxy groups. The double bond of the side chain may have an *E*- or *Z*-configuration. *E*- and *Z*-cinnamic acids are easily interconvertible.

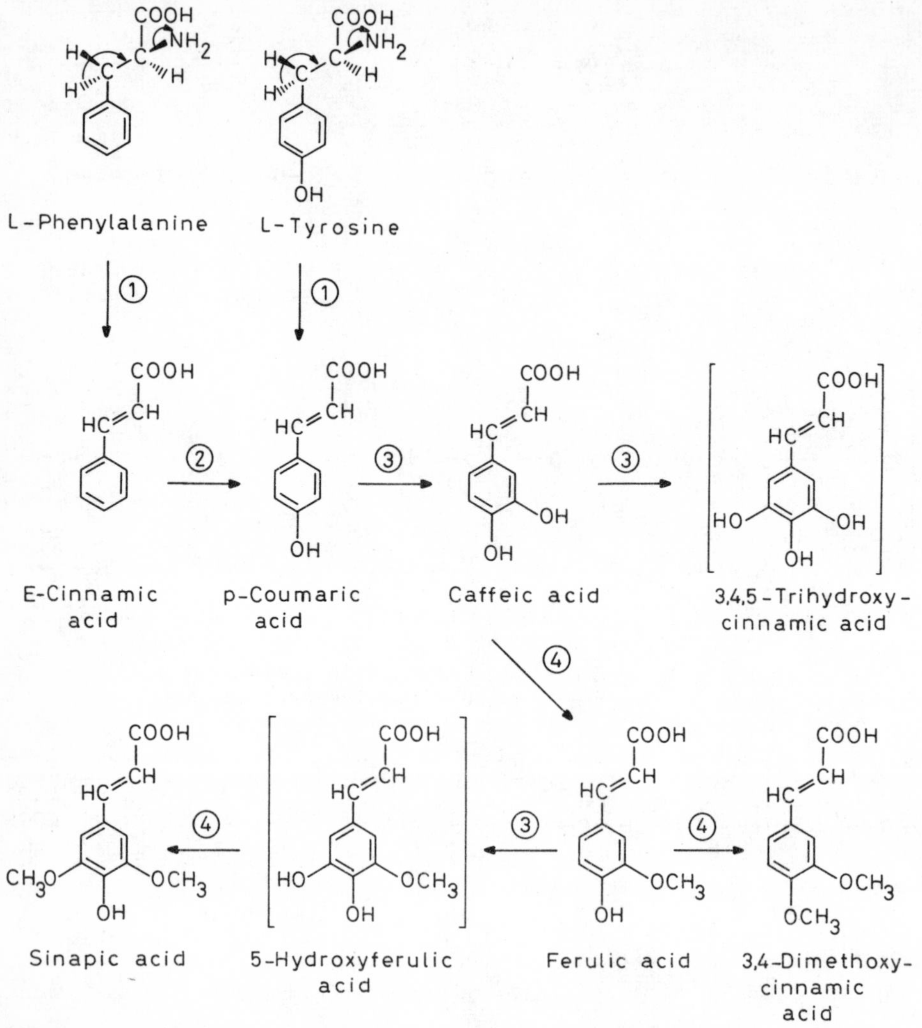

Fig. 296. Formation of cinnamic acids
1 Phenylalanine ammonia-lyase; *2* cinnamate 4-monooxygenase; *3* phenol oxidase; *4* O-methyltransferases, e. g., caffeate O-methyltransferase

Biosynthesis

In plants and microorganisms cinnamic acid is formed from L-phenylalanine by phenylalanine ammonia-lyase (PAL). This enzyme catalyzes the antiperiplanar elimination of the pro 3(S)-hydrogen atom and the NH_2-group to yield E-cinnamic acid (Fig. 296). Most PAL preparations deaminate also L-tyrosine, but to a smaller extent. In some organisms, however, a special tyrosine ammonia-lyase exists.

In animals L-tyrosine may be transformed to p-coumaric acid on the pathway shown in Fig. 297.

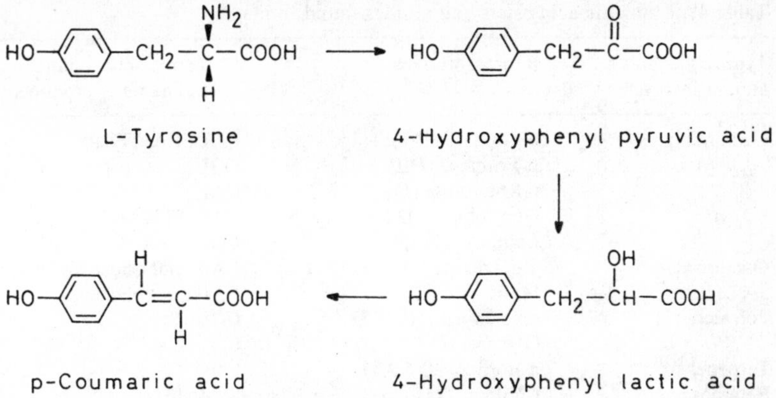

$$HO-\langle\text{phenyl}\rangle-CH_2-\overset{NH_2}{\underset{H}{C}}-COOH \longrightarrow HO-\langle\text{phenyl}\rangle-CH_2-\overset{O}{C}-COOH$$

L-Tyrosine   4-Hydroxyphenyl pyruvic acid

$$HO-\langle\text{phenyl}\rangle-\overset{H}{\underset{H}{C}}=C-COOH \longleftarrow HO-\langle\text{phenyl}\rangle-CH_2-\overset{OH}{CH}-COOH$$

p-Coumaric acid   4-Hydroxyphenyl lactic acid

Fig. 297. Biosynthesis of p-coumaric acid in animals

Cinnamic acid is the precursor of a group of hydroxylated and methoxylated derivatives (Fig. 296). In the formation of p-coumaric acid the p-hydrogen undergoes an NIH-shift to the m-position. (In the hydroxylation of p-coumaric acid to caffeic acid, however, no NIH-shift proceeds, C 6.2.5.)

Significance

E-Cinnamic acid causes self-inhibition of growth of the roots of *Parthenium argentatum*. *Z*-3,4-Dimethoxycinnamic acid inhibits spore germination of rusts. p-Coumaric acid esters are cofactors of ethylene biosynthesis in cauliflower (E 2.2). o- and p-Coumaric acids, ferulic acid, and syringic acid are involved in plant-plant interactions (E 5.3.).

Transformation

a) Esters and amides of cinnamic acids are widespread in higher plants (Table 48). They are formed on two independent pathways. On the one hand, CoA esters are built via AMP derivatives (C 2.1) by CoA ligases, e. g., cinnamate CoA ligase, caffeate CoA ligase, and ferulate CoA ligase. These CoA esters react with hydroxylated compounds and amines. On the other hand, 1-*O*-acyl glucosides are built from UDP-D-glucose (C 6) and free cinnamic acids. The cinnamic acid residue may then be transferred from these compounds to other acceptor molecules by transacylation. Sinapine, for instance, a characteristic constituent of Brassicaceis is formed by the reaction of 1-*O*-sinapoyl-*β*-D-glucose with choline (D 10) catalyzed by 1-*O*-sinapoyl-*β*-D-glucose: choline sinapoyltransferase.

Ferulic acid is frequently associated in plants with proteins as well as with cellulose (D 1.4.1), lignins (D 23.2.3), and other constituents of the cell wall. The methylesters of *Z*-3,4-dimethoxycinnamic acid and *Z*-ferulic acid inhibit the germination of rust spores (E 2.2). Chlorogenic acid acts as a fungitoxin, e.g., in potato tubers (E 5.4).

Table 48. Cinnamic acid esters and amides found in plants

Type of alcohols and amines involved	Representatives	Cinnamic acids bound[a] (trivial name of products)
Carbohydrates	D-Glucose (D 1)	COU, CAF, FER
	D-Fructose (D 1)	COU
	L-Rhamnose (D 1.1)	CAF
	Gentiobiose (D 1.4.1)	CAF, FER
	Cellulose (D 1.4.1)	COU, FER
Organic acids	Tartaric acid (D 1.2)	CAF (cichoric acid)
	Malic acid (D 4)	CAF (phaselic acid)
Polyalcohols	myo-Inositol (D 1.3)	COU
	Glycerol (D 2)	COU, CAF
Terpenes	d-Borneol (D 6.2.1)	COU
Betaines	Choline (D 10)	SIN (sinapine)
Amines	Spermidine (D 12.1)	COU, CAF, FER
	Putrescine (D 17)	COU, CAF (paucine), FER
	Piperidine (D 18)	TRI (piperlongumine)
	Agmatine (D 19)	COU (hordatines)
	Tryptamine (D 22.1)	FER
Alkaloids	Lupinine (D 18)	COU
Cyclohexane carboxylic acids	Quinic acid (D 21)	CAF (chlorogenic acid, cynarin)
	Shikimic acid (D 21)	COU, CAF
Amino acids	L-Tryptophan (D 22)	CAF
	L-DOPA (D 23)	CAF (clovamide)
Flavonoids	Kaempferol-3-glucoside (D 23.3.3)	COU (tiliroside)
	Cyanidin-3-glucoside (D 23.3.3)	COU (hyacinthin)

[a] COU p-coumaric acid, CAF caffeic acid, FER ferulic acid, SIN sinapic acid, TRI thrimethoxy cinnamic acid.

b) Dihydrocinnamic acids (phenylpropionic acids) are formed from cinnamic acids by hydrogenation of the side chain. They are precursors of benzoic acids (D 23.2.5). Melilotic acids (dihydro-o-coumaric acid) and dihydrocaffeic acid are characteristic constituents of higher plants. Melilotic acid is a precursor of dicoumarol (D 23.2.2). Phloretamide is a constituent of the xylem sap of apple trees.

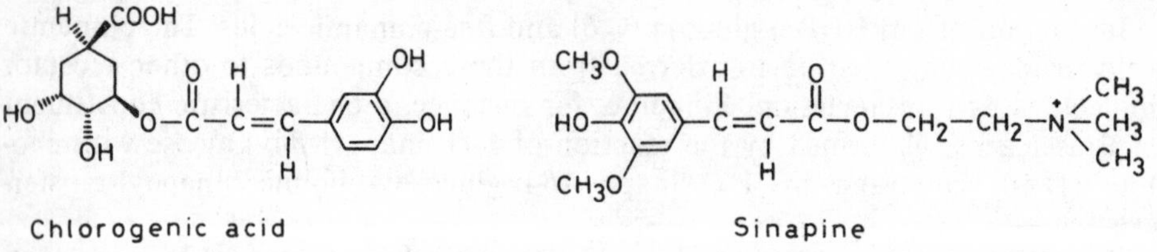

Chlorogenic acid

Sinapine

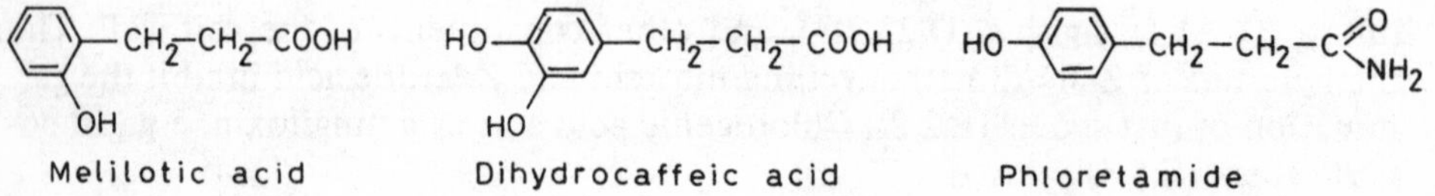

Melilotic acid

Dihydrocaffeic acid

Phloretamide

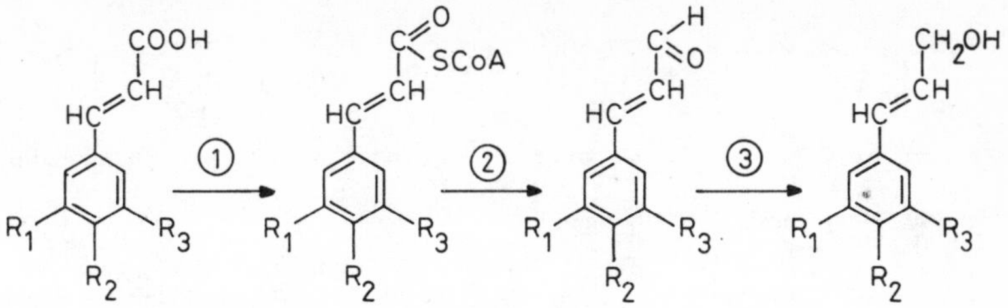

Fig. 298. Activation and reduction of cinnamic acids
1 Cinnamate CoA ligase, p-coumarate CoA ligase; *2* cinnamoyl CoA reductase; *3* cinnamoyl alcohol dehydrogenase

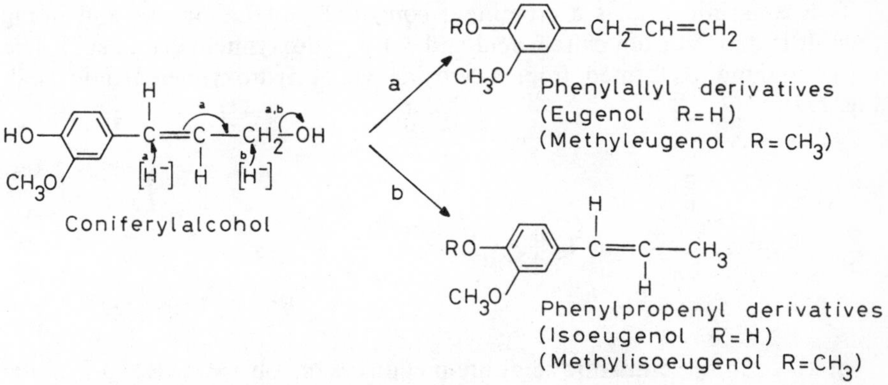

Fig. 299. Probable pathway in the biosynthesis of phenylpropene derivatives

c) Cinnamic alcohols are formed from the CoA-esters of cinnamic acids via cinnamic aldehydes (Fig. 298). The cinnamic alcohols are intermediates in the biosynthesis of lignins (D 23.2.3) and probably of phenylallyl and phenylpropene derivatives (Fig. 299). The latter compounds are constitutents of essential oils. Eugenol and methyleugenol are dominant components of the flower scents of *Eugenia* and *Cassia* sp. (E 5.5.1). Cinnamic aldehyde, myristicin, apiol, and eugenol dominate the smell of several spices (F 1). Methylisoeugenol attracts the carrot root fly (E 5.5.3). Eugenol is used as a dental analgesic (F 2). Dihydroconiferyl alcohol acts synergistically with the plant hormones gibberellic acid, indole-3-acetic acid, and kinetin (E 2.2).

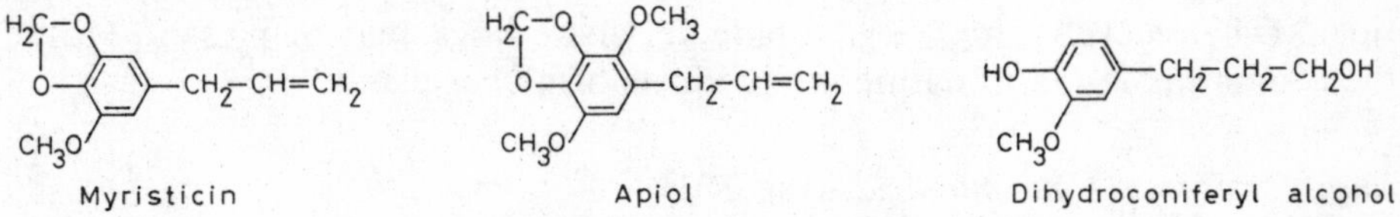

Myristicin   Apiol   Dihydroconiferyl alcohol

d) 6(*S*)-Gingerol is synthesized from one molecule each of ferulic acid, acetate/malonate [with loss of the carboxy group(s)], and hexanoic acid (Fig. 300). It is the most abundant constituent of the flavoring fraction of ginger, the rhizome of *Zingiber officinale* (F 1).

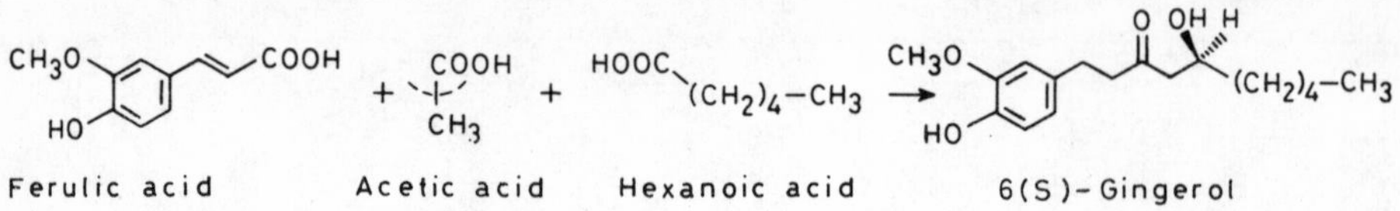

Ferulic acid   Acetic acid   Hexanoic acid   6(S)-Gingerol

Fig. 300. Biosynthesis of 6*(S)*-gingerol

e) Curcumin, a pigment from the rhizomes of *Curcuma* sp., is probably derived from two molecules of ferulic acid and one acetate/malonate unit on a pathway resembling that of gingerol biosynthesis. Powdered *Curcuma* rhizome is the main constituent of curry (F 1).

f) Rosmarinic acid is a prominent constituent of Lamiaceae and Boraginaceae. It is derived from caffeic acid and 3,4-dihydroxyphenyllactic acid. The latter compound is formed from L-tyrosine via 4-hydroxyphenyllactic acid (cf. Fig. 297).

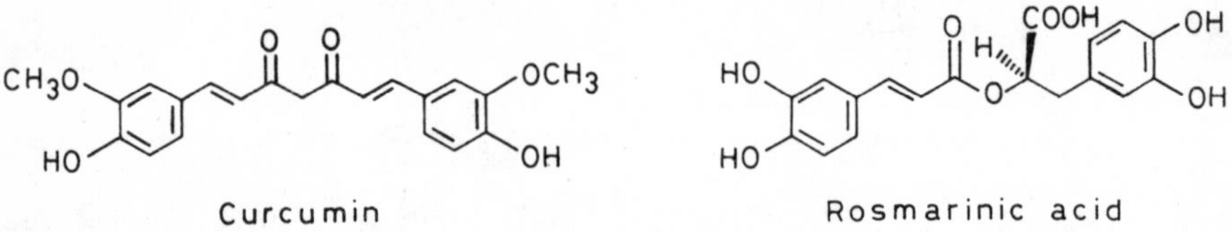

Curcumin   Rosmarinic acid

g) In addition, cinnamic acids are precursors of coumarins (D 23.2.2), lignins (D 23.2.3), lignans (D 23.2.4), benzoic acids, and benzylalcohols (D 23.2.5), of polyketides, e.g., stilbenes and flavonoids (D 23.3.), as well as of hydroquinone and benzoquinone derivatives (D 23.4.).

References for Further Reading

143, 190, 297, 321, 324, 355, 400

D 23.2.2 Coumarins

Chemistry and Distribution

Coumarins are benzopyrane derivatives with a lactone group. The aromatic ring may be substituted, e.g., by hydroxy, methoxy, and isopentenyl groups. Additional *O*-heterocyclic rings, e.g., furane or pyrane rings, may be present. Most of the coumarins found in nature are products of higher plants.

Biosynthesis from Cinnamic Acids (Fig. 301)

(cf. also the formation of polyketidic and flavonoid coumarins, D 5.3.7 and D 23.3.3)

Cinnamic acids which may carry hydroxy and methoxy groups, are hydroxylated in the o-position to give the corresponding o-coumaric acid derivatives

Table 49. Naturally occurring coumarins and their cinnamic acid precursors

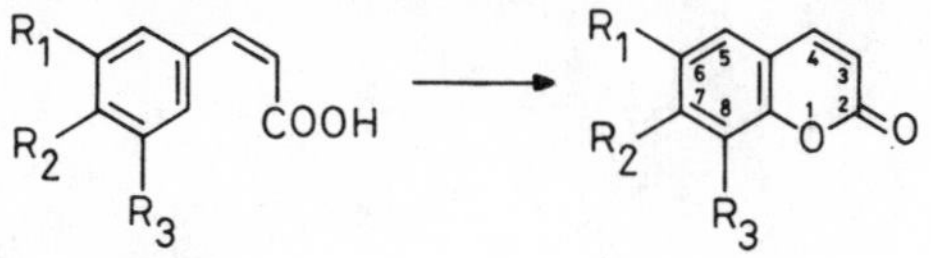

Cinnamic acids	Coumarins	R_1	R_2	R_3
Z-Cinnamic acid	Coumarin	H	H	H
Z-p-Coumaric acid	Umbelliferone	H	OH	H
	Herniarin	H	OCH_3	H
Z-Caffeic acid	Aesculetin	OH	OH	H
Z-Ferulic acid	Scopoletin	OCH_3	OH	H
	Scopolin	OCH_3	O-β-D-Glc-p	H
Z-5-Hydroxyferulic acid	Fraxetin	OCH_3	OH	OH

which may cyclize to the coumarins shown in Table 49. An important step in this biosynthetic pathway is a light-dependent, nonenzymatic *E-Z*-isomerization of the cinnamic acid side chain. Glucosylated derivatives are important intermediates. D-Glucosides of coumarinic acid derivatives predominate in fresh plants. Cleavage of D-glucose and formation of coumarins is usually the result of the breakdown of compartmentalization, e. g., by mechanical injury or wilting. Hence, plants like woodruff (*Asperula odorata*), sweet clover (*Melilotus officinalis*), and *Anthoxanthum odoratum* develop the characteristic smell of coumarin during drying.

In some cases the coumarins formed may be further hydroxylated, methylated, or substituted by isopentenyl residues. Furanocoumarins, e. g., pimpinellin, psoralen, and xanthotoxin (Fig. 302), and dihydropyranocoumarins, e. g.,

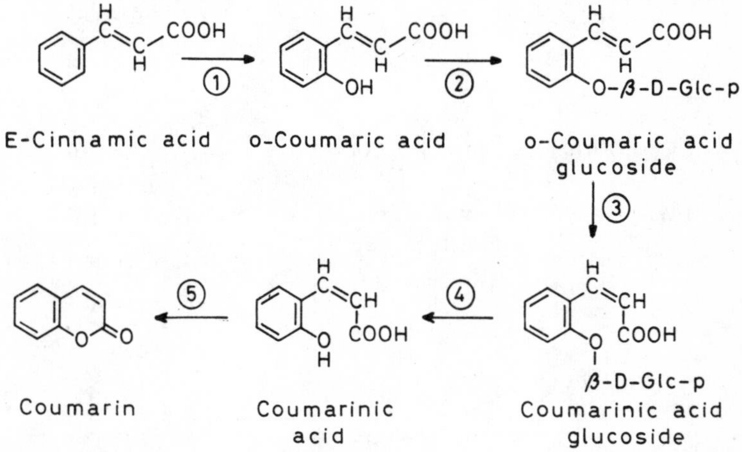

Fig. 301. Formation of coumarin
1 Cinnamate 2-monooxygenase; *2* glucosyltransferase; *3* light-dependent isomerization; *4* glucosidase; *5* spontaneous cyclization

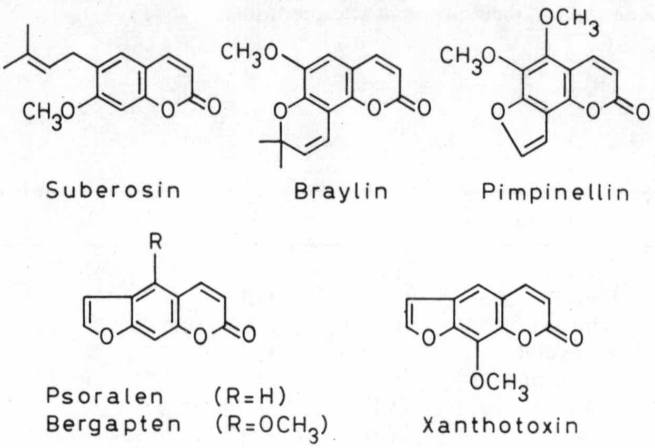

CH_3O
Suberosin
CH_3O
Braylin
OCH_3
CH_3O
Pimpinellin
R
Psoralen (R=H)
Bergapten (R=OCH_3)
OCH_3
Xanthotoxin

Fig. 302. Structural formulas of some isopentenylcoumarins and furanocoumarins

braylin, are derived from isoprenylated coumarins. The additional furane ring may be formed by the series of reactions outlined in section D 21.4.2 (biosynthesis of furoquinolines).

Significance

Psoralen is an inhibitor of seed germination in *Psoralea subacaulis* (E 2.2) and a phototoxic compound responsible, for instance, for the photosensitization of sheep feeding on spring parsley *Cymopterus watsonii* (E 5.5.3). Psoralen and bergapten are used medically in enhancing skin pigmentation (F 2).

Xanthotoxin and psoralen are phytoalexins (E 5.4). Scopolin (scopoletin-7-glucoside) is a fungitoxin formed in potatoes (E 5.4). Coumarin itself attracts certain specialized insects feeding on coumarinic acid glucoside-containing plants (E 5.5.3).

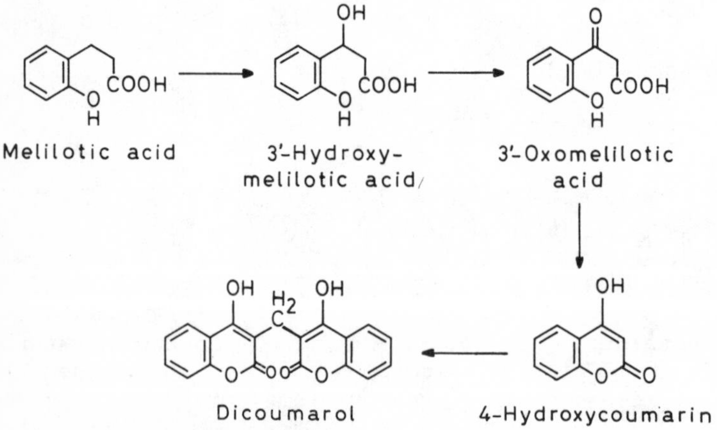

COOH
Melilotic acid
OH
COOH
3'-Hydroxy-melilotic acid
COOH
3'-Oxomelilotic acid
OH
Dicoumarol
OH
4-Hydroxycoumarin

Fig. 303. Formation of dicoumarol

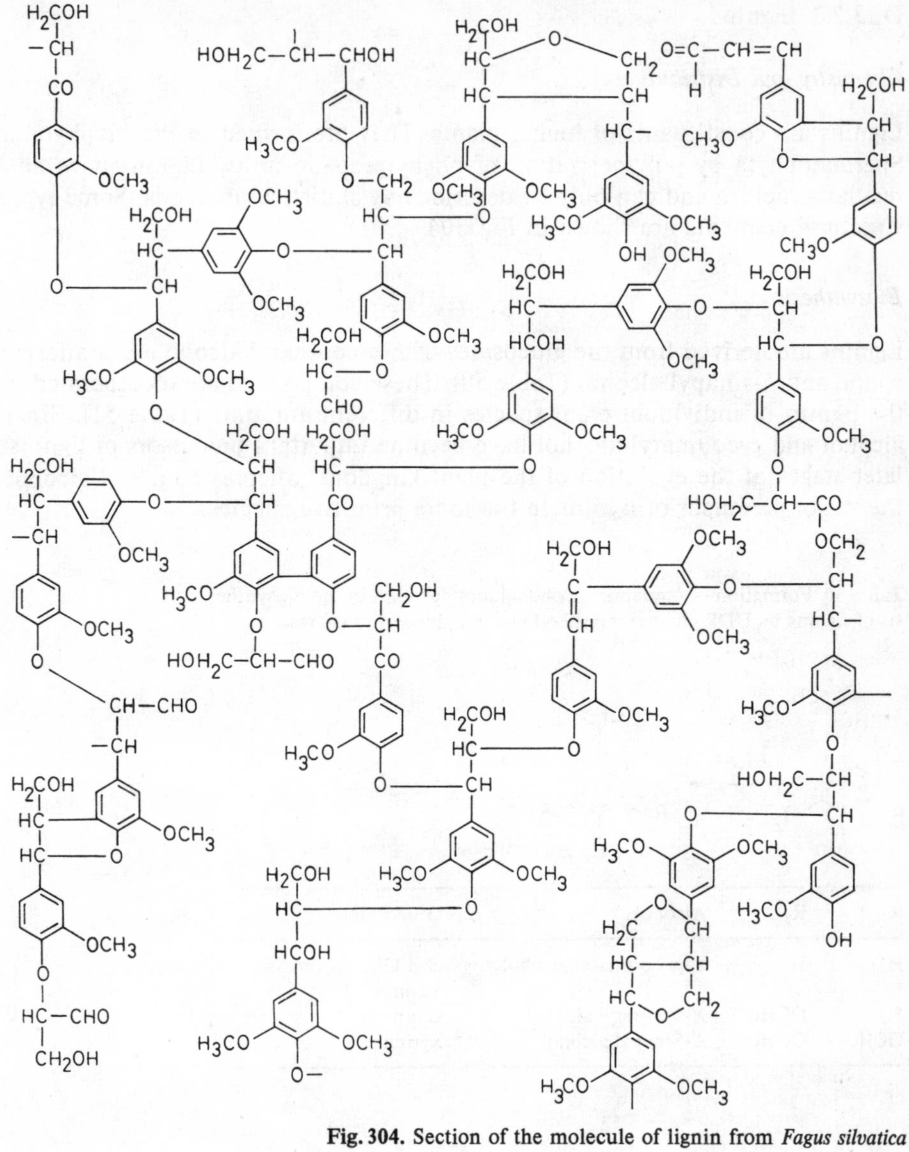

Fig. 304. Section of the molecule of lignin from *Fagus silvatica*

Transformation

Dicoumarol is formed from melilotic acid (Fig. 303) by some microorganisms, e. g., *Aspergillus fumigatus*. The carbon of the methylene bridge may originate from a formaldehyde equivalent. Dicoumarol is the toxic factor of decaying sweet clover *(Melilotus officinalis)*. It inhibits blood coagulation.

References for Further Reading

111, 112, 150, 190, 205, 548

D 23.2.3 Lignins

Chemistry and Distribution

Lignins are constituents of higher plants. They are formed in Pteridophyta and Spermatophyta by polymerization of phenylpropane units. Lignins have an irregular structure and can only be described by statistical methods. Some typical structural elements are shown in Fig. 304.

Biosynthesis

Lignins are derived from the glucosides of *E*-p-coumaryl alcohol, *E*-coniferyl alcohol, and *E*-sinapyl alcohol (Table 50). These compounds are incorporated into the lignins of individual plant species in different amounts (Table 51). Sinapyl alcohol and p-coumaryl alcohol have become important precursors of lignins at later stages of the evolution of the plant kingdom, whereas coniferyl alcohol is the major precursor of lignins in the more primitive species.

Table 50. Formation of cinnamic alcohol glucosides used in the biosynthesis of lignins by UDP-glucose: coniferyl alcohol glucosyltransferase

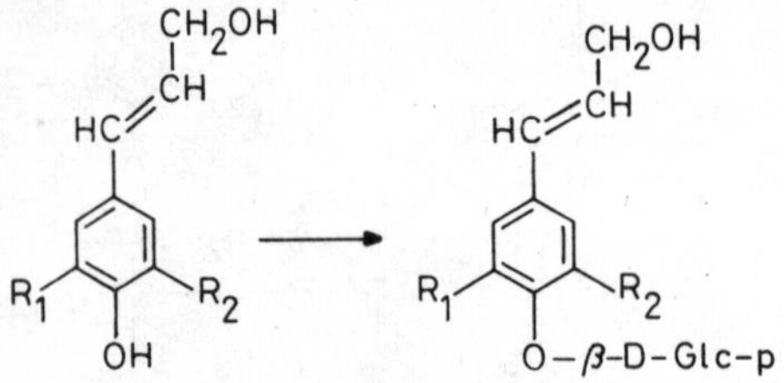

R_1	R_2	Alcohols	Glucosides
H	H	*E*-p-Coumaryl alcohol	*E*-p-Glucocoumaryl alcohol
H	OCH$_3$	*E*-Coniferyl alcohol	Coniferin
OCH$_3$	OCH$_3$	*E*-Sinapyl alcohol	Syringin

Table 51. Lignin precursors in different groups of plants

Systematic unit	Precursors	Systematic unit	Precursors
Gymnosperms	Mostly *E*-coniferyl alcohol, but some *E*-p-coumaryl alcohol and *E*-sinapyl alcohol	Dicotyledons	Mostly *E*-coniferyl alcohol and *E*-sinapyl alcohol, less *E*-p-coumaryl alcohol
		Monocotyledons	*E*-Coniferyl alcohol, *E*-sinapyl alcohol, and *E*-p-coumaryl alcohol

Five processes may be distinguished in lignin formation from cinnamic alcohol glucosides:

a) Alcohols are formed from the glucosides by β-glucosidases (C 6, Fig. 305).

b) The free alcohols are attacket at the unsubstituted 4-hydroxyl group by peroxidase (C 2.4.1) forming an *O*-radical (a) which is in equilibrium with the mesomeric *C*-radicals (b-d, Fig. 305).

The H_2O_2 required in this reaction is formed by the peroxidase from molecular oxygen and NAD(P)H:

$$O_2 + NAD(P)H + H^+ \rightarrow H_2O_2 + NAD(P)^+.$$

The NAD(P) is regenerated by dehydrogenation of L-malic acid catalyzed by malate dehydrogenase (D 4):

$$\text{Malic acid} + NAD(P)^+ \rightarrow \text{Oxaloacetic acid} + NAD(P)H + H^+.$$

Probably a malate-oxaloacetate shuttle across the plasmalemma exists, which allows the regeneration of malic acid and the reuse of oxaloacetic acid.

c) The radicals are able to add unsaturated compounds such as other molecules of cinnamic alcohols (Fig. 306). In this reaction the radicalic structure is retained and further unsaturated compounds may be added. Hence, a chain of reactions is initiated, which leads to the formation of high molecular weight polymeric substances.

In addition, radical-radical interactions take place. In vitro experiments showed, for instance, the formation of dihydrodiconiferyl alcohol and pinoresinol (Fig. 307).

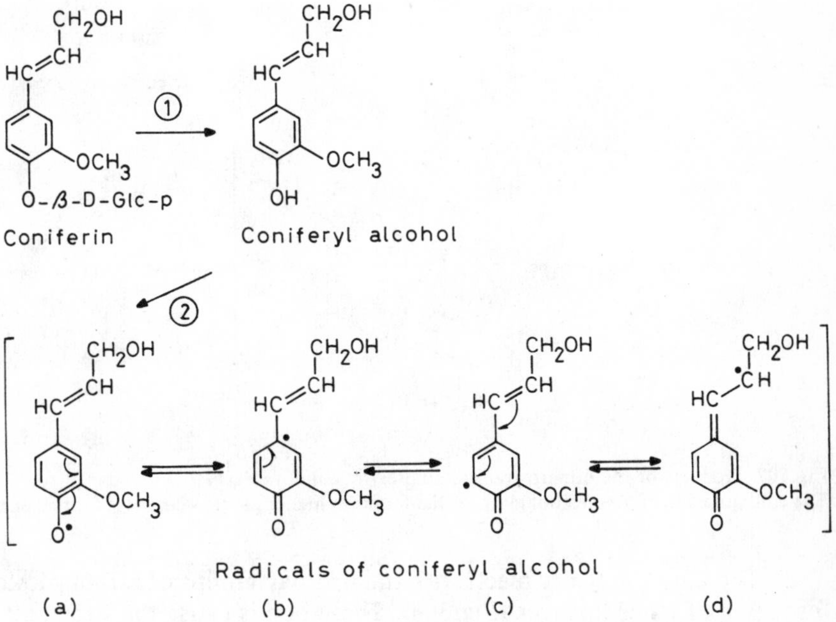

Fig. 305. Formation of coniferyl alcohol radicals
1 β-Glucosidase; *2* peroxidase

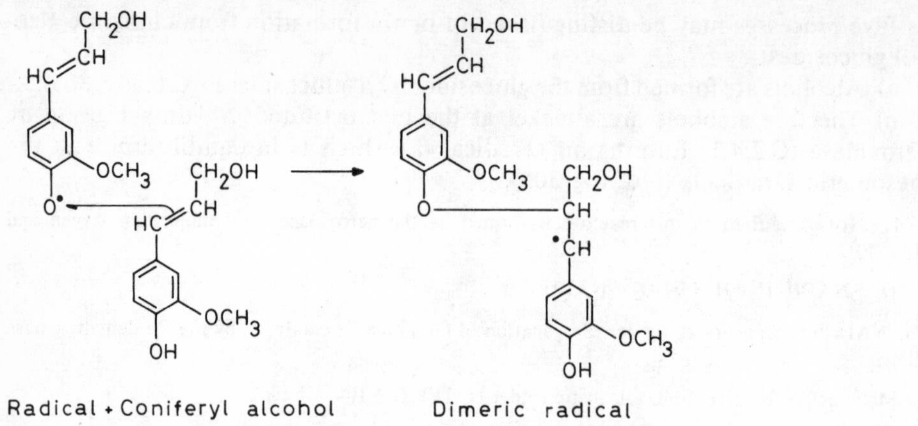

Radical + Coniferyl alcohol   Dimeric radical

Fig. 306. Addition of a molecule of coniferyl alcohol to a coniferyl alcohol radical

Radicals (d) + (c) ⟶

Dihydrodiconiferyl alcohol

Radicals (d) + (d) ⟶

Pinoresinol

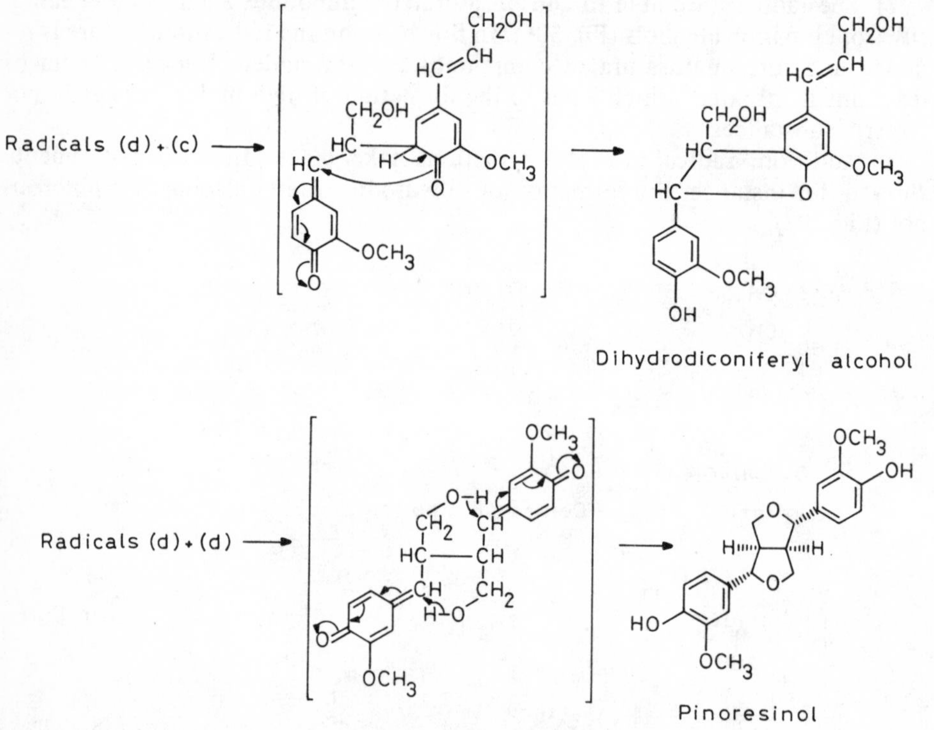

Fig. 307. Products of the dimerization of coniferyl alcohol radicals
The configuration of pinoresinol given is that found in higher plants where it is formed enzymatically

d) The lignin polymer reacts with the hydroxy groups of carbohydrates by the formation of ether linkages (Fig. 308). These bonds cause the strong attachment of lignin to cellulose which originates the characteristic properties of wood (see below).

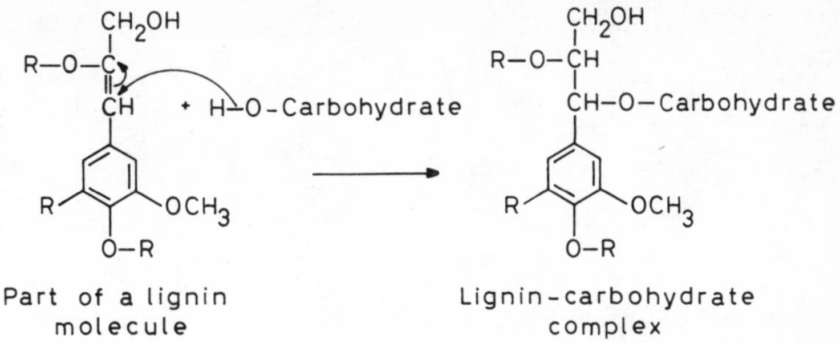

CH$_2$OH
|
R—O—C
‖
CH   + H—O-Carbohydrate

CH$_2$OH
|
R—O—CH
|
CH—O—Carbohydrate

R   OCH$_3$
O—R

R   OCH$_3$
O—R

Part of a lignin
molecule

Lignin-carbohydrate
complex

Fig. 308. Binding of lignin to carbohydrates

e) Free hydroxy groups may be esterified with hydroxylated and methoxylated cinnamic acids (D 23.2.1) and benzoic acids (D 23.2.5) or methylated with *S*-adenosyl-L-methionine (D 12.1).

Significance

Lignins are constituents of the cell wall of various cell types of plants, e.g., wood fibers, vessels, and tracheids. They constitute 20–30 % of the weight of wood. Lignins encrust as an amorphic mass the cellulose fibers which gives the lignified cell wall high mechanical strength (E 2.2) and increases the resistance to microbial degradation (E 5.4). Wood is an important raw material in may fields of technology (F 5).

References for Further Reading

190, 228, 288, 298, 356

D 23.2.4 Lignans

Chemistry and Distribution

Lignans consist of two phenylpropane units linked by a C—C-bond between the carbon atoms 2′ and 2″ of the side chains. The presence of further *O*-heterocyclic rings is common (cf. the formulas of cubebin, β-peltatin, podophyllotoxin, enterolactone, *Aegilops* lignan, and olivil, Fig. 309).

Lignans are widespread in higher plants. Cubebin is found in large quantities in the fruits of *Piper cubeba* (cubebs), olivil constitutes 50 % of the resinous excretions of the olive tree *(Olea europaea)*, and guaiaretic acid is an important constituent of guaiacum resin (obtained from *Guajacum officinale*). Lignans are formed, however, also in certain mushrooms.

Biosynthesis (Fig. 310)

Lignans are formed by the reductive dimerization of cinnamic alcohols or cinnamic acids. During the formation of the C—C-bond between the C-atoms 2′

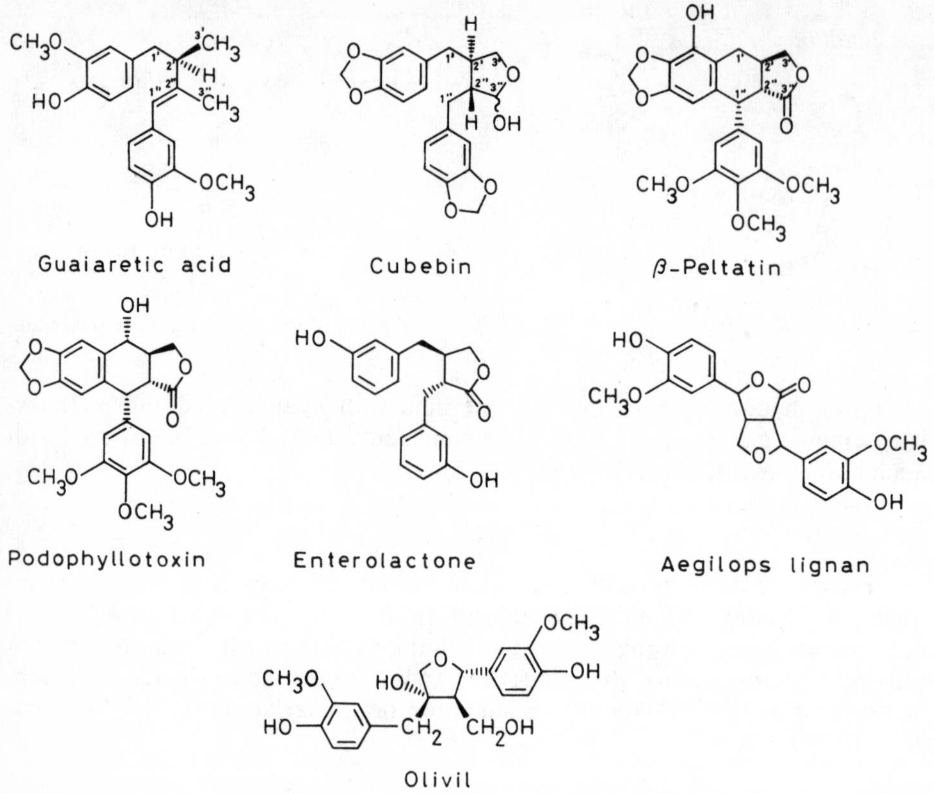

Fig. 309. Structural formulas of lignans

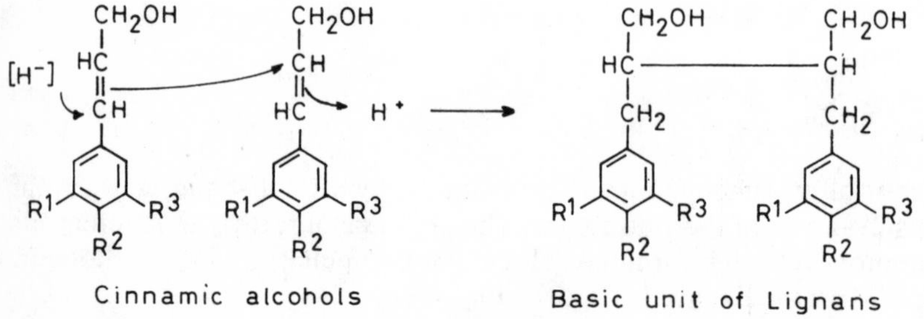

Fig. 310. Reductive dimerization of cinnamic alcohols

and 2″ two new centers of asymmetry are formed. The additional rings present in several lignans are built after the dimerization.

Significance

Aegilops lignan inhibits the germination of seeds of *Aegilops orata* (E 2.2). Podophyllotoxin is cytotoxic and a powerful irritant. It is of significance in repelling potential predators (E 5.5.3). Enterolactin has been identified in the urine of animals and humans (E 1). The amount excreted probably depends on the function of the gonads and ovaries.

References for Further Reading

150, 595, 625, 703, 811

D 23.2.5 Benzoic Acid Derivatives

Chemistry and Distribution

Benzoic acid (phenylcarboxylic acid) and several hydroxylated and methoxylated derivatives of benzoic acid are formed in microorganisms, plants, and animals (Table 52). They are the mother substances of reduced compounds (derivatives of benzaldehyde and benzylalcohol), of decarboxylated products (hydroquinone-derived compounds), but also of more complex substances, like the gallotannins and ellagitannins.

Biosynthesis

In plants and animals dihydrocinnamic acids (D 23.2.1) are the most important precursors of benzoic acids (for the formation of benzoic acid derivatives from shikimic acid and chorismic acid of D 21 and D 21.1). The dihydrocinnamic acids are cleaved with the formation of a two-carbon fragment and an aromatic

Table 52. Benzoic acid derivatives and their cinnamic acid precursors

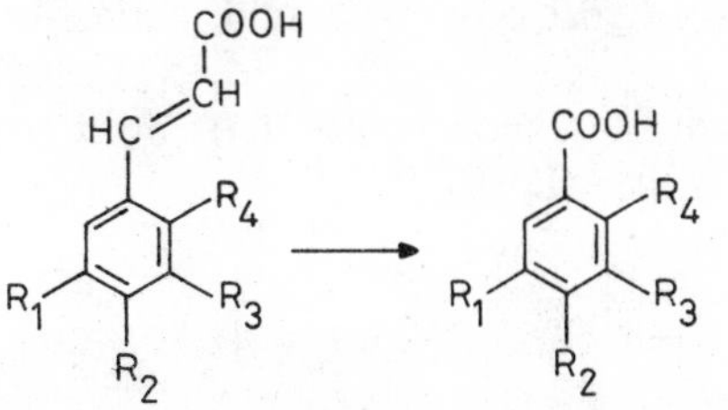

R_1	R_2	R_3	R_4		
H	H	H	H	Cinnamic acid	Benzoic acid
H	OH	H	H	p-Coumaric acid	4-Hydroxybenzoic acid
OH	OH	H	H	Caffeic acid	Protocatechuic acid
OCH_3	OH	H	H	Ferulic acid	Vanillic acid
OCH_3	OH	OCH_3	H	Sinapic acid	Syringic acid
OH	OH	OH	H	3,4,5-Trihydroxycinnamic acid	Gallic acid
H	H	H	OH	o-Coumaric acid	Salicylic acid
OH	H	H	OH	2,5-Dihydroxycinnamic acid	Gentisic acid

aldehyde, which may be oxidized to the corresponding acids or reduced to the alcohols (Fig. 311). Table 52 lists some benzoic acids with their respective cinnamic acid precursors. Activated benzoic acid derivatives, e. g., the CoA esters, may be reduced to benzaldehyde and benzoic alcohol derivatives.

Significance

Gentisic acid glycosides act as leaf movement factors in *Mimosa* sp. (E 2.2). Gallic acid is an inhibitor of flowering in *Kalanchoe blossfeldiana* (E 2.2). Benzoic acid, 4-hydroxybenzoic acid, as well as protocatechuic acid methyl and ethyl esters, are secreted from the pygidial glands of water beetles as antiseptics and feeding deterrents (E 5.1).

4-Hydroxybenzoic, vanillic, and salicylic acid are involved in the interactions of plants (E 5.3). Benzoic acid is a phytoalexin of *Malus pumila*, protocatechuic aldehyde, protocatechuic acid, and catechol are precursors of fungitoxic o-quinones (E 5.4). Vanillin is the dominant constituent of flower scents, e. g., of *Vanilla* and *Orchis* sp. (E 5.5.1). It dominates the smell of vanilla fruits used as spices (F 1). Acetosyringenone and related phenolic compounds trigger gene expression in *Agrobacterium tumefaciens* (A 2.1, E 5.4).

Transformation

a) L-Ephedrine found in various species of *Ephedra* and probably L-norpseudoephedrine are formed from a benzaldehyde-derived C_6—C_1 unit and an unknown C_2N fragment. L-Ephedrine is toxic to potential predators and acts as feeding deterrent (E 5.5.3). It is used in therapeutics (F 2). L-Norpseudoephedrine is the active constituent of khat (Abyssinian tea), i. e., the leaves of *Catha edulis*, which are chewed in East African and Arab countries as a stimulant (F 3).

b) Ubiquinones (D 21.3) are derived from 4-hydroxybenzoic acid, which in plants is built from 4-hydroxycoumaric acid.

c) 4-Hydroxybenzoylcholine is a constituent of several plants. It is built by the reaction of 4-hydroxybenzoyl CoA with choline (D 10).

L-Ephedrine   L-Norpseudoephedrine   4-Hydroxybenzoylcholine

d) Gallotannins (hydrolyzable tannins) ar constituents of many higher plants and some green algae. They are derived from gallic acid and a sugar, usually D-glucose, but in the case of β-hamamelis tannins, for instance, also D-hamamelose (D 1.1). Usually several molecules of gallic acid are bound to one sugar molecule. Gallic acid moieties may be linked together by ester groupings (depsidic bonds) formed between the carboxylic group of one molecule and the hydroxy group of another molecule (cf. the formula of 1,3,4,6-tetra-*O*-galloyl-2-*O*-trigalloyl-glucose). The galls produced on the leaves of *Rhus semialata* may contain more than 60 % gallotannins.

Ellagitannins, another group of plant constituents, are derived from hexahydroxydiphenic acid, which is formed by oxidative coupling (C 2.3.1) from two molecules of gallic acid. Ellagic acid is built by spontaneous cyclization from sugar esters of hexahydroxydiphenic acid, e.g., geraniin, after hydrolysis.

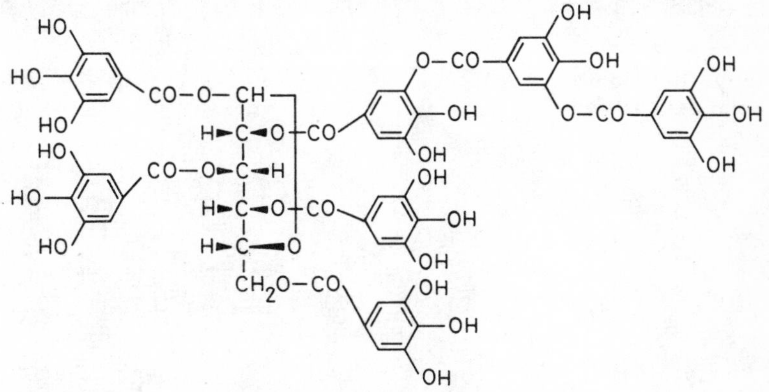

1,3,4,6-Tetra-O-galloyl-2-O-trigalloyl-glucose

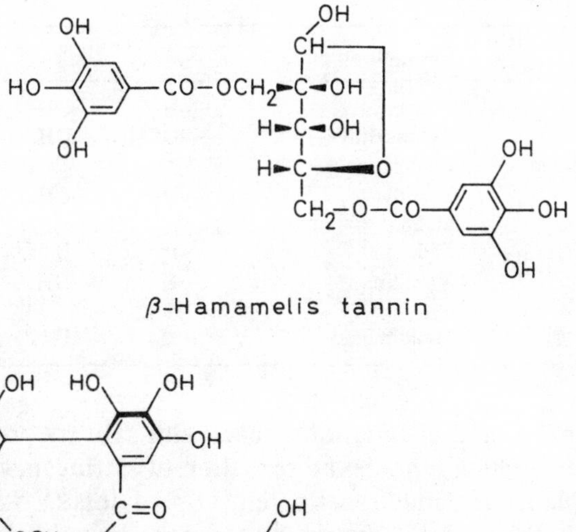

β-Hamamelis tannin

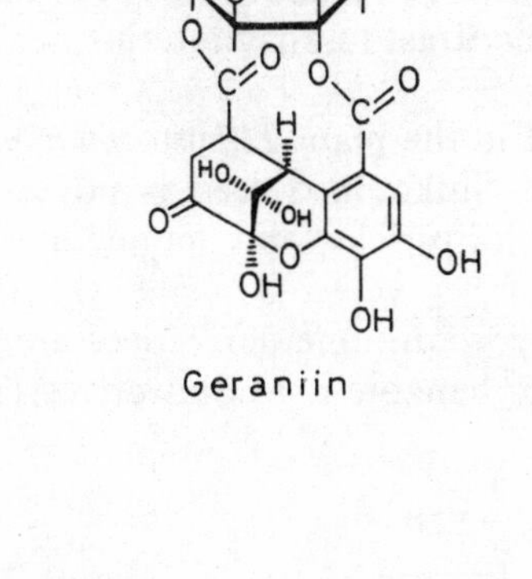

Geraniin

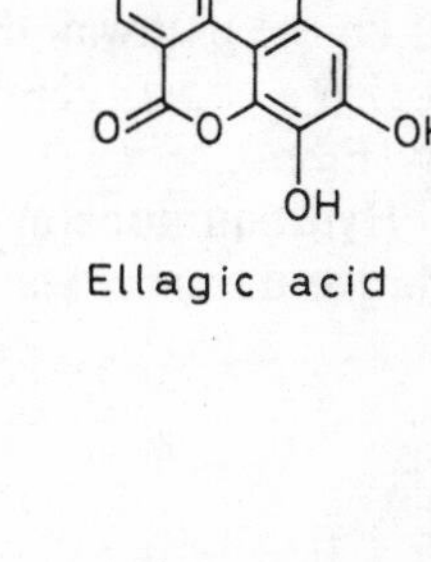

Ellagic acid

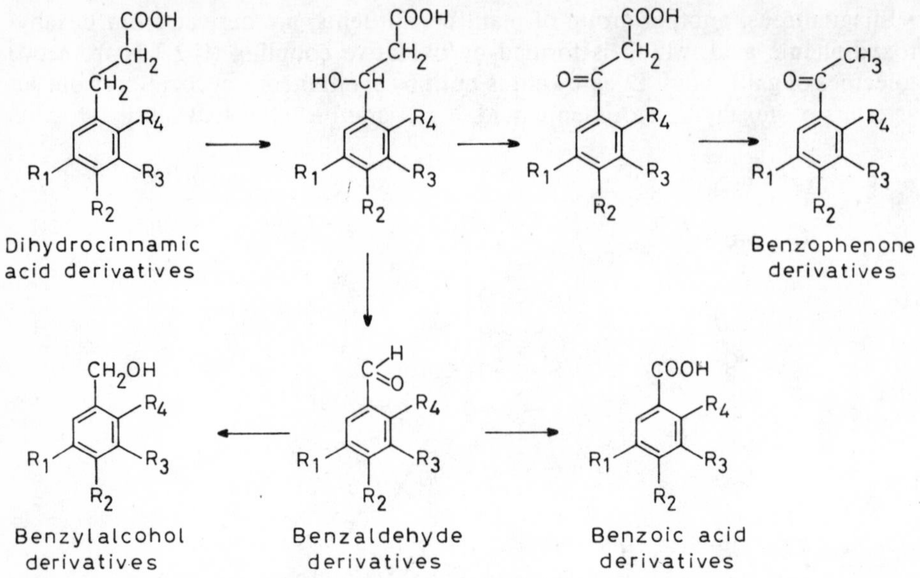

Fig. 311. Biosynthesis of benzophenone, benzylalcohol, benzaldehyde, and benzoic acid derivatives from dihydrocinnamic acids

Group	Compound	R_1	R_2	R_3	R_4
Benzo-phenone derivatives	Pungenin	—H	—OH	—OH	—H
	Acetosyringe-none	—OCH$_3$	—OH	—OCH$_3$	—H
Benzyl-alcohol derivatives	Vanillyl-alcohol	—H	—OH	—OCH$_3$	—H
	Salicin	—H	—H	—H	—O-β-D-Glc-p
Benz-aldehyde derivatives	Vanillin	—H	—OH	—OCH$_3$	—H
	Helicin	—H	—H	—H	—O-β-D-Glc-p

Gallotannins and ellagitannins react with salivary proteins and glycoproteins in the mouth, which renders the sensation of astringency. They reduce the palatability of plants and may repel potential predators (E 5.5.3). Like other types of polyphenols they inhibit the growth of nonadapted plants after being released from decaying cells and bound to the soil (E 5.3). Both types of tannins are used as astringents in therapeutics (F 2) and in the production of leather (F 5). Gallotannins are relatively rarely found in contrast to ellagitannins, which are abundant in plants.

e) Shikonin, a red pigment formed in the plant *Lithospermum erythrorhizon*, is built on the pathway shown in Fig. 312. Shikonin is used as a dyestuff for fabrics. It is also a constituent of cosmetics (biolipsticks) and applied as an antimycotic drug (F 2).

f) Hydroquinone and substituted hydroquinone derivatives are formed by the oxidative decarboxylation of 4-hydroxybenzoic acid derivatives (Table 53). Hy-

Table 53. Formation of some hydroquinone derivatives

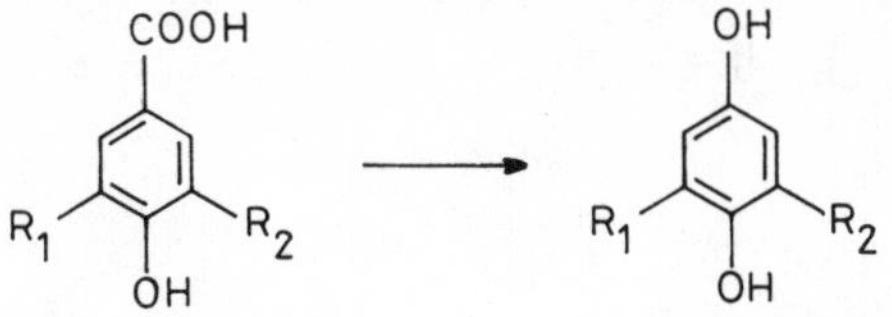

Substituents		Benzoic acid derivatives	Hydroquinone derivatives
R_1	R_2		
H	H	4-Hydroxybenzoic acid	Hydroquinone
OCH_3	H	Ferulic acid	2-Methoxyhydroquinone
OCH_3	OCH_3	Syringic acid	2,6-Dimethoxyhydroquinone

droquinone is an important constituent of the defense secretions of cockroaches and other insects. It has antimicrobial activity (E 5.1). In plants it is a precursor of arbutin, the active constituent of the leaves of *Arctostaphylos uva-ursi* used medically as a diuretic and urinary disinfectant (F 2).

All hydroquinone derivatives are reversibly oxidized to the corresponding p-quinones. Primin, which probably is derived from 2-methoxyhydroquinone, is the contact allergen of *Primula obconica* (E 5.5.3).

g) The urushiols may be formed from catechol, the decarboxylation product of protocatechuic acid and an unknown side chain precursor, probably a fatty acid derivative (D 5.2). They are contact allergens of the plant *Rhus toxicodendron* (E 5.5.3).

h) Gentisein, a xanthone derivative from *Gentiana* sp., is probably a polyketide formed from 4-hydroxybenzoic acid as starter and three molecules of malonyl CoA (D 5.3).

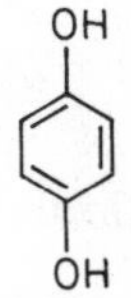

Hydroquinone 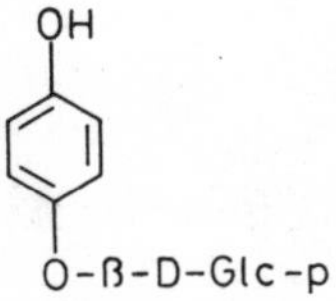   Arbutin 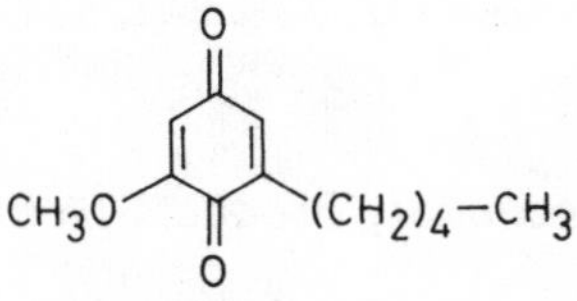   Primin

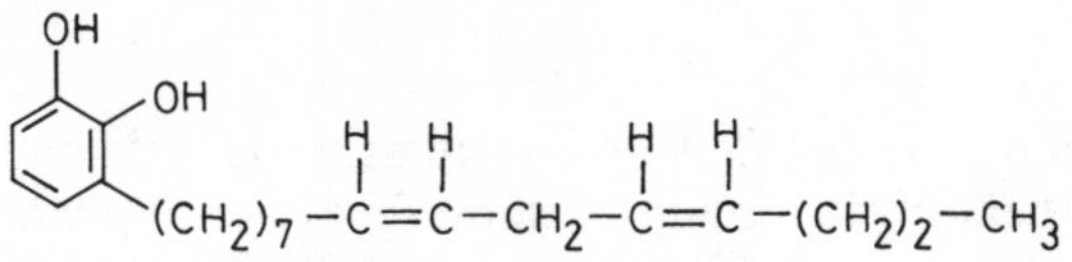

Urushiol 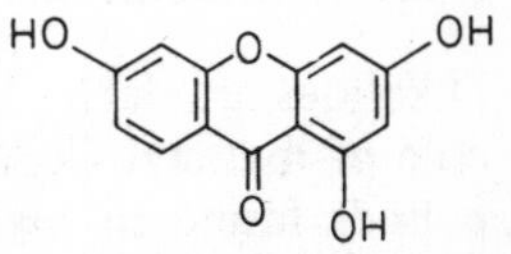   Gentisein

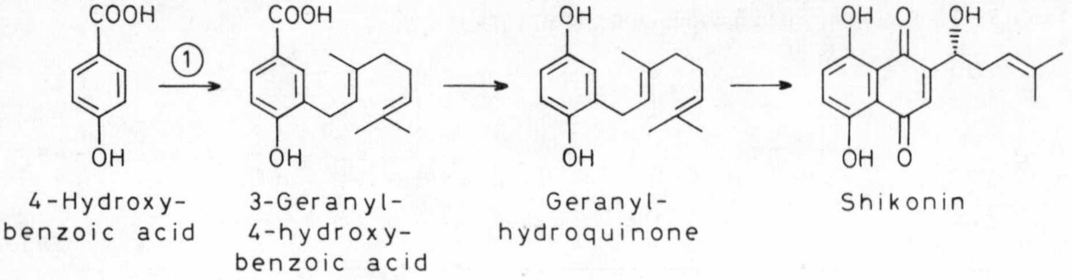

4-Hydroxy-   3-Geranyl-   Geranyl-   Shikonin
benzoic acid   4-hydroxy-   hydroquinone
                  benzoic acid

Fig. 312. Pathway of shikonin biosynthesis
1 Geranyl diphosphate: 4-hydroxybenzoate geranyltransferase

i) Sclerotins form the exoskeleton of insects as well as the egg capsules and cocoons of these animals. They consist of scleroproteins interlinked by dihydroxylated benzoic acid and benzylalcohol derivatives, by *N*-acetyldopamine (D 23.1.1), or 3-hydroxyanthranilic acid (D 21.4.1).

Two glands participate in the biosynthesis of the threads of cocoons and the formation of the egg capsules (Fig. 313). The larger gland synthesizes the scleroproteins as well as 3,4-dihydroxybenzoic acid glucoside, or 3,4-dihydroxybenzylalcohol glucoside and an oxidase. The scleroproteins are contained in the gland as sizeable globules.

In the smaller gland a glucosidase (C 6) is produced which is able to split the glucosides formed in the larger gland. During secretion, the contents of the large and the small gland are mixed, forming 3,4-dihydroxybenzoic acid or 3,4-dihydroxybenzylalcohol. These compounds are susceptible to the action of the oxidase and are transformed to 4-carboxy-o-benzoquinone or 4-hydroxymethyl-o-benzoquinone. These quinones are capable of polymerization and condensation with the peptide chains of the scleroproteins by reaction with free amino groups, e.g., of L-lysine moieties. Subsequent oxidation to the quinoid form yields brown quinone and quinhydrone pigments which give sclerotin its dark color.

References for Further Reading

190, 297, 324, 337, 338, 341, 637

D 23.3 Secondary Products Derived from Cinnamic Acids and Malonate

A special group of polyketides (D 5.3) is formed with the CoA-esters of cinnamic acids as starter and one, two, or three molecules of malonate.

D 23.3.1 Diketides

Diketides are derived from a cinnamoyl CoA derivative as starter and one molecule of malonyl CoA. Members of this group of substances are aliphatic compounds found in higher plants, like piperine, the pungent principle of pepper (F 1), and α-pyrones, like paracotoin.

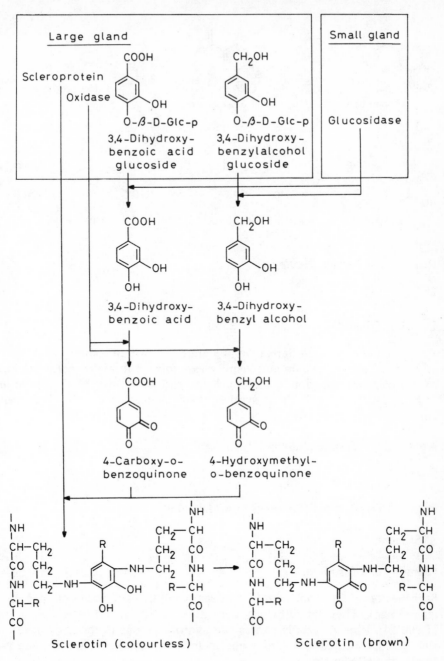

Fig. 313. Formation of sclerotin in insects

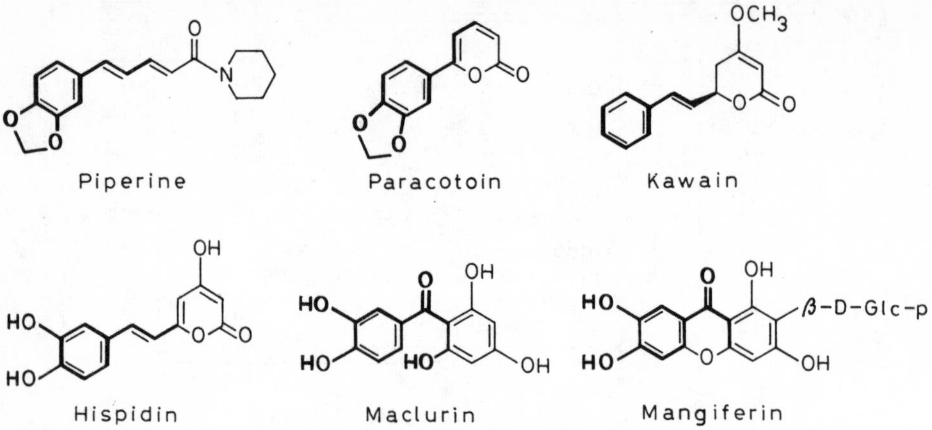

Piperine   Paracotoin   Kawain

Hispidin   Maclurin   Mangiferin

Reference for Further Reading

297

D 23.3.2 Triketides

From a cinnamoyl CoA derivative as starter and two molecules of malonyl CoA styrylpyrones, e. g., kawain and hispidin, are formed in higher plants and microorganisms. Benzophenones, like maclurin, and xanthones, like magniferin, are built in plants. Magniferin is synthesized from maclurin via *C*-glucosylmaclurin.

References for Further Reading

5, 297

D 23.3.3 Tetraketides (Stilbenes and Flavonoids)

A Stilbenes

Chemistry and Distribution

Stilbenes occur in several families of spermatophytes (Gymnospermae and Angiospermae). They are ethylene derivatives substituted by two phenyl rings (Table 54). Ring A usually carries two hydroxy groups in the m-position, while ring B is substituted by hydroxy and methoxy groups in the o-, m-, and p-position to the ethylene core.

Biosynthesis

Stilbenes are formed by stilbene synthase, a polyfunctional protein (A 3.1.1) from one molecule of a cinnamoyl CoA derivative and three molecules of malo-

Table 54. Formation of stilbenes from cinnamic acid derivatives

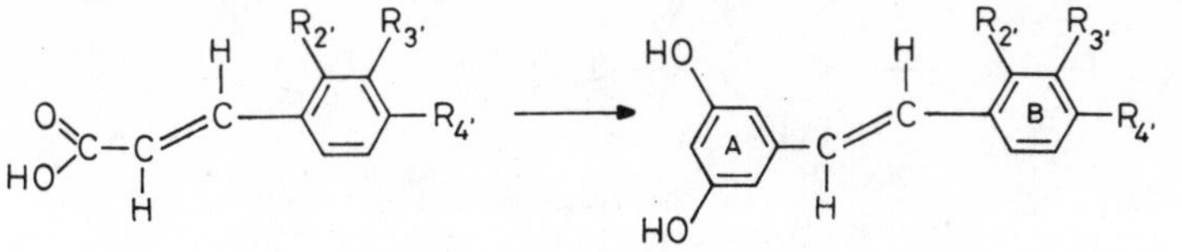

$R_{2'}$	$R_{3'}$	$R_{4'}$	Cinnamic acid	Stilbene
H	H	H	Cinnamic acid	Pinosylvin
H	H	OH	p-Coumaric acid	Resveratrol
OH	H	OH	2',4'-Dihydroxy cinnamic acid	Hydroxyresveratrol
H	OH	OH	Caffeic acid	Piceatannol
H	OH	OCH$_3$	Isoferulic acid	Rhapontigenin

nyl CoA (Fig. 314). The substitution pattern of the cinnamic acid determines that of ring B of the stilbenes (Table 54). The stilbene synthase from *Rheum rhaponticum*, for instance, forms resveratrol from p-coumaryl CoA and malonyl CoA. As with other polyketides the synthesis probably proceeds via an unstable polyoxo acid. The first stable product ist stilbene carboxylic acid.

Significance

Pinosylvin and resveratrol are involved in the resistance of plants to microbial infections (E 5.4).

Transformation

a) Dihydrostilbene carboxylic acids, like lunularic acid, and dihydrostilbenes, like lunularin and batatasin III occur in plants. Lunularic acid is derived from

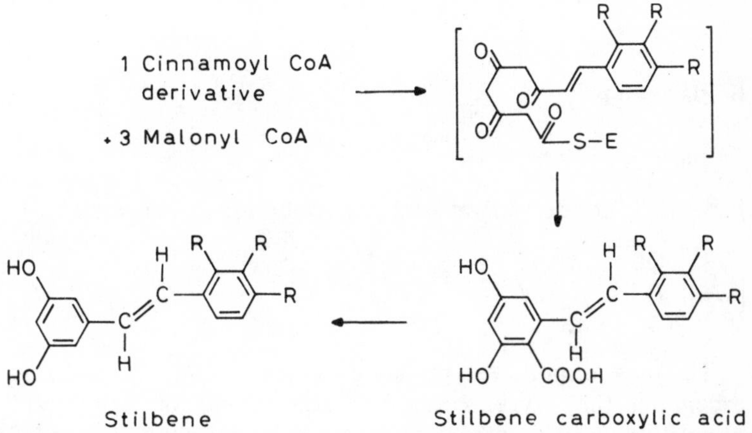

Fig. 314. Biosynthesis of stilbenes

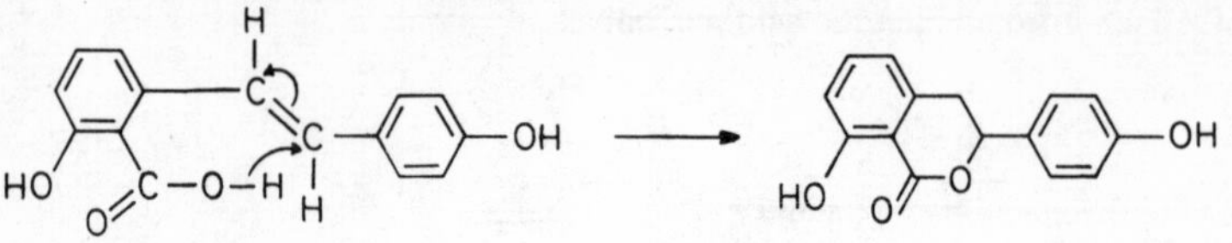

Hydrangic acid — Hydrangenol

Fig. 315. Biosynthesis of hydrangenol

prelunularic acid. It has hormone activity in liverworts (E 3.1). Batatasin III induces dormancy in the bulbils of yam (E 2.2).

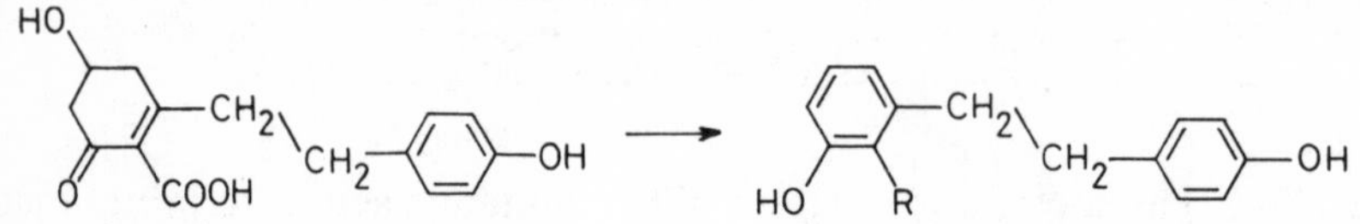

Prelunularic acid — Lunularic acid (R = COOH), Lunularin (R = H)

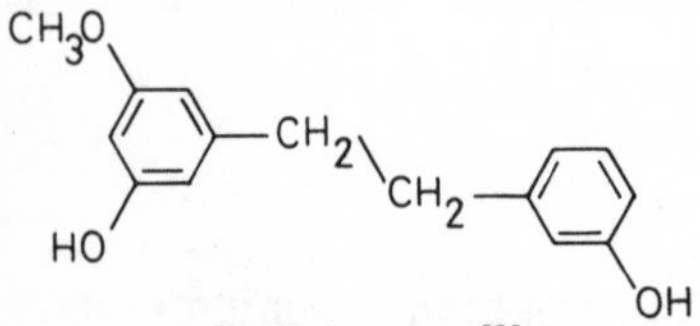

Batatasin III

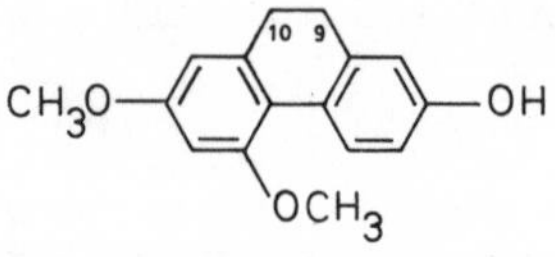

Orchinol

b) 9,10-Dihydrophenanthrenes, e. g., orchinol, are derived from dihydrostilbenes by radicalic coupling (C 2.3.1). Orchinol is a phytoalexin (E 5.4).

c) Hydrangenol, a constituent of *Hydrangea macrophylla*, seems to be formed by the addition of the carboxyl group of hydrangic acid to the ethylene double bond (Fig. 315).

B Flavonoids

Chemistry and Distribution

Flavonoids are derivatives of 2-phenylchromane (flavane):

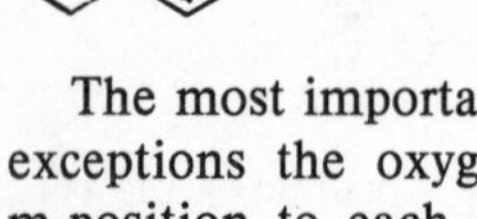

The most important types are collected in Table 55 and Fig. 316. With a few exceptions the oxygen-containing substituents of ring A are located in the m-position to each other, while the substitution pattern of ring C resembles that of cinnamic acid derivatives.

Table 55. Widespread flavonoid aglycones (formulas cf. Fig. 316)

Substituents			Cinnamoyl CoA derivatives	Chalcones	Flavanones[a]	Dihydroflavonols[b]	Flavones
R_1	R_2	R_3					
H	OH	H	p-Coumaryl CoA	2', 4,4', 6' Tetrahydroxychalcone	Naringenin	Dihydrokaempferol (Aromadendrin)	Apigenin
OH	OH	H	Caffeoyl CoA	2', 3,4,4', 6' Pentahydroxychalcone	Eriodictyol	Dihydroquercetin (Taxifolin)	Luteolin
OCH₃	OH	H	Feruloyl CoA				
OH	OH	OH	Trihydroxy cinnamoyl CoA				
OCH₃	OH	OH	5-Hydroxyferuloyl CoA				
OCH₃	OH	OCH₃	Sinapoyl CoA				

Substituents			Flavonols	Procyanidins (flavane-3,4-diols)	Catechins	Anthocyanidins
R_1	R_2	R_3				
H	OH	H	Keampferol			Pelargonidin
OH	OH	H	Quercetin	(+)-Mollis acacidin[c] (−)-Leucofisetinidin[d]	Catechin[e] Epicatechin[f]	Cyanidin
OCH₃	OH	H	Isorhamnetin			Peonidin
OH	OH	OH	Myricetin		Gallocatechin[e] Epigallocatechin[f]	Delphinidin
OCH₃	OH	OH				Petunidin
OCH₃	OH	OCH₃				Malvidin

[a] 2S-Configuration, shown in Fig. 316.
[b] 2R,3R-Configuration, shown in Fig. 316.
[c] 2R,3S,4R-Configuration, shown in Fig. 316.
[d] 2S,3R,4R-Configuration.
[e] 2R,3S-Configuration, shown in Fig. 316.
[f] 2R,3R-Configuration.

More than 3000 flavonoids are known today. Most are formed in higher plants (including mosses and ferns). Only a few are synthesized in microorganisms, e.g., *Aspergillus candidus*. Animals are unable to build flavonoids. The flavonoids present, for instance, in the wings of butterflies are derived from compounds ingested with the diet.

Biosynthesis

The flavane ring system is formed by the polyfunctional protein chalcone synthase (A 3.1.1) from one molecule of a cinnamoyl CoA ester and three molecules of malonyl CoA. The enzyme isolated from parsley *(Petroselinum hortense)* reacts with p-coumaroyl CoA and caffeoyl CoA as starter molecules. It contains neither an acyl carrier domain nor pantetheine.

The mutual interconversion of flavanones and chalcones is catalyzed by chalcone isomerase, an enzyme forming 2(S)-flavanones.

Also spontaneous cyclization of chalcones may occur. This reaction is favored if ring A carries hydroxyl groups at position 2′ and 6′. Hydrogen bonding to the carbonyl group in this case fixes the molecule in a spatial position which favors the addition of the second hydroxyl group to the double bond. Cyclization of chalcones is hindered if one of the hydroxyl groups in the position 2′ and 6′ is substituted.

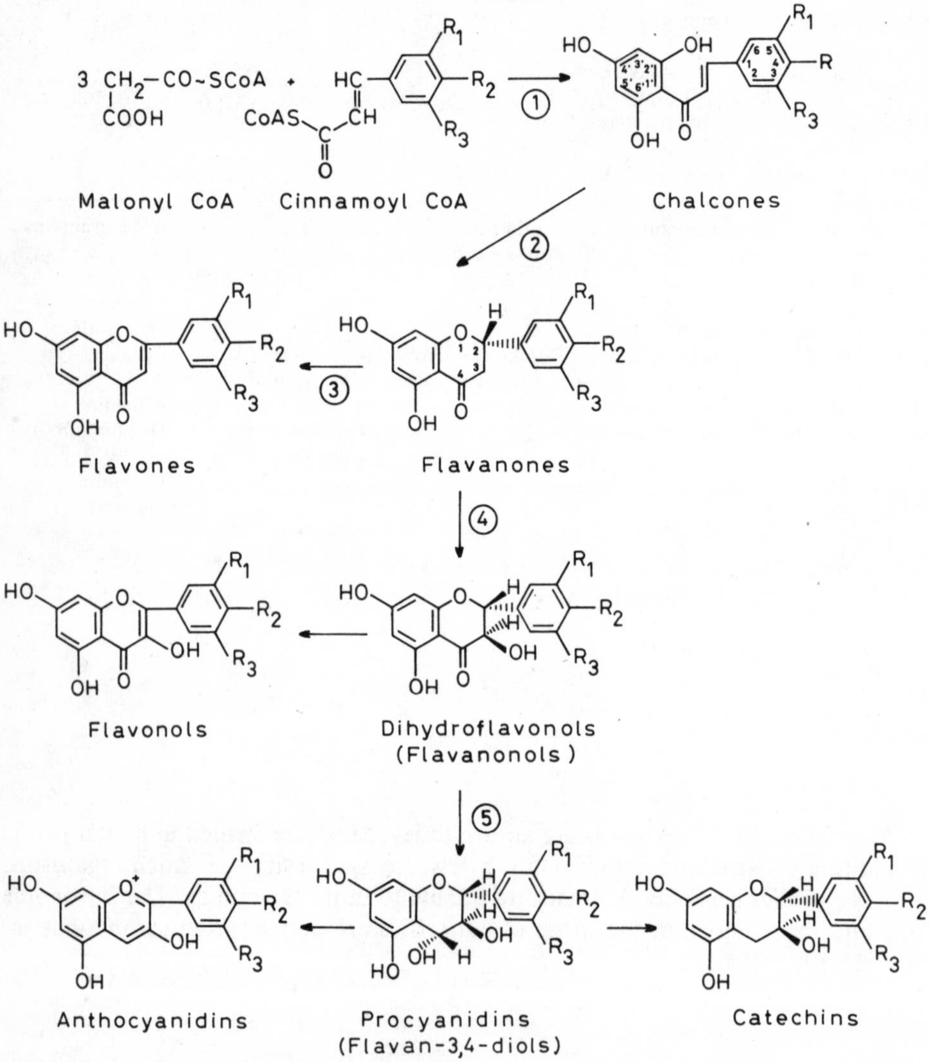

Fig. 316. Formation of the most important groups of flavonoids
1 Chalcone synthase; 2 chalcone isomerase; 3 flavanone oxidase; 4 flavanone hydroxylase (intermolecular dioxygenase, C 2.5); 5 dihydroflavonol 4-hydrogenase

The flavonanes are intermediates in the formation of the other flavonoids shown in Fig. 316.

The different types of flavonoids may undergo further modifications which lead to a large number of individual compounds. In this respect the following reactions are of significance:

a) Hydroxylation. Flavonoids bearing additional hydroxyl groups in rings A and C are common. These hydroxyl groups may be introduced by phenol oxidases (C 2.3.1) or peroxidases (C 4).

b) Methylation. Highly methoxylated flavonoids (cf. the formula of nobiletin) occur frequently in the exsudates of buds and in other lipophilic secretions. The methylations are catalyzed by specific S-adenosyl-L-methionine-dependent methyltransferases (C 3.3).

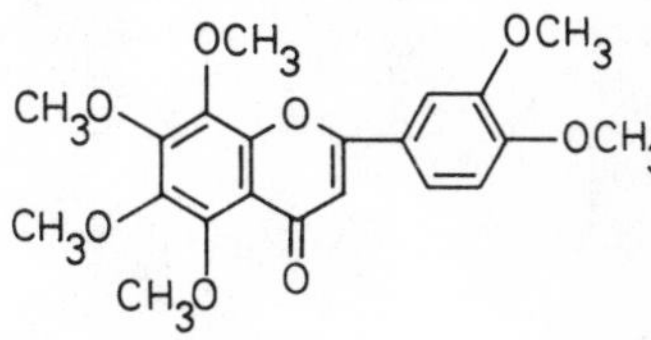

Nobiletin

c) Acylation. The hydroxy groups may react with different acid CoA esters, e. g., malonyl CoA, or with PAPS (sulfatation, D 11). These reactions are catalyzed by special acyltransferases, e. g., malonyltransferases and sulfotransferases.

d) Glycosylation. Most flavonoids accumulating in vacuoles or hydrophilic secretions are O-glycosides (cf. the formulas of rutin, naringin, and apiin). Glycosylation is catalyzed by glycosyltransferases (C 6), e. g., glucosyl, rhamnosyl, and apiosyltransferases. It often proceeds in positions 3 and 7. Also C-glycosylation may occur at electronegatively charged C-atoms (cf. the formula of vitexin). The substrate specificity of the glycosyltransferases with respect to the flavonoids is relatively high.

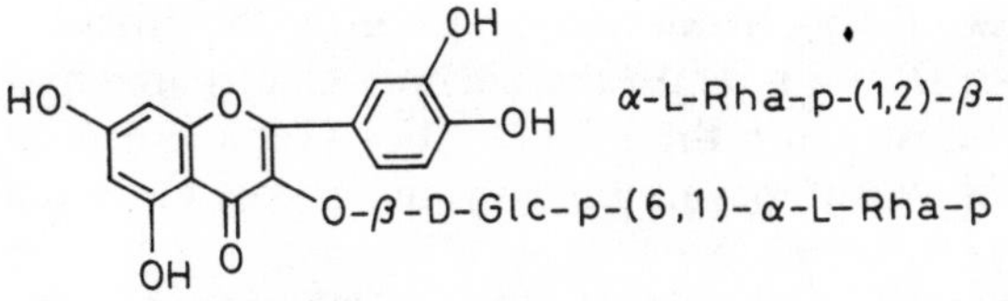

Rutin

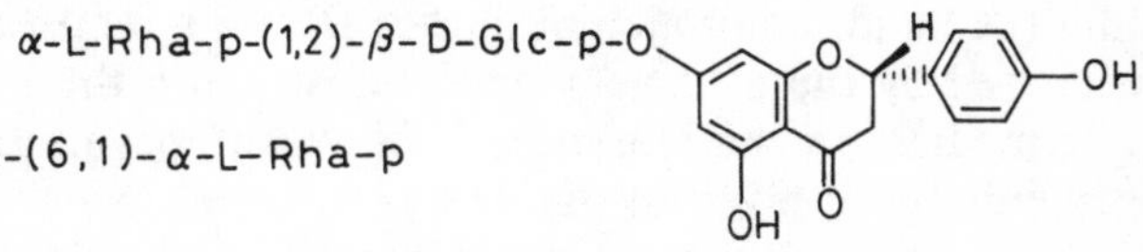

Naringin

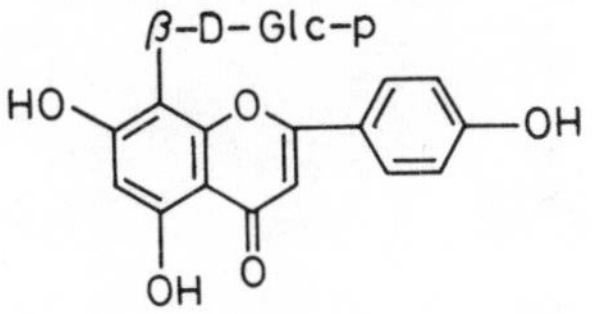

Vitexin

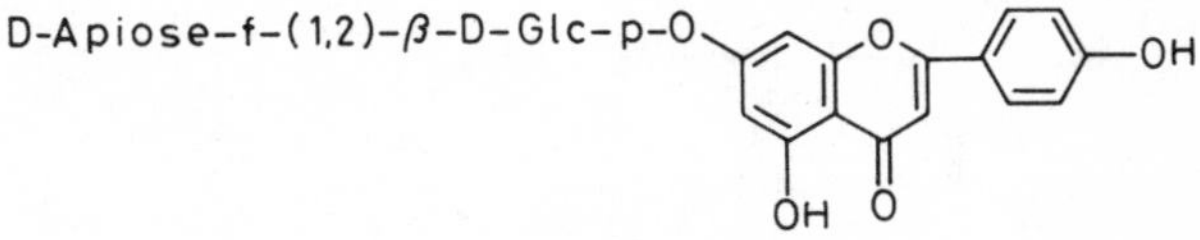

Apiin

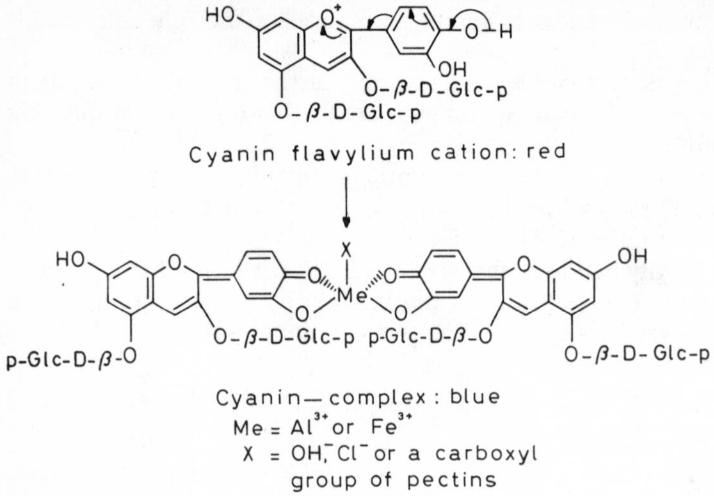

Cyanin flavylium cation: red

Cyanin—complex : blue
Me = Al^{3+} or Fe^{3+}
X = OH,$^-$ Cl$^-$ or a carboxyl
group of pectins

Fig. 317. Formation of cyanin metal complexes

e) *C*-Alkylation. Occurrence of *C*-methylated or *C*-prenylated flavonoids is relatively rare. *C*-Alkylation is more commonly found in ring A than in ring C, probably due to the greater nucleophilicity of certain C-atoms. Alkylation increases the lipophilicity of flavonoids.

f) Binding to polysaccharides. Anthocyanins (glycosides of anthocyanidins), which carry hydroxy groups at positions 3' and 4' frequently form chelates with Fe^{3+} or Al^{3+} ions linked to pectinelike polysaccharides (Fig. 317). Anthocyans, which are not part of such complexes, exist as red oxonium (flavylium) ions at physiological pH values, whereas the complexes which have quinoidal structures under these conditions are deep blue.

g) Polymerization. Proanthocyanidin polymers, also called nonhydrolyzable or condensed tannins, are the most important products of flavonoid polymerization. They exist as chains of C-4—C-8 (or C-6) linked flavan-3-ol units (Fig. 318). The monomer unit may be based on either of two stereochemistries designated as cis and trans and on either of two B-ring hydroxylation patterns, the procyanidin (PC) and the prodelphinidin (PD) units. Thus, the polymer chains are characterized by the average stereochemistry and the PC:PD ratios (Table 56). Key intermediate in the biosynthesis of proanthocyanidin polymers seems to be the carbenium ion given in Fig. 318.

In addition, dimeric flavones occur in some plants species. One of these compounds is amentoflavone found, e.g., in *Viburnum prunifolium*.

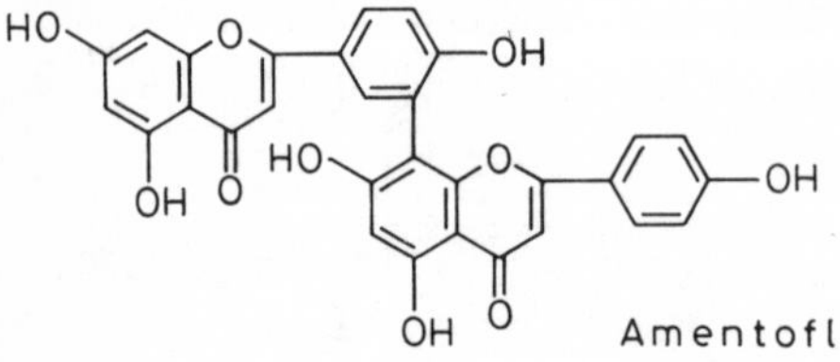

Amentoflavone

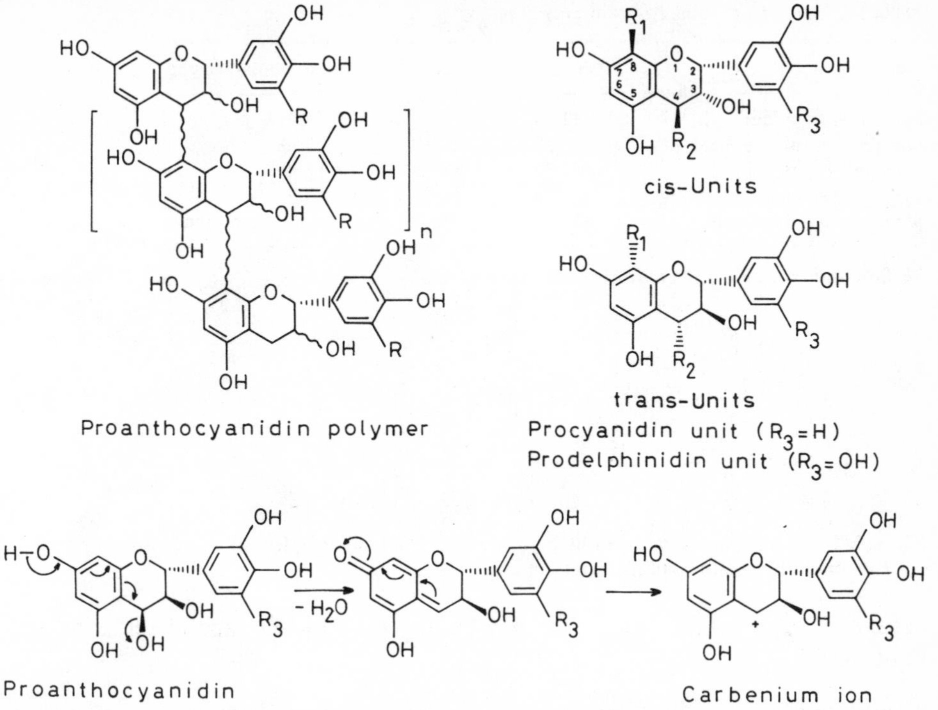

Fig. 318. Structural formulas of condensed tannins and their building blocks
The designations cis and trans relative to the procyanidin and prodelphinidin units indicate the position of the substituents at C-2 and C-3

Degradation of Flavonoids

Most groups of flavonoids, e.g., chalcones, dihydrochalcones, flavanones, flavonols, aurones, and isoflavones undergo a continuous turnover and may be degraded to products of primary metabolism (A 5). The degradation proceeds in two subsequent steps:
   a) formation of the aglycones by glycosidases (C 6) and
   b) degradation of the flavonoid skeleton by peroxidases (C 2.4.1).
   It includes the following types of reactions (Fig. 319):
— Hydroxylations,
— Fission of the skeleton to form phenylpropane, phenylethane, and phenylmethane derivatives. Hence, cinnamic acids and benzoic acids may be derived from rings B and C and phloroglucinol carboxylic acid from ring A. The cinnamic acids built are in part reused in flavonoid biosynthesis.
— Elimination of ring C as a hydroquinone derivative with the formation of chromones from rings A and B. This type of fission is most prominent with flavonoids whose A-ring hydroxy groups are blocked, e.g., by methylation.
   The degradation probably starts with the formation of a phenoxy radical at position 4', which is stabilized by several resonance forms (Fig. 305). These forms may react with OH radicals or water and may undergo further oxidation. As a

Table 56. Data of proanthocyanidin polymers

Plant, organ	PC:PD	cis:trans	Molecular weight
Pinus radiata, phloem	52:48	74:26	2 300
Pinus radiata, middle bark	90:10	41:59	1 750
Betula alba, catkins	84:16	82:18	1 800
Ribes nigrum, leaf	6:94	14:86	4 300
Malus pumila, ripe fruit	100: 0	93: 7	Not determined

PC Procyanidin units; *PD* prodelphinidin units

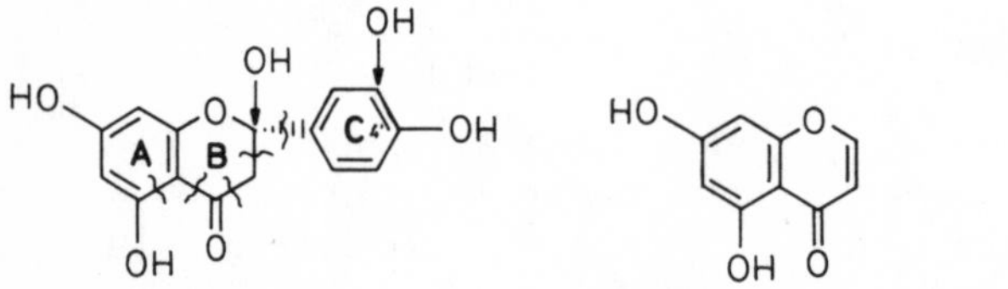

Naringenin   5,7-Dihydroxychromone

Fig. 319. Degradation of naringenin by peroxidases: → OH: hydroxylations; ~ : fissions 5,7-Dihydroxychromone is an important product

consequence, the whole molecule may be degraded to CO_2, acetic acid, and succinic acid.

Significance

Different types of flavonoids are light screens protecting plants from UV and short wavelength visible radiation (E 2.2). They are involved in the regulation of the hormone metabolism of plants by inhibition or augmentation of indoleacetic acid oxidase activity (E 2.2). Certain plant flavonoids are involved in the regulation of gene expression in *Rhizobium* (E 5.4). Naringenin is a gibberellin antagonist in some plants (E 2.2). Highly methoxylated flavonoids have frequently been found in bud exsudates and other lipophilic secretions. Like nobiletin, a constituent of Citrus leaves, they are fungitoxic (E 5.4). Flavones, flavonols, anthocyanins, and other groups of flavonoids are flower pigments attracting pollinators (E 5.5.1). Other flavonoids repel predators. They are, however, feeding stimulants for certain specialized insects (E 5.5.3). Naringin is the bitter principle of *Citrus* fruits (F 1). Rutin and other flavonoids reduce capillary fragility. They are constituents of many herbs used in medicine (F 2).

Condensed tannins interact with glycoproteins of the salivary and the mouth mucosa causing the sensation of astringency. They reduce the palatability of plant proteins and repel potential predators (E 5.5.3). After being set free from decaying plant material they reduce the viability of nonadapted plants in habitats like heaths and moorlands (E 5.3). Condensed tannins are used in therapeutics as astringents (F 2) and in technology as tanning agents (F 5).

Transformation

a) The dihydrochalcone phloridzin is a characteristic constituent of *Malus* sp. It is formed from the corresponding chalcone by reduction (Fig. 320).

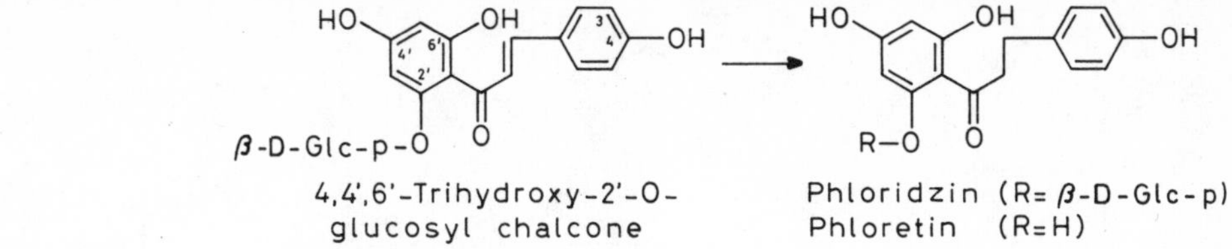

HO⟶⟨OH⟩⟨OH⟩ ⟶ HO⟶⟨OH⟩⟨OH⟩

β-D-Glc-p-O O   R-O O

4,4',6'-Trihydroxy-2'-O-   Phloridzin (R= β-D-Glc-p)
glucosyl chalcone   Phloretin (R=H)

Fig. 320. Biosynthesis of dihydrochalcones

During degradation phloridzin is cleaved to glucose and phloretin which either
— is degraded to phloroglucinol and 4-hydroxydihydrocinnamic acid by a c-acylhydrolase or
— is hydroxylated in position 3 and dehydrogenated by phenol oxidase to the corresponding o-quinone, which has fungitoxic properties (E 5.4). The o-quinone easily forms insoluble polymers.

b) Aurones are derived from chalcones by the formation of a five-membered O-heterocyclic ring. Both groups of substances frequently occur together in plants.

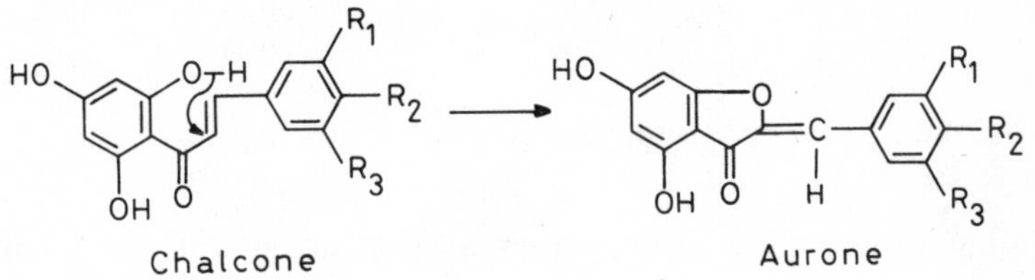

Chalcone   Aurone

c) Isoflavones (3-phenylchromanes) are frequent constituents of Leguminosae. They have been found, however, also in culture filtrates of mushrooms, *Aspergillus* and *Streptomyces* sp. Isoflavones are formed by a shift of the phenyl group from C-2 to C-3 of the flavane ring system. Immediate precursor is a dihydroflavone (Fig. 321). Certain isoflavones, e.g., genistein, daidzein, and formononetin possess estrogenic activity in mammals and birds (E 5.5.3).

Luteone and 2'-deoxyluteone inhibit the development of phytotoxic fungi (E 5.4).

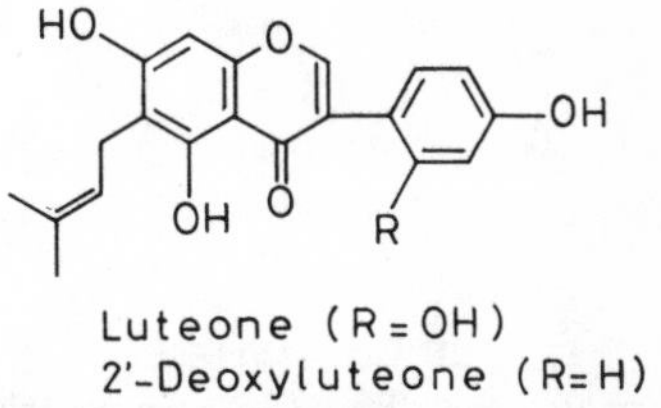

Luteone (R = OH)
2'-Deoxyluteone (R= H)

d) Pterocarpans are derived from isoflavones by the formation of a new o-heterocyclic ring with participation of the oxygen atom at position 4 (Fig. 322). Medicarpin, pisatin, phaseollin, and 6-hydroxyphaseollin act as phytoalexins (E 5.4.). A special group of pterocarpans are the 3-phenylcoumarins which have a lactone structure (Fig. 322). Coumestrol possesses estrogenic activity in animals (E 5.5.3).

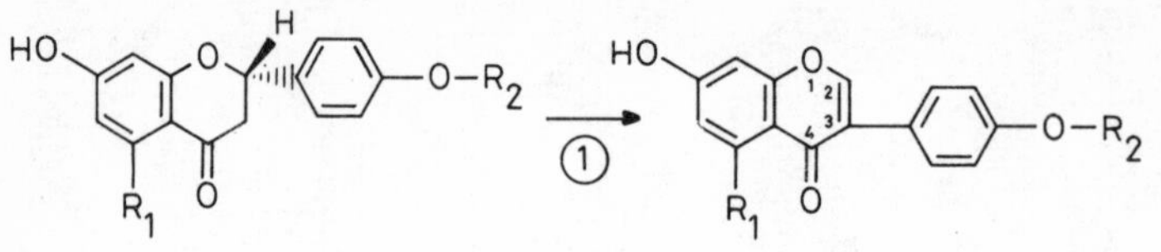

Flavanone

Isoflavone	R_1	R_2
Genistein	OH	H
Daidzein	H	H
Formononetin	H	CH_3
Biochanin A	OH	CH_3

Fig. 321. Biosynthesis of isoflavones from flavanones by isoflavone synthase

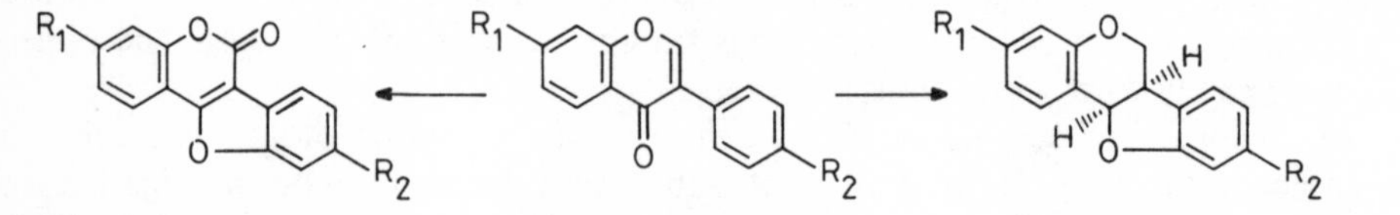

3-Phenylcoumarin
Coumestrol (R_1=OH, R_2=OH)

Isoflavone
Daidzein (R_1=OH, R_2=OH)

Pterocarpan
Medicarpin (R_1=OH, R_2=OCH$_3$)

Fig. 322. Formation of pterocarpans and phenylcoumarins

e) Rotenone occurs together with isoflavones in the plant *Derris elliptica*. It is a toxic compound repelling potential predators (E 5.5.3). Rotenone inhibits mitochondrial respiration. The carbon atom marked by ● originates from the 2-methoxy group of an isoflavonone type intermediate.

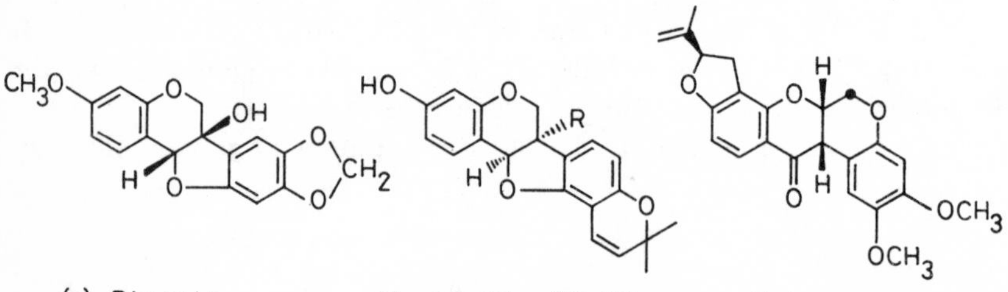

(+)-Pisatin

Phaseollin (R=H)
6-Hydroxyphaseollin (R=OH)

Rotenone

f) Homoisoflavones, e. g., Z-eucomin, are formed from chalcone-type intermediates. The C-atom marked by "●" is derived from a methoxyl carbon probably after dehydrogenation (Fig. 323).

g) Retroflavonoids, i.e., compounds in which the aromatic ring of the chromane system is derived from cinnamic acid and the substituting phenyl ring from malonic acid, occur for instance in *Glycyrrhiza* sp. An example is licoricone. Retroflavonoids are probably derived from retrochalcones which may be formed from chalcones by the set of reactions shown in Fig. 324.

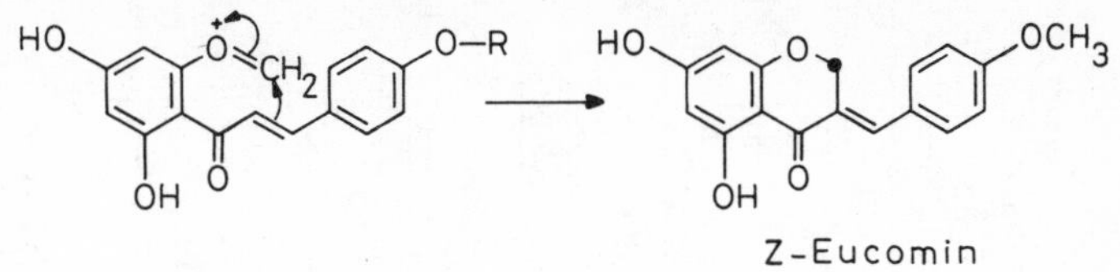

Fig. 323. Assumed key reaction in homoisoflavone biosynthesis

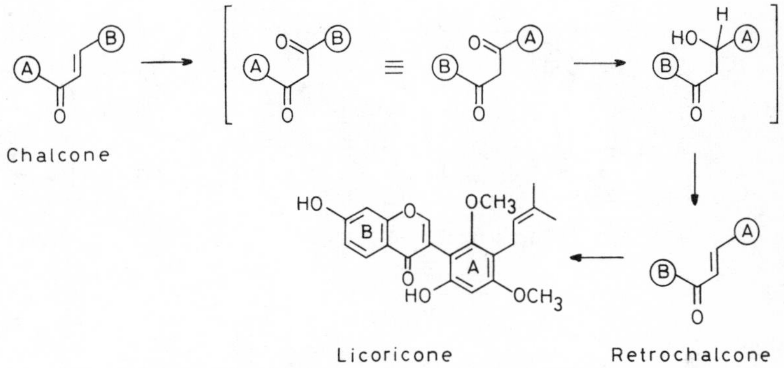

Fig. 324. Probable pathway of retroflavonoid biosynthesis
Ⓐ and Ⓑ: A- and B-ring of the precursor chalcone, respectively

References for Further Reading

20, 32, 150, 166, 190, 196, 225, 276, 277, 314, 324, 327, 328, 337, 351, 380, 608, 692, 727, 826, 842, 887

D 23.4 Homogentisic Acid-Derived Hydroquinone and Benzoquinone Derivatives

Chemistry and Distribution

The hydroquinones and benzoquinones formed from homogentisic acid carry methyl or ethyl groups attached to the aromatic ring. In addition, most of them are joined with an isoprenoid side chain, which may be more or less saturated (cf. the formulas of plastoquinols/plastoquinones and tocopherols/tocopherylquinones).

Plastoquinols and plastoquinones are formed in plants, tocopherols and tocopherylquinones in plants and microorganisms. 2-Methyl- and 2-ethylbenzoquinones are synthesized in animals.

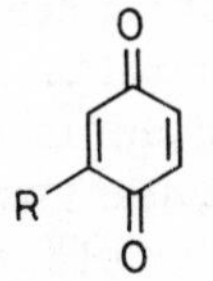

2-Methylbenzoquinone
$(R = CH_3-)$

2-Ethylbenzoquinone
$(R = CH_3-CH_2-)$

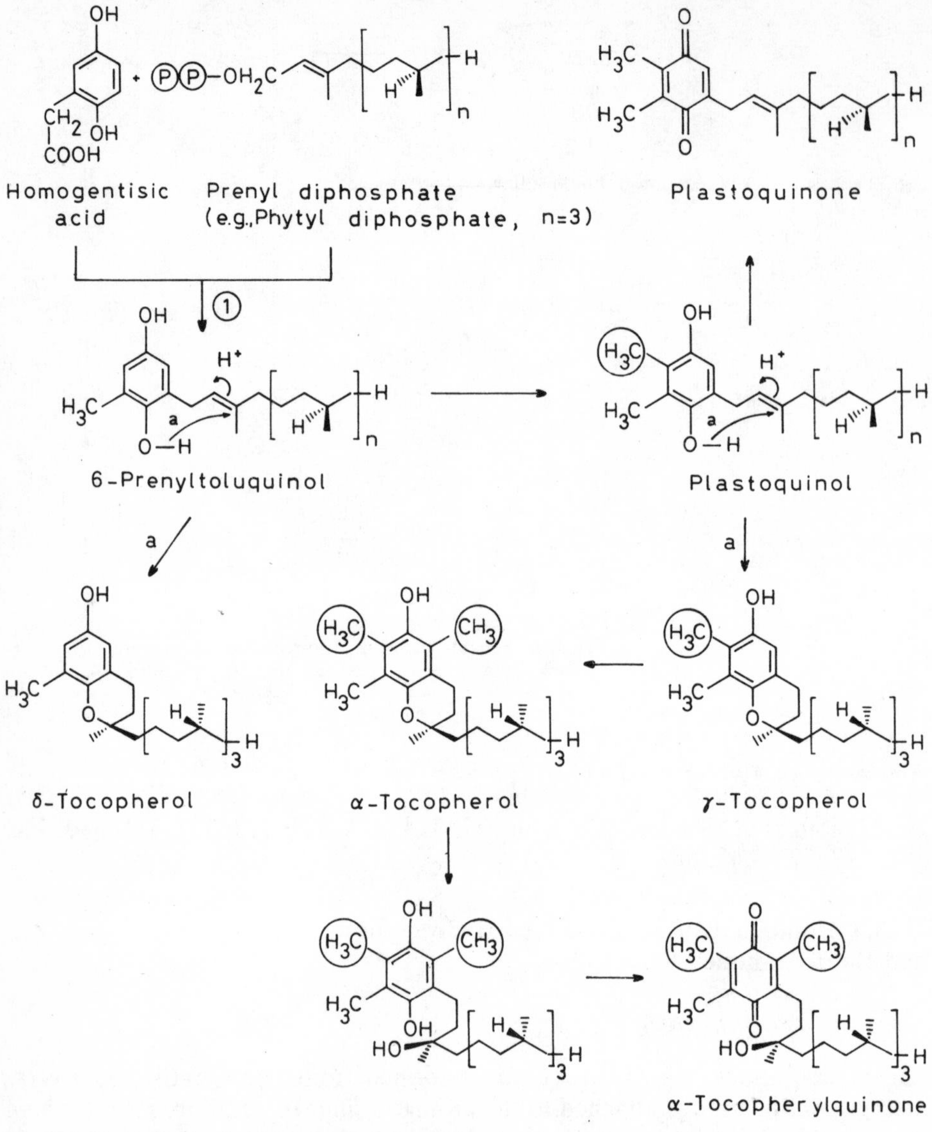

Fig. 325. Biosynthesis of plastoquinols and tocopherols
The *encircled* methyl groups are derived from L-methionine. *1* Homogentisate prenylase

Biosynthesis (Fig. 325)

The aromatic nucleus and one of the methyl groups of the quinols and quinones are derived from homogentisic acid, a product of L-tyrosine degradation (D 23). Homogentisic acid reacts with prenyldiphosphates of different chain length (D 6). The quinols formed are reversibly dehydrogenated to the corresponding quinones, e.g., plastoquinones. Tocopherylquinones are formed by dehydrogenation after hydrolytic splitting of the *O*-heterocyclic ring of tocopherols.

Homogentisic acid is probably also the precursor of 2-methyl- and 2-ethyl-benzoquinones synthesized in some beetles.

Significance

Plastoquinols/plastoquinones and tocopherols/tocopherylquinones serve as redox systems in cell respiration and photosynthesis (E 2.2). They are formed in plastids (A 3.1.3). Tocopherols and tocopherylquinones are vitamins for humans and higher animals (vitamin E, E 2.1). 2-Methyl- and 2-ethylbenzoquinone are constituents of the defense secretions of certain beetles (E 5.1).

References for Further Reading

490, 763

D 24 Porphyrin-Derived Secondary Products

Chemistry and Distribution

Porphyrins contain a cycle of four pyrrole rings interlinked by methine groups. The pyrrole rings are substituted with acetyl, propionyl, methyl, vinyl, and ethyl groups. Most porphyrins form a tetravalent complex with a central metal ion (Mg^{2+}, Fe^{2+}, Fe^{3+}) chelated by covalent and noncovalent bindings. Most important porphyrins are heme (ferroprotoporphyrin IX), containing Fe^{2+}, hemin (ferri-protoporphyrin IX, hematin), containing Fe^{3+}, and chlorophyll a, containing Mg^{2+}, as central atoms.

Heme possesses a planar conformation in which two coordination bonds of the Fe^{2+} atom are perpendicular to the plane of the porphyrin ring system. In heme proteins, e. g., hemoglobin and myoglobin, the fifth coordination position of Fe^{2+} is occupied by the imidazole moiety of a protein-bound histidine residue. The sixth binding position is either free or may be occupied by oxygen (oxygenated form). In the cytochromes the central Fe^{2+} ion may reversibly change to Fe^{3+}. In these porphyrin proteins both perpendicular binding sites of the iron atom are occupied by specific amino acids, a structure which prevents oxygenation. In cytochrome f the porphyrin ring system is covalently bound to protein by addition of the SH-group of an L-cysteine residue to one of the vinyl groups.

Biosynthesis of Porphyrin

The fundamental building block of porphyrins is porphobilinogen, which originates in bacteria, yeasts, and animals from succinyl CoA and glycine on the pathway given in Fig. 326. In higher plants and algae, however, 5-aminolevulinic acid is derived from L-glutamic acid (Fig. 327).

Porphobilinogen polymerizes to the porphyrin ring system with the elimination of ammonia. Of special significance is the formation of uroporphyrinogen III, which differs from the symmetrical uroporphyrinogen I, which is of minor importance, by reversion of the acetyl and the propionyl group in ring D.

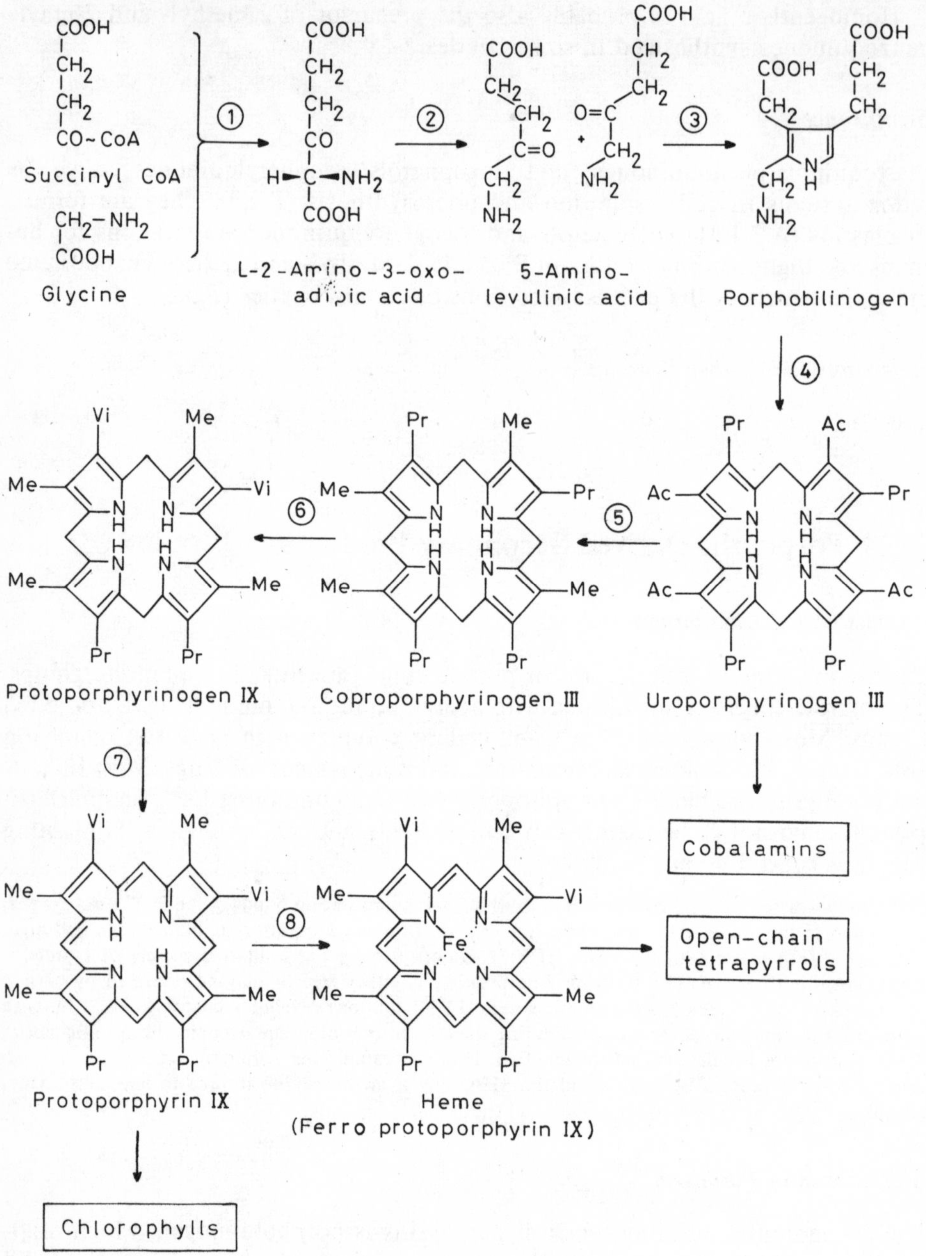

Fig. 326. Formation of heme

1 5-Aminolevulinate synthase; *2* spontaneous reaction; *3* porphobilinogen synthase; *4* uroporphyrinogen synthase, uroporphyrinogen co-synthase; *5* uroporphyrinogen decarboxylase; *6* coproporphyrinogen oxidase; *7* protoporphyrinogen oxidase; *8* ferrochelatase

Ac acetyl (—CH₂COOH); *Et* ethyl (—CH₂CH₃); *Me* methyl (—CH₃); *Pr* propionyl (—CH₂CH₂COOH); *Vi* vinyl (—CH=CH₂)

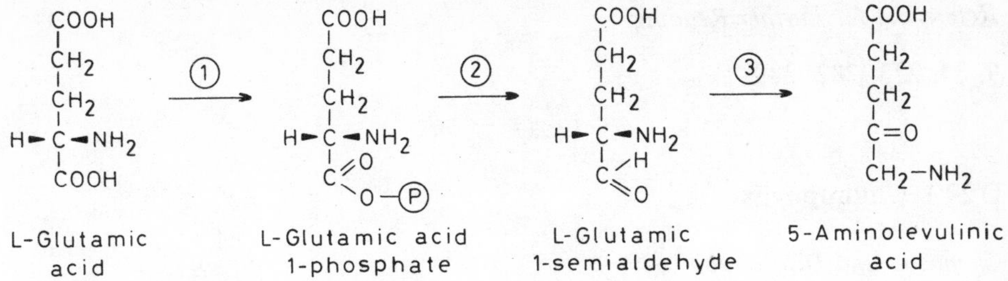

Fig. 327. Formation of 5-aminolevulinic acid from glutamic acid
1 ATP: L-glutamate 1-phosphotransferase; *2* L-glutamate 1-semialdehyde: NADP$^+$ oxidoreductase (phosphorylating); *3* L-glutamate 1-semialdehyde aminotransferase (catalyzing an intramolecular transfer of the amino group)

In the biosynthesis of coproporphyrinogen III the carboxylic groups of the acetate residues are removed, leaving methyl groups in the corresponding positions.

During the formation of protoporphyrinogen IX the propionic acid side chains are decarboxylated at positions 2 and 4 and converted to vinyl groups by dehydrogenation. A further oxidation step yields protoporphyrin IX which is the starting material in the synthesis of the iron-containing porphyrins of the heme and hemine types.

Heme is the prosthetic group of hemoglobin and myoglobin which reversibly bind oxygen and are involved in the oxygen transport in animals (E 2.2). Heme and heme derivatives also mediate the electron transport by cytochromes which is accompanied with a reversible change of Fe^{2+} and Fe^{3+} (E 2.2). In addition, hemin is the prosthetic group of peroxidases (C 2.4.) and catalases, and heme that of dioxygenases (C 2.5).

Secondary Products Built from Porphyrins

Secondary products are formed from the porphyrins of primary metabolism in microorganisms, plants, and animals (Fig. 326). The most important groups are chlorophylls (D 24.1), open-chain tetrapyrrols (D 24.2), cobalamins (D 24.3), and factor 430. The latter is the prosthetic group of methylcoenzym M reductase (C 3.2). It is one of the few secondary products containing nickel.

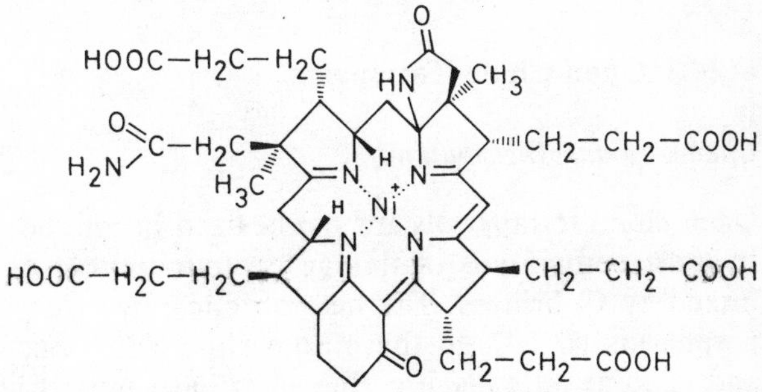

Factor 430

References for Further Reading

9, 34, 223, 282, 889

D 24.1 Chlorophylls

Chemistry and Distribution

Chlorophylls are porphyrins with Mg^{2+} as the central atom and a phytyl side chain. The most important representatives are the chlorophylls a and b. Chlorophylls are synthesized in the "green" tissues of higher plants as well as in algae and some bacteria.

Biosynthesis

The chlorophylls are derived from protoporphyrin IX (Fig. 328). The Mg^{2+} ion is introduced by a specific chelatase. The propionyl side chain of ring C is methylated and a new five-membered ring is formed (protochlorophyllide a). Reduction of ring D and ester formation with geranylgeranyl diphosphate (D 6.4) yields geranylgeranylchlorophyllide a, which is subsequently reduced (formation of the phytyl residue) to chlorophyll a. Chlorophyll b is formed from chlorophyll a by the oxidation of a methyl group to an aldehyde group.

During the senescence of leaves or the ripening of fruits the chlorophylls are degraded to compounds of primary metabolism. In green leaves they undergo a rapid turnover (A 5).

Significance

The chlorophylls a and b, as well as a number of structurally related compounds, are involved in the electron transport in photosynthesis (E 2.2).

References for Further Reading

34, 40, 370, 454, 610, 855, 889

D 24.2 Open-Chain Tetrapyrrols

Chemistry and Distribution

Open-chain tetrapyrrols are synthesized in microorganisms, plants, and animals. In most compounds of this group four pyrrole or pyrroline residues are interlinked by C_1-bridges. The heterocyclic rings carry the substituents typically for porphyrins (D 24). In the biliproteins, which occur in plants, especially algae, and in some animals, e.g., in caterpillars, butterflies, and mammals, open-chain tetrapyrrol residues are linked noncovalently or covalently to proteins.

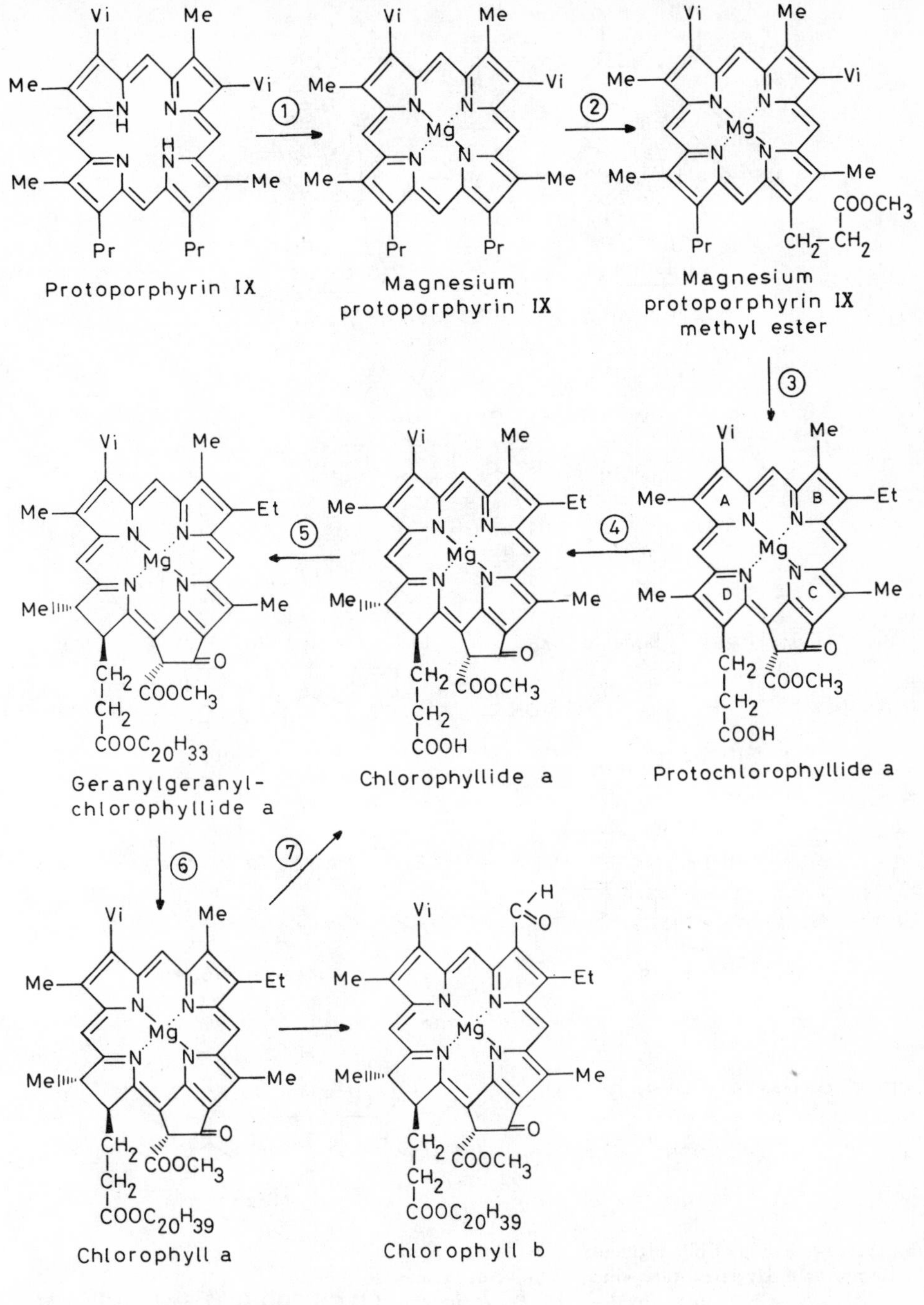

Fig. 328. Biosynthesis of chlorophyll a

1 Magnesium chelatase; *2* magnesium protoporphyrin methyltransferase; *3* NADH-dehydrogenase, other enzymes not fully elucidated; *4* membrane-bound dehydrogenase, in higher plants light requiring; *5* chlorophyllide a phytylating enzyme; *6* reductase; *7* chlorophyllase

Et ethyl (—CH$_2$CH$_3$); *Me* methyl (—CH$_3$); *Pr* propionyl (—CH$_2$CH$_2$COOH); *Vi* vinyl (—CH=CH$_2$)

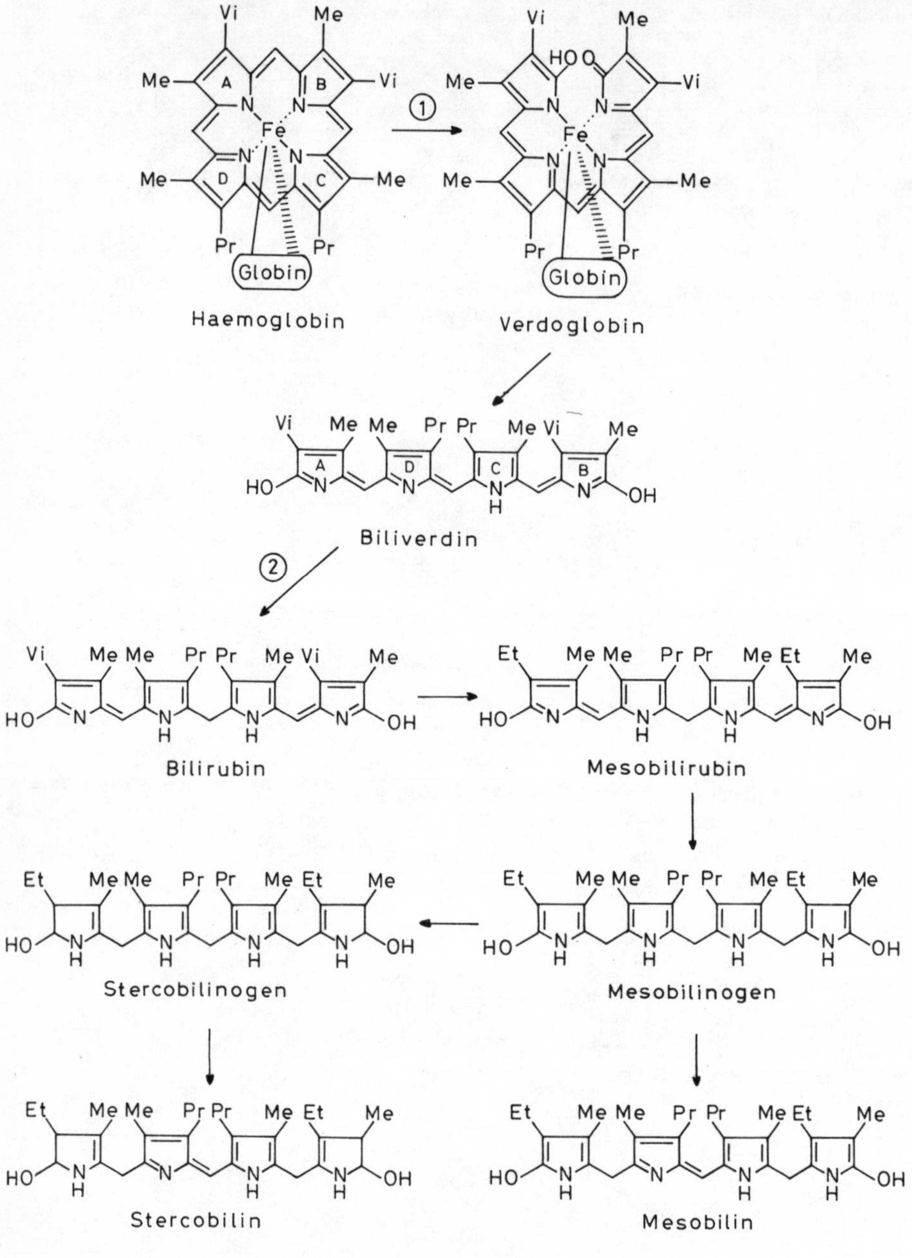

Fig. 329. Formation of bile pigments
1 Hemoglobin oxygenase (decycling); *2* biliverdin reductase
Et ethyl (—CH₂CH₃); *Me* methyl (—CH₃); *Pr* propionyl (—CH₂CH₂COOH); *Vi* vinyl (—CH=CH₂)

Biosynthesis

In higher animals and plants the open-chain tetrapyrrols are formed by cleavage of hemoglobin and other protoporphyrin IX derivatives between rings A and B. In butterflies the porphyrin ring is opened between rings C and D.

In humans and higher animals hemoglobin is degraded to the so-called bile pigments (Fig. 329). In the first step it is converted to the green pigment choleglobin (verdoglobin). From verdoglobin the blue-green compound, biliverdin, is formed by the loss of iron and the protein component globin. Biliverdin is easily reduced to the orange-red pigment bilirubin. Albumin-bound bilirubin occurs in small quantities in the bloodstream. This bilirubin-protein complex is decomposed in the liver and bilirubin, in the free state or glycosidically linked with D-glucuronic acid (D 1.2), is excreted into the bile (E 1).

Bilirubin and bilirubin glucuronide are the most important bile pigments. They impart a yellow color to the feces, which is retained if chyme passes rapidly through the intestine. Usually, however, bilirubin is transformed by enzymes of the intestine and by the action of intestinal bacteria. Mesobilirubin is formed by reduction of the vinyl groups. Further reduction yields mesobilirubinogen and stercobilinogen. In these compounds the pyrrol rings are linked by CH_2-groups. The dark brown pigments of the feces, mesobilin and stercobilin, originate from the latter compounds by spontaneous, nonenzymatic oxidation. Also compounds containing only two pyrrol moieties may be formed.

Significance

Open-chain tetrapyrrols have ecological significance as pigments in microorganisms, plants, and animals. They participate in the coloration of Cyanophyceae, algae, caterpillars, and butterflies and form the green to brown shades and spots in the shell of bird eggs (E 5.6). The formation of bile pigments enables the elimination of compounds with a porphyrin ring system from the body of higher animals and human beings (E 1). In algae certain phycobiliproteins, e.g., the intense red phycoerythrins, participate as accessory pigments in photosynthesis. In the dinoflagellate *Pyrocystis lunula*, causing phosphorescence of the sea, an open-chain tetrapyrrol acts as luciferin (C 2.5). In higher plants an open-chain tetrapyrrol is the prosthetic group of phytochrome, a chromoprotein regulating many light-dependent processes (E 2.2), including the formation of secondary products (A 4.4.2).

PHYTOCHROME

P_r
(inactive)

P_{fr}
(active)

The linkages between the tetrapyrrol moiety and the protein involve a thioester bond and a covalent bond to one of the propionic acid side chains. The nature of the substituents R_1, R_2, and R_3 in P_{fr} is not clarified as yet.

Phytochrome exists in two spectral forms, the chemical structure of which is still unknown. These forms are interconvertible by means of irradiation. Irradiation with far red light (maximum at 730 nm) leads to the formation of an inactive form (P_r), whereas a physiologically active form (P_{fr}) is produced by the action of red light (maximum 660 nm). Biophysical experiments suggest that this photoconversion proceeds via a series of instable intermediates whose life time at 0 °C is in the region of milliseconds to seconds.

References for Further Reading

29, 40, 269, 454, 660

D 24.3 Cobalamins

Chemistry and Distribution

Corrin, the porphyrinlike ring system of the cobalamins resembles uroporphyrinogen III (D 24) relative to the arrangement and the structure of the side chains. In contrast to the porphyrins, however, the pyrrol rings are reduced to a higher degree and two of them are directly linked to one another. The central atom of the corrin ring system is Co^{2+}, to which covalently a —CN, —CH_3, —OH, or 5'-deoxyadenosyl residue may be bound. The corrin ring is linked via a 1-amino-2-propanol unit and D-ribose with 5,6-dimethylbenzimidazol, which is α-glycosidically connected with the D-ribose.

Cobalamins are built in microorganisms and plants.

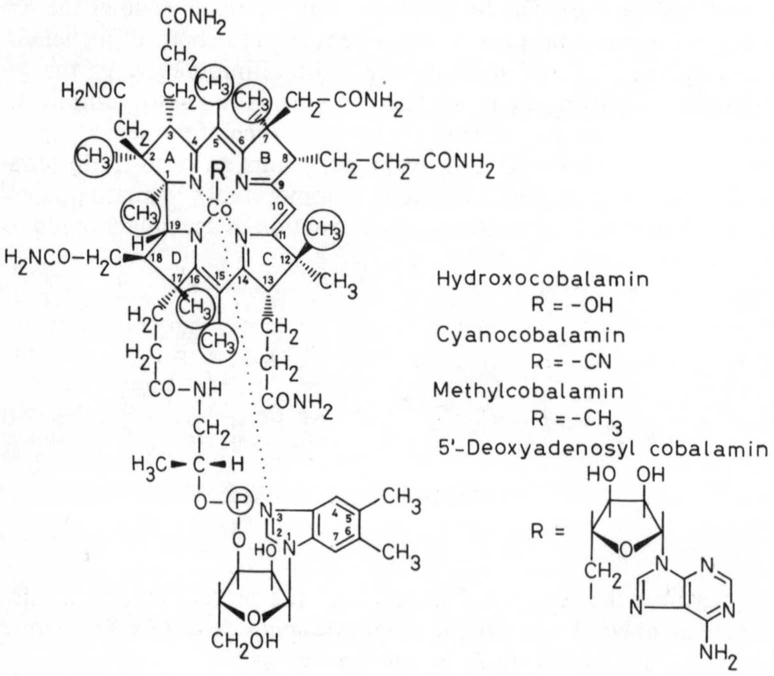

Hydroxocobalamin
R = -OH
Cyanocobalamin
R = -CN
Methylcobalamin
R = -CH$_3$
5'-Deoxyadenosyl cobalamin

Biosynthesis

The corrin ring system is formed from uroporphyrinogen III (D 24) by reductive contraction of the porphyrin ring. All seven extra methyl groups (encircled in the formula given above) are derived from *S*-adenosyl-L-methionine.

The 5,6-dimethylbenzimidazole residue is derived from riboflavin (D 25.4). The 1-amino-2-propanol unit is formed from L-threonine by decarboxylation (D 8.2).

Significance

Cobalamins are vitamins for higher animals and humans (vitamins B_{12}, E 2.1). They are transformed into two types of coenzymes:

a) Aquocobalamin, in which the sixth coordination site of Co^{2+} is occupied by a water molecule. This compound is involved, for instance, in the biosynthesis of L-methionine and in the formation of methane (C 3.2). Methylcobalamin is a transient intermediate in these reactions.

b) 5'-Deoxyadenosyl cobalamin, which is involved in the intramolecular shift of hydrogen coupled with a 1,2-shift of some other groups:

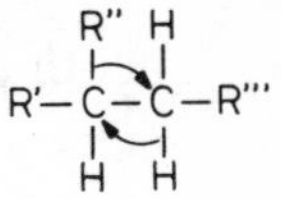

$$R'-\underset{\underset{H}{|}}{\overset{\overset{R''}{|}}{C}}-\underset{\underset{H}{|}}{\overset{\overset{H}{|}}{C}}-R'''$$

An example is the transformation of methylmalonyl CoA to succinyl CoA (D 3).

References for Further Reading

9, 34, 35, 454, 684, 888

D 25 Purine Derivatives

Purines are heterocyclic compounds with the following basic skeleton:

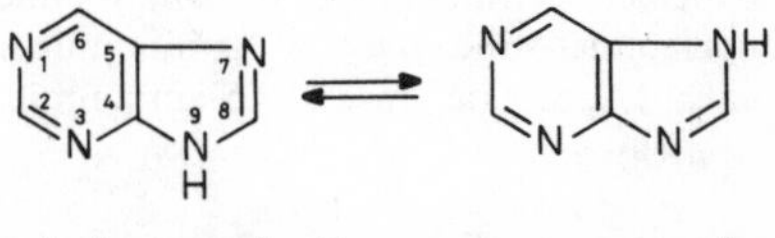

Biosynthesis of the purine ring system proceeds in nearly the same manner in all groups of organisms. Starting with ribose-5-phosphate the purine ring is formed from a set of small pieces by the reactions shown in Fig. 330. It is of interest that 5-aminoimidazole ribonucleotide is also a precursor of thiamine (D 25.6) and that 5-aminoimidazole-4-carboxamide ribonucleotide is an intermediate in L-histidine biosynthesis.

The first purine derivative being formed is inosine-5-monophosphate (inosinic acid), which holds a central position in primary purine metabolism. It is the

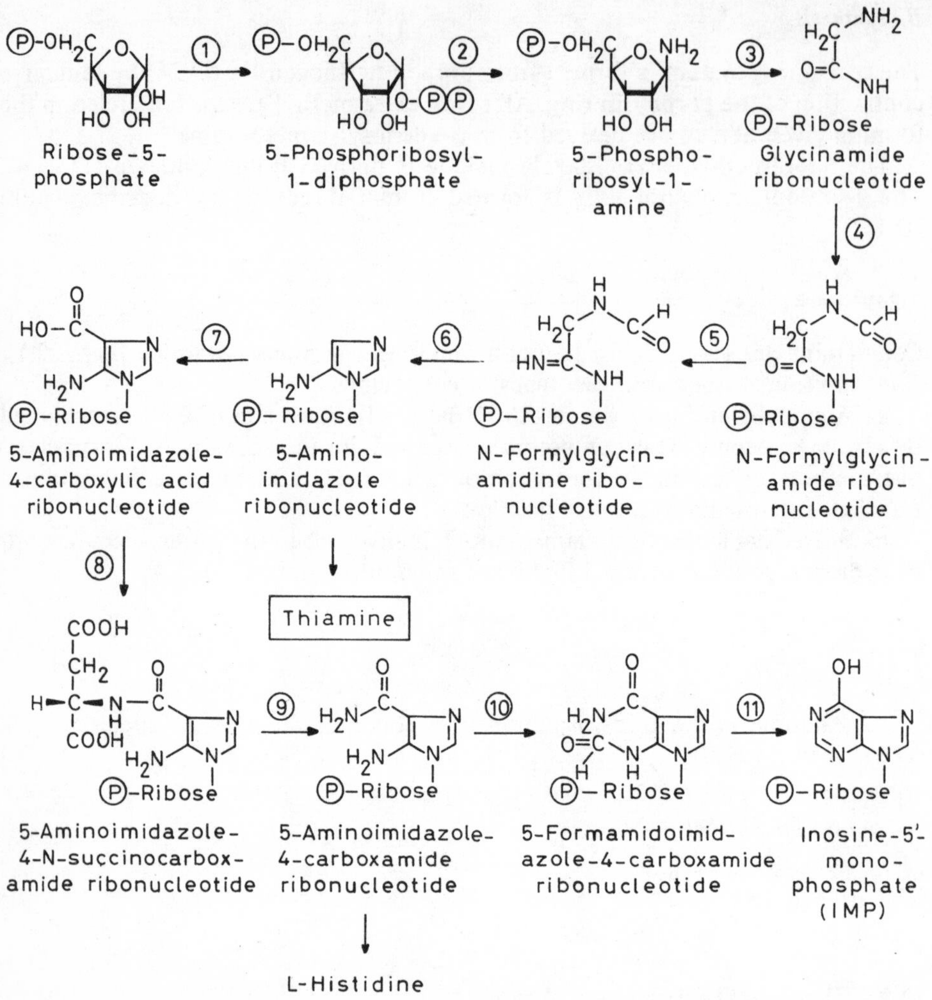

Ribose-5-phosphate

①

5-Phosphoribosyl-1-diphosphate

②

5-Phospho-ribosyl-1-amine

③

Glycinamide ribonucleotide

④

N-Formylglycin-amide ribo-nucleotide

⑤

N-Formylglycin-amidine ribo-nucleotide

⑥

5-Amino-imidazole ribonucleotide

⑦

5-Aminoimidazole-4-carboxylic acid ribonucleotide

⑧

Thiamine

5-Aminoimidazole-4-N-succinocarbox-amide ribonucleotide

⑨

5-Aminoimidazole-4-carboxamide ribonucleotide

⑩

5-Formamidoimid-azole-4-carboxamide ribonucleotide

⑪

Inosine-5'-mono-phosphate (IMP)

L-Histidine

Fig. 330. Synthesis of the purine ring system
1 5-Phosphoribosyl-1-diphosphate synthetase; *2* aminophosphoribosyltransferase; *3* phosphoribosyl-glycinamide synthetase; *4* glycinamide ribonucleotide transformylase; *5* N-formylglycinamidine ribonucleotide amidoligase; *6* 5-aminoimidazole ribonucleotide synthetase; *7* 5-aminoimidazole ribonucleotide carboxylase; *8* 5-aminoimidazole-4-N-succinocarboxamide ribonucleotide synthetase; *9* adenylosuccinate lyase; *10* 5-aminoimidazole-4-carboxamide ribonucleotide transformylase; *11* inosinicase

precursor of adenosine-5'-monophosphate (adenosinic acid) as well as of guanosine-5'-monophosphate (guanosinic acid), compounds which may also be converted back to inosine-5'-monophosphate (Fig. 331).

From the 5'-monophosphates the corresponding 5'-diphosphates and 5'-triphosphates, e. g., adenosine triphosphate (ATP) and guanosine triphosphate (GTP), the unphosphorylated ribosides inosine, adenosine, and guanosine, as well as the free purines hypoxanthine, adenine, and guanine are formed. Guani-

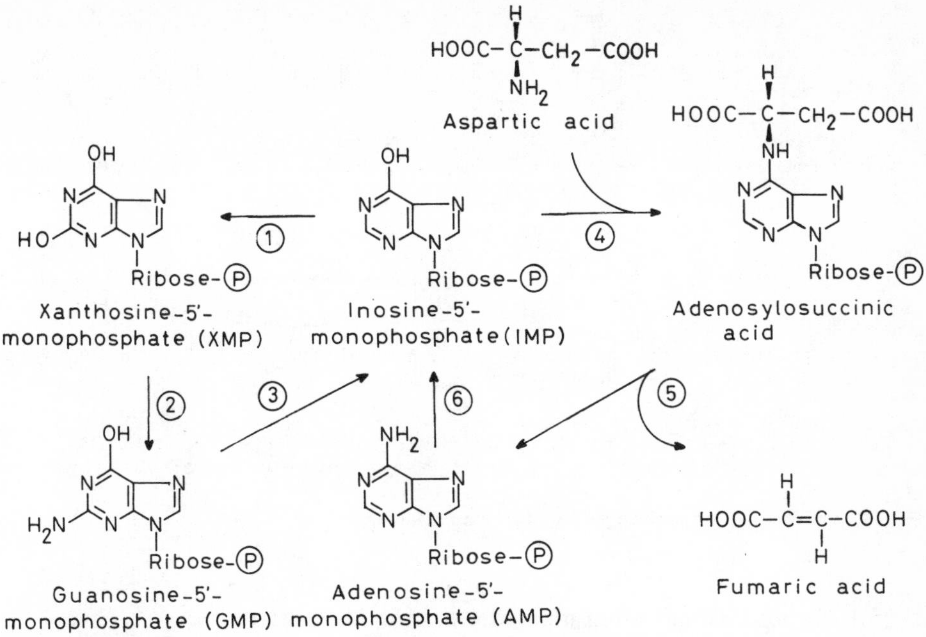

$$\text{HOOC}-\overset{\overset{\displaystyle H}{|}}{\underset{\underset{\displaystyle NH_2}{|}}{C}}-CH_2-COOH$$

Aspartic acid

$$\text{HOOC}-\overset{\overset{\displaystyle H}{|}}{\underset{\underset{\displaystyle NH}{|}}{C}}-CH_2-COOH$$

Xanthosine-5'-
monophosphate (XMP)

Inosine-5'-
monophosphate (IMP)

Adenosylosuccinic
acid

Guanosine-5'-
monophosphate (GMP)

Adenosine-5'-
monophosphate (AMP)

$$\text{HOOC}-\overset{\overset{\displaystyle H}{|}}{C}=\overset{\overset{\displaystyle}{|}}{\underset{\underset{\displaystyle H}{|}}{C}}-COOH$$

Fumaric acid

Fig. 331. Interconversion of inosine, adenosine, and guanosine monophosphates
1 Inosine monophosphate dehydrogenase; *2* guanosine monophosphate synthase; *3* guanosine monophosphate reductase; *4* adenylosuccinate synthetase; *5* adenylosuccinate lyase; *6* adenosine (phosphate) deaminase

dine crystals in the scales of fish cause light reflection and the characteristic silver glitter (E 5.6). The purine glycosides adenosine and guanosine are constituents of the nucleic acids. Their phosphate esters play an important role as energy-rich compounds (C 1.1).

Methylated purines (and pyrimidines, cf. D 26) have been found in the nucleic acid species of all groups of organisms. Rich in methylated purines is tRNA which contains, in addition, also purines with other substituents (Fig. 332).

The "unusual" purine derivatives which occur in the nucleic acids originate from adenine and guanine (in some cases also from hypoxanthine) residues already present in the nucleic acid chain. S-Adenosyl-L-methionine (C 3.3) and 3,3-dimethylallyl diphosphate (D 6) act as donors of the methyl and isopentenyl groups, respectively.

Chemistry and Distribution

The most important groups of secondary products derived from the purines of primary metabolism still show either the purine skeleton (D 25.1) or that basic structure is modified by loss of a nitrogen atom (D 25.2), by enlargement with a carbon atom (D 25.3), or the presence of an additional carbocyclic ring (D 25.4).

References for Further Reading

677, 803

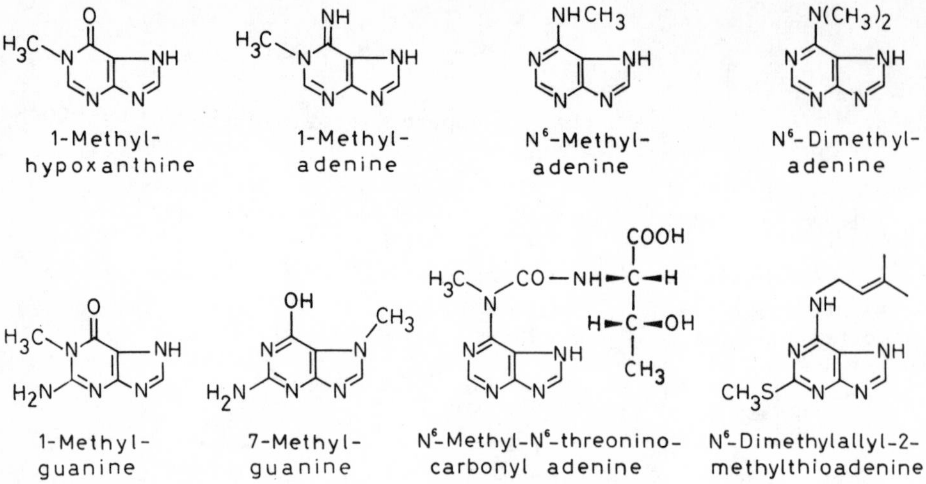

Fig. 332. Unusual purine derivatives found in the nucleic acids

D 25.1 Hypoxanthine, Adenine, and Guanine Derivatives

Chemistry and Distribution

Several types of secondary products occur in microorganisms, plants, and animals formed from the purines of primary metabolism by derivation of the purine nucleus, the D-ribose/D-deoxyribose moiety, or the phosphate residues.

Groups of Secondary Products

a) Cyclic phosphate esters of purine glycosides. Of special significance are adenosine-3':5'-phosphate (cyclic AMP, cAMP) and guanosine-3':5'-phosphate (cyclic GMP, cGMP) which are formed from ATP and GTP by adenylate cyclase and guanylate cyclase, respectively (Fig. 333). cAMP and cGMP are degraded to AMP and GMP by specific phosphodiesterases. cAMP mediates as second messenger the action of many animal hormones (E 3.1). It is involved in catabolite repression in bacteria (A 4.4.1, E 3.1) and is an aggregation pheromone in the slime mold *Dictyostelium* (E 4). cGMP participates in the regulation of gene expression in prokaryotes and eukaryotes.

b) Polyphosphorylated purines. Adenosine as well as guanosinetetraphosphates and pentaphosphates occur in bacteria, which have significance as hormonelike signals (E 3.1). They are formed and degraded according to Fig. 334. In addition, diadenosine and diguanosinetetraphosphates ($A^{5'}$—ⓅⓅⓅⓅ—$^{5'}A$, $G^{5'}$—ⓅⓅⓅⓅ—$^{5'}G$) have been found which are formed as side products during the activation of amino acids by certain tRNA synthetases (Fig. 335). The compounds act as alarmones signaling, e. g., heat shock or oxidative stress, and affect DNA synthesis and gene expression in prokaryotic and eukaryotic cells. They are degraded by specific hydrolases either to 2 ADP or to ATP + AMP.

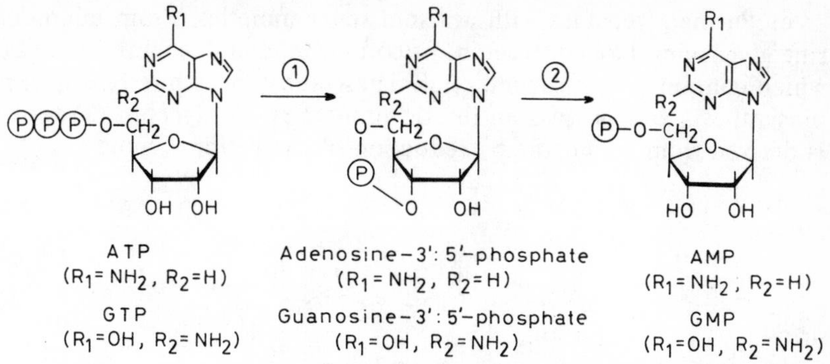

ATP
($R_1 = NH_2$, $R_2 = H$)

GTP
($R_1 = OH$, $R_2 = NH_2$)

Adenosine–3′: 5′-phosphate
($R_1 = NH_2$, $R_2 = H$)

Guanosine–3′: 5′-phosphate
($R_1 = OH$, $R_2 = NH_2$)

AMP
($R_1 = NH_2$, $R_2 = H$)

GMP
($R_1 = OH$, $R_2 = NH_2$)

Fig. 333. Biosynthesis and degradation of cyclic nucleotides
1 Adenylate cyclase, guanylate cyclase; *2* phosphodiesterase

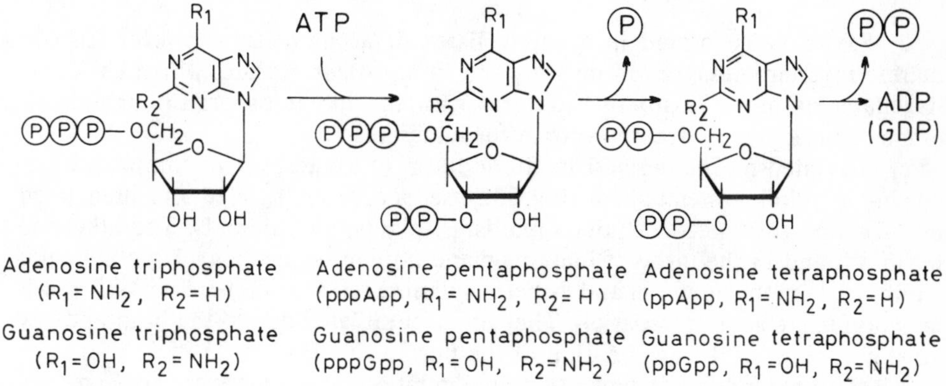

Adenosine triphosphate
($R_1 = NH_2$, $R_2 = H$)

Guanosine triphosphate
($R_1 = OH$, $R_2 = NH_2$)

Adenosine pentaphosphate
(pppApp, $R_1 = NH_2$, $R_2 = H$)

Guanosine pentaphosphate
(pppGpp, $R_1 = OH$, $R_2 = NH_2$)

Adenosine tetraphosphate
(ppApp, $R_1 = NH_2$, $R_2 = H$)

Guanosine tetraphosphate
(ppGpp, $R_1 = OH$, $R_2 = NH_2$)

Fig. 334. Biosynthesis and degradation of ppApp and ppGpp

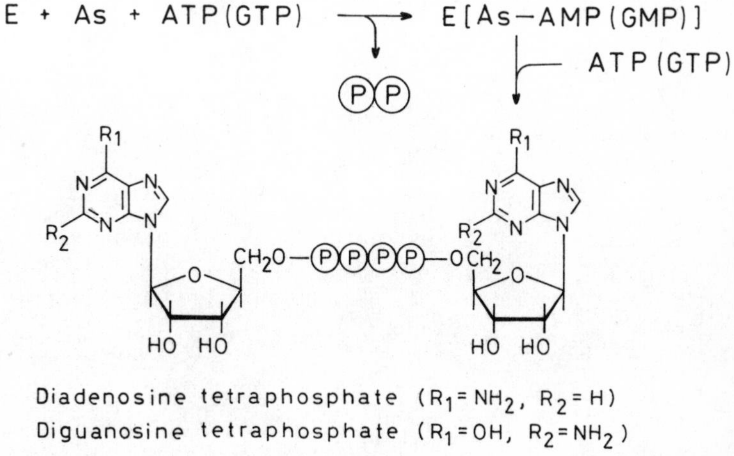

E + As + ATP(GTP) ⟶ E[As–AMP(GMP)]
                                    ⟶ ATP(GTP)

Diadenosine tetraphosphate   ($R_1 = NH_2$, $R_2 = H$)
Diguanosine tetraphosphate   ($R_1 = OH$, $R_2 = NH_2$)

Fig. 335. Synthesis of $A^{5'}$-ⓅⓅⓅⓅ-$^{5'}A$ and $G^{5'}$-ⓅⓅⓅⓅ-$^{5'}G$

c) Purine glycosides with unusual sugar moieties. From microorganisms purine glycosides, like cordycepin, psicofuranin, and decoyinin, have been isolated which contain rare sugar moieties. These substances interfere with nucleic acid biosynthesis and possess antibiotic properties. Cordycepin (3′-deoxyadenosine) is derived from adenosine by reduction of the D-ribose moiety.

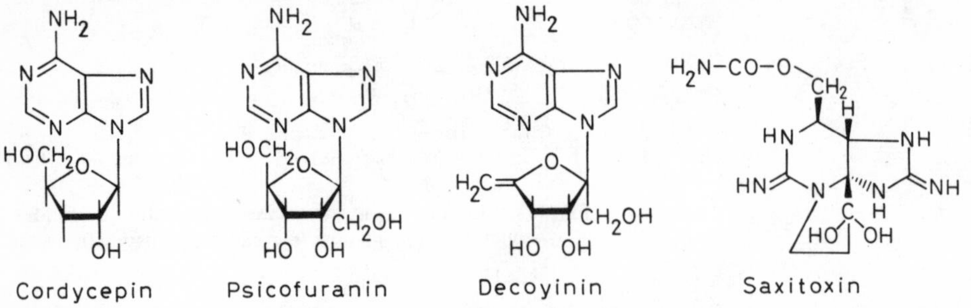

Cordycepin   Psicofuranin   Decoyinin   Saxitoxin

d) Saxitoxin is formed in dinoflagellates. It accumulates, however, also in marine invertebrates, e.g., in mussels and in the Alaskan clam, which use dinoflagellates as prey. Saxitoxin interferes with the ion transport through membranes. It is a very active neurotoxin for mammals.

e) Alkylated purine derivatives. In addition to the respective compounds occurring in primary metabolism (Fig. 332), secondary methylated and prenylated purines are synthesized in microorganisms, plants, and animals. The alkylated purines found in the urine of mammals and human beings, e.g., 1-methylhypoxanthine, 1-methyladenine, and 7-methylguanine (cf. Fig. 332) are possibly products of nucleic acid degradation. They are accumulated due to the hindrance by alkyl groupings of the degradation of the purine ring system (D 25.5).

Alkylations and other types of transformation may, however, also proceed with free purines. In plants, for instance, condensation of AMP with dimethylallyl diphosphate is catalyzed by AMP isopentenyltransferase. The N^6-dimethylallyl-AMP formed is dephosphorylated to N^6-dimethylallyladenosine which is hy-

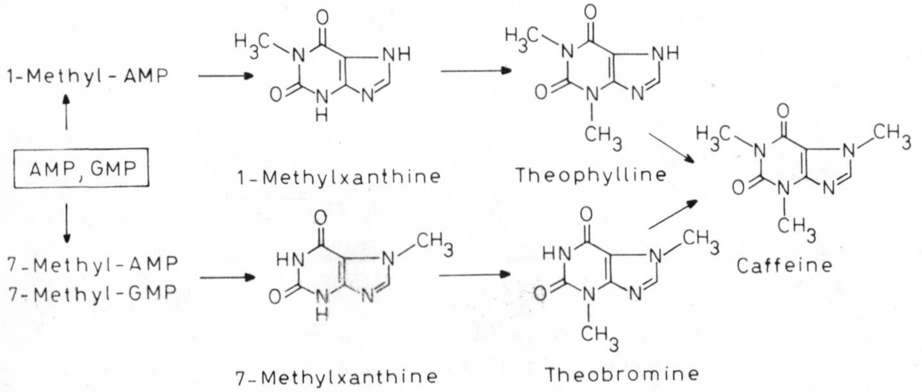

Fig. 336. Biosynthesis of theophylline, theobromine, and caffeine

drolyzed to N^6-dimethylallyladenine. In mammalian tissue 3-methyladenine is formed from adenine, and in plants the alkaloids theobromine and caffeine are derived from the pool of free purines participating in nucleic acid biosynthesis (Fig. 336).

Alkylated purines are degraded in metabolism. N^6-Dimethylallyladenine and similar compounds are split to adenine and an isopentenylaldehyde by cytokinin oxygenase. Methylated purines may be demethylated and the purine ring system may be degraded as shown in D 25.5.

Significance

1-Methyladenine is a hormone of Echinodermata. Zeatin and other N^6-substi-tuted adenine derivatives show cytokinin activity in higher plants (E 3.1). They are synthesized also by certain plant pathogens (E 5.4). Discadenine inhibits the germination of the spores of *Dictyostelium discoideum* (E 2.2). Caffeine, theobro-mine, and theophylline are plant constituents toxic to fungi and insects (E 5.5.3). They have a stimulating effect on humans (F 3).

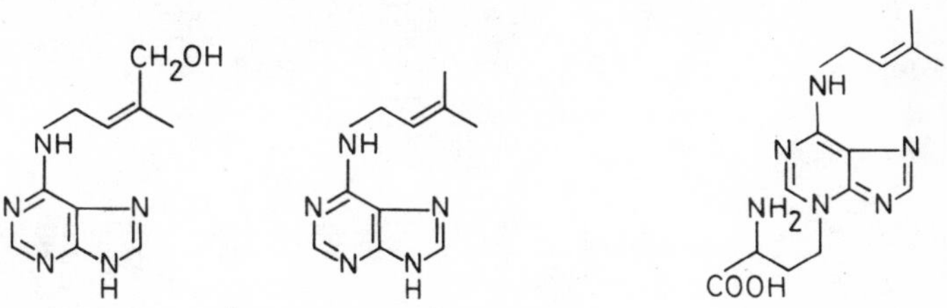

Zeatin N^6-Dimethylallyladenine   Discadenine

References for Further Reading

8, 474, 574, 677

D 25.2 Purine Analogs

Chemistry and Distribution

The most important group of purine analogs are the pyrrolopyrimidines (deaza-purines). Representatives are tubercidin and toyocamycin formed in bacteria, and queuine, a constituent of the tRNA of several groups of organisms.

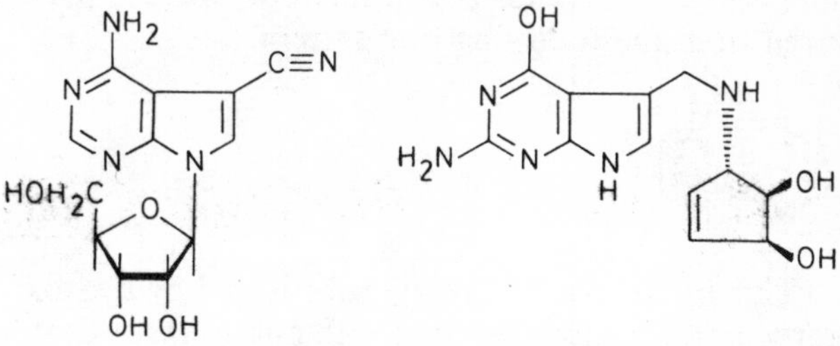

Toyocamycin   Queuine

Biosynthesis

Tubercidin and toyocamycin are derived from adenosine. Presumably the carbon atoms 1 and 2 of D-ribose are converted to the carbon atoms 8 and 7 of tubercidin (Fig. 337). Opening of the purine ring and use of D-ribose in the formation of the new ring resemble the biosynthesis of pteridines (D 25.3). The C-atoms 1, 2, and 3 of D-ribose also serve in the formation of the C-atoms 8, 7, and the cyano carbon of toyocamycin.

The pyrrolopyrimidines are rapidly phosphorylated to the corresponding mono-, di-, and triphosphates. They inhibit the biosynthesis of nucleic acids and act as antibiotics (E 5.2).

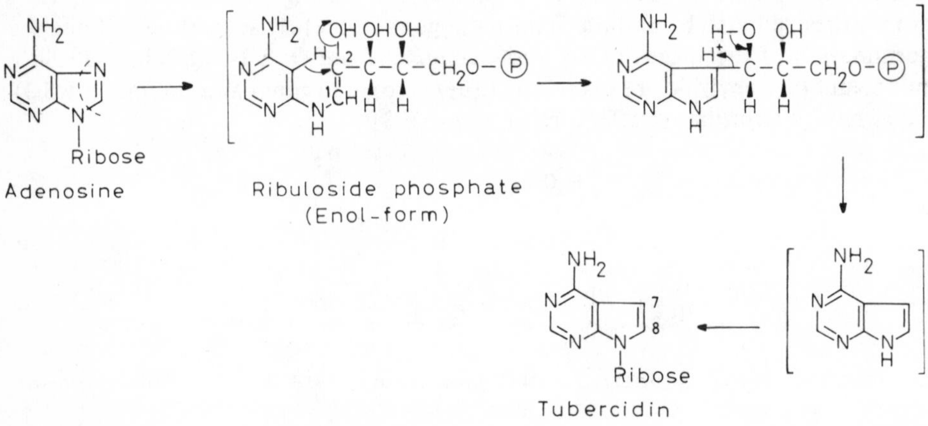

Fig. 337. Probable pathway of tubercidin formation from adenosine

Reference for Further Reading

677

D 25.3 Pteridines

Chemistry and Distribution

Pteridines are synthesized in microorganisms, plants, and animals. They are derived from the following ring system:

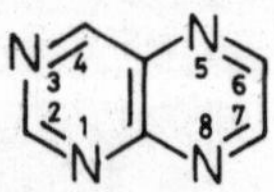

This nucleus is frequently substituted by hydroxy and amino groups and may carry an aliphatic side chain with one or three C-atoms. 2-Amino-4-hydroxypteridines are called pterins.

Biosynthesis (Fig. 338)

Pteridines are formed from guanosine triphosphate. C-Atom 8 of the guanine residue is lost as formic acid during this reaction (see also Fig. 339). The D-ribose moiety is transformed to a D-ribulose moiety by an Amadori rearrangement (cf. also the formation of L-histidine, D 20, and of L-tryptophan, D 22). The new six-membered ring is closed by formation of an azomethine group.

7,8-Dihydroneopterin is a branch point in pteridine metabolism. In the formation of sepiapterin, isosepiapterin, and tetrahydrobiopterin the three-carbon side chain is retained. In the biosynthesis of folic acid derivatives two carbon atoms are eliminated, whereas in the formation of xanthopterin and leucopterin the side chain is completely lost.

During the biosynthesis of tetrahydrofolic acid 6-hydroxymethyl-7,8-dihydropterin diphosphate condenses with 4-aminobenzoic acid (D 21.1) to dihydropteroic acid. This compound reacts with L-glutamine in the presence of ATP to dihydrofolic acid from which folic acid as well as tetrahydrofolic acid may be formed.

Precursors of tetrahydrofolic acid are also pteroylpoly-L-glutamic acids (D 9.2) from which the excess L-glutamic acid residues may be removed by hydrolytic cleavage.

Significance

Tetrahydrobiopterin and tetrahydrofolic acid are coenzymes of monooxygenases (C 2.6.1) and of enzymes of C_1-metabolism, respectively (C 3.2). Folic acid is a vitamin for humans. It is reduced in the human body to tetrahydrofolic acid (E 2.1). Pterins, e.g., xanthopterin and leucopterin, sepiapterin, and isosepiapterin are pigments of insects. They occur, for instance, in the wings of butterflies and flies. Similar compounds are skin pigments of fishes, amphibians, and reptiles and in this respect are of ecological significance (E 5.6).

References for Further Reading

71, 107, 153, 568, 677, 789, 836, 846, 870, 872, 879

D 25.4 Benzopteridines

Chemistry and Distribution

The benzopteridines possess the following nucleus:

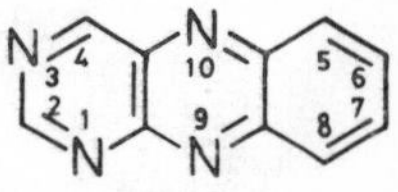

This ring system is substituted in positions 2 and 4 by hydroxy groups and in positions 6 and 7 by methyl groups. The 2,4-dihydroxylated benzopteridines are called isoalloxazines. Benzopteridines are synthesized in microorganisms, plants, and animals.

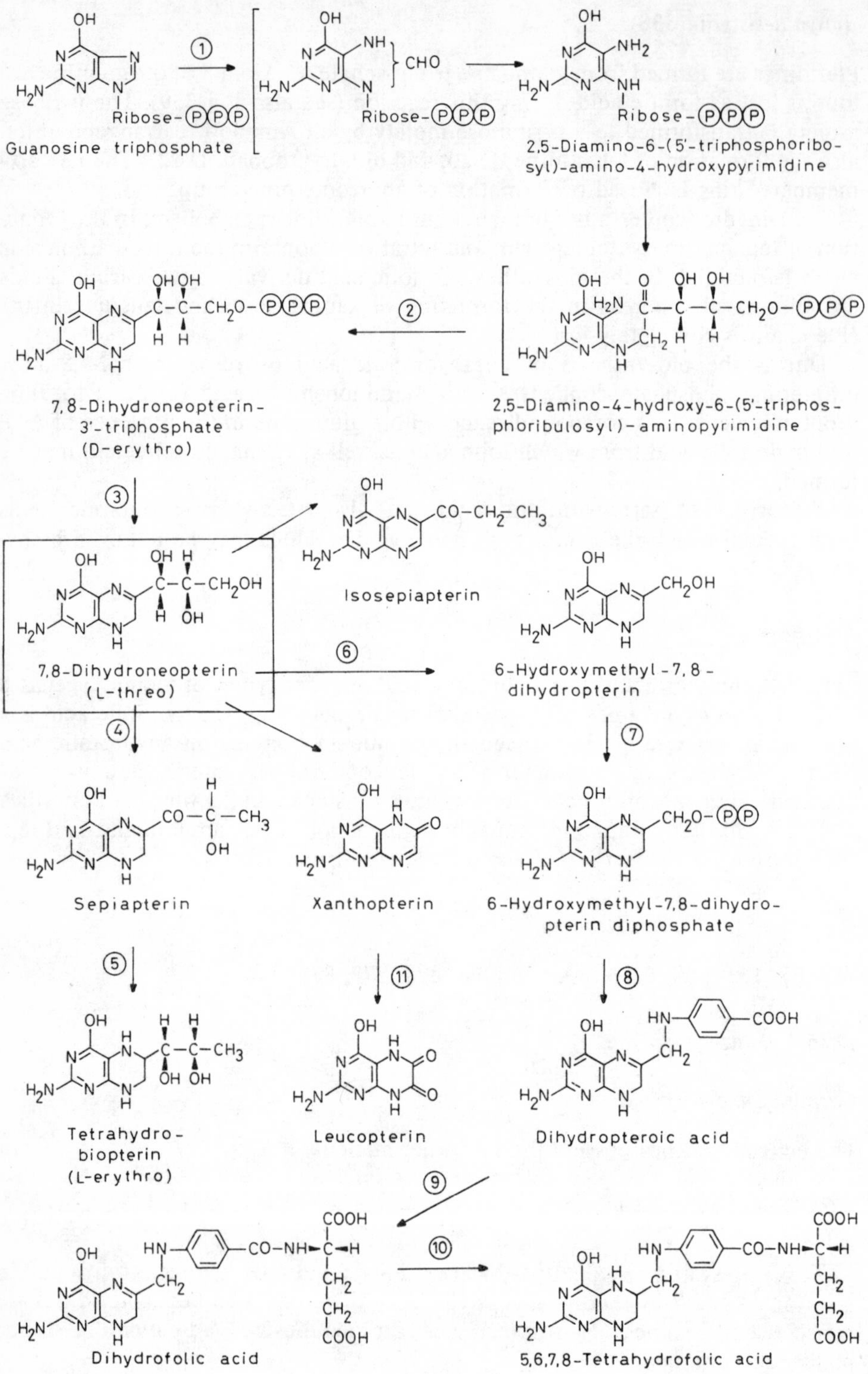

Guanosine triphosphate

2,5-Diamino-6-(5'-triphosphoribo-
syl)-amino-4-hydroxypyrimidine

7,8-Dihydroneopterin-
3'-triphosphate
(D-erythro)

2,5-Diamino-4-hydroxy-6-(5'-triphos-
phoribulosyl)-aminopyrimidine

7,8-Dihydroneopterin
(L-threo)

Isosepiapterin

6-Hydroxymethyl-7,8-
dihydropterin

Sepiapterin

Xanthopterin

6-Hydroxymethyl-7,8-dihydro-
pterin diphosphate

Tetrahydro-
biopterin
(L-erythro)

Leucopterin

Dihydropteroic acid

Dihydrofolic acid

5,6,7,8-Tetrahydrofolic acid

Biosynthesis

The benzopteridines are derived from guanosine triphosphate (Fig. 339). In the first step the C-atom 8 of the guanine moiety is lost (cf. also Fig. 338). The ribityl side chain is formed from the D-ribosyl residue of 5-amino-2,6-dihydroxy-4-ribo-sylaminopyrimidine phosphate. 6,7-Dimethyl-8-ribityllumazine is built from 4-ribitylamino-5-aminourazil phosphate and tetrolose phosphate which is derived from a pentosephosphate by loss of C-4. From two molecules of 6,7-dime-thyl-8-ribityllumazine one molecule of riboflavin is formed. In this reaction the carbon atoms 6 and 7, together with the attached methyl groups of one molecule are transferred to the other molecule in the reversed sequence. As a second product 4-ribitylamino-5-aminouracil is built, which after phosphorylation again condenses with the tetrolose phosphate.

Riboflavin may react with ATP to riboflavin-5'-phosphate (flavin mononucle-otide, FMN). With another molecule of ATP FMN yields riboflavin adenine di-nucleotide (FAD). FAD may be covalently bound to enzyme proteins at one of the methyl groups. Acceptors are one of the ring N-atoms of a L-histidinyl resi-due or the S-atom of an L-cysteinyl residue:

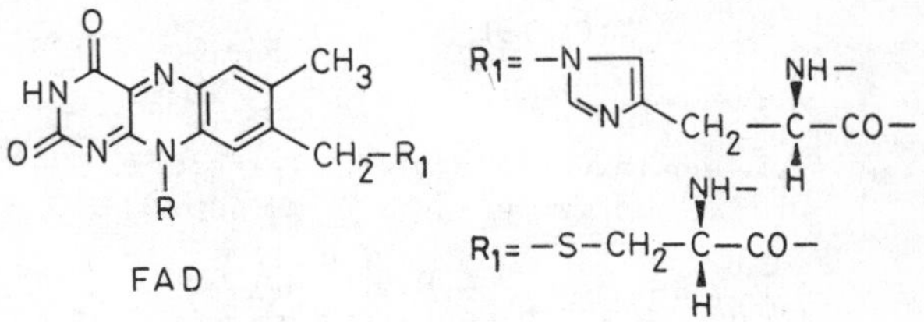

FAD

Significance

Riboflavin is a vitamin for human beings (vitamin B_2, E 2.1). In microorganisms it may be involved in the transport of iron (E 2.2). Riboflavin is found in rela-tively large quantities in milk and milk products. FMN and FAD are coenzymes of oxidoreductases (C 2.2). A unique type of flavin-dependent monooxygenases are bacterial luciferases. These enzymes catalyze the oxidation of $FMNH_2$ and the oxygenation of aliphatic aldehydes with a chain length above eight carbons (Fig. 340). The light-emitting molecule is an enzyme-bound excited flavin deriv-ative. It forms FMN with the elimination of water. Light-emitting bacteria live as symbionts in special organs of higher organisms, such as fishes and squids, but also in free sea water, in the gut of fishes, or in saprophytic or parasitic modes.

◀

Fig. 338. Formation of pteridine derivatives
1 Guanosine triphosphate cyclohydrolase I; *2* dihydroneopterintriphosphate synthetase; *3* diphos-phorylase, phosphatase; *4* dihydropterin synthetase; *5* sepiapterin reductase, dihydropterin reduc-tase; *6* dihydroneopterin aldolase; *7* hydroxymethyldihydropterin diphosphokinase; *8* dihydropter-oate synthase; *9* dihydrofolate synthetase; *10* dihydrofolate reductase; *11* xanthine oxidase

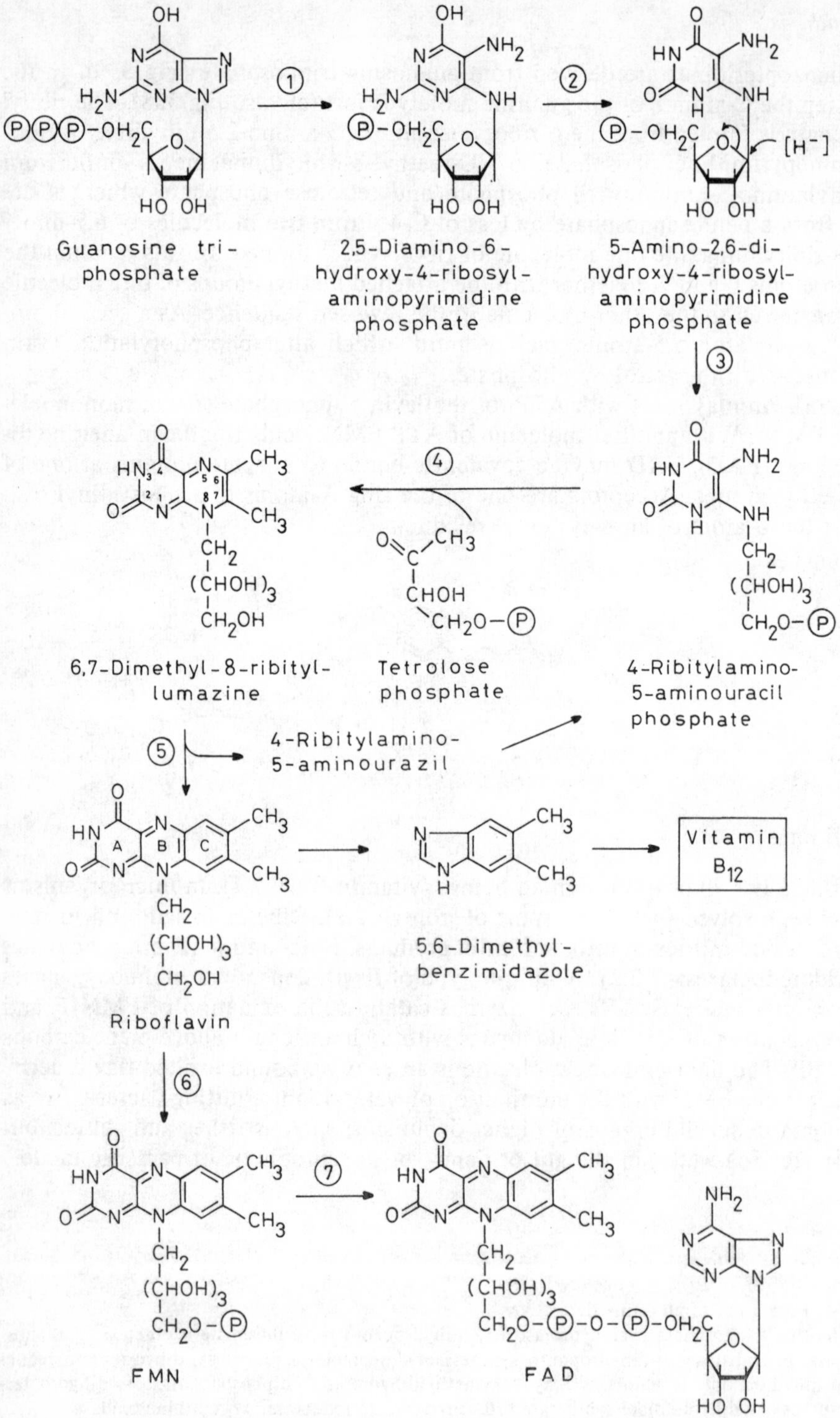

Guanosine tri-
phosphate

2,5-Diamino-6-
hydroxy-4-ribosyl-
aminopyrimidine
phosphate

5-Amino-2,6-di-
hydroxy-4-ribosyl-
aminopyrimidine
phosphate

6,7-Dimethyl-8-ribityl-
lumazine

Tetrolose
phosphate

4-Ribitylamino-
5-aminouracil
phosphate

4-Ribitylamino-
5-aminourazil

Riboflavin

5,6-Dimethyl-
benzimidazole

Vitamin
B₁₂

FMN

FAD

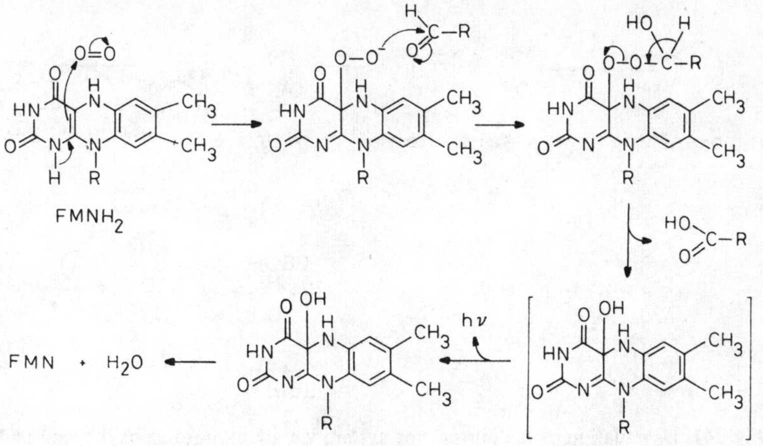

Fig. 340. Postulated mechanism of the reaction of bacterial luciferase

Transformation

a) Russopteridines have been isolated from certain higher fungi, e.g., *Russula sp.* These pigments obviously have an ecological function (E 5.6).

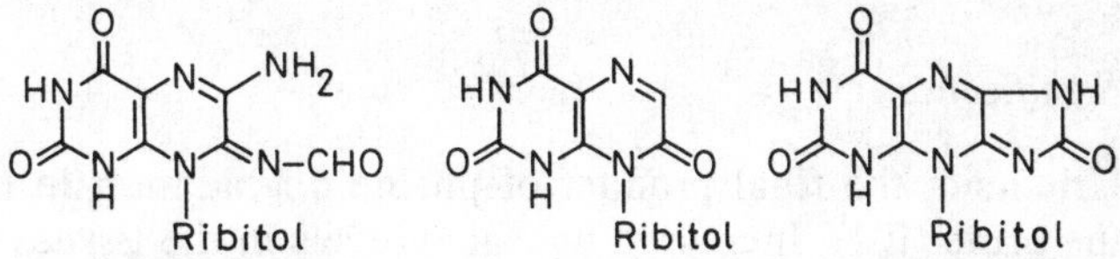

b) 5,6-Dimethylbenzimidazole (Fig. 339), is a constituent of vitamin B_{12} (D 24.3).

References for Further Reading

237, 344, 677, 836

D 25.5 Products of Purine Degradation

Distribution and Biosynthesis

The degradation of purines proceeds in most organisms by the reactions given in Fig. 341. While human beings, higher monkeys, birds, terrestrial reptiles, and most insects excrete uric acid as the end product of purine metabolism, almost

◀

Fig. 339. Biosynthesis of riboflavin, FMN, and FAD
1 Guanosine triphosphate cyclohydrolase II; *2* deaminase; *3* reductase; *4* dimethylribityl lumazine synthase; *5* riboflavin synthase; *6* riboflavin kinase; *7* flavinmononucleotide adenylyltransferase

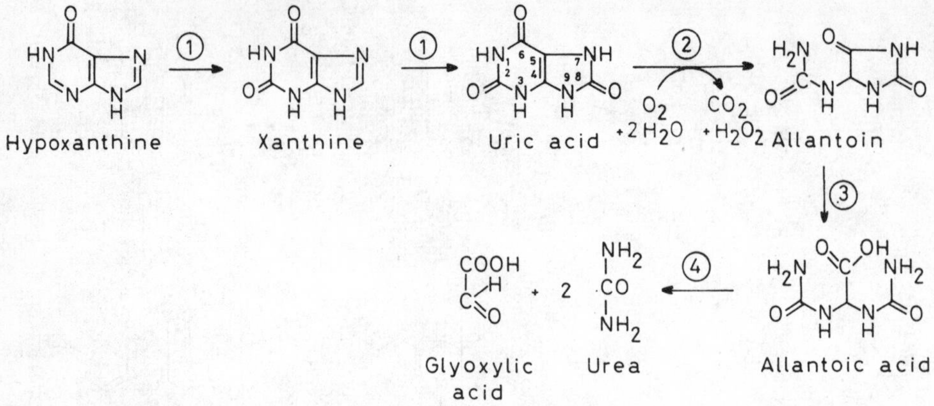

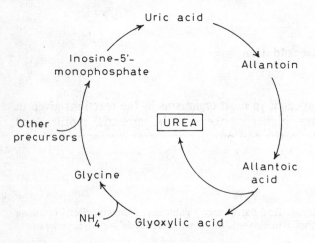

Hypoxanthine → Xanthine → Uric acid → Allantoin

Glyoxylic acid + 2 Urea ← Allantoic acid

Fig. 341. Degradation of the purine ring system via allantoin to glyoxylic acid and urea
1 Xanthine oxidase or xanthine dehydrogenase; *2* urate oxidase; *3* allantoinase; *4* allantoicase

all other animals, as well as plants and microorganisms, degrade the purine ring system to allantoin and allantoic acid, or via these compounds to urea and glyoxylic acid, NH_4^+ and CO_2. Before degradation adenine and guanine are transformed to hypoxanthine and xanthine, respectively, by hydrolytic cleavage of the amino groups.

Significance

Uric acid, the final product of purine degradation in humans, is excreted with the urine (E 1). Increased uric acid levels in the tissues of man, probably as a result of increased purine biosynthesis, may lead to gout and chronic arthritis. In the course of these diseases uric acid crystals are deposited in joints and may form stones in the kidneys.

The so-called ureide plants use allantoin and allantoic acid as a nitrogen store from which ammonia may be liberated by further degradation (E 2.2). In the

Uric acid

Inosine-5'-
monophosphate

Allantoin

Other precursors

UREA

Glycine

Allantoic acid

NH_4^+

Glyoxylic acid

Fig. 342. Urea formation by the glycine-allantoin cycle in lung fish

liver of lungfish a glycine-allantoin cycle (Fig. 342) causes the formation of urea (cf. the formation of urea via L-arginine, D 19).

References for Further Reading

677, 758

D 25.6 Thiamine

Chemistry and Distribution

Thiamine consists of a substituted pyrimidine ring linked via a CH_2-group to a thiazole moiety. It is formed in microorganisms and plants.

Biosynthesis

Thiamine monophosphate is built from 2-methyl-4-amino-5-hydroxymethyl pyrimidine diphosphate and 4-methyl-5-(2'-hydroxyethyl)-thiazole (Fig. 343). The

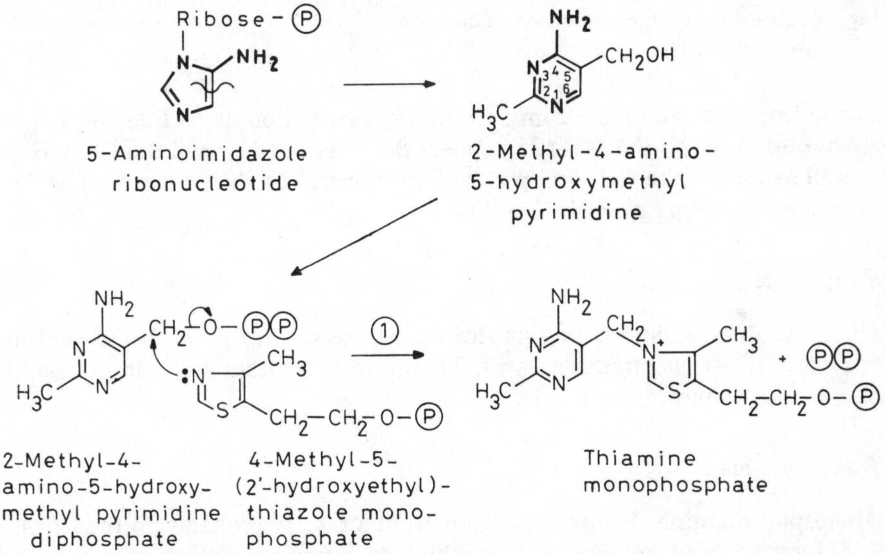

Fig. 343. Formation of thiamine monophosphate *1* Thiamine monophosphate synthase

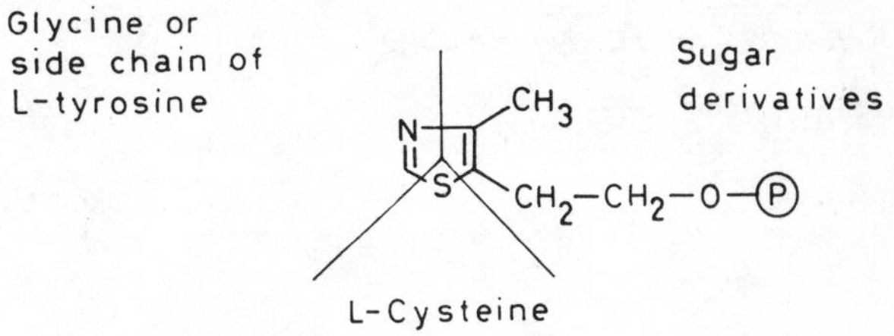

Fig. 344. Precursors of 4-methyl-5-(2'-hydroxyethyl)-thiazole monophosphate

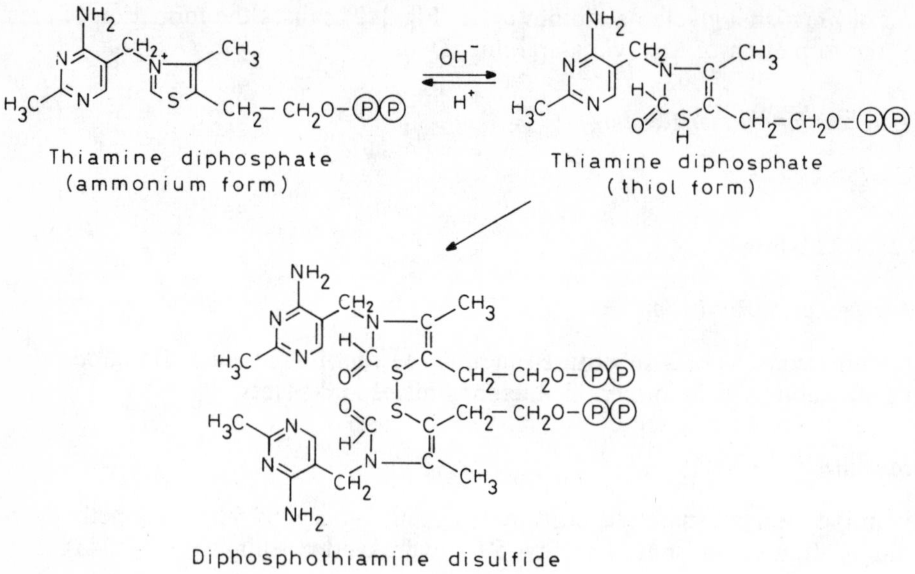

Fig. 345. Formation of diphosphothiamine disulfide

pyrimidine part is derived from 5-aminoimidazole ribonucleotide, an intermediate of purine biosynthesis (D 25). As yet the origin of C-5 and the CH_2OH-group as well as that of the CH_3-group is still unknown. The thiazole moiety is derived from the precursors given in Fig. 344.

Significance

Thiamine diphosphate is of significance as the coenzyme of 2-oxo acid decarboxylases (C 4) and transketolases. Thiamine is a vitamin for animals and human beings (vitamin B_1, E 2.1).

Transformation

Diphosphothiamine disulfide isolated from yeast is probably formed from the thiol form of thiamine diphosphate which in aqueous solution is in equilibrium with the ammonium form (Fig. 345). As yet it is not clear whether diphosphothiamine disulfide possesses physiological significance.

References for Further Reading

95, 107, 678, 836

D 26 Pyrimidine Derivatives

Biosynthesis and Degradation of Precursors

The pyrimidines of primary metabolism are formed from carbamyl phosphate and L-aspartic acid (Fig. 346). Orotic acid is the key intermediate. It is transformed to uracil, thymine, cytosine, and 5-methylcytosine. The biosynthesis of these compounds takes place at the level of the ribosides and ribotides. The amino group of cytosine is derived from L-glutamine. The methyl donor in the biosynthesis of thymidine and 5-methylcytosine is 5-methyltetrahydrofolic acid (C 3.2), that in the formation of 5-hydroxymethylcytosine is 5,10-methylenetetra-hydrofolic acid (C 3.2). In preformed tRNA and DNA chains, however, thymine is also built by *S*-adenosyl-L-methionine-dependent methyltransferases (C 3.3).

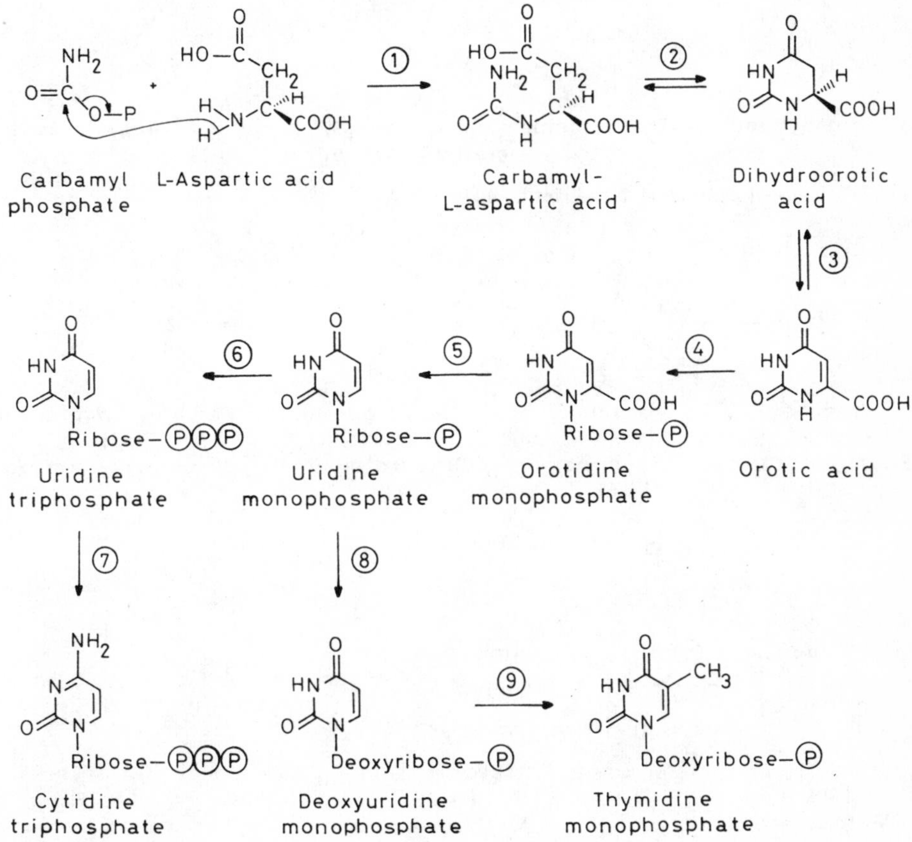

Fig. 346. Formation of orotic acid, uridine, and cytidine derivatives
1 Aspartate carbamyltransferase; *2* dihydroorotase; *3* orotate reductase; *4* orotate phosphoribosyl-transferase; *5* orotidine-5'-phosphate decarboxylase; *6* cytidylate kinase, nucleotide diphosphate kinase; *7* cytidine triphosphate synthetase; *8* nucleoside monophosphate kinase, ribonucleoside diphosphate reductase, phosphatase; *9* thymidylate synthase

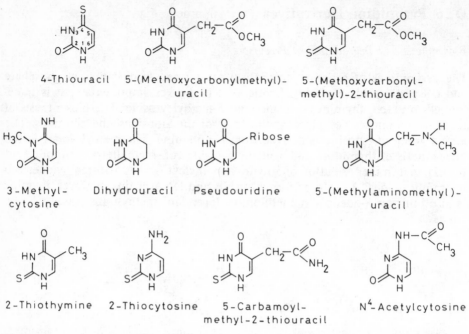

Fig. 347. Unusual pyrimidine derivatives found in tRNA

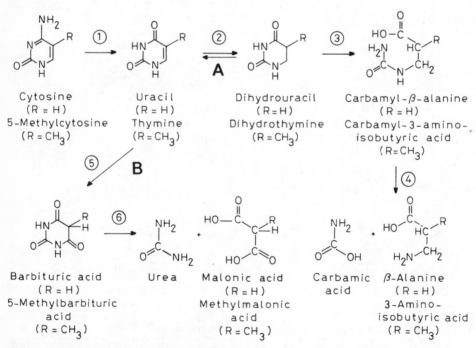

Fig. 348. Degradation of pyrimidine derivatives
1 Cytosine deaminase; *2* dihydrouracil dehydrogenase (uracil reductase); *3* dihydropyrimidinase; *4* β-ureidopropionase; *5* uracil dehydrogenase; *6* barbiturase (barbituric acid amino hydrolase)

2- and 4-Thiouracil and 2- and 4-thiocytosine derivatives are formed by thiol transferases from L-cysteine and uracil and cytosine, respectively. Pseudouridine is derived from uridine by shift of the D-ribose residue.

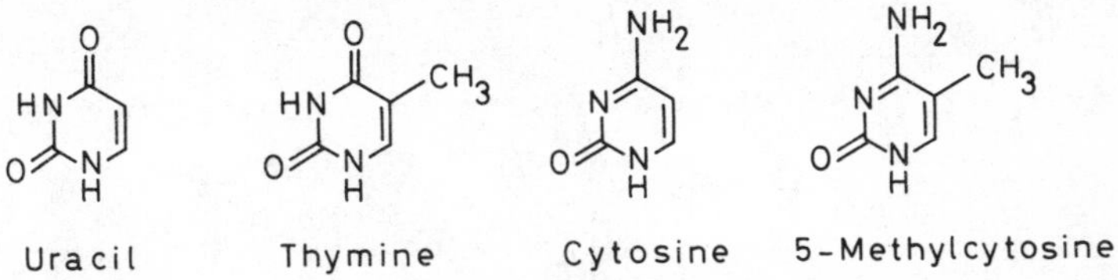

Uracil   Thymine   Cytosine   5-Methylcytosine

Uracil, thymine, cytosine, and 5-methylcytosine are constituents of nucleic acids and in the form of their nucleotides play an important role as energy-rich compounds, e.g., in the metabolism of sugars (C 6) and phospholipids (D 5.2.4). The pyrimidines shown in Fig. 347 are constituents of tRNA. Together with substituted hypoxanthine, adenine, and guanidine derivatives (D 25.1), they are of significance for the spatial structure of the different tRNA species and their specific interaction with amino acyl tRNA synthetase and the ribosomes during protein biosynthesis.

Degradation of the pyrimidines follows in most organisms route A of Fig. 348. The final products are carbamic acid (which spontaneously hydrolyzes to CO_2 and NH_3) β-alanine (D 16), and 3-aminoisobutyric acid. In some bacteria, however, a second pathway, route B, exists which yields urea and malonic/methylmalonic acids.

Groups of Secondary Compounds

Secondary pyrimidines carrying unusual substituents at the heterocyclic ring are formed in microorganisms and plants.

a) Lathyrine is a nonprotein amino acid found in plants of the genus *Lathyrus*. It is formed from uracil and *O*-acetyl-L-serine (Fig. 349). Lathyrine may be transformed to erythro-4-hydroxy-L-homoarginine and threo-4-hydroxy-L-homoarginine found in Leguminosae, marine cucumbers, and sea anemones. L-Homoarginine, accompanying its hydroxylated derivatives, acts as an L-arginine antagonist in certain insects (E 5.5.3).

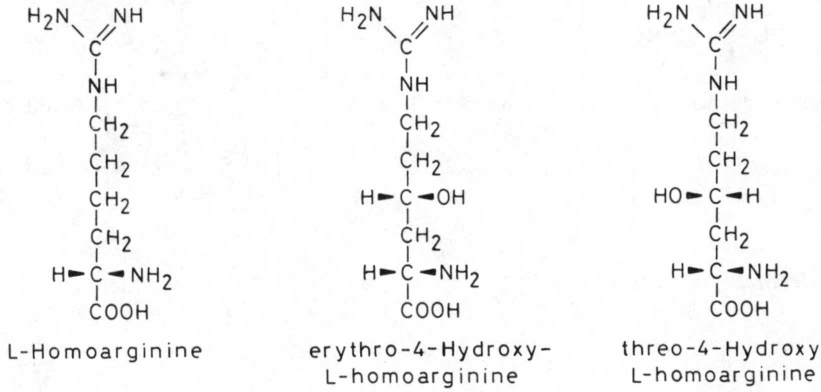

L-Homoarginine   erythro-4-Hydroxy-   threo-4-Hydroxy-
                        L-homoarginine   L-homoarginine

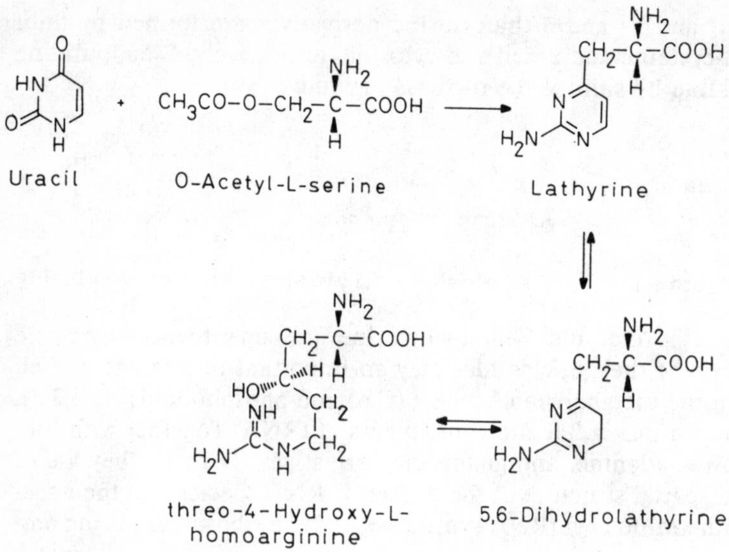

Fig. 349. Biosynthesis of lathyrine

b) Willardiine and isowillardiine are nonprotein amino acids from the seeds of certain *Acacia*, *Mimosa*, *Fagus*, and *Pisum* sp. Both compounds are derived from uracil and *O*-acetyl-L-serine by willardiine/isowillardiine synthase in a reaction resembling the biosynthesis of lathyrine (Fig. 349).

c) Blasticidin S is a product of *Streptomyces griseochromogenes*. It is used as an antifungal antibiotic in agriculture (E 5.2).

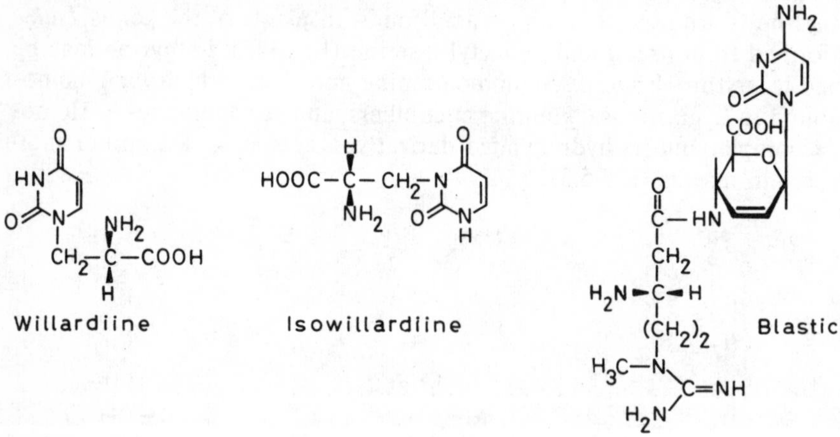

Willardiine   Isowillardiine   Blasticidin S

References for Further Reading

8, 287

E The Significance of Secondary Metabolism and Secondary Products for the Producer Organisms

Two features treated in Chapter A indicate that secondary metabolism is of profit for the producer organisms:

a) secondary products are formed although their synthesis drains resources from primary metabolism and therefore may reduce growth and reproduction, and

b) secondary metabolism is characterized by a high degree of order, cf. the precise regulation of enzyme amount and activity in secondary metabolic pathways, the strict compartmentalization and channeling of enzymes, precursors, intermediates, and products, and the integration of secondary metabolism in the programs of cell specialization (A 4), which need a continuous selection pressure to be maintained.

Indeed, numerous functions for different types of secondary products and the processes of their formation have been detected:
— secondary products may be formed as the result of detoxication of substances accumulating in primary metabolism (E 1)
— secondary products may possess physiological significance, e.g., as coenzymes or cosubstrates, they may increase membrane rigidity, are involved in light perception and ATP regeneration, in the storage of nitrogen, etc. (E 2)
— secondary products may be used as chemical signals in the coordination of cell metabolism in multicellular organisms (hormones, neurotransmitters, E 3)
— secondary products may coordinate the activity of different organisms of the same species (pheromones) (E 4), and
— secondary products may be involved in the ecological relations between the different groups of organisms (E 5).

On the one hand, secondary products meet specific needs of individual producer organisms and, on the other hand, they solve more general problems.

Examples for specific actions of secondary products are:
— the inhibition of the germination of seeds of the producer plants by geniposide or psoralen (E 2.2)
— the attraction of individuals of only the producer organisms by pheromones (E 4), and
— the increase of pollinator constancy by compounds present in the nectar of certain plants (E 5.5.1). These substances are toxic to most animals with the exception of the pollinators.

Examples for a general significance of secondary products are:
— the use of lignin as a supportive and strengthener of the cell wall in many plants (E 2.2)
— the use of chlorophylls in harvesting light energy in different types of prokaryotic and eukaryotic photoautotrophic organisms (E 2.2), and
— the action of the different types of hormones as chemical signals in large groups of plants or animals (E 3.1).

In microorganisms and plants the accumulation of secondary products is obviously an important means for survival, which in several respects is equivalent to the mobility of most animals: it deters potential predators, attracts pollinators (corresponding to sexual mates), discourages competing species, etc. (E 5).

However, there is no long-term project realized in the evolution of these functions and the significance of secondary products in the different groups of organisms. The secondary substances rather form a fund which has been used in manifold ways, depending on the opportunities. Hence, as in other fields, evolution has worked on secondary metabolism "like a tinkerer—a tinkerer who does not know exactly what he is going to produce, but uses whatever he finds around him ..." [385].

The nonexistence of a preconceived plan regarding the different aspects in the use of secondary products may be shown by the following examples:

a) Different groups of secondary products may serve the same function in different organisms as long as the necessary physical or chemical properties are met.

Examples are:
— the red-violet flower pigments which in most plants are anthocyanins, but in Centrospermae are betacyanins (E 5.5.1)
— the use in membrane strengthening of hopanoids in certain prokaryotes instead of sterols and of phytanyl glycerol ethers instead of carotenoids, which have the same function in most other groups of organisms (E 2.2)
— the use of different lipids for surface coatings (E 2.2), e.g., alkanes, alkenes, waxes, suberin, cutin, and sporopollenin
— the significance of enterobactin, riboflavin, dioxopiperazine N-oxides, and sideramines, but also of citric acid in the solubilization and uptake of Fe^{3+} ions in microorganisms (E 2.2)
— the use of phytic acid, creatine phosphate, and L-arginine phosphate in addition to nucleoside phosphates, like ATP, as energy stores in different groups of organisms (E 2.2).

b) Secondary products which possess a certain kind of significance in one group of organisms may be devoid of this or have deviating uses in other organisms.

Typical examples are secondary products used as hormones or neurotransmitters (E 3). Table 63 shows that nearly all secondary products with hormone activity in plants or animals are also synthesized in organisms where a comparable physiological activity is absent. Hence, it is the development of an appropriate target mechanism rather than any special chemical structure which makes a secondary product act as a hormone.

c) Certain groups of secondary products may be useful for the producer organisms in different ways.

Examples are:
— carotenoids which, on the one hand, strengthen membranes and, on the other hand, percept light (E 2.2), and
— open-chain tetrapyrrols which, for instance, in man are simple degradation products of porphyrins without any physiological use, whereas in higher plants they form the prosthetic group of phytochrome, a chromoprotein involved in light responses, and in some algae are accessory pigments in photosynthetic light absorption (E 2.2).

However, even if one takes into account the restricted knowledge in the fields of use of secondary products, not all compounds seem to be of significance for the producing organisms. It is doubtful, for instance, whether all the individual members of the families of structurally related secondary products, which occur in most producer organisms (A 2.2), are of significance. It has been suggested that the modifications of the fundamental chemical structures will overcome feedback inhibition of the early enzymes of secondary pathways (A 2.2) and that the occurrence of groups of slightly different compounds enhances the difficulties in adaptation of predators or other enemies to secondary products used as allelochemicals (E 5.5.3). It seems to be more likely, however, that the many details of the chemical structures of most secondary products cause neither advantages nor disadvantages for the producer organism.[6] Indeed, it is difficult to understand that each of the more than 100 indole alkaloids produced in the plant *Catharanthus roseus*, or each of the several hundreds of compounds, which may be present in a single essential oil, really create a specific advantage for the producer. Obviously a relatively low selection pressure rests on the details of the chemical structures of most secondary products, preserving the results of random mutations. Hence, in many instances these details may reflect what has been called the *"Spiel der Natur"* [542] or the biochemical playground of evolution [838].

In this respect there are striking similarities to morphological characteristics. It is, for example, under most circumstances obviously advantageous for a higher plant to spread the chlorophyll-containing cells by formation of a leaf lamina. But in most cases there is no rationale for the details of the shape of the lamina, which in the different plant species may be smooth-edged, toothed, serrate, divided, lobed, etc.

Several secondary products may be relics which have had significance in former periods of evolution. There are no figures available giving an idea how fast the formation of a secondary product disappears, if it is no longer needed. With microorganisms it has been demonstrated that the ability to form "unnecessary" secondary products vanishes within a short time. Microbial strains synthesizing secondary products which may be of significance as allelochemicals (E 5), e. g., as antibiotics (E 5.2), in the natural habitat, are easily overgrown by spontaneously arising, nonproducing cell lines if developing in pure cultures, i. e., under conditions in which allelochemicals are unnecessary. This instability is a problem of great economic significance in the fermentation industry.

It has not yet been investigated thoroughly whether in other classes of organisms unused secondary products also disappear rapidly. However, regarding the conservatism in the formation of certain morphological structures, which have become unnecessary in later stages of evolution, it seems likely that in some cases the biosynthesis of secondary products might be preserved for shorter or longer periods even if the secondary products have lost their significance for the producer organism.

The disappearance of secondary product biosynthesis during evolution does not necessarily proceed in a single step, and the following features may be regarded as intermediate steps of the decay:

a) The escape of enzymes from the economy in regulation. Examples might be enzymes whose amount and activity are regulated independently from those

---

[6] Exceptions are the chemical details of compounds acting via specific target mechanisms, like hormones, neurotransmitters, and pheromones.

of other enzymes belonging to the same metabolic chain though no acceptable reason is detectable for this diversity.

Cyclopenase, for instance, an enzyme involved in alkaloid metabolism in *Penicillium cyclopium* (D 21.4.2)
— is restricted to the conidiospores of this mold, whereas the other enzymes of alkaloid biosynthesis are present in hyphae and conidiospores
— shows a regulation which is independent of that of the other enzymes of alkaloid metabolism. Cyclopenase activity appears during conidiospore maturation, whereas the other enzymes are constitutive proteins of the spores (Fig. 6), and
— in contrast to the very high cyclopenase activity after cell disintegration, the enzyme is nearly inactive in living cells.

b) The formation of one or several enzymes of pathways which in toto are not expressed in a given organism.

Cell cultures of *Digitalis* sp., for instance, hydroxylate the steroid ring system of cardiac glycosides in the 12β- or 16β-positions or glucosylate these compounds at the terminal deoxysugar residue. These specific transformations occur though the cell cultures are unable to form cardiac glycosides de novo. In the cells grown in vitro, thus the respective monooxygenases and glucosyltransferases, are expressed in contrast to the early enzymes of cardenolide biosynthesis.

c) The nonexpression of secondary metabolic enzymes under normal physiological conditions, although the corresponding genetic material is present. Probably a large part of the genes of most organisms remain dormant during the whole lifetime. They may be expressed, however, under unusual physiological conditions, e.g., if cells are grown in vitro or after administration of unphysiological amounts of precursors or intermediates. In plant cell cultures, for instance, secondary products may be synthesized, which are not formed in the plants from which these cultures were derived.

Examples in this respect are:
— the formation of sesquiterpenes of the paniculide B-type (D 6.3) in cell cultures of *Andrographis paniculata*
— the biosynthesis of *N*-norsanguinarine (cf. the formula of sanguinarine, Fig. 285) in cell cultures of *Papaver somniferum*
— the formation of harman alkaloids (D 22.3) after the administration of large amounts of tryptophan in cell cultures of *Phaseolus vulgaris*, a plant unable to built any indole alkaloids under normal conditions, and
— the biosynthesis of quinolizidine alkaloids (D 18) in stressed cell cultures of *Atropa belladonna*, *Chenopodium rubrum*, *Conium maculatum*, *Daucus carota*, *Spinacea oleracea*, and *Symphytum officinale*, i.e., plants which produce in the native state either other types of alkaloids or no alkaloids at all.

The unexpressed genetic material, the unused secondary metabolic enzymes, and the secondary products without definite function obviously form a reservoir of chemical potential which may be activated if a respective need is developed in the course of further evolution of the producer organisms. It probably allows the establishment of new intercellular or interorganismic signals, of new ecological relations, etc. in a quicker and easier manner than de novo evolution and may in this respect be of significance.

References for Further Reading

385, 477, 542, 825, 838

E 1 Formation and Transformation of Secondary Products — a Process of Detoxication

Certain groups of secondary substances are synthesized as part of reactions detoxifying excess or waste products. Of significance in this respect are:

a) Secondary products formed from primary metabolic substances whose complete degradation is impossible.

Man, like most vertebrates, for instance, is unable to degrade the aromatic amino acids L-histidine and L-tryptophan, the skeletons of purines and pyrimidines, methyl groups, etc. After transformation to secondary products these substances and groupings are removed with the urine and in feces and sweat (Table 57). Most plants also cannot degrade certain aromatic amino acids. Hence, they store large amounts of secondary products, e.g., derived from L-tryptophan (D 22) and L-phenylalanine/L-tyrosine (D 23).

b) Secondary products synthesized under special metabolic conditions, which make the complete degradation of certain primary products impossible.

Examples are the transformation of pyruvic acid under anaerobic conditions to ethanol, e.g., in yeast, or to lactic acid, e.g., in the muscles of animals (D 2), and the transformation of acetyl CoA to acetoacetic acid or acetone (D 5.1), compounds which are excreted with the urine of diabetic persons (Table 57).

c) Secondary products built from primary metabolic substances whose complete degradation would result in the formation of toxic compounds.

Examples are the excretion of sulfur in the form of taurine and hypotaurine from the animal body, but not as sulfate which is strongly acidic (D 11.1), and of nitrogen as urea (D 19), but not as the strongly basic ammonium ion.

d) Secondary products formed from other secondary compounds which undergo a turnover, but cannot be excreted or degraded to primary metabolic compounds.

In animals, for instance, steroid hormones (D 6.6.5—6.6.7) and epinephrine (D 23.1.1) are transformed to other secondary products which may be excreted more easily and are removed from the body with urine or bile (Table 57).

In addition, there are many groups of secondary products which can be handled in the producer organism only if detoxified by further metabolic reactions, e.g., by methylation of "conjugation", i.e., coupling with primary or other secondary compounds, e.g., with sugars, amino acids, uronic acids, sugar alcohols, amines, etc. There are a great variety of conjugating compounds (Table 58) and different types of conjugates may be formed from a particular mother compound. Aliphatic and phenolic hydroxyl groups, mercapto groupings, carboxyl functions, and amino groups of the mother substances may be involved in the biosynthesis of conjugates.

The formation of conjugates is a characteristic feature of many groups of secondary products, e. g., of fatty acid derivatives (D 5.2.4), anthraquinones (D 5.3.6), saponins (D 6.6 and D 6.6.2), cardiac glycosides (D 6.6.8), cyanogenic glycosides (D 8.3), glucosinolates (D 8.4), coumarins (D 23.2.2), and flavonoids (D 23.3.3), but is found seldom with isoprenoids (D 6) and alkaloids (D 8.5). Methylations and conjugations are characteristics of the late stages of secondary

Table 57. Secondary products found in urine, bile, and sweat of mammals (including man)

Urine
D-Glucaric acid (D 1.2)
Acetoacetic acid, L-3-hydroxybutyric acid, acetone (D 5.1)
Progesterone, corticosteroid, androstane and estrane derivatives (D 6.6.5–D 6.6.7)
D-Amino acids, methylated amino acids (D 8.1)
Glutathione conjugates (D 9.2)
Glycine conjugates (D 10)
Creatinine (D 10)[a]
Taurine, hypotaurine (D 11.1)
1-Methylnicotinamide, 1-methyl-5-carboxamidopyridone-(2) (D 16.1)
N^2-substituted L-glutamyl and glutaminyl derivatives (D 17)
L-Pipecolic acid (D 18)
Urea, N^{guanyl}-hydroxymethyl-L-arginine (D 19)
Urocanic acid, hydantoin propionic acid, methylated imidazole derivatives, imidazole acetic acid riboside (D 20)
Indole-3-acetic acid, indoxyl, urinary indican, indigo, indirubin (D 22)
1,2,3,4-Tetrahydro-β-carboline (D 22.3)
Kynurenic acid and related quinolines (D 22.5)
Vanillic acid D 23.1.1)[b]
3′,4′-Deoxynorlaudanosoline carboxylic acid (D 23.1.2)
L-Thyroxine (D 23.1.5)
Enterolactone (D 23.2.4)
Stilbenes (D 23.3.3)
7-Methylguanine, 1-methyladenine and other alkylated purines (D 25.1)
Biopterin (D 25.3)
Uric acid (D 25.5.)[c]

Bile[d]
Cholesterol (D 6.6.1)[e]
Bile alcohols and acids (D 6.6.4)
Corticosteroid, androstane and estrane derivatives (D 6.6.5–D 6.6.7)
L-Thyroxine (D 23.1.5)
Bile pigments (D 24.2)[e]

Sweat
Lactic acid (D 2)
Androstane derivatives (D 6.6.6)
Choline (D 4)
Urea (D 19)
Histamine, urocanic acid (D 20)

---

[a] Removing excess methyl groups.
[b] Derived from epinephrine and norepinephrine.
[c] Constituent of urinary stones.
[d] The compounds found in bile may be reabsorbed in the intestine (enterohepatic cycle).
[e] Cholesterol and calcium bilirubinate are constituents of human gall stones.

Table 58. Compounds participating in the conjugation of secondary products

---

Sugars (D 1.1)
D-Glucose	D-Ribose
D-Fructose	L-Arabinose
D-Galactose	D-Xylose
L-Rhamnose	D-Glucosamine
D-Mannose	D-Apiose and other secondary monosaccharides

Holosides (D 1.4.1)
Rutinose	Glucosylxylose
Gentiobiose	Rhamnosylarabinose
Primverose	Apiosylglucose
Glucosylarabinose	Xylosylgalactose
Arabinosylglucose	Oligosaccharides

Sugar carboxylic acids (D 1.2)
D-Glucuronic acid	D-Galacturonic acid

Alcohols
Glycerol (D 1.3)	Alkanols (D 5.2.3)
myo-Inositol (D 1.3)	3,3-Dimethylallyl alcohol (D 6.1)

Aliphatic acids
Tartaric acid (D 1.2)	Acetic acid (D 5)
Citric acid (D 4)	Malonic acid (D 5)
L-Malic acid (D 4)	Fatty acids (D 5.2)

Amines
Putrescine (D 8.2)	Spermine (D 12.1)
Cadaverine (D 8.2)	Tryptamine (D 22.1)
Choline (D 10)	Tyramine (D 23.1.1)
Spermidine (D 12.1)	

Amino acids
Glycine (D 10)	L-Glutamine (D 17)
L-Alanine (D 10)	L-Ornithine (D 17)
L-Valine (D 13)	L-Tryptophan (D 22)
L-Aspartic acid (D 16)	L-Phenylalanine (D 23)
L-Glutamic acid (D 17)	L-DOPA (D 23)

Peptides
Glutathione (D 9.2)	

Hydroaromatic acids
Qinic acid (D 21)	Shikimic acid (D 21)

Cinnamic acids (D 23.2.1)
p-Coumaric acid	Ferulic acid
Caffeic acid	Sinapic acid

Benzoic acids (D 23.2.5)
Benzoic acid	Protocatechuic acid
4-Hydroxybenzoic acid	Gallic acid

Inorganic acids
Sulfuric acid (D 11)	Phosphoric acid

---

metabolic pathways. They resemble the reactions involved in detoxication of xenobiotics, e.g., drugs, herbicides, and pesticides, but also secondary products, which are ingested in the body of animals with the food (E 5.5.3 and F 1) or are released from pathogenic microorganisms into the body of plants (E 5.4). They block chemically and biologically active groupings, alter the chemical and physiological properties, e.g., solubility, pK, biological activity, etc. and in this way change the accumulation, transformation, and degradation of secondary products, as well as their significance for the producer organisms.

Animals possess effective mechanisms for the excretion of excess or waste products. In most of their tissues and body fluids, therefore, only small amounts of secondary products occur.

Exceptions with respect to the excretion of secondary compounds are, however, several insects, e.g. *Collembola* sp., which are devoid of excretory mechanisms during their whole lifetime and pulmonate land snails, like *Helix pomata*, which during winter or periods of drought reside inside their shells and close the openings with covers of lime or hardened secretions. These organisms accumulate secondary products, e.g., uric acid, within the nephridial cells (snails) or the urate cells of the fat body (Collembola), a process termed "storage excretion".

High concentrations of secondary products may also be present in certain specialized cells and tissues of other groups of animals, e.g., in the hormone glands of mammals (E 3.1), in the exocrine signal glands of insects (E 4), in the defense glands of certain beetles, in the skin glands of salamanders and toads (E 5.1), as well as in hairs and feathers (melanins, D 23.1.3), in the guanophores of fish hypodermis (guanine, D 25), and in the scales of butterflies (pteridines, D 25.3).

In contrast to animals, higher plants lack efficient mechanisms by which waste products are eliminated from the organism. Instead, they accumulate secondary products within their tissues. Frequently these substances are poisonous. Principles by which this toxicity is overcome are therefore of importance. The most widespread are:

a) Segregation of toxic compounds by compartmentation

Most toxic secondary plant products are accumulated beyond the metabolically active sites, e.g., in vacuoles and the cell wall, which are typical constituents of plant cells (storage excretion, A 3.2).

Due to the separation of storage compartments from plasm, compounds which are able to denaturate proteins and other compounds necessary for life can be accumulated within the plant body. The phenolics, for instance, piled up in the sap of vacuoles or in the cell wall are very dangerous to proteins after cell disintegration, i.e., after breaking down compartmentation (B 1.2).

b) Accumulation of protoxins

In several plants secondary products with low physiological activity are accumulated, which are transformed to toxic compounds if cells are damaged or infected. Examples are the cyanogenic glycosides (D 8.3) and the glucosinolates (D 8.4), but also many phenolic glycosides. These substances release toxic genins on the action of glycosidases, which in the native cell are accumulated separately from their substrates.

c) Metabolic adaptation

The toxicity of secondary products interacting as substrate analogs with enzymes of primary metabolism may be overcome by a high substrate specificity of the corresponding enzymes in the producer and storage cells. Several plants are able, for instance, to accumulate toxic nonprotein amino acids which interact with the enzymes of protein biosynthesis (D 8.1).

One of these amino acids is azetidine-2-carboxylic acid (D 12.1), which in most organisms in accepted as a substrate of prolyl-tRNA synthetase, an enzyme obligatory for protein biosynthesis. Azetidine-2-carboxylic acid is incorporated in proteins instead of L-proline, yielding protein molecules with reduced biological efficiency. Prolyl-tRNA in plants which store azetidine-2-carboxylic acid, however, has a much higher substrate specificity and does not react with this compound. Thus, even in the presence of azetidine-2-carboxylic acid normal proteins will be formed.

d) Degradation of toxic compounds

Many groups of secondary products undergo synthesis and degradation in the producer organisms (A 5). Transformation and degradation may play an important role, if toxic secondary products escape into sensitive areas of metabolism.

References for Further Reading

389, 540, 649, 656, 714, 724, 731, 820, 853

E 2 Physiologically Active Secondary Products

E 2.1 Coenzymes and Cosubstrates

Coenzymes and cosubstrates are necessary in all fields of metabolism in catalytic amounts (Table 59). The synthesis of these compounds, however, is restricted within the kingdom of living beings. Humans and animals, for instance, are unable to build most of them. They have to take up these products (or close precursors) as so-called vitamins with the diet. Also most types of plant cells cultivated in vitro and some microorganisms need supplementation with certain vitamins for growth or optimum cell division. Thiamine, for instance, is produced only in the leaves of higher plants and from these organs is transported to other tissues. Hence, it is not formed in plant cells cultivated in vitro which have another state of cell specialization than the leaf tissues. Though with respect to their synthesis, the compounds used as cosubstrates and coenzymes show typical features of secondary products, in contrast to most other secondary substances they are of significance also for the cells in which they are formed and not only for the producer organisms as a whole. In this respect they resemble compounds of primary metabolism. They may therefore be classified as intermediates between primary and secondary metabolic compounds.

References for Further Reading

141, 700, 876

E 2.2 Secondary Products with Special Physiological Functions

Several groups of secondary compounds and individual secondary substances serve the producer organisms in the physiology of growth, development, and reproduction. In Table 60 a few of the many diverse functions are compiled. This

Table 59. Secondary products used as coenzymes and cosubstrates

Secondary products	Mode of action
L-Ascorbic acid (D 1.2)	Cosubstrate of monooxygenases (C 2.6), vitamin for humans (vitamin C)
Pyridoxal-5'-phosphate (D 2)	Coenzyme of transaminases and amino acid decarboxylases (C 5), in humans formed from pyridoxin, pyridoxal and pyridoxamine (vitamins B_6)
Lipoic acid (D 5.2)	Cosubstrate of oxidoreductases, transfer of acyl groups (C 2.7)
Linolenic acid (D 5.2)	Precursor of the eicosanoids (D 5.2.6), vitamin for humans (vitamin F)
Glutathione (D 9.2)	Coenzyme, e.g., of glyoxalase, formaldehyde dehydrogenase, indolepyruvic acid-oxoenoltautomerase and maleylacetoacetic acid isomerase
Biotin (D 10)	Conezyme of carboxylases and carboxyltransferases (C 3.1), vitamin for humans (vitamin H)
Coenzyme A (D 11)	Cosubstrate of carboxylic acid activating enzymes; pantothenic acid, a constituent of coenzyme A, is a vitamin for animals
S-Adenosyl-L-methionine (D 12.1)	Cosubstrate of methyltransferases (C 3.3). L-Methionine is an essential amino acid for humans
Nicotinamide adenine dinucleotide (NAD$^+$) and nicotinamide adenine dinucleotide phosphate (NADP$^+$) (D 16.1)	Coenzymes and cosubstrates of oxidoreductases (C 2.1), in humans synthesized from nicotinic acid and nicotinamide. Both substances belong to the vitamin B complex
Isoprenoid naphthoquinones (D 21.2)	Coenzymes of glutamic acid carboxylase (D 17), vitamins for humans (vitamins K_1 and K_2)
Ubiquinones (D 21.3), plastoquinones, tocopherylquinones (D 23.4)	Constituents of the respiratory and photosynthetic electron transport chains, for humans tocopherol is a vitamin (vitamin E)
Heme (D 24)	Prosthetic group of dioxygenases (C 2.5) and of cytochromes, i.e., proteins participating in the respiratory electron transport and the electron transport in certain monooxygenases (C 2.6.1)
Hemin (D 24)	Coenzyme of peroxidases (C 2.4) and catalase
Factor 430 (D 24)	Prosthetic group of methylcoenzyme M reductase (C 3.2) in methane producing bacteria
Aquocobalamin, methylcobalamin, and 5'-deoxyadenosylcobalamin (D 24.3)	Coenzymes of tetrahydrofolate-dependent methyltransferases (C 3.2) and isomerases (D 24.3), vitamins for humans (vitamin B_{12})
Tetrahydrobiopterin, tetrahydrofolic acid (D 25.3)	Coenzymes of monooxygenases (C 2.6.1), formyltransferases, and certain methyltransferases (C 3.2), in humans synthesized from the vitamin folic acid or from pteroylpoly-L-glutamic acids
Flavinmononucleotide (FMN), flavinadenine dinucleotide (FAD) (D 25.4.)	Coenzymes and cosubstrates of oxidoreductases (C 2.2), in humans formed from riboflavin (vitamin B_2)
Thiamine diphosphate (D 25.6.)	Coenzyme of 2-oxo acid decarboxylases (C 4) and transketolases, in humans formed from thiamine (vitamin B_1)

list is, however, far from being complete. Many other functions may be added, such as the elicitation of bioluminescence mediated by the action of luciferases on different types of luciferins (C 2.5, D 11, D 24.2, D 25.4), the emulsification of dietary fats in the intestine of animals brought about by bile alcohols and bile acids (D 6.6.4), the use of volatile compounds, like ethylene (D 12.1), as "chemical radar" which allows hyphae of molds to pass obstacles without touching, the participation of gentisic acid glycosides (D 23.2.5) in the leaf movement of *Mimosa*, etc.

Table 60. Secondary products with physiological activity

Physiological aktivity/examples

Influence on membranes
Hopanoids (D 6.6) increase the membrane rigidity in certain prokaryotes and replace there the sterols (D 6.6.1) and carotenoids (D 6.7) which have the same function in most other groups of organisms. Phytanylglycerol ethers (D 6.4) substitute in archaebacteria the normal phospholipids (D 5.4.2) as membrane constituents.

Support of mechanical strength
Cellulose and hemicellulosic polysaccharides (D 1.4.1) form the cell wall, i. e., the "skeleton" of higher plants. Lignins (D 23.2.3) stabilize that wall, e. g., in the tracheas in which transpiration causes negative pressure. Wall forming compounds also cover and protect the cells of molds (chitin, D 1.4.1), bacteria (murein, D 9.2.4), as well as the pollen grains of higher plants and the spores of microorganisms (sporopollenins D 6.7). Chitin (D 1.4.1) and sklerotins (D 23.2.5) form the exoskeleton of insects.

Improvement of resistance
Surface lipids like alkanes and alkenes (D 5.2.2), waxes, cutin, and suberin (D 5.2.4) improve the physical resistance of plants and the resistance to microbial attack. Gums (D 1.4.1) form a barrier at lesions hindering the invasion of microorganisms.

Photosynthesis and light perception
Chlorophylls (D 24.1) play a key role in photosynthesis. Carotenoids (D 6.7) and open-chain tetrapyrrols (D 24.2) act as accessory pigments. Retinal (D 6.7) is involved in light perception in microorganisms and animals. An open-chain tetrapyrrol (D 24.2) is the prosthetic group of phytochrome, the most important light-sensitive pigment in plants. Secondary products may act, however, also as simple light screens in all groups of organisms. Examples are urocanic acid (D 20), ommochromes (D 22), melanins (D 23.1.3), and flavonoids (D 23.3.3).

Solubilization and uptake of ions
Siderophores, like sepedonin (D 5.3.3), pulcherriminic acid, neoaspergillic and neohydroxyaspergillic acids (D 9.2.1), aerobactin and ferrichrome (D 9.2.2), enterobactin (D 10), nicotianamine (D 12.1), but also citric acid (D 4) and riboflavin (D 25.4) chelate and transport Fe^{3+}. Bacitracin A is involved in the translocation of Mn^{2+} and valinomycin in that of K^+. The phytochelatins (D 9.2) detoxicate in plants heavy metal ions by complexation.

Storage of nutrients
Secondary compounds participate in the storage and transport of carbon in microorganisms, plants, and animals. Starch, glycogen, levans, and inulins (D 1.4.1), polyhydroxybutyric acid (D 5.1), fats, and fatty oils (D 5.2.4) participate in the storage and transport of carbon in microorganisms, plants, and animals. 4-Methylene-L-glutamine (D 14), L-asparagine (D 16), L-glutamine, N-acetyl-L-proline, and N^5-acetyl-L-ornithine (D 17), N^2-acetyl-L-arginine, N^2-acetyl-L-citrulline, and L-canavaline (D 19), serotonin (D 22.1), allantoin, and allantoic acid (D 25.5) are stores and transport forms of nitrogen. Calcium oxalate (D 4) is a store for Ca^{2+} ions in microorganisms and plants.

Group transfer
Phytic acid (D 1.3), creatine phosphate (D 10), and L-arginine phosphate (D 19) act as phosphate donors. Galactinol (D 1.3) is a glycosyl donor in plants. The dolichols (D 6.8) are involved in transfer of sugar residues in the formation of fungal glycans.

Promotion and inhibition of growth and cell division
myo-Inositol (D 1.3) acts as growth promotor in molds. 1-Triacontanol (D 5.2.3) acts similarly in higher plants. Trigonelline (D 16.1) arrests cells at the G-2 stage of the mitotic cell cycle.

Influence on hormone metabolism
Several secondary plant products act on the biosynthesis, activity, and degradation of plant hormones. Flavonoids (D 23.3.3), for instance, influence the degradation of the auxin indole-3-acetic acid (D 22). Flavonoids with an o-hydroxylated C-ring act as inhibitors of indole-3-acetic acid oxidase, in contrast to flavonoids with a monohydroxylated C-ring, which augment the activity of this

enzyme. Dihydroconiferyl alcohol (D 23.2.1) acts synergistically with gibberellins and other hormones, whereas naringenin (D 23.3.3) is an antagonist of gibberellins in dormant peach buds. In many cases, however, only the influence of secondary products on metabolism and development of plants has been observed. To the group of substances whose mode of action is unknown belong 2-E-6-E-farnesol (D 6.3), which influences the opening of the stomata, 4-hydroxyphenyl acetic acid (D 23.1.1), which acts as growth regulator, psoralen (D 23.2.2) and *Aegilops* lignan (D 23.2.4), which inhibit seed germination in *Psoralea subacaulis* and *Aegilops odorata*, respectively, gallic acid (D 23.2.5), which inhibits flowering of *Kalanchoe blossfeldiana*, and batatasin III (D 23.3.3) which induces dormancy in the bulbils of yam, *Dioscorea batatis*.

Activation and depression of germination
Gramicidin S (D 9.2) inhibits the germination of bacterial spores, methyl Z-3,4-dimethoxycinnamic acid, and methyl Z-ferulic acid (D 23.2.1) that of rust spores. Discadenine, (D 25.1) depresses the germination of spores of *Dictyostelium discoideum*. Geniposide (D 6.2.2) hinders the germination of the seeds of *Gardenia jasmonoides*, psoralen (D 23.2.2) that of *Psoralea subacaulis*. 6-Methylhept-5-en-2-one (D 5.2.2) and nonanal (D 5.2.3) stimulate the outgrowth of rust spores.

Influence on the osmotic value
Glycerol and D-sorbitol (D 1.3) facilitate supercooling in insects and in this way increase cold hardiness. Glycerol (D 1.3) is accumulated in halophilic algae, and betaine (D 8.1) is accumulated in NaCl-adapted plants to resist the osmotic pressure of the surrounding medium. 2-E-6-E-Farnesol (D 6.3) governs the opening of the stomata of higher plants.

All the given examples show that secondary products fulfill these different functions only in restricted systematic groups of organisms, or in certain organs and cells and may be substituted by other compounds with similar properties in other places. This shows the variability and manifoldness in the special fields of metabolism, which reflect the peculiarities of living beings, in contrast to the much higher uniformity in the basic area of primary metabolic reactions.

References for Further Reading

128, 295, 460, 492, 559, 561, 564, 674, 676, 700, 877, 880

E 3 Secondary Products as Intraorganismic Signals

E 3.1 Hormones

Chemical signals are an important means of regulating the cooperation of the specialized cells of multicellular organisms with their different patterns of metabolic reactions. In higher plants and animals hormones play in this respect an outstanding role. Many of them are secondary products (Tables 61 and 62). They share the variability of chemical structures and the restricted distribution within the different groups of living beings with other types of secondary substances. It is of special interest that secondary products acting in one group of organisms as hormones, in other organisms show no comparable physiological activity

Table 61: Secondary products acting as hormones in animals

Hormone (principle source of biosynthesis)	Main biological activity
Prostaglandins, prostacyclins, thromboxanes, leucotrienes (D 5.2.6); vertebrates (several tissues)	Participation in inflammation and blood platelet aggregation, in humans formed from linolenic acid (vitamin F)
Juvenile hormone III (D 6.3); insects (corpora allata)	Triggers the outcome of molts
Ecdysone, ecdysterone (D 6.6.1); insects (prothoracic gland), crustacea (y-organs)	Molting hormones
Progesterone (D 6.6.5); vertebrates (ovary, placenta)	Maintenance of pregnancy
Corticosteroids (D 6.6.5); vertebrates (cortex of adrenal glands)	Regulation of energy utilization and electrolyte balance
Androstanes (D 6.6.6); vertebrates (testes)	Development of male sex characteristics
Estrogens (D 6.6.7); vertebrates (ovary)	Development of female sex characteristics
Calcitriol (D 6.6.9); vertebrates (skin, liver, kidneys)	Control of the Ca^{2+}-content of blood
Retinoic acid epoxide (D 6.7); vertebrates (unknown)	Differentiation of epithelia
Juvenile hormones I and II (D 7); insects (corpora allata)	Determination of the outcome of molts
Angiotensin II (D 9.1); vertebrates (several tissues)	Increase of blood pressure
Bradykinin (D 9.1); vertebrates (several tissues)	Decrease of blood pressure
Thyrotropin-releasing hormone (D 9.1); vertebrates (hypothalamus)	Stimulation of thyrotropin release from pituitary gland
LH/FSA-releasing hormone (D 9.1); vertebrates (hypothalamus)	Stimulation of LH/FSH release from pituitary gland
GH-releasing hormone (D 9.1); vertebrates (pituitary gland)	Stimulation of growth hormone (somatotropin) release from pituitary gland
Oxytocin (D 9.1); vertebrates (pituitary gland)	Stimulation of milk ejection in mammals
Vasopressin (D 9.1); vertebrates (pituitary gland)	Inhibition of water resorption in the kidney during urine concentration, increase of blood pressure
Corticotropin (D 9.1); vertebrates (pituitary gland)	Stimulation of steroid hormone release from the adrenocortex of mammals
β-Melanotropin (D 9.1); vertebrates (pituitary gland)	Cause of redistribution of melanins in melanophores of fish and amphibia
Insulin (D 9.1); vertebrates (pancreas)	Regulation of carbohydrate metabolism
Histamine (D 20); vertebrates (several tissues)	Regulation of blood flow in tissues and secretion in the stomach
Serotonin (D 22.1); vertebrates (several tissues)	Irritation of smooth muscles
Melatonin (D 22.1); vertebrates (pineal gland)	Cause of seasonal and diurnal rhythms

Hormone (principle source of biosynthesis)	Main biological activity
Epinephrine, norepinephrine (D 23.1.1); vertebrates (medulla of the adrenal glands)	Sympaticomimetics, vasoconstrictors, increase of catabolic reactions
L-Thyroxine, L-triiodothyronine (D 23.1.5) vertebrates (thyroid gland)	Affection of body growth and metabolic activity, in amphibia acceleration of metamorphosis
Lignans (D 23.2.4); vertebrates (gonads, ovary)	Involvement in the ovulation cycle?
cAMP (D 25.1); vertebrates (several tissues)	Second messenger, mediating the action, e.g., of nonsteroid hormones
1-Methyladenine (D 25.1); Echinodermata (ovary, interstitial cells)	Shedding of gametes and oocyte maturation

Table 62. Secondary products used as hormones in plants

Hormone (principle source of biosynthesis)	Main biological activity	Hormone (principle source of biosynthesis)	Main biological activity
(−)-Jasmonic acid (D 5.2.1) (unripe fruits, e.g., of *Vicia faba*)	Promotion of senescence, inhibition of seed germination and the growth of seedlings	Brassinolide (D 6.6.1) (unknown)	Promotion of cell division and elongation
(+)-Abscisic acid, phaseic acid, xanthoxin (D 6.3) (leaves, root tip)	Dormancy hormones, inhibition of growth and metabolism, closure of stomata	Ethylene (D 12.1) (several tissues)	Promotion of defoliation and fruit ripening
GA₁ and probably other gibberellins (D 6.4) (unripe fruits, young leaves, roots)	Stimulation of shoot elongation	Indole-3-acetic acid (D 22) (apical meristems of stem)	Regulation of various growth and developmental processes (auxin activity)
		Lunularic acid (D 23.3.3) (whole plant)	Dormancy hormone
		Cytokinins, e.g., zeatin (D 25.1) (predominantly roots)	Delay of senescence and stimulation of cell division
Antheridiogens (D 6.4) (prothalli)	Triggering of antheridia formation		

(Table 63). This may be of significance in ecology, as was demonstrated with animal hormones built in plants (E 5.5.3) and plant hormones synthesized in animals.

An example in this respect is the excretion of cytokinins (D 25.1) from the labiae of larvae of the moth *Stigmella argentipedella* living as mining caterpillars in leaves of the birch *Betula verrucosa*. These cytokinins cause the attraction of nutrients from other parts of the plant and their accumulation in the cells near the insect which are used as prey. The accumulation of nutrients originates the survival of these cells in senescent leaves (formation of "green islands" in the yellow-brown autumn leaves).

In animals hormones are produced either in special glands (cf. Table 61) or they are formed in tissues in which the production of hormones is a biochemical side activity (so-called tissue hormones). In plants special hormone-producing glands are absent, but also in these organisms hormone synthesis is unequal in

Table 63. Occurrence (+) within the kingdom of living beings of secondary products which serve as hormones (H) or neurotransmitters (N) in plants or animals

Compounds	Synthesis in		
	Animals	Plants	Microorganisms
(−)-Jasmonic acid (D 5.2.1)		+ (H)	+
Prostaglandins (D 5.2.6)	+ (H)[a]	+	
(+)-Abscisic acid (D 6.3)		+ (H)	+
Gibberellins (D 6.4)		+ (H)	+
Ecdysone, ecdysterone (D 6.6.1)	+ (H)	+	
Progesterone (D 6.6.5)	+ (H)	+	
Testosterone, androstenedione (D 6.6.6)	+ (H)	+	
Estrone, estradiol, estriol (D 6.6.7)	+ (H)[a]	+	
Calcitriol (1α, 25-dihydroxy-vitamin D_3, D 6.6.9)	+ (H)	+	
Acetylcholine (D 10)	+ (N)	+	
Ethylene (D 12.1)		+ (H)	+
Histamine (D 20)	+ (H)	+	+
Indole-3-acetic acid (D 22)	+	+ (H)	+
Serotonin (D 22.1)	+ (H, N)	+	+
Dopamine (D 23.1.1)	+ (N)	+	
Norepinephrine (D 23.1.1)	+ (N, H)	+	
cAMP (D 25.1)	+ (H)[b]	+	+
Cytokinins (D 25.1)	+	+ (H)	+

[a]   Hormones in vertebrates, but also formed in invertebrates where the compounds have no hormone activity.
[b]   Second messenger, mediating the action of most nonsteroid hormones.

the different types of cells. Cytokinins, for instance, are produced predominantly in roots and indole-3-acetic acid in apical meristems.

In plants the compounds acting as hormones are all synthesized de novo from primary metabolic substances. In animals hormones are either formed from substances of primary metabolism, e.g., steroid hormones, epinephrine, tyroxin, or they are built from special secondary products which have to be taken up with the diet, e.g., unsaturated fatty acids (vitamins F) which are used as precursors of the eicosanoids (D 5.2.6); carotenoids and vitamin A, which are transformed to retinoic acid epoxide (D 6.7); and under certain conditions also the vitamins D, which are the precursors of hydroxylated vitamin D derivatives (D 6.6.9).

Most plant or animal hormones have to be transported from the producer cells to the target cells. This transport proceeds via the blood or similar liquids in animals and the xylem or phloem sap in plants (humoral regulation). The target cells have very fine and sensitive systems of hormone recognition. In animals these may be located within the cells (receptors of most steroid hormones, D 6.6.5–D 6.6.7) or at the cell surface (receptors of most other hormones, especially peptide hormones, D 9.1). The action of the latter group of hormones is brought about by cAMP (D 25.1) as the so-called second messenger which inside the target cells influences the activity of protein kinases. The molecular action of the plant hormones as yet is still more or less unknown.

In microorganisms hormonelike regulatory effectors are rarely found. Examples are:

— butyrolactones, e. g., the so-called A-factor, which in Actinomycetes influences the expression of idiophase characteristics, e. g., the formation of antibiotics and the sporulation

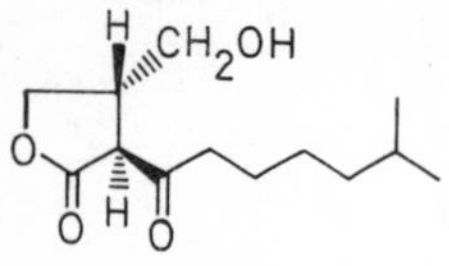

A-factor

— cAMP (D 25.1), which in bacteria mediates the repression of special catabolic enzymes, and
— polyphosphorylated purine glycosides, e. g., ppApp, ppGpp, and ApppppA (D 25.1), which in bacteria affect DNA synthesis and gene expression.

References for Further Reading

283, 363, 410, 423, 570, 607, 753

E 3.2 Neurotransmitters

In addition to humoral regulation (E 3.1), higher animals have a second signal system, the nervous system. Different secondary products, called neurotransmitters, are involved in its functioning (Table 64). Neurotransmitters are released by electric exitation at the synapses, the motoric end plates, or nerve endings, and transmit in a chemical way information from one nerve cell to the next or to the target organ. In the peripheral nerve system acetylcholine or norepinephrine/epinephrine serve as transmitters between the nerve cells and at the neuromuscular junctions. In the central nerve system the spectrum of transmitters is much larger. Acetylcholine, taurine, GABA, histamine, serotonin, 5-methoxytryptamine, dopamine, and norepinephrine as well as the enkephalins and endorphins play a specific role. The enkephalins and endorphins react with the so-called op-

Table 64. Secondary products used as neurotransmitters

Secondary products	Organisms	Secondary products	Organisms
Enkephalins, e.g., met-enkephalin, and endorphins, e.g., β-endorphin (D 9.1)	Vertebrates	Histamine (D 20)	Vertebrates
		Serotonin, 5-methoxy-tryptamine (D 22.1)	Vertebrates and invertebrates
Acetylcholine (D 10)	Vertebrates	Dopamine (D 23.1.1)	Vertebrates
Taurine (D 11.1)	Vertebrates	Epinephrine and norepinephrine (D 23.1.1)	Vertebrates
4-Aminobutyric acid (GABA) (D 17)	Vertebrates, crustacea, insects		

iate receptors in the brain, and as morphine and similar compounds (D 23.1.2), reduce the perception of pain and cause euphoria (F 3).

References for Further Reading

176, 383, 813

E 4 Secondary Products Involved in Intraspecific Communication (Pheromones)

An interesting group of secondary products, the pheromones, is involved in the communication between the individual organisms of animal and microbial species and between certain types of plant cells. Pheromones are used to give signals, e. g.,
— in reproduction (sex pheromones, attracting and arousing sex partners),
— in protection (alarm pheromones, "tagging" other organisms as aggressor), and
— in guidance to food sources (trail pheromones, marking a path which other members of the species can follow), etc. (Table 65).

Table 65. Secondary products used as pheromones

Group of organisms, secondary substances	Mode of action
Microorganisms	
Sex pheromones	
Sirenin (D 6.3)	Sex pheromone of *Allomyces* (water mold)
Antheridiol, oogoniols (D 6.6.1)	Sex pheromones of the aquatic fungus *Achlya*
Trisporic acids (D 6.7)	Sex pheromones of *Mucorales* (fungi)
Mating pheromones	
α-Factor, rhodotorucine A, tremerogen A 10 (D 9.1)	Mating pheromones of the yeasts *Saccharomyces cerevisiae, Rhodosporidium toruloides*, and *Tremella mesenterica*, respectively
Aggregation pheromones	
cAMP (D 25.1)	Aggregation pheromone of the slime mold *Dictyostelium*
Plants	
Sex pheromones	
Fucoserraten (D 5.2.2)	Sex pheromone of the ♀ gametes of *Fucus* sp.
Ectocarpen (D 5.2.2)	Sex pheromone of the ♀ gametes of the brown alga *Ectocarpus siliculosus*
Multifiden (D 5.2.2)	Sex pheromone of the ♀ gametes of brown algae, e. g., *Cutleria* and *Desmarestia* sp.

Group of organisms, secondary substances	Mode of action

Insects

Sex pheromones

Valeric acid (D 5.2.1)	Sex pheromone of ♀ sugar beet wireworm *(Limonius cakfornicus)*
9-Oxo-2-*E*-decenoic acid (D 5.2.1)	Sex pheromone of the honey bee queen *(Apis mellifera)*
7,8-Epoxy-2-methyl-octadecane (D 5.2.2)	Sex pheromone of ♀ *Lymantria dispar*
Bombykol, bombykal (D 5.2.3)	Sex pheromones of ♀ *Bombyx mori*
11-*E*- and 11-*Z*-Tetradecenyl acetate (D 5.2.4)	Sex pheromone of ♀ oak leaf rollermoth *(Archips semiferanus)* and other insects, taken up from the food source
Hexadecanyl acetate (D 5.2.4)	Sex pheromone of the ♂ butterfly *Lycorea ceres ceres*
Myrcene (D 6.2.1)	Sex pheromone of *Dendroctonus brevicomis*, sequestered from pine oleoresin together with two bicyclic ketals synthesized in the beetle
Geranyl and 2-*E*-6-*E*-farnesyl esters (D 6.2.1 and 6.3)	Sex pheromones of ♂ bees of the genus *Andrena* (cf. Table 70)
Pyrrolizidine aldehydes and ketones (D 17.1)	Sex pheromones of ♂ *Danaid* butterflies, produced from pyrrolizidine ester alkaloids taken up from food source *(Senecio* sp.)

Alarm pheromones

Formic acid (C 3.2)	Alarm pheromone (and defense substance) of ants *(Formica* sp.)
Undecane, tridecane, pentadecane (D 5.2.2)	Alarm pheromone of ants *(Formica* sp.)
Isoamyl acetate (D 6.1)	Alarm pheromone of the honey bee *(Apis mellifera)*
Terpinolene (D 6.2.1)	Alarm pheromone of termites *(Amitermes* sp.)

Trail pheromones

3-*Z*-6-*Z*-8-*E*-Dodecatriene-1-ol (D 5.2.3)	Trail pheromone of the termite *Reticulitermes virginicus*, probably taken up from food source
Geraniol (D 6.2.1)	Trail pheromone of the honey bee, probably collected from flower scents
Cembrene A (D 6.4)	Trail pheromone of the termite *Nasutitermes exitisosis*
4-Methylpyrrole-2-carboxylic acid methyl ester (D 17)	Trail pheromone of leaf-cutting ants *(Atta* sp.)

Mammals

Sex pheromones

Propionic acid (D 3), acetic acid (D 5), butyric acid (D 5.2), etc.	Sex pheromones of ♀ *Rhesus* monkey and ♀ *Homo sapiens*
Civetone (D 5.2.2)	Sex pheromone of civet cat *(Viverra civetta)*
Muscone (D 5.2.2)	Sex pheromone of musk deer *(Moschus moschiferus)*
5α-Androst-16-en-3-one and 5α-androst-16-en-3α-ol (D 6.6.6)	Sex pheromones of the boar *(Sus scrofa)*

Alarm pheromones

Crotylmercaptane CH$_3$—CH=CH—CH$_2$SH, isopentylmercaptane (CH$_3$)$_2$—CH—CH$_2$—CH$_2$SH	Alarm pheromones (and defense substances) of the skunk *(Mephitis mephitis)*

Territorial marker

Phenylacetic acid (D 23.1.1)	Territorial marker of the Mongolian gerbil *(Meriones unguiculatus)*

Pheromones are usually mixtures of secondary products. Most widespread are pheromones in the kingdom of insects. Several groups of pheromones are synthesized in and secreted from different glands of these animals and in the course of evolution very sensitive receptor mechanisms have been developed which need only a few molecules for response.

A female silkworm moth, for instance, releases enough bombykol and bombykal to attract male moths in a downwind corridor several kilometers long and over 100 m in diameter. The males begin to react if there are only about 100 molecules of the pheromone per milliliter air.

The existence of pheromones in man is still under debate. Short chain fatty acids found in the vaginal secretions of female humans and 5α-androst-16-en-3-one occurring in the sweat of males have been claimed as female and male sex pheromones, respectively. Recent experiments, however, have made these results questionable.

Most pheromones are synthesized de novo in the animal body. Some, however, are taken up from plant sources and are used directly or in a modified form (cf. the pyrrolizidines secreted as sex pheromones from male *Danaid* butterflies and myrcene used as sex pheromone of *Dendroctonus brevicomis*). Pyrrolizidine alkaloids ingested with the diet by larvae of the moth *Creatonotos* act as pheromone precursors and show hormonelike activity in the control of the morphogenesis of the scent organs.

Pheromones are applied in agriculture. The sex pheromones of boars are used in arousing sows in order to increase the rate of pregnancy at artificial insemination. Some sex pheromones of insects may be used in pest control. On the one hand, synthetically prepared pheromones applied in "traps" may attract insects (point application), on the other hand, whole areas in which insect pests did develop are sprayed. This treatment results in a high general pheromone content in the air, which prevents orientation of insect males to the source of the natural pheromones released by insect females and in this way arrests mating. In addition, chemicals may be sprayed which mask the effect of the natural pheromones and interfere with these signals.

References for Further Reading

25, 69, 83, 91, 92, 323, 325, 388, 406, 411, 431, 446, 521, 619, 620, 681, 683, 781, 859

E 5 Secondary Products (Allelochemicals) Mediating Ecological Relations

Microorganisms, plants, and animals live in a world of chemical signals. These signals, in most cases secondary products, are of special importance in ecology, i.e., the mutual relations between the different groups of organisms, which cause the formation of the characteristic communities of living beings in the forests, meadows, and deserts, in pools, rivers, and the open sea, in coral riffs, ant hills, etc. In these communities many kinds of microorganisms, plants, and animals promote and inhibit each other in a delicate balance. In a process called coevolution step-by-step signal and target mechanisms have emerged, mechanisms of response, adaptation, dependence, and mimicry have developed resulting in a complex network of relations interlinking the organisms of a habitat.

Our present-day knowledge of ecological relations is still unrepresentative. However, ecological significance seems to offer a rational and satisfying explanation for the existence of a good part of the enormous amount of secondary products synthesized in the living world.

References for Further Reading

246, 263, 323, 484, 719, 748, 812

E 5.1 Animal Secondary Products Used in Defense

Certain groups of secondary products protect animals against predation and microbial attack (Table 66). Some allow masking, e. g., the melanins (D 23.1.3), which give the "grey mouse" its characteristic inconspicuous color. Others have an unpleasant smell, like the mercaptane derivatives, produced in special glands of the skunk. Several compounds, like cardenolides and alkaloids, are toxic. Most of these substances have a broad spectrum of physiological activity against different kinds of animals and microorganisms. Animals producing toxins often have a warning coloration. It signals their toxicity to potential predators. This

Table 66. Secondary products which are animal defense substances

Group of organisms, secondary products	Mode of action
Corals Prostaglandin A_2, 15(R)-prostaglandin A_2 (D 5.2.6)	Formed in certain species of corals, deterrent to predatory fish
Cephalopods Eumelanin (D 23.1.3)	Darkening of the water
Fish Tetrodotoxin (D 19)	Accumulated in bowl fish, several amphibia, and cephalopods, inhibits Na^+ transport through cell membranes
Amphibia Batrachotoxins (D 6.6.5)	Accumulated in skin secretions of the poison-dart frog *Phyllobates aurotaenia*, feeding deterrent
Samandarin, cycloneosamandione (D 6.6.6)	Synthesized in skin glands of salamanders, protection against microbial infections, feeding deterrents
Bufotoxin (D 6.6.8)	Synthesized in skin glands of the common toad *Bufo vulgaris*, feeding deterrent
Chimonanthine, calycanthine (D 22)	Accumulated in skin secretions of the poison-dart frog *Phyllobates terribilis*, feeding deterrent
Insects Formic acid (C 3.2)	Synthesized in glands of ants, irritant

Group of organisms, secondary products	Mode of action
Myrmicacin (D 5.2.1)	Synthesized in the metathoracic glands of leaf-cutting ants (*Atta* sp.), prevents germination of undesirable fungal spores in the nest
Citronellal, citral (D 6.2.1)	Synthesized in glands of the ant *Acanthomyops claviger*, feeding deterrent
α- and β-Pinene, abietic acid, etc. (D 6.2.1 and 6.4)	Taken up by larvae of the sawfly *Neodipridion setifer* form food source *(Pinus sylvestris)* and stored in pouches of the foregut, discharged when approached by a predator
Iridodial, dolichodial (D 6.2.2)	Synthesized in glands of the devil's coach horse beetle *Staphylinus olens*, and stick insects (*Anisomorpha* sp.), feeding deterrents
Cantharidin (D 6.3)	Synthesized in the meloid beetle *Lytta vesicatoria*, present in the blood, released by reflex bleeding from joints, irritant
Cortexone, 12-hydroxy-4,6-pregnadien-3,20-dione (D 6.6.5)	Synthesized in the prothoracic glands of the water beetles *Cybister limbatus* and *Cybister tripunctatus*, feeding deterrents
Cardenolide glycosides (D 6.6.8)	Synthesized in certain chrysomelid beetles, feeding deterrents
Calotropin (D 6.6.8)	Formed in the grasshopper *Poekilocerus bufonius* and in larvae of the monarch butterfly *Danaus plexippus* from cardenolides present in the food source (*Asclepias* sp.), stored in the body, deters birds and other predators
Linamarin, lotaustralin (D 8.3)	Releases HCN after degradation which inhibits cell respiration by interaction with the cytochrome system, formed in insects (Zygaenidae and Heliconiini)
Methacrylic acid (D 13) and tiglic acid (D 15)	Synthesized in carabids, constituents of defense secretions
Senecionine (D 17.1)	Taken up by larvae of the moths *Arctia caja* and *Tyria jacobaeae* from food source (*Senecio* sp.), stored in the body, deters birds
Lycopsamine (D 17.1)	Taken up by adult Ithomiini butterflies as a constituent of the nectar they use as prey (E 5.5.1), deters predators, e. g., spiders
Glomerin, homoglomerin (D 21.4)	Synthesized in the millipede *Glomeris marginata*, feeding deterrents
8-Hydroxyquinaldic acid methyl ester (D 22.5)	Synthesized in glands of the water beetle *Ilybius fenestratus*, feeding deterrent, antiseptic
Phenylacetic acid (D 23.1.1)	Secreted from the metathoracic glands of leaf-cutting ants (*Atta* sp.) to keep the nest free of bacterial infections
Aristolochic acid I (D 23.1.2)	Taken up by larvae of the butterfly *Pachlioptera aristolochiae* from food source (Aristolochiaceae), feeding deterrent
Benzoic acid, 4-hydroxybenzoic acid, protocatechuic acid, methyl and ethyl esters, hydroquinone (D 23.2.5)	Secreted from the pygidial glands of the water beetles *Dytiscus* sp., feeding deterrents, antiseptics
Hydroquinone (D 23.2.5), H_2O_2	Synthesized in glands of the bombardier beetles *Brachynus* sp. (Fig. 350)
2-Methyl and 2-ethylbenzoquinones (D 23.4)	Synthesized in several beetles
Mammals	
Isopentylmercaptane $(CH_3)_2CH—CH_2—CH_2SH$, crotylmercaptane $CH_3—CH=CH—CH_2SH$, methylcrotyldisulfide $CH_3—CH=CH—CH_2—S—S—CH_3$	Synthesized in and released from the anal glands of the skunk *Mephitis mephitis*, if frightened, feeding deterrent

warning coloration may be mimicked by nontoxic animals which in this case are also avoided by predators. An example in this respect is the yellow-black coloration of several species of hover-flies *(Syrphidae)* which resembles that of wasps and bees.

Most toxic substances protecting animals are synthesized de novo in the animal body. Some are taken up with the diet and are used directly or in a modified form, e.g., iridoids (D 6.2.2), cardiac glycosides (D 6.6.8), pyrrolizidine alkaloids (D 17.1), and saxitoxin (D 25.1). Some compounds are built, stored, and secreted in special glands, e.g., the defense glands of insects and the skin glands of salamanders and toads. Others are constituents of blood or gut.

Most of the toxins have a broad spectrum of activity against different kinds of predators or microorganisms. There are substances interfering with ion transport (batrachotoxins, cardiac glycosides and esters, tetrodotoxin) and cell respiration (hydrocyanic acid set free from cyanogenic glycosides). In most cases, however, the mode of action is still unknown.

Passive and active defense may be distinguished. In passively toxic animals the toxic compounds are simply accumulated in the body, like the cardiac glycoside, calotropin, in the monarch butterfly and tetrodotoxin (D 19), a guanidine derivative found in bowl fish, several amphibia, and arthropods. In actively toxic animals the toxins may be discharged or actively expelled.

Examples are
— cantharidin, a constituent of the hemolymph of meloid beetles, which is actively excreted by reflex bleeding if the insects are attacked,
— the hormones acetylcholine, serotonin, dopamine, and norepinephrine (E 3.1) which are constituents of the toxins of stinging insects and snakes, and
— the irritating compounds, like formic acid and quinones, ejected from special glands of ants and beetles, respectively (Fig. 350).

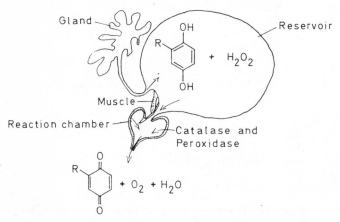

Fig. 350. Operation of the defense gland of the bombardier beetle *Brachynus sp.*
From the defense glands of bombardier beetles a mixture of hydroquinones and hydrogen peroxide is discharged into an extracellular reservoir. There it is accumulated and, if the animal is irritated, is released in the so-called reaction chamber, containing catalase and peroxidase. A sudden reaction takes place, the mixture heats up to 100 °C and is ejected by the evolved oxygen gas with a blow. The secreted liquid is very corrosive. It terrifies and injures the aggressor.

Secondary products with antimicrobial activity are of special significance for animals living submerged or in wet habitats. Salamanders, for instance, synthesize alkaloids in skin glands as protection against microbial infections. Certain water beetles inhibit growth of microorganisms on their chitin coverings by hydroquinone, 4-hydroxybenzoic, and protocatechuic acid derivatives, and leaf-cutting ants prevent bacterial infections in their moisture-saturated nests by phenylacetic acid.

References for Further Reading

76, 108, 323, 590, 618, 630, 675

E 5.2 Antibiotics, Microbial Defense Products

Antibiotics are microbial secondary products which inhibit the growth of other microorganisms or cause the death of bacterial, fungal, and amoebal cells occurring in the neighborhood of the producers. In addition, several antibiotics are active against animals, e. g., mites, which live on microbial cells. Antibiotics are important drugs used in medicine. They allow the effective control of many human and animal microbial pathogens (F 2). Large amounts are used in agriculture in the treatment of plant diseases and are fed to farm animals as anabolics

Table 67. Mode of action of secondary products with antibiotic activity

---

Inhibition of nucleic acid biosynthesis
Cordycepin, psicofuranin, decoyinin (D 25.1), tubercidin, toyocamycin (D 25.2)

Inhibition of DNA biosynthesis
Daunomycin, adriamycin (D 3), edeine A (D 9.2), actinomycins (D 21.4.1)

Inhibition of mRNA biosynthesis
Rifamycin B (D 3), actinomycins (D 21.4.1)

Inhibition of mRNA translation
Streptomycin, neomycin C, gentamycin, kanamycin A, kasugamycin (D 1.3), erythromycins (D 3), tetracyclines, cycloheximide (D 5.3.8), fusidic acid (D 6.6.1), chloramphenicol (D 21.1), blasticidin S (D 26)

Amino acid antagonists
L-2,5-Dihydrophenylalanine (D 23)

Inhibition of cell wall biosynthesis of bacteria
Bacitracin A (D 9.2), penicillins, cephalosporins (D 9.2.3), D-cycloserine (D 10)

Increase of membrane permeability
Nystatin, amphotericin B (D 3), gramicidin S, tyrocidine A, polymyxin B_1 (D 9.2)

Facilitation of ion transport through membranes
Lasalocid A, monensin A (D 3)

Inhibition of uptake and transport of ions
Valinomycin (K^+, D 9.2), nocardamine (Fe^{3+}, D 9.2.2)

Unknown
Griseofulvin (D 5.3.5), pyrrolnitrin (D 22), xanthocillin X (D 23.1.1)

30*

to accelerate weight increase. Most antibiotics influence nucleic acid and protein biosynthesis (Table 67). Others act on cell wall formation, membrane permeability, or on the availability and uptake of ions.

Substances with antibiotic properties can be easily detected by appropriate screening procedures. Many of the microbial secondary products have been examined with these methods and more than 5000 compounds with antibiotic properties have been found. Approximately 3500 of them are derived from different Actinomycetales, about 90% originate from *Streptomyces* sp. About 1000 compounds with antibiotic properties come from different fungi and about 600 are formed from various bacterial strains (including Pseudomonales). Under natural conditions, however, obviously only a few of these substances play an ecological role in microbial defense. This is indicated by the following facts:

— many organisms producing antibiotics are just as sensitive to their own antibiotics as are their competitors,
— only in a few cases have antibiotics been found in the natural habitat of producer strains and the spectrum of activity of the antibiotics usually does not correspond to the spectrum of the naturally occurring competing microorganisms, and

Table 68. Secondary plant products inhibiting growth and development of plants in the neighborhood of the producer

Secondary products	Mode of action
1,8-Cineole, d-camphor and other monoterpenes (D 6.2)	Released from leaves of the sagebrush, *Salvia leucophylla*, and of *Artemisia californica* and absorbed in the soil, inhibition of germination of the seeds of annual plants
3-Aminopropionitrile (D 8.3)	Released from seedlings of *Lathyrus odorata*, toxic to *Lathyrus aphaca* seedlings, but not to seedlings of lettuce
Quinolizidine alkaloids (D 18)	Released from roots, inhibition of seed germination and plant growth
L-Canavanine (D 19)	Released from germinating seeds and seedlings of *Neonotonia wightii*, inhibits growth of other plant seedlings
3-Acetyl-6-methoxybenzaldehyde (D 21.1)	Released during decomposition of leaves of *Encelia farinosa*, inhibits growth of annual plants
Juglone (D 21.1)	The leachate of *Juglans* leaves contains 1,4,5-trihydroxynaphthalene-4-glucoside which after hydrolysis and oxidation yields juglone. Juglone inhibits seed germination and development of many plants
o- and p-Coumaric acids, ferulic acid, syringic acid (D 23.2.1), 4-hydroxybenzoic acid, vanillic acid, salicylic acid (D 23.2.5)	Leached, e. g., from leaves of *Adenostoma fasciculatum*, *Arctostaphylos glandulosa*, and *Quercus falcata*, inhibits development of many other plants
Tannins (D 23.2.5 and D 23.3.3)	Set free, e. g., from decaying parts of heather *(Calluna vulgaris)* and bracken *(Pteridium aquilinum)*, reacting with minerals and organic soil matter to complexes resistant to microbial attack. These complexes exclude many plant genera, e. g., from the growth on heath and moorlands

— most producers form the antibiotics during the phase of cell specialization, i. e., after completion of the phase of rapid growth (A 4.1).

A more detailed analysis of the action of antibiotics has demonstrated that many of these compounds are of significance in the physiology of the producer organism, like bacitracin A in the transport of ions, and gramicidin S in the inhibition of spore outgrowth (E 2.2), rather than in ecology. Knowledge in this field is, however, still scarce.

References for Further Reading

58, 124, 155, 164, 186, 255, 278, 315, 374, 448, 526, 838, 903

E 5.3 Plant-Plant Interactions (Allelopathy) Caused by Secondary Products

Plants compete with each other for moisture, light, and nutrients. Part of this competition is chemical interaction (allelopathy), which in most cases is brought about by secondary products (Table 68). The substances involved occur in leaves, stems, and roots.

They are either leached (wet climates) or volatilized (arid climates) from living organs or set free during the decay of plant material. Often the substances are stored in the living plant cells in a form which needs activation before it exerts the biological action (cf. the transformation of 1,4,5-trihydroxynaphthalene-4-glucoside to juglone). Whether secondary products which show allelopathic activity in vitro are active also in nature depends to a great extent on their resistance to microbial degradation.

References for Further Reading

323, 325, 529, 624, 636

E 5.4 Secondary Products Involved in the Relations between Plants and Microorganisms

Plant Secondary Products Mediating Defense

Resistance of higher plants to microbial infections includes secondary products as part of the general defense system. In this respect the following principles are of significance (Table 69):

a) Accumulation of secondary products acting as mechanical barriers.

Most of these substances are constituents of the different protective layers covering plants (wax, cuticle, cork, cell wall, etc.)

b) Accumulation of secondary compounds toxic to invading microorganisms in the extracellular space, e. g., the atmosphere near the plants (constituents of essential oils), in the moisture film covering the surface of leaves or other over-

Table 69. Secondary products participating in the protection of plants against microbial infections

Secondary products	Mode of action
Preinfectious barriers	
Cell wall polysaccharides (D 1.4.1)	Mechanical barrier to different types of microorganisms, more or less resistant to microbial degradation because of bound phenolic compounds, e. g., ferulic acid (D 23.2.1), or unusual branching of the carbohydrate chain
Alkanes, alkenes (D 5.2.2), alkanols (D 5.2.3), wax, cutin, suberin (D 5.2.4)	Mechanical barrier to different types of microorganisms, resistant to microbial degradation, often forming ridges and spikes at the surface of the aerial parts of plants, favoring maintenance of a moisture film which contains many types of toxic secondary products
Lignins (D 23.2.3)	Mechanical and chemical barrier to most cell wall degrading microorganisms
Preinfectious toxins	
Phenylheptatriyne (D 5.2.5)	Fungitoxin, disturbs membrane functions
Citral, thymol (D 6.2.1)	Released into the atmosphere. The compounds may reach considerable concentrations near the plant surface
Allicin-type compounds (D 11.2)	Antimicrobial compounds, e. g., of Liliaceae
Quinolizidine alkaloids (D 18)	Constituents of several Leguminosae
Hordatines (D 19)	Fungitoxins of *Hordeum vulgare* seedlings
Berberine (D 23.1.2)	Fungitoxin of *Mahonia trifoliata* leaves
Gallotannins, ellagitannins, and condensed tannins (D 23.2.5, D 23.3.3)	Tannins form multiple hydrogen bonds to proteins and inactivate the exoenzymes of invading microorganisms
Pinosylvin (D 23.3.3)	Fungitoxin present in the heartwood of Pinaceae
Resveratrol (D 23.3.3)	Fungitoxin of the roots of *Veratrum grandiflorum*
Nobiletin (D 23.3.3)	Fungitoxin of *Citrus* leaves
Luteone, 2'-deoxyluteone (D 23.3.3)	Toxins of *Lupinus* leaves, inhibit spore germination and germ tube development of *Helminthosporium*
Caffeine, theobromine, theophylline (D 25)	Fungitoxins
Prototoxins	
Tuliposides (D 6.1)	Tulipalins formed from tuliposides by hydrolysis and cyclization are fungitoxins of tulip bulbs
Aucubin (D 6.2.2)	The aglycone formed by glucosidases has antimicrobial activity
Linamarin (D 8.3)	Hydrocyanic acid, a fungitoxin, is released by hydrolysis of linamarin in infected *Lotus corniculatus* cells
Sinigrin (D 8.4)	Allyl isothiocyanate, a fungitoxin, is released from sinigrin by hydrolytic cleavage in infected cells of *Brassica*
Protocatechuic aldehyde, protocatechuic acid, catechol (D 23.2.5)	Constituents of onion bulbs and bananas, oxidized after infection to the corresponding fungitoxic o-quinones
Arbutin (D 23.2.5)	Hydroquinone formed from arbutin by glycosidases has antimicrobial activity
Phloridzin (D 23.3.3)	In infected apple leaves phloridzin is hydrolyzed to phloretin which is hydroxylated to 3-hydroxyphloretin and oxidized to the corresponding fungitoxic o-quinone
Phytoalexins	
Safynol, wyerone (D 5.2.5)	Antimicrobial compounds from *Carthamus tinctoria* and *Vicia faba*, respectively
Rishitin, ipomeamarone (D 6.3)	Phytoalexins from *Solanum tuberosum* and *Ipomea batatis*, respectively

Secondary products	Mode of action
Dianthramides, dianthalexin, and avenalumin I (D 21.4)	Phytoalexins of *Dianthus caryophyllus* and oat, respectively
Chlorogenic acid (D 23.2.1)	Fungitoxin formed in potato tubers in the neighborhood of *Phytophthora* infections
Xanthotoxin, psoralen (D 23.2.2)	Phytoalexins of *Pastinaca sativa* and *Petroselinum crispum*, respectively
Scopolin (D 23.2.2)	Fungitoxin formed in potato tubers in the neighborhood of *Phytophthora* infections
Orchinol (D 23.3.3)	Phytoalexin of *Orchis militaris*
Pterocarpans, e. g., medicarpin, pisatin, phaseollin, 6-hydroxy-phaseollin (D 23.3.3)	Phytoalexins of Leguminosae

ground parts, in the epidermis of the cell wall, but also in the cells themselves (preinfectious chemical barriers).

Cultivated plants, which have a reduced level of secondary products due to selection (F 6), suffer from microbial infections more frequently than their wild-type ancestors. The many milder-flavored cultivars of *Brassica*, for instance, with low glucosinolate accumulation, are frequently infected by mildew, in contrast to the related wild-type species with high glucosinolate content.

c) Formation of toxic compounds from nontoxic substances (protoxins) after breakdown of compartmentation in plant cells during the infection.

Of significance in this respect are glycosidases (C 6) and/or oxidases (D 2.3.1, C 2.4, etc.)

d) De novo synthesis or increase of the rate of synthesis of certain secondary products, so-called phytoalexins, in the infected tissue or its immediate neighborhood, thus warding off ("alexos") invading microorganisms.

In higher plants synthesis of phytoalexins is an important protecting device against microbial infections. Because defense is costly, it should be advantageous for plants to be less well defended when enemies are absent and to increase commitment to defense when attacked. Synthesis of phytoalexins is triggered by certain peptide and polysaccharide fractions from the pathogens or the plant cell wall (elicitors, D 1.4.1), but also by chemical or physical stress, e. g., treatment with inorganic salts, UV radiation, temperature shocks, wounding, etc. Usually several phytoalexins are formed at different speed with differing physiological activities.

Most of the secondary products participating in chemical defense are phenolics with a relatively high degree of lipophilicity due to masking of OH-groups by methylation or formation of methylenedioxy rings.

Microbial Secondary Products Poisonous to Host Plants

In addition, the infecting microorganisms employ a variety of chemicals which may damage or destroy higher plants (Table 70). Of special importance are the following approaches:

a) Synthesis of secondary products which possess hormone activity in higher plants, i.e., may disregulate the metabolism, cause abnormal growth, etc., and

Table 70. Secondary products with phytotoxic activity produced in microorganisms

Secondary products	Mode of action
E-Epoxysuccinic acid (D 4)	Blighting factor in almond leaves, formed after *Rhizopus* infection of the hulls, which causes overproduction of fumaric acid
Zinniol (D 5.3.2)	Nonhost-specific toxin of *Alternaria* sp., induces chlorophyll retention and necrosis
Ceratocystis toxins (D 5.3.3)	Products of *Ceratocystis ulmi*, causing Dutch elm disease
Helminthosporal (D 6.3)	Product of *Helminthosporium sativum*, causing cereal necrosis
Helminthosporium sacchari (HS) toxin A (helminthosporoside) (D 6.2)	Host-selective toxin acting on sugarcane
Gibberellins (D 6.4)	Products of *Fusarium moniliforme (Gibberella fujikuroi)* and *Sphaceloma manihoticola*, causing elongation of internodes of rice and cassava plants, respectively, due to interference with the hormone metabolism of the host
Tentoxin (D 9.2)	Product of *Alternaria tenuis*, causing chlorosis in cotton plants
Phaseolotoxin (D 9.2)	Formed in *Pseudomonas syringae* pv. *phaseolicola*, causing the halo blight disease of beans
Tabtoxin (D 9.2)	Produced in *Pseudomonas tabaci*, originating wildfire disease of tobacco
Helminthosporium carbonum race 1 (HC) toxin (D 9.2)	Host-selective toxin acting on maize
Alternaria mali (AM) toxin I (D 9.2)	Host-selective toxin causing leaf spot disease of apple
Norcoronatine (D 13), coronatine (D 15)	Produced by *Pseudomonas syringae* pvs. *atropurpurea* and *glycinea*, causing chlorosis in young leaves
Fusaric acid (D 16)	Product of *Fusarium* sp., causing wilting, e. g., of tomatoes
Picolinic acid (D 16.1)	Product of *Pyricularia oryzae*, causing rice blast disease
Phaseotoxin (D 17)	Formed in *Pseudomonas phaseolicola*, causing the halo blight disease of beans
Indole-3-acetic acid (D 22)	Produced by *Pseudomonas syringae* pv. *sevastanoi*, or by Ri- and Ti-transformed plant cells (A 2.1, D 22), causes rhizogenesis and together with cytokinins uncontrolled cell division (tumor formation) in the host plants
Tryptophol (D 22)	Produced by *Drechslera nodulosum*, causes necrotic lesions in leaves of *Eleusine indica*
N^6-Dimethylallyladenine, zeatin, and derivatives of these compounds (D 25.1)	Products of *Corynebacterium fascians* and *Pseudomonas syringae* pv. *savastanoi*, causing fasciation, e. g., in pea plants, due to interference with the hormone (cytokinin) metabolism of the host

b) Phytotoxin formation, i. e., synthesis of products poisonous to plant cells. Phytotoxins include proteins, e. g., enzymes which degrade plant cell walls and give rise to the formation of the fragile and easily decaying protoplasts (pectinases, hemicellulases, cellulases, etc.), as well as toxic secondary products. Host-selective and nonspecific toxins with a general toxicity to plants may be distinguished. The host-selective compounds are often toxic to a few susceptible genotypes of the host species only.

In addition, infecting microorganisms may use secondary plant products for tracing host plants. The volatile polyacetylenes 1-tridecene-3,5,7,9,11-pentayne and *E*- and *Z*-trideca-1,11-diene-3,5,7,9-tetrayne (D 5.2.5), for instance, stimulate the germination of rust spores infecting safflower.

The defense compounds produced in higher plants, as well as the toxic microbial products may be detoxified by hydroxylation, conjugation, etc. (E 1) in the invading microorganisms and the infected plants, respectively. Even degradation to substances of primary metabolism occurs. This complicated field of metabolism, however, is poorly understood as yet.

Plant Secondary Products Influencing Gene Expression in Microorganisms

A special field is the interaction of plants and microorganisms during colonization and symbiosis. Acetosyringone and related phenolics (D 23.2.5), occurring in the exsudates of wounded or metabolically active plant cells, trigger the expression of the virulence genes of *Agrobacterium tumefaciens*. The activation of these genes initiates the transfer and integration of T-DNA in the genome of the plant cells (A 2.1). Certain flavones and flavanones (D 23.3.3) released from legume roots induce transcription of the nod genes of *Rhizobium*, whereas isoflavones and flavonol glycosides turn off this process.

References for Further Reading

10, 24, 33, 102, 126, 129, 145, 154, 168, 209, 251, 291, 323, 325, 381, 418, 486, 491, 597, 658, 670, 738, 849, 895, 902

E 5.5 Secondary Products and the Interactions of Plants and Animals

Secondary plant products play an outstanding role in the mutual connections between plant and animals. They are involved, for instance, in such general fields as predation, pollination, and dispersal of seeds (D 5.5.1–D 5.5.3). In addition, they participate in special relations, like coniine (D 5.3.2), which is released by the pitcher traps of the carnivorous plant *Sarracenia flava*, and, on the one hand, entices insects into the trap and, on the other hand, paralyzes them so that they are digested more easily.

E 5.5.1 The Attraction of Plant Pollinators by Secondary Products

The coevolution of plants and insects, birds, and bats with respect to the pollination of flowers has given rise to the formation of special morphological features and chemical characteristics. The main reason why pollinators visit flowers is to gather nectar and pollen as nutrients which contain carbohydrates, fats, and proteins. With respect to secondary products

— nectar may comprise a large spectrum of substances causing the individual flavor. In addition, special nutrients like 3-acetoxy fatty acids (D 5.2.4, several families of Angiospermae) and toxic compounds, such as catalpol and catalposide (D 6.2.2, *Catalpa speciosa*), grayanotoxins (D 6.4, Ericaceae), pyrrolizidine alkaloids (D 17.1, e. g., lycopsamine in *Trichogonia gardneri*), quinolizidine alkaloids (D 18, *Sophora microphylla*), and arbutin (D 23.2.5, *Arbutus unedo*) may be present

Table 71. Characteristic constituents of flower scent volatiles

Group of compounds	Individual substances (occurrence)
Monoterpenes	Limonene *(Citrus)*, geraniol (roses) (D 6.2.1)
Sesquiterpenes	β-Ionone (violets), bisabolol (orange), (−)-δ-cadinene (orchids)[a] (D 6.3)
Amines	Monoamines *(Arum)* (D 8.2)
Phenylpropanes	Eugenol *(Eugenia)*, methyleugenol *(Cassia)* (D 23.2.1)
Benzoic acid derivatives	Vanillin *(Vanilla*, orchids) (D 23.2.5)

[a]   (−)-δ-Cadinene, a constituent of the flower scent of orchids of the genus *Ophrys*, attracts and arouses male bees of the genus *Andrena*. It mimics the sex pheromones, *E*-farnesyl and geranyl esters (E 4), of the female bees.

— pollen may contain pigments, such as carotenoids (D 6.7) and flavonoids (D 23.3.3), cinnamic acid derivatives (D 23.2), and cell wall constituents such as carbohydrates (D 1.4.1) and sporopollenins (D 6.7).

These substances serve in the improvement of the flower constancy of adapted pollinators, which reduces the waste of nectar and pollen by visiting animals not carrying conspecific pollen.

At a greater distance pollinators may be lured by flower scents. These flower volatiles cover a wide range of chemical structures (Table 71). They are usually complex mixtures possessing hundreds of constituents including simple aliphatic alcohols, ketones, and esters as well as aromatic, hydroaromatic, and heterocyclic compounds. Scents are of special significance for pollination of flowers by night-flying animals. Bat- and moth-pollinated, but also many bee-pollinated flowers have a strong smell. Pollinators which are close to the flowers are attracted by secondary products absorbing visible light or UV radiation (Table 72). These pigments facilitate the recognition of flowers among green leaves.

Table 72. Flower pigments attracting plant pollinators

Coloration of flowers, secondary products used as pigments	Pollinators attracted
White: Flavones, e.g., luteolin; quercetin (D 23.3.3)	Bees and other insects (recognizing UV absorption), bats (mostly color blind),
Yellow: Carotenoids (D 6.7), betaxanthins (D 23.1.4), yellow flavonols (D 23.3.3)	Bees, butterflies
Red, violet, blue: Betacyanins (D 23.1.4); anthocyanins (D 23.3.3)	Bees (recognizing intense blue colors, insensitive to red), birds (attracted by vivid scarlets, sensitive to red), butterflies attracted by bright colors), moths (attracted by dull and drab colors)

The significance of flower pigments for pollination may easily be seen with mutants of bluebell (*Scilla non-scripta*) and bellflowers (*Campanula*) having unpigmented (white) flowers. These mutants occasionally appear in the natural population possessing blue flowers. Their seed set, however, is very poor, due to the more or less complete lack of pollinators. The mutants are therefore unable to maintain themselves in the natural habitat.

References for Further Reading

519, 891

E 5.5.2 The Influence of Secondary Products on Seed Dispersal

Color, smell, and taste of ripe fruits indicate to animals that they are suitable as food. The animals swallow the fruits (including the seeds they contain), digest the flesh, and scatter the seeds together with the feces.

The smell of fruits is caused, e. g., by mono- and sesquiterpenes, i. e., the components of essential oils. The sweet taste usually is brought about by different monosaccharides (D 1.1), sugar alcohols (D 1.3), or oligosaccharides (D 1.4.1). Disappearance of toxic compounds, e. g., alkaloids during ripening of fleshy fruits (cf. the degradation of α-tomatine, D 6.6.3, in maturating tomatoes, as well as degradation or binding by soluble polysaccharides of astringent polyphenols, D 23.2.5, D 23.3.3) ensure the palatability of fruits to animals and the dispersal of seeds at an appropriate stage of seed development. On the other hand, the retention of toxic principles even in ripe fruits may exclude nonadapted "low quality" dispersers.

In most angiosperm fruit coloration is caused either by flavonoids or carotenoids. Fruits colored by anthocyanins are, for instance, cherries, strawberries, red and black current, plums, apples, and aubergines (skin), pomegranate, and blood orange (juice). Fruits colored by carotenoids are capsicum, tomatoes, and hips. The wax layer of glaucous fruits, e. g., grapes and the fruits of *Prunus* sp., juniper, and figs, increases UV reflection and hence their visibility for birds (D 5.2.4).

References for Further Reading

116, 299, 323, 584

E 5.5.3 Secondary Plant Products Attracting or Repelling Predators

The feeding behavior of herbivores is strongly affected by the secondary products which determine the texture, taste, and flavor of plants. The influence on animals of secondary products ingested with the diet is complex. It may include the alteration of orientation, metabolic efficiency, growth, mating, etc., and secondary products may even induce death. Deterring and toxic compounds can be distinguished.

Deterring substances may cause mechanical or chemical irritation

Examples are:
— the needles of calcium oxalate (D 4) found in many plants,
— protoanemonin formed from ranunculin, e. g., in *Ranunculus* and *Pulsatilla* sp.,

β–D–Glc–p–OCH₂   Ranunculin   →   Protoanemonin

- the mustard oils built from glucosinolates (D 8.4), and
- capsaicin occurring in the fruits of red pepper (D 23.1.1),

and may sensitize the skin to UV radiation,

Examples are:
- hypericin (D 5.3.6) and the furanocoumarins (D 23.2.2) occurring in *Hypericum* sp. and Ammiaceae, respectively

or they may act as allergens, which after sensibilization by further contact cause dramatic skin affects

Examples are:
- sesquiterpene lactones (D 6.3) found in Asteraceae
- urushiol (D 23.2.5) present in *Rhus toxicodendron*, and
- benzoquinone and naphthoquinone derivatives like primin, a constituent of *Primula obconica* (D 23.2.5).

Many of the repellent secondary products reduce the fitness of animals if ingested (Table 73). Often they have a warning taste, like cucurbitacins (D 6.6), cardiac glycosides (D 6.6.8), strychnine (D 22.3), and quinine (D 22.4), which are bitter, or the tanning substances (D 23.2.5, D 23.3.3), which are astringent. This facilitates the effect of learning in the potential predators. It is of interest that the bitter taste is mimicked by certain groups of secondary products which are without a prominent biological activity, like iridoids (D 6.2.2), certain sesquiterpenes (D 6.3), and diterpenes (D 6.4).

The fitness-reducing secondary products may be divided into two groups:

a) Compounds with a Relatively Low Physiological Activity. The compounds of this group, such as calcium oxalate crystals (D 4), which cause mechanical irritation in the mouth of potential predators, and resins (D 6.4), lignins (D 23.3.3), and tannins (D 23.2.5 and D 23.3.3), which reduce the digestibility of plant materials, act in a dosage-dependent fashion. They are often present in high concentrations (quantitative defense). Tannins may reach concentrations up to 60 % of the dry weight. They form more or less indigestible complexes with proteins, reducing the rate of assimilation of dietary nitrogen and the activity of digestive enzymes. Essential oils and resins inhibit rumen microbial enzyme activity. Substances of quantitative defense provide protection against all classes of enemies (herbivores, pathogens) that acquire nutrients by the breakdown of plant tissues.

b) Compounds with High Toxicity. On the other hand, highly toxic compounds may be present in relatively small concentrations (qualitative defense, cf. Table 73). These substances interact with the central nerve system, like the alkaloids ibotenic acid, pantherine, L-hyoscyamine, L-scopolamine, psilocine, psilocybine, harmine, mescaline, morphine, or the peripheral nerves of animals, like arecoline, nicotine, L-hyoscyamine, cocaine, cytisine, physostigmine, ergotamine, reserpine, yohimbine, L-ephedrine. They may influence ion transport

Table 73. Secondary plant products toxic to animals

Secondary products	Mechanism of toxicity
Fluoroacetic acid (D 5)	Inhibits citrate synthase (D 4)
Trifluoroacetic acid (D 5)	Mechanism of toxicity unknown
Malonic acid (D 5)	Inhibits succinate dehydrogenase (D 4)
Sterculic acid (D 5.2.1.)	Mechanism of toxicity unknown
Acetylenic compounds (D 5.2.5)	Phytotoxic to insects and nematodes
Coniine (D 5.3.2)	Paralyzes the spiral marrow and respiration
Anthraquinone and anthrone derivatives (D 5.3.6)	Laxatives
Hypericin (D 5.3.6.)	Causes photosensitization followed by skin necrosis, etc. in mammals
Pyrethrins (D 6)	Insecticidal activity, mechanism unknown
Saponins (D 6.6 and 6.6.2)	Increase of the permeability of cell membranes
Solanum alkaloids, e.g., demissine (D 6.6.3)	Binding to membrane sterols
Veratrum alkaloids, e.g., veracevine (D 6.6.3)	Affects DNA stability
Cardiac glycosides (D 6.6.8)	Inhibitors of Na^+, K^+-ATPase
Vitamin D_3 (D 6.6.9)	Calcification of soft tissues (vertebrates)
L-Selenocysteine and L-selenomethionine (D 8.1)	L-Cysteine and L-methionine analogs, incorporated in proteins, altering the biological properties
Cyanogenic glycosides (D 8.3)	Releases HCN which inhibits cell respiration by interaction with the cytochrome system
3-Aminopropionitrile, 2,4-diaminobutyric acid (D 8.3)	Causes lathyrism (vertebrates)
Glucosinolates (D 8.4)	Releases isothiocyanates which in high doses cause diarrhea and act as goitrogens in mammals, including humans
α-Amanitin (D 9.2)	Inhibitor of RNA-polymerase II
syn-Propanethial S-oxide (D 11.2)	Lachrymatory irritant of onions
Azetidine-2-carboxylic acid (D 12.1)	L-Proline analog, incorporated in proteins
4-Substituted L-glutamic acids (D 14)	L-Glutamic acid analogs, incorporated in proteins
Arecoline (D 16.1)	Interacts with the peripheral nerve system
Hypoglycine A (D 16)	Inhibitor of the β-oxidation of fatty acids
Miserotoxin (D 16)	Releases nitrite after hydrolysis
Ricinine (D 16.1)	Mechanism of toxicity unknown
Nicotine (D 16.1)	Inhibits acetylcholine receptors in the peripheral nerve system
Pantherine (and ibotenic acid?) (D 17)	Interacting with the central nerve system
Pyrrolizidines (D 17.1)	Toxins of kidneys and nerve system
Cocaine (D 17.2)	Interacting with the peripheral nerve system
L-Hyoscyamine, L-scopolamine (D 17.2)	Paralyzes certain cerebral and medullary centers of the nerve system
Mimosine, mimoside (D 18)	Inhibitors of pyridoxal-5′-phosphate-containing enzymes
Lobeline (D 18)	Emetic
Sedamine (D 18)	Mechanism of toxicity unknown
Quinolizidines (D 18)	Oxytocic compounds
Cytisine (D 18)	Interacts with the peripheral nerve system
L-Canavanine (D 19)	L-Arginine analog, incorporated into proteins
L-Canaline (D 19)	L-Ornithine analog, inhibitor of pyridoxal-5′-phosphate-containing enzymes
Pilocarpine (D 20)	Inhibitor of acetylcholine esterase

Secondary products	Mechanism of toxicity
L-3-Hydroxymethylphenylalanine, L-4-hydroxy-3-hydroxymethyl-phenylalanine (D 21.1)	Inhibitors of L-phenylalanine uptake
Juglone (D 21.2)	Inactivates essential proteins by reaction with free amino groups
Hypaphorine (D 22)	Probably L-tryptophan analog
Physostigmine (D 22)	Inhibitor of acetylcholine esterase
5-Hydroxy-L-tryptophan (D 22.1)	Precursor of serotonin, interfering with hormone metabolism
Psilocine, psilocybine (D 22.1)	Interacts with central nerve system
Ergoline alkaloids, e.g., ergot-amine (D 22.2)	Vasoconstricting compounds, causing necrosis (ergotismus gangrenosus), also affecting the central nervous system (ergotismus convulsivus)
Harmine (D 22.3)	Interacts with the central nerve system
Yohimbine (D 22.3)	Interacts with the peripheral nerve system
Reserpine (D 22.3)	Central inhibitor of vegetative functions
Strychnine (D 22.3)	Central spasmic
Quinine (D 22.4)	Inhibition of nucleic acid biosynthesis by complexation with DNA
Quinidine (D 22.4)	Cardiac depressant
L-DOPA (D 23)	Interferes with tyrosinase, an enzyme of cuticle synthesis in insects (D 23.2.5)
Mescaline (D 23.1.1)	Interacts with the central nerve system
Papaverine (D 23.1.2)	Decreases muscle tonus
Morphine, codeine (D 23.1.2)	Hypnotics, narcotic analgesics
Colchicine (D 23.1.2)	Causes polyploidy
Emetine (D 23.1.2)	Emetic
Berberine (D 23.1.2)	Interacts with DNA
L-Ephedrine (D 23.2.5)	Interacts with the peripheral nerve system
Psoralen (D 23.2.2)	Photosensitizing compound
Podophyllotoxin (D 23.2.4)	Cytotoxic compound
Primin (D 23.2.5)	Contact allergen
Urushiols (D 23.2.5)	Contact allergens
Rotenone (D 23.3.3)	Insecticide, inhibits NADH-dependent dehydrogenase in the respiratory chain
Theobromine, theophylline, caffeine (D 25.1)	Inhibits phosphodiesterase

through membranes (cardiac glycosides), block the energy formation in the respiratory chain (hydrocyanic acid released from cyanogenic glycosides), interact with protein synthesis (nonprotein amino acids) or RNA formation (α-amanitin), etc. These toxins are less costly to plant metabolism than the substances causing quantitative defense. They bring about, however, a less general protection due to the development of potent mechanisms of tolerance or detoxication by specialized herbivores. Substances causing qualitative defense are frequently present in short-lived plants, whereas a high percentage of woody plant species contains substances originating quantitative defense.

The domesticated herbivores, e.g., cattle, tolerate only small amounts of toxic secondary products. Plant species or cultivars with low toxin content, like the sweet lupins, which have a reduced level of quinolizidine alkaloids (D 18), therefore, have to be used as fodder. Under the wild animals, however, species exist that are adapted to feed on plants containing large amounts of highly toxic secondary compounds.

Examples are:
— the polyphagous larvae of the tiger moth, *Arctia caja*, which freely ingest the overground parts of *Digitalis* and *Convallaria* sp. (containing cardenolides, D 6.6.8) as well as the leaves of *Solanum*, *Aconitum*, and *Senecio* sp. (containing alkaloids, D 6.6.3 and D 17.1)
— the tobacco hornworm, the tobacco bud worm, and the green peach aphid feeding on tobacco plants, containing nicotine (D 16.1) known as an effective insecticide, and
— roes and gray squirrels, which ingest without intoxication mushrooms like *Amanita phalloides* containing the highly dangerous polypeptides, phalloidin and α-amanitin (D 9.2).

The mechanisms causing resistance to toxic secondary products may be different. Resorption in the intestine may be reduced, as is the case in insects tolerating phytoecdysones (D 6.6.1), or the toxic compounds may simply be sequestered in certain regions of the body without adverse effects. In a few animals sophisticated mechanisms of adaptation have evolved.

The bruchid beetle *Caryedes brasiliensis*, for instance, a specialized insect living on L-canavanine-containing seeds, possesses an arginyl-tRNA synthetase that, in contrast to the enzymes of most other organisms, discriminates against L-canavanine (D 19). Moreover, the beetle is able to degrade L-canavanine to L-canaline and urea and to hydrolyze the latter compound to ammonia.
In many instances the secondary products ingested, like other "xenobiotics", are transformed to less poisonous compounds. Of importance in this respect are monooxygenases (C 2.6), which introduce into many secondary products functional groups capable of further conjugation (E 1). In other cases special detoxifying enzymes occur, like tropinesterase (D 17.2). These mechanisms of detoxication are, however, outside the scope of this book.

Practically every group of secondary products may be tolerated by certain specialized animals. This tolerance originates ecological niches in which the "specialists" can feed without competition with most of the other potential predators. To facilitate the finding of the respective niches the tolerated secondary products

Table 74. Secondary plant products attracting feeding insects

Secondary products	Biological action
Terpene mixture of pines *(Pinus ponderata)* (D 6.2.1)	Attracts the western pine beetle *(Dendroctonus brevicomis)*. The female beetles attract males by sequestering myrcene, a constituent of pine oleoresin (E 4)
Terpinyl acetate, linaloyl acetate, linalool (D 6.2.1)	Feeding attractant of *Bombyx mori* larvae (silkworm)
Iridoid glycosides (D 6.2.2)	Feeding attractant of *Ceratomia catalpae* (moth)
Cucurbitacins (D 6.6)	Feeding attractant of *Diabrotica undecimpunctata* (beetle)
Δ^7-Stigmastenol (D 6.6.1)	Attractant of *Drosophila* sp., essential precursor for ecdysone, a hormone of the insects (E 3.1)
Sinigrin (D 8.4)	Attractant of *Pieris brassicae* and *P. rapae* (cabbage butterfly), *Brevicoryne brassicae* (aphid), and *Plutella maculipennis* (diamondback moth)
Sparteine (D 18)	Feeding attractant of *Acyrthrosiphon spartii* (aphid)
Methylisoeugenol (D 23.2.1)	Oviposition stimulant of the carrot root fly
Coumarin (D 23.2.2)	Feeding attractant of *Sitonia cylindricollis* (weevils)
Flavonoids (D 23.3.3)	Feeding stimulants of *Scolytus mediterraneus* (beetle), *Bombyx mori* larvae (silkworm), and many other insects

Table 75. Secondary plant products which mimic the action of anmial hormones or show antihormone activity

Secondary products	Biological action
Ageratochromene (D 6.1)	Antijuvenile hormone activity in insects
Juvabione, dehydrojuvabione (D 6.3)	Juvenile hormone activity in certain insects
Cyasterone, ponasterone A (D 6.6.1)	Molting hormone activity in insects
Genistein, daidzein, formononetin (D 23.3.3)	Estrogenic activity in mammals and birds
Coumestrol (D 23.3.3)	Estrogenic activity in mammals and birds

may be used as cues to aid in the location of host plants not used as food resources by most other species (Table 74). Often these specialized animals start feeding at those parts of the plant which have the highest secondary product content.

In addition, plant toxins may be used by animals for their own protection (E 5.1) or as pheromones (E 4).

The monarch butterfly *(Danaus plexippus)*, for instance, contains cardiac glycosides (D 6.6.8), taken up with the food during the larval stage from milkweeds *(Asclepias* sp.), the cinnabar moth *(Tyria jacobaeae)* holds pyrrolizidine alkaloids (D 17.1), ingested with the larval diet from ragwort *(Senecio jacobaea)*, and the butterfly *Pachlioptera aristolochiae* preserves aristolochic acids if the larvae have fed on *Aristolochia* sp. (D 23.1.2). These insects are avoided by predators, e. g., birds, due to these stored secondary products. Also butterflies which have a similar pigment pattern resembling that of toxic species, e.g., the monarch, are rejected by birds, even if they do not contain toxins (mimicry).

Of special interest is the fact that certain plants synthesize secondary products which in animals possess hormone or antihormone activity. Either true animal hormones are formed (Table 63) or substances are built which mimic the animal hormones (Table 75). Both types of compounds may cause metabolic and developmental abnormalities in the predators, reduce reproduction, or, in the case of the trichomes of the stinging nettel *(Urtica dioica)*, containing histamine (D 20) and serotonin (D 22.1), after injection cause irritation of the skin which may repel predators.

Rather large quantities of the hormones may be produced, as is shown by the fact that from 1 g of dried rhizomes of the fern *Polypodium vulgare*, 10 mg ecdysterone have been isolated, in striking contrast to the 0.33 mg obtained from a ton of silkworms.

Defense by secondary products withdraws compounds and energy from primary metabolism which may reduce growth and reproduction. Plants synthesize secondary defense products therefore in direct proportion to their risk.

Acyanogenic morphs of *Trifolium repens*, for instance, have a higher vegetative and sexual reproductive vigor in the absence of herbivores than cyanogenic morphs, whereas in the presence of herbivores, which avoid the cyanogenic morphs, the reverse holds true. The distribution of acyanogenic and cyanogenic morphs thus directly reflects the activity of predators.

Within the producer organism secondary defense products are allocated in direct proportion to the risk of suffering detriment of the particular organs and tissues. Hence, the pattern of distribution is influenced by:

— the relative costs to the plant of herbivore-influenced damage or loss of the parts in question

— the relative probability of discovery and successful herbivore attack of different plant parts in the absence of chemical defense

— physiological constraints of synthesis and/or storage of toxic compounds in different kinds of tissues, and

— the distribution of other defense mechanisms within the plant.

The external tissues of plants are usually the first to come into contact with herbivores. It is therefore not surprising that toxic secondary products are present in relatively high concentrations in these tissues, e. g., in the bark and in trichomes at the surface of leaves, stem, etc.

Plants having several lines of chemical defense are obviously favored. Heterogeneity may facilitate protection against a broader spectrum of herbivores and pathogens, because it may be more difficult to evolve resistance to multiple defenses. Furthermore, different types of secondary products may act synergistically. The steroid saponins (D 6.6.2) present in *Digitalis* leaves, for instance, increase the resorption of cardenolides (D 6.6.8) in the animal intestine. The series of structurally related compounds found in many plants may increase the repellent or toxic response by analog synergism, as is known, for instance, for the purgative effect of the structurally related anthraquinone derivatives (D 5.3.6) of rhubarb, cassia leaves, etc.

References for Further Reading

41, 46, 47, 82, 100, 137, 154, 171, 210, 214, 246, 285, 323, 325, 331, 342, 496, 506, 514, 524, 528, 534, 635, 657, 682, 710, 728, 748, 784, 797

E 5.6 Colored Secondary Products Used as Signals

The colors of living beings are brought about either by pigments or by physical means (reflection or scattering of light, as on the surface of the feathers of birds or on proteins in the "blue" eyes of the scarcely pigmented North Europeans). In several instances pigmentation and physical means also cause mixed colors.

Examples in this respect are:
— the green color of frogs, which is caused by the combined effect of a yellow pigment and the scattering of blue light on skin proteins, and
— the green color of human eyes which is brought about by a yellow phaeomelanin-pigmented iris (D 23.1.3) and a layer of blue light scattering proteins above it.

Most pigments are secondary products. With the exception of a few, e. g., the heme-type porphyrins (D 24, red) or the chlorophylls (D 24.1, green), they have their main or exclusive function in ecology (Table 76).

Colors originate visual signals which participate in the interactions between animals and between animals and other organisms. Colored areas on the surface of animals

— may attract sexual partners

Examples are the bright colored wings of butterflies and other insects.

— may facilitate hiding

Table 76. Light absorbing or reflecting secondary products used as signals in the establishment of ecological relations

Color/Pigment	Occurrence
Red	
Carotenoids, e. g., capsanthin and astaxanthin (D 6.7)	Fruits, goldfish, salmon, flamingo, ladybird beetle
Ommochromes (D 22)	Eyes of insects
Phaeomelanins (D 23.1.3)	Fur of the red fox, red human hairs
Betacyanins, betanin (D 23.1.4)	Flowers, fruits, mushrooms
Anthocyanins (D 23.3.3)	Flowers, fruits
Hemoglobin (D 24)	Comb of cock, cheeks of humans
Bilirubin (D 24.2)	Egg shell of birds
Yellow	
Anthraquinones, e. g., physcion (D 5.3.6)	Fruit bodies of mushrooms
Carotenoids, e. g., xanthophylls (D 6.7)	Flowers, fruits, fish
Betaxanthins, e. g., indicaxanthin (D 23.1.4)	Flowers, fruits, mushrooms
Muscaflavin (D 23.1.4)	Fly agaric
Styrylpyrones, e. g., hispidin (D 23.3.2)	Mushrooms
Flavonoids, e. g., flavonols (D 23.3.3)	Flowers, fruits
Xanthopterin (D 25.3)	Wings of butterflies
Russopteridines (D 25.4)	Mushrooms (*Russula* sp.)
Brown	
Ommochromes, e. g., xanthommatin (D 22)	Skin and wings of insects
Phaeomelanins (D 23.1.3)	Hair and skin of animals and human beings
Quinones, quinhydrones (D 23.2.5)	Exoskeleton of insects
Green	
Carotenoid protein complexes, chromophore: β-carotene (D 6.7)	Insects, e. g., locust integuments
Chlorophylls (D 24.1)	Green parts of plants
Open-chain tetrapyrrols, e. g., biliverdin (D 24.2)	Egg shells of birds; caterpillars, insects
Blue	
Carotenoid-protein complexes, chromophore: astaxanthin (D 6.7)	Insects, e. g., *Oedipoda* hind wings
Anthocyanin complexes, e. g., cyanin (D 23.3.3)	Flowers, fruits
White	
Betulin (D 6.6)	Bark of birch
Flavones, flavonols, e. g., luteolin and quercetin (D 23.3.3)	Flowers
Leucopterin (D 25.3)	Wings of butterflies
Black	
Eumelanins (D 23.1.3)	Hair and skin of animals and humans
Reflection of visible light	
Guanidine crystals (D 25)	Scales of fish
Reflection of UV light	
Waxes (D 5.2.4)	Fruits

Caterpillars, aphids, and grasshoppers feeding on plant shoots are frequently green, animals living in forests have a dull brown-gray color, whereas the fur of the arctic hare and other nordic animals is unpigmented (white).

— signal a warning

Monarch butterflies which contain bitter and vomitory cardenolides (D 6.6.8) have a bright coloration. They are avoided by other animals if they have been once tested. The same happens with the yellow-black colored wasps, or the red and black colored fire salamanders and ladybird beetles which also contain toxic substances.

Colored parts and organs of plants may attract pollinators or may facilitate seed dispersal (E 5.5.1 and E 5.5.2). In addition, they may signal toxicity, as is the case with the red skin of the fly agaric, *Amanita muscaria*, containing ibotenic acid (D 17) or the yellow-green skin of the bulbous agaric, *Amanita phalloides*, containing phalloidin and α-amanitin (D 9.2).

References for Further Reading

100, 265, 299, 556, 857

F Secondary Products in Human Life

Man was confronted with and has used secondary products since ancient times. When gathering his food he was attracted by certain secondary products and repelled by others (F 1). He used secondary substances to cure diseases (F 2), to enjoy stimulants and hallucinogens (F 3) and, to poison animals (and human beings; F 4). He processed cellulosic fibers to make clothes and he used wood, a mixture of cellulose and lignins, for building homes and cooking meals (F 5).

Much has changed since those times, but even now man is surrounded by secondary products. We enjoy the taste and flavor of tea (in addition to caffein containing a large number of secondary products causing the aroma). We use medical remedies which in many cases contain secondary products as active principles. We read the newspaper (made from cellulose) and have pleasure in the scent and color of flowers (pigmented with flavonoids and carotenoids and releasing a large variety of volatile secondary substances). We use perfume (containing terpenoids and other secondary scent components), we drink cognac (containing ethanol, tannins, and the secondary products originating its characteristic flavor), and we smoke cigarettes (with the secondary product nicotine as the most important physiologically active compound). In many instances man has influenced the formation of secondary compounds in microorganisms, plants, and animals to meet his needs (F 6). Recently the modern genetic techniques have opened a completely new prospect in this field.

F 1 The Influence of Secondary Products on the Food Preferences of Man

Taste, flavor, and color of foodstuffs and beverages of humans are dominated by secondary products. They attract and repel human beings in the same way as other vertebrates (E 5.5.3). The most attractive qualities for humans are sweetness and aroma, repellent are sharpness, bitterness, and astringency. But attractiveness or repellence is strongly dose-dependent, as may be seen with the flavoring sulfur compounds of garlic, which are attractive in trace amounts, but repellent if present in higher quantities, or with beverages which are especially attractive if possessing a suitable balance of sweet, acidic, and astringent flavors, as present in good quality wines.

In Table 77 some secondary products are compiled which make foodstuffs and beverages attractive to humans. In some cases the characteristic flavor is

Table 77. Secondary products dominating taste and smell of food stuffs and beverages used in human nutrition

Secondary products	Quality of taste or smell
Mono- and oligosaccharides (D 1.1, 1.4.1), sugar alcohols (D 1.3)	Sweet taste of fruits
Diacetyl, acetoin (D 5)	Flavor of butter
α-Nonalactone (D 5.2.1)	Smell of coconut
Undecalactone (D 5.2.1)	Smell of peach
2-E-6-Z-Nonadienal (D 5.2.3)	Smell of cucumber
Ethyl 2-E-4-Z-decadienoate (D 5.2.4)	Smell of pear
Ethyl 2-methylbutyrate (D 5.2.4)	Smell of apple
Amyl acetate and amyl propionate (D 5.2.4), eugenol (D 23.2.1)	Smell of banana
Humulone, lupulone (D 5.3.1)	Bitter taste of beer
Citral (D 6.2.1)	Smell of lemon
$\acute{\alpha}$-Carvone (D 6.2.1)	Smell of caraway
Iridoids (D 6.2.2)	Bitter taste of liquors (gentian)
ar-Turmerone (D 6.3)	Pungent principle of curcuma
Isothiocyanates (D 8.4)	Pungent principle of mustard, cabbage, radish, horseradish
syn-Propanethial S-oxide (D 11.2)	Smell of onion
Diallyldisulfide (D 11.2)	Smell of garlic
Lenthionine	Smell of the mushroom *Lentinus edodes*
Methyl N-methyl anthranilate (D 21.4) and thymol (D 6.2.1)	Smell of mandarin orange
Capsaicin (D 23.1.1)	Pungent principle of *Capsicum* fruits
Cinnamic aldehyde (D 23.2.1)	Smell of cinnamon
Myristicin (D 23.2.1)	Smell of nutmeg
Apiol (D 23.2.1)	Smell of parsley
Eugenol (D 23.2.1)	Smell of cloves
6(S)-Gingerol (D 23.2.1)	Flavor of ginger
Vanillin (D 23.2.5)	Smell of vanilla fruits
Piperine (D 23.3.1)	Pungent principle of pepper
Naringin (D 23.3.3)	Bitter principle of *Citrus* fruits

dominated by only one main component, as with apple, peach, or coconut. Usually, however, complex mixtures of secondary products influence man's choice of food. (In the aroma of wines, for instance, more than 800 components have been shown to exist.) There is no simple relationship between the chemical structure and the taste or flavor of secondary products. Small changes in chemistry may alter these properties completely (cf. the chemical structures of the flavoring principles of peach and coconut).

Humans are very sensitive to some types of secondary products. The bitter taste of the alkaloid brucine (D 22.3) may be recognized in concentrations of $10^{-4}\%$, the odor of cucumber in a concentration of $10^{-8}\%$. The threshold concentrations are different for different persons and depend on the physiological state even in one and the same person.

Today there are only about a dozen major crop plants used in human nutrition: wheat, rice, corn, potato, barley, sweet potato, cassava, grapes, soybean, oats, sugarcane, sugar beet, and sorghum. In these plants the concentration of nonnutritional secondary products has been reduced by plant breeding and selection in comparison with the wild progenitors (F 6.1).

Examples in this respect are the fruits of Cucurbitaceae, containing less cucurbitacins (D 6.6), potatoes, containing less triterpene alkaloids (D 6.6.3), *Manihot* (cassava roots) containing less cyanogenic glycosides (D 8.3), and cabbage containing less glucosinolates (D 8.4).

Humans become adapted to the reduced amounts of secondary products present in crop plants. People living on cassava receive a daily dose of about 35 mg HCN, which is half the lethal dose for unadapted persons, and men living on potatoes tolerate the small amounts of steroid alkaloids present.

Due to the decreased content of secondary products, pest resistance of cultivated plants and probably also the ability for competition with weeds, is reduced. Hence, man needs synthetic chemicals, such as pesticides and herbicides, to protect his crops. This has an unfavorable impact on ecology, which has become increasingly obvious in the last years.

In some instances crop plants were also selected for a higher content of certain types of secondary products. Cultivated apples, pears, and other fruits have a better aroma, i.e., contain more volatile secondary substances, and in addition to a higher sugar content are much more attractively colored than their wild progenitors. The pigments (E 5.5.2) and scent volatiles are usually nontoxic for humans, in contrast, for example, to the tanning substances which are present in many of the wild forms of our cultivated fruits.

References for Further Reading

37, 60, 299, 323, 402, 425, 503, 623, 764, 859, 898

F 2 Secondary Products as Active Principles of Drugs

Since antiquity man has used secondary products for the alleviation of diseases, and even today the active principles of many drugs are secondary products of microbial, plant, and animal origin (Table 78). Three groups of substances are significant in this respect:

a) compounds with physiological activity, formed in the human body itself, like many hormones (E 3.1), the neurotransmitters (E 3.2), the bile acids (D 6.6.4), etc., if their rate of formation is insufficient

b) secondary products which have to be taken up with the diet, e.g., the vitamins (E 2.1), if the composition of food and beverages do not meet the needs, and

c) physiologically active secondary products formed in other groups of organisms which are able to normalize the metabolism of human beings (or animals) by certain diseases.

Most of the latter products have emerged during the course of evolution as plant or microbial defense substances attacking microorganisms, deterring preda-

Table 78. Secondary plant products used in medicine

Secondary products	Pharmacological activity
D-Sorbitol (D 1.3)	Sweetener for diabetics
Lactose (D 1.4.1)	Mild laxative for children
Castor oil (D 5.2.4)	Cathartic
Linseed oil (D 5.2.4)	Demulcent and laxative, protective if applied externally
Spermaceti, beeswax (D 5.2.4)	Ingredients of ointments, cerates, and plasters
Prostaglandins (D 5.2.6)	Disruption of pregnancy, initiation of delivery, synchronization of rut
Anthraquinone and anthrone derivatives (D 5.3.6)	Cathartics
l-Menthol (D 6.2.1)	Local antipruritic, counterirritant
d-Camphor (D 6.2.1)	Irritant
d-Carvone (D 6.2.1)	Carminative
Thymol (D 6.2.1)	Antifungal and antibacterial agent
d-Fenchone (D 6.2.1)	Carminative
Gentiopicroside (D 6.2.2)	Bitter
Valepotriates (D 6.2.2)	Sedative
Triterpene saponins (D 6.6)	Antitussives, treatment of ulcus cruris, etc.
Cardiac glycosides (D 6.6.8)	Cardiotonics
L-Hyoscyamine, atropine (D 17.2)	Parasympathetic depressants, spasmolytics
L-Scopolamine (D 17.2)	Cerebral sedative
Cocaine (D 17.2)	Local anesthetic
Lobeline (D 18)	Respiratory stimulant
Sparteine (D 18)	Oxytocic, cardiac antiarrhythmic
Pilocarpine (D 20)	Cholinergic, used in ophthalmology
Physostigmine (D 22)	Inhibitor of acetylcholine esterase (D 10), used in ophthalmology
Ergotamine and derivatives (D 22.2)	Uterus contraction, analgesic in migraine
Reserpine (D 22.3)	Antihypertensive
Rescinnamine (D 22.3)	Antihypertensive
Yohimbine (D 22.3)	Peripheral vasodilatans
Vinblastine, vincristine (D 22.3)	Treatment of cancer
Strychnine (D 22.3)	Central stimulant
Quinine (D 22.4)	Antimalarial, treatment of colds, tonic
Quinidine (D 22.4)	Cardiac antiarrhythmic
Emetine (D 23.1.2)	Antiamebic, expectorant, and emetic
Papaverine (D 23.1.2)	Spasmolytic
Codeine (D 23.1.2)	Antitussive
Morphine (D 23.1.2)	Narcotic analgesic
Tubocurarine (D 23.1.2)	Skeletal muscle relaxant
Eugenol (D 23.2.1)	Dental analgesic
L-Ephedrine (D 23.2.5)	Adrenergic, bronchodilator
Psoralen, bergapten (D 23.2.2)	Compounds enhancing skin pigmentation
Arbutin (D 23.2.5)	Diuretic, disinfectant
Shikonin (D 23.2.5)	Antimycotic
Tannins (D 23.2.5 and 23.3.3)	Astringents
Rutin (D 23.3.3)	Compound reducing capillary fragility

tors, or reducing their fitness. These substances have been selected for strong physiological activity on animals and/or microorganisms (E 5.2 and E 5.5.3). Due to the similarity of primary metabolism in all living beings (A) they influence also the physiology of humans and that of pathogenic microorganisms which in the natural habitat may have no relation to the producer organisms. As with other drugs the toxicity of these substances depends on the dosage, and sublethal amounts may have a curative effect. Examples are the cardenolides (D 6.6.8) which in high doses are toxic, but in lower (curative) doses support the efficiency of the heart muscle. Secondary defense substances often possess a warning bitter taste (E 5.5.3). In the traditional and folk medicines, which frequently use drugs containing secondary defense products produced in plants as active principles, the bitter taste of a remedy is therefore often synonymous with high quality.

Important secondary products of plant origin are steroids (obtained semisynthetically, e.g., from sterols, D 6.6.1, diosgenin, D 6.6.2, or solasodine, D 6.6.3), alkaloids, like L-hyoscyamine, atropine, and scopolamine, pilocarpine, codeine, reserpine, quinidine, and L-ephedrine, as well as the cardiac glycosides, like digoxin and digitoxin (Table 78). In addition, many crude drugs (herbs) contain secondary products with weak physiological activity, e.g., anthraquinones, essential oils (with main constituents like l-menthol, d-carvone, thymol, and d-fenchone), saponins, tannins, or flavonoids, like rutin. The use of crude drugs with weak activity has shown a renaissance in the last years.

The most significant microbial secondary compounds used in the treatment of diseases are the antibiotics (E 5.2) of which some 100 products are on the market. In 1980 worldwide antibiotic production was estimated at about 25 000 tons, including 17 000 tons of penicillins (D 9.2.3), 5 000 tons of tetracyclines (D 5.3.8), 1 200 tons of cephalosporins (D 9.2.3), and 800 tons of erythromycins (D 3). The search for new antibiotics continues because of the frequent development of resistant microbial strains and the need for cheaper, safer, and more active products. Chemical modification of natural antibiotics is of increasing significance in this respect. Antibiotics are not only used in human or veterinary medicine, but on a large scale also for growth promotion of farm animals (ergotropics).

In several cases the physiological action of secondary products in the course of evolution has become so effective that man could not prepare synthetically more active substances with similar structures. Examples in this respect are the cardiac glycosides and certain antibiotics, like streptomycin and chloramphenicol. In other cases the secondary products have been used as leaders in the chemical synthesis of new drugs with better properties. Examination of the relations between structural elements and physiological action has allowed the elucidation of the so-called pharmacophoric groups and their use and further modification in the synthesis of better drugs.

Of significance in the improvement of secondary products have been:

a) The Alteration of Physicochemical Properties. In several instances solubility, lipophilicity, the stability in acidic or alkaline solutions, etc. of secondary products are not suitable for medical use.

The antibiotics of the penicillin group (D 9.2.3), for instance, during evolution have been optimized to act in the natural habitat, but not to pass the human stomach with its low pH value. As a result their stability to hydrolysis is not sufficient. Hence, natural penicillins cannot be used orally, e.g., in tablets, a drawback which in the meantime has been overcome by the synthesis of acid-stable derivatives.

b) Prolongation of the Physiological Activity by Delay of Metabolization. In the producer organism the actual concentration of many secondary compounds is governed by synthesis and degradation (A 5). Thus, many secondary products contain chemical groups which can be attacked by transforming and degrading enzymes, including those of other organisms. Blockage of these groups with suitable substituents introduced by chemical reactions has lowered in several cases the rate of transformation/degradation and has yielded products with stronger or prolonged activity.

An example is the extension of the physiological action of acetylcholine derivatives. Acetylcholine, a neurotransmitter in higher animals and human beings (E 3.2), is rapidly degraded by choline esterase (D 10). Hydrolysis of the synthetically prepared ester of choline with carbamic acid (carbachol) is, however, much slower, and the physiological action of carbachol therefore lasts much longer than that of acetylcholine.

c) Simplification of Chemical Structures. In several instances the molecule of secondary products can be simplified considerably without loss of its physiological activity.

For the muscle-relaxing properties of tubocurarine (D 23.1.2), for instance, only part of the complex structure of this alkaloid is necessary: the presence (in the protonated form) of two quarternary N-atoms at a certain distance from each other. This structure is mimicked in the pharmacologically similar synthetically prepared compounds of the decamethonium and suxamethonium type, in which both the quarternary N-atoms are separated either by a simple saturated carbon chain or a succinic acid bischoline ester residue.

The unnecessary complexity of the chemical structures of many secondary products has originated by the fact that in the course of evolution already existing substances have been modified to meet new needs (E). Hence, old chemical structures have been overlayed by new features, and in many instances not the simplest solution for a given problem has been realized.

References for Further Reading

59, 208, 441, 467, 517, 686, 742, 794, 860

F 3 Secondary Products with Stimulating and Narcotic Properties

The most widespread human stimulants are of plant origin (Table 79). Alcoholic beverages, caffeine-containing drinks, and nicotine-containing tobacco goods are consumed in most countries. In addition, there are products in local use, like betel, khat, and coca leaves.

Table 79. Secondary plant products used as stimulants and hallucinogens

Secondary products	Biological activity
Ethanol (D 2)	Stimulant and hallucinogen, used worldwide
$(-)$-Δ^1-Tetrahydrocannabinol (D 6.2.1)	Hallucinogen, formed in *Cannabis sativa*, constituent of hashish and marijuana, used in many countries
Nicotine (D 16.1)	Stimulant, constituent of tobacco, used worldwide
Arecaidine (D 16.1)	Stimulant, formed from arecoline, a constituent of the seeds of *Areca catechu* (betel nuts), used in Southeast Asia
Pantherine (and ibotenic acid?) (D 17)	Hallucinogens, constituents of fly agaric *(Amanita muscaria)*, used in North America, Europe, and Asia
Cocaine (D 17.2)	Hallucinogen, constituent of the leaves of *Erythroxylon coca*, chewed as stimulant in South America
L-Scopolamine and probably other tropane alkaloids (D 17.2)	Hallucinogens if used in high doses. Main constituents of the medieval witch beverages and ointments produced from Solanaceae (e. g., *Hyoscyamus, Datura,* and *Atropa* sp.)
Psilocybine and psilocine (D 22.1)	Hallucinogens, constituents of mushrooms of the genus *Psilocybe* and *Stropharia*, used in Mexico
N, N-Dimethyltryptamine, bufotenine and 5-methoxy-N, N-dimethyltryptamine (D 22.1), 2-methyl and 1,2-dimethyl-6-methoxy-tetrahydro-β-carboline (D 22.3)	Hallucinogens, constituents of snuffs and decoctions derived from *Virola* and *Adenanthera* sp. used in South America
Lysergic acid amide, lysergic acid α-hydroxyethylamide (D 22.2)	Hallucinogens, constituents of the seeds of *Ipomea* sp., used in Mexico
Harmine, harmaline, and tetrahydroharmine (D 22.3)	Hallucinogens, constituents of the plant *Banisteriopsis caapi*, used in the northern part of South America
Ibogaine (D 22.3)	Hallucinogen, constituent of the plant *Tabernanthe iboga*, used in the western part of central Africa
Mescaline (D 23.1.1)	Hallucinogen, constituent of *Lophophora williamsii* and *Trichocereus pachanoi*, used in North and South America
Morphine (D 23.1.2)	Hallucinogen, constituent of opium, i. e., the dried latex of *Papaver somniferum*
L-Norpseudoephedrine (D 23.2.5)	Stimulant, constituent of *Catha edulis* (khat), used in the Middle East
Caffeine, theobromine, theophylline (D 25.1)	Stimulants, constituents of cofee, tea, cocoa, used worldwide

Secondary products are also the active constituents of hallucinogens, which had and still have a central place in the magic ceremonies of nearly all primitive human societies. Hallucinogenic beverages and snuffs have been ingested by priests and in some cases also by the whole community "to contact" ancestors, gods, or ghosts, or to "see" past and future. Different plants have been used as ingredients in different areas of the world in the preparation of hallucinogens (Table 79). Most of them contain alkaloids, such as scopolamine, psilocybine, dimethyltryptamine, 2-methyl-6-methoxytetrahydro-β-carboline, lysergic acid amide, harmine, ibogaine, and mescaline. In the Middle Ages in Europe hallucinogens based on the tropane alkaloids occurring in Solanaceae (*Hyoscyamus, Datura, Atropa,* and *Mandragora* sp.) have played a dominant role in the brews of witches, causing visions of flying, dancing, and even sexual intercourse with the devil.

In modern times in the countries of Western Europe and North America hallucinogens have become abused by certain groups of people as a possibility of escaping the disliked daily routine, stress, and frustration. People usually start with "soft" drugs, like hashish. After some time many change to the ingestion of the "hard" natural products, cocaine and morphine, or to the use of chemically modified compounds like LSD (lysergic acid dietylamide, derived from ergoline alkaloids, D 22.2) and heroin (diacetylmorphine, derived from morphine, D 23.1.2). Heroin is the most powerful and most dangerous hallucinogen known today. It is estimated that per year about 2 000 tons of opium are produced illegally for the production of heroin, which is worth about 100 000 million dollars.

Many of the hallucinogenic substances show structural relations to secondary products acting as neurotransmitters (E 3.2). They interfere with the receptors, and the biosynthesis and/or degradation of these substances.

The hallucinogenic indole alkaloids, for instance, are related to serotonin and 5-methoxytryptamine. Mescaline is similar to dopamine, epinephrine, and norepinephrin, and morphine and heroin are antagonists of the enkephalins and endorphins, peptides regulating the perception of pain (D 9.1).

References for Further Reading

1, 213, 581, 696

F 4 Toxic Secondary Products in Human Life

Many microorganisms, plants, and animals contain secondary products toxic to human beings. Primarily toxic organisms form the poisonous substances within their own body, whereas secondarily toxic organisms import them from outside, e.g., take up the compounds with the food. Actively toxic animals are able to attack human beings directly. Passively toxic organisms, like plants and microorganisms, usually act poisonous only after ingestion or skin contact.

Humans have known the toxicity of certain microbial cultures, plants, and animals throughout the ages. Acute toxicity, i. e., the immediate action of the poisonous compounds on the human body, could easily be traced by trial and error. Chronically acting toxins, however, whose desastrous effects become obvious only after long periods of time, frequently escaped recognition and remained a risk even in modern times. Poisoning of human beings by secondary products has been a prominent and widespread problem especially in periods of famine, when the usual foodstuffs have to be substituted by materials otherwise classified as unsuitable.

Examples are:
— the poisoning of humans with ergot alkaloids (D 22.2) synthesized in the mold *Claviceps purpurea*, a parasite on ray. In the Middle Ages flouers produced from ray grains containing the sclerotia of *Claviceps* caused repeatedly in France and Germany mass poisoning (convulsive ergotism; gangrenous ergotism, St. Anthony's fire)

— and the poisoning of men and cattle in India with seeds of certain *Lathyrus* sp. containing L-3-cyanoalanine and 3-aminopropionitrile (D 8.3). In periods of famine these seeds are used as food. They cause, however, the degeneration of the nerve system (neurolathyrism) as well as anomalies of bone and mesenchymal tissue (osteolathyrism).

A few toxic secondary products have been and are used in human societies in everyday life. African and South American aborigines have smeared arrow heads and darts with preparations containing toxic quarternary indole and isoquinoline alkaloids (curare alkaloids, D 22.3 and D 23.1.2) obtained from plants or frogs to paralyze animals hunted as prey, and they have used rotenone (D 23.3.3), a compound which causes lysis of gill epithelial cells in killing fish. Modern man uses pyrethrins (D 6) and scillaren A (D 6.6.8) in killing insects and rodents, respectively, and applies strychnine (D 22.3) for poisoning foxes and other beasts of prey in the forests.

In addition, in all periods of history toxic secondary products have been misused in killing humans. A well-known victim was Sokrates who was put to death by a brew from *Conium maculatum* (hemlock) containing coniine (D 5.3.2). In modern times criminal poisoning has become relatively seldom. Nowadays more than two-thirds of the cases of acute poisoning occur accidentally due to the reduced knowledge of modern man on toxic organisms in his surroundings. Some plants and mushrooms, which in this respect are a risk, are listed in Table 80. Especially children exploring their neighborhood and people interested in "natural food" (e.g., mushrooms, as well as wild growing fruits and vegetables) may ingest parts of these organisms since even some of the widespread garden plants, like *Delphinium sp.*, *Convallaria majalis*, *Digitalis purpurea*, *Laburnum anagyroides*, and *Sarothamnus scoparius* are toxic.

$$H_3C-CH=N-N\begin{matrix} \diagup CH_3 \\ \diagdown C\diagdown H \\ \diagdown O \end{matrix} \qquad\qquad HO\cdots\begin{matrix} \\ \end{matrix}\quad H_3C-\overset{O}{\underset{}{\bigcirc}}-CH_2-\overset{+}{N}\begin{matrix}\diagup CH_3 \\ -CH_3 \\ \diagdown CH_3\end{matrix}$$

Gyromitrin   Muscarine

A few special dishes are occasionally poisonous, e. g., improperly prepared mushrooms or bowl fish, *Sphaeroides rubipes*, whose inner organs contain tetrodotoxin (D 19). Furthermore, comparison of Tables 80 and 78 shows that many secondary products used in medicine, and widespread stimulants, like nicotine, in overdoses act as toxins and indeed there are no clear frontiers between healing and toxic compounds.[7]

Special secondary substances with a general toxicity to all groups of organisms are formed as microbial defense products. Most important in this respect are the so-called mycotoxins (Table 81). Wheat, peanut, beens, and other foodstuffs colonized by mold strains producing these compounds have repeatedly caused disastrous losses of farm animals (and probably human beings).

---

[7] "Was ist das nit gifft ist? alle ding sind gifft/vnd nichts ohn gifft/Allein die dosis macht das ein ding kein gifft ist." (Everything is poison and nothing is without poison. Only the dose makes it that a thing is no poison); Theophrastus Bombastus von Hohenheim, called Paracelsus (1493–1541).

Table 80. Plants and mushrooms accidentally poisonous to human beings

Species	Most dangerous part	Toxic secondary product(s)
Aconitum napellus	Leaves, seeds	Aconitine (D 6.4)
Amanita muscaria,	Mushroom	Ibotenic acid (D 17),
A. pantherina		muscarine
A. phalloides	Mushroom	Phalloin, α-amanitine (D 9.2)
Atropa belladonna	Berries	L-Hyoscyamine, L-scopolamine (D 17.2)
Colchicum autumnale	Seeds	Colchicine (D 23.1.2)
Convallaria majalis	Berries	Cardiac glycosides (D 6.6.8)
Datura stramonium	Seeds	L-Hyoscyamine, L-scopolamine (D 17.2)
Delphinium sp.	Leaves	Diterpene alkaloids (D 6.4)
Digitalis purpurea	Leaves, seeds	Cardiac glycosides (D 6.6.8)
Gyromitra (Helvella) esculenta	Mushroom	Gyromitrin
Hyoscyamus niger	Seeds	L-Hyoscyamine, L-scopolamine (D 17.2)
Inocybe sp.	Mushroom	Muscarine
Laburnum anagyroides	Flowers, leaves	Cytisine (D 18)
Nicotiana tabacum	Fermented leaves (cigarettes, cigars)	Nicotine (D 16.1)
Sarothamnus scoparius	Flowers, seeds	Sparteine (D 18)
Solanum dulcamara	Berries	C_{27}-Steroid alkaloids (D 6.6.3)

Table 81. Toxic compounds (mycotoxins) produced in fungi

Substance	Mode of action	Substance	Mode of action
Patulin (D 5.3.2)	Carcinogenic, alkylating proteins and nucleic acids	Trichothecenes (D 6.3)	Inhibitors of protein synthesis by interaction with the ribosomes
Citrinin (D 5.3.3)	Nephrotoxic, carcinogenic	Erythroskyrin (D 13)	Hepatotoxic
Aflatoxins (D 5.3.7)	Hepatotoxic, carcinogenic	α-Cyclopiazonic acid (D 22)	Nephrotoxic, hepatotoxic

The way toxins act on humans (and related animals) is diverse. Frequently they inhibit enzymes, increase the permeability of membranes, or interact with the nerve system (cf. Tables 66, 73, and 81).

References for Further Reading

64, 96, 144, 249, 252, 254, 310, 312, 358, 417, 440, 527, 528, 583, 627, 712, 733, 757, 770, 814, 851

F 5 Secondary Products as Raw Materials in Technology

Several secondary products used in manufacture are listed in Table 82. Of great significance are cellulose, wood, rubber, and tannins. In the future probably so-called botanochemicals, i. e., energy-rich, mostly secondary plant products, useful as substituents of petrochemicals, may gain importance as renewable resources.

Cellulose and Wood

Today cellulose and the combination of cellulose and lignin, i. e., wood, are of special importance. Billions of tons of CO_2 fixed by the green plants every year are transformed to these products and whole branches of industry are involved in their processing. Natural cellulose fibers (cotton, flax, jute) or fibers produced from the cellulose of wood play an important role in making thread, cloth, etc. Cellulose of fast-growing trees in the raw material of paper, and wood is even today one of the most important materials in constructing houses, furniture, etc.

Terpentine Oil/Rosin

Terpentine oil is derived either by steam distillation from terpentine, the solution of rosin in terpentine oil harvested from pine trees, or is obtained as a by-product during the processing of fir or pine wood to cellulose. The world production of terpentine oil is about 300 000 t per year. The main constituents are α- and β-pinene, car-3-ene, and limonene (D 6.2.1).

Rubber

Natural rubber is a constituent of the latex of *Hevea brasiliensis*, a tree grown in the tropics. The latex is present in laticiferous vessels and is collected after incision of the bark. It contains about 30 % rubber in the form of small droplets and coagulates after the addition of acids. The coagulate is processed to bandlike sheets of raw rubber (smoked sheets, crepes) which in industry are heated with sulfur (vulcanization) to harden it. At the present time about half of the rubber used in industry is prepared from Hevea latex, the other half is of synthetic origin.

Tannins

Tannins are involved in the production of leather, a process based on hydrogen bond formation between the phenolic hydroxy groups of the tannins and the peptide groups of the amorphous regions of the collagen fibrils of hides. Furthermore, quinones formed by the oxidation of o- or p-phenolic groups give rise to covalent bonds with the free amino groups of basic amino acids (D 8.5). Both types of bindings prevent the proteins of the hides from microbial attack, and increase flexibility and mechanical strength. Even though synthetic substitutes

Table 82. Secondary products used as raw materials in technology

Secondary product	Usage
Carbohydrates	
Cellulose (D 1.4.1)	Preparation of paper and textile fibers, explosives, cellulose acetate
Wood (cellulose, D 1.4.1 and lignin, D 23.2.3)	Construction of houses, furniture, etc.
Starch (D 1.4.1)	Paper and cloth sizing, laundry starching
Agar (D 1.4.1)	Gel-forming agent (food industry)
Alginate (D 1.4.1)	Thickener (food industry, cosmetics)
Gum arabic, gum tragacanth (D 1.4.1)	Demulcent and emollient in cosmetics, cloth printing, pharmaceutical technology
Dextran (D 1.4.1)	Osmotically active compound in plasma expanders (after partial hydrolysis)
Xanthan (D 1.4.1)	Thickener, gum
Glucose degradation	
Ethanol (D 2)	Fuel
Acids and acid derivatives	
Methane (C 3.2)	Fuel
Polyhydroxybutyric acid (D 5.1)	Raw material for plastics
Linseed oil (D 5.2.4)	Use in paints, production of linoleum
Castor oil (D 5.2.4)	Manufacture of soaps, lubricant for combustion engines
Waxes, e.g., beeswax and carnauba wax (D 5.2.4)	Constituents of polishes, manufacture of candle wax, varnishes, etc.
Suberin (D 5.2.4)	Constituent of cork, production of filters, stoppers, and insulating materials
Terpenoids	
Monoterpenes, sesquiterpenes (D 6.2.1 and D 6.3)	Essential oils (parfumery, cosmetics), solvents
d-Camphor (D 6.2.1)	Production of plastics
α-Pinene (D 6.2.1)	Synthesis of d-camphor
Rosin (colophony, D 6.4)	Manufacture of varnishes, varnish and paint dryers, printing ink, soap, scaling wax
Ergosterol (D 6.6.1)	Formation of vitamin D_2
Diosgenin (D 6.6.2), solasodine (D 6.6.3)	Preparation of steroid hormones
Rubber (D 6.8)	Rubber industry
Polyphenols	
Tannins (D 23.2.5 and 23.3.3)	Leather industry

play an increasing role, leather prepared with tanning agents from plants is still of great economical importance.

References for Further Reading

38, 114, 117, 228, 270, 357, 361, 419, 420, 432, 531, 562, 629, 777, 827

F 6 Manipulation of Secondary Product Formation

F 6.1 Improvement and Depression of the Biosynthesis of Secondary Compounds in Natural Producers

Since the beginning of domestication man has used the genetic variation of plants and animals as a fund for the selection of suitable genotypes. With respect to the formation of secondary products aims were the enhancement of biosynthesis as well as its depression. The latter was most important with crop plants. It resulted in the diminuation of nearly all "toxic" secondary products formed and stored as allelochemicals in the wild-type ancestors, like polyphenols, glucosinolates, etc. (F 1).

Specimens with increased rates of secondary product biosynthesis were selected from populations of microorganisms producing antibiotics (E 5.2), of plants synthesizing secondary products used in medicine (F 2), and of trees grown, e.g., for cellulose or rubber (F 5). In addition, plants were selected which, due to an increased content of secondary products, had more colorful and more flavored fruits (apples, pears, etc.), roots *(Daucus carota)*, or other edible parts (F 1).

Since ancient times genetic variation as the basis of selection of suitable genotypes has been broadened by cross-breeding and recently by the artificial introduction of mutations (use of UV, or X-rays as well as by chemicals like N-methyl-N'-nitro-N-nitrosoguanidine). Whereas cross-breeding, due to well-developed sexuality, has been frequently used with plants and animals, the artificial introduction of mutations is most suitable for microorganisms (B 2). It has been effectively applied, for instance, on the producers of antibiotics.

Examples are the *Penicillium sp.* which synthesize penicillin G (D 9.2.3). A low producing strain of *Penicillium notatum* was found accidentally by Fleming 1928. The commercial production began in the last years of World War II with a few milligrams per liter culture medium. In the meantime this amount could be increased by repeated mutation and selection of *Penicillium chrysogenum* strains to more than $40 \, g \, l^{-1}$, i.e., several 1000-fold. In this multistep process each mutation widened one of the bottle necks in precursor metabolism, biosynthesis of penicillin from precursors, resistance of the cells to penicillin, etc. The strains obtained by this procedure must be regarded as metabolic cripples forming the large amounts of penicillin as pathophysiological products. They are unstable and cannot survive in the natural habitat.

In the last years attempts have been made to use accumulating knowledge on the means regulating the fluxes of precursors (A 3) and on the regulation of secondary metabolic pathways (A 4) in the improvement of the production of secondary compounds. Mutants have been selected which overproduce the precursors in the presence of precursor analogs or show no further depression by the products (A 4.3) and excess nutrients (A 4.4.1). In general, however, this type of experiment has not been very successful and as yet the selection of high-yielding producer organisms is still a more or less empirical approach.

Reference for Further Reading

782

F 6.2 The Biosynthesis of Secondary Products in Artificial Systems

F 6.2.1 Formation of Secondary Compounds in Cell Cultures of Plants and Animals

Specialized cells of animals and plants may form secondary products if culti-vated in vitro, i. e., outside the intact organism. Most specialized animal cells capable of synthesizing secondary substances, however, cannot divide and due to their limited life span survive only in vitro for limited periods. Exceptions are cells of appropriate tumors combining the ability for growth in vitro with the for-mation of secondary products. Cell lines derived from neuroblastoma tumors, for instance, form acetylcholine (D 10). Suitable cell strains may also be obtained by fusion of nonmalignant, specialized cells synthesizing secondary compounds with rapidly dividing tumor cells which do not form secondary products (hybri-dome technique). As yet, however, no commercial production of secondary pro-ducts with animal cell cultures has been realized.

Plant cells grow easily in vitro and many cell lines synthesize secondary com-pounds (Table 83). In some cases the formation of secondary products with plant cell cultures has been developed up to industrial scale. Examples are the produc-tion of ginsenosides (D 6.6), berberine (D 23.1.2), and shikonin (D 23.2.5) by cell cultures of *Panax ginseng*, *Coptis japonica*, and *Lithospermum erythrorhizon*, re-spectively. In addition, plant cell cultures may synthesize secondary products not built in the intact plant. An example is the biosynthesis of paniculide B (D 6.3) and related compounds in cell cultures of *Andrographis paniculata*, which are ab-sent in the total plant. These substances are a fund for the selection of new, valu-able, secondary products.

Plant cell cultures are able to transform added secondary compounds stereo-specifically by oxidation, reduction, hydroxylation, glycosylation, etc., even if the cultures do not built secondary products de novo. Examples are the glucosy-

Table 83. Some groups of secondary products formed in plant cell cultures

Secondary products	Producer cell culture	Secondary products	Producer cell culture
Diosgenin gluco-sides (D 6.6.2)	*Dioscorea deltoides*	Capsaicin (D 23.1.1)	*Capsicum fructescens*
Cardiac glycosides (D 6.6.8)	*Digitalis lanata*	Berberine (D 23.1.2)	*Berberis stolonifera, Coptis japonica*
Nicotine (D 16.1)	*Nicotiana tabacum*	Rosmarinic acid (D 23.2.1)	*Coleus blumei*
Anthraquinones (D 21.2)	*Morinda citrifolia, Galium molugo*	Shikonin (D 23.2.5)	*Lithospermum erythrorhizon*
Acridine alkaloids (D 21.4.2)	*Ruta graveolens*	Ubiquinones (D 21.3)	*Nicotiana tabacum*
Ginsenosides (D 6.6)	*Panax ginseng*	Flavonoids (D 23.3.3)	*Petroselinum hortense*
Ajmaline, serpentine (D 22.3)	*Catharanthus roseus*	Caffeine (D 25.1)	*Coffea arabica*

lation and 12β-hydroxylation of digitoxin (D 6.6.8) and related compounds in cell cultures of *Digitalis sp.*, which are unable to form cardenolides. It may be expected that reactions of this type will be applied in industry in the future as part of sophisticated synthetic processes.

References for Further Reading

42, 62, 152, 230, 245, 536, 652, 749, 892

F 6.2.2 Formation of Known Secondary Products in Nonproducers
After Transfer of Genetic Material

For the biotechnological production of secondary compounds some organisms are more suitable than others because of a better growth rate, simple cultivation, etc. Most appropriate are bacteria, like *Escherichia coli*, *Bacillus* sp., or in special cases eukaryotic microorganisms, like yeasts. In the last years the transfer of genetic material from nonsatisfactory producers to these organisms, i. e., the genetic transformation of these organisms, has become possible. The developed techniques (genetic engineering) allow the transfer into the acceptor organisms of the structural genes for the desired secondary products, in combination with appropriate regulatory sequences. The latter make possible an effective expression of the structural genes in the acceptor organisms and are not necessarily those governing the expression of the secondary genetic material in the donor organism. The transferred genetic material may be located on plasmids or may be integrated in the genome of the acceptor organism, thus achieving a more stable product formation.

In the last years the transfer and expression of human genetic material in microorganisms has become of economical importance. This includes DNA encoding secondary products, especially peptide hormones, because these compounds are closely related to the proteins derived from the transferred DNA (D 9.1). Commercially produced are e. g., the A and B chains of insulin, and human growth hormone. A probable next step will be the transfer of gene clusters of prokaryotes encoding secondary products (A 2.1) to organisms which are biotechnologically easier to handle than the original producers. The transfer of genetic material which is scattered over the genome, as is the secondary genetic material of typical euarotes, however, does not seem to be possible in the near future.

F 6.2.3 The Construction of New Secondary Pathways and the
Creation of New Secondary Compounds

The selection of new genotypes of plants and animals has created qualitative alterations in secondary metabolism, in addition to the quantitative changes mentioned in F 1, e.g., new colors in the flowers of garden plants produced by flavonoids not occurring in the ancestors, and new flavors of fruits or vegetables, spices, and herbs. How successfully this selection has been carried out may easily be seen by visiting an exhibition of ornamental plants, fruits, vegetables, and agricultural products.

Recently, the cross-breeding methods used for the construction of new genotypes have been completed by procedures of genetic engineering, like protoplast fusion, and have been successfully applied to nonsexual organisms. As one of the first results new antibiotics of the rifamycin and anthracycline types (D 3) have been obtained by fusion of *Nocardia* and *Streptomyces* strains, respectively.

In addition, new secondary products (in most cases antibiotics) or at least an altered pattern of secondary compounds have been obtained by the following techniques:

Interrupted Biosynthesis

In these attempts the metabolic chains leading to secondary compounds have been blocked by mutations or metabolic inhibitors.

Precursor-directed biosynthesis

Analogs of natural precursors have been incorporated into secondary products due to the relatively low substrate specificity of certain enzymes involved. The addition of branched-chain amino acids, L-proline and its analogs, pipecolic acid, sarcosine, glycine, or azetidine-2-carboxylic acid to *Streptomyces* sp., for instance, enhanced the formation of special actinomycins (D 21.4.1), normally produced only in trace amounts, as well as the synthesis of novel representatives of these secondary compounds.

Mutasynthesis

This technique is a combination of the methods mentioned above, in which analogs of natural precursors and intermediates are added to organisms with secondary pathways blocked by mutations.

Reference for Further Reading

376

References

1 Abel EL (1980) Marihuana, the first twelve thousand years. Plenum Press, New York
2 Abraham EP (1986) Enzymes involved in penicillin and cephalosporin formation. In: Klein-kauf H, von Döhren H, Dornauer H, Nesemann G (eds.) Regulation of secondary metabolite formation. VCH Verlagsgesellschaft, Weinheim, pp 115-132
3 Achenbach H, König F (1972) Die Frage der biogenetischen Gleichwertigkeit der beiden Xan-thocillin-Hälften. Chem. Ber. 105: 784-793
4 Adam G, Marquardt V (1986) Brassinosteroids. Phytochemistry 25: 1787-1799
5 Adityachaudhury N, Das AK (1979) Recent advances in the chemistry of naturally occurring 2-pyrone derivatives. J Sci Industr Res 38: 265-277
6 Adler G, Hiller K (1985) Bisdesmosidische Triterpensaponine. Pharmazie 40: 676-693
7 Aharonowitz Y (1980) Nitrogen metabolite regulation of antibiotic biosynthesis. Ann Rev Microbiol 34: 209-233
8 Ajitkumar P, Cherayil JD (1988) Thionucleotides in transfer ribonucleic acid: diversity, structure, biosynthesis, and function. Microbiol Reviews 52: 103-113
9 Akhtar M, Jordan PM (1979) Porphyrin, chlorophyll, and corrin biosynthesis. Comprehensive organic chemistry, vol 5, Biological compounds. Pergamon Press, Oxford, pp 1121-1166
10 Albone ES (1984) Mammalian semiochemistry: the investigation of chemical signals between mammals. Wiley, Chichester
11 Alibert G, Ranjeva R, Boudet AM (1977) Organisation subcellulaire des voies de synthèse des compóses phénoliques. Physiol Veg 15: 279-301
12 Amino acids and peptides, a specialist periodical report (1969ff) Royal Society of Chemistry, London
13 Aoki H, Okuhara M (1980) Natural β-lactam antibiotics. Ann Rev Microbiol 34: 159-181
14 Applebaum SW, Birk Y (1979) Saponins. In: Rosenthal GA, Janzen DH (eds) Herbivores, their interaction with secondary plant metabolites. Academic Press, New York, pp 539-566
15 Archer BL (1980) Polyisoprene. In: Bell EA, Charlwood BV (eds) Encyclopedia of plant physiology, new series, vol 8, secondary plant products. Springer, Berlin Heidelberg New York, pp 309-327
16 Asakawa Y (1982) Chemical constituents of the Hepaticae. Prog Chem Org Nat Prod 42: 1-285
17 Aspinall GO (ed) (1982-1985) The polysaccharides, vols 1-3. Academic Press, Orlando
18 Atkins EDT (ed) (1985) Polysaccharides, topics in structure and morphology. Macmillan, Basingstoke
19 Atta-Ur-Rahman, Basha A (1983) Biosynthesis of indole alkaloids. Claredon Press, Oxford
20 Ayaba S, Furuja T (1982) Studies on plant tissue cultures, part 36, biosynthesis of a retrochalcone, echinatin, and other flavonoids in the cultured cells of Glycyrrhiza echinata. A new route to a chalcone with transposed A- and B-rings. J Chem Soc Perkin Trans I: 2725-2734
21 Baas WJ (1985) Naturally occurring seco-ring-A-triterpenoids and their possible biological significance. Phytochemistry 24: 1875-1889
22 Baddiley J (1984) Aspects of cell wall formation in bacteria. In: Biologically active principles of natural products. Thieme, Stuttgart, pp 286-293
23 Bader G, Hiller K (1987) Neue Ergebnisse zur Struktur und Wirkungsweise von Triterpensaponinen. Pharmazie 42: 577-597
24 Bailey JA, Mansfield JW (eds) (1982) Phytoalexins. Blackie, Glasgow London
25 Baker R, Herbert RH (1984) Insect pheromones and related natural products. Nat Product Reports 1: 299-318

26 Banthorpe DV, Branch SA (1984-1987) The biosynthesis of C_5-C_{20} terpenoid compounds. Nat Product Reports 1: 443-449, 2: 513-524, 4: 157-173

27 Banthorpe DV, Charlwood BV (1980) The terpenoids. In: Bell EA, Charlwood BV (eds) Encyclopedia of plant physiology, new series, vol 8, secondary plant products. Springer, Berlin Heidelberg New York, pp 185-220

28 Barber J (1987) Composition, organization, and dynamics of the thylakoid membrane in relation to its function. In: Hatch MD, Boardman NK (eds) The biochemistry of plants, vol 10, photosynthesis. Academic Press, San Diego, pp 75-130

29 Barbier M (1981) The status of blue-green bile pigments of butterflies and their phototransformations. Experientia 36: 1060-1062

30 Baron ML, Bothroyd CM, Rogers GI, Staffa A, Rae ID (1987) Detection and measurement of fluoroacetate in plant extracts by ^{19}F NMR. Phytochemistry 26: 2293-2295

31 Barrett GC (ed) (1985) Chemistry and biochemistry of the amino acids. Chapman and Hall, London

32 Barz W, Köster J (1981) Turnover and degradation of secondary (natural) products. In: Conn EE (ed) The biochemistry of plants, vol 7, secondary plant products. Academic Press, New York, pp 35-84

33 Barz W, Willeke U, Weltring KM (1980) Microbial degradation of phytoalexins and related compounds. Ann Phytopathol 12: 435-452

34 Battersby AR (1987) Nature's pathways to the pigments of life. Nat Product Reports 4: 77-87

35 Battersby AR (1984) Recent research on the biosynthesis of vitamin B_{12}. In: Voelter W, Daves DG (eds) Biologically active principles of natural products. Thieme, Stuttgart, pp 31-37

36 Bauer K (1985) Degradation and biological inactivation of neuropeptides. In: Hamprecht B, Neuhoff V (eds) Selected topics of neurobiochemistry. Springer, Berlin Heidelberg New York, pp 43-54

37 Bauer K, Garbe D (1985) Common fragrance and flavor materials. VCH Verlagsgesellschaft, Weinheim

38 Baumann H, Bühler M, Fochem H, Hirsinger F, Zöbelein H, Falbe J (1988) Natural fats and oils—renewable raw materials for the chemical industry. Angew Chem Int Ed Engl 27: 41-62

39 Baumgarten G (1963) Die herzwirksamen Glykoside. Thieme, Leipzig

40 Beale SI (1984) Biosynthesis of photosynthetic pigments. In: Baker NR, Barber J (eds) Chloroplast biogenesis. Elsevier, Amsterdam, pp 133-205

41 Beart JE, Lilley TH, Haslam E (1985) Plant polyphenols—secondary metabolism and chemical defense: some observations. Phytochemistry 24: 33-38

42 Becker H (1987) Regulation of secondary metabolism in plant cell cultures. In: Green CE, Somers DA, Hackett WP, Biesboer DD (eds) Plant tissue and cell culture. Alan R. Liss, Inc., New York, pp 199-212

43 Becker H (1984) Inhaltsstoffe der Erdkröte (Bufo bufo). Pharmazie in unserer Zeit 13: 129-136

44 Beker JT (1974) Tyrischer Purpur, ein antiker Farbstoff, ein modernes Problem. Endevour 33: 11-17

45 Bell EA (1980) Nonprotein amino acids. In: Bell EA, Charlwood BV (eds) Encyclopedia of plant physiology, new series, vol 8, secondary plant products. Springer, Berlin Heidelberg New York, pp 403-432

46 Bell EA (1980) The possible significance of secondary products in plants. In: Bell EA, Charlwood BV (eds) Encyclopedia of plant physiology, new series, vol 8, secondary plant products. Springer, Berlin Heidelberg New York, pp 11-21

47 Bell WE, Carde RT (1984) Chemical ecology of insects. Chapman and Hall, London

48 Bender DA (1985) Amino acid metabolism. Wiley, Chichester

49 Benkovic SJ (1980) On the mechanism of action of folate- and biopterin-requiring enzymes. Ann Rev Biochem 49: 227-251

50 Bennet JW (1983) Differentiation and secondary metabolism in mycelial fungi. In: Bennet JW, Ciegler A (eds) Secondary metabolism and differentiation in fungi. Marcell Dekker, New York, pp 1-32

51 Bennet JW (1983) Secondary metabolism as differentiation. J Food Safety 5: 1-11

52 Bennet JW, Bentley R What's in a name?—Microbial secondary metabolism. Adv Appl Microbiol, in press

53 Bennet JW, Christensen SB (1983) New perspectives on aflatoxin biosynthesis. Adv Appl Microbiol 28: 53-92

54 Bentley KW (1984-1988) β-Phenylethylamines and the isoquinoline alkaloids. Nat Product Reports 1: 355-370, 2: 81-96, 3: 153-169, 5: 265-292

55 Bentley R, Meganathan R (1987) Biosynthesis of isoprenoid quinones: ubiquinone and menaquinone. In: Neidhard FC (ed) *Escherichia coli* and *Salmonella typhimurium*—cellular and molecular biology, ASM Publications, pp 12-520

56 Bentley R, Meganathan R (1983) Vitamin K biosynthesis in bacteria—precursors, intermediates, enzymes, and genes. J Nat Prod 46: 44-59

57 Benveniste P (1986) Sterol biosynthesis. Ann Rev Plant Physiol 37: 275-308

58 Berdy J (1980-82) CRC Handbook of antibiotic compounds, vol 1-10. CRC Press, Boca Raton

59 Berger F (1949-67) Handbuch der Drogenkunde, vols 1-7. W. Maudrich, Wien

60 Bergner KG (1984) Weinqualität aus chemischer Sicht. Deutsche Apotheker Ztg 124: 2179-2183

61 Berkeley RCW, Gooday GW, Ellwood DC (1979) Microbial polysaccharides and polysaccharidases. Academic Press, London

62 Berlin J (1986) Secondary products from plant cell cultures. In: Pape H, Rehm HJ (eds) Biotechnology, vol 4. VCH Verlagsgesellschaft Weinheim, pp 629-658

63 Berridge MJ (1987) Inositol triphosphate and diacylglycerols: two interacting secondary messengers. Ann Rev Biochem 56: 159-193

64 Betina V (ed) (1984) Mycotoxins—production, isolation, separation and purification. Elsevier, Amsterdam

65 Beyer TA, Sadler JE, Rearick JI, Paulson JC, Hill RL (1981) Glycosyltransferases and their use in assessing oligosaccharide structure and structure-function relationships. Adv Enzymol 52: 23-175

66 Bhakuni DS, Jain S (1986) Protoberberine alkaloids. In: Brossi A (ed) The alkaloids, vol 28. Academic Press, New York, pp 95-181

67 Bhattacharyya P, Chakraborty DP (1987) Carbazole alkaloids. Prog Chem Org Nat Prod 52: 159-209

68 Birch AJ (1962) Some pathways in biosynthesis. Proc Chem Soc, pp 3-13

69 Birch MC (ed) (1974) Pheromones. North Holland, Amsterdam

70 Black SD (1987) P_{450} cytochromes: structure and function. Adv Enzymol 60: 35-87

71 Blakley RL, Benkovic SJ (eds) (1985) Folates and pterins, vols 1 and 2, chemistry and biochemistry of pterins. Wiley, New York

72 Bleasdale JE, Eichberg J, Hauser G (eds) (1985) Inositol and phosphoinositides. Humana Press, Clifton

73 Bloch K (1977) Control mechanisms in the synthesis of saturated fatty acids. Ann Rev Biochem 46: 263-298

74 Bloch K (1982) Sterols and membranes. In: Martonosi AN (ed) Membranes and transport, vol 1. Plenum Press, New York, pp 25-35

75 Block E (1985) The chemistry of garlic and onion. Sci American 252: 114-119

76 Blum MS (1981) Chemical defenses of arthropods. Academic Press, New York

77 Bode R, Böttcher F, Birnbaum D, Samsonova IA (1975) Der Kurzschlußweg—einziger Weg zur de novo-Biosynthese von Nicotinsäure bei *Hansulena henricii*. Z Allg Mikrobiol 15: 149-155

78 Böhm H (1985) The biochemical genetics of alkaloids. In: Mothes K, Schütte HR, Luckner M (eds) Biochemistry of alkaloids. Deutscher Verlag der Wissenschaften, Berlin and Verlag Chemie, Weinheim, pp 25-36

79 Bohlmann F, Burkhardt T, Zdero C (1973) Naturally occurring acetylenes. Academic Press, London

80 Bonner J (1950) Plant biochemistry. Academic Press, New York, p 537

81 Bonner J, Galston AW (1952) Principles of plant physiology. Freeman, San Francisco

82 Boppre M (1986) Insects pharmacophagously utilizing defensive plant chemicals (pyrrolizidine alkaloids). Naturwiss 73: 17-26

83 Boppre M, Schneider D (1985) Pyrrolizidine alkaloids quantitatively regulate both scent morphogenesis and pheromone biosynthesis in male *Creatonotos* moths (Lepidoptera: Arctiidae) J Comp Physiol A 157: 569-577

84 Boyer PD (ed) (1977) The enzymes, vol 13, dehydrogenation, oxidases, hydrogen peroxide cleavage. Academic Press, New York

85 Boyer PD (ed) (1973) The enzymes, vol 8, group transfer, part A. Academic Press, New York

86 Boyer PD (ed) (1976) The enzymes, vol 12, oxygenation—reduction, part B, electron transfer, oxygenases, oxidases. Academic Press, New York

87 Boyer PD (ed) (1983) The enzymes, vol 16, lipid enzymology. Academic Press, New York

88 Boyer PD (ed) (1973) The enzymes, vol 9, group transfer, part B, phosphoryl transfer, one-carbon group transfer, glycosyl transfer, amino group transfer, other transferases. Academic Press, New York

89 Boyer PD (ed) (1972) The enzymes, vol 6, carboxylation and decarboxylation (nonoxidative), isomerization. Academic Press, New York

90 Boyer PD, Chance B, Ernster L, Mitchell P, Racker E, Slatter EC (1977) Oxidative phosphorylation and photophosphorylation. Ann Rev Biochem 46: 955-1026

91 Brahmachary RL (1986) Ecology and chemistry of mammalian pheromones. Endeavour New Series 10: 65-68

92 Brand JM, Young JC, Silverstein RM (1979) Insect pheromones: a critical review of recent advances in their chemistry, biology and application. Prog Chem Org Nat Prod 36: 1-190

93 Brattsten LB (1979) Biochemical defense mechanisms in herbivores against plant allelochemicals. In: Rosenthal GA, Janzen DH (eds) Herbivores, their interaction with secondary plant metabolites, Academic Press, New York, pp 199-270

94 Braunstein AE, Goryachenkova EV (1984) The β-replacement specific pyridoxal-P-dependent lyases. Adv Enzymol 56: 1-89

95 Bray RC, Engel PC, Mayhev SG (eds) (1984) Flavins and flavoproteins. de Gruyter, Berlin

96 Bresinsky A, Besl H (1985) Giftpilze. Wissenschaftliche Verlagsgesellschaft Stuttgart

97 Brewin NS, Davies DD, Robins RJ (1987) Immunochemistry for enzymology. In: Davies DD (ed) The biochemistry of plants, vol 13, methodology. Academic Press, San Diego, pp 1-31

98 Brian PW (1957) The ecological significance of antibiotic production. Soc Gen Microbiol Symp 7: 168-188

99 Bringmann G (1979) Chemische Mechanismen der Alkohòl-Wirkung. Naturwissenschaften 66: 22-27

100 Britton G. (1983) The biochemistry of natural pigments. University Press, Cambridge

101 Britton G (1984-1986) Carotenoids and polyterpenoids. Nat Product Reports 1: 67-85, 2: 349-387, 3: 591-614

102 Brooks CJW, Watson DG (1985) Phytoalexins. Nat Product Reports 2: 427-459

103 Brooks J, Grant PR, Muir M, van Gijzel P, Shaw G (ed) (1971) Sporopollenin, Academic Press, London

104 Brossi A (ed of the last vols) (1950ff) The alkaloids. Academic Press, San Diego

105 Brown AG, Roberts SM (eds) (1985) Recent advances in the chemistry of β-lactam antibiotics. Royal Society of Chemistry, Letchworth, England

106 Brown EG, Flayeh KAM, Gallon RJ (1982) The biosynthetic origin of the pyrazole moiety of β-pyrazol-1-yl-L-alanine. Phytochemistry 21: 863-867

107 Brown GM, Williamson JM (1982) Biosynthesis of riboflavin, folic acid, thiamine, and pantothenic acid. Adv Enzymol 53: 345-381

108 Brown KS (1984) Adult-obtained pyrrolizidine alkaloids defend ithomiine butterflies against a spider predator. Nature 309: 707-709

109 Brown R (ed) (1982) Cellulose and other natural polymer systems, biogenesis, structure and degradation. Plenum Press, New York

110 Brown SA (1972) Methodology. Specialist periodical report, biosynthesis 1: 1-40

111 Brown SA (1981) Coumarins. In: Conn EE (ed) The biochemistry of plants, vol 7, secondary plant products. Academic Press, New York, pp 269-300

112 Brown SA (1986) Biochemistry of plant coumarins. Rec Adv Phytochem 20: 287-316

113 Bryan JK (1980) Synthesis of the aspartate family and branched-chain amino acids. In: Miflin BJ (ed) The biochemistry of plants, vol 5, amino acids and derivatives. Academic Press, New York, pp 403-452

114 Buchanan RA, Otey FH, Bagby MO (1980). Botanochemicals. Rec Adv Phytochem 14: 1-22

115 Buck KT (1987) The bisbenzylisoquinoline alkaloids. In: Brossi A (ed) The alkaloids, Vol 30, Academic Press, New York, pp 1-222

504   References

116 Burkhardt D (1982) Birds, berries and UV. Naturwiss 69: 153-157
117 Bushel ME (1983) Microbial polysaccharides. Progr Ind Microbiol 18
118 Butt VS, Lamb CJ (1981) Oxygenases and the metabolism of plant products. In: Conn EE (ed)
    The biochemistry of plants, vol 7, secondary plant products. Academic Press, New York,
    pp 627-665
119 Bu'Lock JD (1975) The two-faced microbiologist: contributions of pure and applied microbiol-
    ogy to good research. Dev Ind Mircrobiol 66: 11-19
120 Bu'Lock JD (1985) The biosynthesis of natural products. McGraw-Hill, London
121 Bu'Lock JD (1961) Intermediary metabolism and antibiotic synthesis. Adv Appl Microbiol 3:
    293-342
122 Bu'Lock JD, Hamilton D, Hulme MA, Powell AJ, Smalley HM, Shepherd D, Smith GN (1965)
    Metabolic development and secondary biosynthesis in *Penicillium urticae*. Can J Microbiol 11:
    765-778
123 Bu'Lock JD, Kilgour WJ, Knauseder F, Demnerova K, Steinbuchel A (1982) The azasterols of
    Geotrichum flaveobrunneum—a problem in autotoxicity, resistance, adaptation and selection. In:
    Krumphanzl V, Sikyta B, Vanek Z (eds) Overproduction of microbial products. Academic Press,
    London, pp 77-87
124 Bycroft BW (1988) Dictionary of antibiotics and related compounds. Chapman and Hall, Lon-
    don
125 Cabib E (1987) The synthesis and degradation of chitin. Adv Enzymol 59: 59-101
126 Callow JA (1983) Biochemical plant pathology. Wiley, Chichester
127 Campbell IM (1983) Correlation in secondary metabolism and differentiation. In: Bennet JW,
    Ciegler A (eds) Secondary metabolism and differentiation in fungi, Marcell Dekker, New York,
    pp 55-71
128 Campbell IM (1984) Secondary metabolism and microbial physiology. Adv Microbial Phy-
    siol 25: 1-60
129 Cangelosi GA, Nester EW (1988) Initial interaction between plant cells and *Agrobacterium tume-
    faciens* in crown gall tumor formation. In: Conn EE (ed) Opportunities for phytochemistry in
    plant biotechnology. Plenum Press, New York, pp 99-126
130 Cannon M (1987) Antimicrobial peptides: a family of wound healers. Nature 328: 478
131 Cantoni GL (1975) Biological methylation: selected aspects. Ann Rev Biochem 44: 435-451
132 Caspi E (1978) Nonoxidative cyclization of squalene by *Tetrahymena pyriformis*. The incorpora-
    tion of a 3β-hydrogen (deuterium) atom into tetrahymanol. Proc 11th Int Symp Chem Nat Pro-
    ducts, vol 4. Bulg Acad Sci, Sofia, pp 166-181
133 Cattel L, Balliano G, Caputo O, Viola F (1981) Biosynthesis of cucurbitacins in *Bryonia dioica*
    seedlings. Planta medica 41: 328-336
134 Cavallini D, Gaull GE, Zappia V (eds) (1980) Natural sulfur compounds. Plenum Press, New
    York
135 Chandel RS, Rastogi RP (1980) Triterpenoid saponins and sapogenins: 1973-1978. Phytochem-
    istry 19: 1889-1908
136 Chang C, Huang AHC (1981) Metabolism of glyoxalate in isolated spinach leaf peroxysomes.
    Kinetics of glyoxalate, oxalate, carbon dioxide, and glycine formation. Plant Physiol 67:
    1003-1006
137 Chapman RF, Bernays EA, Stoffolano JG (1987) Perspectives in chemoreception and behaviour.
    Springer, Berlin Heidelberg New York
138 Chawia AS, Jackson AH (1984-1986) Erythrina and related alkaloids. Nat Product Reports 1:
    371-373, 3: 555-564
139 Chen Z, Zhu D (1987) Aristolochia alkaloids. In: Brossi A (ed) The alkaloids, vol 31, Academic
    Press, New York, pp 29-65
140 Christie WW (1984, 1987) Simple and complex lipids: their occurrence, chemistry, and bio-
    chemistry. Nat Product Reports 1: 499-511; 4: 113-128
141 Chytil F, McCormick DB (eds) (1986) Vitamins and coenzymes, parts G and H. Methods in en-
    zymology, vol 122 and 123. Academic Press, New York
142 Clarke S (1985) Protein carboxyl methyltransferases: two distinct classes of enzymes. Ann Rev
    Biochem 54: 479-506
143 Clausen S, Olsen O, Sorensen H (1982) 4-Hydroxybenzoylcholine, a natural product present in
    Sinapis alba. Phytochemistry 21: 917-922

144 Cole RJ, Cox RH (1981) Handbook of toxic fungal metabolites. Academic Press, New York
145 Collinge DB, Slusarenko AJ (1987) Plant gene expression in response to pathogens. Plant Molecular Biology 9: 389-410
146 Colowick SP, Kaplan NO (eds) (1975) Methods in enzymology, vol 35, lipids, part B. Academic Press, New York
147 Colowick SP, Kaplan NO (eds) (1975) Methods in enzymology, vol 43, antibiotics. Academic Press, New York
148 Conn EE (1980) Cyanogenic glycosides. In: Bell EA, Charlwood BV (eds) Encyclopedia of plant physiology, new series, vol 8, secondary plant products. Springer, Berlin Heidelberg New York, pp 461-492
149 Conn EE (1988) A biosynthetic relationship between cyanogenic glycosides, glucosinolates and nitro compounds. Symp Am Chem Soc, in press
150 Conn EE (ed) (1986) The shikimic acid pathway. Plenum Press, New York
151 Conolly JD, Hill RA (1984-1986) Triterpenoids. Nat Product Reports 1: 53-65, 2: 1-17, 3: 421-442
152 Constabel F, Vasil IK (eds) (1988) Phytochemicals in plant cell culture. Academic Press, New York
153 Cooper BA, Whitehead VM (eds) (1986) Chemistry and biology of pteridines 1986. de Gruyter, Berlin
154 Cooper-Driver GA, Swain T, Conn EE (eds) (1985) Chemically mediated interactions between plants and other organisms. Plenum Press, New York
155 Corcoran JW (ed) (1981) Antibiotics, vol 4, biosynthesis. Springer, Berlin Heidelberg New York
156 Cordell GA (1977) The sesterterpenes, a rare group of natural products. Prog Phytochem 4: 209-256
157 Cordell GA (1981) Introduction to alkaloids, a biogenetic approach. Wiley-Interscience, New York
158 Cori MO (1983) Enzymic aspects of the biosynthesis of monoterpenes in plants. Phytochemistry 22: 331-341
159 Cormier MJ, Lee J, Wampler JE (1975) Bioluminescence: recent advances. Ann Rev Biochem 44: 255-275
160 Cosgrove DJ (1980) The inositol phosphates. Elsevier/North Holland, Amsterdam
161 Cossins EA (1987) Folate biochemistry and the metabolism of one-carbon units. In: Davies DD (ed) The biochemistry of plants, vol 11, biochemistry of metabolism. Academic Press, San Diego, pp 317-353
162 Cossins EA (1980) One-carbon metabolism. In: Davies DD (ed) The biochemistry of plants, vol 2, metabolism and respiration. Academic Press, New York, pp 365-418
163 Costello M, Lesh DH, Karol MH, Gottlieb FJ, Campbell IM (1989) Cross reactivity of plant and fungal phenylalanine ammonia-lyases. Have fungi been stealing plant genes? Phytochemistry, in press
164 Crandall LW, Hamill RL (1986) Antibiotics produced by Streptomycetes. Major structural classes. In: The bacteria, vol 9. Academic Press, New York, pp 355-401
165 Crews P, Naylor S (1985) Sesterterpenes: an emerging group of metabolites from marine and terrestrial organisms. Prog Chem Org Nat Prod 48: 203-269
166 Crombie L (1984) Rotenoids and their biosynthesis. Nat Product Reports 1: 3-19
167 Croteau R (1987) Biosynthesis and metabolism of monoterpenoids. Chem Rev 87: 929-954
168 Cutler HG (1988) Unusual plant-growth regulators from microorganisms. CRC Crit Rev Plant Sci 6: 323-343
169 Czapek F (1921) Biochemie der Pflanzen, vol 3, p 220. Fischer, Jena
170 Daehne W von, Godtfredsen WO, Rasmussen PR (1979) Structure-activity relationships in fusidic acid-type antibiotics. Adv Appl Microbiol 25: 95-146
171 Daly JM, Deverall BJ (1983) Toxins and plant pathogenesis. Academic Press, Sidney
172 Daly JW (1982) Alkaloids of neotropical poison frogs (Dendrobatidae). Prog Chem Org Nat Prod 41: 205-340
173 Danielsson H, Sjövall J (eds) (1985) Sterols and bile acids. Elsevier, Amsterdam
174 Darvill AG, Albersheim P (1984) Phytoalexins and their elicitors—a defense against microbial infection in plants. Ann Rev Plant Physiol 35: 243-298
175 Darvill A, McNeil M, Albersheim P, Delmer DP (1980) The primary cell wall of flowering

plants. In: Tolbert NE (ed) The biochemistry of plants, vol 1, the plant cell. Academic Press, New York, pp 91-162

176 Davidson N (1976) Neurotransmitter amino acids. Academic Press, London

177 Davies BH (1980, 1981) Carotenoid biosynthesis. In: Czygan FC (ed) Pigments in plants. Fischer, Stuttgart and Akademie Verlag, Berlin, pp 31-56

178 Davies BH (1985) Carotenoid metabolism in animals—a biochemists review. Pure Appl Chem 57: 679-684

179 Davies HG, Green RH (1986) Avermectins and milbemycins. Nat Product Reports 3: 87-121

180 De Rosa M, Gambacorta A, Nicolaus B, Sodano S (1982) Incorporation of labelled glycerols into ether lipids in *Caldariella acidophila*. Phytochemistry 21: 595-599

181 Dellaporta SL, Chomet PS (1986) The activation of maize controlling elements. Progr Chem Nat Prod 50: 169-216

182 Delmer DP (1987) Cellulose biosynthesis. Ann Rev Plant Physiol 38: 259-345

183 DeLuca HF (1983) The metabolism of vitamin A and its functions. In: Nutritional factors in the induction and maintenance of malignancy. Academic Press, New York, pp 149-167

184 Demain AL (1982) Catabolite regulation in industrial microbiology. In: Krumphanzl V, Sikyta B, Vanek Z (eds) Overproduction of microbial products. Academic Press, London, pp 3-20

185 Demain AL (1976) Genetic regulation of fermentation organisms. Stadler Genet Symp 8: 41-55

186 Demain AL (1980) Do antibiotics function in nature? Search 11: 148-151

187 Deus-Neumann B, Zenk MH (1986) Accumulation of alkaloids in plant vacuoles does not involve an ion-trap mechanism. Planta 167: 44-53

188 Dev S (ed) (1985, 1986) Handbook of terpenoids: diterpenoids, vols 1-4. CRC Press, Boca Raton

189 Dev S, Narula ASP, Vadav JS (1982) Handbook of terpenoids, monoterpenoids, vols 1 and 2. CRC Press, Boca Raton

190 Dewick PM (1984-1988) The biosynthesis of shikimate metabolites. Nat Product Reports 1: 451-469, 2: 495-511, 3: 565-585, 5: 73-98

191 Dewick PM (1984) The biosynthesis of cyanogenic glycosides and glucosinolates. Nat Product Reports 1: 545-549

192 Dey PM (1980) Biochemistry of α-D-glycosidic linkages in the plant kingdom. Adv Carbohydr Chem Biochem 37: 283-372

193 Dey PM, Brinson K (1984) Plant cell walls. Adv Carbohydr Chem Biochem 42: 265-382

194 Dey PM, del Campillo E (1984) Biochemistry of the multiple forms of glycosidases in plants. Adv Enzymol 56: 141-249

195 Dey PM, Dixon RA (eds) (1985) Biochemistry of storage carbohydrates in plants. Academic Press, Orlando

196 Dhar DN (1981) The chemistry of chalcones and related compounds. Wiley, New York

197 Die aktivierte Essigsäure und ihre Folgen: Autobiographische Beiträge von Schülern und Freunden Feodor Lynens; gewidmet von den Autoren anläßlich seines 65. Geburtstages. (1976) de Gruyter, Berlin

198 Dimroth F, Walter H, Lynen F (1970) Biosynthese von 6-Methylsalicylsäure. Eur J Biochem 13: 98-110

199 Diplock AT (ed) (1985) Fat soluble vitamins: their biochemistry and applications. Heinemann, London

200 Dirscherl W (1970) Biochemie der Vanillinsäure. Arzneimittelforschung 20: 405-409

201 Döring HP (1985) Plant transposable elements. Bio Essays 3: 164-171

202 Döring HP, Starlinger P (1986) Molecular genetics of transposable elements in plants. Ann Rev Genetics 20: 175-200

203 Dolphin D, Avramovic O, Poulson R (1987) Pyridine nucleotide coenzymes: chemical, biochemical, and medical aspects, parts A and B. Wiley, New York

204 Dolphin D, Poulsen R, Avramovic O (eds) (1986) Vitamin B_6, pyridoxal phosphate: chemical, biochemical, and medical aspects. Wiley, New York

205 Dreyer DL (1986) Some structural and stereochemical aspects of coumarin biosynthesis. Rec Adv Phytochem 20: 317-337

206 Dürckheimer W, Blumbach J, Lattrell R, Scheunemann KH (1985) Recent developments in the field of β-lactam antibiotics. Angew Chem Int Ed Engl 24: 180-202

207 Dufy-Barbe L (1985) Hypothalamic hormones. Endeavour New Series 9: 42-51

208 Duke JA (1985) CRC Handbook of medicinal herbs. CRC Press, Boca Raton

209 Durbin RD (1981) Toxins in plant disease. Academic Press, New York
210 Ehrlich PR, Raven PH (1965) Butterflies and plants: a study in co-evolution. Evolution 18: 586–608
211 Eibl H (1984) Phospholipids as functional constituents of biomembranes. Angew Chem Int Ed Engl 23: 257–271
212 Elix JA, Whitton AA (1984) Recent progress in the chemistry of lichen substances. Prog Chem Org Nat Prod 45: 103–234
213 Emboden WA (1979) Narcotic plants of the world. Macmillan, New York
214 Engelbrecht L (1976) Phytohormone excretion by insects. Nova Acta Leopoldina, Suppl 7: 503–505
215 Epstein WW, Poulter CD (1973) A survey of some irregular monoterpenes and their biogenetic analogies to presqualene alcohol. Phytochemistry 12: 737–747
216 Erman WF (1985) Chemistry of the monoterpenes—an encyclopedic handbook. Marcell Dekker, New York
217 Etten CH van, Tookey HL (1979) Chemistry and biological effects of glucosinolates. In: Rosenthal GA, Janzen DH (eds) Herbivores, their interaction with secondary plant metabolites. Academic Press, New York, pp 471–500
218 Evans EA (1974) Tritium and its compounds. Butterworth, London
219 Evans EA, Muramatsu M (eds) (1977) Radiotracer techniques and application. Marcell Dekker, New York
220 Evans LS, Tramontano WA (1984) Trigonelline and promotion of cell arrest in G2 of various legumes. Phytochemistry 23: 1837–1840
221 Everse J, Anderson B, You K (eds) (1982) The pyridine nucleotide enzymes. Academic Press, New York
222 Eyre DR, Paz MA, Gallop PM (1984) Cross-linking in collagen and elastin. Ann Rev Biochem 53: 717–748
223 Fässler A, Kobelt A, Pfaltz A, Eschenmoser A, Bladon C, Battersby AR, Thauer EK (1985) Zur Kenntnis des Faktors F_{430} aus methanogenen Bakterien: absolute Konfiguration. Helv Chim Acta 68: 2287–2298
224 Fangmeier N, Leistner E (1981) Conversion of D-lysine into L-lysine via L-pipecolic acid in *Nicotiana glauca* L. plants and cell-suspension cultures. J Chem Soc Perkin I: 1769–1772
225 Farkas L, Gabor M, Kallay F (eds) (1986) Flavonoids and bioflavonoids 1985. Elsevier, Amsterdam
226 Faulkner DJ (1984–1987) Marine natural products. Nat Product Reports 1: 251–280, 551–598, 3: 1–33, 4: 539–576
227 Feingold DS, Avigad G (1980) Sugar nucleotide transformation in plants. In: Preiss J (ed) The biochemistry of plants, vol 3, carbohydrates, structure and function. Academic Press, New York, pp 101–170
228 Fengel D, Wegener G (1984) Wood—chemistry, ultrastructure, reactions. de Gruyter, Berlin
229 Fischer NH, Olivier EJ, Fischer HD (1979) The biogenesis and chemistry of sesquiterpene lactones. Prog Chem Org Nat Prod 38: 47–390
230 Flores HE, Hoy MW, Pickard JJ (1987) Secondary metabolites from root cultures. Trends Biotechnology 5: 64–69
231 Florkin M, Scheer BJ (1967–79) Chemical zoology, vol 1–11. Academic Press, New York
232 Florkin M, Stotz EH (eds) (1975) Comprehensive biochemistry, vol 31, a history of biochemistry part III. Elsevier/North Holland, Amsterdam
233 Floss HG (1980) The biosynthesis of ergot alkaloids (or the story of the unexpected). In: Phillipson JD, Zenk MH (eds) Indole and biogenetically related alkaloids. Academic Press, New York, pp 249–270
234 Floss HG (1981) Stereochemical aspects of natural products biosynthesis. In: Conn EE (ed) The biochemistry of plants, vol 7, secondary plant products. Academic Press, New York, pp 177–214
235 Floss HG (1972) Instrumental methods in biosynthetic studies. Lloydia 35: 399–417
236 Floss HG, Anderson JA (1980) Biosynthesis of ergot toxins. In: Steyn PS (ed) The biosynthesis of mycotoxins. Academic Press, New York, pp 17–67
237 Floss HG, Keller PJ, Le Van Q, Nielsen P, Neuberger G, Bacher A (1986) Biosynthesis of riboflavin. In: Atta-ur-Rahmann (ed) Natural product chemistry. Springer, Berlin Heidelberg New York, pp 108–120

238 Floss HG, Robbers JE, Heinstein PF (1974) Regulatory control mechanisms in alkaloid biosynthesis. Recent Adv Phytochem 8: 141-175

239 Fodor GB (1980) Alkaloids derived from phenylalanine and tyrosine. In: Bell EA, Charlwood BV (eds) Encyclopedia of plant physiology, new series, vol 8, secondary plant products. Springer, Berlin Heidelberg New York, pp 92-127

240 Fodor GB (1980) Alkaloids derived from histidine and other precursors. In: Bell EA, Charlwood BV (eds) Encyclopedia of plant physiology, new series, vol 8, secondary plant products. Springer, Berlin Heidelberg New York, pp 160-166

241 Fodor G, Dharanipragada R (1984-1988) Tropane alkaloids. Nat Product Reports 1: 231-234, 2: 221-225, 3: 181-184, 5: 67-72

242 Forest JC (1978) Properties of plant aminotransferases. Phytochemistry 17: 1455-1471

243 Foster JW (1949) Chemical activities of the fungi. Academic Press, New York

244 Fowden L (1981) Nonprotein amino acids. In: Conn EE (ed) The biochemistry of plants, vol 7, secondary plant products. Academic Press, New York, pp 215-247

245 Fowler MW (1986) Interactions and interrelationships between primary and secondary metabolism. In: Morris P, Scragg AH, Stafford A, Fowler MW (eds) Secondary metabolism in plant cell cultures. Cambridge University Press, New York Cambridge, pp 103-107

246 Fraenkel GS (1959) The raison d'être of secondary plant substances. Science 129: 1466-1470

247 Frage BM (1984-1987) Natural sesquiterpenoids. Nat Product Reports 1: 105-169, 2: 147-161, 3: 273-296, 4: 473-498

248 Franck B (1980) The biosynthesis of ergochromes. In: Steyn PS (ed) The biosynthesis of mycotoxins. Academic Press, New York, pp 157-191

249 Franck B (1984) Mycotoxins from mold fungi—weapons of uninvited fellow-boarders of man and animal: structures, biological activity, biosynthesis, and precautions. Angew Chem Int Ed Engl 23: 493-505

250 Frenkel RA, McGarry JD (eds) (1980) Carnitine biosynthesis, metabolism and functions. Academic Press, New York

251 Friend J (1981) Plant phenolics, lignification and plant disease. Prog Phytochem 7: 197-261

252 Frohne D, Pfänder HJ (1986) Giftpflanzen. Wissenschaftliche Verlagsgesellschaft, Stuttgart

253 Fry SC (1986) Cross-linking of matrix polymers in the growing cell walls of Angiosperms. Ann Rev Plant Physiol 37: 165-186

254 Fuller JG (1968) The day of St. Anthony's fire. Macmillan, New York

255 Gale EF, Cundliffe E, Reynolds PE, Richmond MH, Waring MJ (1981) The molecular basis of antibiotic action. Wiley, London

256 Galliard T (1980) Degradation of acyl lipids: hydrolytic and oxidative enzymes. In: Stumpf PK (ed) The biochemistry of plants, vol 4, lipids: structure and function. Academic Press, New York, pp 85-116

257 Galliard T, Chan HWS (1980) Lipoxygenases. In: Stumpf PK (ed) The biochemistry of plants, vol 4, lipids: structure and function. Academic Press, New York, pp 131-161

258 Galneder E, Rueffer M, Wanner G, Tabata M, Zenk MH (1988) Alternative final steps in berberine biosynthesis in Coptis japonica cell cultures. Plant Cell Reports 7: 1-4

259 Geike F (1971) Zur Biochemie der Phosphonoaminosäuren. Naturwiss Rundschau 24: 335-340

260 Gersch D (1980) Metabolic regulation by cyclic AMP in macrolide antibiotic producing strains of *Streptomyces hygroscopicus*. Process Biochem 15: 21-25

261 Gheysen G, Dhaese P, Van Montagu M, Schell J (1985) DNA flux across genetic barriers: the grown gall phenomenon. In: Hohn B, Dennis ES (eds) Genetic flux in plants (advances in plant gene research, vol 2). Springer, Wien, pp 11-47

262 Gibbons GF, Mitropoulos KA, Myant NB (1982) Biochemistry of cholesterol. Elsevier Biomedical Press, Amsterdam

263 Gilbert LE, Raven PH (eds) (1975) Coevolution of animals and plants. University of Texas Press, Austin–London

264 Gilchrist DG, Kosuge T (1980) Aromatic amino acid biosynthesis and its regulation. In: Miflin BJ (ed) The biochemistry of plants, vol 5, amino acids and derivatives. Academic Press, New York, pp 507-531

265 Gill M, Steglich W (1987) Pigments of fungi (Macromycetes). Prog Chem Org Nat Prod 51: 1-317

266 Giovanelli J, Mudd SH, Datko AH (1980) Sulfur amino acids in plants. In: Miflin BJ (ed) The

biochemistry of plants, vol 5, amino acids and derivatives. Academic Press, New York, pp 453-505

267 Givan CV, Harwood JL (1976) Biosynthesis of small molecules in chloroplasts of higher plants. Biol Rev 51: 365-406

268 Glasby JS (1977) Encyclopedia of the alkaloids, vol 1-3. Plenum Press, New York

269 Glazer AN (1982) Phycobilisomes: structure and dynamics. Ann Rev Microbiol 36: 173-198

270 Goldstein IS (1981) Organic chemicals from biomass. CRC Press, Boca Raton

271 Goldsworthy GJ, Wheeler CH (1985) Neurosecretory hormones in insects. Endeavour New Series 9: 139-143

272 Golebiewski WM, Spenser ID (1984) ^{2}H NMR Spectroscopy as a probe of the stereochemistry of biosynthetic reactions: the biosynthesis of lupinine. J Am Chem Soc 106: 1441-1442

273 Gooday GW (1983) Hormones and sexuality in fungi. In: Bennet JW, Ciegler A (eds) Secondary metabolism and differentiation in fungi. Marcell Dekker, New York, pp 239-266

274 Goodwin TW (1986) Metabolism, nutrition, and function of carotenoids. Ann Rev Nutr 6: 273-297

275 Goodwin TW (1980, 1984) The biochemistry of the carotenoids, vols 1 and 2. Chapman and Hall, London

276 Gorham J (1980) The stilbenoids. Prog Phytochem 6: 203-252

277 Goto T (1987) Structure, stability and color variation of natural anthocyanins. Prog Chem Org Nat Prod 52: 113-158

278 Gottlieb D (1976) The production and the role of antibiotics in soil. J Antibiot 29: 987-991

279 Graebe JE (1980) GA-biosynthesis: the development and application of cell-free systems for biosynthetic studies. In: Skoog F (ed) Plant growth substances. Springer, Berlin Heidelberg New York, pp 180-187

280 Graebe JE (1987) Gibberellin biosynthesis and control. Ann Rev Plant Physiol 38: 419-465

281 Gräfe U, Schlegel R, Bergholz M (1984) Polyether antibiotica. Pharmazie 39: 661-670

282 Granick S, Beale SI (1978) Hemes, chlorophylls, and related compounds: biosynthesis and metabolic regulation. Adv Enzymol 46: 33-203

283 Gray CH (ed) (1979) Hormones in blood, vol 1-3. Academic Press, New York

284 Grayson DH (1984-1987) Monoterpenoids. Nat Product Reports 1: 319-337, 3: 251-272, 4: 377-397

285 Green MB, Hedin PA (eds) (1986) Natural resistance of plants to pests — roles of allelochemicals. American Chemical Society, Washington

286 Grein A (1987) Antitumor anthracyclines produced by *Streptomyces pseucetius*. Adv Appl Microbiol 32: 203-214

287 Griffith OW (1986) β-Amino acids: mammalian metabolism and utility as α-amino acid analogs. Ann Rev Biochem 55: 855-878

288 Grisebach H (1981) Lignins. In: Conn EE (ed) The biochemistry of plants, vol 7, secondary plant products. Academic Press, New York, pp 457-478

289 Grisebach H (1980) Branched-chain sugars: occurrence and biosynthesis. In: Preiss J (ed) The biochemistry of plants, vol 3, carbohydrates: structure and function. Academic Press, New York, pp 171-197

290 Grisebach H (1980) Biochemische Genetik der Flavonoide. Biologie in unserer Zeit 10: 52-56

291 Grisebach H (1983) Phytoalexins. In: Bennet JW, Ciegler A (eds) Secondary metabolism and differentiation in fungi. Marcell Dekker, New York, pp 377-428

292 Gröger D (1980) Alkaloids derived from tryptophan and anthranilic acid. In: Bell EA, Charlwood BV (eds) Encyclopedia of plant physiology, new series, vol 8, secondary plant products. Springer, Berlin Heidelberg New York, pp 128-159

293 Gröger D (1985) Alkaloids derived from tryptophan. In: Mothes K, Schütte HR, Luckner M (eds) Biochemistry of alkaloids. Deutscher Verlag der Wissenschaften, Berlin and Verlag Chemie, Weinheim, pp 272-313

294 Gross D (1985) Alkaloids derived from nicotinic acid. In: Mothes K, Schütte HR, Luckner M (eds) Biochemistry of alkaloids. Deutscher Verlag der Wissenschaften, Berlin and Verlag Chemie, Weinheim, pp 163-187

295 Gross D (1980) Wachstumsregulatorisch wirksame Pflanzeninhaltsstoffe. Z Chemie 20: 397-406

296 Gross D, Schütte HR, Schreiber K (1985) Isoprenoid alkaloids. In: Mothes K, Schütte HR,

Luckner M (eds) Biochemistry of alkaloids. Deutscher Verlag der Wissenschaften, Berlin and Verlag Chemie, Weinheim, pp 354-384

297 Gross GG (1981) Phenolic acids. In: Conn EE (ed) The biochemistry of plants, vol 7, secondary plant products. Academic Press, New York, pp 301-316

298 Gross GG (1979) Recent advances in the chemistry and biochemistry of lignin. Rec Adv Phytochem 12: 177-220

299 Gross J (1987) Pigments in fruits. Academic Press, New York

300 Große W (1982) Function of serotonin in seeds of walnuts. Phytochemistry 21: 819-822

301 Grue-Sorensen G, Spenser ID (1983) Deuterium nuclear magnetic resonance spectroscopy as a probe of the stereochemistry of biosynthetic reactions: the biosynthesis of retronecine. J Am Chem Soc 105: 7401-7404

302 Grundon MF (1984–1987) Indolizidine and quinolizidine alkaloids. Nat Product Reports 1: 349-353, 2: 235-243, 4: 415-422

303 Grundon MF (1984-1988) Quinoline, quinazoline, and acridone alkaloids. Nat Product Reports 1: 195-200, 2: 393-400, 4: 225-236, 5: 293-308

304 Grundon MF (1984-1987) Amaryllidaceae alkaloids. Nat Product Reports 1: 247-250, 2: 249-251, 4: 89-94

305 Grunwald C (1980) Steroids. In: Bell EA, Charlwood BV (eds) Encyclopedia of plant physiology, new series, vol 8, secondary plant products. Springer, Berlin Heidelberg New York, pp 221-256

306 Guinaudeau H (1984) Dimeric aporphinoid alkaloids. J Nat Products 47: 565–580

307 Gunsalus IC, Pederson TC, Sliger SG (1975) Oxygenase-catalyzed biological hydroxylations. Ann Rev Biochem 44: 377-388

308 Gunstone FD (1984, 1987) Fatty acids and glycerides. Nat Product Reports 1: 483–497, 4: 95-112

309 Gunstone FD, Harwood JL, Padley FB (eds) (1986) The lipid handbook. Chapman and Hall, London

310 Habermehl G (1987) Gift-Tiere und ihre Waffen. Springer Verlag, Berlin Heidelberg New York Tokyo

311 Habermehl G (1981) Venomous animals and their toxins. Springer, Berlin Heidelberg New York

312 Habermehl G (1985) Giftpflanzen und Pflanzengifte in Mitteleuropa. Springer Verlag, Berlin Heidelberg New York Tokyo

313 Hager A (1980, 1981) The reversible, light-induced conversions of xanthophylls in the chloroplast. In: Czygan FC (ed) Pigments in plants. Fischer, Stuttgart and Akademie Verlag, Berlin

314 Hahlbrock K (1981) Flavonoids. In: Conn EE (ed) The biochemistry of plants, vol 7, secondary plant products. Academic Press, New York, pp 425-456

315 Hahn FE (ed) (1979) Antibiotics, vol 5, mechanism of action. Springer, Berlin Heidelberg New York

316 Hammond SM, Lambert PA, Rycroft AN (1984) The bacterial cell surface. Croom Helm, London

317 Hanahan DJ (1986) Platelet activating factor: a biologically active phosphoglyceride. Ann Rev Biochem 55: 483-509

318 Hanson JR (1984) The biosynthesis of C_5-C_{20} terpenoid compounds. Natural Product Reports 1: 443-449

319 Hanson JR (1986) Sesterterpenoids. Nat Product Reports 3: 123-152

320 Hanson JR (1984-1988) Diterpenoids. Nat Product Reports 1: 171-180, 339-348, 533-544, 3: 307-322, 4: 399-413, 5: 211-228

321 Hanson KR, Havir EA (1981) Phenylalanine ammonia-lyase. In: Conn EE (ed) The biochemistry of plants, vol 7, secondary plant products. Academic Press, New York, pp 577-625

322 Hanson KR, Havir EA (1979) An introduction to the enzymology of phenylpropanoid biosynthesis. Recent Adv Biochem 12: 91-137

323 Harborne JB (1988) Introduction to ecological biochemistry. Academic Press, New York

324 Harborne JB (1980) Plant phenolics. In: Bell EA, Charlwood BV (eds) Encyclopedia of plant physiology, new series, vol 8, secondary plant products. Springer, Berlin Heidelberg New York, pp 329-402

325 Harborne JB (1986) Recent advances in chemical ecology. Nat Product Reports 3: 323-344

326 Harborne JB (1977) Flavonoid sulphates: a new class of natural products of ecological significance in plants. Prog Phytochem 4: 189-208

327 Harborne JB (ed) (1988) The flavonoids, Chapman and Hall, London
328 Harborne JB, Mabry TJ, Mabry H (eds) (1975) The flavonoids. Chapman and Hall, London
329 Harborne JB, Turner BL (1984) Plant chemosystematics. Academic Press, New York
330 Harding RW, Shropshire W (1980) Photocontrol of carotenoid biosynthesis. Ann Rev Plant Physiol 31: 217-238
331 Harlin MM (1987) Allelochemistry in marine macroalgae. CRC Crit Rev Plant Sci 5: 237-249
332 Harrison DM (1984, 1986) Steroidal alkaloids. Nat Product Reports 1: 219-224, 3: 443-449
333 Harrison DM (1985) The biosynthesis of triterpenoids and steroids. Nat Product Reports 2: 525-560
334 Harrison DM (1986) The biosynthesis of carotenoids. Nat Product Reports 3: 205-215
335 Harwood JL (1988) Fatty acid metabolism. Ann Rev Plant Physiol, Plant Mol Biol 39: 101-138
336 Hasapis X, Mac Leod AJ (1982) Benzylglucosinolate degradation in heat-treated *Lepidium sativum* seeds and detection of a thiocyanate-forming factor. Phytochemistry 21: 1009-1013
337 Haslam E (1981) Vegetable tannins. In: Conn EE (ed) The biochemistry of plants, vol 7, secondary plant products. Academic Press, New York, pp 527-556
338 Haslam E (1986) Hydroxybenzoic acids and the enigma of gallic acid. Rec Adv Phytochem 20: 163-200
339 Haslam E (1985) Metabolites and metabolism – a commentary on secondary metabolism. Claredon Press, Oxford
340 Haslam E (1986) Secondary metabolism – fact and fiction. Nat Product Reports 3: 217-249
341 Haslam E (1982) The metabolism of gallic acid and hexahydroxydiphenic acid in higher plants. Prog Chem Org Nat Prod 41: 1-46
342 Haslam E, Lilley TH (1986) Polyphenol complexation. In: Farkas L, Gabor M, Kallay F (eds) Flavonoids and bioflavonoids 1985. Elsevier, Amsterdam, pp 113-138
343 Hastings JW (1983) Biological diversity, chemical mechanism, and the evolutionary origins of bioluminescent systems. J Mol Evol 19: 309-321
344 Hastings JW, Nealson KH (1977) Bacterial luminescence. Ann Rev Microbiol 31: 549-595
345 Hayashi N, Komae H (1980) Components of the ant secretions. Biochem Syst Ecol 8: 293-295
346 Heathcote JG (1978) Aflatoxin: chemical and biological aspects. Elsevier, Amsterdam
347 Hegnauer R (1986) Phytochemistry and plant taxonomy – an essay on the chemotaxonomy of higher plants. Phytochemistry 25: 1519-1535
348 Hegnauer R (1962-1973) Chemotaxonomie der Pflanzen, vols 1-6. Birkhäuser, Basel
349 Heide L, Arendt S, Leistner E (1982) Enzymatic synthesis, characterization, and metabolism of the coenzyme A ester of o-succinylbenzoic acid, an intermediate in menaquinone (vitamin K_2) biosynthesis. J Biol Chem 257: 7396-7400
350 Heinemeyer W, Buchmann I, Tonge DW, Windass JD, Alt-Moerbe J, Weiler EW, Botz T, Schröder J (1987) Two *Agrobacterium tumefaciens* genes for cytokinin biosynthesis: Ti plasmid-coded isopentenyltransferases adapted for function in prokaryotic or eukaryotic cells. Mol Gen Genet 210: 156-164
351 Heller W, Tamm C (1981) Homoisoflavones and biogenetically related compounds. Prog Chem Org Nat Prod 40: 105-152
352 Hemmerich P (1976) The present status of flavin and flavocoenzyme chemistry. Prog Chem Org Nat Prod 33: 451-517
353 Herbert E (1981) Discovery of pro-opiomelanocortin a cellular polyprotein. TIBS, pp 184-188
354 Herbert RB (1984-1987) The biosynthesis of plant alkaloids and nitrogenous microbial metabolites. Nat Product Reports 1: 181-193, 2: 163-179, 3: 185-203, 4: 423-440
355 Herrmann K (1978) Hydroxyzimtsäuren und Hydroxybenzoesäuren enthaltende Naturstoffe in Pflanzen. Prog Chem Org Nat Prod 35: 73-132
356 Higuchi T (ed) (1985) Biosynthesis and biodegradation of wood components. Academic Press, Orlando
357 Hill AF (1952) Economic botany, a textbook of useful plants and plant products. McGraw Hill, New York
358 Hirono I (ed) (1987) Naturally occurring carcinogens of plant origin. Elsevier, Amsterdam
359 Hlavka JJ, Boothe JH (1985) The tetracyclines. Springer, Berlin Heidelberg New York
360 Hodkinson A (1977) Oxalic acid in biology and medicine. Academic Press, London
361 Hölzl J, Bancher E (1965) Bau und Eigenschaften der organischen Naturstoffe; Einführung in die organische Rohstofflehre. Springer, Wien

362 Hösel W (1981) Glycosylation and glycosidases. In: Conn EE (ed) The biochemistry of plants, vol 7, secondary plant products. Academic Press, New York, pp 725-753
363 Hoffmann J (ed) (1984) Biosynthesis, metabolism and mode of action of invertebrate hormones. Springer, Berlin Heidelberg New York
364 Hohn B, Dennis ES (eds) (1985) Genetic flux in plants. Springer, Wien
365 Hohn T, Schell J (eds) (1987) Plant DNA infectious agents. Springer, Wien New York
366 Holloway PJ (1983) Some variations in the composition of suberin from the cork layers of higher plants. Phytochemistry 22: 495-502
367 Hollstein U, Mock DL, Sibbitt RR, Roisch U, Lingens F (1978) Incorporation of shikimic acid into iodinin. Tetrahedron Lett 2987-2990
368 Holtzer H, Bischoff R, Chacko S (1969) Activities of the cell surface during myogenesis and chondrogenesis. In: Smith RT, Good RA (eds) Cellular recognition developmental immunology workshop, Appleton-Century-Crofts, New York, pp 19-25
369 Holzapfel CW (1980) The biosynthesis of cyclopiazonic acid and related tetramic acids. In: Steyn PS (ed) The biosynthesis of mycotoxins. Academic Press, New York, pp 327-355
370 Hoober JK (1987) The molecular basis of chloroplast development. In: Hatch MD, Boardman NK (eds) The biochemistry of plants, vol 10, photosynthesis, Academic Press, San Diego, pp 1-74
371 Hooykaas PJJ, Schilperoort RA (1984) The molecular genetics of crown gall tumorigenesis. Adv Genetics 22: 209-283
372 Hopwood DA (1981) Genetic studies of antibiotics and other secondary metabolites. In: Glover SW, Hopwood DA (eds) Genetics as a tool in microbiology. Cambridge University Press, Cambridge, pp 187-218
373 Hürlimann H, Cherbuliez E (1981, 1985) Konstitution und Vorkommen der organischen Pflanzenstoffe (exclusive Alkaloide), Ergänzungsband 2, Teile 1 and 2. Birkhäuser, Basel
374 Hütter R, Leisinger T, Nüesch J, Wehrli W (1978) Antibiotics and other secondary metabolites. Academic Press, London
375 Husson HP (1985) Simple indole alkaloids including β-carbolines and carbazoles. In: Brossi A (ed) The alkaloids, Vol 26, Academic Press, New York, pp 1-51
376 Hutchinson CR (1986) Biological methods for studying the biosynthesis of natural products. Nat Product Reports 3: 133-152
377 Hutchinson CR (1980) Biosynthetic studies of antitumor indole alkaloids. In: Phillipson JD, Zenk MH (eds) Indole and biogenetically related alkaloids. Academic Press, New York, pp 143-158
378 Huxtable R, Barbeau A (eds) (1976) Taurine. Raven Press, New York
379 Ikegami F, Kaneko M, Lambein F, Kuo YH, Murakoshi I (1987) Difference between uracilylalanine synthases and cysteine synthases in Pisum sativum. Phytochemistry 26: 2699-2704
380 Ingham JL (1983) Naturally occurring isoflavonoids. Prog Chem Org Nat Prod 43: 1-266
381 Ingham JL (1973) Disease resistance in plants: the concept of pre-infectional and post-infectional resistance. Phytopathol Z 78: 314-335
382 Inouye H, Uesato S (1986) Biosynthesis of iridoids and secoiridoids. Prog Chem Org Nat Prod 50: 169-236
383 Iversen LL, Goodman EC (eds) (1986) Fast and slow chemical signalling in the nervous system. Oxford University Press, Oxford
384 Iwanow A, Hill RE, Sayer BG, Spenser ID (1984) Biosynthesis of vitamin B_6: incorporation of a C-N unit derived from glycine. J Am Chem Soc 106: 1840-1841
385 Jacob F (1977) Evolution and tinkering. Science 196: 1161-1166
386 Jacob J (1977) Bürzeldrüsenlipide. Prog Chem Nat Prod 34: 373-438
387 Jacoby WB, Griffith OW (eds) (1987) Methods in enzymology, vol 143, sulfur and sulfur amino acids. Academic Press, Orlando
388 Jaenicke L, Boland W (1982) Signalstoffe und ihre Reception im Sexualcyclus mariner Braunalgen. Angew Chem 94: 659-670
389 Jakoby WB (1978) The glutathione S-transferases: a group of multifunctional detoxification proteins. Adv Enzymol 46: 383-414
390 Jakubke HD, Jeschkeit H (1982) Aminosäuren, Peptide, Proteine. Akademie Verlag, Berlin
391 Jaroszewski JW, Ettlinger MG (1981) Ring cleavage of phenols in higher plants: biosynthesis of triglochinin. Phytochemistry 20: 819-821

392 Jeffery J (ed) (1980) Dehydrogenases. Birkhäuser, Basel
393 Jefford CW, Cadby PA (1981) Molecular mechanisms of enzyme-catalyzed dioxygenation. Prog Chem Org Nat Prod 40: 191-265
394 Jente R, Olatunji A, Bosold F (1981) Formation of natural thiophene derivatives from acetylenes by *Tagetes patula*. Phytochemistry 20: 2169-2175
395 Johne S (1984) The quinazoline alkaloids. Prog Chem Org Nat Prod 46:159-229
396 Johne S (1986) Quinazoline alkaloids. In: Brossi A (ed) The alkaloids, vol 29. Academic Press,New York, pp 99-140
397 Johnson M, Caray F, McMillan RM (1983) Alternative pathways of arachidonate metabolism: prostaglandins, thromboxanes and leucotrienes. Essays Biochem 19: 40-141
398 Johnson RG, Carlson NJ, Scappa A (1978) ΔpH and catecholamine distribution in isolated chromaffin granules. J Biol Chem 253: 1512-1521
399 Jones BJ, Porter JW (1986) Biosynthesis of carotenes in higher plants. CRC Crit Rev Plant Sci 3: 295-324
400 Jones DH (1984) Phenylalanine ammonia-lyase: regulation of its induction and its role in plant development. Phytochemistry 23: 1349-1359
401 Jones H, Rasmusson GH (1979) Recent advances in the biology and chemistry of vitamin D. Prog Chem Org Nat Prod 39: 63-121
402 Jones KL, Jones SE (1984) Fermentations involved in the production of cocoa, coffee, and tea. Progr Ind Microbiol 19: 411-456
403 Jones RCF (1984) Macrocyclic microbial metabolites. Nat Product Reports 1: 87-103
404 Journal of natural products (Lloydia) (1938ff) American Society of Pharmacognosy and the Lloyd Library and Museum
405 Kametani T, Honda T (1985) Aporphine alkaloids. In: Brossi A (ed) The alkaloids, vol 24. Academic Press, New York, pp 153-251
406 Kamiya Y, Sakurai A (1981) Mating pheromones of heterobasidiomycetous yeasts. Naturwissenschaften 68: 128-133
407 Kaneda M, Mizutani K, Tanaka K (1982) Lilioside C, a glyerol glucoside from Lilium lancifolium. Phytochemistry 21: 891-893
408 Kanfer JN, Hakomori S (1983) Handbook of lipid research, vol 3: sphingolipid biochemistry. Plenum Press, New York, pp 485
409 Karlson P (1976) Animal hormones as secondary plant products and their ecological relevance. In: Luckner M, Mothes K, Nover L (eds) Secondary metabolism and coevolution. Nova acta Leopoldina, Suppl 7: 423-432
410 Karlson P (1986) Warum sind so viele Hormone Steroide? Leopoldina (R. 3) 30. 1984: 97-106
411 Karlson P, Lüscher M (1959) Pheromone. Ein Nomenklaturvorschlag für eine Wirkstoffklasse. Naturwissenschaften 46: 63-64
412 Karrer W (1958) Konstitution and Vorkommen der organischen Pflanzenstoffe (exclusive Alkaloide). Birkhäuser, Basel
413 Karrer W, Cherbuliez E, Eugster CH (1977) Konstitution und Vorkommen der organischen Pflanzenstoffe (exclusive Alkaloide), Ergänzungsband 1. Birkhäuser, Basel
414 Kasai T, Larsen PO (1980) Chemistry and biochemistry of γ-glutamyl derivatives from plants including mushrooms (Basidiomycetes). Prog Chem Org Nat Prod 39: 173-285
415 Kaufman S (ed) (1987) Methods in enzymology, Vol 142: metabolism of aromatic amino acids and amines. Academic Press, Orlando
416 Kean EA, Lewis CE (1981) Biosynthesis of L-β-(methylenecyclopropyl)-alanine (hypoglycine) in *Blighia sapida*. Phytochemistry 20, 2161-2164
417 Keeler RF, Tu AT (1983, 1984) Handbook of natural toxins, vols 1 and 2. Marcell Dekker, New York
418 Kemp MS, Burden RS (1986) Phytoalexins and stress metabolites in the sapwood of trees. Phytochemistry 25: 1261-1269
419 Kennedy JF, Bradshaw IJ (1984) Production, properties and application of xanthan. Progr Ind Microbiol 19: 319-371
420 Kennedy JF, Phillips GO, Williams PA (eds) (1987) Wood and cellulosics: industrial utilization, biotechnology, structure, and properties. Halsted Press, New York
421 Kennedy JF, White CA (1983) Bioactive carbohydrates. Ellis Harwood, Chichester

514   References

422 Keys AJ (1980) Synthesis and interconversion of glycine and serine. In: Miflin BJ (ed) The biochemistry of plants, vol 5, amino acids and derivatives. Academic Press, New York, pp 359-374
423 Khokhlov AS (1982) Low molecular weight microbial bioregulators of secondary metabolism. In: Krumphanzl V, Sikyta B, Vanek Z (eds) Overproduction of microbial products. Academic Press, London, pp 98-109
424 Kindl H (1979) Compartments and microcompartments channelling intermediates in phenylpropanoid metabolism. In: Luckner M, Schreiber K (eds) Proc 12th FEBS Meeting, vol 55, regulation of secondary product and plant hormone metabolism. Pergamon Press, Oxford
425 Kinghorn AD, Soejarto DD (1986) Sweetening agents of plant origin. CRC Crit Rev Plant Sci 4: 79-120
426 Kirby GW, Robins DJ (1980) The biosynthesis of gliotoxin and related epipolythiodioxopiperazines. In: Steyn PS (ed) The biosynthesis of mycotoxins. Academic Press, New York, pp 301-326
427 Kleinkauf H, von Döhren H (1986) Enzyme systems synthesizing peptide antibiotics. In: Kleinkauf H, von Döhren H, Dornauer H, Nesemann G (eds) Regulation of secondary metabolite formation. VCH Verlagsgesellschaft, Weinheim, pp 173-207
428 Kleinkauf H, von Döhren H (1987) Biosynthesis of peptide antibiotics. Ann Rev Microbiol 41: 259-289
429 Kleinkauf H, von Döhren H (eds) (1982) Peptide antibiotics, biosynthesis and functions. de Gruyter, Berlin
430 Koch M, Giregson RP (1984) Brominated phlorethols and nonhalogenated phlorotannins from the brown alga *Cystophora congesta*. Phytochemistry 23: 2633-2637
431 Kochert G (1978) Sexual pheromones in algae and fungi. Ann Rev Plant Physiol 29: 461-486
432 Kollmann FFP, Cote WA (1984) Principles of wood science and technology. Springer Verlag, Berlin Heidelberg New York
433 Korpela T, Christen P (eds) (1987) Vitamin B_6 catalysis. Birkhäuser, Boston
434 Kossel A (1891) Über die chemische Zusammensetzung der Zelle. Arch Physiol, p 181
435 Kosterlitz AW (1987) Biosynthesis of morphine in the animal kingdom. Nature 330: 606
436 Kosuge T, Sanger M (1986) Indolacetic acid, its synthesis and regulation: a basis for tumorigenicity in plant disease. Rec Adv Phytochem 20: 147-161
437 Kramer KJ, Koga D (1986) Insect chitin, physical state, synthesis, degradation, and metabolic regulation. Insect Biochem 16: 851-877
438 Krebs HC (1986) Recent developments in the field of marine natural products with emphasis on biologically active compounds. Prog Chem Org Nat Prod 49: 151-363
439 Krieger DT, Barnstein MJ, Martin JB (1983) Brain peptides. Wiley, Chichester
440 Krogh P (ed) (1987) Mycotoxins in food. Academic Press, New York
441 Krogsgaard-Larsen P, Christensen SB, Kofod H (eds) (1984) Natural products and drug development. Munksgaard, Copenhagen
442 Kühn H, Schewe T, Rapoport SM (1986) The stereochemistry of the reactions of lipoxygenases and their metabolites. Proposed nomenclature of lipoxygenases and related enzymes. Adv Enzymol 58: 273-311
443 Kumar R (ed) (1984) Vitamin D: basic and clinical aspects. Martinus Nijhoff Publ., Boston
444 Kutney JP (1987) Studies in plant tissue culture. The synthesis and biosynthesis of indole alkaloids. Heterocycles 25: 617-640
445 Kwan-sa You (1985) Stereospecificity for nicotinamide nucleotides in enzymatic and chemical hydride transfer reactions. CRC Critical Reviews in Biochem 17: 313-451
446 Kydonieus AF, Beroza M (1982) Insect suppression with controlled release pheromone systems. CRC Press, Boca Raton
447 Lai SMF, Manley PW (1984) Prostaglandins, thromboxanes, leucotrienes, and related arachidonic acid metabolites. Nat Product Reports 1: 409-441
448 Lancini G, Parenti F (1982) Antibiotics. Springer, New York
449 Lands WEM, Smith WL (eds) (1982) Prostaglandins and arachidonate metabolites. Methods Enzymol, vol 86
450 Larsen PO (1981) Glucosinolates. In: Conn EE (ed) The biochemistry of plants, vol 7, secondary plant products. Academic Press, New York, pp 501-525
451 Larsen PO, Wieczorkowska E (1978) Transformation of 3-(3-carboxyl-phenyl)alanine in *Iris* species. Biochim Biophys Acta 542: 253-262

452 Laskin AI, Lechevalier HA (eds) (1981-1987) CRC Handbook of microbiology, vol 3, microbial composition: amino acids, proteins, and nucleic acids; vol 4, microbial composition: carbohydrates, lipids, and minerals; vol 5, microbial products; vol 8, toxins and enzymes. CRC Press, Boca Raton

453 Le Quesne PW, Purdy RH (1986) The oxydative metabolism of the aromatic A-ring of steroidal estrogens. In: Atta-ur-Rahmann (ed) Natural Product Chemistry, Springer, Berlin Heidelberg New York, pp 213-226

454 Leeper FJ (1985, 1987) The biosynthesis of porphyrins, chlorophylls, and vitamin B_{12}. Nat Product Reports 2: 19-47, 2: 561-580, 4: 441-469

455 Leete E (1980) Alkaloids derived from ornithine, lysine and nicotinic acid. In: Bell EA, Charlwood BV (eds) Encyclopedia of plant physiology, new series, vol 8, secondary plant products. Springer, Berlin Heidelberg New York, pp 65-91

456 Leete E, Bjorklund JA, Kim SH (1988) A new hypothesis for the biosynthesis of cocaine. Abstracts 16th Int Symp Chem Nat Prod, May 29-June 3, 1988, Kyoto

457 Leistner E (1981) Biosynthesis of plant quinones. In: Conn EE (ed) The biochemistry of plants, vol 7, secondary plant products. Academic Press, New York, pp 403-423

458 Leistner E (1986) Biosynthesis of iso-chorismate-derived quinones. Rec Adv Phytochem 20: 243-261

459 Lennarz WJ (ed) (1981) The biochemistry of glycoproteins and proteoglycans. Plenum Press, New York

460 Letham DS, Goodwin TW, Higgins TJV (eds) (1978) Phytohormones and related compounds – a comprehensive treatise, vol 2. Elsevier/North Holland, Amsterdam

461 Lewis DH (ed) (1984) Storage carbohydrates in vascular plants: distribution, physiology, and metabolism. Cambridge University Press, Cambridge

462 Lewis JR (1984-1986) Muscarine, imidazole, and peptide alkaloids, and other miscellaneous alkaloids. Nat Product Reports 1: 387-389, 2: 245-248, 3: 587-590

463 Liebig J (1843) Handbuch der Chemie mit Rücksicht auf Pharmacie. Heidelberg

464 Liebisch HW, Schütte HR (1985) Alkaloids derived from ornithine. In: Mothes K, Schütte HR, Luckner M (eds) Biochemistry of alkaloids. Deutscher Verlag der Wissenschaften, Berlin and Verlag Chemie, Weinheim, pp 106-127

465 Liebisch HW, Schütte HR (1985) Lysine-derived alkaloids. In: Mothes K, Schütte HR, Luckner M (eds) Biochemistry of alkaloids. Deutscher Verlag der Wissenschaften, Berlin and Verlag Chemie Weinheim, pp 128-162

466 Lindahl U, Feingold DS, Roden L (1986) Biosynthesis of heparin. Trends Biochem Sci 11: 221-225

467 List PH, Hörhammer L (eds) (1967-1980) Hagers Handbuch der Pharmazeutischen Praxis. Springer, Berlin Heidelberg New York

468 Loewus FA (1980) L-Ascorbic acid: metabolism, biosynthesis, function. In: Preiss J (ed) The biochemistry of plants, vol 3, carbohydrates: structure and function. Academic Press, New York, pp 77-99

469 Loewus FA, Loewus MW (1980) Myo-inositol: biosynthesis and metabolism. In: Preiss J (ed) The biochemistry of plants, vol 3, carbohydrates: structur and function. Academic Press, New York, pp 43-76

470 Loewus FA, Loewus MW (1987) Biosynthesis and metabolism of ascorbic acid in plants. CRC Crit Rev Plant Sci 5: 101-119

471 Loewus FA, Tanner W (eds) (1982) Plant carbohydrates I, intracellular carbohydrates (Encyclopedia of plant physiology, new series, vol 13 A). Springer, Berlin Heidelberg New York

472 Loomis WD, Croteau R (1980) Biochemistry of terpenoids. In: Stumpf PK (ed) The biochemistry of plants, vol 4, lipids: structure and function. Academic Press, New York, pp 363-418

473 Loomis WF (1967) Skin pigment regulation of vitamin D biosynthesis in man. Science 156: 501-506

474 Looser E, Baumann TW, Wanner H (1974) The biosynthesis of caffeine in the coffee plant. Phytochemistry 13: 2515-2518

475 Luckner M (1969) Chinoline. In: Mothes K, Schütte HR (eds) Biosynthese der Alkaloide. Deutscher Verlag der Wissenschaften, Berlin, pp 510-550

476 Luckner M (1985) Alkaloids formed from histidine. In: Mothes K, Schütte HR, Luckner M (eds)

516   References

Biochemistry of alkaloids. Deutscher Verlag der Wissenschaften, Berlin and Verlag Chemie, Weinheim, pp 351-353

477 Luckner M (1980) Expression and control of secondary metabolism. In: Bell EA, Charlwood BV (eds) Encyclopedia of plant physiology, new series, vol 8, secondary plant products. Springer, Berlin Heidelberg New York, pp 23-63

478 Luckner M (1982) The expression of secondary metabolism—an aspect of cell specialization. In: Nover L, Luckner M, Parthier B (eds) Cell differentiation. Fischer, Jena and Springer, Berlin, pp 408-425

479 Luckner M (1969) Der Sekundärstoffwechsel in Pflanze und Tier. Fischer, Jena

480 Luckner M (1971) Was ist Sekundärstoffwechsel? Pharmazie 26: 717-724

481 Luckner M (1980) Alkaloid biosynthesis in *Penicillium cyclopium*—does it reflect general features of secondary metabolism? J Nat Prod 43: 21-40

482 Luckner M, Diettrich B, Lerbs W (1980) Cellular compartmentation and channelling of secondary metabolism in microorganisms and higher plants. Progr Phytochem 6: 103-142

483 Luckner M, Johne S (1985) Alkaloids derived from anthranilic acid. In: Mothes K, Schütte HR, Luckner M (eds) Biochemistry of alkaloids. Deutscher Verlag der Wissenschaften, Berlin and Verlag Chemie Weinheim, pp 314-337

484 Luckner M, Mothes K, Nover L (eds) (1976) Secondary metabolism and coevolution: cellular, intercellular and interorganismic aspects. Nova Acta Leopoldina, Suppl 7. Barth, Leipzig

485 Luckner M, Nover L, Böhm H (1977) Secondary metabolism and cell differentiation (molecular biology, biochemistry and biophysics, vol 23). Springer, Berlin Heidelberg New York

486 Lugtenberg B (ed) (1987) Recognition in microbe-plant symbiotic and pathogenic interactions. Springer, Berlin Heidelberg New York

487 Lynch DR, Snyder SH (1986) Neuropeptides: multiple molecular forms, metabolic pathways, and receptors. Ann Rev Biochem 55: 773-799

488 Mabry TJ (1980) Betalains. In: Bell EA, Charlwood BV (eds) Encyclopedia of plant physiology, new series, vol 8, secondary plant products. Springer Berlin Heidelberg New York, pp 513-533

489 Mac Leod AJ, Rossiter JT (1985) The occurrence and activity of epithiospecifier protein in some Crucifereae seeds. Phytochemistry 24: 1895-1898

490 Machlin LJ (ed) (1980) Vitamin E: a comprehensive treatise. Marcell Dekker, New York

491 Macko V (1983) Structural aspects of toxins. In: Daly JM, Deverall BJ (eds) Toxins and plant pathogenesis. Academic Press, New York, pp 41-80

492 Macko V (1981) Inhibitors and stimulants of spore germination and infection structure formation in fungi. In: Turian G, Hohl HR (eds) The fungal spore: morphogenetic controls. Academic Press, New York, pp 565-584

493 Mah RA, Ward DM, Baresi L, Glass TL (1977) Biogenesis of methane. Ann Rev Microbiol 31: 309-314

494 Mahato SB, Ganguly AN, Sahu NP (1982) Steroid saponins. Phytochemistry 21: 959-978

495 Malik VS (1982) Genetics and biochemistry of secondary metabolism. Adv Appl Microbiol 28: 27-115

496 Mandava NB (1985) Handbook of natural pesticides: methods. CRC Press, Boca Raton

497 Mandelstam J (1976) Bacterial sporulation: a problem in the biochemistry and genetics of a primitive developmental system. Proc Roy Soc London B 193: 89-106

498 Mangold HK, Paltauf F (eds) (1983) Ether lipids: biochemical and biomedical aspects. Academic Press, New York

499 Mangold KH, Spener F (1980) Biosynthesis of cyclic fatty acids. In: Stumpf PK (ed) The biochemistry of plants, vol 4, lipids: structure and function. Academic Press, New York, pp 647-663

500 Manitto P (1981) Biosynthesis of natural products. E. Horwood, Chichester

501 Mann J (1987) Secondary metabolism. Oxford University Press, Oxford

502 Margna U (1977) Control at the level of substrate supply—an alternative in the regulation of phenylpropanoid accumulation in plant cells. Phytochemistry 16: 419-426

503 Markakis P (1982) Anthocyanins as food colors. Academic Press, New York

504 Marshall VP, Wiley PF (1986) Biomodification of antibiotics by Streptomyces. In: The Bacteria, vol 9. Academic Press, New York, pp 323-353

505 Martell AE (1982) Reaction pathways and mechanisms of pyridoxal catalysis. Adv Enzymol 53: 163-199

506 Martin RA, Lynch SP (1988) Cardenolide content and thin-layer chromatography profiles of monarch butterflies, *Danaus plexippus* L., and their larval host-plant milkweed, *Asclepias asperula* subsp. *capricornu*, in North Central Texas. J Chem Ecology 14: 295-318

507 Martin SF (1987) The Amaryllidaceae alkaloids. In: Brossi A (ed) The alkaloids, vol 30. Academic Press, New York, pp 251-376

508 Massiot G, Delaude C (1986) Pyrrolidine alkaloids. In: Brossi A (ed) The alkaloids, vol 27. Academic Press, New York, pp 269-322

509 Matile P (1980) Die Senfölbombe: Zur Kompartimentierung des Myrosinasesystems. Biochem Physiol Pflanzen 165: 722-731

510 Matile P (1984) Das toxische Kompartiment der Planzenzelle. Naturwissenschaften 71: 18-24

511 Mattocks AR (1986) Chemistry and toxicology of pyrrolizidine alkaloids. Academic Press, London

512 Mazliak P, Benveniste P, Costes C, Douce R (1980) Biogenesis and function of plant lipids. Elevier/North Holland, Amsterdam

513 McCourt P, Somerville CR (1987) The use of mutants for the study of plant metabolism. In: Davies DD (ed) The biochemistry of plants, vol 13, methodology. Academic Press, San Diego, pp 33-64

514 McLachlan D, Arnason T, Lam J (1986) Structure-function relationships in the phototoxicity of acetylenes from the Asteraceae. Biochem Systematics Ecology 14: 17-23

515 McNeil M, Darvill AG, Fry SC, Albersheim P (1984) Structure and function of the primary cell wall of plants. Ann Rev Biochem 53: 625-663

516 Mead JF, Alfin-Slater RB, Howton DR, Popjak G (1986) Lipids: chemistry, biochemistry and nutrition. Plenum Press, New York

517 Medicinal and aromatic plant abstracts (1979 ff) Publications and informations directorate, New Delhi

518 Menachery MD, Lavanier GL, Wetherly ML, Guinaudeau H, Shamma M (1986) Simple isoquinoline alkaloids. J Nat Prod 49: 745-778

519 Metcalf RL (1987) Plant volatiles as insect attractants. CRC Crit Rev Plant Sci 5: 251-301

520 Meyer AM, Harel E (1979) Polyphenol oxidases in plants. Phytochemistry 18: 193-215

521 Michael RP, Bonsell RW, Warner P (1974) Human vaginal secretions: volatile fatty acid content. Science 186: 1217-1219

522 Miflin BJ (1980) Histidine biosynthesis. In: Miflin BJ (ed) The biochemistry of plants, vol 5, amino acids and derivatives. Academic Press, New York, pp 533-539

523 Mikolajczak KL (1977) Cyanolipids. Prog Chem Fats Other Lipids 15: 97-130

524 Miller JR (1986) Insect-plant interactions. Springer, Berlin Heidelberg New York

525 Mirocha CJ (1972) Phytotoxins and metabolism. In: Wood RKS, Ballio A, Graniti A (eds) Phytotoxins in plant diseases. Academic Press, London, pp 191-209

526 Misato M, Ko K, Yamaguchi I (1977) Use of antibiotics in agriculture. Adv Appl Microbiol 21: 53-88

527 Mislivec PB (1981) Mycotoxin production by conidial fungi. In: Cole GT, Kendrick B (eds) Biology of conidial fungi, vol 2. Academic Press, New York, pp 37-74

528 Mitchell J, Rook A (1979) Botanical dermatology. Greenglass LTD, Vancouver

529 Molisch H (1937) Der Einfluß einer Pflanze auf die andere—Allelopathy. Fischer, Jena

530 Mondovi B (ed) (1985) Structure and functions of amine oxidases. CRC Press, Boca Raton

531 Money T (1985) Camphor: a chiral starting material in natural product synthesis. Nat Product Reports 2: 253-289

532 Moore TS (1982) Phospholipid biosynthesis. Ann Rev Plant Physiol 33: 235-259

533 Mori A, Cohen BD, Lowenthal A (eds) (1985) Guanidines: historical, biological, biochemical, and clinical aspects of the naturally occurring guanidino compounds. Plenum Press, New York

534 Mori A, Nishino C, Enoki N, Tawata S (1988) Cytotoxicity of plant flavonoids against HeLa cells. Phytochemistry 27: 1017-1020

535 Morin RB, Gorman M (eds) (1982) Chemistry and biology of β-lactam antibiotics. Academic Press, New York

536 Morris P, Scragg AH, Stafford A, Fowler MW (eds) (1986) Secondary metabolism in plant cell cultures. Cambridge University Press, Cambridge

518   References

537 Morris RO (1986) Genes specifying auxin and cytokinin biosynthesis in phytopathogens. Ann
     Rev Plant Physiol 37: 509-538
538 Morrison JF, Heyde E (1972) Enzymic phosphoryl group transfer. Ann Rev Biochem 41:
     29-54
539 Mothes K (1980) Historical introduction. In: Bell EA, Charlwood BV (eds) Encyclopedia of
     plant physiology, new series, vol 8, secondary plant products. Springer, Berlin Heidelberg New
     York, pp 1-10
540 Mothes K (1966) Zur Problematik der metabolischen Exkretion bei Pflanzen. Naturwissenschaf-
     ten 53: 317-323
541 Mothes K (1973) Pflanze und Tier; ein Vergleich auf der Ebene des Sekundärstoffwechsels. Sit-
     zungsber Österr Akad Wiss Math Naturwiss Klasse, Sonderheft Abt 1, 181: 1-37
542 Mothes K (1976) Secondary plant substances as materials for chemical high quality breeding in
     higher plants. In: Wallace JW, Mansell RL (eds) Biochemical interactions between plants and
     insects. Plenum Press, New York, pp 385-405
543 Mothes K, Luckner M (1985) Historical introduction. In: Mothes K, Schütte HR, Luckner M
     (eds) Biochemistry of alkaloids, Deutscher Verlag der Wissenschaften, Berlin and Verlag Che-
     mie, Weinheim, pp 15-20
544 Mothes K, Schütte HR, Luckner M (eds) (1985) Biochemistry of alkaloids. Deutscher Verlag der
     Wissenschaften, Berlin and Verlag Chemie, Weinheim
545 Muccino RR (1986) Synthesis and application of isotopically labelled compounds. Elsevier, Am-
     sterdam
546 Mudd JB (1980) Sterol interconversions. In: Stumpf PK (ed) The biochemistry of plants, vol 4,
     lipids: structure and function. Academic Press, New York, pp 509-534
547 Murakoshi I, Kaneko M, Koide C, Ikegami F (1986) Enzymatic synthesis of the neuroexcitatory
     amino acid quisqualic acid by cysteine synthase. Phytochemistry 25: 2759-2863
548 Murray RDH, Mendez J, Brown SA (1982) The natural coumarins. Wiley, Chichester
549 Musso H (1982) Über die Farbstoffe des Fliegenpilzes. Naturwissenschaften 69: 326-331
550 Muzarelli R, Jeuniaux C, Gooday GW (eds) (1986) Chitin in nature and technology. Plenum
     Press, New York
551 Nahrstedt A (1982) Strukturelle Beziehungen zwischen pflanzlichen und tierischen Sekundär-
     stoffen. Planta medica 44: 2-14
552 Naqui A, Chance B (1986) Reactive oxygen intermediates in biochemistry. Ann Rev Biochem
     55: 137-166
553 Natural product reports (1984 ff) Royal Society of Chemistry, London
554 Natural product updates (1987 ff) Royal Society of Chemistry, Nottingham
555 Nebert TW, Gonzalez FJ (1987) P_{450} genes: structure, evolution, regulation. Ann Rev Biochem
     56: 945-993
556 Needham AE (1974) The significance of zoochromes. Springer, Berlin Heidelberg New York
557 Neidleman SL, Geigert J (1987) Biological halogenation: role in nature, potential in industry.
     Endeavour New Series 11: 5-15
558 Neidleman SL, Geigert J (1986) Biohalogenation: principles, basic roles and applications. Ellis
     Horwood Limited, Chichester
559 Neilands JB (1981) Microbial iron compounds. Ann Rev Biochem 50: 715-731
560 Neilands JB (1982) Microbial envelope proteins related to iron. Ann Rev Microbiol 36:
     285-309
561 Neilands JB, Leong SA (1986) Siderophores in relation to plant growth and disease. Ann Rev
     Plant Physiol 37: 187-208
562 Nemethy EK (1984) Biochemicals as an energy resource. CRC Crit Rev Plant Sci 2: 117-129
563 Nes AW, Fuller G, Tsai LS (eds) (1984) Terpenoids in plants: biochemistry and function. Mar-
     cell Dekker, New York
564 Nes WD, Heftmann E (1981) A comparison of triterpenoids with steroids as membrane compo-
     nents. J Nat Prod (Lloydia) 44: 377-400
565 Nester EW, Gordon MP, Amasino RM, Yanofsky MF (1984) Crown gall: molecular and physio-
     logical analysis. Ann Rev Plant Physiol 35: 387-413
566 Neuberger A (1981) The metabolism of glycine and serine. Compr Biochem 19 A: 257-303
567 Neumann D (1985) Storage of alkaloids. In: Mothes K, Schütte HR, Luckner M (eds) Biochem-

istry of alkaloids. Deutscher Verlag der Wissenschaften, Berlin and Verlag Chemie, Weinheim, pp 49–55

568 Nichol CA, Smith GK, Duch DS (1985) Biosynthesis and metabolism of tetrahydrobiopterin and molybdopterin. Ann Rev Biochem 54: 729–764

569 Niedleham P, Turk J, Jadschik BA, Morrison AR, Lefkowith JB (1986) Arachidonic acid metabolism. Ann Rev Biochem 55: 69–102

570 Norman AW, Litwack G (1987) Hormones. Academic Press, New York

571 Norman AW, Schaefer K, Grigoleit HG, von Herrath D (eds) (1985) Vitamin D: chemical, biochemical and clinical update. de Gruyter, Berlin

572 Nover L (1982) Molecular basis of cell differentiation. In: Nover L, Luckner M, Parthier B (eds) Cell differentiation. Fischer, Jena and Springer, Berlin Heidelberg New York, pp 99–254

573 Nover L, Luckner M, Parthier B (eds) (1982) Cell differentiation. Fischer, Jena and Springer, Berlin Heidelberg New York

574 Nover L, Reinbothe H (1982) Biochemistry of gene expression. In: Nover L, Luckner M, Parthier B (eds) Cell differentiation, molecular basis and problems. Fischer, Jena and Springer, Berlin Heidelberg New York, pp 23–74

575 Nozaki M (1979) Oxygenases and dioxygenases. Top Curr Chem 68: 145–186

576 Nüesch J, Heim J, Treichler HJ (1987) The biosynthesis of sulfur-containing β-lactam antibiotics. Ann Rev Microbiol 41: 51–75

577 Omura S (ed) (1984) Macrolide antibiotics: chemistry, biology, and practice. Academic Press, Orlando

578 Ondetti MA, Cushman DW (1982) Enzymes of the renin-angiotensin system and their inhibitors. Ann Rev Biochem 51: 283–308

579 Ortiz de Montellano PR (1986) Cytochrome P_{450}: structure, mechanism, and biochemistry. Plenum Press, New York

580 Ott DG (1981) Synthesis with stable isotopes. Wiley, New York

581 Ott J (1976) Hallucinogenic plants of North America. Wingbow Press, Berkeley

582 Ourisson G, Rohmer M, Poralla K (1987) Procariotic hopanoids and other polyterpenoid sterol surrogates. Ann Rev Microbiol 41: 301–333

583 Ovchinnikov YA (1984) Bioorganic chemistry of polypeptide neurotoxins. Pure Appl Chem 56: 1049–1068

584 Ozawa T, Lilley TH, Haslam E (1987) Polyphenol interactions: astringency and the loss of astringency in ripening fruit. Phytochemistry 26: 2937–2942

585 Pace-Asciak C, Graustrom E (eds) (1983) Prostaglandins and related substances. Elsevier, Amsterdam

586 Paech K (1950) Biochemie and Physiologie der sekundären Pflanzenstoffe. Springer, Berlin Heidelberg New York

587 Panossian AG (1987) Search for prostaglandins and related compounds in plants. Prostaglandins 33: 363–381

588 Parker WL, O'Sullivan J, Sykes RB (1986) Naturally occurring monobactams. Adv Appl Microbiol 31: 181–205

589 Partington JR (1962) A history of chemistry, pp 231–233. Macmillan, London

590 Pasteels JM, Rowell-Rahier M, Braekman JC, Daloze D (1984) Chemical defenses in leaf beetles and their larvae: the ecological, evolutionary and taxonomic significance. Biochem System Ecology 12: 395–406

591 Patel AV, Blunder G, Crabb TA, Sauvaire Y, Baccou YC (1987) A review of naturrally occurring steroidal sapogenins. Fitoterapia 58: 67–107

592 Peiser GD, Wang T-T, Hoffmann NE, Yang SF, Liu H, Walsh CT (1984) Formation of cyanide from carbon 1 of 1-aminocyclopropane 1-carboxylic acid during its conversion to ethylene. Proc Nat Acad Sci USA 81: 3059–3061

593 Pelletier SW (1983–1987) Alkaloids: chemical and biological perspectives, Vols 1–5. Wiley-Interscience, New York

594 Pelletier SW, Page SW (1984, 1986) Diterpenoid alkaloids. Nat Product Reports 1: 375–386, 3: 451–464

595 Pelter A (1986) Lignans: some properties and syntheses. Rec Adv Phytochem 20: 201–241

596 Peter MG, Waespe HR, Woggon WD, Schmid H (1977) Einbauversuche mit (^{3}H und ^{14}C)-doppeltmarkiertem Farnesol in Cantharidin. Helv Chim Acta 60: 1262–1272

597 Peters NK, Long SR (1988) The role of plant compounds in the regulation of *Rhizobium* nodulations genes. In: Conn EE (ed) Opportunities for phytochemistry in plant biotechnology. Plenum Press, New York, pp 83–97

598 Peterson PA (1987) Mobile elements in plants. CRC Crit Rev Plant Sci 6: 105–208

599 Petrosky RJ, Tookey HL (1982) Interactions of thioglucoside glucohydrolase and epithiospecifier protein of cruciferous plants to form 1-cyanoepithioalkanes. Phytochemistry 21: 1903–1905

600 Pfander H (ed) (1987) Key to carotenoids. Birkhäuser, Boston

601 Pfeffer W (1897) Pflanzenphysiologie, vol 1. Engelmann, Leipzig, p 991

602 Philipsborn W, von, Müller R (1986) ^{15}N NMR spectroscopy—new methods and applications. Angew Chem Int Ed Engl 25: 383–413

603 Phillipson JD, Roberts MF, Zenk MH (eds) (1985) The chemistry and biology of isoquinoline alkaloids. Springer, Berlin Heidelberg New York

604 Piattelli M (1981) The betalains, structure, biosynthesis and chemical taxonomy. In: Conn EE (ed) The biochemistry of plants, vol 7, secondary plant products. Academic Press, New York, pp 557–575

605 Pinder AR (1984–1987) Pyrrolidine, piperidine, and pyridine alkaloids. Nat Product Reports 1: 225–230, 2: 181–187, 3: 171–180, 4: 527–537

606 Piper PJ (1981) SRS-A and leucotrienes. Wiley, New York

607 Pirson A, Zimmermann MH (eds) (1980–85) Encyclopedia of plant physiology, new series, vols 9–11: hormonal regulation of development. Springer, Berlin Heidelberg New York

608 Platt RV, Opie CT, Haslam E (1984) Biosynthesis of flavan-3-ols and other secondary plant products from (2S)-phenylalanine. Phytochemistry 23: 2211–2217

609 Ponchet M, Favre-Bonvin J, Hauteville M, Ricci P (1988) Dianthramides (N-benzoyl and N-paracoumarylanthranilic acid derivatives) from elicited tissues of *Dianthus caryophyllus*. Phytochemistry 27: 725–730

610 Porra RJ, Meisch HU (1984) The biosynthesis of chlorophyll. Trends Biochem Sci 9: 99–104

611 Porter JW, Spurgeon SL (eds) (1983) Biosynthesis of isoprenoid compounds. Wiley, New York

612 Poston JM (1986) β-Leucine and the β-keto pathway of leucine metabolism. Adv Enzymol 58: 173–189

613 Poulton JE (1981) Transmethylation and demethylation reactions in the metabolism of secondary plant products. In: Conn EE (ed) The biochemistry of plants, vol 7, secondary plant products. Academic Press, New York, pp 667–723

614 Preiss J (1984) Bacterial glycogen synthesis and its regulation. Ann Rev Microbiol 38: 419–458

615 Preiss J (1982) Regulation of the biosynthesis and degradation of starch. Ann Rev Plant Physiol 33: 431–454

616 Preiss J (ed) (1988) Carbohydrates (The Biochemistry of plants, vol 14). Academic Press, San Diego

617 Preiss J (ed) (1980) Carbohydrates, structure and function (Biochemistry of plants, vol 3). Academic Press, New York

618 Prestwich GD (1986) Chemical defense and selfdefense in termites. In: Atta-ur-Rahmann (ed) Natural Product Chemistry. Springer, Berlin Heidelberg New York, pp 318–329

619 Prestwich GD (1985) Communication in insects. II. Molecular communication in insects. Quart Rev Biol 60: 437–456

620 Prestwich GD, Blomquist GJ (eds) (1987) Pheromone biochemistry. Academic Press, New York

621 Prince RC (1987) Hopanoids: the worlds most abundant biomolecules. Trends Biochem Sci 12: 455–456

622 Progress in the chemistry of organic natural products (1938 ff) Springer, Wien

623 Purseglove JW, Brown EG, Green CL, Robbins SRJ (eds) (1981) Spices. Longman, London

624 Putnam AR, Tang CS (eds) (1986) The science of allelopathy. Wiley, Chichester

625 Rao CBS (1978) The chemistry of lignans. Andhra University Press

626 Raven JA (1987) The application of mass spectrometry to biochemical and physiological studies. In: Davies DD (ed) The biochemistry of plants, vol 13, methodology. Academic Press, San Diego, pp 127–180

627 Rechcigl M (ed) (1983) CRC Handbook of naturally occurring food toxicants. CRC Press, Boca Raton

628 Recsei PA, Snell EE (1984) Pyruvyl enzymes. Ann Rev Biochem 53: 357-387

629 Regel C von (ed) (1962) Die Rohstoffe des Pflanzenreiches. Cramer, Weinheim

630 Reichstein T (1967) Cardenolide (herzwirksame Glykoside) als Abwehrstoffe bei Insekten. Naturwiss Rundschau 20: 499-511

631 Reinbothe H, Miersch J, Mothes K (1981) Special problems of nitrogen metabolism in plants. Comprehensive Biochem 19 A: 51-163

632 Reinbothe H, Miersch J (1985) Metabolism of amino acids related to alkaloid biosynthesis. In: Mothes K, Schütte HR, Luckner M (eds) Biochemistry of alkaloids. Deutscher Verlag der Wissenschaften, Berlin and Verlag Chemie, Weinheim, pp 65-87

633 Rennenberg H (1982) Glutathione metabolism and possible biological roles in higher plants. Phytochemistry 21: 2771-2781

634 Reznik H (1980, 1981) Betalains. In: Czygan FC (ed) Pigments in plants. Fischer, Stuttgart and Akademie Verlag, Berlin, pp 370-392

635 Rhoades DF (1985) Pheromonal communication between plants. In: Cooper-Driver GA, Swain T, Conn EE (eds) Chemically mediated interactions between plants and other organisms. Plenum Press, New York, pp 195-218

636 Rice EL (1983) Allelopathy. Academic Press, London

637 Richards AG (1978) The chemistry of insect cuticle. In: Rockstein M (ed) Biochemistry of insects. Academic Press, New York

638 Riddiford LM, Truman JW (1978) Biochemistry of insect hormones and insect growth regulators. In: Rockstein M (ed) Biochemistry of insects. Academic Press, New York, pp 307-357

639 Rivier L, Bruhn JG (eds) (1981) Coca and cocaine. J. Ethnopharmacol 3, No. 23

640 Roberts JA, Tucker GA (1985) Ethylene and plant development. Butterworths, London

641 Robins DJ (1982) The pyrrolizidine alkaloids. Prog Chem Org Nat Prod 41: 115-203

642 Robins DJ (1984-1987) Pyrrolizidine alkaloids. Nat Product Reports 1: 235-243, 2: 213-220, 3: 297-305, 4: 577-590

643 Robinson JA, Gani D (1985) Enzymology in biosynthesis: mechanistic and stereochemical studies of β-lactam biosynthesis and the shikimate pathway. Nat Product Reports 2: 293-319

644 Robinson R (1955) The structural relations of natural products. Clarendon Press, Oxford

645 Robinson R (1917) A synthesis of tropinone. J Chem Soc 111: 762-768

646 Robinson T (1974) Metabolism and function of alkaloids in plants. Science 184: 430-435

647 Robinson T (1981) Biochemistry of alkaloids. Springer, Berlin Heidelberg New York

648 Roddick JG (1980) Isoprenoid alkaloids. In: Bell EA, Charlwood BV (eds) Encyclopedia of plant physiology, new series, vol 8, secondary plant products, Springer, Berlin Heidelberg New York, pp 167-184

649 Rodgers A, Spector M (1981) Human stones. Endeavour, new series 5: 119-126

650 Rönsch H (1986) Rhoeadine alkaloids. In: Brossi A (ed) The alkaloids, vol 28. Academic Press, New York, pp 1-93

651 Roeske RW, Kennedy SJ (1983) Ion transporting peptides. In: Weinstein B (ed) Chemistry and biochemistry of amino acids, peptides, and proteins, vol 7. Marcell Dekker, New York, pp 205-266

652 Rohdes MJC, Robins RJ (1987) The use of plant cell cultures in studies of metabolism. In: Davis DD (ed) The biochemistry of plants, vol 13, methodology. Academic Press, San Diego, pp 65-125

653 Roos W (1985) Compartmentation and channeling in alkaloid biosynthesis. In: Mothes K, Schütte HR, Luckner M (eds) Biochemistry of alkaloids. Deutscher Verlag der Wissenschaften, Berlin and Verlag Chemie, Weinheim, pp 42-48

654 Roos W, Luckner M (1986) The spatial organization of secondary metabolism in microbial and plant cells. In: Subramanian TAV (ed) Cell metabolism, growth and environment. CRC Press, Boca Raton

655 Rosenthal GA (1982) Plant nonprotein amino and imino acids. Academic Press, New York

656 Rosenthal GA, Bell EA (1979) Naturally occurring, toxic nonprotein amino acids. In: Rosenthal GA, Janzen DH (eds) Herbivores, their interaction with secondary plant metabolites. Academic Press, New York, pp 353-385

522   References

657 Rosenthal GA, Janzen DH (eds) (1979) Herbivores, their interaction with secondary plant metabolites. Academic Press, New York
658 Rossen L, Davis EO, Johnston AWB (1987) Plant induced expression of *Rhizobium* genes involved in host specificity and early stages of nodulation. Trends Biochem Sci 12: 430–433
659 Ruddat M, Garber D (1983) Biochemistry, physiology and genetics of carotenogenesis in fungi. In: Bennet JW, Ciegler A (eds) Secondary metabolism and differentiation in fungi. Marcell Dekker, New York, pp 95–151
660 Rüdiger W (1980, 1981) Plant biliproteins. In: Czygan FC (ed) Pigments in plants. Fischer, Stuttgart and Akademie Verlag, Berlin
661 Ruzicka L (1959) History of the isoprene rule. Proc Chem Soc pp 341–360
662 Ruzicka L (1953) The isoprene rule and the biogenesis of terpenic compounds. Experientia 9: 357–367
663 Sachs J (1868) Lehrbuch der Botanik. Engelmann, Leipzig, p 541
664 Sachs J (1882) Vorlesungen über Pflanzenphysiologie. Engelmann, Leipzig, pp 203–222
665 Sandford PA (1979) Exocellular microbial polysaccharides. Adv Carbohydr Chem Biochem 36: 265–313
666 Sangameswaran L, Fales HM, Friedrich P, De Blas AL (1986) Purification of a benzodiazepine from bovine brain and detection of benzodiazepine-like immunoreactivity in human brain. Proc Nat Acad Sci USA 83: 9236–9240
667 Sankawa U (1980) The biosynthesis of anthraquinoid mycotoxins from *Penicillium islandicum* Sopp and related fungi. In: Steyn PS (ed) The biosynthesis of mycotoxins. Academic Press, New York, pp 357–394
668 Saxton JE (1984–1986) Recent progress in the chemistry of indole alkaloids and mould metabolites. Nat Product Reports 1: 21–51, 2: 49–80, 3: 353–394
669 Schauer R (1982) Sialic acids—chemistry, metabolism and function, Springer, Wien
670 Scheffer RP, Livingston RS (1984) Host-selective toxins and their role in plant diseases. Science 223: 17–21
671 Scheuer PJ (ed) (1987) Bioorganic marine chemistry, vol 1. Springer, Berlin Heidelberg New York London Paris Tokyo
672 Schiff JA (1983) Reduction and other metabolic reactions of sulfate. In: Läuchli A, Bieleski RL (eds) Encyclopedia of plant physiology, new series, vol 15, inorganic plant nutrition. Springer, Berlin Heidelberg New York, pp 401–421
673 Schiff PL (1987) Bisbenzylisoquinoline alkaloids. J Nat Products 50: 529–599
674 Schildknecht H (1986) A chemical power of movement in higher plants. In: Atta-ur-Rahmann (ed) Natural product chemistry. Springer, Berlin Heidelberg New York, pp 358–382
675 Schildknecht H (1970) Die Wehrchemie von Land- und Wasserkäfern. Angew Chem 82: 17–25
676 Schildknecht H (1983) Turgorins, hormones of the endogenous daily rhythms of higher organized plants—detection, isolation, structure, synthesis, and activity. Angew Chem Int Ed Engl 22: 695–710
677 Schlee D (1985) Alkaloids derived from purines. In: Mothes K, Schütte HR, Luckner M (eds) Biochemistry of alkaloids. Deutscher Verlag der Wissenschaften, Berlin and Verlag Chemie, Weinheim, pp 338–350
678 Schlee D, Straube G (1984) Physiology and biochemistry of riboflavin formation. Pharmazie 39: 805–811
679 Schliemann W (1987) β-D-Glucosidasen von Pflanzen und pflanzlichen Zellkulturen. Pharmazie 42: 225–239
680 Schmidt U, Häusler J, Öhler E, Poisel H (1979) Dehydroamino acids, α-hydroxy-α-amino acids and α-mercapto-α-amino acids. Prog Chem Org Nat Prod 37: 251–327
681 Schneider D (1987) The strange fate of pyrrolizidine alkaloids. In: Chapman RF, Bernays EA, Stoffolano JG (eds) Perspectives in chemoreception and behavior. Springer, New York Berlin Heidelberg London Paris Tokyo, pp 123–142
682 Schneider D (1987) Plant recognition by insects: a challenge for neuro-ethological research. In: Labeyrie V, Fabres G, Lachaise D (eds) Insects—plants. Dr. W. Junk Publishers, Dordrecht, pp 117–123
683 Schneider D, Boppre M, Zweig J, Horsley SB, Bell TW, Meinwald J, Hansen K, Diehl EW

(1982) Scent organ development in *Creatonotos* moths: regulation by pyrrolizidine alkaloids. Science 215: 1264-1265

684 Schneider Z, Stroinski A (1987) Comprehensive B_{12}: chemistry, biochemistry, nutrition, ecology, medicine. de Gruyter, Berlin

685 Schnepf E (1976) Morphology and cytology of storage spaces. In: Luckner M, Mothes K, Nover L (eds) Secondary metabolism and coevolution. Nova Acta Leopoldina, suppl 7: 23-44

686 Schönbeck F, Weltzien HC, Wilbert H (1976) Phytomedizin. Paray, Berlin Hamburg

687 Schöpf C (1937) Die Synthese von Naturstoffen, insbesondere von Alkaloiden, unter physiologischen Bedingungen und ihre Bedeutung für die Frage der Entstehung einiger pflanzlicher Naturstoffe in der Zelle. Angew Chem 50: 779-787, 797-805

688 Schroepfer GJ (1982) Sterol biosynthesis. Ann Rev Biochem 50: 585-621, 51: 555-585

689 Schütte HR (1966) Radioaktive Isotope in der organischen Chemie und Biochemie. Deutscher Verlag der Wissenschaften, Berlin

690 Schütte HR (1985) Simple amines. In: Mothes K, Schütte HR, Luckner M (eds) Biochemistry of alkaloids. Deutscher Verlag der Wissenschaften, Berlin and Verlag Chemie, Weinheim, pp 99-105

691 Schütte HR (1986) Monoterpenoid indole alkaloids. Prog Bot 48: 151-166

692 Schütte HR (1985) Special topics of flavonoid metabolism. Prog Bot 47: 118-141

693 Schütte HR (1984) Monoterpenes. Prog Bot 46: 119-139

694 Schütte HR (1987) Aspects of steroid biosynthesis. Prog Bot 49: 117-136

695 Schütte HR, Liebisch HW (1985) Alkaloids derived from tyrosine and phenylalanine. In: Mothes K, Schütte HR, Luckner M (eds) Biochemistry of alkaloids. Deutscher Verlag der Wissenschaften, Berlin and Verlag Chemie, Weinheim, pp 128-271

696 Schultes RE, Hofmann A (1980) Plants of the gods. McGraw-Hill, New York

697 Scogin R, Freeman CE (1987) Floral anthocyanins of the genus Penstemon: correlations with taxonomy and pollination. Biochem Systematics and Ecology 15: 355-360

698 Segal SJ (ed) (1985) The unique action of gossypol, a potential contraceptive for men. Plenum Press, New York

699 Seigler DS (1981) Secondary metabolites and plant systematics. In: Conn EE (ed) The biochemistry of plants, vol 7, secondary plant products. Academic Press, New York, pp 139-176

700 Seigler DS (1977) Primary roles for secondary compounds. Biochem Syst Ecol 5: 195-199

701 Seltmann G (1982) Die bakterielle Zellwand. Fischer, Jena

702 Sermonti G, Lanfaloni L (1982) Antibiotic genes—their assemblage and localization in *Streptomyces*. In: Krumphanzl V, Sikyta B, Vanek Z (eds) Overproduction of microbial products. Academic Press, London, pp 485-497

703 Setchell KDR, Lawson AM, Mitchell FL, Adlercreutz H, Kirk DN, Axelson M (1980) Lignans in man and animal species. Nature 287: 740-742

704 Shamma M (1984-1986) Aporphinoid alkaloids. Nat Product Reports 1: 201-207, 2: 227-233, 3: 345-351

705 Shockman GD, Barret JF (1983) Structure, function and assembly of cell walls of gram-positive bacteria. Ann Rev Microbiol 37: 501-527

706 Siegenthaler PA, Eichenberger W (eds) (1984) Structure, function and metabolism of plant lipids. Elsevier, Amsterdam 1984

707 Sigel H (ed) (1985) Metal ions in biological systems, Vol 19: antibiotics and their complexes. Marcell Dekker, New York

708 Simpson TJ (1984-1987) The biosynthesis of polyketides. Nat Product Reports 1: 281-297, 2: 321-347, 4: 339-376

709 Singh H, Bhardway TR (1986) Brassinosteroids. Ind J Chem 25B: 989-998

710 Slama K (1980) Animal hormones and antihormones in plants. Biochem Physiol Pflanzen 175: 177-193

711 Smith CW (ed) (1987) The peptides, vol 8, chemistry, biology, and medicine of neurophyseal hormones and their analogs. Academic Press, Orlando

712 Smith JE, Moss MO (1985) Mycotoxins—formation, analysis and significance. Wiley, Chichester

713 Smith LW, Culvenor CCJ (1981) Plant sources of hepatotoxic pyrrolizidine alkaloids. J Nat Prod 44: 129-152

714 Smith RL (1973) The excretory function of bile. Chapman and Hall, London

715 Smith TA (1985) Polyamines. Ann Rev Plant Physiol 36: 117-143
716 Smith TA (1981) Amines. In: Conn EE (ed) The biochemistry of plants, vol 7, secondary plant products. Academic Press, New York, pp 249-268
717 Smith TA (1980) Plant amines. In: Bell EA, Charlwood BV (eds) Encyclopedia of plant physiology, new series, vol 8, secondary plant products. Springer, Berlin Heidelberg New York, pp 433-454
718 Smith TW (ed) (1986) Digitalis glycosides. Grune and Stratton, Orlando
719 Sondheimer E, Simeone JB (eds) (1970) Chemical ecology. Academic Press, New York
720 Southgate R, Elson S (1985) Naturally occurring β-lactams. Prog Chem Org Nat Prod 47: 1-106
721 Southon I, Buckingham J (1989) The dictionary of alkaloids. Chapman and Hall, London
722 Specialist periodical reports, aliphatic and related natural product chemistry (1984ff). Royal Society of Chemistry, London
723 Spenser ID (1985) Stereochemical aspects of the biosynthetic routes leading to the pyrrolizidine and quinolizidine alkaloids. Pure Appl Chem 57: 453-470
724 Spiteller G (1985) Combination of chromatographic separation methods with mass spectrometry—a modern technique for studying metabolism. Angew Chem Int Ed Engl 24: 451-465
725 Spurgeon DL, Porter JW (1980) Carotenoids. In: Stumpf PK (ed) The biochemistry of plants, vol 4, lipids: structure and function. Academic Press, New York, pp 419-483
726 Stafford HA (1981) Compartmentation in natural product biosynthesis by multienzyme complexes. In: Conn EE (ed) The biochemistry of plants, vol 7, secondary plant products. Academic Press, New York, pp 117-137
727 Stafford HA (1988) Proanthocyanidins and the lignin connection. Phytochemistry 27: 1-6
728 Stahl E (1888) Pflanzen und Schnecken, biologische Studie über die Schutzmittel der Pflanzen gegen Schneckenfraß. Jenaische Z Naturwiss 22: 657-684
729 Steiner DF, Chan SJ, Welsh JM, Kwok SMC (1985) Structure and evolution of the insulin gene. Ann Rev Genetics 19: 463-484
730 Stephen AM (1980) Plant carbohydrates. In: Bell EA, Charlwood BV (eds) Encyclopedia of plant physiology, new series, vol 8, secondary plant products. Springer, Berlin Heidelberg New York, pp 555-584
731 Sterba G, Wolf G (1976) Excretion storage and its functional integration in animals. Nova Acta Leopoldina Suppl 7: 13-22
732 Stewart GR, Mann AF, Fentem PA (1980) Enzymes of glutamate formation: glutamate dehydrogenase, glutamine synthetase, and glutamate synthase. In: Miflin JB (ed) The biochemistry of plants, vol 5, amino acids and derivatives. Academic Press, New York, pp 271-327
733 Steyn PS, Vleggaar R (1985) Tremorgenic mycotoxins. Progr Chem Org Nat Prod 48: 1-80
734 Stoddart RW (1984) The biosynthesis of polysaccharides. Croom Helm, London
735 Stoeckenius W, Bogomolni RA (1982) Bacteriorhodopsin and related pigments from halobacteria. Ann Rev Biochem 51: 587-616
736 Stöckigt J (1980) The biosynthesis of heteroyohimbine-type alkaloids. In: Phillipson JD, Zenk MH (eds) Indole and biogenetically related alkaloids. Academic Press, New York, pp 113-141
737 Strauss DG (1987) Structure of anthracycline antitumor agents. Pharmazie 42: 289-303
738 Strobel GA (1982) Phytotoxins. Ann Rev Biochem 51: 309-333
739 Strunz GM, Findlay JA (1985) Pyridine and piperidine alkaloids. In: Brossi A (ed) The alkaloids, vol 26. Academic Press, New York, pp 89-183
740 Stumpf PK (ed) (1987) Lipids: structure and function (The biochemistry of plants, vol 9). Academic Press, New York
741 Stumpf PK, Conn EE (eds) (1987) The biochemistry of plants, vol 9, lipids: structure and function. Academic Press, New York
742 Suffness M, Cordell GA (1985) Antitumor alkaloids. In: Brossi A (ed) The alkaloids, vol 25, Academic Press, New York, pp 1-355
743 Sugawara F, Strobel GA (1987) Tryptophol, a phytotoxin produced by *Drechslera nodulosum*. Phytochemistry 26: 1349-1351
744 Sutherland IW (1985) Biosynthesis and composition of gram-negative bacterial extracellular and wall polysaccharides. Ann Rev Microbiol 39: 243-270
745 Suttie JW (1985) Vitamin K-dependent carboxylases. Ann Rev Biochem 54: 459-477

746 Suttie JW (ed) (1988) Current advances in vitamin K research. Elsevier, New York
747 Swain T (1965) Methods used in the study of biosynthesis. In: Pridham JB, Swain T (eds) Biosynthetic pathways in higher plants. Academic Press, London, pp 9-36
748 Swain T (1977) Secondary compounds as protective agents. Ann Rev Plant Physiol 28: 479-501
749 Tabata M, Umetani Y, Ooya M, Tanaka S (1988) Glucosylation of phenolic compounds by plant cell cultures. Phytochemistry 27: 809-813
750 Tabor CW, Tabor H (1984) Methionine adenosyltransferase (S-adenosylmethionine synthetase) and S-adenosylmethionine decarboxylase. Adv Enzymol 56: 251-282
751 Tabor CW, Tabor H (1985) Polyamines in microorganisms. Microbiol Reviews 49: 81-99
752 Tabor CW, Tabor H (1984) Polyamines. Ann Rev Biochem 53: 749-790
753 Takahashi N (ed) (1986) Chemistry of plant hormones. CRC Press, Boca Raton
754 Tanner W, Loewus FA (eds) (1981, 1982) Plant carbohydrates. In: Encyclopedia of plant physiology, vols 13 A and 13 B. Springer, Berlin Heidelberg New York
755 Taylor GT (1982) The methanogenic bacteria. Prog Ind Microbiol 16: 231-329
756 Taylor RF (1984) Bacterial triterpenoids. Microbial Reviews 48: 181-198
757 Teuscher E, Lindequist U (1987) Biogene Gifte. Akademie Verlag, Berlin
758 Thomas RJ, Schrader LE (1981) Ureide metabolism in higher plants. Phytochemistry 20: 361-371
759 Thompson GA (1980) Plant lipids of taxonomic significance. In: Bell EA, Charlwood BV (eds) Encyclopedia of plant physiology, new series, vol 8, secondary plant products. Springer, Berlin Heidelberg New York, pp 535-551
760 Thompson JF (1980) Arginine synthesis, proline synthesis, and related processes. In: Miflin JB (ed) The biochemistry of plants, vol 5, amino acids and derivatives. Academic Press, New York, pp 375-402
761 Thomson RH (1987) Naturally occurring quinones: recent advances. Chapman & Hall, London
762 Thomson RH (1985) The chemistry of natural products. Blackie, Glasgow
763 Threlfall DR (1980) Polyprenols and terpenoid quinones and chromanols. In: Bell EA, Charlwood BV (eds) Encyclopedia of plant physiology, new series, vol 8, secondary plant products. Springer, Berlin Heidelberg New York, pp 288-308
764 Timberlake CF, Henry BS (1986) Plant pigments as natural food colors. Endeavour New Series 10: 31-36
765 Torsell K (1983) Natural product chemistry: a mechanistic and biosynthetic approach to secondary metabolism. Wiley, Chichester
766 Träger L (1977) Steroidhormone – Biosynthese, Stoffwechsel, Wirkung. Springer, Berlin Heidelberg New York
767 Trier G (1912) Über einfache Pflanzenbasen und ihre Beziehungen zum Aufbau der Eiweißstoffe und Lecithin. Bornträger Verlag, Berlin
768 Trudinger PA, Loughlin RE (1981) Metabolism of simple sulphur compounds. Compr Biochem 19 A: 165-256
769 Tsuda M (1987) Photoreception and phototransduction in invertebrate photoreceptors. Photochem Photobiol 45: 915-931
770 Tu AT (ed) (1983ff) Handbook of natural toxins, vols 1-5. Marcell Decker, New York
771 Turk J, Jakschik BA, Morrison AR, Lefkowith JB (1986) Arachidonic acid metabolism. Ann Rev Biochem 55: 69-102
772 Turner WB (1971) Fungal metabolites. Academic Press, London
773 Turner WB, Aldridge DC (1983) Fungal metabolites II. Academic Press, London
774 Udenfriend S, Meienhofer J (eds) (1984) The peptides, analysis, synthesis, biology, vols 1-7. Academic Press, Orlando
775 Udsin E, Borchardt RT, Creveling (eds) (1978) Transmethylation. Elsevier/North Holland, Amsterdam
776 Ueno Y (1983) Trichothecenes. Elsevier, Amsterdam
777 Ullmanns Enzyklopädie der technischen Chemie, vols 1-25. (1972ff) Verlag Chemie, Weinheim/Bergstr.
778 Umezawa H (ed) (1982) Aminoglycoside antibiotics. Springer, Berlin Heidelberg New York
779 Underhill EW (1980) Glucosinolates. In: Bell EA, Charlwood BV (eds) Encyclopedia of plant

physiology, new series, vol 8, secondary plant products. Springer, Berlin Heidelberg New York, pp 493-511

780 van Oycke S, Braekman JC, Daloze D, Pasteels JM (1987) Cardenolide biosynthesis in chrysomelid beetles. Experientia 43: 460-462

781 Vandenbergh JG (ed) (1983) Pheromones and reproduction in mammals. Academic Press, New York

782 Vanek Z, Blumauerova M (1986) Physiology and pathophysiology of secondary metabolite production. In: Vanek Z, Hostalek Z (eds) Overproduction of microbial metabolites. Butterworths, Boston

783 Vanek Z, Mateju J, Cudlin J, Blumauerova M, Sedmera P, Jizba J, Kralovcova E, Tax J, Gauze GF (1982) Biosynthesis of daunomycin-related anthracyclines. In: Krumphanzl V, Sikyta B, Vanek Z (eds) Overproduction of microbial products. Academic Press, London, pp 283-299

784 Vane-Wright RI, Ackery RP (1984) The biology of butterflies. Academic Press, London

785 Vederas JC (1987) The use of stable isotopes in biosynthetic studies. Nat Product Reports 4: 277-337

786 Vennesland B, Conn EE, Knowles CJ, Westley J, Wissing F (1981) Cyanide in biology. Academic Press, London

787 Verma DPS, Hohn T (eds) (1984) Genes involved in microbe-plant interactions. Springer, Wien

788 Virtanen AI (1962) Organische Schwefelverbindungen in Gemüse- und Futterpflanzen. Angew Chem 74: 374-382

789 Visconti, M (1984) L-Biopterin, a history beginning from *Drosophila melanogaster* and ending with the treatment of an atypical phenylketonuria. In: Voelter W, Daves DG (eds) Biologically active principles of natural products. Thieme, Stuttgart, pp 23-30

790 Vleggar R, Steyn PS (1980) The biosynthesis of some miscellanous mycotoxins. In: Steyn PS (ed) The biosynthesis of mycotoxins. Academic Press, New York, pp 395-422

791 Vogel R (1984) Natürliche Enzyminhibitoren. Thieme, Stuttgart New York

792 Vogel S (1974) Ölblumen und ölsammelnde Bienen. Steiner, Wiesbaden

793 Voigt G, Hiller K (1987) Advanves in the chemistry and biology of steroid saponins. Sci Pharm 55: 201-227

794 Wagner H, Hikino H, Farnthworth NR (eds) (1985) Economic and medicinal plant research, vol 1. Academic Press, London

795 Walker JB (1979) Creatine: biosynthesis, regulation and function. Adv Enzymol 50: 177-242

796 Walker JB (1974) Biosynthesis of the monoguanidinated inositol moiety of bluensomycin, a possible evolutionary precursor of streptomycin. J Biol Chem 249: 2397-2404

797 Wallace JW, Mansell RL (eds) (1976) Biochemical interactions between plants and insects. Plenum Press, New York

798 Waller GR, Dermer OC (1981) Enzymology of alkaloid metabolism in plants and microorganisms. In: Conn EE (ed) The biochemistry of plants, vol 7, secondary plant products. Academic Press, New York, pp 317-402

799 Waller GR, Nowacki EK (1978) Alkaloid biology and metabolism in plants. Plenum Press, New York

800 Wallis M (1981) The molecular evolution of pituitary growth hormone, prolactin and placental lactogen: a protein family showing variable rates of evolution. J Mol Evol 16: 10-18

801 Wallis M, Howell SL, Taylor KW (1985) Biochemistry of the polypeptide hormones. Wiley, Chichester

802 Walton DC (1980) Biochemistry and physiology of abscisic acid. Ann Rev Plant Physiol 31: 453-489

803 Wasternack C (1982) Metabolism of pyrimidines and purines. In: Parthier B, Boulter D (eds) Encyclopedia of plant physiology, new series, vol 14 b, nucleic acids and proteins in plants. Springer, Berlin Heidelberg New York, pp 263-301

804 Waterman PG, Gray AI (1987) Chemical systematics. Nat Product Reports 4: 175-203

805 Weiler EW, Schröder J (1987) Hormone genes and crown gall disease. Trends Biochem Sci 12: 271-275

806 Weiner M, van Eys J (1983) Nicotinic acid: nutrient, cofactor, drug. Marcel Dekker, New York

807 Weinstein B (1971-1983) Chemistry and biochemistry of amino acids, peptides, and proteins, vols 1-7. Marcell Dekker, New York

808 Weiss U, Edwards JM (1980) The biosynthesis of aromatic compounds. Wiley-Interscience, New York

809 Wenger RM (1986) Cyclosporine and analogues—isolation and synthesis—mechanism of action and structural requirements for pharmacological activity. Prog Chem Org Nat Prod 50: 123-168

810 Werner C, Matile P (1985) Accumulation of coumarylglucosides in vacuoles of barley mesophyll protoplasts. J Plant Physiol 118: 237-249

811 Whiting DA (1985, 1987) Lignans, neolignans, and related compounds. Nat Product Reports 2: 191-211, 4: 499-525

812 Whittaker RH, Feeney PP (1971) Allelochemics: chemical interactions between species. Science 171: 757-770

813 Wied D (1980) Neuropeptides and psychopathology. Endevour 4: 154-159

814 Wieland T (1987) 50 Jahre Phalloidin—seine Entdeckung, Charakterisierung sowie gegenwärtige und zukünftige Anwendung in der Zellforschung. Naturwissenschaften 74: 367-373

815 Wieland T (1986) Peptides of poisonous Amanita mushrooms. Springer, Berlin Heidelberg New York

816 Wiermann R (1981) Secondary plant products and cell and tissue differentiation. In: Conn EE (ed) The biochemistry of plants, vol 7, secondary plant products. Academic Press, New York, pp 85-137

817 Wightman F, Forest JC (1978) Properties of plant aminotransferases. Phytochemistry 17: 1455-1471

818 Wijnsma R, Verpoorte R (1986) Anthraquinones in Rubiaceae. Prog Chem Org Nat Prod 49: 79-149

819 Willenbrink J (1987) Die pflanzliche Vakuole als Speicher. Naturwissenschaften 74: 22-29

820 Williams RT (1967) The biogenesis of conjugation and detoxication products. In: Bernfeld P (ed) Biogenesis of natural compounds. Pergamon Press, Oxford, pp 589-639

821 Wimmer MJ, Rose IA (1978) Mechanisms of enzyme-catalyzed group transfer reactions. Ann Rev Biochem 46: 1331-1378

822 Wink M (1987) Quinolizidine alkaloids: biochemistry, metabolism, and function in plants and cell suspension cultures. Planta Med 53: 509-514

823 Wink M (1984) Biochemistry and chemical ecology of lupin alkaloids. Proc 3rd International Lupin Congress, Lal Rochelle (France), 4.-8.6.1984, pp 325-343

824 Wink M, Hartmann T (1984) Enzymology of quinolizidine alkaloid biosynthesis. Zawalewski RI, Skolik JJ (eds) Natural products chemistry, Elsevier, Amsterdam, pp 511-520

825 Wink M, Witte L (1983) Evidence for a wide-spread occurrence of the genes of quinolizidine alkaloid biosynthesis. FEBS Letters 159: 196-200

826 Wollenweber E (1984) The systematic implication of flavonoids secreted by plants. In: Rodriguez E, Healey P, Mehta I (eds) Biology and chemistry of plant trichomes. Plenum Press, New York, pp 53-69

827 Wolters B (1985) Steroidhormone, Synthesen mit Hilfe von Mikroorganismen. D Apotheker Ztg 125: 643-647

828 Wood HG, Barden RE (1977) Biotin enzymes. Ann Rev Biochem 46: 385-413

829 Wood WA (ed) (1982) Carbohydrate metabolism, parts D and E. Methods Enzymol, vols 89 and 90

830 Wright CE, Tallan HH, Lin YY (1986) Taurine biological update. Ann Rev Biochem 55: 427-453

831 Wrobel JT (1985) Pyrrolizidine alkaloids. In: Brossi A (ed) The alkaloids, vol 26. Academic Press, New York, pp 327-384

832 Yamazaki M (1980) The biosynthesis of neurotropic mycotoxins. In: Steyn PS (ed) The biosynthesis of mycotoxins. Academic Press, New York, pp 193-222

833 Yang SF, Hoffman NE (1984) Ethylene biosynthesis and its regulation in higher plants. Ann Rev Plant Physiol 35: 155-189

834 Yeaman SJ (1986) The mammalian 2-oxoacid dehydrogenases: a complex family. Trends Biochem Sci 11: 293-296

835 You K (1982) Stereospecificities of the pyridine nucleotide-linked enzymes. Methods Enzymol 87: 101-126

836 Young DW (1986) The biosynthesis of the vitamins thiamin, riboflavin, and folic acid. Nat Product Reports 3: 395-419

837 Young RA, Rowell RM (eds) (1986) Cellulose: structure, modification, and hydrolysis. John Wiley, Chichester

838 Zähner H (1979) What are secondary metabolites? Folia Microbiol 24: 435-443

839 Zähner H, Anke H, Anke T (1983) Evolution and secondary pathways. In: Bennet JW, Ciegler A (eds) Secondary metabolism and differentiation in fungi. Marcell Dekker, New York, pp 153-171

840 Zähner H, Drautz H, Weber W (1982) Novel approaches to metabolite screening. In: Bu'Lock JD, Nisbet LJ, Winstanley DJ (eds) Bioactive microbial products: search and discovery. Spec Publ Soc Gen Microbiol 6, Academic Press, New York, pp 1-70

841 Zamir LO (1980) The biosynthesis of patulin and penicillic acid. In: Steyn PS (ed) The biosynthesis of mycotoxins. Academic Press, New York, pp 223-268

842 Zaprometov MN (1976) Dual function of some phenolic compounds in the metabolism of higher plants. In: Luckner M, Mothes K, Nover L (eds) Secondary metabolism and coevolution. Nova Acta Leopoldina, suppl 7: 497-498

843 Zaprometov MN (1977) Metabolism of phenolic compounds in plants (Russ.). Biochimija 42: 3-20

844 Zenk MH (1967) Biochemie und Physiologie sekundärer Pflanzenstoffe. Ber Dtsch Bot Ges 80: 573-591

845 Zenk MH, Rüffer M, Amman M, Deus-Neumann B (1985) Benzylisoquinoline biosynthesis by cultivated plant cells and isolated enzymes. J Nat Prod 48: 725-738

846 Ziegler I (1987) Pterine: Pigmente, Cofaktoren und Signalverbindungen bei der zellulären Interaktion. Naturwissenschaften 74: 563-572

848 Nahrstedt A (1987) Recent developments in chemistry, distribution and biology of the cyanogenic glycosides. In: Hostettmann K, Lea PJ (eds) Biologically active natural products. Claredon Press, Oxford, pp 213-234

849 Harborne JB (1987) Natural fungitoxins. In: Hostettmann K, Lea PJ (eds) Biologically active natural products. Claredon Press, Oxford, pp 195-211

850 Hiller K (1987) New results on the structure and biological activity of triterpenoid saponins. In: Hostettmann K, Lea PJ (eds) Biologically active natural products. Claredon Press, Oxford, pp 167-184

851 Harris JB (ed) (1986) Natural toxins. Claredon Press, Oxford

852 Morgan DML (1987) Polyamines. Essays Biochemistry 23: 82-115

853 Pickett CB (1987) Structure and regulation of glutathione S-transferase genes. Essays Biochemistry 23: 116-143

854 Turner AJ (1986) Processing and metabolism of neuropeptides. Essays Biochemistry 22: 69-119

855 Hendry GAF, Houghton JD, Brown SB (1987) The degradation of chlorophyll—a biological enigma. New Phytol 107: 255-302

856 Stahl-Biskup E (1987) Monoterpene glycosides, state-of-the-art. Flavour and Fragrance J 2: 75-82

857 Kerkut GA, Gilbert LI (eds) (1985) Comprehensive insect physiology, biochemistry, and pharmacology, vol 10, biochemistry. Pergamon Press, Oxford

858 Spencer CM, Cai Y, Martin R, Gaffney SH, Goulding PN, Magnolato D, Lilley TH, Haslam E (1988) Polyphenol complexation—some thoughts and observations. Phytochemistry 27: 2397-2409

859 Acree TE, Soderlund DM (eds) (1985) Semiochemistry, flavors and pheromones. de Gruyter, Berlin

860 Zepernick B, Langhammer L, Lüdecke JBP (eds) (1984) Lexikon der offizinellen Arzneipflanzen. de Gruyter, Berlin

861 Vining CL (1986) Secondary metabolism. In: Pape H, Rehm HJ (eds) Biotechnology, vol 4, microbial products II. VCH Verlagsgesellschaft, Weinheim, pp 19-38

862 Vanek Z, Hostalek Z, Blumauerova M, Mikulik K, Podojil M, Behal V, Jechova V (1973) The biosynthesis of tetracyclines. Pure Appl Chem. 34: 463-486

863 Kuhn DN, Chappell J, Boudet A, Hahlbrock E (1984) Induction of phenylalanine ammonia-ly-ase and 4-coumarate: CoA ligase mRNAs in cultured plant cells by UV light or fungal elicitor. Proc Nat Acad Sci USA 81: 1102-1106

864 Murakoshi I, Ikegami F, Hinuma Y, Hanma Y (1984) Purification and characterization of β-(pyrazol-1-yl)-L-alanine synthase from *Citrullus vulgaris*. Phytochemistry 23: 973-977

865 Miyazaki JH, Yang SF (1987) Inhibition of the methionine cycle enzymes. Phytochemistry 26: 2635-2660

866 Mayne RG, Kende H (1986) Ethylene biosynthesis in isolated vacuoles of *Vicia faba* L.—requirement for membrane integrity. Planta 167: 159-165

867 Malik VS (1986) Genetics of secondary metabolism. In: Pape H, Rehm HJ (eds) Biotechnology, vol 4, microbial products II. VCH Verlagsgesellschaft, Weinheim, pp 39-68

868 Schellenberger A (1967) Structure and mechanism of action of the active centre of yeast pyruvate decarboxylase. Angew Chem Int Ed Engl 6: 1024-1035

869 Ullrich J, Ostrovsky YM, Eyzaguirre J, Holzer H (1971) Thiamine pyrophosphate-catalyzed enzymatic decarboxylation of α-oxo acids. Vitam Horm 28: 365-398

870 Curtius HC, Blau N, Levine RA (eds) (1987) Unconjugated pterins and related biogenic amines. de Gruyter, Berlin

871 Seidl PH, Schleifer KH (eds) (1985) Biological properties of peptidoglycan. de Gruyter, Berlin

872 Pfleiderer W, Wachter H, Blair JA (eds) (1987) Biochemical and clinical aspects of pteridines, vol 5. de Gruyter, Berlin

873 Gräfe U (1982) Relevance of microbial nitrogen metabolism to production of secondary metabolites. In: Krumphanzl V, Sikyta B, Vanek Z (eds) Overproduction of microbial products. Academic Press, London, pp 63-75

874 Hostalek Z (1980) Catabolite regulation of antibiotic biosynthesis. Folia Microbiol 25: 445-450

875 Mothes K (1969) Biologie der Alkaloide. In: Mothes K, Schütte HR (eds) Biosynthese der Alkaloide. Deutscher Verlag der Wissenschaften. Berlin, pp 1-20

876 Friedrich W (1988) Vitamins. de Gruyter, Berlin

877 Winkelmann G, van der Helm D, Neilands JB (1987) Ion transport in microbes, plants, and animals. VCH Verlagsgesellschaft, Weinheim

878 O'Neill HJ, Gordon SM, Krotoszynski B, Kavin H, Szidon JP (1987) Identification of isoprenoid-type components in human expired air: a possible shunt pathway in sterol metabolism. Biomed Chromatography 2: 66-69

879 Pteridines (1988 ff) de Gruyter, Berlin

880 Meeuse BJD, Raskin I (1988) Sexual reproduction in the arum lily family, with emphasis on thermogenicity. Sex Plant Reprod 1: 3-15

881 Schell J (1986) The T-DNA genes of *Agrobacterium* plasmids appear to be of a complex evolutionary origin. In: Gustafson JP, Stebbins GL, Ayala FJ (eds) Genetics, development, and evolution. Plenum Press, New York, pp 343-353

882 Bender DA, Joseph MH, Kochen W, Steinhart H (eds) (1986) Progress in tryptophan and serotonin research 1986. de Gruyter, Berlin

883 Theodoropoulos D (ed) (1986) Peptides 1986. de Gruyter, Berlin

884 Nicolaus RA (1986) Melanins. Hermann, Paris

885 Swan GA (1974) Structure, chemistry and biosynthesis of melanins. Prog Chem Org Nat Prod 31: 521-582

886 Baldwin JE, Abraham E (1988) The biosynthesis of penicillins and cephalosporins. Nat Product Reports 5: 129-146

887 Barron D, Varin L, Ibrahim RK, Harborne JB, Williams CA (1988) Sulphated flavonoids—an update. Phytochemistry 27: 2375-2395

888 Battersby AR (1988) Synthetic and biosynthetic studies on vitamin B 12. J Nat Prod 51: 643-661

889 Battersby AR (1988) Biosynthesis of the pigments of life. J Nat Prod 51: 629-642

890 Diamond JM (1989) Hot sex in voodoo lilies. Nature 339: 258-259.

891 Edenborough MS, Herbert RB (1988) Naturally occurring isocyanides. Nat Product Reports 5: 229-246

892 Fowler MW, Scragg AH (1988) Natural products from higher plants and plant cell cultures. In: Pais MSS, Mavituna F, Novais JM (eds) Plant cell biotechnology. Springer Berlin Heidelberg New York London Paris Tokyo, pp 165-177

893 Harrison DM (1988) The biosynthesis of triterpenoids, steroids, and carotenoids. Nat Product Reports 5: 387-416

894 Lewis JR (1988) Imidazole, oxazole, and peptide alkaloids and other miscelaneous alkaloids. Nat Product Reports 5: 351-362

895 Mayer AM (1989) Plant-fungal interactions: a plant physiologist's viewpoint. Phytochemistry 28: 311-317

896 Niemeyer HM (1988) Hydroxamic acids (4-hydroxy-1,4-benzoxazin-3-ones), defense chemicals in the Gramineae. Phytochemistry 27: 3349-3358

897 Schmidt G (1986) Recent developments in the field of biologically active peptides. Topics in Current Chemistry 136: 109-166

898 Spencer CM, Cai Y, Martin R, Gaffney SH, Goulding PN, Magnolato D, Lilley TH, Haslam E (1988) Polyphenol complexation—some thoughts and observations. Phytochemistry 27: 2397-2409

899 Stöckigt J, Schübel H (1988) Cultivated plant cells: an enzyme source for alkaloid formation. In: Pais MSS, Mavituna F, Novais JM (eds) Plant cell biotechnology. Springer Berlin Heidelberg New York London Paris Tokyo, pp 251-264

900 Weissenberg M, Levy A, Wasserman RH (1989) Distribution of calcitriol activity in Solanum glaucophyllum plants and cell cultures. Phytochemistry 28: 795-798

901 Weitz CJ, Lowney LI, Faull KF, Feistner G, Goldstein A (1988) 6-Acetylmorphine: a natural product present in mammalian brain. Proc Natl Acad Sci USA 85: 5335-5338

902 Wink M (1988) Plant breeding: importance of plant secondary metabolites for protection against pathogenes and herbivores. Theor Appl Genet 75: 225-233

903 Worthington PA (1988) Antibiotics with antifungal and antibacterial activity against plant diseases. Nat Product Reports 5: 47-66

Index

An asterik (*)indicates that a structural formula is given on this page.

M